T0388488

ADVANCES IN PHYTOCHEMISTRY, TEXTILE AND RENEWABLE ENERGY RESEARCH FOR INDUSTRIAL GROWTH

PROCEEDINGS OF THE INTERNATIONAL CONFERENCE OF PHYTOCHEMISTRY, TEXTILE AND RENEWABLE ENERGY FOR SUSTAINABLE DEVELOPMENT (ICPTRE 2020), AUGUST 12–14, ELDORET, KENYA

Advances in Phytochemistry, Textile and Renewable Energy Research for Industrial Growth

Edited by

Dr. Charles Nzila
School of Engineering, Moi University, Kenya

Dr. Nyamwala Oluoch
Department of Mathematics, Physics & Computing, Moi University, Kenya

Prof. Ambrose Kiprop
Africa Center of Excellence in Phytochemicals, Textiles and Renewable Energy (ACEII-PTRE)/Chemistry and Biochemistry, Moi University, Kenya

Dr. Rose Ramkat
Africa Center of Excellence in Phytochemicals, Textiles and Renewable Energy (ACEII-PTRE/Department of Biological Sciences, School of Biological & Physical Sciences, Moi University, Kenya

Prof. Isaac S. Kosgey
Vice Chancellor, Moi University, Kenya

CRC Press/Balkema is an imprint of the Taylor & Francis Group, an informa business

Typeset in Times New Roman by MPS Limited, Chennai, India

Library of Congress Cataloging-in-Publication Data
A catalog record has been requested for this book

First published 2022

Published by: CRC Press/Balkema
Schipholweg 107C, 2316 XC Leiden, The Netherlands
e-mail: enquiries@taylorandfrancis.com
www.routledge.com – www.taylorandfrancis.com

ISBN: 978-1-032-11871-0 (hbk)
ISBN: 978-1-032-11877-2 (pbk)
ISBN: 978-1-003-22196-8 (ebk)
DOI: 10.1201/9781003221968

Advances in Phytochemistry, Textile and Renewable Energy Research for Industrial Growth – Nzila et al. (Eds)
© 2022 Copyright the Editor(s), ISBN: 978-1-032-11871-0
Open Access: www.taylorfrancis.com, CC BY-NC-ND 4.0 license

Table of contents

Section 2: Insights in progressive textiles, transformative industrialization & phytochemistry

Section 3: Renewable energy for economic and industrial growth

Advances in Phytochemistry, Textile and Renewable Energy Research for Industrial Growth – Nzila et al. (Eds)
© 2022 Copyright the Editor(s), ISBN: 978-1-032-11871-0
Open Access: www.taylorfrancis.com, CC BY-NC-ND 4.0 license

Preface

This book is an outcome from an international conference by the Africa Centre of Excellence in Phytochemicals, Textile and Renewable Energy (ACEII-PTRE); a World Bank funded project under the auspices of the Africa Centers of Excellence (ACE II) that seeks to contribute to addressing critical gaps in technical, scientific and research skills. The ACEII-PTRE aims at providing high quality training and research in the Eastern and Southern African region in Phytochemicals, Textile and Renewable energy. It seeks to build regional specialization by concentrating on postgraduate students and faculty, generating knowledge, developing skills to support industry and private sector demand, and to generate research for the betterment of the community and the region at large.

Our aim in writing this book is to prepare a text which is comprehensive and a stimulating discourse on research for sustainable industrial growth. The context of the book is situated on the premise that the economic growth and industrial investments have boosted demand for greater technological skills and applied research in some of the priority economic sectors like Manufacturing, Agriculture, Energy, Health and others in many developing economies. Prior to Covid-19, many regions of the world experienced remarkable economic growth especially in the last decade. Outstanding skills and continuous vigorous research are therefore needed, particularly in Science, Technology, Engineering and Mathematics (STEM) so as to sustain the economic growth and transformation. However, there is a new global paradigm shift that seeks to consider STEAM instead of STEM – by adding the 'Arts' to the STEM since Arts play an important role in areas related to STEM. Notwithstanding, ethics in research is also a key area of focus in STEAM.

The theme of the book: "Advancing Science, Technology and Innovation for Industrial Growth" and the thematic areas of Phytochemistry, Progressive Textiles, Transformative Industrialization, Renewable and Alternative Energy are relevant in realizing Sustainable Development Goals (SDGs), which is a universal call of action to end poverty, protect the planet and ensure people enjoy peace and prosperity by the year 2030. A better world requires an interdisciplinary approach, hence the incorporation of social science perspectives rendering this book a high-level academic discourse with cross-cutting disciplines.

Advancing research and well-integrated national development strategies can help raise productivity, improve industrial competitiveness, support faster growth, create jobs and promote resilient communities. Therefore, leveraging on the research arena to advance industrial growth is a central goal of this book. The book captures various knowledge creation and development aspects through the presentation of case specific, basic and applied research and innovation from the perspective of multiple contributors. The text is presented in 42 short chapters which are organized in three sections. Section 1 of the book focuses on advances in Science, Technology, Engineering, Arts & Mathematics for responsible consumption and production while section 2 presents insights in progressive textiles, transformative industrialization and phytochemistry. In section 3, various aspects of renewable energy for economic and industrial growth are explained in great detail. Owing to the diversity of topics covered, the purpose of the book is not to train specialists but to present short and comprehensive concepts and techniques necessary to stimulate satisfying discourse and understanding in the thematic areas. The book is recommended as a reference text for graduate courses such as those in STEAM, Phytochemistry, Textile and Industrial Engineering as well as Renewable and Alternative Energy Technologies. We hope that the book will also remain of interest as a valuable supporting text to researchers and industrial practitioners.

Advances in Phytochemistry, Textile and Renewable Energy Research for Industrial Growth – Nzila et al. (Eds)
© 2022 Copyright the Editor(s), ISBN: 978-1-032-11871-0
Open Access: www.taylorfrancis.com, CC BY-NC-ND 4.0 license

Editorial board

PROF. PAUL WAMBUA (*Session Chair ICPTRE2020*)
Professor of Materials Engineering, Moi University

PROF. SAMUEL ROTICH (*Session Chair ICPTRE2020*)
Thematic Leader, Renewable Energy – Physics option, ACEII-PTRE, Moi University.

PROF. JOHN GITHAIGA (*Session Chair ICPTRE2020*)
Thematic Leader, Textile Engineering

PROF. XIAOHONG QIN (*Session Chair ICPTRE2020*)
Professor, Donghua University, (China)

PROF. ZACHARY SIAGI (*Session Chair ICPTRE2020*)
Thematic Leader, Renewable Energy, ACEII-PTRE, Moi University.

PROF. KUN ZHANG (*Session Chair ICPTRE2020*)
Professor, Donghua University, (China)

ENG. PROF. SIMIYU SITATI (*Session Chair ICPTRE2020*)
Dean, School of Engineering, Moi University

PROF. KIRIMI KIRIAMITI (*Session Chair ICPTRE2020*)
Professor, Moi University

PROF. SIMEON MINING (*Session Chair & Panellist ICPTRE2020*)
Director of Research & Coordinator – Intellectual Property, Moi University

PROF. JULIUS KIPKEMBOI (*Session Chair ICPTRE2020*)
Professor, Egerton University

DR. ISAAC KOWINO (*Session Chair ICPTRE2020*)
Senior Lecturer, Department of pure and applied Chemistry, Masinde Muliro University of Science and Technology

PROF. FREDRICK KENGARA (*Session Chair ICPTRE2020*)
Associate Prof., Department of Physical and Biological Sciences, Bomet University College

MIFTAH FEKADU KEDIR (*Session Chair ICPTRE2020*)
Researcher, Central Ethiopian Environment and Forest Research Center, (Ethiopia)

GEOFFREY SSEBABI MUTUMBA (*Session Chair ICPTRE2020*)
Faculty, Department of Economics and Statistics, Kyambogo University, (Uganda)

DR. SEBASTIAN WAITA (*Session Chair ICPTRE2020*)
Professor, University of Nairobi

PROF. MATHEW KOSGEI (*Session Chair ICPTRE2020*)
Professor, Moi University

DR. SHENGYUAN YANG (*Session Chair ICPTRE2020*)
Professor, Donghua University, (China)

PROF. JOEL KIBIIY (*Session Chair ICPTRE2020*)
Professor, Moi University

MR. PNENWIX MUSONYE (*Session Chair ICPTRE2020*)
EPRA/AEPEA

PROF. FUJUN XU (*Session Chair ICPTRE2020*)
Professor, Donghua University, (China)

DR. DAVID NJUGUNA (*Session Chair ICPTRE2020*)
Coordinator, Textile and Industrial Engineering, ACEII-PTRE, Moi University

DR. MARTHA INDULI (*Session Chair ICPTRE2020*)
Deputy Director, KIRDI

DR. JACQUELINE MAKATIANI (*Session Chair ICPTRE2020*)
Coordinator, Industrial linkages, ACEII-PTRE, Moi University

DR. ERIC OYONDI (*Session Chair ICPTRE2020*)
HoD, Manufacturing, Industrial & Textile Engineering, Moi University

BRENDA AKANKUNDA (*Session Chair ICPTRE2020*)
Lecturer, Makerere University, (Uganda)

DR. KORIR KIPTIEMOI (*Session Chair ICPTRE2020*)
Coordinator, Renewable Energy, ACEII-PTRE, Moi University
PROF. ENG. AUGUSTINE MAKOKHA (*Session Chair ICPTRE2020*)
Professor, Moi University

PROF. BIN SHEN (*Session Chair and Keynote Speaker ICPTRE2020*)
Associate Professor of supply chain management, operations-marketing interface, and fashion industry, Donghua University (China)

DR. ROSE RAMKAT (*Session Chair and Panellist ICPTRE2020*)
HoD, Biological Science & Deputy Centre Leader, ACEII-PTRE, Moi University

DR. LYNN KISEMBE (*Session Chair ICPTRE2020*)
Coordinator, Social Safeguards, ACEII-PTRE, Moi University

DR. KEFA CHEPKWONY (*Session Chair ICPTRE2020*)
Manager, Incubation Centre, Moi University

DR. MINGWEI ZHAO (*Session Chair ICPTRE2020*)
Director, International Cooperation Office, Donghua University, (China)

PROF. PHILIPPE GERARDIN, (*Keynote Speaker ICPTRE2020*)
Professor of Natural Products Chemistry, Chemistry, Université de Lorraine (France)

DR. MARCUS PERREIRA PESSOA, (*Keynote Speaker ICPTRE2020*)
Asst. Professor of Product Design and Development, University of Twente, (Netherlands)

DR. ROGER PETRY. (*Keynote Speaker, Session Chair & Panellist ICPTRE2020*)
Professor of Philosophy, Co-coordinator, RCE Saskatchewan & Cluster Co-Chair, SDG 12, Luther College at the University of Regina, (Canada).

DR. BEATRICE MUGANDA (*Keynote Speaker ICPTRE2020*)
Director of Higher Education, Partnership for African Social and Governance Research (PASGR), Kenya

PAVEL ROBERT OIMEKE, EBS (*Keynote Speaker ICPTRE2020*)
Director General – EPRA

PROF. LIEVA VAN LANGEHHOVE, (*Keynote Speaker ICPTRE2020*)
Professor of Functional Properties of Materials, Ghent University, (Belgium)

ENG. HARNESS MUKHONGO (*Panel Moderator ICPTRE2020*)
AEPEA, Kenya

DR. KENETH CHELULE (*Panel Moderator ICPTRE2020*)
Researcher, KIRDI, Kenya

DR. ZINAIDA FADEEVA (*Panel Moderator ICPTRE2020*)
Visiting Professor, Nalanda University (India), Universiti Sains Malaysia (Malaysia)

PROF. SAMSON RWAHWIRE (*Panellist ICPTRE2020*)
Director Graduate Studies, Busitema University, (Uganda)

DR. SIMON GITHUKU (*Panellist ICPTRE2020*)
In-charge of Research, KAM

VESA KORHONEN (*Panellist ICPTRE2020*)
CEO, Nocart Ltd.

DOMINIC WANJIHIA (*Panellist ICPTRE2020*)
CEO, Biogas International

PROF. DAVID TUIGONG (*Panellist ICPTRE2020*)
CEO, KIRDI

DR. ALICE M. MWEETWA (*Panellist ICPTRE2020*)
Senior Lecturer, University of Zambia, (Zambia)

DETLEV LINDAU-BANK (*Panellist ICPTRE2020*)
Researcher for Education & Social Work and Chair of RCE Oldeburger Munsterland, University of Vechta (Germany)

PROF. DR. MARGIT STEIN (*Panellist ICPTRE2020*)
Professor of Education, University of Vechta (Germany)

PROF. KENNETH OCHOA (*Panellist ICPTRE2020*)
Director of Environmental Engineering, El Bosque University (Colombia)

SCIENTIFIC COMMITTEE

Prof. Ambrose Kiprop
Prof. Charles Lagat
Prof. Henry Kiriamiti
Prof. Samuel Rotich
Prof. Simeon Mining
Prof. Stanley Simiyu
Prof. Zachary Siagi
Prof. Augustine Makokha
Dr. Rose Ramkat
Dr. Fredrick Nyamwala
Dr. Charles Nzila
Dr. David Njuguna
Dr. Lynn Kisembe
Dr. Jacqueline Makatiani
Dr. Korir Kiptiemoi
Dr. Cleophas Achisa
Dr. Jackson Cherutoi
Dr. Stephen Talai
Dr. Sarah Chepkwony
Dr. Milton Arimi
Dr. Peter Chemweno
Dr. Eric Oyondi

ICT COMMITTEE

Dr. Joyce Komen
Ms. Suzzanne Kalume
Mr. George Bateta
Ms. Rachel Cheptumo
Mr. Sammy Goin
Mr. Victor Siele
Ms. Zipporah Boto
Mr. Julius Koech
Mr. Godfrey Rono
Ms. Caroline Jepkogei
Mr. Moses Kirong
Ms. Philister Yator
Mr. Gilbert Baigok

Advances in Phytochemistry, Textile and Renewable Energy Research for Industrial Growth – Nzila et al. (Eds)
© 2022 Copyright the Editor(s), ISBN: 978-1-032-11871-0
Open Access: www.taylorfrancis.com, CC BY-NC-ND 4.0 license

Acknowledgements

We are grateful to the entire management of Moi University for their steadfast support during the preparation of the book. We also appreciate all the various authors who contributed different chapters of the book.

On behalf of ACEII-PTRE we would also like to thank the following for their valuable remarks during the virtual international conference which was the basis of this book: Ambassador Simon Nabukwesi, (Principal Secretary, State Department for University Education and Research), Ms. Ruth Charo (Senior Education Specialist, Education Global Practice, World Bank – Kenya), Dr. Pam Fredman (President, International Association of Universities), Prof. Jianyong Yu (President, Donghua University) and Mr. Zhou Pingjian (Chinese Ambassador to Kenya).

Finally, we wish to thank the World Bank for the financial support to ACEII-PTRE through a credit facility to the government of Kenya.

Dr. Charles Nzila
Dr. Nyamwala Oluoch
Prof. Ambrose Kiprop
Dr. Rose Ramkat
Prof. Isaac S. Kosgey

Advances in Phytochemistry, Textile and Renewable Energy Research for Industrial Growth – Nzila et al. (Eds)
ISBN: 978-1-032-11871-0

Messages

MESSAGE FROM THE VICE-CHANCELLOR, MOI UNIVERSITY

On behalf of Moi University Management, Staff, Students and, indeed, on my own behalf, I warmly welcome you all to this maiden virtual ACE II Conference jointly hosted by Moi University-Kenya, and Donghua University-The People's Republic of China. I must state that, this is the first virtual International Conference that Moi University is hosting and, therefore, will go in the annals of history of our University. It also demonstrates our capability in terms of human resource on information communication technology and gives us more impetus to improve our infrastructure in information communication technology.

The theme of the Conference is "*Advancing Science, Technology and Innovation for Industrial Growth*," which strongly resonates with Moi University's Vision *to be a University of choice in nurturing innovation and talent in science, technology and development*. Initially, the Conference was to be held physically at Moi University, but due to the outbreak of Covid-19 pandemic, we resorted to hold it virtually. We thank the World Bank, through the government of the Republic of Kenya, which is the main funding entity of ACEII-PTRE for allowing us the opportunity to hold the Conference virtually. This is actually one opportunity for us to see ourselves as a 'global village'. During the Conference, we will have participants from Africa, Asia, Europe and Northern America, and we intend to have over hundred papers for presentations during this two-day Conference.

Brief of Moi University

Moi University was established in 1984 as the second public University in Kenya. It started with one faculty and, over the years, has expanded into fourteen Schools and four Directorates offering diverse academic programmes and involved in various research activities. We have our Main Campus located in Kesses, Uasin Gishu County, and five other Campuses, that is, three in Eldoret, and one in Nairobi and Mombasa respectively. Additionally, the University has three Institutes and two subsidiary companies – Rivatex East Africa Limited and Innovation Firm Limited. The University also prides itself as a host of three Centers of Excellence in: (i)Education Research in East and Southern Africa – CERMESA, (ii)Phytochemicals, Textile and Renewable Energy (PTRE) and (iii)African Studies.

About the ACEII-PTRE Centre

The Africa Centre of Excellence in Phytochemicals, Textile and Renewable Energy (ACEII-PTRE) based at Moi University, Eldoret, Kenya, was established in the year 2016. The objective of the ACE II Project is to strengthen selected Eastern and Southern Africa higher education institutions to deliver quality postgraduate education and build collaborative research capacity in the regional priority areas. The ACEs are expected to address specific development challenges and skill gaps facing the region through graduate training in Masters', Doctorate (Ph.D.) and short-term courses and applied research in the form of partnerships and collaborations with other institutions as well as the private sector.

Impact of the ACE II-PTRE Centre

The Centre, since its inception, has made great impact on resource mobilization to the University for infrastructural development such as lecture rooms, laboratories and equipment among others. It has also contributed in capacity building by offering scholarships for staff and students, and staff exchange programmes. Such contributions have enabled Moi University to enhance its global visibility in research, innovation and teaching. I take this opportunity to thank the World Bank for its noble idea to establish Centres of Excellence across the Africa, and the Republic of Kenya in particular through the Ministry of Education, and for facilitating Moi University to be a beneficiary of the same. The local community has also greatly benefitted from the ACE II Project through outreach services in short course training on self-sustainability through making detergents and soap, making of natural dyes, and training on biogas production for surrounding high schools, just to mention a few.

Coping and Containment Mechanisms of Covid-19

Moi University is well aware of the impact of the Covid-19 pandemic and swung into action to be part of the solution through inter-disciplinary research in combating the scourge. We are also involved in Covid-19 containment measures such as mass production of masks for public use and also medical suits through our Rivatex Textile Factory at subsidized costs. Additionally, plans are underway for the production of subsidized hand sanitizers through ACE II PTRE Project hosted in the School of Sciences and Aerospace Studies for local consumption and commercial purposes eventually.

Conclusion

I would like to conclude by once more thanking the World Bank for their continued support as our Sponsor, Donghua University – China for jointly hosting this Conference, and the Inter-University Council of East Africa, Regional Universities Forum for Capacity Building in Agriculture, International Association of Universities, and all our other partners for their effort in being in the forefront of ensuring that Moi University achieves its vision and mandate. I end by thanking the participants, presenters, moderators, rapporteurs, and the Organizing Committee led by the Centre Leader, for the great effort they have made to ensure the virtual Conference is a success. *'Asanteni Sana'.* I wish all the participants the very best during this Conference. Thank you and welcome to the Conference.

Prof. Isaac Sanga Kosgey,
Vice-Chancellor

MESSAGE FROM MS. RUTH CHARO, SENIOR EDUCATION SPECIALIST, WORLD BANK

We do congratulate Moi University for moving forward with this conference. I am aware there was a lot of dialogue on whether to proceed with this conference (or not). But we must adopt to the new norm, the new reality in the COVID 19 context and ensure that we are making progress in what we committed to implement at the beginning of the year. We most sincerely thank Moi University for taking up this challenge-to hold this conference virtually. We do believe the success of this virtual conference will inform African Centers of Excellence (ACEs) planning in Kenya and in the Region.

There is context to this international conference. As we are aware, the economic growth and foreign investment have boosted demand for greater technological skills and applied research in some of the priority economic sectors like Manufacturing, Agriculture, Energy, Heath and others in Kenya and in the region. Prior to Covid-19, this country, Kenya, and the region, were experiencing remarkable economic growth in the last decade or so. What does this mean for higher education? To sustain economic growth and transformation of the economy, means that high order skills and research are needed, particularly in Science, Technology, Engineering and Mathematics, which we commonly refer to as the STEM areas. There is a new global trend to consider STEAM instead of STEM- adding the 'Arts' to the STEM. Arts play an important role in areas related to STEM. For example, development of soft skills or the 21st century skills; entrepreneurship, interior design and other areas need to be meaningfully mainstreamed in the STEM curriculum.

The World Bank, working closely with the Government, is contributing to addressing critical skills gaps in technical, scientific and research skills. As you are aware, the support to the ACEs and other higher education operations, is mainly through credits, which means that the Government, and in this case the Government of Kenya, is committed and taking lead to ensure improvements in development of the most needed high order skills that could support priority economic sectors, and research.

The ACE approach, as you have been informed by the Vice Chancellor and Prof. Ambrose Kiprop, the ACE Centre leader at Moi University, aims to build regional specialization among the ACEs by concentrating on limited top-level faculty, generating knowledge, and developing skill to support industry and private sector demand for skills, and to generate research which can be of immediate use or uptake to the development challenges of the community. The ACE project emphasizes the need for effective linkages between universities and communities – the ACEs must serve the community within which they are situated. I am encouraged that the Vice Chancellor's remarks on the efforts by Moi university to partner with the community, including schools in areas such as biogas and the textiles industry.

The Vice Chancellor has already mentioned that Moi University is part of the network of Universities participating in the Africa Centers of Excellence (ACE II) regional project and has outlined what Moi University has achieved under the ACE II project. therefore, I will not be going into details of the ACE II regional project. Additional information on the overall ACE II project is available online.

Experiences from East Asia and Latin America show how profound, long-term and continued investment and reform efforts into training in high-quality skills and scientists for very concrete industries can lead to complete transformation of the targeted economic sectors and emergence of new competitive and higher value-added production. I am aware that Moi university is engaged in interventions related to value addition including in the textiles and biogas areas. Ideally, these are the kind of linkages that we need to demonstrate the relevance of higher education and training to the marketplace as well as in contributing to productivity of the Country, and job creation.

We cannot overemphasize the need to strengthen and consolidate the overall higher education eco-system. I am glad we are joined by our Chief guest today. We have been having dialogue on the need to review and align the overall governance of the higher education sub sector specifically in quality assurance, accreditation, student financing, research and innovation areas. Unless we have a solid higher education ecosystem, it becomes very challenging to build a foundation for excellence in higher education training, as well as collaborative research. We are optimistic moving forward, we will work more towards a cohesive higher education sub-sector, including in the management of research and innovation funds. We need to strive for excellence in higher education training and research. This vision will ensure University education is contributing to development of quality and relevant skills and cutting-edge research-and uptake of this research to inform innovative solutions aimed at addressing development challenges not only in Kenya but also in the region.

Once more, we are very happy that Moi University is hosting this conference. We cannot over emphasize the essence of networking and knowledge sharing. We have continued to encourage the ACEs to network and to pursue partnerships in the country, region and internationally, because this approach will enable the ACEs to synthesize training and research output and uptake of the same either at the country level or in the region. We have also been highlighting the importance of the ACEs working closely with relevant line ministries to influence policies, spending and investments. These linkages will enable the training and research outputs, and innovations from the Centers to inform and or contribute to core development agenda of the Country. Unless we can create these linkages effectively, it becomes very challenging to justify 'the value addition' of investing in the ACEs, and in the overall higher education subsector. Higher education may be considered an investment for the betterment of the individual pursuing 'it', but higher education should contribute to a higher order or higher-level development objective.

We acknowledge ongoing efforts by Moi University to promoting increased women participation in the postgraduate programme, specifically in the STEM fields. This is commendable. The ACE Centre at Moi University is supporting women through scholarships; improved conditions for accommodation; and mentorship. Through these efforts, over the past three years, the Centre at Moi University has been increasing the number of women being enrolled at postgraduate level, particularly at the PhD level. The university has proven with deliberate efforts, it is possible to attract young females to pursue post graduate training in the STEM fields -by creating a conducive environment and the right incentives for women. It is our hope that these women will be supported to develop their careers as faculty and contribute to higher education sub-sector and the ACE at Moi University.

The COVID-19 situation has created challenges, as well as opportunities. We will continue to work closely with Moi University, as well as other ACEs, to ensure learning continuity for the students during COVID-19. I do not think this situation will 'go away' soon. Kenya has in place relevant technologies and resources for the Centers to implement online based support for the students. We really do not have any excuses as to why the students at the Centre cannot be supported to for example present and 'defend' research proposal and thesis. Probably data collection could be deferred for students needing to be in the field for data collection, but when the COVID-19 situation allows, the students should be supported to get back on track. I am going to end here and thank the Ministry of Education and Moi University for this conference. We look forward to the next sessions, and to learn more about the specific themes for this conference in Science, Technology and Innovation. Thank you very much.

Ms. Ruth Charo,
Senior Education Specialist, Education Global Practice, World Bank

MESSAGE FROM AMBASSADOR SIMON NABUKWESI, PRINCIPAL SECRETARY, STATE DEPARTMENT FOR UNIVERSITY EDUCATION AND RESEARCH, MINISTRY OF EDUCATION

I am indeed delighted to participate in this Virtual International Conference. I thank the Vice Chancellor for the invite. This Conference is coming at a time when the world is going through health, economic and social turmoil due to the impact of Covid-19 pandemic which is affecting the entire world. Such challenges come with lessons to be learned. One such lesson being the reassertion that the world is more networked than ever before. Challenges facing one side of the world affect the entire world and need concerted effort of each and every one of us in the whole world to find possible lasting solutions. The Covid-19 pandemic has made us learn that no nation or individual has monopoly of knowledge, and the world is always a work-in-progress. It has also made us learn the important roles our health professionals play in our lives.

Fighting Covid-19 requires a collaborative scientific and interdisciplinary approach, ranging from efforts in developing a vaccine against it, curative and prevention measures, to social distancing and protective equipment. In these, social, biological and physical sciences are all applied. The Ministry of Education is at the forefront in developing capacity in Science, Technology and Innovation through provision of research funds which are competitively awarded to researchers. However, such funding is never enough and to mitigate this, the Ministry has provided an enabling environment for Universities to seek additional funding from the global community to bridge the funding gap and ensure they are relevant in so far as education and research are concerned.

I take this opportunity to thank the World Bank for the support they have given in the establishment of three (3) Africa Centers of Excellence (ACEs) in Kenya, i.e.

i. Africa Centre of Excellence in Sustainable Agriculture and Agribusiness Management (CESAAM) at Egerton University,
ii. Africa Centre of Excellence in Phytochemicals, Textiles and Renewable Energy (ACEII-PTRE) at Moi University and,
iii. Africa Centre of Excellence in Sustainable Use of Insects as Foods and Feeds (INSEFOODS) at Jaramogi Oginga Odinga University of Science and Technology.

It is from this support that Moi University is able to offer postgraduate teaching and research in Phytochemicals, Textile and Renewable Energy.

The theme of the Conference: **"Advancing Science, Technology and Innovation for Industrial Growth"** and the thematic areas of Phytochemistry, Progressive Textiles, Renewable Energy and Transformative Industrialization are relevant in realizing Sustainable Development Goals (SDGs), which is a universal call of action to end poverty, protect the planet and ensure people enjoy peace and prosperity by the year 2030. A better world requires an interdisciplinary approach by including social sciences, of which this Conference embraces, having incorporated Sino-Africa Culture Exchange (SACE), making it a three-in-one International Conference with high level academic discourse and cross-cutting disciplines. As I conclude, I wish to state that post Covid-19 will require us to build new bridges in scientific cooperation and re-double our efforts in Science, Technology and Innovation with determination and distinction. Universities are expected to lead in this aspect.

It is now my honour to officially declare the Conference open and wish all the participants effective presentations and productive discussions. Thank you.

Ambassador Simon Nabukwesi
Principal Secretary, State Department for University Education and Research, Ministry of Education.

MESSAGE FROM PRESIDENT INTERNATIONAL ASSOCIATION OF UNIVERSITIES

This is the start of a conference with an impressive program on an important topic. It is a great honour for me to represent International Association of Universities, (IAU). The Covid 19 pandemic is still ongoing and with an uncertain development. We are all facing the short term and trying to foresee the long-term consequences of the pandemic, far beyond the health issues. Covid 19 has brought awareness of the interconnection between the SDGs and that global perspectives on a sustainable future, reaching Agenda 2030, are crucial and necessary.

UN has clearly stated that higher education, through research and education, plays a key role in Agenda 2030 towards the realization of the SDGs. The IAU has actively promoted and advocated for Higher Education for Sustainable Development (HESD), since the early 90s. IAU was created under the auspices of UNESCO in 1950 and is a membership-based organization serving the global higher education community through: expertise & trends analysis, publications & portals, advisory services, peer-to-peer learning, events, global advocacy. It is an NGO with over 600 universities and university organizations as members all around the globe.

HESD is one of IAU strategic priorities and the overall objectives are to encourage peer to peer learning, monitor, trends, sharing expertise, fostering whole institutions approach and providing leadership training, capacity building and networking services (https://www.iau-aiu.net/HESD). One important role is to engage in policy discussions, conferences and in policy documents. For example, IAU is key partner in the UNESCO Global Action Program on Education for Sustainability (GAP on HESD) and with its member institutions and organizations active in the High-Level Political Forum 2019 and 2020.

With the aim to foster further support for universities in their role in societal transformation for a global sustainable development IAU in 2018 started a university network for HESD. For each of the SDG, one university, actively working with that specific goal and interconnected goals, was invited as leader institutions and asked to build a network around that goal. Today the network comprises over 80 universities around the world.

The leader institution for SDG12 is the University of Regina and Luther college and they have formed a network with six universities around the world, the host for this meeting, Moi university, and from Malaysia, Sri Lanka, Germany, Colombia and Peru. I am sure that this meeting will bring important knowledge to be shared within the IAU Cluster. Through the network there will be a global voice of the universities actions and needs to fulfill their role in reaching agenda 2030 and the SDGs. For IAU to act in promoting and advocating for HESD the global networking is crucial. Each leading institution and their satellite institutions has a role to interact with other institutions in research and education to extend the sharing and the strength of the voice of higher education to policy makers, governments, public and private funding agencies.

Covid 19 has increased the awareness of the inequalities in resources and capacities to perform research and higher education. Recent surveys including that performed by IAU (https://www.iau-aiu.net/IMG/pdf/iau_covid19_and_he_survey_report_final_may_2020.pdf) on the consequences of Covid 19 on HE pays attention to the risk of increasing inequality. This highlights the need of cohesion and cooperation between universities and between universities and the society built on global engagement and local relevance.

Knowledge creation and development through basic and applied research and innovation needs cooperation on common interests but must be built on respect and trust for each stakeholder's role, goals and legislations. For universities the fundamental principles for Higher Education needs to be respected, academic freedom in research and education, institutional autonomy, education built on science and/or proven experience. These are values that foster skills as critical thinking, analytical competence and creativity. Skills that together with disciplinary knowledge will be bring through students to society and empowering for taking action for a sustainable development.

IAU will continue to actively engage in promoting and advocate for the key role of universities in society for realizing the Agenda 2030. SDG12 "Ensure sustainable consumption and production patterns" safeguards and accommodates all SDGs and the outcome of this conference is of value for them all. I wish you a successful conference.

Dr. Pam Fredman, IAU President

MESSAGE FROM THE PRESIDENT, DONGHUA UNIVERSITY

The Leadership of Kenyan Government, The Leadership of Moi University, Distinguished scholars from Africa, China and the rest of the world, welcome to our joint conference.

On behalf of Donghua University, I'd like to express my sincere appreciation for the joint efforts made by both sides in organizing this great academic event. This event is both a new endeavour and an old tradition at the same time. It is new because this is the first online version of Sino-Africa International Symposium on Textiles and Apparel (SAISTA), and for the first time held alongside with another big event at Moi, namely, the Africa Centre for Excellence in Phytochemicals, Textile and Renewable Energy. It is old because China and Africa have always been a community with a shared future. China is the largest developing country while Africa is the continent with the largest concentration of developing countries. Our peoples have established solid friendship between us.

Africa has a rich textile history, and Donghua University (DHU) distinguishes itself by textiles, so it is the joint textile bond that turns into solid foundation of our cooperation. As a state-key university in China, DHU was selected as a member of "Sino-Africa 20+20 University Cooperation Project" in 2010. By virtue of its advantages in Textile Engineering, Material Science, Design, etc., DHU actively builds up the platform of China-Africa textile and apparel research as well as cultural exchange: in 2015, Confucius Institute at Moi University, the first and so far only Confucius Institute featuring textile engineering and fashion design was officially launched; in the same year, the first Sino-Africa International Symposium on Textile and Apparel (SAISTA) was successfully held at DHU; in 2017, the "Belt & Road" Advanced Seminar for Textile Industry and International Cooperation in Production Capacity opened at DHU; in 2018, university-enterprise jointly established the "Belt & Road" Textile Education and Training Center (Africa) in Ethiopia. Up to 2019, SAISTA has been successfully held 5 times with 5 different themes, which attracted educators and professionals from China, Kenya, Sudan, Zimbabwe, Tanzania, Uganda, and South Africa as well as from all over the world. It's my utmost pleasure to witness, out of our SAISTA and SACEF conferences, fruitful achievements of talent training for African textile and apparel industry as well as great advancement in China-Africa textile science and technology development.

Africa is a geographical and natural extension of the "Belt and Road" initiative as well as an essential participant. Forum on China-Africa Cooperation, a main platform for expanding and deepening bilateral cooperation, provides so many possibilities for Africa ranging from abundant resources and various pathways to huge market and space to diversified development prospects. Although this yar we all face a most challenging time during the outrage of Covid-19, globalization and It can be envisioned that through this Symposium, more and more scholars and entrepreneurs will join hands and explore the connotation and effectiveness of educational collaborations between China and Africa. Together, we will make a difference in textile and apparel education, research and industry liaison between China and Africa. Finally, I hope the joint conference this year a complete success and that everyone has an inspiring online experience.

Prof. Jianyong Yu
President, Donghua University

MESSAGE FROM THE ACEII-PTRE CENTER LEADER

On behalf of the Conference Organizing Team, I take the earliest opportunity to welcome our Chief Guest and all participants to the first International Conference organized virtually by The Africa Centre for Excellence in Phytochemicals, Textile and Renewable Energy (ACEII – PTRE), based at Moi University, Eldoret, Kenya, and Sino-Africa Symposium on Textiles and Apparel (SAISTA), Donghua University, China.

ACEII – PTRE's aim is to advance technology development and innovation in Phytochemicals, Textile and Renewable Energy through delivery of quality post-graduate training and collaborative research in the regional priority areas. Thus, ACEII – PTRE has a special focus on capacity-building. Towards this end, the Centre has equipped more than seven (7) laboratories at the School of Sciences and Aerospace Studies and School of Engineering that support research for both staff, students in the University and partners. As such, students under the Centre partial scholarship program have the great opportunity to present their research papers detailing the results realized in relation to their research during this virtual conference. The conference was supposed to be held through face to face at Moi University, Eldoret, but due to the outbreak of covid-19 pandemic, we resorted to hold it virtually and, on this note, we thank the University Management, World Bank and the Ministry of Education for allowing us the latitude to hold the conference virtually.

More than two hundred papers were submitted for the conference from all over the world. Papers selected for presentations reveals the amazing diversity aligned to several thematic areas that include Phytochemistry, Progressive Textiles, Renewable Energy, and Transformative Industrialization. The conference also captures Cross – cutting topics and research in Sustainable Technology and Innovations and Science, Technology, Engineering, Arts and Mathematics (STEAM) with relevance to the Theme of the Conference which is, "***Advancing Science, Technology and Innovation for Industrial Growth***."

The conference highlights the remarkable contribution which ACEII – PTRE has made under the support of World Bank and provides a valuable opportunity for research scientists, industry specialists and decision-makers to share experiences. Further, we are grateful to the Regional Facilitating Unit: Inter-University Council for East Africa (IUCEA) for continuous facilitation. Besides the World Bank and IUCEA, we also acknowledge the participation of our important partners and supporters; Kenya Industrial Research & Development Institute (KIRDI), RIVATEX East Africa Limited (REAL), University of Lorraine (France), University of Gezira (Sudan), Regional Universities Forum for Capacity Building in Agriculture (RUFORUM), Busitema University (Uganda), Flexi Biogas International, Nocart, Association of Energy Professionals of East Africa (AEPEA), National University of Science and Technology (Zimbabwe), University of Gent (Belgium), University of Linkoping

(Sweden), Kenya Bureau of Standards (KEBS), Elsevier, Ruparelia Consultants Ltd (RCL), Kenya Association of Manufacturers (KAM), Seeding Labs (USA), and SDG 12 cluster working group.

We are grateful to the many experts who have come to share their knowledge in the conference and wish all participants fruitful deliberations in the next two days. Special thanks to the World Bank, partners and supporters, University Council, University Management and the entire Moi University fraternity for their continued support to the Centre. Last but not least we express gratitude to the Conference Organizing Team and Reviewers for their tireless effort in ensuring all the arrangements for the conference are done well and promptly despite the challenges occasioned by Covid-19 pandemic and the same goes to presenters for carrying out research at this time of difficulties. We look forward to more successful deliberations. Thank you.

Prof. Ambrose Kiprop
Centre Leader, Africa Centre for Excellence in Phytochemicals, Textile and Renewable Energy (ACEII – PTRE)

MESSAGE FROM THE CHAIRMAN, ICPTRE2020

Introduction

Science, technology and innovation (ST&I) are the seedbed for development. In addition, advancing competitiveness in ST&I is a key prerequisite to sustained industrial growth and increased standards of living. Research around ST&I must therefore continuously realign with the global dynamics and demanding needs for industrial growth. Industries around the world continue to face different afflictions elicited by several factors including social conflicts, economic, health and energy crises leading to suboptimal operations. Since the beginning of the year, the world has been thrust deep into uncharted territory and very challenging circumstances occasioned by the emergence of COVID-19. Notwithstanding these very difficult times, the ACE II-PTRE in partnership with SAISTA provides selected participants from around the world with an opportunity to showcase their research results in the inaugural icptre2020 under the theme of *Advancing Science, Technology and Innovation for Industrial Growth*.

Advancing Science, Technology and Innovation

The world has experienced a variety of disruptive shocks over time. While the disruption to people and livelihoods in many developing countries is certainly not a new phenomenon, the developments in building rapid resilience and diverse capabilities in equally disruptive technologies are worth being celebrated. Therefore, ST&I have a critical role to play in all global afflictions.

The experience from developed partners and some of the most successful countries globally show that advancing ST&I through research and well-integrated national development strategies can help raise productivity, improve industrial competitiveness, support faster growth, create jobs and promote resilient communities. Therefore, advancing industrial growth and fortifying our position in the research arena is one of the central goals of our time. Notwithstanding the prevailing circumstances, our shared resolutions and strategic partnerships are well positioned to deliver our mutual aspirations. Today, through the ACE II-PTRE and SAISTA partnership, we are moving forward very much aware more than ever before that any meaningful and sustainable progress in industrial growth requires humanity coming together as a global community of researchers.

The conference structure

The icptre2020 is structured to offer vibrant presentations and deliberations on key ST&I research. The conference thematic areas include Phytochemistry, Progressive textiles, Renewable Energy, Transformative Industrialization, Sustainable Technology and Innovations, in addition to Science, Technology, Engineering, Arts and Mathematics presentations that contribute to the aforementioned thematic areas. The conference also brings on board the responsible consumption and production (SDG 12) cluster-working group of the International Association of Universities (IAU) Higher Education for Sustainable Development (HESD) Cluster. The Cluster promotes the role that Higher Education Institutions globally have to fulfil in order to achieve the Sustainable Development Goals (SDGs) and Agenda 2030. *Universities are to address the SDGs, which themselves impact on and transform universities*. The Cluster encourages a holistic approach to the SDGs, focusing specifically on the whole institution approach.

In the course of the next three days, over 200 participants from around the world shall actively engage on research findings along the theme of the conference. The conference presentations will tackle diverse topics including issues pertaining to sustainable exploitation of bio-based resources, sustainable technologies, transformative nanomaterials, accelerating the transition to a renewables-based energy system among other topics. Besides, 9 keynote speakers will share their insights on key thematic topical areas. In addition, 13 selected panellists representing different regions of the globe and sectors of the economy shall engage with the audience and deliberate on a number of topical areas including contract researching, research and development, ethical issues, developing and implementing collaborative research, resilience and collaborations in pandemic times, intellectual property issues in collaborative research as well as sustainable production and consumption.

Conclusion

In summary, the conference presentations avail a unique opportunity to share knowledge, promote the actualization of several SDGs while creating new opportunities as well as contributing to the enhancement of livelihoods and global resilience to adverse disruptions. The world is currently united in the commitment to realize this opportunity. On behalf of the entire conference organizing team, it is my humble duty to wish all the participants the most rewarding, thought provoking and vibrant deliberations.

Dr. Charles Nzila (PhD)
Chairman – ICPTRE2020 Organizing Committee.

Section 1: Advancing science, technology, engineering, arts & mathematics for responsible consumption and production

Advances in Phytochemistry, Textile and Renewable Energy Research for Industrial Growth – Nzila et al. (Eds)

A survey of engineers' and engineering students' perceptions on ethical behaviour

Emmanuel C. Kipkorir*, Sum Kipyego & Rioba Oteki
Department of Civil and Structural Engineering, School of Engineering, Moi University, Kenya

ABSTRACT: The engineering industry is known for its low ethical performance. Professional ethics instruction in Kenyan engineering faculties is commonly conducted by examining case studies in light of the code of ethics and conduct for engineers. Although the tenets of a code of ethics may leave a lasting impression, students generally gain their professional identity from relatives, colleagues, and practicing engineers. Their engineering professional ethics tend to be mostly an extension of their personal ethics. Instruction on ethics during training generally serves only to reinforce students' inclination to act ethically and encourages them to act on these beliefs. This study based on survey on engineering ethics adopted moral awareness which is one of the Rest model's four processes. The survey was conducted (n = 164) to examine the personal ethical perceptions of engineering students (n = 120, consisting of n = 79 for 1st year and n = 41 for 5th year) and practicing engineers (n = 44). The survey consisted of 16 acts that challenged respondents by examining their personal ethical beliefs in light of the professional ethics requirements of the Engineers Board of Kenya (EBK) code. The survey measured how respondents perceive their own ethical beliefs and how they perceive the ethical beliefs and actions of their peers. After familiarization with the EBK code, respondents were also invited to comment regarding their beliefs regarding adherence to the code. Results indicate that, although generally the engineers and engineering students sampled agreed that the acts listed were unethical, several items raised concern. In particular, the item concerning "continuous professional development" was rated as one of the least unethical behaviours. This result points strongly to the need to further reinforce the need for relevant lifelong learning for engineers both during training and practice. Also, results indicated that there is evidence of self-perception-versus-other disparity. For six unethical acts for students and four acts for engineers, in the surveyed list, the means of data for self-perception and colleague perceptions were statistically significantly different at the alpha level of 0.05. When the act was perceived as more unethical, both engineers and students tended to rate themselves as more ethical compared to their peers. Action research through mentorship is recommended as part of the solution to addressing ethical issues in engineering practice.

Keywords: identity, ethics, engineering practice, perceptions, student, professional, development

1 INTRODUCTION

The engineering community has experienced numerous scandals involving unethical and illegal engineering practices; many of them are committed by large and well-known engineering companies and government agencies (Bairaktarova & Woodcock 2017). According to Bowen, Akintoye, Pearl, and Edwards (2007), ethical codes and codes of conduct, bribery and corruption, and favouritism have a negative implication on engineering processes in any organization and may lead to decreased performance and service delivery. This occurs at a time when professional engineers have a personal and professional obligation to society to act ethically (Passino 1998). To address the challenge, the Engineers Board of Kenya (EBK) has developed a code of conduct and ethics which forms the basis and framework for responsible professional practice in Kenya, as it prescribes the standards of conduct to be observed by all engineers. The code is based on broad tenets of truth, honesty, trustworthiness, respect for human life and welfare, fairness, openness, competence, accountability, engineering excellence, protection of the environment, and sustainable development (EBK 2016).

Development of ethical judgment skills in future engineers is a key competency for engineering schools as engineering ethics is part of the engineering thinking, identity, and professional practice of engineers (Harris, Davis, Pritchard, & Rabins, 1996). A study by Loui (2005) on ethical and moral development revealed that the greatest benefit of professional engineering ethics education is to reinforce students' inclination to act ethically. Therefore, instruction on moral reasoning frameworks and professional codes of conduct to engineering students encourages them

*Corresponding author

DOI 10.1201/9781003221968-1

to act on the personal ethical and moral convictions already held (Stappenbelt 2012). Naturally then, the question arises regarding the ethical inclination of engineering freshmen at the onset of their engineering education and what can be done to promote and encourage further personal ethical development as they progress with their studies, and later when they start practicing in the industry after graduating. The present study aims to assess the personal ethical perceptions of engineering students and practicing engineers by examining their personal ethical beliefs in light of the professional ethics requirements of the EBK code.

2 LITERATURE REVIEW

Ethics is the philosophical discipline of studying what contributes to good and bad conduct, including related actions and values (Barry 1979). It is concerned with the nature of specific decisions made and the goodness or badness of those decisions in terms of the consequences of those decisions (Chonko 1995). Thompson (2005) defines ethics as the study of right and wrong; of the moral choices people make and how they seek to justify them. In the engineering industry, engineers are required to keep up with a fast-paced, constantly changing environment, which makes it even more important for engineers to be taught ethics and professionalism (Li & Fu 2012). Harris, Pritchard, Rabins, James, and Englehardt (2013) mention that ethics education in engineering programmes is important in preparing the engineering undergraduate students – the future engineers – to carry out their duties professionally with a sense of responsibility towards society and the biosphere. Consequently, engineering ethics education has been made a compulsory subject as part of the conditions for accreditation in many countries. Despite its importance, ethics is not much investigated in engineering education (Sethy 2017).

Engineering ethics is an important topic in the engineering education curriculum therefore 'professional ethics' is offered as a compulsory course to undergraduate engineering students in many countries including India and Kenya. In teaching engineering ethics courses, various pedagogical approaches have recorded a positive impact on students' attitudes towards ethics, opening up a new dimension in ethics education which highlights the importance of teaching strategies in developing the attitude towards engineering ethical issues (Balakrishnan, Azman, Indartono 2020). Therefore, it has been recommended that for engineering institutions to "strengthen their students' ethical development, they should consider tracking students' exposure to these issues, identifying where and how this learning takes place" (Colby & Sullivan 2008).

Engineering solutions through design are humanity's way of facing the continuous stream of various global challenges (Barakat 2015). Design shares a broadly common understanding of design ethics and the main difference is in the scope, complexity, and the human interface. However, certain phases of the design process appear to prompt consideration of specific principles; students' interactions with users and project partners stimulate the most reflection on their ethical decision-making (Humphries-Smith, Blount, & Powell 2014).

The present study adopts Rest's model that consists of four processes, namely moral awareness, moral judgment making ability, moral intention, and moral character (Rest 1986). In this study, the moral awareness process is very important as it triggers one's thought and cognition before one makes judgment and the act is considered. According to the model, in the awareness process, an individual generally has the ability to recognize that there is a moral issue in a situation, which serves as a kind of an activating mechanism that initiates the ethical decision-making process (Sparks & Merenski 2000). Moral awareness involves an individual undertaking role-taking, but the person must realize that violating some moral norm, or allowing unethical situations to occur, can "affect the needs, interest, welfare, and expectation of others." However, Rest (1986) asserts that not everyone can interpret situations or be sensitive to unethical situations. Thus, a disparity exists in how sensitive an individual is to a particular moral situation.

Codes of ethics are widely understood to act as a mechanism for facilitating and ensuring ethical behaviours within organizations (Yallop 2012). Codes of ethics may be variously described as codes of conduct, codes of practice, ethical codes, corporate ethical codes, ethical guidelines, business conduct, codes of professional behaviour, operating principles, and so on (Fisher 2001; Marnburg 2000). Internationally, several studies stress the importance of codes of ethics as a necessary tool for creating and establishing an ethical environment within organizations (Ferrell & Skinner 1988; Hunt, Chonko, & Wilcox 1984; Oliver, Kearins, & McGhee 2005; Schlegelmilch & Houston 1989; Segal & Giacobbe 2007; Ziegenfuss & Martinson 2002). Therefore, a code of ethics develops standards by which a leader can judge the effects that different behaviours have on one another (Hickman 1998). In summary, ethics comes down to a choice to influence self-perception and others into doing the right thing.

In the present study, a survey was used to measure how three sets of participants (first-year and final (fifth) year engineering students, and practicing engineers) perceive their own ethical beliefs and how they perceive the ethical beliefs and actions of their peers. And after familiarization with the EBK code, respondents were also invited to comment regarding their beliefs regarding adherence to the code.

3 METHODS

This study used a questionnaire to conduct a survey on the perception of personal ethical behaviour of engineers. A pilot survey involving 30 randomly sampled respondents was conducted to determine the effectiveness of the questionnaire in meeting the objectives of the study, and whether the questions selected represent the personal ethical perception under study, and to ascertain whether the target respondents could understand and interpret the questions. The respondents were requested to be honest when filling in the survey and were informed that they had the right to rescind their participation in the survey at any time. They were assured of the confidentiality of their answers, and that their responses would be used for research purposes only. The obtained pilot survey results indicated that the questionnaire could assure validity, trustworthiness, and reliability of the results.

After the pilot survey was completed, the actual survey was conducted (n = 164) examining the personal ethical perception of the first-year and fifth-year Moi University, School of Engineering students, and practicing engineers. The students were all enrolled in respective courses during a semester where professional ethics was taught. The first sample response gathered was 79 out of a population of 260 first-year students following the Introduction to Engineering Profession & Safety – a common course offered in all the six engineering programs during the first semester of the 2019/2020 academic year. The second sample response was gathered of 41 out of a population of 45 fifth-year students following the Law, Ethics & Professional Practice course during the second and the final semester in the 2019/2020 academic year in the Civil and Structural Engineering program. In the fifth-year course case studies one pedagogical approach is used in delivery. The third sample response of 44 practicing engineers who attended a seminar on continuous professional development (CPD) was sampled from the Institution of Engineers of Kenya (IEK), western branch data base. The sample size of 164 respondents was adopted based on an estimated large population of 5,000 engineering students and practicing engineers in the IEK western branch, to allow the study to determine the personal ethical perception with a confidence interval of ±7.5%.

The questionnaire used was developed in Google Form (Kuczenski 2013) and consisted of a list of sixteen acts that were developed by (Stappenbelt 2012) by adopting a set of twelve from the study by (O'Clock & Okleshen 1993). Whereby respondents examined their personal ethical beliefs in light of the professional ethics requirements of the EBK code (EBK 2016). The sixteen acts were as follows:

1. Accepting gifts or favours in exchange for preferential treatment
2. Undertaking work in an area in which you are not competent
3. Passing blame for errors to an innocent colleague
4. Not supporting a colleague who is trying to do the right thing
5. Giving gifts or favours in exchange for preferential treatment
6. Claiming credit for someone else's work
7. Not reporting others' violation of organization policies
8. Divulging confidential information
9. Withholding relevant information from a colleague or client
10. Calling in sick to take a day off
11. Pilfering organization material and supplies
12. Doing personal business on organization time
13. Not keeping up to date with the latest developments in your area
14. Concealing one's errors
15. Taking extra personal time (lunch hour, breaks, early departure)
16. Using organization services for personal use

The survey measured how respondents perceive their own ethical beliefs and how they perceive the ethical beliefs and actions of their peers. In the survey, participants were to rank the sixteen unethical acts on a five-point Likert scale with (1) being very unethical and (5) not being unethical at all. The ranking, given in the list below, was both in terms of their personal beliefs and their perceptions towards the actions of their peers.

1. Very unethical
2. Basically unethical
3. Somewhat unethical
4. Not particularly unethical
5. Not at all unethical

After familiarization with the EBK code, respondents were also invited to comment regarding their beliefs in adherence to the code. Also, data relating to personal details were collected in the survey: age, gender, religion, and professional class.

The survey was presented to the three sets of respondents after a presentation and discussion of ethics in the classroom/seminar. The intentions of the study was first introduced to the students, they were then asked to respond anonymously, and were assured that the responses would only be taken en masse and could not affect their grades. The practicing engineers were briefed on the study after a seminar on CPD organized by IEK western branch held on 29th February 2020 in Eldoret, Kenya. The same survey that was set to receive only one response per respondent was given online via Google Form (Kuczenski 2013) to the three sets of samples separately.

The respective student's class representative circulated the survey questionnaire through their respective class social media and e-mail platforms and the students had a maximum of two days after class to fill and submit the questioner online. The survey questionnaire was circulated to the practicing engineers using their respective e-mails captured during registration for attendance of the seminar. The procedure used ensured that all the respondents were surveyed anonymously with only time stamp information collected. Although the questionnaire consisted of many questions, every respondent participated willingly as the subject seemed to be of wide interest. However, a general limitation attributed to survey method used in this study is oversimplification of social reality, by an arbitrary design of the questionnaire and adopting multiple-choice questions with pre-conceived categories with an overly simple view of reality.

Data were summarized using means and standard deviation for continuous data, while frequency and percentages were used for categorical data. The t-test was used to compare ratings between groups, while ANOVA was used to compare ratings across the three groups.

4 RESULTS AND DISCUSSIONS

4.1 *Basic data*

4.1.1 *Gender of participants*

The study participants were 79.9% (131) males and 20.1% (33) females (Table 1). Though the observations made from the responses indicate that engineering is still a male-dominated field, the proportion of females in Year 5 (26.8%) that responded is more than in Year 1 (19.0%). This may indicate a moderate increase (7.8%) in persistence and active participation of female students in activities in undergraduate engineering programs in recent years. However, more needs to be done to address the gender imbalance in engineering practice since the ratio is still lower than the recommended one third for either gender.

Table 1. Gender of participants per category.

Participant category	Male (N)	Male (%)	Female (N)	Female (%)
First-year students	64	81.0	15	19.0
Fifth-year students	30	73.2	11	26.8
Practicing engineers	37	84.1	7	15.9
Total	131	79.9	33	20.1

4.1.2 *Age of participants*

Table 2 shows the distribution of respondents by age for the three sets of participants.

Table 2. Age distribution of participants per category.

Age of participants (years)	First-year students (%)	Fifth-year students (%)	Practicing engineers (%)
15–19	67.1		
20–24	31.6	70.8	
25–29	1.3	29.2	2.3
30–34			22.7
35–39			27.3
40–44			15.9
45–49			4.5
50–54			11.4
55–59			13.6
60–64			0.0
>=65			2.3

Two thirds (67.1%) of first-year students were older teenagers of age group 15–19 years, while the majority (70.8%) of fifth-year students were emerging adults of age group 20–24 years. On the other hand, half (50%) of the practicing engineers were adults aged 30–39 years.

4.1.3 *Recent ethical perception of Kenyans*

The data for the survey question: *in the last 6–10 years in Kenya, have people in general become*; less ethical, remain the same, or become more ethical, to the three sets of 164 participants is presented in Table 3. The IEK membership classes of practicing engineering participants consisted of Graduate Engineers (56.8%) and Corporate Members (43.2%).

Table 3. Engineers recent ethical perception of people in Kenya.

Question: *in the last 6–10 years in Kenya have people in general*	First-year students		Fifth-year students		Practicing engineers		Total
	N	(%)	N	(%)	N	(%)	(%)
Become less ethical	67	84.5	32	79.5	32	72.7	79.9
Remain the same	1	1.9	5	11.4	7	15.9	7.9
Become more ethical	11	13.6	4	9.1	5	11.4	12.2

Results indicate overall that the perception of the majority (79.9%) was that people have become less ethical, 7.9% indicated that people have remained the same, and 12.2% indicated that people have become more ethical, however, 98% of the responses indicate that they are religious. This response is interesting given the religious inclination, because particular groups of people were not specified, just people in general, and it is not clear what might be driving this response. One possible reason could be an increase in ethics perception due to the prevalence of similar issues in recent Kenyan media coverage or increased awareness due to ethics instruction after the lecture/presentation before this study was carried out.

4.2 *Ethical awareness and conduct of participants*

The results of the survey for the three sets of participants are presented in Figures 1–3 for the three sets of participants self-perception and their perception of colleagues' beliefs regarding unethical behaviour. The mean ratings in the three Figures 1–3 indicate that the participants generally understood all the sixteen acts or behaviours listed in the survey to be unethical to

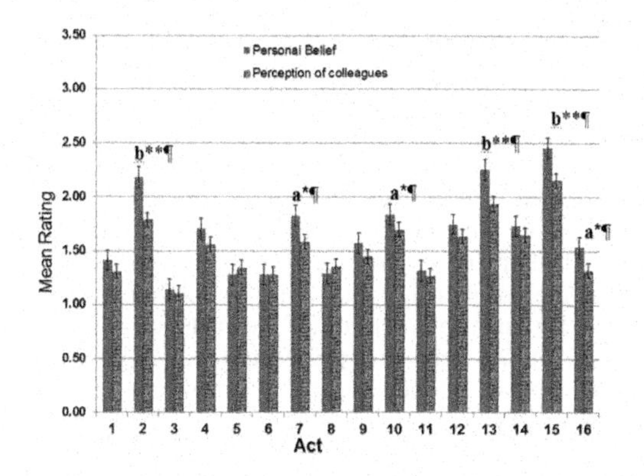

Figure 1. Mean rating of first-year engineering students' self-perception and their perception of colleagues' beliefs regarding unethical behaviour (statistically significant difference in means between self-perception and colleague ethical perception rating a* at $\alpha = 0.05$ level; b** at $\alpha = 0.005$ level).

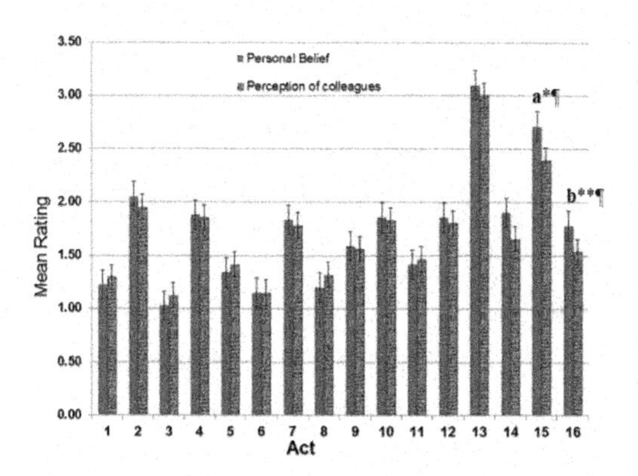

Figure 2. Mean rating of fifth-year engineering students' self-perception and their perception of colleagues' beliefs regarding unethical behaviour (statistically significant difference in means between self-perception and colleague ethical perception rating a* at $\alpha = 0.05$ level; b** at $\alpha = 0.005$ level).

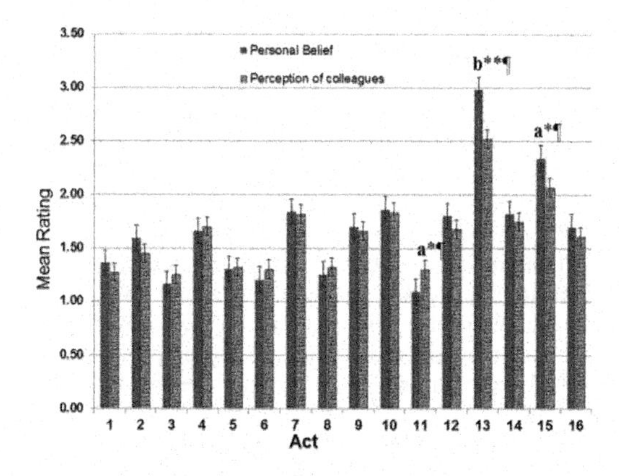

Figure 3. Mean rating of practicing engineers' self-perception and their perception of colleagues' beliefs regarding unethical behaviour (statistically significant difference in means between self-perception and colleague ethical perception rating a* at $\alpha = 0.05$ level; b** at $\alpha = 0.005$ level).

some degree, ranging from very unethical, basically unethical, or somewhat unethical. The most unethical acts rated by the three sets of participants are six acts: 1, 3, 5, 6, 8, 11, that they rated as very unethical. This was closely followed by nine acts: 2, 4, 7, 9, 10, 12, 14, 15, 16, that they rated as basically unethical. The major difference in the results for the three sets of participants was in act 13: *not keeping up to date with the latest developments in your area*, where fifth-year students and practicing engineers both rated it least unethical (somewhat unethical) while the first-year students rated the act basically unethical.

The rating of act 13: *not keeping up to date with the latest developments in your area* as the least unethical is somewhat disturbing. The rating of act 13 by the final year students and practicing engineers as the least unethical behaviour is similar to the findings of Stappenbelt (2012) and this points strongly to the need to reinforce the relevant lifelong learning mainly in the final year of study before graduating from university and continuously during engineering practice in academia or the industry based on CPD policy (EBK 2017). Demonstration of the dynamic nature of engineering knowledge through periodic review in the teaching curriculum will also alert the engineering students on the need to stay updated while in the industry. This is because it is generally accepted that a person's ability to maintain high levels of professional competence is achieved by continually upgrading his/her skills and knowledge (EBK 2017).

From the comparison of data (Figures 1–3), for each of the three sets of participants self-perception and their perception of colleagues' beliefs regarding unethical behaviour, it is inferred that there is evidence of self-perception -versus-colleague disparity. For three unethical acts (13, 15, and 16) in the survey list, the null hypothesis that the means of data for self-perception and colleagues perceptions are statistically significantly different is accepted at the levels ($\alpha = 0.005$ and $\alpha = 0.05$) indicated in Figures 1–3. Results indicate that when the act was perceived as more unethical, participants, irrespective of their level, tended to rate themselves consistently and significantly as more ethical compared with their peers. This can be attributed to the fact that people generally view themselves as morally superior to others (Sezer, Gino, & Bazerman, 2015).

We also compared the total scores from the 16 acts in the questionnaire by the three sets of groups as well as by personal beliefs versus perceptions of peers. This was done using a minimum total score of 16 for very unethical and a maximum of 80 for not at all unethical. We observed that the perception of own personal belief was not statistically significantly different from the perception of their peers ($\alpha > 0.05$). However, when we compared across the three groups there was a statistically significant difference in the scores for the perception of peers with scores higher among the fifth-year students compared to the first-year students and practicing engineers ($\alpha < 0.05$). The results are shown in Table 4.

Table 4. Comparison of total scores for the three sets of groups.

	First-year	Fifth-year	Practicing engineers	Total	P-value*
Personal beliefs	26.5 (7.9)	29.5 (6.8)	26.7 (5.8)	27.3 (7.2)	0.081
Perception of peers	24.4 (8.5)	28.7 (7.5)	25.9 (7.6)	25.9 (8.2)	0.023
P-value**	0.101	0.615	0.583		

*ANOVA by comparing by the three groups.
**t-test by comparing personal versus peer perception.

These results suggest that when students join an engineering program, they are very concerned with unethical issues among their peers compared to their final or fifth-year students. Overall there seems to be a consensus that unethical acts affect us in our day-to-day engineering practices and a lot needs to be done to improve the ethical code of conduct as stipulated in the EBK code of conduct.

The survey results for engineering participants' perception regarding adherence to the EBK code of conduct are presented in Table 5. The breakdown of responses to the question *Do you believe you always act in accordance with the tenets of the EBK code of conduct?* indicated that the majority of respondents (56% fifth-year students, and 79.5% practicing engineers) indicated Yes as a response. This is consistent with the personal ethical beliefs reported by the three sets of participants in this study.

Table 5. Engineers perception regarding adherence to EBK code of conduct.

Question/ Responses	Fifth-year students N	Fifth-year students (%)	Practicing engineers N	Practicing engineers (%)
Question: Do you believe you always act in accordance with the tenets of the EBK code of conduct?				
Yes	23	56.1	35	79.5
No	5	12.2	2	4.5
Unsure	13	31.7	7	15.9
Question: Do you believe that most practicing engineers always abide by the EBK code of conduct?				
Yes	10	24.3	10	22.7
No	22	53.7	23	52.3
Unsure	9	22.0	11	25.0
Question: Do you believe that professional engineers can realistically be expected to abide at all times by the EBK code of conduct?				
Yes	30	73.1	38	86.4
No	9	22.0	3	6.8
Unsure	2	4.9	3	6.8

That more than half the number of fifth-year students (53.7%) and practicing engineers (52.3%) do not believe practicing engineers always act ethically is very worrying (Table 5). This result is similar to the earlier finding of the perception of the majority (79.9%) of participants in this study that people have become less ethical in the recent past in Kenya. On a positive note, the majority of the fifth-year students (73.1%) and practicing engineers (86.4%) surveyed stated that they believe that professional engineers can realistically be expected to abide by the EBK code at all times. The awareness that has been created so far regarding abiding by the code of conduct needs to be stepped up, starting from self-realization that the engineering profession will be better positioned in society and industry with ethical practices. Part of the solution to addressing these worrying results is providing mentorship to engineers based on action research, which is a values-based approach to researching one's professional work, contributing to the ongoing development of the engineer and, potentially, the engineering industry. Through action research in engineering practice, it enables one to look at what they are doing from a critical point of view and reflect on how they do their work, to foster engineering professional growth.

5 CONCLUSIONS AND RECOMMENDATIONS

The present study through a survey measured how three sets of respondents – first-year and fifth-year engineering students and practicing engineers – perceive their own ethical beliefs and how they perceive the ethical beliefs and actions of their peers. Although generally, the majority of participants sampled in this study agreed that people have become less ethical in the recent past in Kenya and that the acts listed in the study were unethical, the rating of the act *not keeping up to date with the latest developments in your area* as the least unethical raised major concern. It is concluded that there is a need to reinforce the relevant lifelong learning by incorporating action research using mentorship programs during engineering education in the university and continuously during engineering practice both in academia and in industry. Adopting action research in engineering practice will enable one to look at what they are doing from a critical point of view and reflect on how they do their work, to foster professional growth both in academia and engineering industry. The finding that half of both fifth-year students and practicing engineers do not believe practicing engineers always act ethically is of great concern. This suggests that action is urgently required in engineering ethics education and in shaping engineering students' professional identities and enforcement of the code of ethics by EBK. Results from the present study also indicated that when an act was perceived as more unethical, participants, irrespective of their level, tended to rate themselves as more ethical compared with their peers. Specifically, for three unethical acts in the survey list, namely, *not keeping up to date with the latest developments in engineering, taking extra personal time*, and *using organization services for personal use*, a notable statistical significant difference was observed between the ethical perception of the individual and their perception of their colleagues' beliefs. It is concluded that participants in this study rated themselves consistently and significantly more ethical than their peers. The findings from this study will be useful to the following institutions: EBK, IEK, Schools of Engineering, among others. The study recommends the need for engineering students and practicing engineers to be people of integrity to withstand the force to do wrong. Further, there is a need to emphasize ethical lessons in engineering education and cultivate the change in attitude in practicing engineers. Areas for further research identified are promoting professional identity and persistence in engineering.

ACKNOWLEDGMENTS

Many thanks go to the engineering students and practicing engineers involved in this study. The authors are also grateful for the support received from the Associate Dean, School of Engineering, Moi University.

REFERENCES

Bairaktarova, D., & Woodcock, A. (2017). Engineering Student's Ethical Awareness and Behaviour: A New Motivational Model. *Sci Eng Ethics,* 23:1129–1157.

Balakrishnan, B., Azman, M. N.A. & Indartono, S. (2020). Attitude towards Engineering Ethical Issues: A Comparative Study between Malaysian and Indonesian Engineering Undergraduates, *International Journal of Higher Education,* 9(2): 63–69.

Barakat N. (2015). Engineering ethics and professionalism education for a global practice, QScience Proceedings (Engineering Leaders Conference) http://dx.doi.org/10.5339/qproc.2015.elc2014.5.

Barry, V.E., (1979), *Moral issues in business*, Belmont, CA: Wadsworth Publishing Company.

Bowen, P., Akintoye, A., Pearl, R., & Edwards, P. J. (2007). Ethical behaviour in the South African construction industry, *Journal Construction Management and Economics*, 25(6): 631–648.

Chonko, L.B., (1995), *Ethical decision making in marketing*, Newbury Park CA: Sage Publications.

Colby, A., & Sullivan, W. M. (2008). Ethics Teaching in Undergraduate Engineering Education. *Journal of Engineering Education*, 97(3):327–338.

EBK (2016). *Code of Ethics and Conduct for Engineers*, Engineers Board of Kenya, Nairobi. Retrieved from https://ebk.or.ke/download/code-of-ethics-for-engineers/

EBK (2017). *Continuous Professional Development (CPD)*, Guidelines for Professional Engineers, Engineers Board of Kenya, Nairobi. Retrieved from https://ebk.or.ke/download/engineers-continuous-professional-development-cpd-guidelines-for-professional-engineers/

Ferrell, O.C., Skinner, S.J., (1988), Ethical behaviour and bureaucratic structure in marketing research organizations, *Journal of Marketing Research*, 25(1):103–109.

Fisher, C., (2001), Managers' perceptions of ethical codes: dialectics and dynamics, *Business Ethics: A European Review,* 10(2):145–156.

Harris, C. E. Jr., Pritchard, M. S., Rabins, M. J., James, R., & Englehardt, E. (2013). Engineering ethics: Concepts and cases. Cengage Learning.

Harris, C. E. Jr., Davis, M., Pritchard, M. S., & Rabins, M. J. (1996). Engineering ethics: What? Why? How? and When? *Journal of Engineering Education*, 85(2): 93–96.

Hickman, G. R. (1998). *Transactional and transforming leadership leading organizations perspectives for a new era* (First ed.). Thousand Oaks: Sage.

Humphries-Smth, T., Blount, G, & Powell, J. (2014). Ethics – Research, Engineering, Design…They are all the same aren't they? *International Conference on Engineering and Product Design Education,* University of Twente, The Netherlands.

Hunt, S.D., Chonko, L.B., & Wilcox, J.B., (1984), Ethical problems of marketing researchers, *Journal of Marketing Research,* 21 (3): 309–324.

Kuczenski, J.A. (2013). Student Ethics in Engineering: A Comparison of Ethics Survey Results from Undergraduate Engineering Students at Three Different Engineering Programs and Institutions. *120th ASEE Annual Conference & Exposition*, American Society for Engineering Education.

Li, J., & Fu, S. (2012). A Systematic Approach to Engineering Ethics Education. *Science & Engineering Ethics*, 18(2):339–349.

Loui, M.C. (2005). Ethical Development of Professional Identities of Engineering Students. *Journal of Engineering Education*, 94(4): 383–390.

Marnburg, E., (2000), The behavioural effects of corporate ethical codes: Empirical findings and discussion, *Business Ethics: A European Review,* 9 (3): 200–210.

O'Clock, P. & Okleshen, M. (1993). A comparison of ethical perceptions of business and engineering majors. *Journal of Business Ethics*, 12(9):677-687. http://dx.doi.org/10.1007/BF00881382.

Oliver, G., Kearins, K., & McGhee, P., (2005), Professional codes of ethics: a New Zealand study, *New Zealand Journal of Applied Business Research,* 4 (1): 97–116.

Passino, K.M. (1998). Teaching professional and ethical aspects of electrical engineering to a large class. *IEEE Transactions on Education*, 41(4): 273–285.

Rest, J. (1986). *Moral Development: Advances in Research and Theory*. New York, New York, USA: Praeger.

Schlegelmilch, B.B., Houston, J.E., (1989), Corporate codes of ethics in large UK companies: an empirical investigation of use, content and attitudes, *European Journal of Marketing,* 23 (6): 7–24.

Segal, M.N., Giacobbe, R.W., (2007), Ethical issues in Australian marketing research services: An empirical investigation, *Services Marketing Quarterly*, 28 (3): 33–53.

Sethy, S.S. (2017). Undergraduate engineering students' attitudes and perceptions towards 'professional ethics' course: a case study of India, *European Journal of Engineering Education*, 42(6): 987–999.

Sezer, O., Gino, F., & Bazerman, M. H. (2015). Ethical blind spots: Explaining unintentional unethical behaviour. *Current Opinion in Psychology*, 6:77–81.

Sparks, J., R., & J. P. Merenski. (2000). Recognition-Based Measures of Ethical Sensitivity and Reformulated Cognitive Moral Development: An Examination and Evidence of Nomological Validity. *Teaching Business Ethics,* 4(4): 359–377.

Stappenbelt, B. (2012). Ethics in engineering: student perception and their professional identity development. *Journal of Technology and Science Education*, 3(1): 86–93.

Thompson, M., (2005), Ethical theory (access to philosophy) (2nd ed.), London, UK: Hodder & Stoughton.

Yallop, A. C. (2012). The Use and Effectiveness of Codes of Ethics – A Literature Review, *International Conference "Marketing – from information to decision" 5th Edition,* Risoprint Publishing House, 502–514.

Ziegenfuss, D.E., & Martinson, O.B., (2002), The IMA code of ethics and IMA members' ethical perception and judgment, *Managerial Auditing Journal*, 17 (4): 165–173.

Advances in Phytochemistry, Textile and Renewable Energy Research for Industrial Growth – Nzila et al. (Eds)

Research ethics and scientific innovation nexus: Unpacking the essentials

Julius Kipkemboi
UNESCO Chair in Bioethics, Egerton University/UNESCO Regional Bioethics Centre, Egerton University, Njoro, Kenya

Violet Naanyu
School of Arts and Social Sciences, Moi University, Eldoret, Kenya
AMPATH IU-Kenya partnership, Kenya

ABSTRACT: Research on scientific technologies and innovations remains critical in providing solutions to societal challenges. Sustainable development policy decisions must be informed by sound scientific research that adheres to ethical values and practice. The quality of research, innovations and consequently the technologies-arising therein should be within acceptable integrity standards. This implies that ethical lapses in research and innovation development can harm those involved as research participants, researchers, the public and the environment. It also compromises research and innovations outcomes. Besides this, deliberate, dangerous and negligent deviations can damage institutional image. It is therefore vital for researchers to conduct research with integrity and sustain a research environment that fosters veracity. Furthermore, for scientific research, innovations and scholarships to endure, these efforts must be founded on objectivity, clarity, reproducibility and utility. This article emphasises on ethics as an integral part of research on scientific technologies and innovations. It reflects on what ethical research and innovation entails and the integrity expected from researchers. It also discusses importance of complying with ethical requirements with emphasis on informed consent, where human participants are involved. Ethical review of research and innovation protocols by an accredited committee is one way of mitigating the risk of research misconduct. Furthermore, in the case of innovations, one of the key integrity issue is the intellectual property rights. In this paper, we discuss ethical issues in the research and innovation lifecycle. There is no doubt that the quest for new knowledge and innovations is inevitable as society continues to face new challenges, which require scientific, technological and innovative solutions. Nonetheless, if advances in science, technology and innovations are to create meaningful contribution to communities, then ethical considerations are unavoidable.

Keywords: Ethical clearance; Good research practices; Informed consent; Intellectual property rights; Research misconduct; Scientific integrity; Subsidiarity

1 INTRODUCTION

Research and innovations form the cornerstones of human advancement and by extension sustainable development. These two aspects of scientific and technological advancement are intricately related yet independent. The common denominator in both cases is the basis of ideas, thinking and questioning the existing reality in a bid to find solutions to problems affecting humanity. Research is systematic investigation to establish facts about these problems, and possible solutions and expanding conclusions. Innovation is embedded in research because the initial idea can only be translated into a prototype through research (Fassin 2000). Innovations however, include strong components of commercialisation and continuous improvements phases. Research and innovations have converging end points with respect to better societal outcomes, which include new ideas and or products hence by extension and improved quality of life. Research and innovations when put in the context of development are essential ingredients of any industrial growth.

Sustainable development heavily relies on research in science, technology and innovation (Levi, Rakićević & Jovanović 2018). Advances in science, technology and innovation are therefore considered as the toolboxes for industrial advancement. Technological knowledge accrued from scientific research can be applied to solve problems at community, institutional and industrial scale. Innovations can be technical when new ideas emanating from practical knowledge or experience are conceived and tested through a scientific process or industrial when the focus is the development of new products or new processes by the industrial sector (Fassin 2000). Scientific development

DOI 10.1201/9781003221968-2

entails action by humankind to provide solutions to societal problems and nurtures curiosity through organised attempt to conceptualise and understand the processes through hypothesis testing and answering questions. Technology on the other hand uses knowledge generated or existing to produce and improve goods and services. As such science and technology hubs play a pivotal role in driving research and innovation.

Centres of excellence in scientific and technological advancement have a major role in contributing to sustainable development through research and innovations. The quality of any research, whether basic or applied, by and large depends on the process through which the outcomes are generated (Resnik & Shamoo 2011). Institutions can only gain reputation, value and public benefit based on the integrity of their research and innovations. This implies that research and innovation by university faculty and students ought to be conducted within a framework of integrity. Any deliberate, dangerous or negligent deviations negate the core values of the research institution. It is for this reason that ethical aspects of research and innovation are considered paramount not only from the level of individual researchers, but also at the institutional governance level. This underscores the important role of institutional research ethics committees in shaping ethical research innovation and sustainable development. Researchers and innovators should therefore be guided by international, national regulations and guidelines. It also implies that they should have a strong culture of institutional values and individual responsibility.

In this paper, we highlight the importance of looking at research and innovation through the lens of integrity and ethical considerations. While focusing on research ethics in science and technology, we make reference to innovation in broad sense. For the purpose of this article, we limit the innovation scope to research and development phases. We reflect on the importance of carrying out quality research, characteristics of ethical research, the role of research ethics committees, special considerations in research involving human participants, and governance of research and associated outputs.

2 QUALITY OF RESEARCH AND ITS IMPLICATIONS

The public and industry expect researchers and innovators to undertake quality research in an effort to advance science, technology and innovation. Quality research is known to be a precursor to quality evidence, thus, investigators are expected to ensure their study design, specific objectives, and methods are coherent, consistent, and justifiable. The characteristics of good quality and ethical research include use of clear and specific research questions, sampling methods that allow representativeness and reduce bias, sound study design, well-worded tools and consent documents, and well justified data management and analysis plans. The research should also have fully described data collection processes. Evidently, quality research calls for objectivity, internal and external validity, reliability, rigor, open-mindedness, and truthful reporting (FOCUS 2005; Wooding & Grant 2003).

Research integrity is an essential element of quality of research. It is demonstrated through good research practices such as observance of professional standards by generating and documenting results, honesty, and proper storage of primary data (Resnik & Shamoo 2011). Lead scientists in research institutions should strive to develop standards for research and mentorship. This should be done in such a way that it is binding for the heads of the individual scientific working units. For instance, primary data which is the basis for publications should be securely stored for a period defined by institutional policy. The disappearance of primary data from a laboratory is a violation of basic principles of careful scientific practice and creates an impression of dishonesty or gross negligence. Research group leaders are responsible for ensuring that mentees receive adequate training and supportive supervision. Healthy communication within a research group and supportive supervision are best ways to mitigate against poor quality research and adverse outcomes such as research misconduct (GRF 1997).

The implications of poor quality research are far reaching (Schneider 2000). It negatively impacts not only on the researcher, but also on the institutional image. Any compromises made in the course of research affect the expected outcomes with respect to new knowledge, and by extension a waste of resources both human (especially time spent) and financial. It also implies that institutional expectations from research undertakings are at stake when chances of research misconduct arise knowingly or unknowingly. Another effect of poor-quality research is the erosion of public trust in research findings and innovations. Poor research mentorship negates the primary role of knowledge hubs in producing skilled personnel through research and training. Consequently, it curtails the potential effects of research and innovation in driving industrial and economic growth.

3 ETHICAL ISSUES IN SCIENTIFIC AND TECHNOLOGICAL RESEARCH

Advances in science and technology are meant to benefit humanity through returns through social benefits and justice and by extension contribute to peaceful societies (UNESCO 2018). African Development Bank points out that research, science and technology is the promise that can spur industrial growth and development and hence improve the quality of life of the people of Africa (ADB 2017). However, the returns to investment in science and technology research can only be realized if such advancement is viewed in the context not only on the outcomes to the intended beneficiaries and doing the right thing at every stage. There are often concerns when investments in science and technology are done at the expense of human values and without consideration of the entire spectrum of social dimensions and inclusivity. As such the rights of the individuals, essence of justice, and environmental concerns should always be part and parcel of every stage in research.

Research ethics alludes to morals and professional codes applied by researchers as they conduct their research. It is important to adhere to ethical principles in order to protect the dignity, rights and welfare of research participants. All research involving human beings should be reviewed by an ethics committee to ensure that the appropriate ethical standards are being upheld (CIOMS 2016; WHO 2020). These standards should be applied in all global contexts, whether carried out in low income countries (LICs) or high-income countries (HICs). It is in this spirit that the UNESCO Commission for Ethics of Science and Technology recommends that science can be sustained when the actors involved in research consider this noble task as a responsibility (COMEST 2015).

Scientific and technological advancement make positive contribution to humanity when they resonate with national and international agenda, promote enhancement of society, and give provisions for environmental consideration. In this respect scientific research should be conducted in a humane, and responsible manner scientifically, socially and ecologically. In recommending civic and ethical aspects of scientific research, UNESCO advocates for researchers take responsibility for themselves, their institutions, their nation, and global community at large (COMEST 2015). Building on Rawls' principle of difference, as we distribute any resources associated with innovative research endeavors, we need to pay special attention to vulnerable communities and individuals who have experienced injustice and historical disadvantages over the years – and actually support those who need help the most. This would be justice in action and it would embody the principle of subsidiarity, whereby, resources are consciously distributed to those in great need (Rawls 2001).

Science and technology are currently characterised by a process of rapid change. It is notable that cutting-edge converging technologies may escape existing normative frameworks or regulations (UNESCO 2018). A good case in mind is the contemporary big data discussion. With increasing advancement in information and communication technology, data acquisition, storage - including aspects of personal privacy – can easily be compromised by this transformation. Beyond research and innovation, such big data collection arises from how people consume products from industry (Strand & Kaiser 2015). A second example is biobanking and associated innovations. Studies involving the use of biological materials are on the rise in Africa and there is need to conduct research activities following strict ethics. Biobanks are organized collections of human biological materials and associated information stored for research purposes (Zatloukal 2018). Heated debates concerning exportation of biological samples, exploitation of participants, benefit-sharing, secondary use of samples and data over time, and return of results, among others in relation to LICs have been documented (Tindana, Molyneux, Bull, & Parker 2014; Wonkam, Kenfack, Muna, & Ouwe-Missi-Oukem-Boyer 2011). Where research engages sample donors and utilizes the samples in numerous future analyses, there is a need to give the sample donors and their local communities' feedback on findings and any resultant products (CIOMS 2016).

4 ETHICAL ISSUES RELATED TO INNOVATIONS

In this paper we focus on research and development stages of innovation. One of the common ethical issue at the onset is the source of ideas. Innovations often require teamwork inputs, but they can also be enabled by outstanding innovators (Stückelberger & de Waal 2014). Innovative ideas may be derived from diverse sources including indigenous informal, and formal sectors. Once the idea gains momentum for research, ethical test sets in. Sometimes the source of knowledge is not appropriately credited, acknowledged, and compensated. Moreover, the mechanisms to ensure the resulting innovations is transferred back to the source are not well established (UNESCO 2018). This presents a potential ethical lacuna.

Like scientific and technological advancement, the societal benefits of innovation should be placed in an ethical scale both in the short term and long-term. This is because innovations bring numerous benefits but also present some potential risks. The main areas that require attention are the potential negative effects of innovations even with improvements and uncertainties. For example, the development of Dichlorodiphenyltrichloroethane (DDT) was initially considered as breakthrough chemistry innovation but later negative effects on bioaccumulation and biomagnification in the food chain causing harm in birds and other species. Similarly, the development of chlorofluorocarbons (CFCs) used as refrigerants and propellants, was for a long time proved to be a great success until the negative effects on the ozone layer were discovered (Stückelberger & de Waal 2014). Lessons from past innovations call for due diligence, in this case on ethical concerns. It is incumbent upon the scientific community to therefore view all new inventions through an ethical lens so as to avoid "innovation ethical blindness". This calls for a balanced approach in the innovation value chain so that both risks and potential benefits can be critically reviewed. It is also important to bear in mind that along the innovation value chain, ethical concerns at consumption level may arise for instance on how available, accessible, affordable and acceptable the perceived benefits are at commercial level (Bastos de Morais & Stückelberger 2014).

Like basic and applied research, innovations are based on evidence collected to support prototype. Often these evidence comprise of the novel narratives supported by data generated through research (Fassin 2000). One of the common ethical issues that arises in the innovation environment is related to the claim of intellectual property ownership. when improvement is made on an existing undeveloped rudimentary of idea, technology or product. Innovation from institutions may be further complicated by the context in which it has been developed. For instance, in the case of a laboratory set up with multiple persons and efforts, integrity is likely to be at stake. The status of different staff who may contribute directly or indirectly to an innovation – from the professor to laboratory assistants – can create problems. The source of funding if not addressed right from the beginning through formal documentation (such as memoranda of understanding and agreements) can also raise conflicts and even ethical issues. However, all these can be overcome if there is a culture of integrity based on good governance, best practice and institutional support for researchers. Institutions ought to have appropriate policies, and procedures that guide innovations arising from students, staff and collaborators.

If we look at the world as one big global village where resources should be distributed in an egalitarian way, the concepts of distributive and commutative justice become important. It is hoped globalization will result in widespread innovations and improvement in average incomes and LICs will be expected to grow rapidly. 'In this "win-win" perspective, the importance of nation-states fades as the "global village" grows and market integration and prosperity take hold' (Scott 2001). The HICs, where many research initiatives in LICs are incubated and funded, should routinely embrace a humanistic world-view. We should all create and honor common values that will uphold ethical conduct of research across the world. Many LICs were plundered by colonialists and were left in unbalanced treaties that favored the HICs to continue unfair pillage of natural resources from LICs. The sharing of research resources from HICs to LICs shows the universal moral duty to support victims of global economic order and to respond to humanitarian needs while consciously paying attention to those in greatest need (Rawls 2001) Any innovative research or development endeavors being designed or funded globally ought to prioritize topics that will better the underserved populations. Ethical research will help us dignify the lives of vulnerable communities that we commonly engage and prioritize the needs of least advantaged.

Beyond the design of quality scientific proposals and ethical implementation of approved research protocols, ethical issues may also arise during publication (Beisiegel 2010). The same applies to sharing of benefits at the commercialisation stage of innovations. It is therefore also important that beyond ethical issues during design, data collection and analysis, the investigators and ethical review committees should consider any other issues that may arise at manufacturing, marketing, promotion and even during continuous improvement of subsequent innovative products (Miles, Munilla, & Covin 2004).

5 RESEARCH ETHICS COMMITTEES AND THE REVIEW PROCESS

Research Ethics Committees (RECs) are multidisciplinary, independent groups of individuals appointed to review research protocols involving human beings to help ensure in particular that the dignity, fundamental rights, safety, and well-being of research participants are duly respected and protected (Council of Europe 2012; UNESCO 2005). The primary role of a REC is to evaluate research proposals for ethical implications, foreseeable research outcomes, and potential consequences of research results for society. 'Society' can encompass both local and wider contexts and may include the potential interests of future generations. RECs aims to safeguard participants' rights, dignity, safety, and well-being. They also put into consideration value deliberation based on the principle of justice. This requires that the benefits and burdens of research are distributed fairly among all groups and classes in society, taking into account in particular age, gender, economic status, culture and ethnic considerations (Eckstein 1992). Further to this they provide the independent advice to the appointing authority, researchers, funding agencies and professionals on the extent to which proposals for research studies comply with recognized ethical standards (ALLEA 2017; European Union 2013; Scherzinger & Robbert 2017).

The REC review process allows for critical evaluation of the ethical soundness of a data collection process for the purpose of scientific and technological development (Kass, Hyder, Ajuwon, AppiahPoku, & Barsdorf 2007). Although it comes at the initial phase of a project, the review process ideally is not restricted to pre-project phase. It is a dynamic process that should encompass the lifetime of the project from application for funding, through the design stage, to completion of the work, its publication and the application of the results. Throughout this process, every opportunity is taken by the REC to ensure ethical and scientific conduct of research. RECs are responsible for acting primarily in the interest of research participants and concerned communities. Furthermore, they take into account the interests, needs and safety of researchers who are trying to undertake research (Kruger, Ndebele, & Horn 2014).

In Kenya, RECs are accredited by the National Commission for Science Technology and Innovation. During constitution and accreditation, research ethics committee are provided with terms of reference that define their mandate (NACOSTI 2017). Any research involving humans or has direct effect on human life must be approved by an accredited REC in Kenya. Ideally all research and innovations require ethical review. Special consideration is required for conduct of research and innovation involving vulnerable populations, human biology, intrusive procedures, use of human tissues, experimental human psychology, and use of animals in experiments. It may also be new data from human participants e.g. interview, observation, original survey; analysis of secondary data especially where data has not been anonymised, including any record showing date of birth. Furthermore, data whose outcome has implications for an identifiable group of people such as use of human genetic data in estimating genetic potential for disease in an identifiable group of living people is also within the purview of special ethical review.

The ethical review of any research or innovation should be guided by Standard Operating Procedures. Figure 1 outlines the core stages in the ethical review process for all research. The focus of the research ethics committee spans across various elements of a research protocol (Wasunna & Bukusi 2014). From the research background analysis, scientific question(s) and/or a hypothesis should be postulated so as to advance science. Further to this, the research design and methodology is scrutinised for ethical issues. In case of research involving human participants, the REC must review the inclusion and exclusion criteria, study procedures, statistical validation of the sample size, study population, and analytical plan for assessing results. The location(s) where research will be conducted, as well as the duration of the proposed research study is not only important for justification for the site selection but also monitoring. Adequate description of the process of conducting research including the work plan and the resources, such as the funds, equipment and facilities required are part of the quality assessments. Lastly the data management including the description of the statistical analysis plan are essential in supporting the production of statistically valid conclusions.

6 GOVERNANCE ISSUES IN RESEARCH AND INNOVATION

The application of advances in science, technology and innovation after successful research phase is a complex, open-ended and unpredictable process. Although risks and side-effects may not always be anticipated, ethical blindness should always be avoided at all costs. Responsible governance that deals thoroughly and proactively with potential hazards and other ethical concerns, is a good strategy in promoting the positive developments to be expected. Functional governance structure is essential to regulate societal implications of science, technology and innovation. However, this can only work if there is sufficient ethical knowledge and understanding of the content of the said research innovations (Strand & Kaiser 2015).

Guided by core values and responsibility to the public and organizations that support research, knowledge institutions should establish mechanism through which research can be regulated on ethical matters. One of the means through which this can be achieved is through setting up institutional research ethics committees or ethics review boards. Where possible there should be a research integrity office with clear policies and codes of conduct for researchers and innovators. The establishment is such structures will not only protect the image of the institution but also the

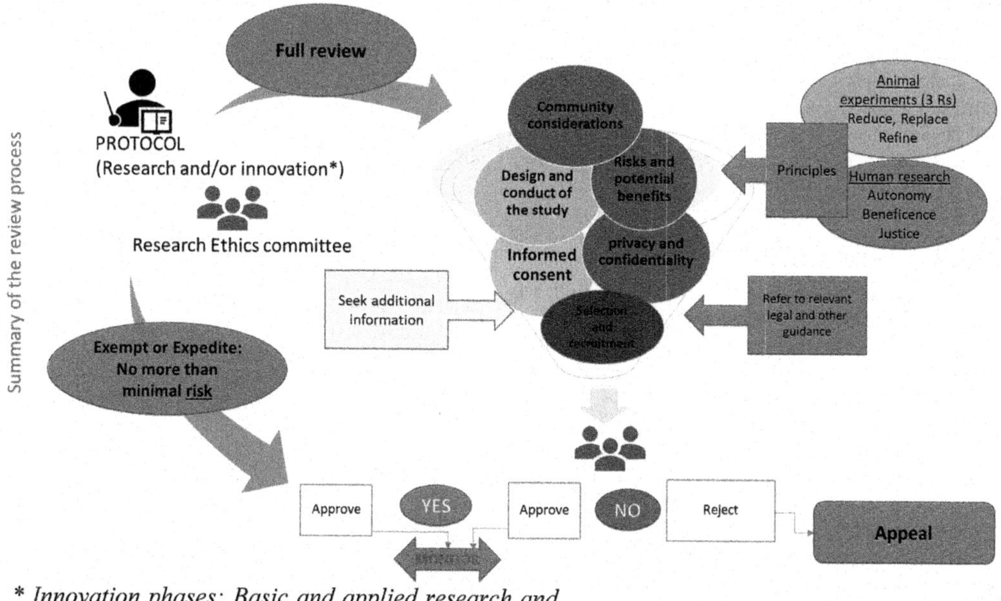

* *Innovation phases: Basic and applied research and development*

Figure 1. The ethical review process of a research or innovation protocol. (Conceptualised from Kruger, et al. 2014).

researchers from possible retribution. The governance of research and innovation must be viewed at local, national and global level so as to be able to competently assess risks if any and avoid systemic injustice and promote equitable benefit sharing. (COMEST 2015). Armin (2011) argues that the issue of responsibility in science, technology and innovation should be viewed in the context of the individual, situation in which responsibility is required, the existing governance structure, and quality of knowledge available with respect to consequences of actions under which scientific and technological knowledge is generated (Armin 2011). Whereas institutions provide a platform from which research and innovations are developed, it is important that they respect the intellectual property rights (IPR) of individual researchers.

IPR comprises the entire landscape of protection of intellect including inventions arising from research and innovations in scientific, artistic, product development domains (Granstrand 2006). The debate of IPR in the context of developing countries is beyond the scope of this article. Ethical issues are inevitable when commercial interests curtail the societal benefits. Tradeoffs and sustainable solutions revolve around the entire governance ecosystem. Like in the case of research and innovation, IPR requires utmost responsibility (Schuijff & Dijkstra 2020). This responsibility should be reflected at all levels of knowledge advancement.

7 CONCLUSION

Ethical issues are inevitable in research and innovation. Nathan (2015) and Lubberink, Blok, van Ophem, and Omta (2017) assert that unless innovation decision-making is based on ethical decision-making framework, its likely to be tainted by undesired impacts and concerns. One of the mitigative approach is the ethical review of every step of research and innovation process. Advances in emerging sciences and technologies that pose serious ethical issues and concerns, on the individual, community and even international level require collective responsibility. Responsible research and innovation is a transparent, interactive process by which societal actors and innovators become mutually responsive to each other with a view to the (ethical) acceptability, sustainability and societal desirability of the innovation process and its marketable products. This will allow proper embedding of scientific and technological advances in our society. Secondly, research sponsors and investigators have the right to study whatever they want, however, ethical reflections suggested that they ought to be morally obligated to routinely include innovative works that may be a matter of life and death in LICs. They should also not withhold any available basic resources that can better the lives of their study participants, and any resources that they distribute should not be used to manipulate vulnerable communities to consent to research participation.

8 RECOMMENDATIONS

Research ethics is currently in its nascent stage in many LICs, therefore many innovative research projects are likely to miss out the microscopic scan on how well autonomy, beneficence, non-maleficence, and justice are addressed by the research teams. As such, research participants and their local communities may be getting very little for contributions made. This then stands out as one research-related injustice and amounts to exploitation of communities, who generally have limited frameworks for defense. Better governance is needed to increase the ethical and social robustness of new and emerging sciences and technologies. This is needed within the individual institutional establishments as well as at national level. There should be overarching and institutionalized framework of regulations, standards, policies, codes, principles to ensure that quality research and innovations are part and parcel of knowledge advancement for industrial growth and sustainable development.

DISCLAIMER

The views, interpretations, and conclusions expressed in this report are entirely those of the author(s) and do not necessarily express the views of the institutions in which the authors are affiliated.

REFERENCES

ADB.2017. Industrialize Africa. Strategies, Policies, Institutions, and Financing. African Development Bank Group.

ALLEA. 2017. The European Code of Conduct for Research Integrity. All European Academies, Berlin Germany.

Armin, G. 2011. Responsible Innovation: Bringing together Technology Assessment, Applied Ethics, and STS research, Enterprise and Work Innovation Studies, 7, IET, pp. 9–31.

Bastos de Morais J-C., & Stückelberger C. (eds.), 2014. Innovation Ethics. African and Global Perspectives. Globethics.net and African Innovation Foundation, Geneva.

Beisiegel, U. 2010. Research integrity and publication ethics. Atherosclerosis, 212 (2) https://doi.org/10.1016/ j.atherosclerosis.2010.01.050

CIOMS. 2016. Council for International Organizations of Medical Sciences (CIOMS). International Ethical Guidelines for Health-related Research Involving Humans. Retrieved July 13, 2020 from https://cioms.ch/wp-content/uploads/2017/01/WEB-CIOMS-Ethical Guidelines.pdf

COMEST. 2015. Ethical Perspective on Science, Technology and Society: A Contribution to the Post-2015 Agenda. Report of COMEST, SHS/YES/COMEST-8EXTR/14/3, UNESCO, Paris.

Council of Europe. 2012. Guide for Research Ethics Committee Members. Steering Committee on Bioethics. https://www.coe.int/en/web/bioethics/guide-for-research-ethics-committees-members.

Eckstein, S. (Ed.). 1992. Manual for Research ethics committees. 6th Edition, Centre of Medical Law, Kings College, London, Cambridge University Press.

Fassin, Y. 2000. Innovation and Ethics Ethical Considerations in the Innovation Business. Journal of Business Ethics, Sep., 2000, Vol. 27, No. 1/2, Business Challenging Business Ethics: New Instruments for Coping with Diversity in International Business: The 12th Annual EBEN Conference, pp. 193–203.

FOCUS. 2005. What Are the Standards for Quality Research? FOCUS. Center on Knowledge Translation for Disability and Rehabilitation Research (KTDRR) Technical Brief Number 9 Focus: A Technical Brief from the National Center for the Dissemination of Disability Research. Southwest Educational Development Laboratory. Retrieved July 13, 2020 from https://ktdrr.org/ktlibrary/articles_pubs/ncddrwork/focus/focus9/Focus9.pdf

German Research Foundation (GRF). 2013. Proposals for Safeguarding Good Scientific Practice Recommendations of the Commission on Professional Self-Regulation in Science Retrieved July 13, 2020 from https://onlinelibrary.wiley.com/doi/pdf/10.1002/9783527679188

Granstrand, O. 2006. Innovation and Intellectual Property Rights. In: Jan Fagerberg and David C. Mowery (Eds). The Oxford Handbook of Innovation. DOI: 10.1093/oxfordhb/9780199286805.003.0010.

Kass, N.E., Hyder, A.A., Ajuwon, A., AppiahPoku, J., & Barsdorf, N. 2007 The structure and function of research ethics committees in Africa: A case study. PLoS Med 4(1): e3. doi:10.1371/journal.pmed.0040003

Kruger, M., Ndebele, P., & Horn, L. 2014. Research Ethics in Africa: A Resource for Research Ethics Committees. SUN MeDIA Stellenbosch.

Levi Jakšić, M., Rakićević, J., & Jovanović, M. 2018. Sustainable Technology and Business Innovation Framework A Comprehensive Approach. Amfiteatru Economic, 20(48), pp. 418–436.

Lubberink, R., Blok, V., van Ophem, J., & Omta. O. 2017. A Framework for Responsible Innovation in the Business Context: Lessons from Responsible-, Social- and Sustainable Innovation. In L. Asveld et al. (eds.), Responsible Innovation, Springer International Publishing, DOI 10.1007/978-3-319-64834-7_11

Miles, M.P., Munilla, L. S., & Covin, J.G. 2004. Innovation, Ethics, and Entrepreneurship. *Journal of Business Ethics,* 54: 97–101.

NACOSTI 2017. Guidelines for Accreditation of Institutional Ethics Review Committees in Kenya. National Commission for Science, Technology and Innovation (NACOSTI).

Nathan, G. 2015. Innovation process and ethics in technology: an approach to ethical (responsible) innovation governance. Journal on Chain and Network Science 2015; 15(2): 119-134 Wageningen Academic Publishers, SPECIAL ISSUE: Responsible innovation in the private sector, ISSN 1875-0931 online, DOI: 10.3920/JCNS2014.x018119

Rawls, J. 2001. A theory of justice. Harvard University Press, Nueva York.

Resnik, D,. & Shamoo, A. 2011. The Singapore Statement on Research Integrity. Accountability in research. 18. 71–5. DOI:10.1080/08989621.2011.557296

Scherzinger, G., & Bobbert, M. (2017). Evaluation of Research Ethics Committees: Criteria for the Ethical Quality of the Review Process. *Accountability in Research*, 24:3, 152–176, DOI: 10.1080/08989621.2016.1273778

Schneider, C. 2000. Safeguarding good Scientific practice: New institutional approaches in Germany. Sci. Eng. Ethics, 6, 49–56. https:doi.org/10.1007/s11948-000-0022-2

Schuijff, M., & Dijkstra, A.M. 2020. Practices of Responsible Research and Innovation: A Review. Sci Eng. Ethics 26, 533–574. https://doi.org/10.1007/s11948-019-00167-3

Scott, B.R. 2001. The Great Divide in the Global Village. Retrieved June 21, 2020. https://www.globalpolicy. org/component/content/article/218/46491.html

Strand, R., & Kaiser M. 2015. Report on Ethical Issues Raised by Emerging Sciences and Technologies. Report written for the Council of Europe, Committee on Bioethics. Centre for the Study of the Sciences and the Humanities, University of Bergen, Norway.

Stückelberger, C., & de Waal, M. 2014. An African Ethical Innovator: Herman Chinery-Hesse. In: Bastos de Morais J-C. and Stückelberger C. (eds.). Innovation Ethics. African and Global Perspectives. Globethics.net and African Innovation Foundation, Geneva.

Tindana, P., Molyneux, C.S., Bull, S., & Parker, M. 2014. Ethical issues in the export, storage and reuse of human biological samples in biomedical research: perspectives of key stakeholders in Ghana and Kenya. BMC Medical, 15(1), 76. https://bmcmedethics.biomedcentral.com/articles/10.1186/1472-6939-15-76

UNESCO 2018. Recommendation on science and scientific researchers. United Nations Educational, Scientific and Cultural Organization, Paris, France.

UNESCO. 2005. Universal Declaration on Bioethics and Human Rights. UNESCO, Paris.

Wasunna, C., & Bukusi, E.A. 2014. A Stepwise Approach to Protocol Review. In: Kruger, M., Ndebele, P., Horn, L. (Eds.), Research Ethics in Africa: A Resource for Research Ethics Committees. SUN MeDIA Stellenbosch.

WHO. 2020. Ethical standards and procedures for research with human beings. Retrieved June 21, 2020. https://www.who.int/ethics/research/en/

Wonkam, A., Kenfack, M.A., Muna, W.F.T., & Ouwe-Missi-Oukem-Boyer, O. 2011. Ethics of human genetic studies in sub-Saharan Africa: The case of Cameroon through a bibliometric analysis. Developing World Bioethics, 11(3). http://onlinelibrary.wiley.com/doi/10.1111/j.1471-8847.2011.00305.x/full

Wooding, S., & Grant, J. 2003. Assessing research: The researchers' view. Cambridge, England: RAND Europe.

Zatloukal, K. 2018. Biobanks in personalized medicine. In Expert Review of Precision *Medicine and Drug Development,* 3:4:265–273.

Advances in Phytochemistry, Textile and Renewable Energy Research for Industrial Growth – Nzila et al. (Eds)

Impact of Covid-19 on businesses in Uasin Gishu county, Kenya

Charles Lagat, Benard Nassiuma, Kefa Chepkwony & Stephen Bitok
Moi University, Eldoret, Kenya

ABSTRACT: The COVID-19 pandemic's announcement by the World Health Organization (WHO), a wide range of measures and guidelines, has been issued by the Government of Kenya through the Ministry of Health MOH). Citizens are being reminded through various communication channels to adhere to MOH containment measures, including social distancing, testing, and isolation. This report assesses the impact of COVID-19 pandemic on the operations of businesses in Uasin Gishu County based on a survey conducted in April 2020. The information reveals how the coronavirus pandemic had already affected businesses and the business owners' suggestions on mitigating the situation. KNCCI, as a business membership organization, includes members from all the sectors of the economy in the county and has more than 700 members in Uasin Gishu Chapter. The study revealed 29% of the businesses had already closed down, several operating at less than 50% at the survey time. Further, 80% of the companies had already sent about 1,400 employees on leave, most of them without pay. This report provides useful information for both the Local and National Governments and other stakeholders in coming up with mitigation measures.

Keywords: Coronavirus, World health organization, Outbreak, Ministry of health

1 INTRODUCTION

The origin of the coronavirus (COVID-19) pandemic has been categorized as a public health emergency globally through the world health organization (WHO). The first incident of COVID-19 disease and it's subsequent named SARS-CoV-2 was first reported by officials in Wuhan City, China, in December 2019 (Ali, Baloch, Ahmed, Ali, & Iqbal 2020; Chinazzi, Davis, Ajelli, Gioannini, Litvinova, Merler, & Viboud 2020). Coronaviruses (CoV) originates from a family of viruses that result from a short illness arising from a common cold to severe diseases. Coronavirus disease has resulted in a new struggle against humanity on the planet. WHO its core mandate to combat other diseases like polio, Ebola, HIV, tuberculosis, cancer, diabetes, and mental health, amongst other diseases and conditions afflicting humankind.

The COVID-19 pandemic challenge presents a universal economic crisis whose extent is still not entirely apparent (Paules, Maston, & Fauci 2020). More than 3 million coronavirus cases globally, more than 243,569 deaths, and only less than 1,104,723 cases had recovered on 1st May 2020. The death and recovery rates are crucial statistics, which might enable the country to identify the magnitude of the risk areas for efficient medical and through stratifying patients based on the extent of attention needed, which may lead to lockdowns (Khafaie & Rahim 2020; Potluri & Lavu 2020). The coronavirus disease epidemic has been labeled a public health crisis. The countries preparedness calls for comprehensive strategies to manage the hospital's space optimally, staff and supplies on service delivery, and prevent further transmission (Wong et al. 2020). In Kenya, testing for the coronavirus has been ongoing. Four hundred eleven confirmed cases were reported with 240 active incidents of the disease, according to an update by the Ministry of Health on 1st May 2020. One victim has been reported in Uasin Gishu County (Daily Nation, 1st May 2020).

On March 27th, 2020, the International Monetary Fund (IMF) indicated an economic recession globally. Their main concerns include bankruptcies and layoffs, which could impact to crisis preventing the recovery process (Beck, Degryse, De Haas, & Van Horen 2018). The lockdown of economies has significantly affected the capital flows and disruptions to the manufacturing industry. According to the World Food Programme (WFP), the East African Community's economies are anticipated to deteriorate owing to inferior internal and external economic chains.

The WFP report on 15th April 2020 projects a gloomy future in terms of a rise in infections in East Africa, which is worrying despite the various measures taken. The prediction of virus trends shows a significant impact on the economies (COVID and Team 2020). WFP estimates that 20 million people worldwide were probably food insecure, and likely to be worsened.

DOI 10.1201/9781003221968-3

The admission of patients into the hospital is to be presumed to have been confirmed. The authors believe several patients had underlying diseases that included diabetes, hypertension, and cardiovascular disease (Huang et al. 2020). The epidemic doubles resulting in a prolonged incubation period making isolation difficult to the suspect. Thus, MOH has come-up with countering the spread through school shutdown, public screenings, and disbanding of mass gatherings (You, Lin, & Zhou 2020; Wu, Leung, & Leung 2020;. Researchers believe that COVID-19 was likely to spread throughout the county via unknown mechanisms, which has resulted from an increased number of infections and death cases globally (Bassetti, Vena, & Giacobbe 2020; Holshue et al. 2020). Therefore, this survey aimed at investigating the impact of Covid 19 on businesses in Uasin Gishu County.

i) *Action by Kenya government*

The ministry of health (MOH) has several containment protocols like social distancing, public testing, and suspects' isolation. To a large extent, Kenya's response to coronavirus has been reassuring. The leadership's continued agility will determine the scope of the crisis's impact and effects and the speed and direction of recovery (Daily Nation Newspaper Editorial, 27th March 2020). Indeed, President Uhuru Kenyatta and his entire leadership have been giving clear guidelines since COVID-19 was a global pandemic. In Kenya, testing for the coronavirus has been going on, and 435 have been reported positive with two from Uasin-Gishu County (Daily Nation Editorial, 1st May 2020).

Recently, the Kenyan parliament enacted laws assented to by the president to cushion the economy's collapse. The amendment bills entailed reviews of the income tax (IT) act (CAP 470), value-added tax (VAT) act of 2013, excise duty (ED) act (2015), tax procedures (TP) act (2015), Miscellaneous Levies and Fees Act (2016), and the retirement benefits (RB) act (1997). Besides the revision of laws, the government has further increased the threshold for turnover tax from the speculative tax by lowering and lessening PAYE tax from 30 to 25 percent to support low waged earners. Scrutiny on corporation tax to 25 percent and value-added tax (VAT) rate was reduced to benefit from the consumers' essential products. The retirement benefits (RB) act (1997) was revised to enable citizen access to affordable housing as propagated in the housing pillar in the big four agenda.

ii) *KNCCI Uasin Gishu chapter*

Kenya has 47 counties, Uasin Gishu county, included and situated in Eldoret town, its main administrative and commercial center. The main economic activities in the County include various sectors like large-scale farming and horticulture, manufacturing, sports, and tourism. There numerous medium and large scale industries, including Kenya Cooperative Creameries (KCC), Doinyo Lessos' Creameries, Rivatex Textiles, Raiply Wood Factory, and Kenya Pipeline Company Ltd. Eldoret Town commercial business district (CBD) is a hub of wholesale and retail businesses as well as financial, healthcare, and transport services.

Uasin Gishu County created a COVID-19 Response Committee composed of representatives from the National Government led by the County Commissioner and County Government led by the Governor. The other members are from the Ministry of Health and representatives of the business community. Uasin Gishu County Covid-19 response Committee has put additional measures to ensure there is no spread of the virus in the County. Among the actions being taken is a requirement for all long-distance public service vehicles to keep a record of all travelers because it makes it easier to track all individuals should any case be reported. According to Governor H. E. Jackson Mandago, the County Government has put up necessary infrastructure and support systems to ensure minimum economic activity disruption in the recent past.

The Kenya National Chamber of Commerce is a Business Membership Organization with membership across the country. The Chamber is the voice representing the member's diverse interests at various levels. They protect, promote, and develop the business community's interests by providing an enabling environment to be competitive. There is a chapter in each of the 47 counties in Kenya consisting of business owners within the respective counties. The Uasin Gishu Chapter has been operational since 2001, and its leadership is made up of Directors representing different sectors of the economy. The chapter is to formalize collaborations with Moi University to support the business community through training, capacity building, supporting agribusiness growth, setting up incubator centers, and advocating sustainable development. This report has been prepared jointly by a team from Moi University and KNCCI Uasin-Gishu Chapter to assess the region's COVID-19 pandemic's impact.

Mr. Willy Kenei, the Chairman KNCCI Uasin Gishu Chapter & Businessman in the Health care industry, the coronavirus is a significant threat to business operations globally. Its impact on the business sector in Uasin Gishu County has been unprecedented and will continue to affect every sector of the economy. The COVID-19 crisis is coming when most businesses have been having challenges with unpredictable weather patterns and late rains in the last two years, affecting the agricultural sector, which is the region's main activity. Most of the other businesses in the Uasin Gishu region are either processing agricultural produce or dealing in farm inputs. The crisis will affect the area's economic performance during the year, and it may take longer for some businesses to recover.

The study findings reveal how they have been affected by the coronavirus crisis. Some of the companies have either closed down, while the majority are operating way below capacity. Most businesses rely on daily sales revenue to run their operations, and some businesses have bank loans, which they regularly service. The chapter appreciates the team from Moi University. They have assisted in compiling the report on the coronavirus outbreak's impact and bringing out suggestions and recommendations for action by the various stakeholders. I also appreciate the support from CIPE, our board, and the Secretariat. The region's growth will require approval from the County Government, National Government, and Non-governmental players.

2 METHODOLOGY AND DATA COLLECTIONS

A survey was conducted between the date 14th and 24th April 2020. There are more than 700 businesses in the County; however, the target respondents were 400 members of the KNCCI Uasin Gishu chapter whose contact details were readily available at KNCCI offices in Eldoret. A survey research method was used to collect primary data from KNCCI members to gain information and insights into various interest topics. A survey research design was chosen to reach a large portion of participants in KNCCI members.

Quantitative data was used because of the standardized procedures that ensure that each respondent answered the questions to avoid biased opinions that could influence the research outcome. The process involved asking for information through online (Weigold, Weigold, and Russell 2013). However, with the arrival of new technologies during the pandemic period, the questionnaire was distributed using digital media such; social networks and respondent email with the help of KNCCI.

3 RESULTS AND DISCUSSION

The members operate a wide range of businesses within Uasin Gishu County. Responses from 145 business owners were received, giving a response rate of 36%. Among the respondents, 74% had operations with Eldoret Town CBD, while 26% operated from premises outside the CBD. The respondents were evenly spread across all the sub-counties: Turbo – 29, Kesses – 28, Kapseret – 24, Ainabkoi – 20, Soy – 18, and 13 from outside the County. The details of the distribution were presented in Figure 1.

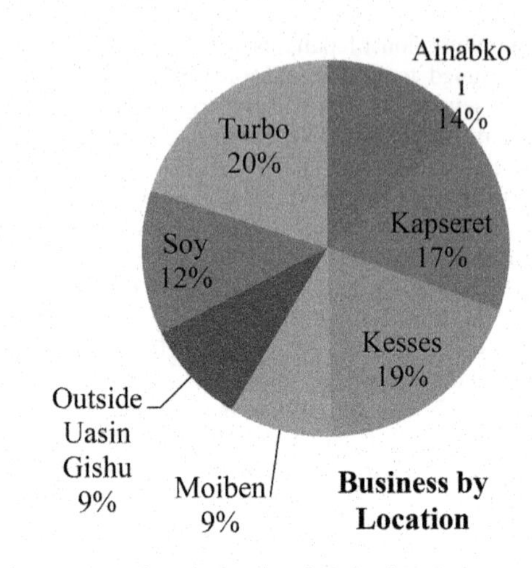

Figure 1. Business enterprises location in UG county.

i) *Size of business and period of operation*
Regarding the size of the businesses, the survey indicated that 45% (65) had less than five employees, 27% (39) had between 6 and 10 employees, 20% (29) had between 11 and 50 employees, and only 8% (11) had more than 51 employees. In this respect, most enterprises' spectrum falls in micro small and medium enterprises (MSME).

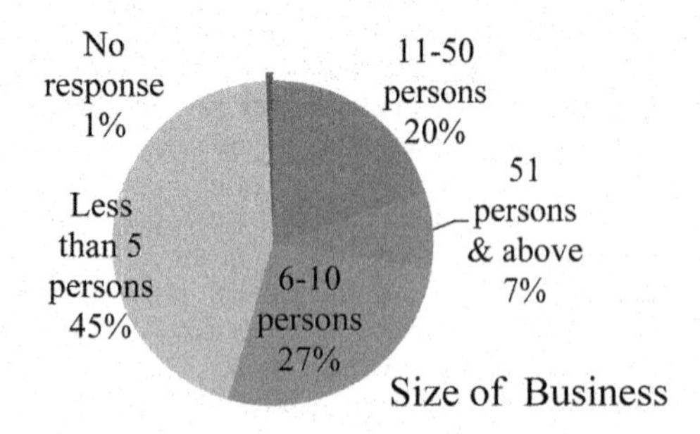

Figure 2. Size of business.

The study revealed that 55% (80) of the businesses had been in operation for more than five (5) years in the operation period. It implies that indicatively that the enterprises have stabilized. Those who had been in operation for less than one year were 8% (12) of the respondents.

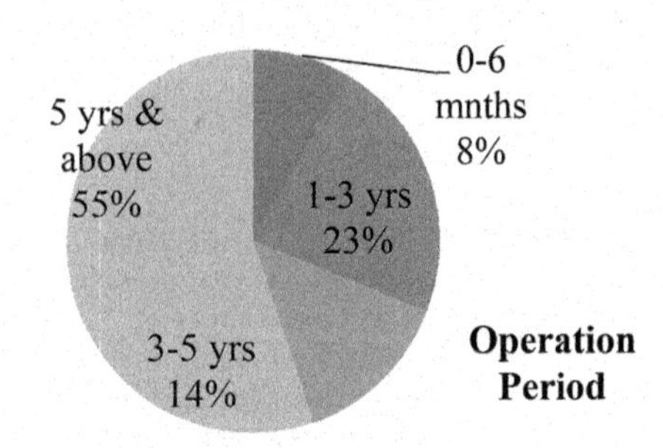

Figure 3. Period of business operation.

ii) *Source of finance*

The survey sought to find out the primary source of finances used for running the business operations. The majority (74%) of 1 07 businesses indicated that they relied on sales revenue (41%) and loans (33%). Only a small percentage indicated that they relied on savings, shares, or equity for running their operations. Therefore, the disruptions caused by travel restrictions and curfew on their everyday business operations have severe implications on the business's survival.

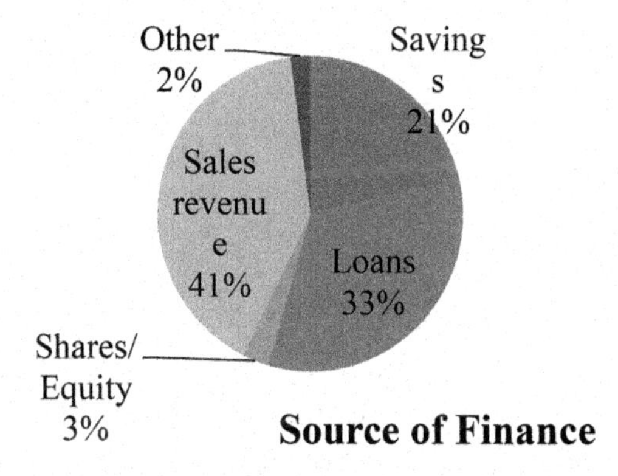

Figure 4. Sources offinances.

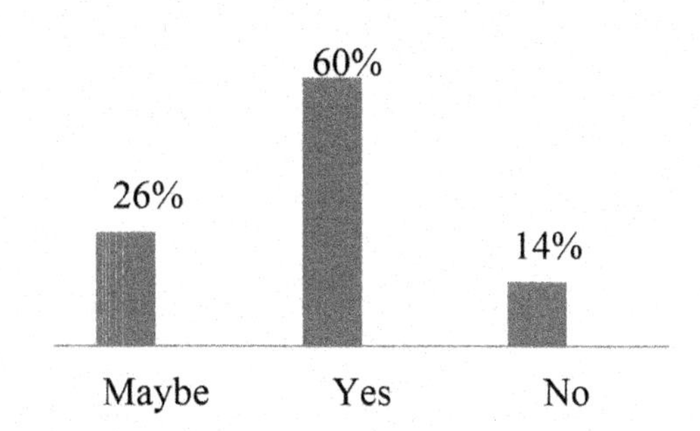

Figure 5. Employee layoff.

iii) *Effect of the outbreak on the businesses operations*

The survey found 42 businesses already closed down completely, while 65 were operating at 25% capacity, 28 at 50% capacity, and only ten are still running at close to 100%. Among the businesses surveyed, 87 have laid off some employees, 37 are currently considering layoffs because of the corona pandemic, while 14 (12%) are still having all the employees. Implying that about 86% of the businesses may lay off employees owing to the pandemic. The number of employees who have either been laid off or on leave by the 145 firms was reported to be 1,416 employees. This indicates that in the 700 businesses in Uasin Gishu, more than 6,000 employees have been affected by the pandemic. The details on business operation and employee layoffs are provided in Figures 5 and 6.

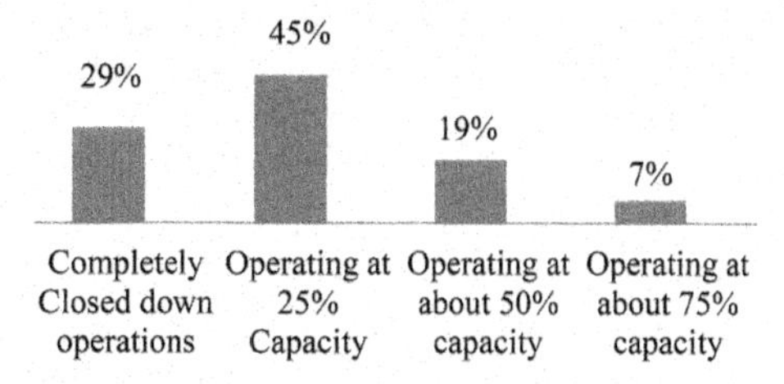

Figure 6. Effect of outbreak.

iv) *Effect of the outbreak on different sectors*

The COVID-19 pandemic had impacted several sectors in the economy; therefore, no industry in all the businesses was operating at 100% operations. Analysis of each sector's effect was conducted to explain the impact of the industry. The most affected sector is the hospitality, which has seen some either closed down or operating at less than 25%. The manufacturing, construction, real estate, and ICT businesses work at varying levels, mostly between 25% and 50%. Some had closed down already. Most of the health care establishments and a few in manufacturing are operating at above 75%.

The retail and transport sectors were partially operational but facing a significant drop in sales and the implementation of COVID-19 containment measures (Chinazzi et al. 2020). CBD's business stalls and microenterprises experienced fewer sales as consumers spend on essentials, non-perishables, and fast-moving consumer goods. Mama Mbogas, particularly those who were well-positioned, are benefiting from sales of fresh fruit and vegetables. However, this depends on the continued supply of their produce from the farmers. In the agricultural sector, those businesses that deal with agribusiness and processing face challenges due to collections of inputs and market access. The details on the impact of the outbreak on different sectors are provided in Figure 7.

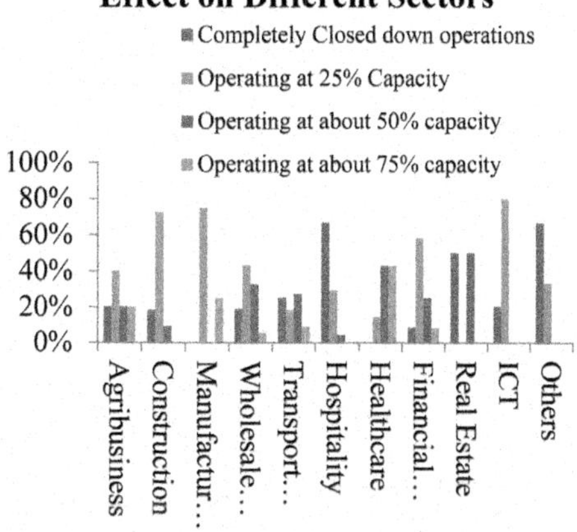

Figure 7. Impact of the pandemic in different sectors.

v) *Challenges faced as a result of COVID-19*
As a result of containment measures, mainly travel restrictions, stay home orders, curfew, and social distancing, businesses indicated they face a wide range of challenges. The study found challenges listed in Table 1 below ranked above 3.92 on a scale of 1-7, above the average of 3.5. The standard deviations were high (2.08–2.39) since the challenges affecting different sectors varied. The most pressing challenge is the travel restrictions followed by the curfew, rent payments, and finance in that order (Wang, Zhang, Zhao, Zhang, & Jiang 2020). The businesses in Kenya source their materials from other regions and hence are facing material delivery delays. These are because the restrictions had significantly reduced their operating hours, with most reporting that they had to close by 4.00 PM. Those in wholesale and retail businesses usually run their businesses operated the whole day but had to close by 6.00 PM. Most of the companies operating from rented/leased premises expected to pay rent at the end of are finding it challenging to meet their rent/lease obligation due to low business. The travel restrictions and requirements for maintaining social distance have also affected the marketing of products. Other challenges that were identified included meeting taxes, levies, and workforce retention and risk management obligations.

Table 1. Ranking of different challenges.

Challenges Facing the Enterprise (scale 1-7)	Mean	Std Dev.
Finance and liquidity	4.71	2.29
Workforce: Human Resource/Staff/Employees	3.92	2.26
Taxes and levies	4.22	2.08
Travel Restrictions	4.82	2.39
Market Access	4.69	2.21
Restrictions on operating time	4.61	2.25
Stock or Material Supplies	4.48	2.26
Marketing and Promotion	4.21	2.27
Risk Management	4.59	2.13
Rent payment	4.74	2.37
Curfew	4.66	2.28

vi) *Expectations of business performance in 2020*
The survey sought to find out from the business owners their expectations of business performance in 2020 compared to 2019. Overall, respondents indicated that they expected their business's poor performance with a decrease of more than 20% during 2020, as shown in Figure 7 below.

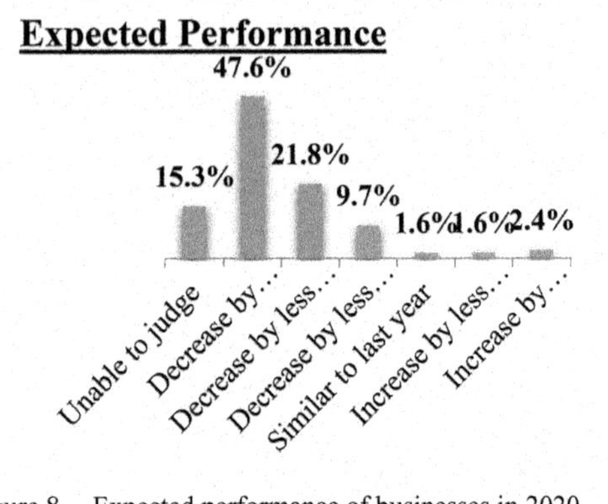

Figure 8. Expected performance of businesses in 2020.

The expectations of business performance during the year varied across sectors and sizes of business. The elephant - speaking truth to power- that workers are likely to be further exposed to job losses and financial insecurity than those in the micro and small enterprise informal sector (Ndii 2020). However, most of the business surveyed were formal and but were already facing experiencing operational challenges.

vii) *Businesses take new measures*
The businesses have been called upon to take preventive measures by respective agencies in response to the outbreak of COVID-19, and which were still in operations reported that they had devised new arrangements in their businesses to mitigate the challenges while complying with government directives, which included online business, 41%, house-to-house deliveries, or take-away services 23% besides other measures 34%.

Table 2. Expected performance and size of business.

Expected Performance Size of Business	Be Similar to last year	Decrease by less than 20%	Decrease by less than 50%	Decrease by more than 50%	Increase by less than 20%	Increase by more than 20%	Unable to judge
11–50 persons	1 3%	5 17%	7 24%	11 38%	1 3%	1 3%	3 10%
51 persons and above	1 9%	0 0%	2 18%	4 36%	0 0%	2 18%	2 18%
6-10 persons	0 0%	3 8%	9 23%	21 54%	1 3%	0 0%	5 13%
Less than five persons	1 2%	7 11%	14 22%	30 46%	1 2%	1 2%	12 18%

The respondents reported that they were utilizing all communication forms to interact with their suppliers and customers and employees who have been required to work from their homes most of the time. It is significate to note that only one respondent had introduced a night shift in their operations. Most of them had reported that they had reduced the amount of credit accorded to their customers.

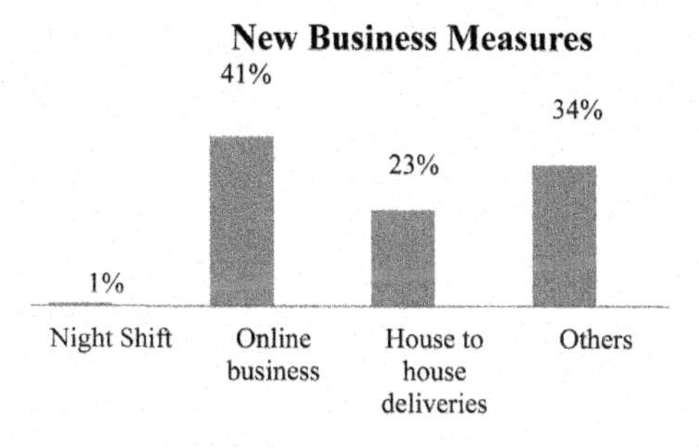

Figure 9. New measures undertaken by businesses.

4 SUGGESTIONS TO COUNTY GOVERNMENT ON THE COVID-19

The respondents suggested the action of the Uasin Gishu County Government to mitigate the impact. The suggestions made included; reduction or eradicating fees and levies; reduction of restrictions on movement/ travel; public awareness and sensitization on COVID-19; actions to cushion businesses on loans; rent payment for premises and other business operations; as well as financing and related support to businesses. Further suggestions were made for the national Government to act.

i) *County government fees and levies*
A summary of the business people's perceptions regarding the county government's actions concerning levies on trading permits, cess on commodities, water bills and payment of rental tax, and land rates are tabulated in Table 3 below.

Table 3. County government fees and levies by frequency and percent.

Action required	Frequency	Percent
To waive/ scrap business license fees	8	33
Reduce license fee	7	29
Clear Pending Bills.	5	21
The government to increased surveillance and security, reduce anxiety.	1	4
Implement procurement plans to give business to the private sector	1	4
Reduce land rates	1	4
Reduce charges such as water bills	1	4
Total	**24**	**100**

ii) *Restrictions on movement/travel*
Responses from the business people on movement are presented in Table 4. The respondents suggested that businesses should operate within the legal framework to access goods from other companies across counties. This will enable them to generate income to enhance their livelihoods and hence be cushioned against price hikes. Also, government departments that provide essential services should be allowed to reopen, such as the judiciary and lands office.

Table 4. Restrictions on movement/travel by frequency and percent.

Action required	Frequency	Percent
Allow the business to open	4	50
Allow movement of basics	2	25
cushion against price hiking	1	13
Access to essential services-lands and judiciary	1	13
Total	**8**	**100**

iii) *Sensitization on COVID-19*
Responses on the actions that should be taken to control the COVID-19 are presented in Table 5. The results show that the majority (13, 54%) want sensitization to occur, while (4, 16%) suggest that provision of testing kits, protective gear such as ventilators, Expansion of ICU bed capacity should be allowed through means such as importation. On random testing and increased surveillance and security to reduce anxiety was considered significant (2-8%) of the respondents. Responses on team working suggest collaboration of all agents involved in the already established disaster management committees and working with village elders to ensure human movement is restricted and ensure that affected counties are locked and to supplement/ support society's

Table 5. Sensitization on COVID 19 actions by frequency and percent.

Action Required	Frequency	Percent
Education and awareness of COVID 19	13	52
Stay at home	1	4
Concentrate on finding a vaccine too	1	4
Avail food to the vulnerable	1	4
Collaboration and coordination committees	2	8
Provide protective gears	4	16
Do random testing,	1	4
Increased surveillance, security, assurance from Government to reduce anxiety	2	8
Total	**25**	**100**

vulnerable members. Responses on other measures such as staying at home, concentrate on searching for the vaccine, avail food to the weak constituted 4 percent each.

iv) *Actions required on loans*
The responses on loans are presented in Table 6. The results show that the majority (11, 52%) of the respondents suggested that businesses should be given unsecured loans to boost the firms after the pandemic is over. 20% of the respondents indicated that a stimulus package should be introduced to support businesses; 10% suggested that the repayment period be extended. In comparison, 5% respectively indicated that tax holidays and reduction in parking fees be offered to business owners.

Table 6. Actions required on loans.

Action Required	Frequency	Percent
Avail unsecured loans to SMEs	11	52
Ease loan payments	2	10
Cushions Businesses through giving tax holidays	1	5
Give stimulus packages	6	29
Reduce fees payable, e.g., parking, licensing	1	5
Total	**21**	**100**

v) *Rent payment for premises*
The responses on rent issues are presented below. The results show that the majority (86%) of the respondents suggest that the government intervene to freeze rent or give more time within which rent should be paid until the economy is stable. Regarding rental properties owned by the county government, the respondents (14%) suggested that rent be waived for the pandemic period. Some landlords indicated that rental income is their only income source, making it difficult for them to either waive or reduce the rent.

vi) *Other business operations*
Responses to other business operations are presented in Table 7. The results show that most (50%) of the respondents suggested that businesses remain open but comply with the regulations on COVID-19. 14% indicated that the provision of essential services be guaranteed during the pandemic period, with some suggesting that they had experienced police harassment even before the curfew time. Respondents indicated that there is a need to reduce the cost of production costs and unfair supplier prices.

5 MITIGATION MEASURES BY THE NATIONAL GOVERNMENT

The covid-19 pandemic outbreak has affected nations on all continents, therefore, calling for mitigation measures (Ebrahim, Ahmed, Gozzer, Schlagenhauf, & Memish 2020). On the tax issue, respondents suggested that a significant reduction of all taxes should be made to enable the businesses to again.

Table 7. Other business operations.

Action Required	Frequency	Percent
Allow businesses to operate at 50 percent just like matatus and hotels	7	50
Stop harassment before curfew times	2	14
Create online jobs or else to employ us on shifts	1	7
Provision Electronic services	2	14
Reduction of production cost for essential products	1	7
Profound control over supplier unfair price increase	1	7
Total	**14**	**100**

i) *Financing and related businesses support*
Responses on the actions that should be undertaken concerning the financing of businesses are presented in Table 8 below. Results on the perception of the respondents (5, 50%) how that they want the government to amend the law to make banks flexible in the lending practices, as a strategy of the government protecting businesses from collapse and mass unemployment, to provide support to the low-income groups and to incite a stimulus program to support businesses. Respondents (2, 20%) suggest that companies should be kept to restock their enterprises. The respondents (3, 30%) indicated that loans should not earn interest else the central bank should reduce interest charges on interest rate charges.

Table 8. Financing & related business support.

Required action	Frequency	Percent
Financing support	2	20
Central bank to consider the reduction of interest rates on loans.	3	30
National Government action on the support	5	50
Total	**10**	**100**

ii) *Perceptions of lockdown*
Responses on the lockdown are presented in Table 9. The results show that most (6, 46%) of the respondents wished that a total lockdown be done to control the virus's spread and manage it quickly

and thoroughly. Some of the respondents (3, 23%) suggested that the curfew should have conditions and laid down procedures then ultimately be lifted. The respondents who didn't want the lockdown were 2, 15%. They argued that adherence to social distancing was enough, and the sale of alcohol should be allowed. Those respondents who suggested that the lockdown is for 21 days were 8 percent while those offering curfews be also extended 8 %.

Table 9. Perceptions of lockdown.

Action required	Frequency	Percent
Lockdown	6	46
Enforce a lockdown for 14 days	1	8
Curfew hour to be extended to 9.	1	8
Lessen the trade restrictions, e.g., curfew	3	23
Let us businesses operate with compliance	2	15
Total	**13**	**100**

iii) *Other measures to be taken*
Responses on other measures were presented in Table 10. The results show that most of the respondents suggested that the government reduce the effects of Covid- 19. Other respondents viewed it as a growth opportunity, felt the police harassment affected supplies. Some also expressed that the government should build program livelihoods to support the vulnerable and underprivileged in society.

Table 10. Action required by frequency and percent.

Action Required	Frequency	Percent
Cushion the business community	6	60
The crisis itself is a growth opportunity	1	10
Police harassment affected supplies	1	10
Mass testing should be conducted	1	10
Support the needy	1	10
Total	**10**	**100**

6 CONCLUSIONS

Based on study findings, concluded the majority of businesses in Uasin Gishu County operated within Eldoret town. Most of them have less than ten (10) employees and can be categorized as micro small and medium-sized enterprises (MSME). Commonly, enterprises have also been in business for more than five (5) years. Indicatively most of the companies rely on sales revenue for operations. Their primary source of capital is savings, debt, and equity.

COVID-19 pandemic has had unique effects on the business environment globally. It has seriously affected the operations and functioning of enterprises to the extent of threatening their very survival. For example, in Uasin Gishu County, the members of the KNCCI have intimated that their sales have dropped significantly due to travel restrictions, stay home, social distancing, and curfew orders, which triggers a reduction of traffic along commercial thoroughfares, drastically sinking consumption of goods and services.

During the period of the pandemic, it was reported that some businesses had closed completely (9%); most of them are operating at 25% capacity; some are operating at 50%, while a few still run at 100% capacity. The hospitality industry is the worst affected, followed by manufacturing, construction, real estate, and ICT. The businesses project a grim performance for the year 2020 compared to 2019 due to the situation created by the COVID-19 pandemic crisis. Some businesses have resorted to carrying out their business online and engaged in-home deliveries. Managing social distancing for customers and employees in their premises, awareness creation about the pandemic, and providing sanitization facilities (Chinazzi et al. 2020).

The study that demand for some products existed. However, it was difficult for businesses to supply them due to the stringent COVID-19 health and safety measures. Some supply chains have been disturbed and rendered limited, while others have become ineffective. Exporters in the agribusiness industry, particularly those dealing with horticulture, have been the worst hit. Despite good production, they cannot deliver their produce to international markets due to the complete lockdown of international air transport besides their international customers' inability to buy the product due to global economic downtime. Arising from this situation, the risk of these businesses failing due to their failure to meet their obligation, and eventually losing their key markets becomes a reality.

Micro and small and medium-sized enterprises' survival seems to be most threatened to owe to their dependence on demography of customers who rely on daily wages. These kinds of customers may have had taken wage-cuts or may even have lost their jobs due to their employers' inability to support them due to the pandemic.

The study established that businesses during the pandemic are unable to honor their obligations owing to low sales. This has affected existing employment relationships. The remuneration of employees has been involved, with some businesses considering salary cuts for its employees. Other companies have resorted to releasing part of its workforce, mostly casual laborers, with others delaying salary payment.

Similarly, some businesses have experienced challenges in complying with their tax obligations. Companies, on the other hand, should develop strategies for mitigating potential tax liabilities. Besides these, businesses have had challenges meeting their rent and leasing obligations.

7 RECOMMENDATIONS

The resulting context requires crafting mechanisms for businesses and government (both County and National) through relevant agencies to co-create solutions and develop best practices that can lessen coronavirus's effects on the business sector. It is also necessary for these agencies to build a support system and mechanisms for resilience and adaptation for vulnerable enterprises across supply and value chains and citizens.

Towards this end, the government should create a disaster stimulus fund to revive businesses primarily affected. Pay daily wages for the elderly, marginalized, and underserved population; a cash transfer system should be developed and executed during the pandemic to spur consumption and therefore buoy up demand for essential goods and services, thus keeping businesses afloat.

Further, it will be necessary to develop and implement emergency policies that will ease the production and consumption of locally manufactured goods and services. This is expected to substitute imports and, through upscaling, make local enterprises build export orientation. Such policies should target the utilization of local resources and easing the process of importation of manufacturing inputs. The result is building local production capacity, creating employment, utilizing local resources, foreign exchange savings, and the Kenyan economy's independence.

As part of the economic stimulus, the government needs to consider tax relief or reduction and possibly waive administrative fees for individual key businesses that provide necessary goods and services. Besides, it will be important for the government to increase subsidies for vital and essential programs to spur economic activities.

The government together with other stakeholders, needs to seek partnerships and agreements to support businesses through the access of long-term low-interest financing and interest-free LPO financing for international markets, which shall enable traders to expand their sourcing markets and have access to capital.

Through relevant training and research, stakeholders must imprint entrepreneurial life skills and knowledge to survive in such tumultuous situations. Further, similar problems will be dealt with in the future by developing and executing risk management strategies that effectively serve their business

. To cushioning businesses from severe effects on its operations and functioning, possible government stimuli must stabilize and maintain the business environment's integrity. It is also crucial for both National and County Governments to create a great atmosphere and opportunities for businesses to manufacture personal protective equipment (PPEs), necessary medical supplies, and equipment such as ventilators to supplement imports. The Jua Kali sector can take centre stage in innovating and production of some medical equipment.

Finally, the requirement for hand washing and sanitization if expensive for small enterprises. This requires support from especially county governments by ensuring a continuous supply of water and subsidizing the cost of water. It is also recommended that there is a need to install UV light and fumigation sanitization systems in strategic locations in urban areas to ensure the containment of the virus and build public confidence.

REFERENCES

Ali, S. A., Baloch, M., Ahmed, N., Ali, A. A., & Iqbal, A. (2020). The outbreak of Coronavirus Disease 2019 (COVID-19)—An emerging global health threat. *Journal of infection and public health*.

Bassetti, M., Vena, A., & Giacobbe, D. R. (2020). The novel COVID-19 (2019-nCoV) infections: Challenges for fighting the storm. *European journal of clinical investigation*, 50(3), e13209.

Beck, T., Degryse, H., De Haas, R., & Van Horen, N. (2018). When arm's length is too far: Relationship banking over the credit cycle. *Journal of Financial Economics*, 127(1), 174–196.

Chinazzi, M., Davis, J. T., Ajelli, M., Gioannini, C., Litvinova, M., Merler, S., . . . & Viboud, C. (2020). The effect of travel restrictions on the spread of the 2019 novel coronavirus (COVID-19) outbreak. *Science*, 368(6489), 395–400.

COVID, T. C., & Team, R. (2020). Severe Outcomes among Patients with Coronavirus Disease 2019 (COVID-19)-United States, February 12-March 16, 2020. *MMWR Morb Mortal Wkly Rep*, 69(12), 343–346.

Ebrahim, S. H., Ahmed, Q. A., Gozzer, E., Schlagenhauf, P., & Memish, Z. A. (2020). Covid-19 and community mitigation strategies in a pandemic.

Holshue, M. L., DeBolt, C., Lindquist, S., Lofy, K. H., Wiesman, J., Bruce, H., . . . & Diaz, G. (2020). First case of 2019 novel coronavirus in the United States. *New England Journal of Medicine*.

Huang, C., Wang, Y., Li, X., Ren, L., Zhao, J., Hu, Y., . . . & Cheng, Z. (2020). Clinical features of patients infected with 2019 novel coronavirus in Wuhan, China. The Lancet, 395(10223), 497–506

Khafaie, M. A., & Rahim, F. (2020). Cross-country comparison of case fatality rates of COVID-19/SARS-COV-2. Osong Public Health and Research Perspectives, 11(2), 74.

Ndii, D. (2020). Thoughts of a pandemic, geoeconomics, and Africa's urban sociology. *The Elephant*, 25.

Paules, C. I., Marston, H. D., & Fauci, A. S. (2020). Coronavirus infections—more than just the common cold. Jama, 323(8), 707–708.

Potluri, R., & Lavu, D. (2020). Making sense of the Global Coronavirus Data: The role of testing rates in

understanding the pandemic and our exit strategy. Available at SSRN 3570304.
Wang, G., Zhang, Y., Zhao, J., Zhang, J., & Jiang, F. (2020). Mitigate the effects of home confinement on children during the COVID-19 outbreak. *The Lancet*, 395(10228), 945–947.
Weigold, A., Weigold, I. K., & Russell, E. J. (2013). Examination of the equivalence of self-report survey-based paper-and-pencil and internet data collection methods. *Psychological methods*, 18(1), 53.
Wong, J., Goh, Q. Y., Tan, Z., Lie, S. A., Tay, Y. C., Ng, S. Y., & Soh, C. R. (2020). Preparing a COVID-19 pandemic: a review of operating room outbreak response measures in a large tertiary hospital in Singapore. *Canadian Journal of Anesthesia/Journal canadien d'anesthésie*, 1–14.
Wu, J. T., Leung, K., & Leung, G. M. (2020). Nowcasting and forecasting the potential domestic and international spread of the 2019-nCoV outbreak originating in Wuhan, China: a modeling study. *The Lancet*, 395(10225), 689–697.
You, C., Lin, Q., & Zhou, X. H. (2020). An estimation of the total number of cases of NCIP (2019-nCoV)—Wuhan, Hubei Province, 2019–2020. China CDC Weekly, 2(6), 87–91.

Advances in Phytochemistry, Textile and Renewable Energy Research for Industrial Growth – Nzila et al. (Eds)

Responsible consumption and production for wholistic transformation of universities: Campus innovation through a sustainable livelihoods approach

Roger A. Petry*
Luther College, University of Regina, Regina, Saskatchewan, Canada

Jocelyn Crivea
Institute for Energy, Environment and Sustainable Communities, University of Regina, Regina, Saskatchewan, Canada

ABSTRACT: Of the 17 UN *Sustainable Development Goals* (SDGs) adopted as global goals from 2015 to 2030, Sustainable Development Goal #12 on *Responsible Consumption and Production* has been described by some as the "heart" of the Goals: our production systems are central to meeting our human needs and aspirations along with important dimensions of human well-being (SDGs #1-5); they shape and organize our economic practices, our choice of material inputs, and their efficient and effective long-term sustainable use (SDGs #6-11); they also ensure whether or not we live within the carrying capacity of our natural ecosystems and whether these systems are healthy and resilient (SDGs #13-15). Yet in addition to these social, economic, and environmental sustainability goals, SDG 12 is central at a deeper *cultural* level. It enables (and is enabled by) peace, justice, and strong institutions (SDG 16), and motivates new productive partnerships between a variety of organizations at all geographic scales (SDG 17 on "Partnerships for the Goals"). This paper explores the cultural dimensions of SDG 12 through its potential for a whole institution approach on university campuses. In particular, its role in reshaping university governance and scholarly identities (particularly around the concept of sustainable livelihoods and political dimensions as discussed by Ian Scoones (2015), understandings of scholarly impact, and greater inclusivity of our disciplinary specializations (through living laboratories and development of transformative technologies) are explored. The University of Regina in Saskatchewan, Canada, is then used as a case study to illustrate these cultural dimensions in the recent development of its strategic plan and living laboratories on campus. The paper demonstrates that a traditional sustainable livelihoods approach to development can be used to create a new *sustainable scholarly livelihood* identity on campuses. This new scholarly identity allows for substantive transformations in university governance to advance campus sustainability, address core livelihood anxieties within communities, and develop key innovations in the scholarship of sustainable livelihoods.

Keywords: sustainable consumption and production, sustainable development goal 12, sustainable livelihoods, cultural sustainability, whole-institution approach, whole-university approach, university governance, living laboratories.

1 INTRODUCTION

In 2015 the United Nations adopted 17 Sustainable Development Goals (SDGs) to advance the global development agenda from 2015 to 2030, replacing the earlier Millennium Development Goals (2000–2015). As "global goals," they focus on both developed and developing countries mobilizing governments at multiple geographic scales including national, state, provincial, regional, and local governments. Other sectors, however, are also willingly contributing to these goals—something central to achieving the SDGs if we recognize the diverse productive capacities held by each organizational type. These organizations include the business sector, non-governmental and other civil society organizations, faith organizations, and professions. Concerns, however, have been raised that the achievement of the SGDs might be substantially impaired due to the impact of COVID-19 that has exposed significant weaknesses and lack of resilience in the global economy. Attempts are being made to get the economy "back to normal," reflecting pre-COVID-19 days, despite the unsustainability of this earlier system. Depressed markets and lack of investment (particularly in least developed countries) have created fragile businesses, heightened government and

*Corresponding author

DOI 10.1201/9781003221968-4

citizen indebtedness, and loss of individual and corporate resources for charitable giving to non-profit organizations and faith communities. These recent developments have undermined what had been seen as organizational pillars in achieving the SDGs. With the impairment of global production systems, SDG 12 on responsible (or sustainable) consumption and production (SCP), viewed, by some as being "at the heart and soul" of the 2030 Agenda (Paul 2018), have taken even greater prominence.

A key organizational sector, however, that perhaps has more resilience due to its traditional countercyclical nature to economic activity and its own organizational autonomy is Higher Education (see Petry & Ramkat 2019). Even prior to the COVID-19 crisis, Higher Education (HE) already had been recognized as having important contributions to make to the SDGs through its knowledge production and dissemination activities. Longstanding global scholarly organizations, such as the United Nations University (UNU, established in 1972) and the International Association of Universities (IAU, established in 1950), for example, had mobilized university-led initiatives to generally advance sustainable development and now, more specifically, the 17 SDGs. Strategic initiatives include the UNU's Regional Centres of Expertise (RCEs) on Education for Sustainable Development (ESD) and the IAU's Global Cluster *Higher Education and Research for Sustainable Development* (HESD), established in 2019 (IAU 2020; UNU 2020). In both cases, integrative approaches have been sought: a "whole-institution approach" for universities in the case of the IAU initiative, and a "regional approach" integrating Higher Education and community partners in the case of RCEs. Such whole-institution HE approaches recognize the significant unknowns associated with how to achieve sustainable development, thereby acknowledging the prudential need for inclusion of the full range of scholarly disciplines. They also recognize the value of participation of entire organizations beyond the domains of traditional scholarship (e.g., staff, administrators, alumni) and the value of partnering with other organizations in effective and efficient resource mobilization. In part, this is again tied to the uncertainty of such investments, both in terms of how to best meet the multiple dimensions of well-being of populations in diverse ecological and social contexts in the present generation, much less for future generations.

But a whole-institution approach within universities faces multiple challenges. These are sufficiently substantial that individual scholars historically have sought to create innovations in scholarly methods *outside* the traditional academy, such as the Royal Society of London (that gave rise to modern science) and the formation of the Trilingual Colleges (that gave rise to the humanities in Northern Europe; see Petry 2014). Some of these academic barriers are historic, such as commitments to disciplinary centered research (vs. inter/multi/trans-disciplinary community engaged research) and traditional scholarly outputs (such as books and journals) that constrain research and teaching from being measured, in part, by its transformation impact on development patterns for sustainability. Some of these academic barriers, however, are historically more recent, such as the corporatization of the university over the last three decades (Polster & Newson 2015). This corporatization has led to the erosion in collegial governance through the centralization and growth of university administrations whose focus includes the commercialization and privatization of knowledge to support particular industrial and high tech sectors through new products and services. Some disciplines have been eroded through the focus of resources on particular disciplines viewed as valuable in the knowledge economy (such as STEM disciplines). The scholarly profession is itself being dismantled. This unbundling includes the hiring of precarious academic labor solely for teaching, research-only faculty, and the assertion of organizational control of intellectual property scholars generate (either by private companies or university administrations) to generate revenue. The reengineering of universities and their scholarly outputs to serve the economic needs of business (including universities seen as businesses) and government for short-term economic growth and the profitability of existing for-profit entities only reinforces the existing patterns of unsustainable development. Global market production does not significantly factor in those principally left out of the market (poor and marginalized individuals and communities who have little money with which to consume), much less not yet existing future generations and non-human species that can effect little (or no) market demand (at least on their own). Yet these three areas are the primary focus of sustainable development. What is also precluded is a scholarly focus on the *best types* of scholarly questions enabled through traditional academic freedom and the investigator-driven pursuit of knowledge. In the case of education for sustainable development (ESD) it would be those questions that lead to the most highly transformative knowledge, innovations, and dissemination techniques for sustainable development—whether or not this occurs through existing market organizations, traditional patterns of investment and employment, or even markets at all. Scholars need to consider that some of the better (or perhaps best) forms of sustainable development may rely on new, not yet existing firms (including small and medium sized enterprises (SMEs) or co-operatives), non-business organizations (such as government, households, or the non-profit sector), or non-market activities (such as volunteerism).

If universities as a whole are to be mobilized for sustainable development, a whole institutional approach must engage the deeper cultural dimensions of the university. It is these cultural dimensions that both unify universities as organizations while protecting diverse forms of scholarship. The foregoing suggests that earlier scholarly identities relying on academic freedom and collegial governance, both informing and modeling ideas of free citizenship within democracies, have been inadequate to resist corporatization and the growth in centralized administrations (particularly entrenched with centralized organizational responses to the COVID-19 crisis). Some new scholarly identity needs to reconcile older scholarly institutions while taking seriously the market forces shaping the corporate university and the social, economic, and ecological imperatives of sustainability. A relatively new identity (from a historic perspective), that of "sustainable livelihoods" (SL), has emerged since the early 1990s (Scoones 2015, p. 5–9). The "sustainable livelihoods approach" (SLA) has been applied to development principally in the non-university context of international development (such as rural areas in developing countries; ibid., p. 11–13). It will be argued that SLs as an identity within the academy would allow universities to embrace their traditions of academic freedom and collegial governance while addressing in a robust way their embeddedness in market norms and corporate institutional frameworks. Case studies will be used to illustrate how a new identity of "sustainable scholarly livelihoods" (SSLs) can create the individual and organizational assets and capabilities needed by scholars to directly engage the more promising avenues of research, teaching, and service to achieve the sustainable development goals.

2 LITERATURE REVIEW AND TERMINOLOGY

As a starting point there are two general research questions that this paper seeks to explore: can a Sustainable Livelihoods Approach (SLA) be used in the context of the corporatized university to alter its institutional arrangements and "production and consumption" of scholarly materials to strengthen the traditional scholarly identity (tied to individual academic freedom and collegial university governance) through a new identity of Sustainable Scholarly Livelihoods (SSLs)? Secondly, how might a university be restructured to advance this shared identity to enable innovation for responsible (sustainable) consumption and production within and outside the university context? A review of the scholarly literature combining the concept of "sustainable livelihoods" with "university governance" and/or "scholarly identity" indicates that the sustainable livelihoods approach has not been employed as a conceptual tool in framing debates in either of these other two areas. This is despite extensive and ongoing publications related to each separately. As will be discussed later, Petry, a co-author of this paper, proposed the merits of such an analysis in the conclusions of a paper in 2018. This paper seeks to advance this current gap in scholarship.

Prior to elaborating on the concept of SLs, it is important to recognize how the terms "institution" and "organization" are being employed. Douglas North (1990) argued that institutions should be understood as "the rules of the game" with organizations as "the players." More generally, this paper follows new institutionalism theory that sees institutions as "sets of rules, decision-making procedures, and programs that define social practices, assign roles to the participants in these practices, and guide interactions among the occupants of individual roles" (Young 2002, p. 5). These institutions or rules are those that are actually practiced (vs. codified rules in documents that are not actually followed). Universities themselves are understood as *organizations*, namely, "material entities with employees, offices, equipment, budgets, and (often) legal personality" (ibid.). New institutionalism has its own evolution as a disciplinary study with shifting emphases on institutions as a means of understanding organizational efficacy in achieving wider social goals, to examinations of institutions in their own right, to a return in focus on the organization as the key concern and level of analysis (Greenwood, Hinings, & Whetten 2014, p. 1207–1211).

This paper chooses not to take sides in this theoretical debate in light of the research questions posed. To the extent we are concerned with responsible (or sustainable) consumption and production, *institutions* as rules are constitutive of and shape the actual scholarly productive practices within universities. This focus on institutions shaping complex and interrelated systems of rules is implicit in UNESCO's (2014) definition of whole-institution approaches: "Whole-institution approaches encompass mainstreaming sustainability into all aspects of the learning environment. This includes embedding sustainability in curriculum and learning processes, facilities and operations, interaction with the surrounding community, governance and capacity-building" (p. 30). On the other hand, the merits of mobilizing the actual resources of the university as organizational wholes by linking its organizational structures or functional units to advance SDG 12 (and the 17 UN SDGs more generally) demands, simultaneously, an *organizational* focus. McMillan and Dyball (2009) discuss the importance of a "whole-of-university approach" to sustainability that "explicitly links the research, educational and operational activities" of the university, "engag[ing] students in each, rather than confining their education solely to the classroom" (p. 56). Both an *institutional* and *organizational* focus are also key elements of a SLA (and as will be seen, a Sustainable *Scholarly* Livelihood Approach (SSLA)). For example, institutions shape the livelihood strategies one might pursue (including one's access and claims to resources and how they can be used) while organizations (through their collective actions and allocation of resources) can act as both barriers to livelihoods or can create mutually reinforcing livelihood strategies. Sustainable livelihood frameworks, such as that of Scoones, commonly have "institutions and organizations" as part of their modeling given how these influence "access to livelihood resources and composition of livelihood strategy portfolio" (2015, p. 36, Figure 3.1).

3 METHODOLOGY

In seeking to advance a whole-institution and whole-university approach to advancing the SDGs in universities this paper chooses to employ the concept of SLs. This choice needs to be defended in light of the range of options one might choose as proposed "drivers of the integrative process" (terminology used by McMillan & Dyball 2009, p. 58) needed for these wholistic approaches. Other drivers could include, for example, the implementation of high-level organizational policies for sustainable development putting in place specific institutional rules (say, around sustainable procurement) or the creation of new organizational structures, such as staffing a campus sustainability office. In an earlier publication, Petry (2018) a co-author of this paper, raised a number of skeptical questions about the necessity and adequacy of campus sustainability policies and organizational resourcing of dedicated sustainability offices through centralized administrations. While these arguments will not be reintroduced here they point to the need for further, alternative drivers of integration. In his concluding remarks Petry introduced the potential of a SLA as a lens for critiquing and reforming specific university policies, especially in response to university corporatization (ibid., p. 9–11). However, this, effectively, was still seeing the needed intervention at purely a policy governance level. This paper seeks to illustrate how a SLA can transform universities at a much deeper, more profound level when employed as part of a values framework that seeks to shift the *entire culture* of modern universities, adjusting scholarly institutions and organizational activities and resourcing to support the ethical concerns implicit within a SLA.

To intervene at this cultural level within a particular kind of organization (in this case universities), we need to understand what is captured by the concept of *culture* more generally. Sutter (2016) citing Worts (2006, 2011), defines culture as "a pervasive and evolving suite of values, beliefs, attitudes and behaviours" (p.3). Sutter then draws upon the work of Dyball and Newell (2015) to talk about the importance of culture in advancing sustainability: "culture can provide an effective perspective for sustainability work, because it is rooted in the values that drive our individual and collective behaviours, and it responds and contributes to the complex systems that govern so much of our increasingly globalized world" (ibid.). This, in turn, necessitates "events and opportunities that encourage cultural reflection...at both individual and collective levels if humanity is to remain in sync with the changing world" (ibid.). It will be argued and illustrated through two avenues that a SLA provides an appropriate set of questions needed for the cultural reflection and organizational and institutional transformation within universities to advance sustainable development at this time. SLs as a new sharable identity for scholars also provides an important means to preserve the important cultural heritage tied to traditional scholarship, including its non-formal, intangible dimensions.

If there is a central contribution of SLs to enriching the culture of universities it is at the level of values, particularly those that are foundational and shared across the academy that inform a shared scholarly identity. For example, the earliest universities embraced the methods of argument of the ancient philosophical academies with the expectation that beliefs would be formulated, supported by, and evaluated against rational standards. Scholars also value reproducible and rigorous scholarly methods (such as the scientific method) that are the basis for accepting new disciplines into the university. To these might be added the value of the skepticism scholars apply when studying and interpreting written texts (something championed by the early humanist scholars). Scholars also share ethical values implicit in the ethical criteria employed when approving research, especially research involving humans and other living organisms. With the advent of the corporate university, core values around academic freedom, investigator-driven pursuit of knowledge, and collective resourcing towards answering "the best questions" (as determined through processes of collegial governance) have received considerable attention, in part, due to their erosion.

What particular value, then, might SLs contribute to the scholarly identity in response to the corporatization of the university, the dominance of market organizations in the global economy, and the call of SDG 12 to advance more responsible and sustainable patterns of consumption and production? For those not familiar with the academic literature around livelihood analysis (including the components of a livelihood) and the normative values implicit in sustainable livelihoods it is worth noting briefly that livelihoods capture the diverse ways one makes a living (for a good overview see Scoones 2015). While government policy makers frequently focus on formal employment generation as the primary end of public policy (and a key way we are to make a living), livelihoods themselves are more than a job and potentially involve a diverse set of market and non-market strategies. A diverse set of market roles are also incorporated into the typical analysis of individual livelihoods. Here one acts as an *owner* and *investor* in one's own assets (including one's human capital), as a *board of directors* setting the livelihood outcomes one chooses to pursue, as a *manager* of one's resources or assets (including one's own labor) in advancing livelihood strategies to achieve these outcomes and mitigate risks, and as an *autonomous corporate entity* that contends with other livelihoods and organizations, while concerning itself with the *total stocks* of natural, human, social, physical, and financial capital upon which one's own livelihood strategies depend. A *sustainable* livelihood generalizes (or universalizes) one's concerns for one's own livelihood improvement to include those of others (especially the most marginalized and future generations), thereby establishing ethical boundaries and opportunities in choosing specific strategies. An early definition of SL by Chambers and Conway (1992) reflects both these pragmatic and normative or ethical concerns: "[a] livelihood is sustainable when it can cope with and recover from stresses and shocks, maintain and enhance its capabilities and assets, and provide sustainable livelihood opportunities for the next generation; and which contribute net benefits to other livelihoods at the local and global levels and in the short and long term" (p. 6).

While it might be sufficient to see a SLA directly engaging the *culture* of markets and corporate entities (including corporatized universities) by employing their roles and conceptual tools this, in itself, does not address the deeper question, namely the motivation(s) that underlie this market culture. Do SLs fulfill or address these deeper motive(s) at the heart of market culture? Here early comments by Aristotle (1946) in his *Politics* are helpful. At a time when markets were gaining prominence, Aristotle was trying to make sense of what motivated some to pursue unlimited acquisition of wealth through retail trade, levels of wealth that went well beyond what they needed to meet the needs of their own households (p. 25–26; 1257b–1258a). For Aristotle merchants engaged in this kind of acquisition were motivated differently from those who transacted as traditional managers of the household, thereby acquiring materials needed to fulfill basic human needs that contributed to well-being—a form of wealth acquisition that had limits. Instead those engaged in retail trade were motivated by "a general anxiety about livelihood" (p. 26; 1257b), an anxiety that lent itself, in Aristotle's view, to limitless acquisition of wealth. While a SLs framework ordinarily incorporates the fulfillment of basic human needs in its set of livelihood outcomes (the concerns of Aristotle's household manager), the framework itself deals much more profoundly with addressing livelihood anxiety generated by a range of uncertainties from one's livelihood context beyond one's control. These include whether there is predictable access to assets (or resources) from which to construct one's livelihood strategies, the hazards to which these assets are exposed, changing capabilities shaping one's viable livelihood strategies tied to shifting institutional and organizational contexts, uncertainties in managing various livelihood activities to produce outputs (e.g., goods or services), whether or not these outputs achieve one's desired livelihood outcomes (e.g. desired components of well-being), and whether, ultimately, one's chosen livelihood outcomes are even well founded. All of these uncertainties compound one's livelihood anxiety. Yet strategies to advance sustainable livelihoods in a wholistic way can realistically relieve all these anxieties, likely more constructively than the unlimited resource acquisition exhibited by Aristotle's merchant class.

That this livelihood anxiety is a central concern of our times, particularly in an area of high levels of joblessness—even before COVID-19—will not be argued for here. Addressing this anxiety is a shared motivation for those engaging in market activity and is reflected in modern, for-profit corporate structures that also seek to amass wealth without limit. Yet even if livelihood anxiety is generally held, can a SLA be successfully integrated into the basic culture *of universities*—especially in light of the relative lack of structural uptake of the SLA elsewhere? Even in areas where it has been employed (such as development agencies seeking to advance rural agrarian livelihoods especially in developing countries—a focus of Scoones (2015, p. 11)), Universities are a very different livelihood context. In order to answer the general research questions of (1) whether a SLA can be used to create a new (and viable) Sustainable Scholarly Livelihood (SSL) identity and (2) how the corporatized university might be restructured to advance this shared identity for innovation for sustainable consumption and production, two avenues will be explored. The first avenue provides general argumentation for the viability of implementing a SLs identity in universities, restructuring their institutions guiding their production processes and the organization itself through changes in governance. This will be done by addressing four core political areas of the SLA identified by Ian Scoones (2015) needing to be addressed to revitalize livelihood analysis in a given context (p. 110). These are "the politics of interests, individuals, knowledge and ecology" (ibid.). The second avenue looks at a particular university case study—the University of Regina (U of R), in Saskatchewan, Canada—to illustrate the universities engagement of sustainability at a cultural level, both at the level of the entire organization and also in a particular area of production and consumption. Here the university's most recent strategic planning process is presented to illustrate how the wholistic mobilization of the general membership of a campus can occur attentive to its shared cultural concerns while addressing the 17 SDGs. A specific goal emerging from this plan, the development of campus living laboratories is then discussed in light of existing U of R programs focused on the sustainable production and consumption of food on campus that sustain diverse livelihoods.

4 RESULTS AND DISCUSSION

4.1 *A sustainable livelihood approach to revitalizing the scholarly identity in universities*

The previous discussion has suggested that a core motivation for pursuing sustainable livelihoods is to address various types of anxiety about livelihood. If we are to implement the ethics of SLs within the cultural context of universities, thereby shaping its production and consumption processes, we need to step back to first ask, what is the core value(s) shaping the current scholarly identity that informs these scholarly institutions and processes? While some of these have already been outlined, one underlying core value that informs the need for investigator driven pursuit of knowledge is that of curiosity or wonder, whether about practical or theoretical questions. This curiosity shapes the research questions that mobilize entire processes of scholarship. The individual quest to satisfy one's burning questions alongside others in an academic community is guarded by scholarly institutions, such as academic freedom and tenure, and organizationally through collegial governance. If this is a core value (or perhaps *the* core value) of universities, the ability to mobilize scholars to embrace SLs as part of their identity will depend on how livelihood anxiety within a scholarly context is currently being generated in relation to this core value satisfying curiosity or wonder is being jeopardized. This anxiety within the scholarly community is then analogous to the grain of sand that irritates the oyster to create a pearl. With this tension in mind we can now address the four core political areas Scoones identifies as needed for revitalized livelihoods analysis (2015, p. 109-116), and consider what would be done differently if campuses shared commitments to "sustainable scholarly livelihoods" (SSLs).

The politics of interests

Scoones' "politics of interests" identifies how livelihood opportunities are shaped by one's interests but that these interests are, in turn, shaped by "the structural features that define our own lives" (p. 110). That curiosity has been a central, shared interest for mobilizing resources within scholarly communities is identified very early on. Aristotle in describing the origins of the pursuit of theoretical knowledge (and philosophy more generally) grounds it in a general desire found in all humans to know things for their own sake (1947, p. 243–249 (980a–983a)).While this shared interest is embedded within the modern university, as Scoones notes one also needs to consider the broader political economy in which livelihoods (in this case scholarly livelihoods) are situated. Scoones points to impacts on livelihoods of "intense globalization under neoliberalism," which includes "the appropriation of resources for livelihoods through commodification and financialization" (p. 111). These forces have led, in part, to features of the corporatized university including the commodification of teaching materials as well as research outputs (through applying intellectual property rights to these outputs and knowledge commercialization). The separation of these ownable materials from the scholar, now understood as a knowledge worker, leads to further unbundling of the scholarly profession.

How might a SLA address this? Up until recently, the integration of the scholarly identity was assumed. All scholars were expected to do research, teaching, and service (whether to the scholarly or wider community). This meant that the scholar was highly portable and self-sustaining. No matter what occurred historically to universities as organizations (whether plagues, fires, or barbarian invasions, etc.), the integration of this identity meant scholars could sustain themselves as individuals and form new scholarly communities as needed while being generally welcomed as a resource by a new community. A SL framework that takes as its start individual livelihoods naturally scales down to that of the traditional "lonely scholar" seeking answers that satisfy his or her own curiosity. This points to the need to reintegrate the individual scholarly identity. The rebundling of the scholarly identity necessitates the rejection of university contractual relationships that remunerate only one element of the scholarly identity (whether it be just research, teaching, or administrative service). In the classroom, such rebundling would also view students as teachers and faculty as learners. Scholars have also traditionally sustained their scholarly activities through shared assets, including their libraries and learning spaces (such as laboratories). In response to corporatization and commodification of the scholarly materials of universities, a sustainable *scholarly* livelihood (SSL) would suggest scholars fully embracing open licensing of research and teaching materials along with open patenting of university innovations (or minimally sharing with scholars in universities that are legally committed to employing all their resources towards scholarly purposes). To further advance campus sharing, other elements of the sharing economy need to be embraced for scholarly purposes. This would include sharing of buildings, lands, tools, and equipment for scholarly ends. For example, university libraries would now need to expand to include campus tool libraries.

To advance the "whole-institution approach" a university as a corporate body could view itself as its own SL writ large, in the service of the many SSLs that comprise it. This would demand a de-bifurcation of the university. Non-academic university administrators and staff serving the academic community would be contracted to share the SSL identity alongside traditional scholars. At the same time, political science and administration departments could explore how the bicameral university's governance model (separating academic functions from financial management) and corporate divisions (between management and labor) could be altered so that scholars themselves control boards of governors and the allocation of material resources of the university. In a university setting this is enabled both by traditions of collegial governance but also by faculty members themselves being, through tenure, effectively treated as long-term (human) capital assets (and therefore investments) of the university. Universities would then become effectively a kind of "worker cooperative" (an observation made by Tirole 2017). The university viewed as its own SSL would also seek to internalize its production and consumption processes, where possible, to minimize risk. This could entail creating its own learning spaces, buildings, and equipment customized to its local ecosystems and social and scholarly preferences. Such campus production and consumption would, by definition, contribute to a circular economy enabling cradle-to-cradle (C2C) production, something seen as central to achieving sustainable consumption and production (SDG 12; Paul

2018). Lastly, if universities each saw the other as a SSL, this would entail a highly cooperative (vs. competitive) relationship between universities. Universities would see themselves as sharing a common planetary fate and the need to preserve the unique knowledges held by each university particularly as it relates to the sustainability of their respective regions.

The politics of individuals

Scoones (2015) notes that in livelihood analysis one needs to also focus on "actor-oriented approaches" focused on "human agency, identity and choice," approaches that drill down "to what individuals think, feel and do"—their "behaviour, emotion, and responses" and the intensely personal dimensions of who we are (p. 111). These are tied, for example, to the "the politics of the body, gender and sexuality" (ibid., p. 112). If we are to incorporate and advance this dimension of the SLA in forming new SSLs, we need to understand the relevance of these features to university scholarly production processes. At one level universities are ideal as these personal dimensions are the object of study of multiple academic disciplines (including entirely new disciplines). Universities also have the qualitative and grounded theoretical methodologies (among others) that can generate this knowledge. What would be the contribution to a SSL? In this case universities would care not only about institutional and organizational features tied to scholars in general (and perhaps replicable across universities) as set out in *The Politics of Interests* (see above). Rather a SSL approach would include concerns of the particular scholars making up a given university campus and how their scholarly livelihoods could be made more sustainable through this study. This would enable a profound self-understanding of scholars that connect these dimensions of self to the areas of traditional livelihood analysis, namely understandings of hazards and risk, assets, institutional and organizational barriers and opportunities, viable livelihood strategies, and desired livelihood outcomes. It is the latter which would be most affected as many outcomes tied to dimensions of well-being are very personalized, tied to a particular individual's subjectivities, self-understandings, and context, and not generalizable. Robust livelihood fulfillment in general, much less the fulfillment of a *scholar's* livelihood would seem almost impossible in the absence of this concern with individuals. Universities as "whole-institutions" seeking to see themselves as their own SSL would require this self-understanding of the particular scholars making up the university and need to put in place supportive institutions and appropriate changes to their own governance models.

The politics of knowledge

Scoones (2015) notes that in livelihood analysis "whose knowledge counts?" (p. 112) is a key question along with the question raised by Robert Chambers (1997) of "whose reality counts?" Both of these questions for Scoones have shaped policy debates about "[w]hich versions of whose livelihood is seen as valid and which is seen as deviant and in need of change" (2015, p. 112). At its root this is a normative debate about "what makes a good livelihood?" (ibid.). In terms of this debate, universities are well equipped to the extent, again, there are entire disciplines dedicated to these questions. In particular, philosophy explores the underpinnings of all three questions in the sub-fields of epistemology, metaphysics, and ethics. These scholarly resources could be readily mobilized to address these questions. To the extent these debates are often tied to marginalized peoples whose livelihoods have historically been undervalued due to, for example, colonialism and racism, again universities have the scholarly resources to critique and re-evaluate earlier erroneous judgments by listening to these "unheard voices" and offering new and multiple forms of assessment—something also of concern for Scoones (p. 113–14). These new, more wholistic forms of livelihood assessment would be especially beneficial at this time in assessing scholarly livelihoods, particularly where scholarly output is measured by a narrow assessment of publications in specific formats (journals and books) without consideration of its broader impact on how these transform the university or the wider society for sustainable development. Importantly, universities have the opportunity in shaping SSLs to validate multiple knowledges, especially those of Indigenous peoples, through recognizing these as contributing important methodologies and discoveries to the scholarly community. These knowledges can provide important insight into more sustainable livelihood strategies pursued both on and off campuses. A SSL identity on campuses can also serve to protect a diversity of ways scholars choose to question, pursue methods to answer these questions, and apply and disseminate knowledge, especially in ways that are optimal for long-term sustainable development. This, then, directly challenges external efforts to direct university research through primarily STEM disciplines for knowledge privatization and product commercialization. Commercialization efforts of university knowledge aimed at sustainable development, particularly that which generates sustainable livelihoods, have been shown to be especially problematic in particular university settings (see Petry 2008).

The politics of ecology

Scoones (2015) begins by emphasizing rapid environmental change impacting livelihoods and the recursive relationship between ecology and politics, with each shaping the other (p. 114). Livelihoods need to be able to rapidly change to respond to ecological shifts (ibid.). The ethics of sustainable livelihoods also presupposes that resulting livelihood burdens and opportunities are maintained "in an equitable and socially just way" including between globally interconnected regions (ibid.). Again, universities are at a distinct advantage at addressing these multiple concerns. Universities have departments of biology and environmental studies, specialized scholars in the area of political ecology and environmental ethics, and are traditionally one of the most globalized organizations—long before economic globalization. From the perspective of individual scholars, the scholarly livelihood is traditionally sustained through an examination of human and non-human natural systems. Natural environments serve as a source of questions, experimental models, inspired solutions, and places for testing hypotheses. For this reasons scholars and campuses are frequently situated in natural settings; symbols of knowledge have historically been tied to plants and animals (e.g., the tree of knowledge in Genesis (2:17)). Unlike non-university environments where environmental science is highly contested (due, in part, to the disruptive implications of findings on traditional industries and interest groups), scholars are trained to accept well-supported, peer reviewed findings and through collegial governance are motivated to conform to environmental limits even when other organizations do not (for example, limiting greenhouse gas emissions contributing to climate change). University campuses are, in themselves, sufficiently large (akin to small towns or even cities) that, in doing so, they can effectively model gradual or rapid changes in livelihood and lifestyle, sustainable energy production and consumption, and sustainable building design for other communities. At the same time, given the typical land area of universities, they have ample room for experimentation with new forms of production that increase overall biomass and natural habitats while generating new natural capitals for campus use.

Against these four criteria of Scoones (i–iv), universities seem to be a credible (if not ideal) location for implementation of the SLA and the creation of a new SSL identity.

4.2 *Case Study of the University of Regina: How a whole-institution, cultural, place-based, approach can promote sustainable livelihoods through campus living laboratories*

The UN's Sustainable Development Goals are a roadmap for sustainability and well-being. As higher education addresses sustainable development and the SDGs, universities are impacted and transformed through living laboratories. A university living lab integrates academic teaching and research into campus operations such as planning, infrastructure, and community engagement. The university's campus is used as a lab to test out new ideas, initiatives, and opportunities that may arise in a local/community context. Students, staff, faculty, alumni, community members, and organizations can come together to collaborate. They also learn, from both successes and failures, and in the latter case it can be argued that failures are just as helpful to developing local strategies of responsible consumption and production.

The paper will now illustrate the application of a whole institution, place-based, living laboratory approach in a case study of a comprehensive, mid-size university, the University of Regina (U of R). The U of R also includes three federated college partners: First Nations University of Canada, Campion College and Luther College (the last two have religious affiliations). Treaties between Indigenous peoples (First Nations) and the British Crown (and later the Government of Canada) are formative documents in the development of Canada as a country. The U of R campuses are located on Treaty 4 and Treaty 6 lands. These are the territories of the *nehiyawak* (Cree), *Anishinapek* (Saulteaux), Dakota, Lakota, and Nakota nations, and the homeland of the Métis/Michif Nation (University of Regina 2018). This connection is mentioned in relation to living labs and what Scoones refers to as "appreciating the complexity of people in places and…understanding of the wider, structural and relational dynamics that shape localities and livelihoods" (Scoones 2015, p.113). The U of R has committed to support truth and reconciliation with Indigenous peoples, which involves responsibly sharing the lands that the campus now sits on, establishing mutually respectful relationships (University of Regina 2018), and the University's role in developing and applying knowledge, space, and maintaining relationships with Indigenous peoples and communities.

University of Regina "high level" livelihood strategy: Institutional Strategic Plan

In June of 2020, the U of R's Board of Governors approved a five-year strategic plan entitled *All Our Relations: kahkiyaw kiwahkomakaninawak.* The plan reflects a broad cultural approach. The title acknowledges the interconnectedness of the University with Indigenous peoples both in the use of Cree, one of the languages of the peoples originally living on the territory before the campus was built, and the centrality of maintaining good and healthy relationships between individuals, communities, and humans and other species and the land that are found in Indigenous worldviews. The plan's development was also culturally grounded in the University's own context, being led by a volunteer team of faculty, staff and students, and engaged stakeholders ("students, faculty, alumni, local communities, Indigenous Elders, industry partners, government representatives" and staff) through broad forms of participation (such as town halls, focus groups, world cafés, and other forms of consultation). The resulting content of the plan reflects what people across the campus said that they wanted, based on participation from more than 1,300 individuals. Such engagement is similar to earlier open processes used by the University for campus planning of operations in the *Campus Master Plan 2016,* and sustainability activities and principles outlined in the President's Advisory Committee on Sustainability's *Strategic Plan for Sustainability: 2015–2020.* These plans reflect varying degrees of a whole institution, participatory approach in generating interests and goals to be shared across the entire university, and the role of volunteerism in both their development and mobilizing participation for their resulting implementation.

The resulting plan has five strategic priorities (these "areas of focus" are "discovery," "truth and reconciliation," "well-being and belonging," "environment and climate action," and "impact and identity") along with specific goals to meet in each area. Sustainability remains an overarching area of emphasis for the University, with a view to making decisions that consider impacts on future generations such as a commitment to responsible stewardship of land and resources (University of Regina 2020). As well, the plan aligns the 17 United Nations Sustainable Development Goals (SDGs) with the U of R's five areas of focus, a nod to the local priorities generated through deep local consultation also reflecting global concerns identified through UN policy processes. The plan has a social impact objective of meeting the community's various social, cultural, economic, environmental, and technological needs. This objective includes specific targets such as reducing greenhouse gas emissions, reducing production of waste and consumption of water, and broadening partnerships and connections "with communities in the pursuit of knowledge and discovery projects" (ibid.), all of which can shape campus generated sustainable livelihood strategies.

The University of Regina and a Living Laboratory Approach through Food

The University of Regina in its new strategic plan made a specific commitment to advance learning through living laboratories (University of Regina 2020). According to *All Our Relations* the University's labs are described as a form of

> *...open innovation through modelling and creating real-life environments sustained by partnerships and collaborations for the creation, prototyping, validating, and testing of new technologies, services, products, and systems in real life contexts...a great opportunity to act, monitor, and model sustainability for communities as a Living lab that pilots ways to achieve net zero emission sustainability...[and may] act as a model for future research in additional areas of focus.* (ibid. p.15)

Living labs are conducive to expanding relationships and generating sustainable livelihood strategies that positively impact society. One specific livelihood area of particular relevance for the U of R and its region is food production and consumption. The University of Regina's main campus is located in the capital city of Regina with approximately 215,000 residents—a small city by global standards. The province of Saskatchewan, located in western Canada, has long-standing roots in a prairie agricultural economy. While the economy has diversified in recent years, agriculture is an important contributor to the provincial economy, both domestically and internationally through primarily agricultural exports. With modern agricultural production in Saskatchewan, however, there are substantial adverse environmental impacts such as water pollution and land degradation from agricultural fertilizers, pesticides, and herbicides, high fossil fuel use through energy intensive, large scale cereal grain farming (contributing to climate change), and destruction of natural habitats (through wetland destruction for increased cropping area, conversion of on-land water storage to rapid drainage, and seeding of crop monocultures). Further unsustainable impacts occur through food processing production. Individuals and households influence these impacts as well through their eating choices and habits of food purchasing and domestic gardening. These affect the environment through food-related energy consumption and waste generation, making them an interesting focus for "study" in a living lab related to responsible consumption and production.

At the U of R a systems approach to food (social, economic, cultural, environmental) has developed, exemplifying a campus living laboratory approach connected to livelihood. The U of R has started to transform the physical landscape of the campus by integrating production and consumption of food through establishing campus fruit and vegetable gardens and an apiary (beehive). These initiatives build on strengths from the campus and local community. The food gardens were started by the Regina Public Interest Research Group (RPIRG), a student-funded and run not-for-profit research organization founded in 2011. The University also contributes funds to the gardens through a campus sustainability initiative called the Sustainability and Community Engagement Fund (University of Regina, n.d.). While the campus gardens are planned and delivered by RPIRG they are primarily volunteer-run due to the many relationships established with all scholars (Petry 2018) including faculty members (mostly from science, social science and humanities disciplines), students, staff (including facilities management) and community organizations. RPIRG organizes tours, cooking nights, and seminars on local food production, harvesting, and reducing waste such as composting that anyone can participate in. Produce grown in the gardens is donated to students and individuals in need, with a portion of the harvest donated to Carmichael Outreach, a local community-based organization that provides services to people "who are experiencing, or at risk of experiencing, homelessness" (Carmichael Outreach, n.d.).

In 2015, the edible campus apiary (beekeeping) project was started with a grant from the Sustainability and Community Engagement Fund to encourage pollination for the fruits, vegetables, and flowering plants on campus. It is part demonstration and part teaching project. The apiary is used in an undergraduate biology class to study the relationship between food production and bee ecology, and for visits from community members. For example, children from the campus daycare can learn about local food production and the importance of bees and pollinators in the food web. Several potential formal research projects have been discussed, but these are only in their early planning stages at the time of writing this article. The apiary also provides positive externalities by helping to pollinate nearby community gardens that are located off-campus in a nearby neighborhood. The apiary was started by a small group of staff, faculty, and students from biology, psychology, and environment and sustainability with the advice of a local beekeeper who shared their knowledge and experience with the group to help them establish the hive. A volunteer manager (a U of R staff member who has beekeeping experience) coordinates and delivers care for the hive, including monitoring the bees, identifying maintenance issues, diseases and pests, and preparing the bees for cold, prairie winters. They group connects with the local bee club for education and technical advice. Then of course there is the bonus of honey for the volunteers in return for their labor.

This case provides an example of sustainable scholarly livelihood opportunities through the increase of scholarly assets and capabilities for teaching and research on the University of Regina campus. The gardens and apiary living lab promotes exploring the interconnections of local food systems and our relationships to and responsibility within these systems. These new spaces for experiential learning can form the basis for more traditional scholarship. This living lab is a hub for education and research, and allows a better understanding of biodiversity and how to grow healthy food on the Canadian prairies while minimizing adverse impacts. It also contributes to understanding how more sustainable patterns of food production can contribute to livelihood outcomes beyond satisfying only basic needs, such as experiences of nature that enhance well-being along with the benefit of forming new social networks between people from different disciplines, and of differing cultures and ages yet having shared interests. Living labs on campuses challenge us to be curious and look at our roles, not just as passive local consumers of food but as local producers as well, in our pursuit of education and knowledge for sustainability. The formation of new local food production and consumption linkages between the university campus and local community share mutually reinforcing and positive feedback loops. This highlights how attempts to advance sustainable scholarly livelihoods provide, at the same time, positive impacts on the livelihood assets, capabilities, strategies, and livelihood outcome achievements of community member livelihoods more generally.

5 CONCLUSIONS AND RECOMMENDATIONS

The paper has attempted to show how a Sustainable Livelihoods Approach (traditionally used in the context of international development projects) can reshape the scholarly identity to transform universities at a deep cultural level. Such a cultural shift could help sustain scholarly production processes and enable new innovations for sustainable livelihoods (SL) and sustainable development. The case was made that there exists a strong congruence between the traditional values and commitments of the university and four political areas identified by Scoones (2015) needed for revitalized livelihood analysis. In reviewing each area, specific institutional and organizational changes to construct Sustainable *Scholarly* Livelihoods (SSL) were proposed. The paper then introduced a case study, the University of Regina in Saskatchewan, Canada, to illustrate how open, highly participatory planning processes and campus-led living laboratories in the area of food production and consumption could both engage an entire campus community at the level of culture while advancing sustainable livelihoods both on and off campus.

The implications of this analysis for non-university organizations, we believe, are many. By transforming universities around a shared SSL identity, universities can model new governance patterns and institutional processes for other organizations. These models promote robust notions of autonomy and self-sufficiency for both the organizations themselves and individuals that participate within them. Such a new SL model reflecting self-sufficient individuals, organizations, and communities, addresses key livelihood anxieties that some have traditionally addressed by pursuing unlimited accumulation of wealth (whether as physical or financial capital). Addressing these shared human anxieties through a shared SL identity within the global economy may halt the current unsustainable use and accumulation of global resources. At the same time sustaining universities as innovators through their institutional restructuring around the concept of SSL is increasingly an imperative—especially when other organizations are in crisis due to COVID-19 creating weaknesses of the global economy. Universities themselves viewed as integrated and self-sufficient SSL (alongside other universities viewed in the same light) have the potential for rapid innovation of technologies and livelihood strategies tied to each university's own ecosystems and the communities they serve within their region. Central to pioneering these livelihood strategies will be the new living laboratories universities construct. A SSL framework provides the ethical and pragmatic norms needed for their construction.

ACKNOWLEDGEMENTS

The authors received no financial support for the research, authorship, and/or publication of this article.

REFERENCES

Aristotle. 1946. *The Politics of Aristotle* (E. Barker, Trans. Oxford: Oxford University Press. Aristotle. 1947. *Metaphysics* (W.D. Ross, Trans.). In *Introduction to Aristotle*. R. McKeon, ed. New York: The Modern Library.

Carmichael Outreach. n.d. Carmichael Outreach. https://carmichaeloutreach.ca (Accessed July 14, 2020).

Chambers, R. 1997. *Whose reality counts? Putting the first last.* London: Intermediate Technology Publications (ITP).

Chambers, R. & Conway, G. 1992. *Sustainable rural livelihoods: practical concepts for the 21st century*. Brighton: Institute of Development Studies.

Dyball, R. & Newell, B. 2015. *Understanding Human Ecology: A Systems Approach to Sustainability.* New York: Routledge.

Greenwood, R., Hinings, C.R., & Whetten, D. 2014. "Rethinking institutions and organizations." *Journal of Management Studies*, 51(7). doi: 10.1111/joms.12070

International Association of Universities. 2020. IAU Global Cluster on HESD. Higher Education and Research for Sustainable Development. https://www.iau-hesd.net/en/contenu/4648-iau-global-cluster-hesd.html

McMillan, J. & Dyball, R. 2009. Developing a whole-of- university approach to educating for sustainability: linking curriculum, research and sustainable campus operations. *Journal of Education for Sustainable Development 3*(1), 55–64. doi: 10.1177/097340820900300011

North, D. 1990. *Institutions, institutional change, and economic performance*. Cambridge: Cambridge University Press.

Paul, D. 2018, July 13. SDG 12 Review at HLPF Calls for Circular Economies, Sustainable Lifestyles. IISD SDG Knowledge Hub. http://sdg.iisd.org/news/sdg-12-review-at-hlpf-calls-for-circular-economies-sustainable-lifestyles/

Petry, R.A. 2008. "The role of free knowledge at universities and its potential impact on the sustainability of the prairie region." [Unpublished doctoral thesis]. University of Regina. https://search-proquest-com.libproxy.uregina.ca/docview/304547745

Petry, R.A. 2014. Advancing ESD through new multisectoral learning partnerships: parallels between the RCE initiative and the earlier rise of humanism and science. In Z. Fadeeva *et al* (Eds.). *Building a resilient future through multi-stakeholder learning and action* (pp. 238–254). Tokyo: UNU-IAS.

Petry, R.A. 2018. Sustainability policies if necessary but not necessarily sustainability policies. *EcoThinking Journal,* 1, 1–13. https://eco-thinking.org/index.php/journal/article/view/992/608

Petry, R.A., & Ramkat, R. C. 2019. From Ivory Tower to Tree of Life: Leading the World through University Self-Transformation. *IAU Horizons*, 24(2), 25–26. https://www.iau-aiu.net/IMG/pdf/iau_horizons_vol.24.2_light.pdf

Polster, C. & Newson, J. 2015. *A penny for your thoughts: how corporatization devalues teaching, research, and public service in Canada's universities.* Ottawa, Ontario: Canadian Centre for Policy Alternatives.

Scoones, I. 2015. *Sustainable Livelihoods and Rural Development.* Rugby, U.K.: Practical Action Publishing.

Tirole, Jean. 2017. *Economics for the Common Good* (S. Rendall, Trans.). Princeton: Princeton University Press.

United Nations Educational, Scientific and Cultural Organization. 2014. *Shaping the future we want: UN Decade of Education for Sustainable Development (2005–2014) final report.* Paris: UNESCO.

University of Regina. 2018. A Guide to Implementing the Truth and Reconciliation Commission of Canada's Calls to Action at the University of Regina. https://www.uregina.ca/president/assets/docs/pdf/Major%20Documents/trc-working-group-guide-12-2018.pdf

University of Regina. 2020. University of Regina Strategic Plan 2020–2025: All Our Relations: kahkiyaw kiwahkomakaninawak. https://www.uregina.ca/strategic-plan/index.html

University of Regina. n.d. Sustainability and Community Engagement Fund. Retrieved from https://www.uregina.ca/president / campus-sustainability/Sustainability-Awards/index.html

United Nations University. 2020. Global RCE Network: Education for Sustainable Development. RCE Network. https://www.rcenetwork.org/portal/

Worts, D. 2006. Measuring museum meaning: a critical assessment framework. *Journal of Museum Education,* 31, 41–49. https://doi.org/10.1080/10598650.2006.11510528

Worts, D. 2011. Culture and museums in the winds of change: the need for cultural indicators. *Culture and Local Governance*, 3, 117–132. https://doi.org/10.18192/clg-cgl.v3i1.190

Young, O. 2002. *The institutional dimensions of environmental change: fit, interplay, and scale.* Cambridge, Massachusetts: The MIT Press.

Advances in Phytochemistry, Textile and Renewable Energy Research for Industrial Growth – Nzila et al. (Eds)

An innovative ergonomic design of classroom furniture based on anthropometric measurements at tertiary institutions

A. Esmaeel
Faculty of Industries Engineering and Technology, University of Gezira, Wad Medani, Sudan

D. Starovoytova & O. Maube
School of Engineering, Moi University, Eldoret, Kenya

R. Asad
Faculty of Industries Engineering and Technology, University of Gezira, Wad Medani, Sudan

ABSTRACT: In many institutions, classroom furniture does not meet any national ergonomic criteria. This study aimed to use the concept of ergonomics to design a classroom desktop–chair for students in Uasin-Gishu County, Kenya. Anthropometric data were collected from a total of 382 students of both genders. The fourteen anthropometric measurements were taken from students with the help of anthropometric tools. The research applied fundamental engineering principles of product design and was carried out in compliance with ISO 7250-1:2017. The data obtained was analysed using Minitab 17.0 statistical package. Using the collected anthropometric data, a students' desktop–chair was proposed. In conclusion, one type of ergonomically suitable classroom desktop–chair design was proposed to improve the match between classroom desktop–chairs' dimensions and students' anthropometric characteristics. It is highly recommended that similar scientific research should be carried out in other countries.

Keywords: Desktop–chair, Classroom environment, MSDs, Awkward position.

1 INTRODUCTION

1.1 *Problem statement, significance, and purpose of the study*

The basic philosophy of ergonomics is to make any design comfortable. Students require well-designed classroom furniture for their comfort in the learning context. This requires that in designing classroom furniture, designers should include anthropometric sciences (Igbokwe et al. 2019b; Taifa & Desai 2017). According to some estimations, about 44 million workers in Europe suffer from occupational musculoskeletal disorders (Yusop et al. 2018). This shows that the ergonomics problem is a major issue that needs to be solved to avoid further suffering in the future. Therefore, there is need for ergonomists to treat the issue of furniture design for students as a necessity, and educational institutes/universities should treat the selection of the right kind of furniture as a social responsibility towards the students' community (Igbokwe et al. 2019). It is very essential for an institution of learning to have their anthropometric measurements regarding students so that they can be used by designers who intend to make ergonomic furniture, for them. This will ensure safety, comfort, adaptability, suitability, and ultimately guarantee user satisfaction, as well as result in the reduction of musculoskeletal disorders (MSDs) (Igbokwe, Osueke, Opara, Ileagu, & Ezeakaibeya 2019a). This research is of paramount importance because it expended current knowledge in the field of anthropometry to provide a database for future research and it is potentially beneficial to all future student. This research, therefore, seeks to use anthropometry for the design of classroom furniture for students to improve physical responses and their performance. The main purpose of the study is to conduct anthropometric measurements of students from four selected tertiary institutions and to design a desktop–chair using the collected anthropometric measurements.

2 LITERATURE REVIEW

Al-Hinai et al. (2018a) noted that the compatibility between classroom furniture dimensions and students' anthropometric characteristics has been identified as a key factor in improving some students' physical responses. Besides, there is a large amount of research worldwide (Castellucci, Gonçalves, & Arezes 2010; Chung & Wong 2007; Saarni et al. 2007) that shows a clear mismatch between anthropometric characteristics and the dimensions of classroom furniture. This mismatch might affect the learning process, even during the most stimulating and interesting lessons,

DOI 10.1201/9781003221968-5

and can produce some MSDs, such as low back pain and neck shoulder pain. This study, therefore, will fill the research gaps by providing an innovative ergonomically desktop–chair design based on students' anthropometric measurements. Thus, it will make a contribution to the existing literature on the compatibility between classroom desktop–chair dimensions and students' anthropometric characteristics, and improve the performance of students in terms of attentiveness while professors or instructors are teaching them.

3 RESEARCH METHODOLOGY

The entire methodology of the study can be divided into the steps that are described in Figure 1.

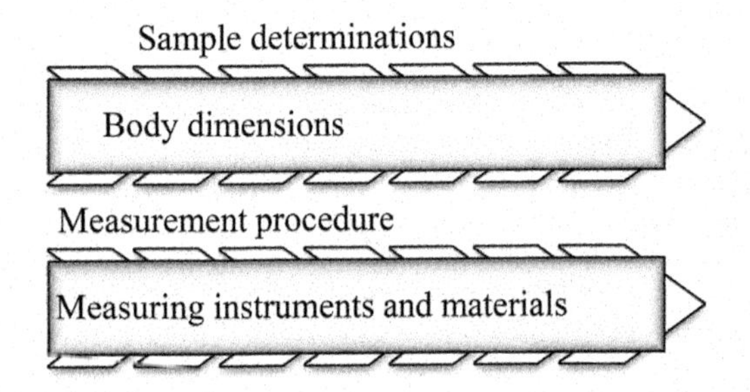

Figure 1. Main steps of the study methods.

3.1 *Sample determination*

Four higher institutions in Uasin Gishu County, Kenya, namely, (i) Moi University (MU), (ii) University of Eldoret (UoE), (iii) Rift Valley Technical Training Institute (RVTTI), and (iv) The Eldoret National Polytechnic (TENP), were selected to participate in the study. The students sample in the study was three hundred and eighty-two, through the use of equations given by Madara (2016):

$$\frac{\text{Students per institutions}}{\text{Total number of students in institutions}} * \text{Sample size}$$

3.2 *Body dimensions*

The design of standard furniture needs the direct involvement of anthropometric measurements. Various researchers (Igbokwe, Osueke, Opara, Ileagu, & Ezeakaibeya 2019b) have recommended the body dimensions which are essential in designing furniture, especially for students. Figure 2 shows all twelve body dimensions that were selected for the study with the addition of weight and forearm–fingertip length as the fourteenth body measurement. Two dimensions were collected while the participant was in the standing position, whereas the remaining twelve dimensions were taken while the participants were seated. The two most relevant anthropometric measurements for chair design are the popliteal height and buttock popliteal length (Igbokwe et al. 2019b). Table 1 indicates the serial number and descriptions of the selected student's body dimensions.

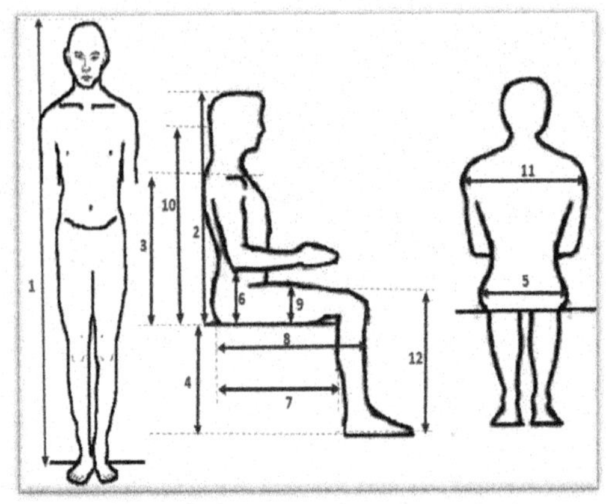

Figure 2. Anthropometric data required in classroom furniture design. *Keys:* (1) Stature, (2) Sitting height, (3) Shoulder height, (4) Popliteal height, (5) Hip breadth, (6) Elbow height, (7) Buttock popliteal length, (8) Buttock knee length, (9) Thigh clearance, (10) Eye height, sitting, (11) Shoulder breadth and (12) Knee height.

3.3 *Measurement procedures*

Three hundred and eighty-two students (191 males and 191 females) were selected (at random) from four selected tertiary institutions. The body size of each student was assessed using standard anthropometric measurement techniques (Esmaeel & Order 2017). The consent of all students was obtained before the commencement of the measurements. In this study, stature (body height) dimensions for each student were taken while they were standing, along with body mass. All other dimensions were measured while they were sitting erect on an adjustable desk with their knees bent at 90°. All anthropometric measurements were taken with the subjects wearing light clothing in a relaxed and erect posture, without shoes and with respect to the local culture. The time taken to measure and record all the dimensions per subject was about 15–20 minutes. Furthermore, measurements were taken every working day for 20 days in February in year 2020. The students' measurements were done in the hostels for each of the four selected tertiary institutions and all measurements were measured in centimetre (cm) except for the body mass (kg).

3.4 *Measurement instruments and materials*

According to ISO 7250-1:2017, the standard measuring instruments recommended are anthropometer, sliding callipers, spreading callipers, weighing scale, and tape measure.

Table 1. Selection of body dimensions to be measured for classroom furniture design.

S/NO. According to ISO 7250	Basic students' body dimensions	Description according to ISO 7250-1:2017
6.1.2	Stature (body height)	The vertical distance from the floor to the highest point of the head (vertex).
6.2.1	Sitting height (erect)	The vertical distance from a horizontal sitting surface to the highest point of the head (vertex).
6.2.4	Shoulder height, sitting	The vertical distance from a horizontal sitting surface to the acromion.
6.2.11	Popliteal height, sitting	The vertical distance from the foot-rest surface to the lower surface of the thigh immediately behind the knee, bent at right angles.
6.2.10	Hip breadth, sitting	The breadth of the body measured across the widest portion of the hips.
6.2.5	Elbow height, sitting	The vertical distance from a horizontal sitting surface to the lowest bony point of the elbow bent at a right angle with the forearm horizontal.
6.4.7	Buttock popliteal length (seat depth)	The horizontal distance from the hollow of the knee to the rearmost point of the buttock.
6.4.8	Buttock knee length	The horizontal distance from the foremost point of the knee-cap to the rearmost point of the buttock.
6.2.12	Thigh clearance	The vertical distance from the sitting surface to the highest point on the thigh.
6.2.2	Eye height, sitting	The vertical distance from a horizontal sitting surface to the outer corner of the eye (ectocanthus).
6.2.8	Shoulder (bideltoid) breadth	The horizontal distance across the maximum lateral protrusions of the right and left deltoid muscles.
6.2.13	Knee height, sitting	The vertical distance from the floor to the highest point of the superior border of the patella (suprapatella, sitting).
6.1.1	Body mass	The total mass (weight) of the body.
6.4.6	Forearm fingertip length	The horizontal distance from olecranon (back of the elbow) to the tip of the middle finger, with the elbow bent at right angles.

Source. (Esmaeel & Order 2017).

4 RESULTS AND DISCUSSION

4.1 *Anthropometric dimension for students*

The results obtained from the four selected tertiary institutions were analysed using Minitab 17.0 statistical package, to get the mean, standard deviation, 5th, 50th, and 95th percentiles. For seat height, the 5th percentile (lower percentile) of the popliteal height of the population is usually recommended so that a larger number of the population is accommodated, thus allowing a short person to use the chair. Similarly, the 95th percentile (larger percentile) of the hip breadth is usually recommended in the design of the seat width to accommodate as many people of the population as possible, thus allowing an overweight person to use the chair. The following results, shown in Table 2, give a summary of the anthropometric measures, based on the average of the collected anthropometric data, that can be used in designing a desktop–chair for students at four selected tertiary institutions in Uasin-Gishu County, Kenya.

After analysing all the anthropometric measurements of the students at the four selected tertiary institutions, the final specification is proposed for the design of the ergonomic desktop–chair that

Table 2. Summary of anthropometric dimension for student of the selected institutions (n = 382).

Variable	Mean	St. Dev	5th Percentile	50th percentile	95th Percentile
Age (Yrs.)	20.51	1.67	18.00	20.00	23.00
Stature	168.38	7.86	155.50	168.00	182.00
Sitting height	81.01	4.01	75.00	81.80	88.00
Shoulder height	54.41	2.76	50.96	54.50	57.80
Popliteal height	44.73	2.78	40.50	44.50	49.60
Hip breadth	33.46	3.49	29.57	32.86	39.36
Elbow height	20.39	1.16	19.11	20.30	22.37
Buttock popliteal length	42.54	2.73	38.10	42.65	46.90
Buttock knee length	51.85	3.05	47.20	51.90	56.70
Thigh clearance	14.54	1.70	11.96	14.43	17.39
Eye height	67.43	3.46	62.88	68.00	72.50
Shoulder breadth	41.78	3.43	37.02	41.47	46.59
Knee height	51.96	3.29	46.80	52.10	57.30
Body mass	60.54	9.09	48.00	59.50	75.50
Forearm fingertip length	47.62	2.70	43.46	47.44	51.88

can cover the maximum number of students. Table 3 shows the recommended dimensions for a new desktop–chair with the criteria for use in four elected tertiary institutions, Uasin-Gishu County, Kenya.

4.2 *Design of the desktop–chair*

After, running the analysis of the recorded data, as shown in Table 3 there is only one type of innovative ergonomically suitable desktop–chair that was identified in the four selected tertiary institutions (the dimensions were the same in the respective institutions), as drawn in Figures 3a and 3b using 3D SolidWorks 2019 software.

Table 3. Recommended dimensions for a new desktop–chair for use in tertiary institutions, Uasin-Gishu County, Kenya.

Seat feature	Anthropo-metric measure	Design dimensions (cm)	Criteria/ Determinant	References
Seat height	Popliteal height	40.95	5th percentile of popliteal height + 0.45 cm shoe heel allowance	(Ismail et al. 2013)
Seat width	Hip breadth	45.26	95th percentile of hip breadth + 15% allowance for clothing	(Musa & Ismaila 2014)
Seat depth	Buttock popliteal length	38.10	5th percentile of buttock popliteal length	(Mohamed et al. 2010)
Desktop height from seat	Elbow height	19.11	5th percentile of elbow height	(Musa et al. 2014)
Backrest height	Shoulder height	50.96	5th percentile of shoulder height	(Mohamed et al. 2010)
Desktop width	–	24.20	Literature review suggestions	(Ismaila et al. 2013)
Desktop length	Forearm fingertip length	47.44	50th percentile of forearm fingertip length	(Ismaila et al. 2013)
Backrest angle	–	109°	Literature review suggestions	(Mohamed et al. 2010)
Desk angle	–	0°	From literature review	(Ansari et al. 2018)
Seat angle	–	110°	From literature review	(Igbokwe et al. 2019)

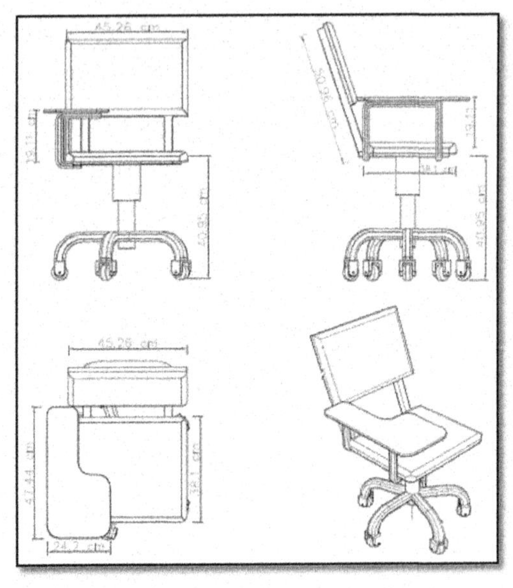

Figure 3a. Sketches of the proposed adjustable students' desktop–chair.

Figure 3b. Complete model of the proposed adjustable students' desktop–chair.

5 CONCLUSION AND RECOMMENDATIONS

5.1 *Conclusions*

From the present study, it is well expected that the determining criteria for an adjustable desktop–chair shown in Table 3 need to be used whenever designers wish to have adjustable classroom furniture (which is ergonomic design desktop–chair) in the four selected tertiary institutions, Uasin-Gishu County, Kenya. In this study, therefore, there is only one type of innovative ergonomically suitable classroom desktop–chair design that was proposed to improve the match between classroom desktop–chairs dimensions and students' anthropometric characteristics.

5.2 *Recommendation*

In this 21st century, it is highly recommended that the analysed anthropometric data set from this study be used for the design of classroom desktop–chairs for students in the four selected tertiary institutions, Uasin-Gishu County, Kenya. Achieving this will ensure safety, comfort, adaptability, suitability, and ultimately guarantee user satisfaction. The authors propose further work on seat chair design for elderly people should be carried out, with ergonomic and anthropometric consideration; this will allow seat chair design that is sustainable, safe, and comfortable as well as helping in health care.

Lastly, the goal is not to be perfect, it is just to be better than before (Pusca & Northwood 2018).

REFERENCES

Al-Hinai, N., Al-Kindi, M., & Shamsuzzoha, A. (2018). An Ergonomic Student Chair Design and Engineering for Classroom Environment. *International Journal of Mechanical Engineering and Robotics Research,* 7(5), 534–543.

Castellucci, I., Gonçalves, A., & Arezes, M. (2010). Ergonomic Design of School Furniture: Challenges for the Portuguese Schools. *Advances in Occupational, Social, and Organizational Ergonomics*, (April 2015), 625–633.

Chung, Y., & Wong, S. (2007). Anthropometric Evaluation for Primary School Furniture Design. *Ergonomics,* 50(3), 323–334.

Esmaeel, A., & Order, S. (2017). International Standard Technological Design —, 2017.

Igbokwe, O., Osueke, O., Opara, V, Ileagu, O., & Ezeakaibeya, U. (2019a). *Considerations of Anthropometrics in the Design of Lecture Hall Furniture,* 7(August), 374–386.

Igbokwe, O., Osueke, O., Opara, V, Ileagu, O., & Ezeakaibeya, U. (2019b). *Considerations of Anthropometrics in the Design of Lecture Hall Furniture,* 7(August), 374–386.

Madara, S. (2016). Consumer-Perception on Polyethylene-Shopping-Bags Consumer-Perception on Polyethylene-Shopping-Bags, (November).

Madara, S., Arusei, K., & Njoroge, N. (2016). Design Simulation and Analysis of Manual Block-Making Machine Design Simulation and Analysis of Manual Block-Making Machine, (August).

Md Yusop, S., Mat, S., Ramli, R., Dullah, R., Khalil, N., & Case, K. (2018). Design of Welding Armrest Based on Ergonomics Analysis: Case Study At Educational Institution in Johor Bahru, Malaysia. *ARPN Journal of Engineering and Applied Sciences,* 13(1), 309–313.

Milanese, S., & Grimmer, K. (2004). School Furniture and the User Population: An Anthropometric Perspective. *Ergonomics,* 47(4), 416–426.

Pusca, D., & Northwood, O. (2018). Design Thinking and Its Application to Problem Solving, (March).

Saarni, L., Nygård, H., Kaukiainen, A., & Rimpelä, A. (2007). Are the Desks and Chairs At School Appropriate? *Ergonomics,* 50(10), 1561–1570.

Taifa, W., & Desai, A. (2017). Anthropometric Measurements for Ergonomic Design of Students' Furniture in India. *Engineering Science and Technology, an International Journal*, 20(1), 232–239.

Advances in Phytochemistry, Textile and Renewable Energy Research for Industrial Growth – Nzila et al. (Eds)

Africa's need for a technological approach to monitor pollutions from mining activities

A. Antwiwaa
All Nations University, Ghana

R. Damoah
Morgan State University, MD, USA

N.K. Gerrar
Cape Coast Technical University, Ghana

J.N. Quansah
ANU-SSTL, Ghana

ABSTRACT: Africa hosts the largest mineral industry in the world and most of these mining countries operate with an unregulated environmental impact. The impact of mining operations touches on the principal elements of the environment affecting the implementation of the SDGs. Air quality is severely degraded by mining through putting a large quantity of dust into the atmosphere. This research seeks to analyze the impact of mining activities on the environment by considering three different mining sites from Ghana, Kenya, and Tanzania by employing the services of the dataset from the NASA Giovanni System. We have used time series analysis of $PM_{2.5}$ from the NASA Giovanni System that spans the years before and during the mining activities at our study sites to determine the pollution levels and the health issues involved. The results obtained show that there was a significant increase in pollution level over time at all the mining sites. This work proposes an Internet of Things (IoT) based solution to help the environmental protection agencies effectively monitor the pollution level from mining activities.

1 INTRODUCTION

Mining plays a significant role in the development of the economy in most countries in Africa. It contributes to about 9.1% of Ghana's gross domestic product (GDP) (Awudi 2002) and provides jobs to about 300,000 people according to the Ghana Statistical Services (2015). It also contributes 1% of the GDP of Kenya (Kenya Mining Investment Handbook 2016) and 3.5% of Tanzania GDP (Ecofin Agency 2020). Alongside the importance of mining to socioeconomic growth, mining activities have a serious effect on the environment and human health. There are two main types of mining, strip mining and underground mining. In Africa, most of the large-scale mining is strip mining. The impact of mining operations in Africa both from the large- and small-scale mining touches on the principal elements of the environment (i.e., land, water, and air). Large tracts of land for farming activities have been acquired by mining companies for large-scale surface mining operations, depriving the communities of their source of livelihood and leading to large-scale deforestation as well as causing atmospheric conditions to deteriorate rapidly over time (Kulkami & Zambare 2018; Parmer, Lakhani, & Chattopadhyay 2017)

The deterioration of the atmospheric conditions leads to ozone layer depletion. Apart from the hazardous effect it has on the environment, pollution has an adverse effect on human health, leading to diseases like lung cancer, stroke, heart diseases, etc. (WHO 2018). Rout, Karuturi and Padmini (2018) stated that about 7 million deaths every year are caused by polluted air and about 90% of the world's population does not have access to clean air. The majority of the pollution is from industry and the mining sector. This research seeks to analyze the impact of mining activities on the environment by considering three different mining sites from Ghana, Kenya, and Tanzania by employing time series analysis of $PM_{2.5}$ derived from the NASA Giovanni System. An area average time series spanning the years before the mining companies were established up to 2020 was used to determine the pollution levels and the health issues involved.

The Environmental Protection Agency in Ghana visits the mining sites quarterly to measure the amount of pollution produced by these sites. Most of these sites are in remote areas. The monitoring is done

DOI 10.1201/9781003221968-6

by using portable hand-held air quality monitoring devices which they carry with them to instantly monitor the levels at the short period of time when they are onsite. This short time measurement will not accurately reveal the actual pollution levels of the area. There is the need for a continuous measurement to determine accurate pollution levels. Therefore we propose an Internet of Things (IoT)-based solution to help the environmental protection agencies to effectively monitor the level of pollution from mining activities.

2 LITERATURE REVIEW

2.1 *Economic impact of mining*

Mining plays a very vital role in the growth of the economy of most countries in Africa. Large-scale mining, which is known as legal mining, generates about 95% of the world's total mineral production and employs approximately 2.5 million people worldwide (Kunanayagam, Mcmahon, Sheldon, Strongman, & Weber-Fahr 2002). Table 1 represents the gold production of Geita gold mining company in Tanzania from 2003 to 2011. (Table 1 was adapted from Annual report archive from 2005 to 2011, AngloGold Ashanti website). Table 2 represents the gold production of Obuasi Gold mines in Ghana from 2002 to 2009. (Table 2 was adapted the AngloGold Ashanti website).

Table 1. Geita gold production from 2003 to 2011 in Tanzania.

YEAR	PRODUCTION	CASH COST PER OUNCE
2003	661,000 ounces	US$ 183
2004	570,000 ounces	US$ 250
2005	613,000 ounces	US$ 298
2006	308,000 ounces	US$ 497
2007	327,000 ounces	US$ 452
2008	264,000 ounces	US$ 728
2009	272,000 ounces	US$ 954
2010	357,000 ounces	US$ 777
2011	494,000 ounces	US$ 536

Table 2. Obuasi gold mines production from 2002 to 2009 in Ghana.

YEAR	PRODUTION	CASH COST PER OUNCE
2002	537,219 ounce	US$ 198
2003	513,163 ounce	US$ 217
2004	255,000 ounce	US$ 305
2005	391,000 ounce	US$ 345
2006	387,000 ounce	US$ 395
2007	360,000 ounce	US$ 459
2008	357,000 ounce	US$ 633
2009	381,000 ounce	US$ 630

The production data of only these two mining sites were available.

2.2 *Effect of pollution from mining in Africa*

Air pollution due to harmful gases released into the atmosphere from various sources has a tremendous adverse effect on the health of humans. Exposure to these harmful gases causes millions of deaths per annum. 6.5 million death in 2012 in middle- and low-income nations were attributed to outdoor and indoor pollution (WHO 2018). According to the World Health Organization, air pollution-related health hazards include stroke, respiratory infections, lung cancer, and cardiovascular disease.

Mining activities contribute to the increasing level of air pollution in most of the low-income nations. Air quality is severely degraded by mining, via the release of a large quantity of dust into the atmosphere through their activities, thereby increasing the aerosol content of the atmosphere. These aerosols or particulate matter range from fine to coarse particles with a diameter between 2.5 and 1.0 micrometers, i.e., $PM_{2.5}$ and PM_{10}, respectively. The ability of particulate matter to enter deep into the lungs and the bloodstream unfiltered makes it the most harmful form of air pollution according to the World Health Organization. They can cause respiratory diseases, heart attacks, and even premature death. According to the special report on global exposure to air pollution in 2018, 4.1 million deaths from heart disease, lung cancer, chronic lung disease, respiratory infections and stroke in 2016 globally were due to exposure to $PM_{2.5}$.

3 METHODOLOGY

This paper reviews some impacts of mining covered in some scientific journals. The second part retrieves the data for $PM_{2.5}$, which is particulate matter released into the atmosphere by the various mining sites, using the NASA GIOVANNI system for monthly and seasonal area-averaged time series of dust column mass density- for a range of years before and after the mining sites $PM_{2.5}$were established. The NASA GIOVANNI (GES-DISC Interactive Online Visualization and analysis Infrastructure) system is a web portal to access satellite-based earth science data. This portal provides access to NASA's earth-observing satellite data that has been accumulated over a period of time. The mining sites under consideration include Geita and Bulyanhulu Gold mines in Tanzania, Obuasi Gold mines and Tarkwa Goldfields in Ghana, and Karebe and Kilimapesa Gold mines in Kenya. The $PM_{2.5}$values from the NASA GIOVANNI portal for the various mining sites are compared with the WHO $PM_{2.5}$ standard monthly and annual mean values. This helps in determining the pollution produced by these mining sites over the years. The final part proposes an Internet of Things (IoT)-based solution to help the environmental protection agencies to effectively monitor the pollution level from mining activities.

4 RESULTS AND DISCUSSION

4.1 *$PM_{2.5}$ emanating from different mining areas in Africa with time*

This section focuses on the level of $PM_{2.5}$ produced from different mining areas in Africa. The mining areas are Geita and Bulyanhulu gold mines in Tanzania, Kilimapesa and Karebe mines in Kenya, Anglo Ashanti Obuasi and Tarkwa Gold field in Ghana.

According to the WHO air quality guidelines (WHO 2005), the annual mean of $PM_{2.5}$ is 10 $\mu g/m^2$ and the monthly mean is 25 $\mu g/m^2$.

Geita gold mine is an open pit gold mine which is operated by Anglo-Gold Ashanti in Tanzania. Anglo-Gold began its operation in the year 2000 and production commenced around 2003. Figure 1a represents the Time Series Area-Averaged of Dust Column Mass Density–$PM_{2.5}$ [MERRA-2 Model M2TMNXAER v5.12.4] for January 1998 to April 2020, and Figure 1b represents the area image of the Geita mining location using the CNES/Airbus Maxar Technology. The $PM_{2.5}$ rate was at the lowest rate during the period of 1998 and 1999 where production had not yet commenced. It began to rise from the year 2000 when production commenced and achieved a higher $PM_{2.5}$ pollution rate of 0.00019 kg m^{-2} around 2004 and 2005. This pollution rate is higher than the WHO $PM_{2.5}$ standard for the annual mean but within the monthly mean standard. It can be observed from Table 1 that Geita gold mine achieved its second highest production of 613,000 ounces within the period. The $PM_{2.5}$ rate for this area in 2020 stood at 0.00010 kg m^{-2}.

Figure 2a represents the Time Series Area-Averaged of Dust Column Mass Density–$PM_{2.5}$ [MERRA-2 Model M2TMNXAER v5.12.4] for January 1990 to April 2020, and Figure 2b represents the area image of Tarkwa Goldfields in Ghana using the CNES/Airbus Maxar Technology. The Tarkwa Goldmine concession right was gained in 1993 by Goldfields Ghana limited. The Tarkwa Goldfields is an open pit mine. The $PM_{2.5}$ of the region in 1993 when Goldfields gained the concession right was 0.0001 kg m^{-2}. The mining activities of the area caused a rise in the pollution level over the years until the highest pollution level was recorded in the year 2008 at 0.00035 kg m^{-2}, which is higher than both the monthly and the annual mean of the WHO standard. The pollution level stood at 0.00025 kg m^{-2} in April 2020.

Figure 3a represents the Time Series Area-Averaged of Dust Column Mass Density–$PM_{2.5}$ [MERRA-2 Model M2TMNXAER v5.12.4] for January 2005 to April 2020 and Figure 2b represents the area image of Karebe gold mine in Kenya using the CNES/Airbus Maxar Technology. Karebe gold mine is an underground mine which was founded in 2008 and started its operation in 2009 in Kenya. The pollution level increased from 0.00012 kg m^{-2} in 2007 to 0.00014 kg m^{-2} in 2008 when the mine was founded. The highest pollution level was obtained in the year 2016 at 0.00019 kg m^{-2} which is an approximate 58% increase from 2007. This increase is within the WHO's monthly mean but above the annual mean of the $PM_{2.5}$ standard

Figure 4 represents the seasonal Area-Averaged of Dust Column Mass Density–$PM_{2.5}$ [MERRA-2 Model M2TMNXAER v5.12.4] for January 1998 to April 2020 of the Bulyanhulu gold mine. Bulyanhulu Gold mine is an underground mine that is currently operated by Acacia mining plc in Tanzania.

This mine was purchased by African Barrick Gold in 1999 and it was opened in 2001. The highest $PM_{2.5}$ value for 1998 before the mine was purchased was 0.00003028 kg m^{-2}. The $PM_{2.5}$ value increased to 0.00006035 kg m^{-2} in 2001 when the mine was opened by African Barrick Gold. It attained a highest pollution level in 2016 at 0.0001644 kg m^{-2}. The difference between the pollution level attained in 1998 and 2016 is 0.00013412 kg m^{-2}, representing a 443% increase in pollution level. The pollution levels were lower for all the years for the months of June, August, and October. The pollution level for April 2020 was 0.0001341 kg m^{-2}.

The pollution level was higher for all the mining sites in February because of the dry and dusty trade winds which occur during that period.

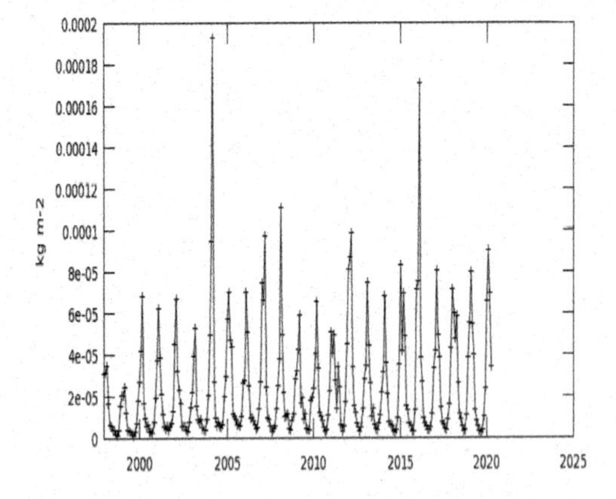

Figure 1a. Time series, area-averaged of dust column mass density–$PM_{2.5}$ monthly 0.5 × 0.625 deg. [MERRA-2 Model M2TMNXAER v5.12.4] kg m^{-2} for January 1998 to April 2020, Region 0E, 2.82S, 32.2073E, 0N.

Figure 1b. Image from Geita Goldmines in Tanzania from Image@2020CNES/Airbus.

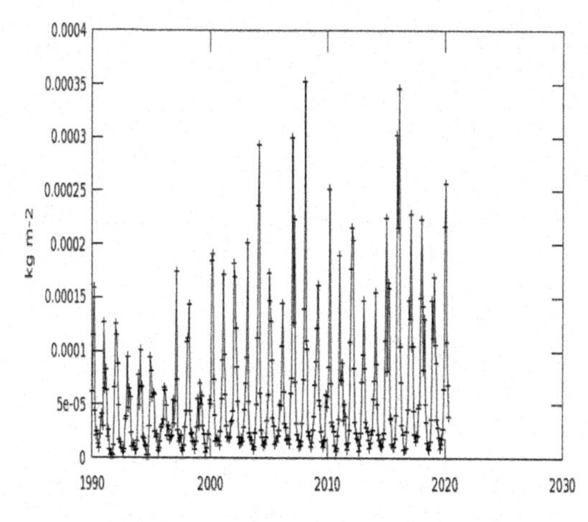

Figure 2a. Time series, area-averaged of dust column mass density–$PM_{2.5}$ monthly 0.5 × 0.625 deg. [MERRA-2 Model M2TMNXAER v5.12.4] kg m^{-2} for January 1990 to April 2020, Region 1.9899W, 0N, 0E, 5.3212N.

Figure 2b. Image from Tarkwa Goldfields in Ghana from Image@2020CNES/Airbus.

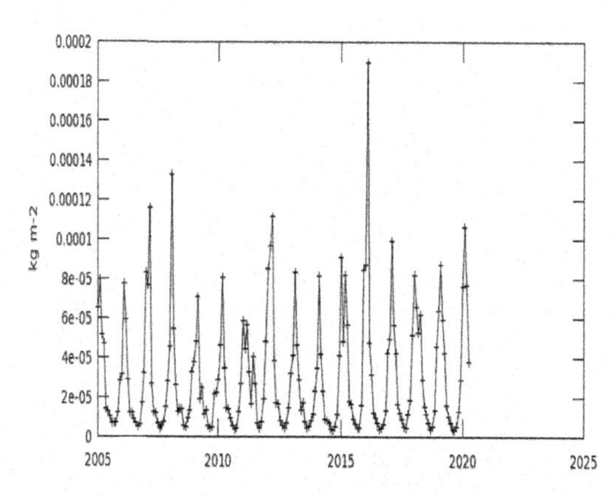

Figure 3a. Time series, area-averaged of dust column mass density–$PM_{2.5}$ monthly 0.5 × 0.625 deg. [MERRA-2 Model M2TMNXAER v5.12.4] kg m^{-2} for January 2005 to April 2020, Region 0E, 0.03175S, 35.0237E, 0N.

Figure 3b. Image from Karebe Gold mine in Kenya from Image@2020CNES/Airbus.

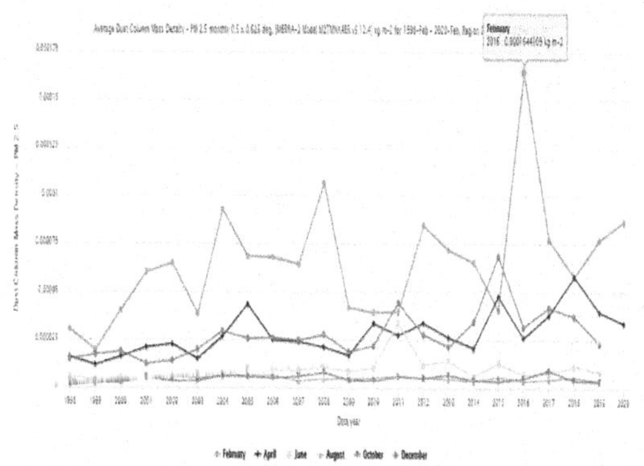

Figure 4. Seasonal area-averaged of dust column mass density–monthly 0.5 × 0.625 deg. [MERRA-2 Model M2TMNXAE $PM_{2.5}$ R v5.12.4] kg m^{-}2 for January 1998 to April 2020, Region 0E, 3.2287S, 32.4837E, 0N (Bulyanhulu Gold Mine, Tanzania).

Figure 5 represents the seasonal Area-Averaged of Dust Column Mass Density–$PM_{2.5}$ [MERRA-2 Model M2TMNXAER v5.12.4] for January 1998 to April 2020 of the Obuasi gold mine in Ghana. This mine was established in 1879 and it was formerly operated by AngloGold Ashanti in 2008. Satellite data was not available for the time the mine was found therefore a seasonal area-averaged of dust column Mass Density–$PM_{2.5}$ was obtained from 1999 to 2020. This covers the duration within which AngloGold commenced its operations. The pollution level from Figure 5 shows that the mine obtained a higher pollution level in 2008 when AngloGold began its operation and 2016 with an area-averaged of 0.0003521 kg m^{-2} and a lower rate of 0.0000050603 kg m^{-2}.This gives a difference of 0.00034704 kg m^{-2} from 1999 to 2008 which is a very large increase in the pollution level of the area.

Kilimapesa Gold mine, situated in Migori Archean Greenstone belt in southwest Kenya, commenced production on January 2009. Figure 6 represents the seasonal Area-Averaged of Dust Column Mass Density–$PM_{2.5}$ [MERRA-2 Model M2TMNXAER v5.12.4] for

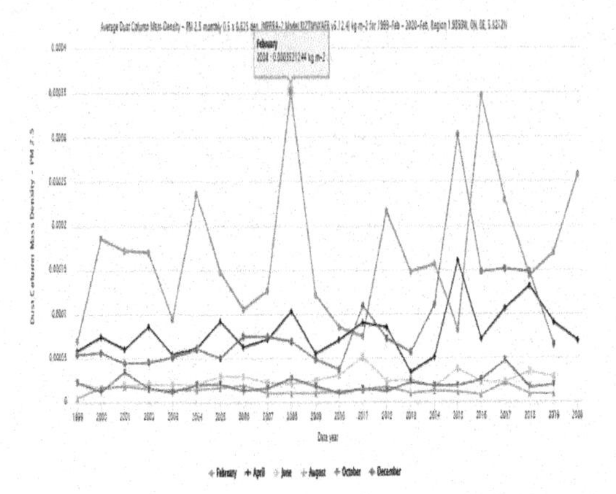

Figure 5. Seasonal area-averaged of dust column mass density–$PM_{2.5}$ monthly 0.5 × 0.625 deg. [MERRA-2 Model M2TMNXAER v5.12.4] kg m^{-2} over January 1999 to April 2020, Region 1.9899W, 0N, 0E, 5.3212N (Obuasi Goldmines, Ghana).

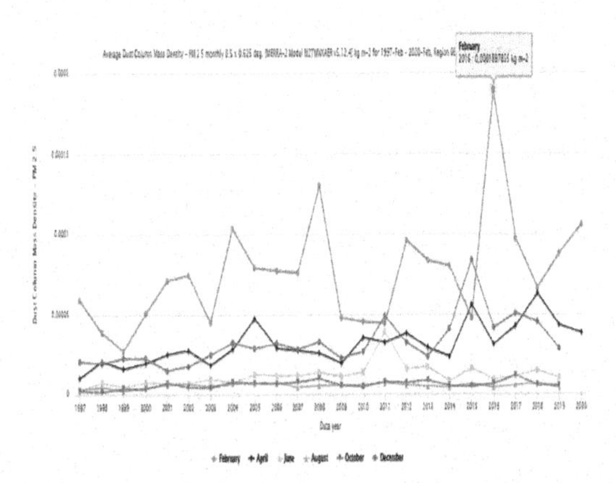

Figure 6. Time series, area-averaged of dust column mass density–$PM_{2.5}$ monthly 0.5 × 0.625 deg. [MERRA-2 Model M2TMNXAER v5.12.4] kg m^{-2} for January 1997 to April 2020, Region 0E, 0.9818S, 34.2509E, 0N (Kilimapesa Gold mine, Kenya).

January 1997 to April 2020 of the Kilimapesa gold mine. The results show that on February 1999, before the mine was established, the pollution level of the area was low as 0.00002706 kg m^{-2}. The mine achieved its highest pollution level of 0.0001898 kg m^{-2} in 2016 which accounts for an approximate 601% increase in the pollution level over 17 years. The pollution levels were lower for all years for the months of June, August, and October.

This shows that mining activities have an effect on the increased level of $PM_{2.5}$ over time. This $PM_{2.5}$ has an adverse effect on our health, therefore there is the need for the regulatory bodies to have a technological approach to provide constant monitoring of these mining sites in Africa.

4.2 *Internet of things for air pollution monitoring around mining sites*

We are in an era where most things have been automated and computerized. All these developments arise from the advancement of technological practices which have improved the standard of living. The Internet nowadays has become a worldwide tool which has been extensively adopted by companies, industries, institutions, and individuals for smooth and fast work and business transactions.

Gone are the days where environment protection agencies embark on on-site environment monitoring activity, which has been the traditional method of taking environment pollution readings.

They visit the mining site with a hand-held monitoring device and take readings of the pollution level. These readings are just a representation of the time during which they are at the mining site. They will not be a true picture of the pollution level in the area. The miners can decide not to perform any active work the day the Environmental Protection Agency is present so as not to increase the pollution readings.

Figure 7 is a typical example which shows an Environment Protection Agency taking a site reading of $PM_{2.5}$ at a mining community in Ghana in 2019. These traditional methods have their limitations. They require a larger workforce and field personnel for the on-site readings. Time becomes an important factor to consider in the traditional approach. It needs more time and effort to obtain data for processing, analysis, and usage. It can be costly and it does not give a true representation of pollution in the area.

With the application of the Internet of things (IoT), the challenges associated with the traditional method of environment pollution data collection can be reduced and eventually eliminated. IoT are normal objects ("things") that are endowed with network capabilities (Uckelmann, Harrison & Michahelles 2011).

Figure 7. Personnel from the Environmental Protection Agency in Ghana taking a manual pollution data from a mining site.

These network capabilities aid the device to transmit and receive information. The IoT system creates a platform for connectivity among computing and digital systems for data transfer over a network without requiring human-to-human/computer interaction.

4.3 *IoT environment pollution monitoring architecture*

An Air Visual Pro monitoring device would be installed at the different mining sites, Ghana, Kenya, and Tanzania for this proposal, as indicated in Figure 8. The AirVisual Pro is an air quality monitoring device that provides real-time measurements of particulate pollution ($PM_{2.5}$), CO_2, humidity, and temperature and displays the data on an easy-to-read color display and over the internet.

The Air Visual Pro sensor device takes field data 24 hours per day. The sensors are connected over the internet via a designated Wi-Fi module onto a cloud server (Air Visual Pro and Ubidots Cloud Server). Most of the mining sites are located in remote areas where GSM (Global System for Mobile communication)-based stations cannot be accessed, therefore a Wi-Fi module will not work in this case. The bandwidth required to transmit the pollution data is also very low, therefore LoRa (Long Range), which is a low-power–wide-area network protocol that allows a long-range transmission of more than 10 km, can be used in the remote areas. Figure 9 represents IoT-based environmental monitoring using LoRa.

An End User Software, Air Visual Pro Software, and an additional developed software using Ubidots would be installed on the Environment Monitoring Agency monitoring station for data analysis and usage. Based on the analyzed data, the agency can effectively monitor and implement the needed measures to reduce the level of pollution and to rightly advise the mining companies on the necessary preventive measures. The monitoring can take place on a 24-hour per day basis.

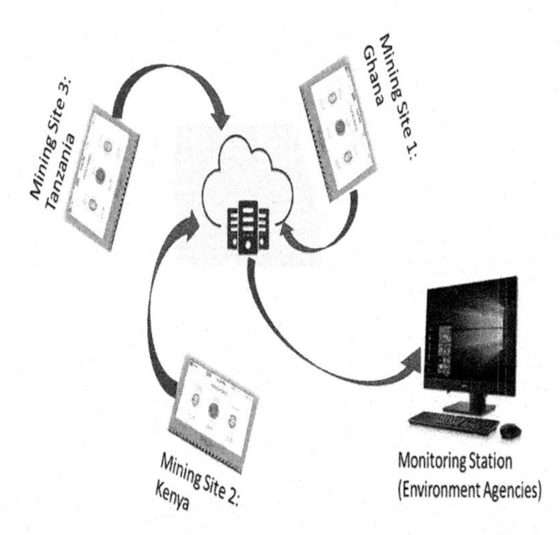

Figure 8. IoT environment pollution monitoring architecture.

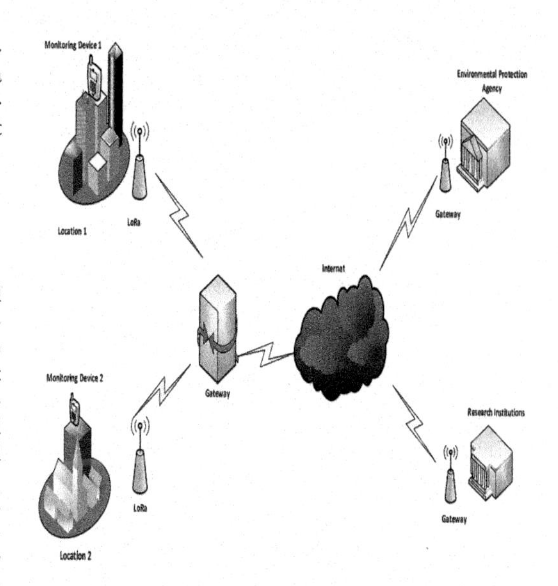

Figure 9. IoT-based environmental monitoring using LoRa.

This approach reduces cost, saves time, and speeds up the monitoring process. There is the surety of high data accuracy and efficiency as human errors in on-site readings are avoided.

5 CONCLUSION

Mining plays an important role in the growth of many developing countries. The GDP growth of most countries depends on mining and it also provides employment to a lot of people. However, there is an adverse effect of mining from both large- and small-scale mining on human health, especially from an increase in $PM_{2.5}$. Results obtained from the $PM_{2.5}$ data retrieved from the NASA GIOVANNI system from all the six mining sites from three different African countries show that the $PM_{2.5}$ values were lower before the mining sites were established. There was a significant increase in pollution level over time at all the mining sites as compared to the WHOs $PM_{2.5}$ standard for monthly and annual mean. Tanzania's Geita goldmine recorded its highest $PM_{2.5}$ value of 0.0019 kg m^{-2} in 1998. Karebe gold mine in Kenya experienced a 58% $PM_{2.5}$ increase from 2007 to 2016. There was a 250% increase in $PM_{2.5}$ from the location of Tarkwa Goldfields in Ghana from 1993 to 2008. Bulyanhulu Gold mine in Tanzania recorded a massive $PM_{2.5}$ increase of 443% from 1998 to 2008. The location for Kilimapesa goldmine in Kenya recorded a 601% increase from 1999 to 2016. Obuasi Gold mines in Ghana recorded the highest $PM_{2.5}$ increase from 0.0000050603 kg m^{-2} in 1999 to 0.0003470 kg m^{-2} in 2008. The outcome of the pollution level from these mining sites can help the regulatory bodies to make policies that will help reduce the pollutants produced by the mining companies, with an effect on human health. There is a need for a technological continuous method for monitoring the pollution levels at the mining sites by the regulatory bodies, therefore an IoT-based solution was proposed to enable continuous real-time monitoring with high data accuracy of the pollution level from the various mining sites.

REFERENCES

AngloGold, A. 2003. *Ashanti Annual Report 2003.* From http://www.anglogold.com/ additional/ annual + rports/ ashanti/annual+2003.htm.

AngloGold, A. 2005. *Annual report 2005.* From http://www.anglogoldashanti.co.za/subwebs/informationforinvestors/annualreport05/report/pdf/ar_report_2005pdf

AngloGold, A. 2006. *Annual Report 2006.* From http://www.anglogold.com/nr/rdonlyres/5f4b53a4-2de4-43ec-87ab-bdb13f5c8c1f/0/ar_report_2006.pdf

AngloGold, A. 2009. *Annual Report 2009.* From http://www.anglogoldashanti.com/subwebs/informationforinvestors/reports09/annualreport09/f/aga_ar09.pdf

AngloGold, A. 2011. *AngloGold Ashanti Annual Financial Statements 2010.* From http://www.an glogoldashanti.com/media/report/aga-annual-financial-statements-2010.pdf

Awudi, G. B. 2002. The role of foreign direct investment (FDI) in the mining sector of Ghana and the environment. *Conference on Foreign Direct Investment and the Environment; 2002* Feb 7-8; Paris, from: http://www.oecd.org/countries/ ghana/1819492.pdf

Ecofin Agency 2020. Tanzania: Mining shares in GDP to reach 10% by 2025, from 3.5% currently. Retrieved 24/02/2020, 18:14 from: http://www.ecofinagency.com/public - management / 2402 - 41024 - tanzania - miming - shares-in-gdp-to-reach-10-by-2025-from-3-5-currently

Ghana Minerals Commission, 2006. Statistical overview of Ghana's mining industry (1990-2004): Accra, Ghana Minerals Commission.

Kenya Mining Investment handbook 2016. Page 16 from https:// www.google.com / search?q = Kenya + Mining + Investment+handbook+2016.+Page+16&oq=Kenya+Mining+Investment+handbook+2016.+Page+16&aqs=chrome..69i57.2375j0j7&sourceid=chrome&ie=UTF-8

Kulkarni K. A., & Zambare, M. S., 2018. The impact study of houseplants in purification of environment using wire less sensor network, *Wireless Sensor Network,* vol. 10, no. 03, pp. 59–69.

Kunanayagam R., Mcmahon G., Sheldon C., Strongman J. E., & Weber-Fahr M. Mining. In: Klugman J, editor.2002 *A sourcebook for poverty reduction strategies.* Vol. 2. Washington, D.C.: The World Bank; 2002, Chapter 25 from: http://documents.worldbank.org/curated/en/ 68165 1468147315119 /pdf/298000v-2.pdf

Parmar G., Lakhani S., & Chattopadhyay M. 2017. An IoT based low cost air pollution monitoring system in 2017 *International Conference on Recent Innovations in Signal processing and Embedded Systems (RISE),* Bhopal, India.

Revised 2014 annual gross domestic product 2015. *Ghana Statistical Service 2015* from: http://www.statsghana.gov.gh/docfiles/GDP/GDP2015/ AnnualGDP2014_template_2014Q4_April%20 2015%20edition_web.pdf

Rout G., Karuturi S., & Padmini T. N., 2018. Pollution monitoring system using IoT, *ARPN Journal of Engineer ing and Applied Sciences,* vol. 13, pp. 2116–2123.

World Health Organization, 2005.WHO air quality guidelines for particulate matter, ozone, Nitrogen dioxide and Sulphur dioxide, global update 2005, summary of risk assessment, from: https://apps.who.int/iris/bitstream/handle/10665/69477/WHO_SDE_PHE_OEH_06.02_eng.pdf;jssionid=520AACA57A12378D3CF63CEAB25B4095?sequence=1

World Health Organization, 2018. Air Pollution and Child Health Prescribing Clean Air, WHO, Geneva, Switzerland, from: https://www.who.int/ceh/publications/Advancecopy-Oct24_18150_Air-Pollution-and-Child-Health-mergedcompressed.pdf.

World Health Organization, 2018. WHO releases country estimates on air pollution exposure and health impact. World Health Organization: Geneva, Switzerland, from: https://www.who.int/news-room/detail/ 27-09-2016-who-releases-country-estimates-on-air-pollution-exposure-andhealth-impact

Uckelmann D., Harrison M., & Michahelles F., 2011. An architectural approach towards the future internet of things, *in Architecting the internet of things,* Springer, pp. 1–24.

UBIDOTS cloud server 2020 from https://ubidots.com/

Advances in Phytochemistry, Textile and Renewable Energy Research for Industrial Growth – Nzila et al. (Eds)

Mitigation of power outages in Rwanda

Boniface Ntambara & Paul M. Wambua
Department of Manufacturing, Industrial and Textile Engineering, School of Engineering, Moi University, Eldoret, Kenya

S. Simiyu Sitati
Department of Electrical and Communication Engineering, School of Engineering, Moi University, Eldoret, Kenya

Jean B. Byiringiro
Siemens Mechatronics Certification Center, Dedan Kimathi University of Technology, Nyeri, Kenya

ABSTRACT: Power outages in Rwanda severely affected most of the Western and Northern grids of Rwanda in 2018, 2019, and 2020. This paper studied the causes and mechanism of power outages and developed the methods and techniques to mitigate the power outages. Two operational elucidations such as a balanced steady state control system and an optimal overcurrent relay settings model for operational HV substation relay coordination have been proposed and developed. The sustainable synchronism between Rwanda Power system regions was becoming difficult over time due to load disturbances/changes. The under and over frequency, and under and over voltage load shedding techniques have been taken to reduce and mitigate the system outages. However, in order to minimize power outages in emergency regions, these systems have been extended in different regions such as Western and Northern grids which need to be maintained. The PID controllers for enhancing and mitigating the power outages have been developed in single and two area power systems. The load disturbance injection of 50MW has been created. Matlab/Simulink 2017a have been used to simulate the frequency load control, overcurrent relay, and power system models. The relay coordination and settings for a 110/15kV substation with a 17156.48A and 13987.10A of maximum and minimum fault currents, plug setting (PS), and actual operating time of the different relay have been ascertained and modeled. The simulation results have been compared with and without PID controller installation in the power system and it has been shown that the frequency response characteristics for single and two area networks in western and northern grids have been minimized to 0.0 Hz, 0.0 Hz, and 2.5 sec for overshoots, steady state errors, and settling times after cascade outages respectively. The overcurrent substation relays have been coordinated with the expected times of 0.0924–0.0622sec, 0.0949–0.0720 sec, and 0.0764–0.0661sec for extremely, very, and standard inverse relay characteristics, respectively.

Keywords: Overcurrent relay; Frequency Load Control; Power Outages; Matlab/Simulink 2017a; PID controllers; power system stability.

1 INTRODUCTION

The power outages in Rwanda clearly demonstrated the relevance of the severe problems of loss of supply and MW loss. The situation is further complicated by the necessity for the power system to survive under competition and uncertain conditions. Based on a detailed power outage analysis, the dangerous overload of the transmission grid is held as a key element initiating the development of many cascading processes (Zalostiba 2020). The outages which occurred in Rwanda from 2018 to June 2020, have mostly proven severe and very significant. It has been reported that during the power outages, about 5 million people have been affected in 5 districts from the Northern region and 5 districts from the Western region, and 70 MW of loads were lost, which is about 27% of the total load. Some other major outages began when lightning and overloading caused the tripping of a major transmission line between Western and Northern grids. Research work in these two regions are aimed at predicting voltage collapse and voltage overload with a view to controlling and reducing its occurrence on power system networks (Weiss 2019).Most of power outages are triggered by distribution circuit failure (Wang L. 2016). Conventionally overcurrent relay settings are provided based on the full load current of power system components. Load frequency control was based on many power control advanced concepts and the dynamic behavior of many industrial plants is heavily influenced

DOI 10.1201/9781003221968-7

by disturbances and in particular, by changes in the operating point (M. Wadi, 2017). The problem of coordinating protective relays in power system networks consists of selecting their suitable settings such that their fundamental protective function is met under the requirements of sensitivity, selectivity, reliability, and speed. In a modern power system, abnormal conditions can frequently occur to cause interruption to the supply, and may damage the equipment connected to the power system, which allows us to note the importance of designing a reliable protective system (Mancer N, 2015). The power system is the interconnection of more than one control areas through tie lines. The generators in a control area always vary their speed together (speed up or slow down) for the maintenance of the frequency and relative power angles to the predefined values in both static and dynamic conditions. If any sudden load change occurs in a control area of the interconnected power system then there will be frequency deviation as well as tie line power deviation (Behera, 2019). The cascade outages that occurred in Rwanda from 2018 to 2020 mostly proved severe and very significant. It has been reported that during the power outages about 5 million people have been affected in 5 districts from the Northern region and 7 districts from the Western region, and 70 MW of load were lost, which was about 27% of the total load. Some other major cascade events began when lightning and overloading caused the tripping of a major transmission line between the Western and Northern grids. Research work in these two regions aimed to predict voltage collapse and voltage overload with a view to controlling and reducing its occurrence on power system networks (Weiss R. 2020). Electrical energy is a primary prerequisite for economic growth. The demand for electrical energy has greatly increased due to large-scale industrialization. A modern power system operates under stressed conditions because of the growth in demand and the deregulation of electric power system. This leads to many problems associated with the operation and control of power systems. The economics of power generation has been a major concern for the power utilities (P. K. Modi, 2006). Power systems are becoming heavily stressed due to the increased loading of the transmission lines and due to the difficulty of constructing new transmission systems as well as the difficulty of building new generating plants near the load centers. All of these problems lead to the voltage stability problem in the power system (Z. Osman, 2006). The effective power system in Rwanda relies heavily on the ability of engineers to ensure a continuous and reliable service in cascade outages. In the ideal case, the load feeding should be at a constant frequency of 50 Hz and voltage. For reasonable operation of consumer devices, voltage and frequency should be maintained and secured within tolerable limits in practical applications. A voltage decrease from 10% to 15% or a reduction in frequency of system can cause stalling of the device loads (Tanwani, 2013). The first requirement of this is the maintenance of parallel operation of the synchronous generator with the necessary ability to grip the load condition. Because, if synchronism between the generator and the systems is missing at any time, it will affect the voltage and current instability and system relays will disconnect the supply at faulty sections (Naresh K. Tanwani, 2014). The load frequency control problems are denoted by regulating the active power output to generate units responding to the disturbances in system frequency and load power interchanges within the prescribed ranges (Congzhi Huang, 2017). The secure operation of power systems with the variation of loads has been a challenge for power system engineers since the 1920s. (Steinmetz, 1920). The voltage and frequency instability in the short term is driven by fast recovering load components that tend to restore power consumption in the time frame of seconds after a voltage drop caused by a contingency. The PID controller and Automatic Voltage Regulator have been shown to be more effective for the enhancement and optimization of power system stability with better damping under small and large disturbances when compared with conventional excitation control (M. J. Hossain, 2009).

2 MATERIAL AND METHODS

2.1 *Balanced state control system*

In this research, the PID controllers for frequency load control have been employed. The parameters and setting values of single and two area generation of hydroelectric power plants like the rated power load of 200 MW, load disturbance of 50 MW, turbine time constant of 0.5 sec, governor time constant of 0.2 sec, generator inertia constant of 5 sec, governor speed regulation of 0.05 p.u, power system time constant of 10 sec, and motor load damping coefficient of 0.6 p.u have been chosen and employed into Ntaruka and Nyabarongo Hydroelectric power plants as shown in Figures 2, 3, 5, and 6 in the Northern and Western grids, which have been controlled at Gikondo National Network Dispatching and Control. The Northern electricity grid has been monitored to check whether there were any cascade blackouts or changing conditions during the system normal operation. The system condition was evaluated by computing the frequency load system using PID controllers. These control schemes have checked the frequency deviation and drips, demand–supply power imbalance, and load change to ensure the reliability and stability of the power system. When the system was determined to be secure (not vulnerable), the monitoring system has continued in operating condition. Otherwise, the vulnerable parts and conditions were identified. As the possible unbalanced frequency and overload problems for those vulnerable conditions have occurred, they were predicted, and suitable and corresponding control strategies to mitigate and prevent the power blackouts were identified and activated whenever needed. The models and simulations of the scenarios have been computed using

Matlab 2017a/Simulink. Figure 1 demonstrates the flowchart of power cascade outages detection and the load–supply power imbalance control from the start of monitoring of the power system to the provided frequency control mechanism.

The governor regulated the turbine speed, load-supply power and the grid frequency by starting the turbine from a still condition and varied the load on the turbo-generator in imbalance conditions. The 0.8% load variation caused 1% frequency change. The rated load active power of 200 MW, nominal frequency of 50 Hz, and a sudden load disturbance of 50 MW have occurred in both area power systems, as shown in Figures 2 and 5 for no PID installed and Figures 3 and 6 for PID installed. The single area power systems without and with PID controllers have been computed and modeled as shown in Figures 2 and 3. The two area power system has been considered in unbalanced frequency load control and the generators have been interconnected in series with the loads fed, and the load disturbances have been applied in area 1 and the two power generations have been synchronized as indicated in the Figures 5 and 6. Figures 4 and 7 show the setting values and parameters of PID controllers in a single area, area 1 and area 2 in two area power generations, respectively. The performance characteristics of power and frequency in two area power generations with and without PID controllers have been observed, recorded and plotted in Figures 10, 13–16.

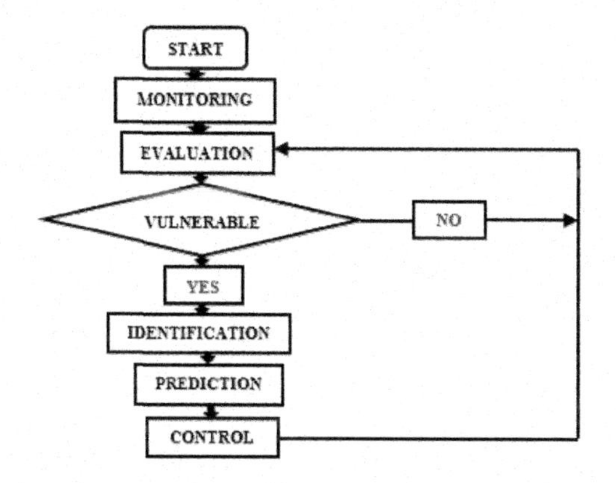

Figure 1. Flowchart of detection for cascade outages minimization in power balance steady system.

2.2 *Optimal overcurrent relay settings for effective substation relay coordination*

The values and parameters of grid and feeder impedance equivalences, phase to phase fault currents, current settings of 110 kV incomer and 15 kV outgoing feeders, and operating times for relay characteristics of the studied Rwanda power system have

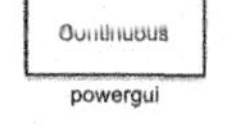

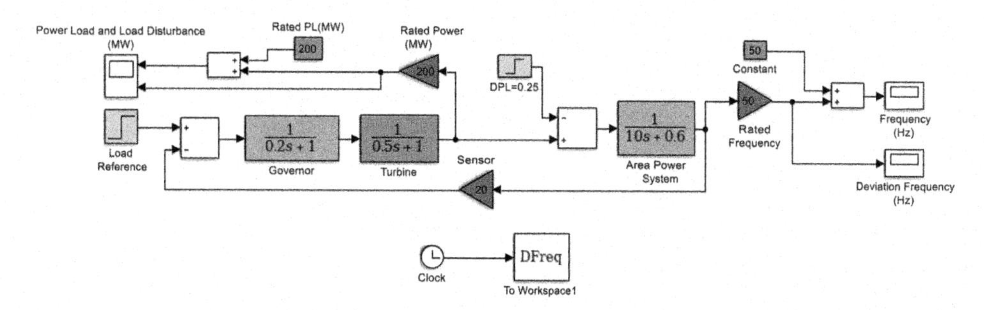

Figure 2. Rwanda Northern grid at Ntaruka hydropower plant without PID controller.

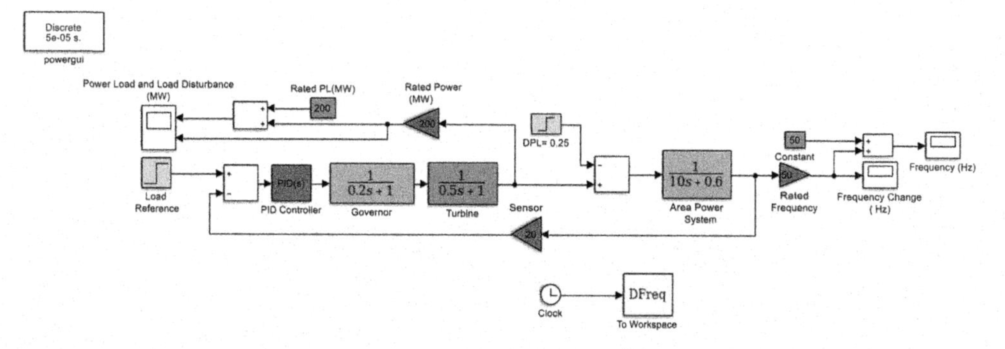

Figure 3. Rwanda Northern electricity grid at Ntaruka hydropower plant with PID controller.

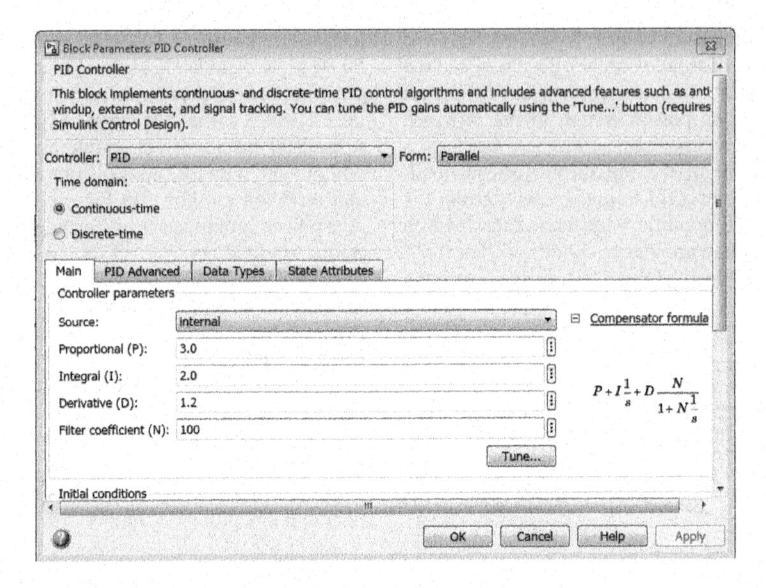

Figure 4. PID setting values of proportional, integral, and derivative.

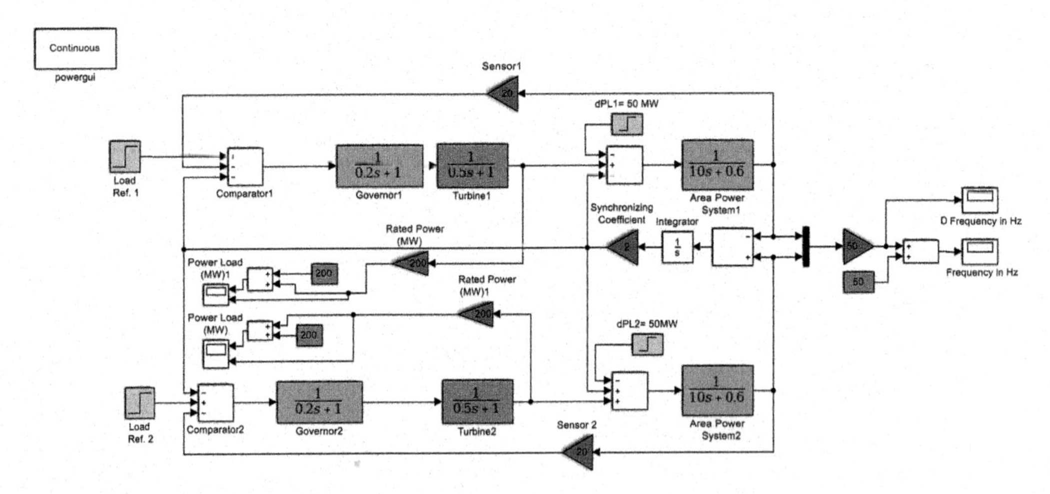

Figure 5. Two area networks without PID controllers.

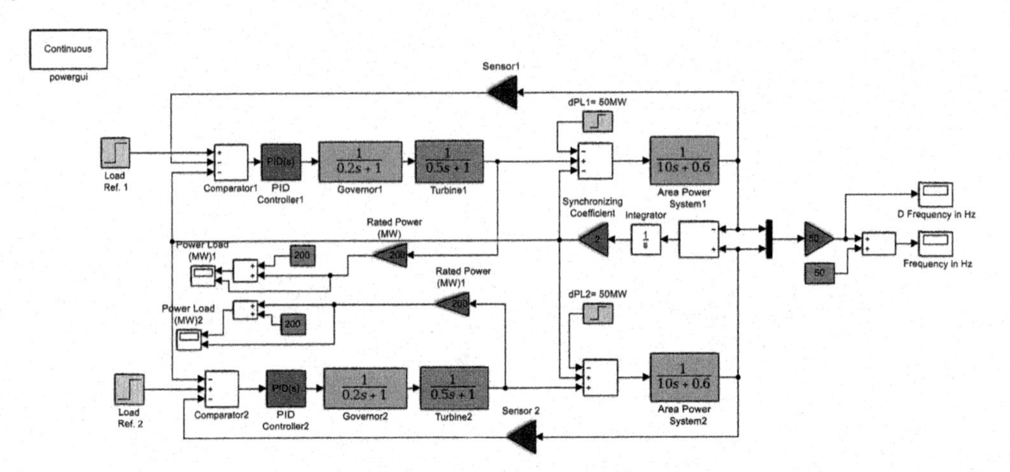

Figure 6. Two area PID controllers of the two area power generation.

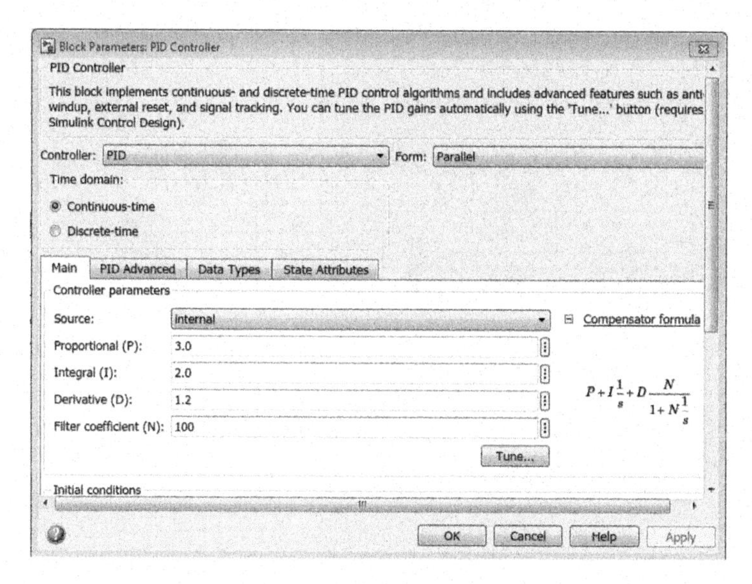

Figure 7. PID controller settings in area 1 and area 2.

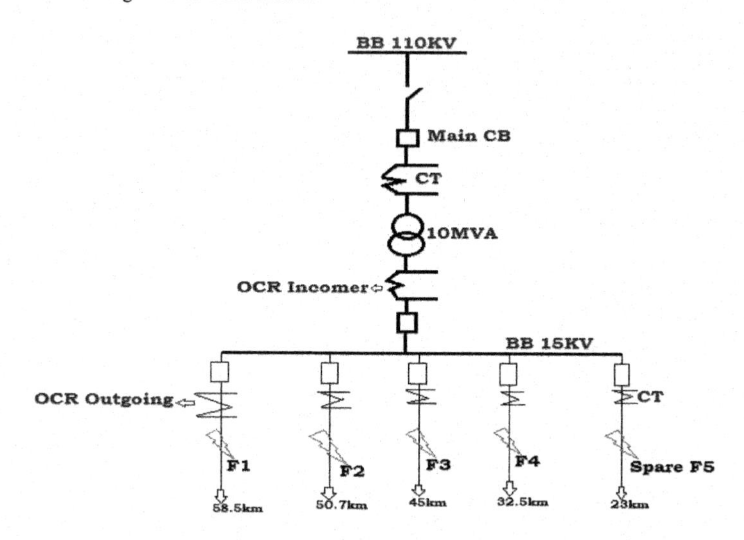

Figure 8. Incoming and outgoing (feeders) for overcurrent relay coordination and protection.

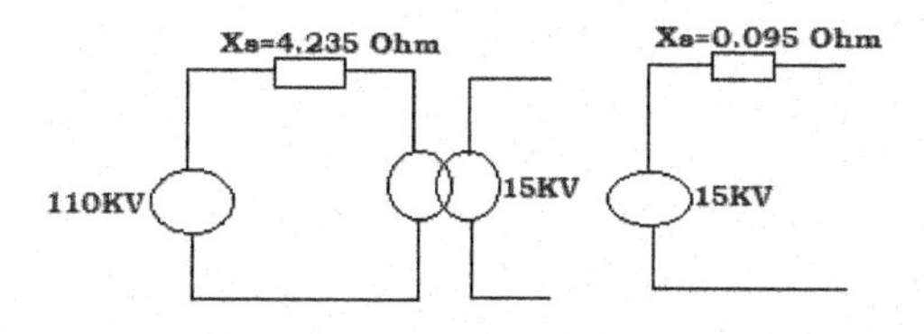

Figure 9. Reference of Xs from BB 15 kV.

been calculated and determined. The maximum and minimum fault currents from the last 3 years were recorded and documented. Figure 12 demonstrates a single line schematic of a substation with transformer 110/15 kV, 10 MVA, distributed feeders F1, F2, F3, F4, and spare F5 considering fault locations. The network MVA short circuit is 2587 MVAsc, both positive and negative sequence impedance of feeder ($Z_1 = Z_2$) is$0.045 + j0.151\Omega/\text{km}$. The power system parameters like (X/R ratio $= 5$, CT ratio $= 2500/4$A, line length $= 100$ Km, positive and zero sequence resistances $= 0.04553$ and $0.1517\Omega/\text{km}$ respectively, positive and zero sequence inductances $= 0.000617$ and $0.001533H/\text{km}$ respectively full load current = 1840.9A, line capacity = 282 MVA, TMS = 0.05, line voltage = 110 kV and CT ratio for outgoing feeders = 500/1A, grid frequency = 50 Hz and real power = 70 MW have been considered and collected from Rwanda Energy Group Ltd under the Energy Utility Corporation Limited subsidiary and introduced into the power system grid to simulate the results from the power system Matlab model (see Figure 10). From the simulation of relay status as indicated in Figure 11, the performance of these relays has been achieved in a reliable and sensitive way and Figure 12 shows how voltages

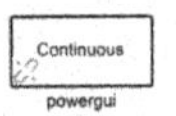

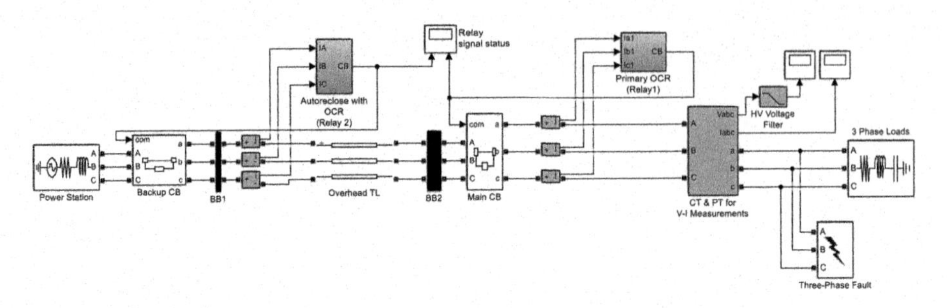

Figure 10. Power system grid with occurred faults model.

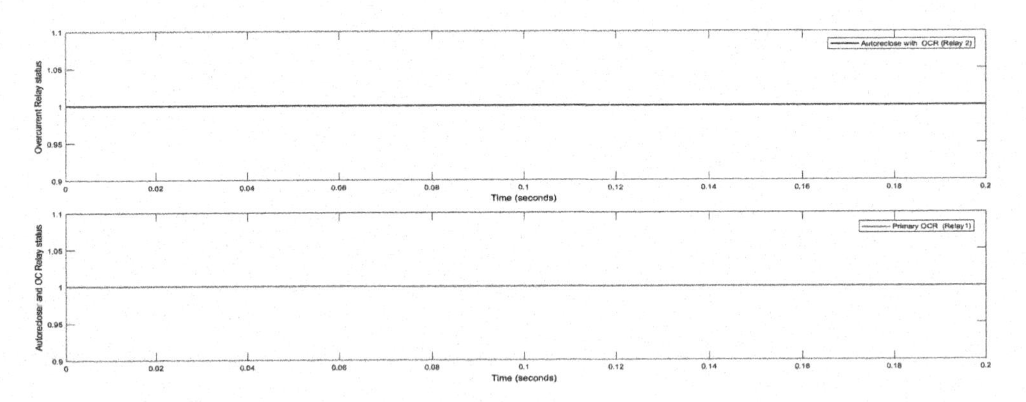

Figure 11. Power network relay status for good operational state.

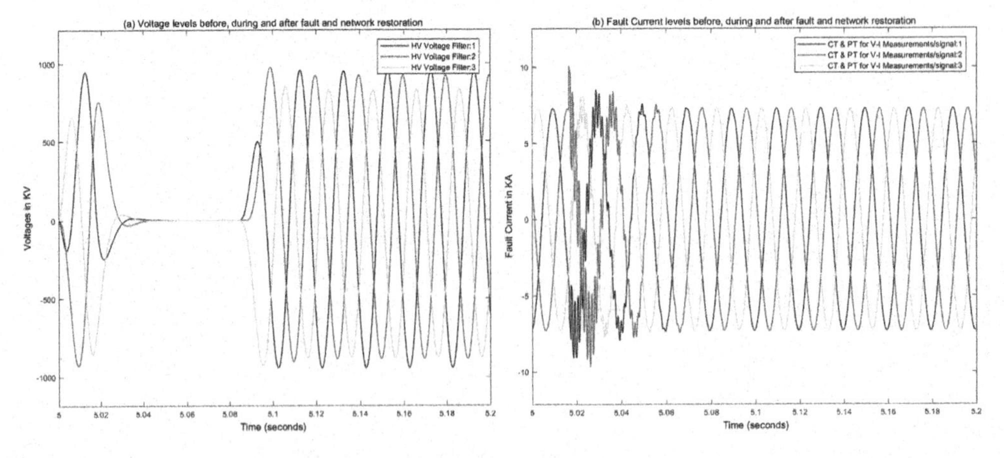

Figure 12. (a) Voltages and (b) fault currents observed during simulation on the power system.

and currents behaved before, during, and after power outages along the network. The relay characteristics in terms of operating times, as shown in Table 5, have been observed during simulation.

2.2.1 *Phase fault currents determination*

The calculation of phase short circuit currents requires the data of grid impedance (X_{s110KV}), transformer impedance (X_t) and feeder impedance (X_f) at fault location.

The impedance of feeder line varied with the feeder distance; the length in consideration was from the 15 kV bus bar to the outgoing feeder fault location. The outgoing of the five feeders, F1 was selected for the purpose of illustration for the short circuit current calculations, since the longest (58.5 km) has

Table 1. Feeder impedance calculation.

% Length	Feeder Impedances
0	0
1	1%*58.5*(0.045 + j0.151)= 0.026+j0.088

been considered. Table 1 shows the calculation of the feeder impedance at a distance of 0% (bus bar 15 kV) and 1% of the length of the F1 feeder. Let $Z_1 = Z_2 = 0.045 + j0.151$.

The grid impedance 110 kV

$$X_{S110KV} = {(KV)^2}/{\text{MVAsc}} = {(110)^2}/{2587} = 4.2352\Omega \tag{1}$$

The grid impedance 15 kV,

$$X_{S15KV} = \frac{(X_{S15KV})^2}{(X_{S110KV})^2} \times (X_{S110KV}) = \frac{15^2}{110^2} \times 4.2352 = 0.095\Omega \tag{2}$$

Transformer impedance

$$(X_t) = \frac{(kV)^2}{MVA} \times 8.67\% = \frac{(15)^2}{10} \times 8.67\% = 1.95\Omega \tag{3}$$

The reduced equivalence impedance diagram is shown in Figure 9.

The equivalent of impedance in the fault location was calculated in series,

$$Z_{1eq} = Z_{2eq} = X_{s(15KV)} + X_t + X_{F1} = 0.045 + j0.151 + X_{F1} X_{F1} = j0.095 + j1.95 = j2.045 \tag{4}$$

Table 2 indicates the feeder impedance equivalence of fault positions 0% and 1% respectively.

Table 2. Feeder impedance equivalence calculation.

% Length	Feeder Impedances
0	j2.045
1	j2.045 + 0.026 + j0.088 = 0.026 + j2.133

Phase to phase short circuit current (I_{SC}) in the feeder is affected by positive and negative sequence of equivalent impedance (Z_{1eq} and Z_{2eq}) at fault location.

$$I_{SC} = \frac{V}{Z1eq + Z_{2eq}} = \frac{V}{2Z_{2eq}} \tag{5}$$

where V is ph–ph voltage. Based on equations (5) and Table 1 for equivalent of impedances following the 25%, 50%, 75%, and 100% feeder locations, by imitating the same procedures, the short circuit currents found after calculation are tabulated in Table 3.

Table 3. Calculated and protected fault current levels based on maximum fault occurrences.

% Length	Calculated Fault level Current (A)	Occurred Fault Level (A)	Protected Fault Current Level (A)
0	3667.48	13489($I_{fault-max}$)	17156.48
1	3515.92		17004.92
25	1833.96		15322.96
50	985.02		14474.02
75	655.76		14144.76
100	498.10	1617($I_{fault-min}$)	13987.10

The current setting (I_{set}) of the OCR relay is 120% of the full load (I_n) equipment installed. The lowest current installed transformation is 1840.9 A. Selected 120% × I_n must be planned for extreme load forbearance.

$$I_{set} = 120\% \times I_n = 120\% \times 1840.9 = 2209.08\text{ A}. \tag{6}$$

The outgoing feeder utilized a current transformer ratio of 500 / 1 A. The setting current (I_{set}) of the OC relay is 120% of the full load current ($I_{full-load}$) of the installed equipments (CT) 500 A.

$$I_{set} = 120\% \times I_{full-load} = 120\% \times 500 = 600A \tag{7}$$

2.2.2 *Calculation of operating time of OCR (t_{OCR})*

The time multiplier setting (TMS) and working time (t_{OCR}) based on the short circuit current on 15 kV bus have been regulated to 0.05 and calculated respectively.

$$\text{SIR (t)} = \frac{0.14}{\text{Psm}^{0.02} - 1} \times \text{TMS}, \tag{8}$$

$$\text{IR (t)} = \frac{13.5}{\text{Psm} - 1} \times \text{TMS} \tag{9}$$

$$\text{EIR (t)} = \frac{80}{\text{Psm}^2 - 1} \times \text{TMS} \tag{10}$$

where

$$\text{Psm} = \frac{I_{SC}}{I_{set}}. \tag{11}$$

By replacing (11) in (8), (9), and (10) we get: standard inverse relay, $\text{TMS} = \frac{\left(\frac{I_{SC}}{I_{set}}\right)^{0.02} - 1}{0.14} \times t_{OCR}$, t_{OCR} is OCR tripping time when fault occur at 15 kV bus bars;

$$t_{OCR} = \frac{0.14}{\left(\frac{I_{SC}}{I_{set}}\right)^{0.02} - 1} \times \text{TMS}; \tag{12}$$

very inverse relay,

$$\begin{aligned} TMS &= \frac{\left(\frac{I_{sc}}{I_{set}}\right) - 1}{13.5} \times t_{OCR}, \\ t_{OCR} &= \frac{13.5}{\left(\frac{I_{sc}}{I_{set}}\right) - 1} \times TMS; \end{aligned} \tag{13}$$

extremely inverse relay,

$$\begin{aligned} TMS &= \frac{\left(\frac{I_{sc}}{I_{set}}\right)^2 - 1}{80} \times t_{OCR}, \\ t_{OCR} &= \frac{80}{\left(\frac{I_{sc}}{I_{set}}\right)^2 - 1} \times TMS; \end{aligned} \tag{14}$$

t_{OCR} is tripping time of overcurrent relay Figure 10 demonstrates the power system model to be set and coordinated using overcurrent relays

3 RESULTS AND DISCUSSIONS

3.1 *Frequency response characteristics*

In this research paper, during the load disturbance conditions, frequency load PID controllers have been developed and performed in balancing steady state condition by considering the values of overshoot, steady state error, and settling time. The frequency has been stabilized and balanced by using PID controllers in single area and two area hydroelectric power plants. The simulation results in Table 4 and in Figures 13–16 for single and two area power systems show that by applying a step load increment of 0.25 p.u MW equivalent to 50 MW before PID controller installation, the generation plants have been collapsed with an overshoot of 0.02 Hz, a steady state error of 0.55 Hz, and a settling time of 10 sec. After PID controller installation in the generation control station and National Dispatching, the power system was recovered to normal

Table 4. Simulation results of frequency response.

SINGLE AREA			TWO AREA	
Frequency Response Characteristics			Frequency Response Characteristics	
	Before PID	After PID	Before PID	After PID
Overshoot (Hz)	**0.02**	**0**	**0.03**	**0**
Settling time (Sec)	**10**	**2.5**	**10**	**2.5**
Steady state error (Hz)	**0.55**	**0**	**0.57**	**0**
Recorded nominal frequency (Hz)	< 49.5	**50**	< 49.5	**50**

state at 0.0 Hz, 0.0 Hz, and 2.0 sec for overshoot, steady state error, and settling time, respectively (see Figure 14 and 16). The generation power plants have been synchronized with a synchronization parameter as shown in Figures 5 and 6, and the simulation results show that by setting the PID controller and install in a power network, the overshoot, steady state error, and settling time have been minimized to 0.0 Hz, 0.0 Hz, and 2.5 sec, respectively, and it has been observed that the frequency has been recovered and remained at 50Hz after load disturbance. Table 2 shows the summarized simulation results of single and two area power systems of the frequency responses before and after PID application in the hydropower station.

Summarizes the single and two area power system frequency results before and after plant perturbations.

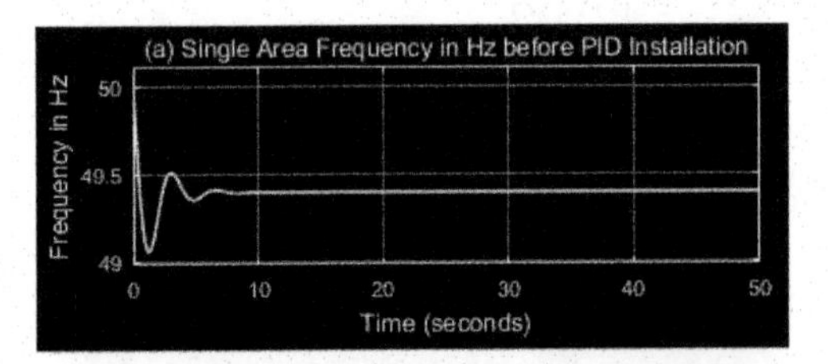

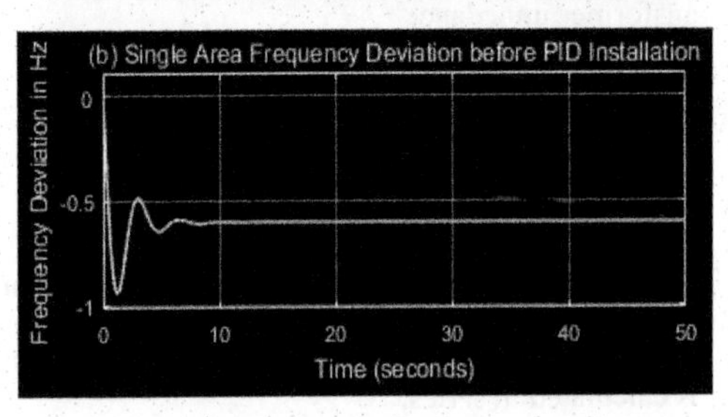

Figure 13. Frequency response without PID controller with load disturbance injection.

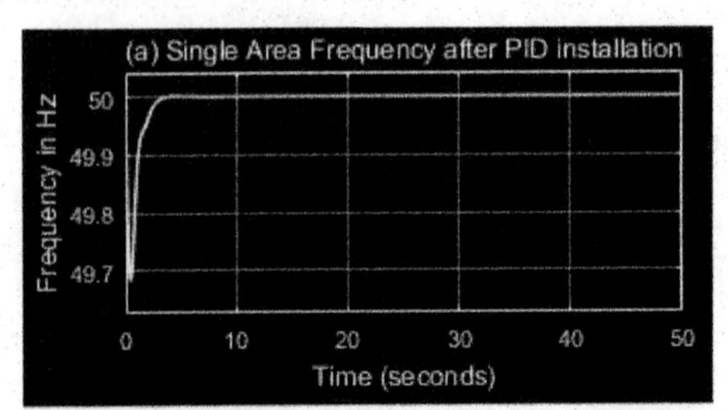

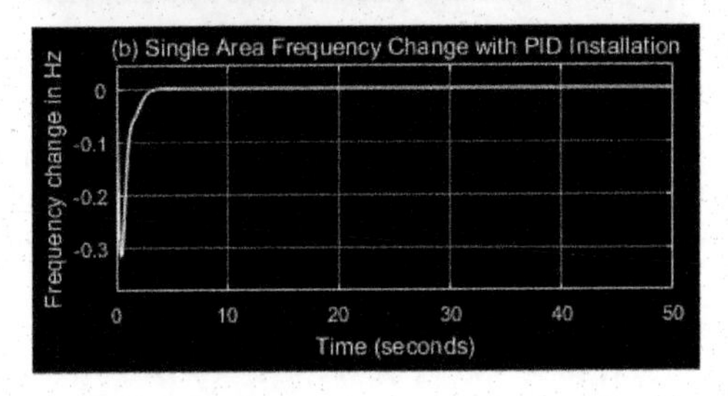

Figure 14. Frequency response with PID controller with load disturbance injection.

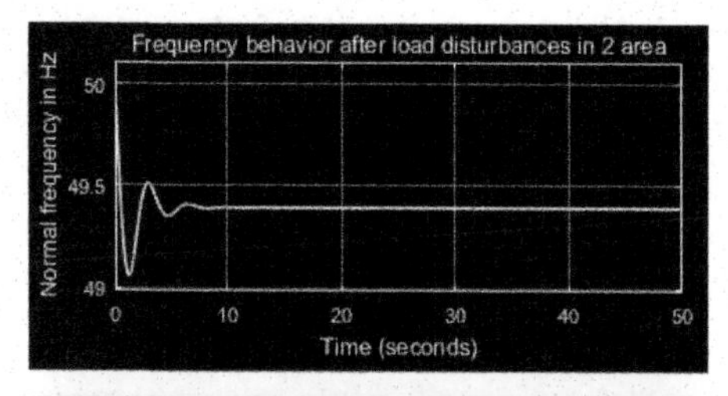

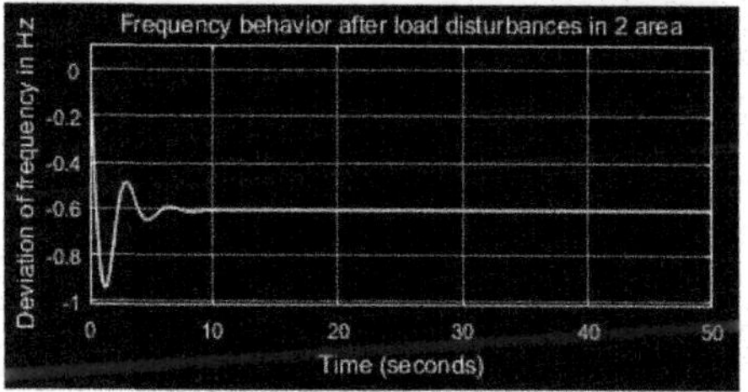

Figure 15. Frequency response of two area power generation without PID controllers.

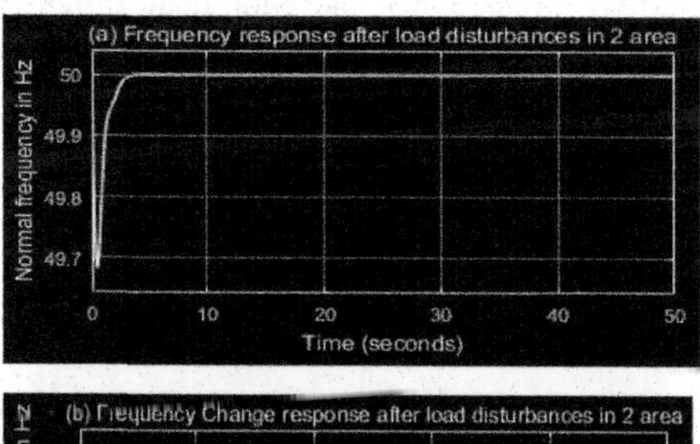

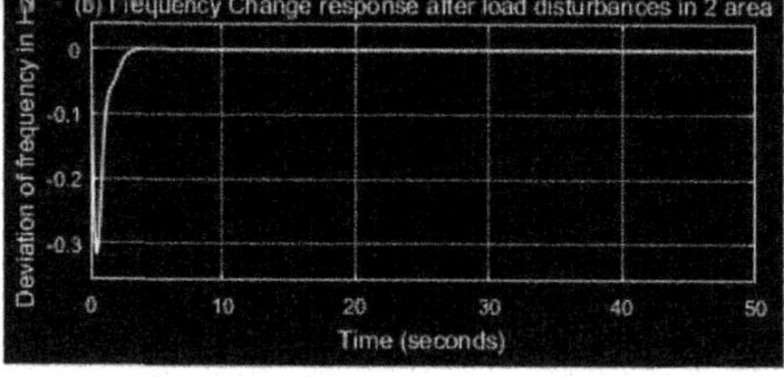

Figure 16. Synchronized frequency response with PID controllers in two areas after disturbances.

3.2 *Overcurrent setting and substation relay coordination*

Results summarized in Table 5 and Figure 17 are based on Table 5 with equations (5), (6), (7), (8), (9), (10), (11), (12), (13), and (14). Calculations were made of operating times of overcurrent relay incoming and outgoing for SIR, VIR, and EIR using I_{set} and I_{SC}.

Table 5. Operating time of incoming and outgoing feeders of OCR characteristics.

Fault Positions (%)	Operating Time (t_{OCR}) in secs					
	OCR INCOMING			OCR OUTGOING FEEDERS		
	EI	VI	SI	EI	VI	SI
0 (15 kV BB)	0.067	0.099	0.167	0.0048	0.0244	0.1009
1	0.068	0.111	0.168	0.0049	0.0246	0.1011
25	0.084	0.113	0.177	0.0061	0.0275	0.1045
50	0.095	0.121	0.182	0.0068	0.0291	0.1064
75	0.100	0.124	0.185	0.0072	0.0299	0.1072
100	0.102	0.126	0.186	0.0073	0.0302	0.1076

Relay coordination and settings were generally based on their characteristics, which indicated the speed of operation The characteristics are: Standard Inverse (SI), Very Inverse (VI), and Extremely Inverse (EI). From Figure 17 and Table 5, it was observed that the extremely inverse, very inverse, and standard inverse were working at 0.067–0.102 sec, 0.099–0.126 sec, 0.167–0.186 sec, and 0.0048–0.0244 sec, 0.1009–0.0073 sec, 0.0302–0.1076 sec for overcurrent incoming and outgoing feeders, respectively. The operating time of overcurrent incoming and outgoing feeders were calculated based on the short circuit currents demonstrated in Table 3, and for a time setting multiplier that was set at 0.05 sec (IEC standards), and

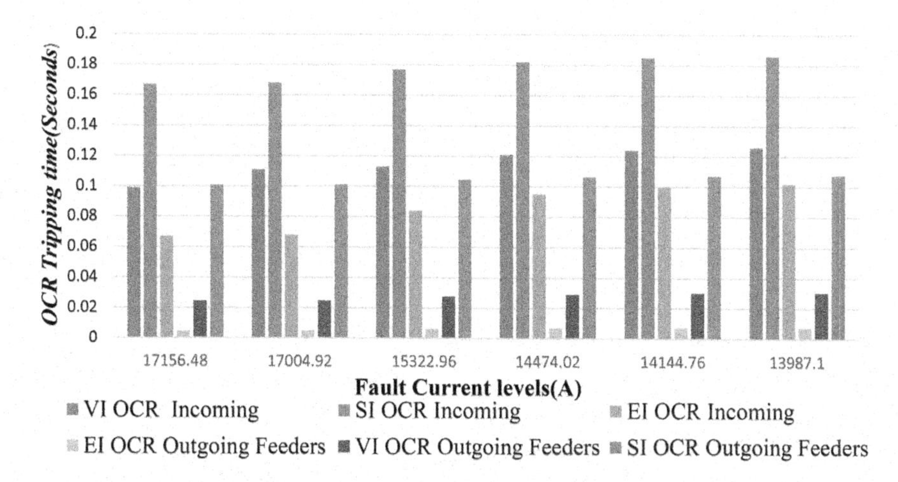

Figure 17. Operating times of overcurrent relay incoming and outgoing feeders.

also based on equations (5), (6), and (7). By application of autorecloser and overcurrent relays and creation of three phase faults in the power network, it has been shown that there was improvement and optimization of voltages and currents waveforms in the transmission line before, during, and after fault occurrences and the outages were cleared (see Figure 12). As the power outages occurred as shown in Figure 10, Relay 1 and Relay 2 were coordinated and sent the tripping signals to Backup CB and Main CB1, respectively, to open and isolate the power system grid and trip the network; the currents increased rapidly, waited for fault clearance restoration signals from relays, and restored quickly. The voltages decreased to zero volts instantly, as shown also in Figure 12. This meant that during the outages, overcurrent and autorecloser relays for extremely inverse tripped at a very short time, as indicated in the Table 5 and Figure 17 normally. Whenever a transient fault has occurred in the system, the autorecloser relay checked the synchronism of the network and the outages were mitigated and prevented. From voltage parts, at the 5th sec, the grid worked normally, after 0.02 sec, an outage occurred, was recorded, and remained in the power system for a period of 0.064 secs, after which the autorecloser with an overcurrent relay commanded the main CB to trip and isolated the faulted part within 0.064 sec; at 5.084 sec it became normal, protected, and optimized. Like the fault current's part, at 5 sec, the system normally worked within 0.01 sec, when the network was faulted during 0.04 secs and restored at 5.05 secs. With reference to Figure 17 and Table 5, the time of operation of overcurrent relays varied, with the extremely inverse relay the smallest, followed by the very inverse and standard inverse for both overcurrent relay incoming and outgoing feeders, respectively. It should be also observed that the three relay characteristics have been considered during the relay settings. The standard inverse characteristic took care of faults within the utility substation. The very inverse characteristic took care of faults at the midpoint of the feeder, while the extremely inverse characteristic took care of faults at the far end of the feeders.

4 CONCLUSIONS AND RECOMMENDATIONS

In this paper, we presented a frequency-based load control scheme to balance demand with supply and regulated frequency in a power system. We set up a load control optimization problem with the objective of minimizing the total frequency variation/error and the constraint of demand–supply power balance. The load frequency PID controllers using local frequency measurements per unit to solve the load disturbances and prove the convergence of the controllers have been designed. Numerical experiments have demonstrated that the proposed PID load control scheme was able to relatively quickly balance the power demand with supply and restore frequency under generation-loss. It can be observed that under steady state condition, incremental turbine power output of single and two area in which disturbance were given, became equal to the load disturbance and that of other area became zero along with zero tie-line power deviation with the controllers used in frequency load control of the case studies. Further research should deal with more than two frequency control areas in a microgrid to ensure the reliability and the security of the generation, transmission, and distribution systems. The overcurrent relay characteristics were developed and modeled in MATLAB/SIMULINK. The performance characteristics of the overcurrent relay were evaluated at a location with three-phase faults. The protection of a 110/15 kV substation with two bus systems and its relay settings has been presented. The optimal coordination of overcurrent with extremely inverse time is better than very inverse and standard inverse characteristics because overcurrent relay feeders as the main protection and overcurrent incoming as backup protection were able to work faster to protect the distribution system from phase-to-phase short circuit currents that occur in the feeders. The information like fault data, transmission line data, load data, power system data, and different relay types, like Mho relay, differential relay, and impedance relay need to be considered for upcoming research. The model can be extended to other categories of relay characteristics and fault types to implement real-time relay modeling of the power system.

ACKNOWLEDGMENT

This work was supported by Moi University, Eldoret, Kenya under Africa Center of Excellence II for Phytochemical, Textile and Renewable Energy (**ACE II PTRE**) through World Bank, Prof. Paul Wambua, Prof. Simiyu Sitati and Prof. Jean Bosco Byiringiro.

REFERENCES

Aimable, N. (2020). Power blackout occured in Rwanda Electricity Grid in 2019 and later 2020. Kigali: Aimable Nsanzimana.

Akbar A., M. P. (2011). Optimal coordination of overcurrent and distance relays by a new particle swarm optimization method. International Journal of Engineering and Advanced Technology, 93–98.

Akbar, M. (2011). Optimal Coordination of Overcurrent and Distance Relays by a New Particle Swarm Optimization Method. International Journal of Engineering and Advanced Technology, 93–98.

Behera, N. (2019). Load Frequency Control of Power System. Rourkela, India: P.K Ray.

Congzhi Huang, J. L. (2017). Linear active disturbance rejection control approach for load frequency control of two-area interconnected power system. Transactions of the Institute of Measurement and Control, 1–9.

Dobson. (2007). Protection relay signals. . Mumbai: Dobson.

G. S. Thakur, A. P. (2014). 'Load frequency in Single area with tradition Ziegler-Nichols PID tuning controller. International Journal of Research in Advanced Technology.

Gerard, E. A. (2020). Total Duration of Power blackouts in Rwanda Electricity Grid Records. Kigali: Energy Utility Corporation Limited.

Gurrala, G., & Sen, I. (2010). Power system stabilizers design for interconnected power systems. IEEE Trans. on Power Systems, XXV(2), 1042–1051.

Hojo, M. O. (2005). Analysis of Load Frequency Control Dynamics Based on Multiple Synchronized Phasor Measurements. 15th PSCC (pp. 22-26). Tokushima: Hojo, M., Ohnishi, K, and Ohnishi, T.

Javad S., V. A. (2019). Optimal coordination ofovercurrent and distance relays with hybrid genetic algorithm. 10th International Conference on Environment and Electrical Engineering. Rome, Italy: Javad S.

Kundur, P. (1994). Power System Stability and Control. New York: McGraw-Hill: P. Kundur.

L. Gao, Y. W. (2016). Transient Stability Enhancement and Voltage Regulation of Power System. . L. Gao, Y.Y Wang.

Limited, E. U. (2018, 2019 and 2020). Total Blackouts in Rwanda Electricity Grid Records. Kigali: EUCL.

M. J. Hossain, H. R. (2009). Excitation Control for Large Disturbances in Power Systems with Dynamic Loads. IEEE Xplore.

M. Wadi, M. B. (2017). Comparison between Open-Ring and Closed-Ring Grids Reliability. 4th International Conference on Electrical and Electronics Engineering. Ankara: M. Wadi, M. Baysal.

Mancer N, M. B. (2015). Optimal Coordination of Directional Overcurrent Relays Using PSO. TVAC Energy Procedia (pp. 1239–1247). Hamed M.

Naresh K. Tanwani, A. P. (2014). Simulation Techniques of Electrical Power System Stability Studies Utilizing Matlab/Simulink. international journal of engineering sciences & research technology, iii(3), 1228-1240.

NEPA. (2007). Basic Protection Course P1"Training and development program. Washington: NEPA.

P. K. Modi, S. P. (2006). Stability Improvement of Power System by Decentralized Energy, Advances in Energy Research. India: P. K. Pradhan.

P. Nelson, S. M. (2017). Frequency regulation free governor mode of operation in power stations. IEEE International Conference on Computational Intelligence and Computing Research, 125–129.

Reza M., H. A. (2010). Optimal relays coordination efficient method in interconnected power systems. Journal of Electrical Engineering , 75–83.

Steinmetz, C. P. (1920). Power control and stability of electric generating stations. AIEE Trans., XXXIX(2), 1215–1287.

Tanwani, N. K. (2013). nalysis and Simulation Techniques of Electrical Power System Stability Studies. Pakistan: Naresh K. Tanwani.

W.B. (2017). frequent power outages in Rwanda. Kigali: World Bank Group.

Wang. (2010). Penetration of automation control and protection in power systems. Delhi: Wang.

Wang, L. (2016). Li Wang 2016 The Fault Causes of Overhead Lines in Distribution Network. 10.1051/ mateconf/20166102017: MATEC Web of conferences 61.

Weiss, R. (2019). Line Voltage Stability Indices Based on Precautionary MeasureApproximation in Smart Grid. Proceedings of the 11th International Conference on Innovation & Management. Kigali, Rwanda: Rwanda Energy Group.

Weiss, R. (2020). Line Voltage Stability Indices Based on Precautionary MeasureApproximation in Smart Grid. Kigali: Rwanda Energy Group.

Weiss, R. (2020). Line Voltage Stability Indices Based on Precautionary MeasureApproximation in Smart Grid. Proceedings of the 11th International Conference on Innovation & Management. Kigali, Rwanda: Rwanda Energy Group.

Z. Osman, M. Y.-D. (2006). Maximum Loadability of Power System Using Hybrid Particle Swarm Optimization. Electric Power System Research,, V(76), 485–492.

Zalostiba. (2020). blackout analysis, the dangerous overload of transmission grid in blackout analysis. Tokyo: Zalostiba.

Advances in Phytochemistry, Textile and Renewable Energy Research for Industrial Growth – Nzila et al. (Eds)

Bayesian inference for simple and generalized linear models: Comparing INLA and McMC

J. Darkwah
Moi University, Eldoret, Kenya

ABSTRACT: Markov chain Monte Carlo (McMC) is a traditional technique in Bayesian inference. Lately, Integrated Nested Laplace Approximations (INLA) has gained popularity as another technique for Bayesian inference. This paper compares the performance of these techniques in terms of accuracy, execution time, and computational burden in simple and generalized linear models. At the end of the simulation study, INLA produced estimates similar to those of the McMC technique. This observation was evident in the estimates of the fixed parameters of the models. Though random effects of the generalized linear model were not considered in this paper, those of the simple linear model were considered and the estimates by the two techniques were found to be closely identical, leading to the conclusion that INLA is as computationally efficient as McMC. Furthermore, INLA took a shorter time in approximating parameters than McMC. Finally, McMC was found to be more computationally intensive than INLA.

1 INTRODUCTION

The two main approaches in inferential statistics are the Frequentist and the Bayesian approaches. Though, if procedures are carefully followed, both approaches produce identical results, they vary mainly in their conceptualization of probability (Fox, 2015). While the Frequentist sees probability as a limiting frequency of an experiment repeated infinitely, the Bayesian, on the other hand, sees it as a subjective quantity which depends on the availability of information (Wakefield, 2013). Applying these approaches in any simulation study requires mathematical models, and in statistics these models are largely classified as either a simple linear model or a generalized linear model. While in simple linear models, the response variable assumes a Gaussian distribution, the response variable of a generalized linear model assumes a non-Gaussian conditional distribution (Fox, 2015). This paper focuses on Bayesian inference for simple and generalized linear models, comparing the traditional Markov chain Monte Carlo (McMC) technique with the relatively new Integrated Nested Laplace Approximation (INLA) technique in terms of computational burden, accuracy and time of execution.

2 LITERATURE REVIEW

Bayesian inference is basically determining parameters of a model by sampling from its posterior marginal and this is accomplished mostly by the McMC technique but with several challenges (Gamerman, D. & Lopes, H.F, 2006). One of the challenges is the convergence of the Markov chain, which mostly takes quite some time. Another challenge is the fact that the McMC process is computationally intensive. These challenges result from the complex form that the posterior distribution takes, making it not easy to sample from (Givens, G. H., & Hoeting, J. A., 2012).

For the past two decades, several software for implementing McMC have been developed. Some of them are OpenBUGS (Lunn, D., Spiegelhalter, D., Thomas, A. & Best, N., 2009), JAGS (Plummer, 2003), and CARBayes (Lee, 2013). While most of these software rely on Monte Carlo integration for the estimation of parameters, INLA, which is another technique in Bayesian inference, computes very accurate approximations of the posterior marginal in a fraction of the time used by McMC and its software, and it does this by numerical integration (Rue, H. & Held, L., 2005) (Rue, H., Martino, & S., Chopin, N., 2009).

In 2010, there was a study titled “Posterior and Cross-validatory Predictive Checks: A Comparison of McMC and INLA in Statistical Modeling and Regression.” This study compared cross-validatory checks between INLA and McMC (Held, L., Schrodle, B. & Rue, H., 2010). (Carroll, R., Lawson, A. B., Faes, C. Kirby, R. S., Aregay, M., & Watjou, K., 2015), and also compared INLA and OpenBUGS for hierarchical Poisson modeling in disease mapping. It concluded that INLA underperformed in estimating the parameters of the random effects when compared with OpenBUGS in disease mapping. This study compares the performance of McMC and INLA in terms of accuracy, time of execution, and computational burden

DOI 10.1201/9781003221968-8

in simple and generalized linear models. The versions of R and INLA used in this simulation study are 4.0 and 20.05.12, respectively.

This paper is organized as follows: the methodology, which consists of the setup and simulation procedures for the models under consideration, results and discussion, and finally, the conclusion, limitations, and recommendations. Emboldened uppercase letters are matrices and emboldened lowercase letters are vectors. Again, emboldened Greek letters are vectors of parameters and non-emboldened Greek letters are parameters.

3 METHODOLOGY

3.1 *McMC formulation for the simple linear model*

3.1.1 *Simple linear model*

A model given by

$$\boldsymbol{y} = \boldsymbol{X\beta} + \boldsymbol{\varepsilon}, \boldsymbol{\varepsilon} \sim \boldsymbol{N}\left(\boldsymbol{0}, \sigma^2 \boldsymbol{I}\right) \tag{1}$$

is a linear model in $\boldsymbol{X}$, where $\boldsymbol{X} =$ design matrix, $\boldsymbol{y} =$ observed data or response vector, $\boldsymbol{\beta} =$ parameter vector, $\boldsymbol{\varepsilon} =$ error vector and $\boldsymbol{I} =$ identity matrix(Bingham, N. J. & Fry, J. M., 2010). Upon the assumption that the error vector, $\boldsymbol{\varepsilon}$, is Normally distributed with mean vector $= 0$ and a diagonal matrix of $\boldsymbol{\sigma}^2$, the variance-covariance matrix, the only unknown parameters to be determined in this model are $\boldsymbol{\beta}$, and $\boldsymbol{\sigma}^2$.

3.1.2 *Setup*

Consider the setup for n observations where the response variable, $y_i, i = 1, 2, \cdots, n$, is normally distributed with mean, $\eta_i = x_i^T \boldsymbol{\beta}$, and variance, τ^{-1} (τ being the precision parameter with $\tau^{-1} = \sigma^2$). With a flat prior distribution on $\boldsymbol{\beta}$ and a Gamma prior on τ, say $\tau \sim Gamma(a, b)$ where a is the shape parameter and b, the rate parameter, the full conditional distributions of β and τ can be deduced. To deduce them, the likelihood and the prior distributions had to be defined, after which the corresponding posterior distribution was determined and thereafter, the full conditionals derived. The details are as follows:

The likelihood of $\boldsymbol{y}$ is given $p\left(\boldsymbol{y}|\boldsymbol{\beta}, \tau^{-1}\right) =$

$$(2\pi)^{-n/2} \tau^{n/2} \exp\left[-\frac{\tau}{2}\left(\boldsymbol{y} - \boldsymbol{X\beta}\right)^T \left(\boldsymbol{y} - \boldsymbol{X\beta}\right)\right] \tag{2}$$

The prior distribution of τ is also given by

$$p(\tau) = \frac{b^a}{\Gamma(a)} \tau^{(a-1)} \exp(-b\tau) \tag{3}$$

Lastly, the flat prior distribution of $\boldsymbol{\beta}$ is given by

$$p(\boldsymbol{\beta}) = 1 \tag{4}$$

The posterior distribution, which is a product of the likelihood in equation (2), the prior distributions of equations (3) and (4), is given by

$$p\left(\boldsymbol{\beta}, \boldsymbol{\tau}|\boldsymbol{y}\right) \propto p\left(\boldsymbol{y}|\boldsymbol{\beta}, \tau^{-1}\right) p(\boldsymbol{\beta})\, p(\tau)$$

Now, the full conditional distribution of $\boldsymbol{\beta}$ is derived as follows:

$$\begin{aligned}
p(\boldsymbol{\beta}|\boldsymbol{y}, \tau) &\propto p\left(\boldsymbol{y}|\boldsymbol{\beta}, \tau^{-1}\right) p(\boldsymbol{\beta}) \\
&= (2\pi)^{-n/2} \tau^{n/2} \exp\left[-\frac{\tau}{2}(\boldsymbol{y} - \boldsymbol{X\beta})^T (\boldsymbol{y} - \boldsymbol{X\beta})\right] \\
&\propto \exp\left[-\frac{\tau}{2}(\boldsymbol{y} - \boldsymbol{X\beta})^T (\boldsymbol{y} - \boldsymbol{X\beta})\right] \\
&= \exp\left[-\frac{\tau}{2}\left(\boldsymbol{y}^T \boldsymbol{y} - 2\boldsymbol{y\beta}^T \boldsymbol{X}^T + \boldsymbol{\beta}^T \boldsymbol{X}^T \boldsymbol{X\beta}\right)\right] \\
&\propto \exp\left[-\frac{\tau}{2}\left(-2\boldsymbol{y\beta}^T \boldsymbol{X}^T + \boldsymbol{\beta}^T \boldsymbol{X}^T \boldsymbol{X\beta}\right)\right] \\
&= \exp\left[-\frac{\tau}{2}\left(\boldsymbol{\beta}^T \boldsymbol{\beta} - 2\boldsymbol{\beta}^T \left(\boldsymbol{X}^T \boldsymbol{X}\right)^{-1} \boldsymbol{X}^T \boldsymbol{y}\right) \right. \\
&\qquad \left. \left(\boldsymbol{X}^T \boldsymbol{X}\right)^{-1}\right] \\
&\propto \exp\left[-\frac{\tau}{2}\left(\boldsymbol{\beta} - \left(\boldsymbol{X}^T \boldsymbol{X}\right)^{-1} \boldsymbol{X}^T \boldsymbol{y}\right)^T \left(\boldsymbol{X}^T \boldsymbol{X}\right)^{-1} \right. \\
&\qquad \left. \left(\boldsymbol{\beta} - \left(\boldsymbol{X}^T \boldsymbol{X}\right)^{-1} \boldsymbol{X}^T \boldsymbol{y}\right)\right] \\
&= \boldsymbol{N}\left(\left(\boldsymbol{X}^T \boldsymbol{X}\right)^{-1} \boldsymbol{X}^T \boldsymbol{y}, \frac{1}{\tau}\left(\boldsymbol{X}^T \boldsymbol{X}\right)^{-1}\right)
\end{aligned} \tag{5}$$

Also, the full conditional distribution of τ is as deduced below:

$$\begin{aligned}
p(\tau \mid \boldsymbol{y}, \boldsymbol{\beta}) &\propto p\left(\boldsymbol{y} \mid \boldsymbol{\beta}, \tau^{-1}\right) p(\tau) \\
&= (2\pi)^{-n/2} \tau^{n/2} \exp\left[-\frac{\tau}{2}(\boldsymbol{y} - \boldsymbol{X\beta})^T (\boldsymbol{y} - \boldsymbol{X\beta})\right] \\
&\qquad \frac{b^a}{\Gamma(a)} \tau^{(a-1)} \exp(-b\tau) \\
&\propto \tau^{\left(a + \frac{n}{2} - 1\right)} \exp\left[-\tau\left(b + \frac{1}{2}(\boldsymbol{y} - \boldsymbol{X\beta})^T (\boldsymbol{y} - \boldsymbol{X\beta})\right)\right] \\
&= Gam\left(a + n/2, b + \frac{1}{2}(\boldsymbol{y} - \boldsymbol{X\beta})^T (\boldsymbol{y} - \boldsymbol{X\beta})\right)
\end{aligned} \tag{6}$$

3.1.3 *Simulation*

The linear model was established as $\boldsymbol{y} = 1 + 2\boldsymbol{x} + \boldsymbol{\varepsilon}$, where $\boldsymbol{\varepsilon} \sim \boldsymbol{N}(\boldsymbol{0}, \boldsymbol{I})$, with mean, $\eta_i = \beta_0 + \beta_1 x_i$. The covariate, x_i, was assigned the standard Normal distribution prior, the parameter, $\beta_i \sim N(0, 1000)$, and $\tau \sim Gam(1, 5e-05)$. These prior distributions were specifically chosen to render them non-informative. By this, their influence on the posterior distribution was minimized to allow a likelihood dominance, thereby rendering the Bayesian procedure objective.(Berger, 2006). With the full conditional distributions of the parameters, $\boldsymbol{\beta}$ and τ, in closed form as indicated in equations (5) and (6), McMC samples were drawn for $\boldsymbol{\beta}$ and τ using the Gibbs sampling algorithm. It is important to note that drawing samples from a full conditional distribution renders a Markov chain stationary (Wakefield, 2013).

3.2 *McMC formulation for the generalized linear model*

3.2.1 *Generalized linear models*

Models with response variable, y_i, having a specific non-Gaussian conditional distribution are said to belong to the family of generalized linear models (Fox, 2015). With such models, the response variable has a distribution that belongs to the exponential family and has a link function (Wakefield, 2013). By the above description, there are several generalized linear models but the Poisson model is what was considered in this study.

3.2.2 *Setup*

Considering a map of n small regions, let y_i denote the observed count and E_i the expected count in each region, i. This setup results in the response variable, y_i, being distributed as $y_i|\eta_i \sim Po(E_i \exp(\eta_i))$, where $\eta_i = X_i^T \boldsymbol{\beta}$ is the relative risk of the Poisson model. $\boldsymbol{\beta} =$ vector of parameters to be determined and $X_i =$ covariates. Let a Normal prior with mean $= \boldsymbol{\beta}_0$ and variance-covariance matrix $= \boldsymbol{\Sigma}_0$, be assigned to $\boldsymbol{\beta}$. Then the posterior distribution of $\boldsymbol{\beta}$ given the data, $\boldsymbol{y}$, is

$$p(\boldsymbol{\beta}|\boldsymbol{y}) \propto p(\boldsymbol{y}|\boldsymbol{\eta})\, p(\boldsymbol{\beta}|\boldsymbol{\beta}_0, \boldsymbol{\Sigma}_0) \tag{7}$$

The terms on the right of equation (7) are

$$p(y|\eta) = \prod_{i=1}^{n} \frac{(E_i \exp(\eta_i))^{y_i}}{y_i!} \exp(-E_i \exp(\eta_i))$$

$$= \exp\left(\sum_{i=1}^{n} (y_i \eta_i - E_i \exp(\eta_i) + y_i ln E_i - \ln(y_i!))\right)$$

$$p(\boldsymbol{\beta}|\boldsymbol{\beta}_0, \boldsymbol{\Sigma}_0) = (2\pi)^{-n/2} |\boldsymbol{\Sigma}_0|^{-1/2} \exp\left[-\frac{1}{2}(\boldsymbol{\beta} - \boldsymbol{\beta}_0)^T \boldsymbol{\Sigma}_0^{-1} (\boldsymbol{\beta} - \boldsymbol{\beta}_0)\right]$$

After ignoring constant terms in equation (7), the posterior distribution becomes

$$p(\boldsymbol{\beta}|\boldsymbol{y}) \propto \exp\left(\sum_{i=1}^{n} (y_i \eta_i - E_i \exp(\eta_i)) - \frac{1}{2}(\boldsymbol{\beta} - \boldsymbol{\beta}_0)^T \boldsymbol{\Sigma}_0^{-1} (\boldsymbol{\beta} - \boldsymbol{\beta}_0)\right)$$

Since the posterior distribution is not in closed form, it is difficult to draw McMC samples from it. To arrive at a proposal distribution that is easy to sample from, a second order Taylor's expansion about a suitable value, say z_i, had to be constructed (Rue, H. & Held, L., 2005). The resulting expression is

$$\tilde{f}(\eta_i) = E_i \exp(z_i)\left(z_i - \frac{1}{2} z_i^2 - 1\right) + (y_i + E_i \exp(z_i)(z_i - 1))\, \eta_i - \frac{1}{2}(E_i \exp(z_i))\, \eta_i^2 \tag{8}$$

Equation (8) is equivalent to

$$\tilde{f}(\eta_i) = a_i + b_i \eta_i - \frac{1}{2} c_i \eta_i^2$$

Now, the full conditional distribution of $\boldsymbol{\beta}$ becomes $p(\boldsymbol{\beta}|\boldsymbol{y}) \propto p(\boldsymbol{y}|\boldsymbol{\beta}) p(\boldsymbol{\beta}|\boldsymbol{\beta}_0, \boldsymbol{\Sigma}_0)$

$$= \prod_{i=1}^{n} \frac{(E_i \exp(\eta_i))^{y_i}}{y_i!} \exp(-E_i \exp(\eta_i)) (2\pi)^{-n/2} |\boldsymbol{\Sigma}_0|^{-1/2} \exp\left[-\frac{1}{2}(\boldsymbol{\beta} - \boldsymbol{\beta}_0)^T \boldsymbol{\Sigma}_0^{-1} (\boldsymbol{\beta} - \boldsymbol{\beta}_0)\right]$$

$$\propto \exp\left(\sum_{i=1}^{n} (y_i \eta_i - E_i \exp(\eta_i))\right) \exp\left[-\frac{1}{2}(\boldsymbol{\beta} - \boldsymbol{\beta}_0)^T \boldsymbol{\Sigma}_0^{-1} (\boldsymbol{\beta} - \boldsymbol{\beta}_0)\right]$$

$$= \exp\left(\sum_{i=1}^{n} (y_i \eta_i - E_i \exp(\eta_i)) - \frac{1}{2}(\boldsymbol{\beta} - \boldsymbol{\beta}_0)^T \boldsymbol{\Sigma}_0^{-1} (\boldsymbol{\beta} - \boldsymbol{\beta}_0)\right)$$

$$= \exp\left(-\frac{1}{2}\boldsymbol{\beta}^T \boldsymbol{\Sigma}_0^{-1} \boldsymbol{\beta} + \boldsymbol{\beta}^T (\boldsymbol{\Sigma}_0^{-1} \boldsymbol{\beta}_0) + \boldsymbol{\eta}^T \boldsymbol{y} - \exp(\boldsymbol{\eta})^T \boldsymbol{E}\right)$$

Let $(\boldsymbol{\eta}) = \boldsymbol{\eta}^T \boldsymbol{y} - \exp(\boldsymbol{\eta})^T \boldsymbol{E}$. Then by the second order Taylor's expansion of $f(\boldsymbol{\eta})$ at $\boldsymbol{z}$, a similar expressions as $\tilde{f}(\boldsymbol{\eta}) \propto \boldsymbol{\eta}^T \boldsymbol{b} - \frac{1}{2} \boldsymbol{\eta}^T \operatorname{diag}(\boldsymbol{c}) \boldsymbol{\eta}$ was arrived at. Hence, replacing $\boldsymbol{\eta}$ by $\boldsymbol{X\beta}$, 'the full conditional of $\boldsymbol{\beta}$ becomes

$$q(\boldsymbol{\beta}|\boldsymbol{y}, \boldsymbol{z}, \boldsymbol{\beta}_0, \boldsymbol{\Sigma}_0) \propto \exp\left(-\frac{1}{2}\boldsymbol{\beta}^T \boldsymbol{\Sigma}_0^{-1} \boldsymbol{\beta} + \boldsymbol{\beta}^T (\boldsymbol{\Sigma}_0^{-1} \boldsymbol{\beta}_0) + (\boldsymbol{X\beta})^T \boldsymbol{b} - \frac{1}{2}(\boldsymbol{X\beta})^T \operatorname{diag}(\boldsymbol{c}) \boldsymbol{X\beta}\right)$$

$$= \exp\left(-\frac{1}{2}(\boldsymbol{\beta}^T \boldsymbol{\Sigma}_0^{-1} \boldsymbol{\beta} - 2\boldsymbol{\beta}^T (\boldsymbol{\Sigma}_0^{-1} \boldsymbol{\beta}_0) - 2(\boldsymbol{X\beta})^T \boldsymbol{b} + (\boldsymbol{X\beta})^T \operatorname{diag}(\boldsymbol{c}) \boldsymbol{X\beta})\right)$$

$$= \exp\left(-\frac{1}{2}(\boldsymbol{\beta}^T (\boldsymbol{\Sigma}_0^{-1} + \boldsymbol{X}^T \operatorname{diag}(\boldsymbol{c}) \boldsymbol{X}) \boldsymbol{\beta} - 2\boldsymbol{\beta}^T (\boldsymbol{\Sigma}_0^{-1} \boldsymbol{\beta}_0 + \boldsymbol{X}^T \boldsymbol{b}))\right)$$

$$= \boldsymbol{N}\left((\boldsymbol{\Sigma}_0^{-1} \boldsymbol{\beta}_0 + \boldsymbol{X}^T \boldsymbol{b})(\boldsymbol{\Sigma}_0^{-1} + \boldsymbol{X}^T \operatorname{diag}(\boldsymbol{c}) \boldsymbol{X})^{-1}, (\boldsymbol{\Sigma}_0^{-1} + \boldsymbol{X}^T \operatorname{diag}(\boldsymbol{c}) \boldsymbol{X})^{-1}\right)$$

The above approximated density is multivariate normal distribution with precision,

$$P = \boldsymbol{\Sigma}_0^{-1} + \boldsymbol{X}^T \operatorname{diag}(\boldsymbol{c}) \boldsymbol{X}.$$

3.2.3 *Simulation*

The expected count for each region, E_i, was fixed at one. About 100 arbitrary regions were considered and the relative risk, η_i, was parametrized with the log relative risk model given by $\log(\eta_i) = 1 + 2C_i$. This parametrization was to aid the modeling of the relationship between the dependent and the independent variables (Wakefield, 2013). The covariate, C_i, was mean centered to assist in the goodness-of-fit of the model. The log of the relative risk of the fitted model was established as $\log \eta_i = \beta_0 + \beta_1 C_i$, where β_0 and β_1, expressed compactly as $\boldsymbol{\beta}$ are the parameters to be estimated. These parameters were assigned a prior distribution of $\beta_i \sim N(0, 1000)$ and $C_i \sim N(0, 1)$. The distribution of β_i was non-informative in order to make sure that the posterior distribution is dominated by the likelihood so as to enhance objectivity in the simulation process. The simulations were carried out using the Metropolis-Hasting algorithm of McMC.

3.3 *INLA formulation for both models*

It was very easy implementing INLA for both models. All that was needed was to establish a relationship between the response variable and the covariates, indicate the required data and then apply the INLA function. This seems not to be the case for the implementation of the McMC algorithms.

4 RESULTS AND DISCUSSION

4.1 *Results in terms of accuracy*

This section states the parameter estimates and other relevant plots of the simple and generalized linear models. The details are as follows:

4.1.1 *Simple linear model*

For the purposes of convergence diagnostics, the trace, and autocorrelation plots of tau, τ, were ascertained. They are as shown in Figure 1. The corresponding convergence diagnostics plots of $\boldsymbol{\beta}$ were similar to those of τ so they were excluded to avoid tautology.

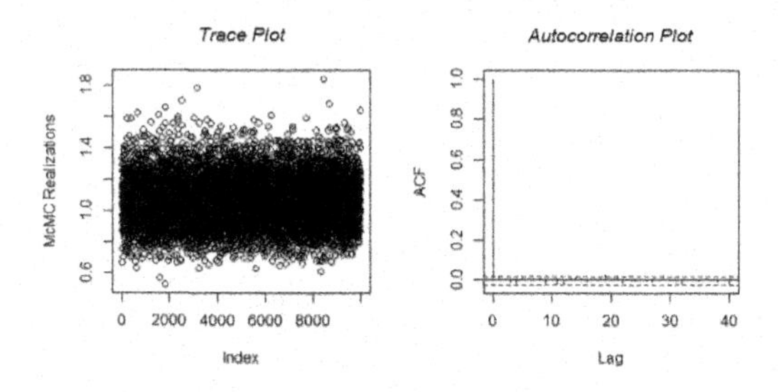

Figure 1. Trace and autocorrelation plots of the McMC realizations of Tau, τ.

Secondly, a table showing the mean values of the parameters estimated by the McMC and INLA techniques, with their respective standard deviations (in brackets) and the true values of the parameters of interest, is as shown in Table 1 above.

Also, graphs of INLA estimates superimposed on the McMC realizations of β_0,β_1, and τ were plotted to complement the values of Table 1. These graphs are as shown in Figure 2 below. The histograms are the McMC realizations while the curvy outlines superimposed on the histograms are the INLA estimates.

Table 1. A table of parameter estimates with their respective standard deviations (in brackets) and the true values of the parameters of interest.

	Parameters		
Technique	β_0	β_1	τ
McMC	1.283(0.11)	2.097(0.12)	0.907(0.13)
INLA	1.281(0.11)	2.098(0.12)	0.907(0.13)
True Values	1.000	2.000	1.000

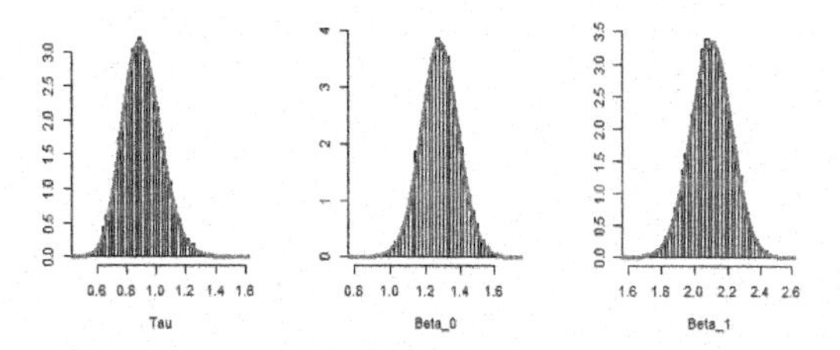

Figure 2. INLA estimates superimposed on McMC realizations for the Simple Linear Model.

4.1.2 *Generalized linear model*

Similarly, Figure 3 shows the trace, and autocorrelation plots of β_1. These plots are to facilitate the convergence diagnostics of the McMC process. Again, similar graphs were plotted for β_0 and they were found to be similar to those of β_1 so they were excluded to avoid the repetition of equivalent plots.

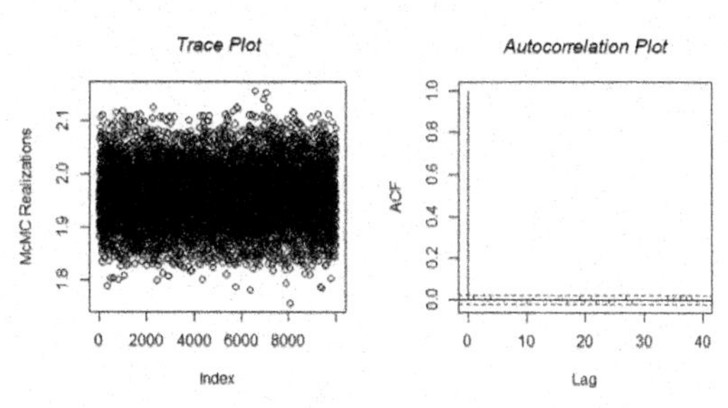

Figure 3. Trace and autocorrelation plots of the McMC realizations for β_1.

Also, the McMC realizations and INLA estimates are as shown in Table 2 below. Table 2 consists of the mean realizations (estimates) with their corresponding standard deviations (in brackets) and the true values of the parameters of $\boldsymbol{\beta}$.

Table 2. Mean realizations of McMC and INLA with their respective standard deviations (in brackets) and their corresponding true values.

	Parameters	
Technique	β_0	β_1
McMC	0.979 (0.06)	2.016 (0.04)
INLA	0.979 (0.06)	2.016 (0.06)
True Values	1.000	2.000

To complement the results in Table 2, graphs of the INLA estimates superimposed on the McMC realizations were plotted. They are as shown in Figure 4 below. The curvy outlines are the INLA estimates while the histograms are the McMC realizations

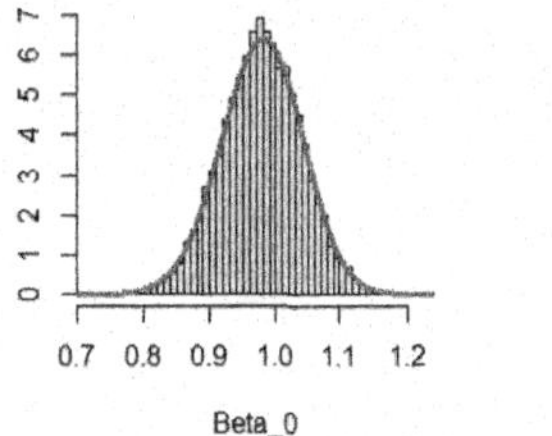

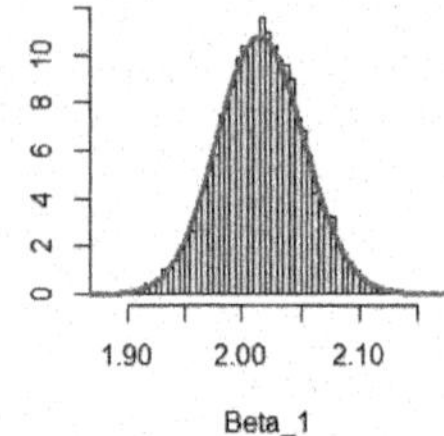

Figure 4. INLA estimates superimposed on the McMC realizations for the Generalized Linear Model.

4.2 *Results in terms of time of execution*

Comparing the time of execution, McMC used 5.14 seconds and 21.87 seconds in the realization of the parameters for the simple and the generalized linear models respectively. On the other hand, INLA used less than 3 seconds in approximating the same parameters for both models.

4.3 *Discussion*

The discussion is in two parts. First, a discussion of the convergence diagnostics of the McMC technique in both models and secondly, a comparison of McMC and INLA in terms of accuracy, time of execution and computational burden. With the results for the simple linear models being relatively identical to that of the generalized linear models, the discussion was clamped together to avoid repetitions. The details are as follows:

4.3.1 *Convergence diagnostics of McMC*

The trace plots of the parameters of interest suggest well-mixed chains in both models. This is typified by how the chains moved away from their initial values rapidly with the sample paths wiggling about vigorously in the area supported by the distributions of the parameters. This phenomenon is also an indication that the chains converged very fast and that the values sampled from the simulations attained stationarity (Nylander, J.A.; Wilgenbusch, J. C.; Warren, D. L. & Swofford, D. L., 2008). The autocorrelation plots of the same parameters for the models also showed the dependence of the realizations in 10 iterates apart. The 10 iterates apart is the lag and as it increased, there was an observed exponential decay, suggesting independence between the iterates (Cowles, M. K. & Carlin, B.P., 1996).

4.3.2 *Comparing McMC and INLA in terms of accuracy*

The McMC realizations and the INLA estimates were relatively equivalent. These are as shown in Tables 1 and 2 with Figures 2 and 4 to complement. By these tables and figures, it can be said that the INLA estimates were as accurate as the McMC realizations.

4.3.3 *Comparing McMC and INLA in terms of time of execution*

To determine the time of execution for the estimation of the parameters, both techniques were timed during their respective executions. It was found out that the McMC process took much more time for its realizations to be ascertained compared to INLA.

4.3.4 *Comparing McMC and INLA in terms of computational burden*

With the determination of the posterior distribution, the corresponding full conditional distributions for the parameters of interest, and the computer intensity of the simulation process, McMC was found to be computationally burdensome, compared to INLA which required the specification of the relationship between the response variable and the covariates in the form of a formula, stating of the data and the application of the INLA function.

5 CONCLUSION, LIMITATION, AND RECOMMENDATION

5.1 *Conclusion*

1. The parameter estimates of INLA were as accurate as the McMC realizations in both the simple and generalized linear models. This implies that INLA is as computationally efficient as McMC.
2. In both models, INLA estimated parameters with lesser time compared to its McMC counterpart.
3. In general, it was computationally burdensome implementing McMC as compared to INLA.

5.2 *Limitation*

1. The random effects of the generalized linear model were not considered in this simulation study.
2. The conclusions drawn in this simulation study are peculiar to the R software.

5.3 *Recommendation*

1. Since there are R packages such as OpenBUGS, JAGS, CARBayes, etc., for implementing McMC in Bayesian inference, it is recommended that a comparative study on the accuracy of these packages with INLA be carried out for simple linear models and generalized linear models.
2. Further work can be carried out on comparing McMC and INLA in generalized linear models with conditional autoregressive priors.
3. A study of this kind may be conducted using other statistical software to validate or otherwise the conclusions drawn in this paper.

ACKNOWLEDGMENT

I would like to acknowledge the immense support and tutelage of Prof. Andrea Riebler.

REFERENCES

Berger, J. (2006). The Case of Objective Bayesian Analysis. *Bayesian Analysis*, 385–402.

Bingham, N. J. & Fry, J. M. (2010). *Regression: Linear Models in Statistics.* London, UK: Springer Science & Business Media, Springer-Verlag.

Carroll, R., Lawson, A. B., Faes, C. Kirby, R. S., Aregay, M., & Watjou, K. (2015). Comparing INLA and OpenBUGS for Hierarchical Poisson Modeling in Disease Mapping. *Spatial and Spatio-temporal Epidemiology*, 45–54.

Cowles, M. K. & Carlin, B.P. (1996). Markov chain Monte Carlo Convergence Diagnostics: A Comparative Review. *Journal of the American Statistical Association*, 883–904.

Fox, J. (2015). *Applied Regression Analysis and Generalized Linear Models.* Los Angeles: Sage Publications.

Gamerman, D. & Lopes, H.F. (2006). *Markov chain Monte Carlo: Stochastic Simulation for Bayesian Inference.* New York: CRC Press.

Givens, G. H., & Hoeting, J. A. (2012). *Computaional Statistics (Vol 703).* N.J.: John Wiley and Sons.

Held, L., Schrodle, B. & Rue, H. (2010). Posterior and Cross-validatory Predictive Checks: A Comparison of McMC and INLA in Statistical Modedling and Regression. *Physica*, 91–110.

Lee, D. (2013). CARBayes: An R package for Bayesian Spatial Modeling with Conditional Autoregressive Priors. *Journal of Statistical Software*, 1–24.

Lunn, D., Spiegelhalter, D., Thomas, A. & Best, N. (2009). The BUGS Project: Evolution, Critique and Future Directions. *Statistics in Medicine*, 3049–3067.

Nylander, J.A.; Wilgenbusch, J. C.; Warren, D. L. & Swofford, D. L. (2008). AWTY (Are We There Yet?): A System for Graphical Exploration of McMC Convergence in Bayesian Phylogenetics. *Bioinformatics*, 581–583.

Plummer, M. (2003). JAGS: A program for Analysis of Bayesian Graphical Models Using Gibbs Sampling . *Statistical Computing*, 1–10.

Rue, H. & Held, L. (2005). *Gaussian Markov Random Fields: Theory and Applications.* Boca Raton, Florida: CRC Press.

Rue, H., Martino, & S., Chopin, N. (2009). Approximate Bayesian Inference for Latent Gaussian Models by Using Integrated Nested Laplace Approximations. *Royal Statistics Society: Series B (Statistical Methodology)*, 319–392.

Wakefield, J. (2013). *Bayesian and Frequentist Regression Methods.* New Yrok: Springer Science and Business.

Advances in Phytochemistry, Textile and Renewable Energy Research for Industrial Growth – Nzila et al. (Eds)

Determination of precursors of acrylamide formation in roasted maize

M.C. Koske
Department of Chemistry, Faculty of Science, Egerton University, Njoro, Kenya
Africa Center of Excellence II in Phytochemicals, Textiles and Renewable Energy (ACE II PTRE), Moi University, Eldoret, Kenya
Department of Chemistry and Biochemistry, School of Sciences and Aerospace Studies, Moi University, Eldoret, Kenya

A. Kiprop
Department of Chemistry and Biochemistry, School of Sciences and Aerospace Studies, Moi University, Eldoret, Kenya
Africa Center of Excellence II in Phytochemicals, Textiles and Renewable Energy (ACE II PTRE), Moi University, Eldoret, Kenya

O.P. Ongoma & S.M. Kariuki
Department of Chemistry, Faculty of Science, Egerton University, Njoro, Kenya

S.M. Kagwanja & J.M. Gichumbi
Department of Chemistry, Faculty of Science, Engineering and Technology, Chuka University, Chuka, Kenya

ABSTRACT: Acrylamide, an organic compound with the formula CH_2=$CHCONH_2$, is a contaminant generated through high-temperature cooking processes as a result of Maillard reactions catalyzed by the presence of reducing sugars and free amino acids in starchy food compounds. Acrylamide and its major metabolite, glycidamide, have been considered probable human carcinogens. In this study, we report on the acrylamide content in roasted maize from some Kenyan markets. Raw maize was purchased from local markets and roasted under laboratory conditions. They were crushed and extracted using water and hexane in a ratio of 2:1. The extract was derivatized with potassium bromate and potassium bromide and further subjected to liquid–liquid extraction using ethyl acetate-hexane (4:1, v/v). The final bromoprop-2-enamide (BPA) analyte was analyzed using gas chromatography–flame ionization detector. Acrylamide was not detected in any of the samples (at limit of detection of 20 μg/kg), which was consistent with reports from other countries.

1 BACKGROUND

1.1 *Acrylamide*

Following the discovery of high concentrations of acrylamides by the Swedish National Food Administration (NFA) and researchers from Stockholm University in April 2002 in food rich in carbohydrates content, it has become a subject of public interest. Acrylamide was found to be carcinogenic in rodents and is classified as a probable human carcinogen (Swedish 2002; Vinci et al. 2012). Food such as French fries, potato crisps, and corn were reported to contain acrylamide when cooked at elevated temperatures (Boroushaki et al. 2010). However, there are no guidelines currently showing the permissible limits of acrylamide in processed food (Hariri et al. 2015). Acrylamide is an organic compound with a formula CH_2=$CHCONH_2$. It is a contaminant formed through the Maillard reaction when starchy foods with appreciable sugar content are heated to elevated temperatures (Gökmen & Şenyuva 2007; Lund & Ray 2017; Mottram et al. 2002). The formation of acrylamide has not been reported in boiled foods or foods that were not heat treated (Ahn et al. 2002). It has become evident that the formation of acrylamide cannot be stopped. However, there is a concerted effort to minimize its presence in human diets and this has called for accelerated research to reduce its formation in foodstuffs (Alam et al. 2018; Fu et al. 2018; Li et al. 2012; Ou et al. 2010; Zeng et al. 2009).

From available reports, the main concern about the possible health effects of acrylamide in food is its probable carcinogenic and genotoxic (DNA-damaging) effects, as evidenced by tumors in laboratory rats (Manière et al. 2005). Since it has been detected in food, detailed research has been done to evaluate

 DOI 10.1201/9781003221968-9

whether acrylamide actually causes cancer in humans, but as yet there is sparse evidence that this is the case. However, acrylamide has been categorized by the International Agency for Research on Cancer (IARC) as a probable human carcinogen (Belkova et al. 2018; Gökmen & Palazoğlu 2008). The biological effect and risks associated with continued consumption of foods with acrylamide have been assessed by many international bodies including the European Food Safety Authority, the Food and Agriculture Organization of the United Nations (FAO), and the World Health Organization (WHO).

A Norwegian exposure assessment reported dietary acrylamide exposure with the mean and median exposure in adolescents and adults ranging between 0.3–0.5 µg/kg bodyweight per day. These estimates are in the same range as the mean daily exposures estimated by the European Food and Safety Authority (EFSA) for adolescents (0.4–0.9 µg/kg) body weight and adults (0.4–0.5 µg/kg body weight. Consumption patterns and dietary intake vary among people of different cultures and backgrounds (Normandin et al. 2013; Wyka et al. 2015).

Some foods analyzed for acrylamide, including infant powdered formula, coffee and chocolate powders, corn snacks, bakery products, and tuber-, meat-, and vegetable-based foods, showed that the levels of acrylamide present were variable among different foods and within different brands of the same food, as reported by European Union Authority (Authority 2012; Pacetti et al. 2015; Wilson et al. 2006). In a toxicological evaluation of acrylamide carried out by the Joint FAO/WHO Expert Committee on Food Additives (JECFA) in February 2005, it was noted that no data or limited information from Latin America and Africa were submitted. It was recommended that for useful assessment and mitigation of effects of acrylamide to reduce human exposure there was a need to have occurrence data on acrylamide in the foods consumed in developing countries (Arisseto et al. 2007).

The mechanism of formation of acrylamide in starchy foods is illustrated in Figure 1 (Krishnakumar & Visvanathan 2014).

Over the last few years various studies have reported the formation of acrylamide in foods to be assisted by precursors such as reducing sugars: fructose and glucose (Elmore et al. 2005; Mesias et al. 2018). Various methods have been employed for analysis, including high-performance liquid chromatography (HPLC-DAD) coupled to ultraviolet–visible (UV) detection (at 195 nm) with the limit of detection (LOD) of 10 µg/L in aqueous matrices (Ghiasvand & Hajipour 2016), and gas chromatography—an electron capture technique on the basis of the bromination of the acrylamide double bond has been developed (Zhang et al. 2006).

The aim of our study was to evaluate the levels of precursors of acrylamide in roasted green maize using gas chromatography-flame ionization detection using a new modified method (Geng et al. 2011) and monitoring precursors (glucose and fructose) using polarimetry. The new method introduced refluxing in place of ultrasonic shakers in the process of extraction. Therefore the simplicity of this method makes it possible to investigate many samples in any laboratory setup.

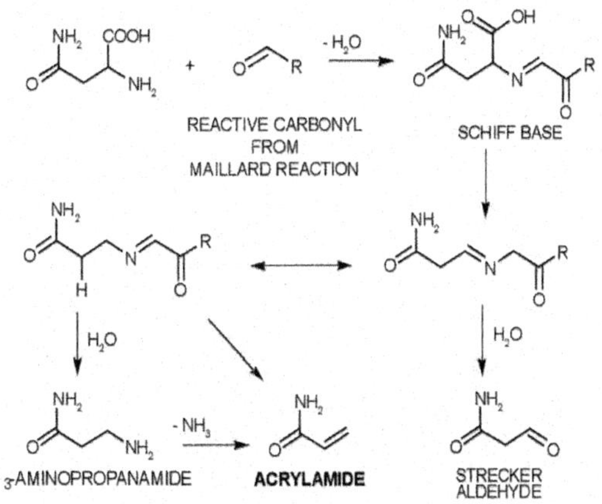

Figure 1. Proposed mechanism for the formation of acrylamide in heat-treated foods. Source: (Krishnakumar & Visvanathan 2014).

2 MATERIALS AND METHODS

2.1 *Chemicals*

Acrylamide (99%), potassium bromate ($KBrO_3$), and potassium bromide (KBr), were purchased from Sigma Aldrich; n-hexane and ethyl acetate were redistilled before use; all the other reagents used were of analytical grade.

2.2 *Equipment/apparatus*

Experiments were done with a Varian 3400 CX chromatograph equipped with a flame ionization detector and a splitless injector. Separations were conducted on a 5, and 30 m × 0.25mi.d PTE capillary column.

2.3 *Sampling procedures and roasting of maize*

For the purpose of this study, 24 raw maize samples were sampled and bought from the local market and transported to the laboratory where roasting was immediately done while they were still fresh. Both raw and roasted maize were crushed, homogenized, and samples analyzed separately.

2.4 *Preparation of roasted maize*

The roasting of maize was done using a laboratory procedure similar to the setups for roasting maize found in homes and on the streets in towns and cities, as illustrated in Figure 2. Raw maize samples collected from the local market were subjected to high-temperature roasting, and then cooled, stored in polyethylene bags, and frozen in the refrigerator to enable the preparation of analysis.

Figure 2. Roasting of maize.

2.5 *Sample preparation*

Maize samples were crushed using a pestle and mortar to obtain a uniform mixture. The extraction of the analyte followed a modified protocol developed by Geng and others (Geng et al. 2011). A measured 5.0 g of homogenized maize sample was weighed and placed in a 100 mL round bottom flask, extracted using 70 mL of distilled water, and refluxed for 50 minutes at 80°C. The defatting process was accomplished using hexane to allow for fatty components to remain in organic phase by adding redistilled hexane (20 mL), shaking, and allowing the mixture to settle. The fatty components remained in the organic layer. Ten milliliters of the lower aqueous layer and 0.6 mL sulfuric acid (10 %v/v) were sequentially added into brown glass tubes. The tubes were then placed into refrigerating cabins for precooling (4°C, 15 min). 0.1 mL of 0.1 M of derivatization reactants, including potassium bromate ($KBrO_3$) and 2.5 g of potassium bromide (KBr) powder, were added to the precooled solution. The tubes were briefly shaken with a vortex mixer and the reaction mixture was allowed to stand for 45 min at 4°C. The derivatization reaction was finalized by adding 1 mL of 0.1 mol/L sodium thiosulfate solution. The mixture was transferred to a 100 mL separatory funnel and extracted with 15 mL of ethyl acetate-hexane (4:1, v/v). The organic phase was filtered into a 100-mL round bottomed flask using a filter paper size (whatman 70 mm Cat No 1004-070) covered with 2 g of calcinated sodium sulfate. The separating funnel and the filter were rinsed twice with 5 mL aliquots of ethyl acetate-hexane (4:1,v/v). Pooled fractions were evaporated to dryness on a rotary evaporator (40, 140 mbar).The residue was then dissolved in 5 mL of hexane prior to analysis by GC-FID (Zhang et al. 2006).

2.6 *Determination of moisture content*

The moisture content of the green raw maize and the roasted maize was determined following the method described in the Association of Official Analytical Chemists (AOAC) 922.6. Maize was crushed using a pestle and mortar until it was a uniform mixture, from which 5.0 g of the sample was heated at 105 ± 2°C in an oven (DSO-500D) for 5 hrs. It was then cooled in a desiccator and weighed. The process of heating and cooling was repeated until a constant mass of the dried maize sample was obtained. The determination was run in duplicate.

2.7 *Determination of glucose and fructose content*

Raw green and roasted green maize samples were crushed separately and homogenized. To this paste in 100 mL round bottomed flask was added 10 mL 0.1% hydrochloric acid solution. The mixture was thoroughly mixed on a mechanical shaker for 20 minutes and then refluxed for 2 hours. It was cooled, filtered, and analyzed for glucose and fructose using a polarimeter (ADP 600 Series).

2.8 *Bromination of calibration standards*

Acrylamide stock solution (1000 mg L^{-1}) was prepared by dissolving 0.1000 g acrylamide solid in acetonitrile in a 100 mL volumetric flask from which working standards in the range of 5–200 μL of acrylamide were prepared and kept in brown glass vials, followed by consecutive addition of distilled water and 0.6 mL sulfuric acid, respectively. Bromination of the standards was done in the same manner as the sample bromination was done. The final solutions were subjected to SPE cleanup procedures and kept in vials under refrigeration at +4°C as preparations for analysis by GC-FID were made.

2.9 *Gas chromatograph conditions*

Acrylamide standards were analyzed on a GC-FID 3400 CX Varian model, Supelcowax capillary 60 m length, 0.25 rom i.d., 0.25 /m film thickness, column used. The injector temperature was maintained at 260°C and the nitrogen carrier gas linear velocity was maintained at 62 cm/sec at 100°C, the oven temperature was held at 100°C for 0.5 min before it was allowed to increase at a rate of 15°C/min to attain a final temperature of 200°C.

3 RESULTS

In this study, foods were sampled in the Njoro area of Kenya. From the results it was observed that the moisture content of raw maize had a mean of 44% whereas the moisture content of roasted maize had a mean of 35% moisture content, indicating a loss of water during the roasting process at elevated temperatures.

Both the raw and roasted samples were extracted with water then analyzed by a polarimeter and the specific angle of rotation was recorded. Acid hydrolysis is done with (0.1 % HCl) and then the analyte was subjected to polarimetry, and the specific angle of rotation was also recorded.

3.1 *Quantification of acrylamide*

Working standard solutions for the standard curve as well as the sample recovered in hexane were analyzed using the GC-FID chromatographic system.

4 DISCUSSION

A study of the levels of precursors responsible for the formation acrylamide (AA) in roasted maize at high temperature is reported. Although the amounts of acrylamide in the roasted maize samples were not quantified, the glucose–fructose ratios explain sufficiently the overall trend of acrylamide content in the samples. It was observed that the moisture content of raw maize was higher when compared to that of roasted maize, indicating a loss of water when samples were roasted at elevated temperature. However, the loss of moisture content after roasting was not significantly high. This implies that the moisture content in roasted maize is still high enough to suppress the formation of acrylamide, which accounts for the low limit of quantification (LOQ) factor observed in this study. This finding is in agreement with observations made by Elmore and Zhang that formation of acrylamide is related to moisture content and it only forms when it falls below 5% for some foods (Elmore et al. 2005; Zhang et al. 2009). The specific angle of rotation recorded from water-based extracts obtained when raw and roasted maize samples were extracted with water proved to be levorotatory, as shown in Figure 3.

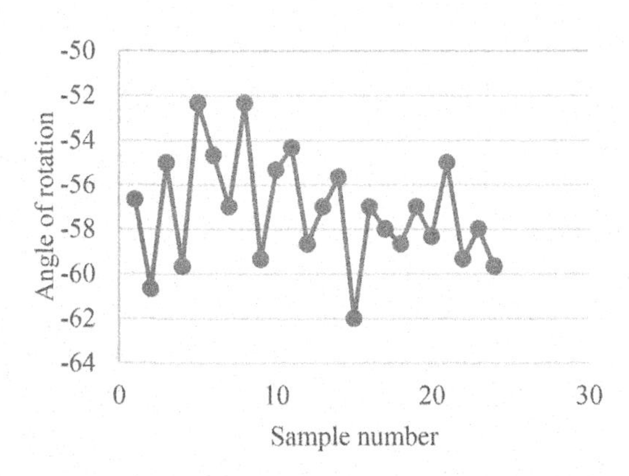

Figure 3. Angle of rotation of water-based extracts obtained from raw maize samples.

The angle of rotation for sucrose in maize or any other starchy food is expected to be dextrorotatory, but observations indicating the angle of rotation to be levorotatory in nature is due to the presence of a mixture of glucose and fructose in a sample. Indeed levorotatory fructose has a greater molar rotation than the dextrorotatory glucose (Panpae et al. 2008).

From our study, and as shown in Figure 4, it is clear that the angles of rotation measured from water-based extracts obtained from roasted maize samples were lower than those observed in raw maize samples—an indication of decreased amounts of fructose in roasted maize, probably due to the Maillard reaction.

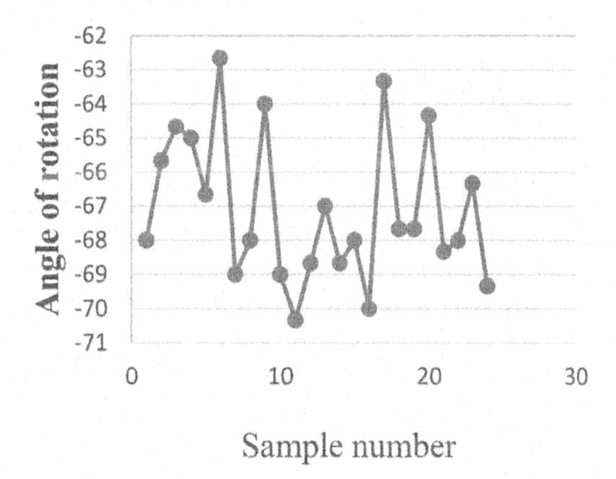

Figure 4. Angle of rotation of water-based extracts obtained from roasted maize samples.

Acid hydrolysis of the sample extracts resulted in a decrease in sucrose content due to its inversion to glucose with subsequent formation of a glucose–fructose mixture. It may be concluded that the more the glucose is formed through inversion process, the more dextrorotatory behavior the extract exhibits, as shown in Figure 5.

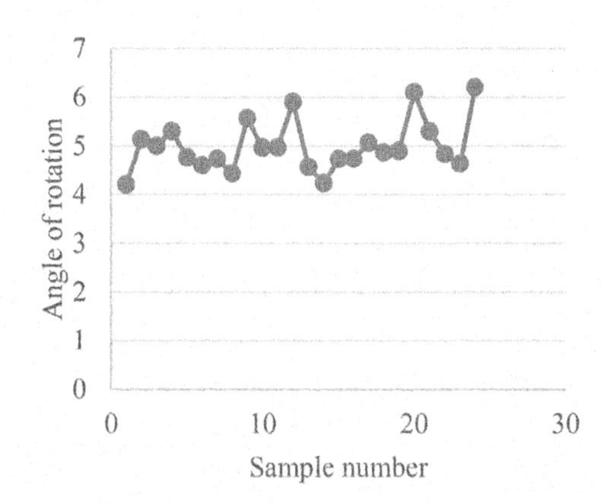

Figure 5. Angle of rotation of raw maize acid hydrolyzed extracts.

The angle of rotation of roasted maize is reduced when maize is roasted suggesting the depletion of glucose occurs when raw maize is roasted, as observed in Figure 6.

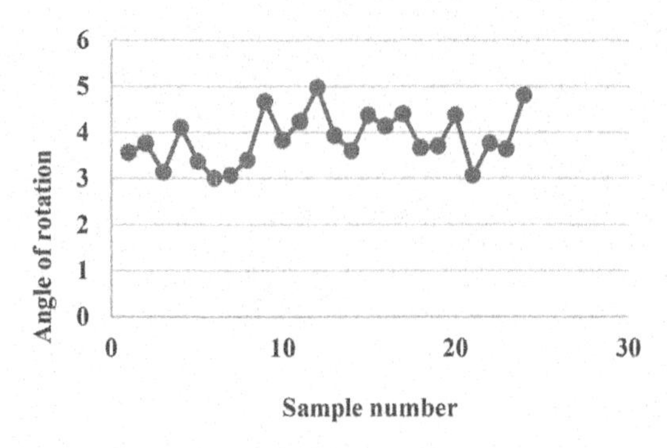

Figure 6. Angle of rotation of roasted maize acid hydrolyzed.

The ratio of fructose to glucose in raw maize samples is higher when compared to that in the roasted maize samples (Figure 7). There is a loss of the amount of glucose/fructose when roasting is done, probably due to the formation of acrylamide; this loss is related to the observed difference in angle of rotation. Reducing sugars such as glucose and fructose are the major contributors to acrylamides (Pedreschi et al. 2014).

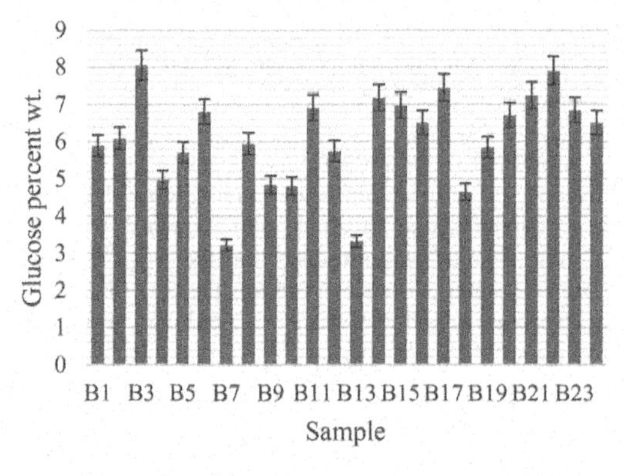

Figure 7. Glucose expressed percent weight in roasted samples.

The decrease in fructose/glucose ratio in raw maize compared to roasted maize could probably explain their role in the formation of acrylamides. However, in our study, acrylamide levels were below LOQ, which is consistent with other studies involving maize-based products. In a study involving quantification of acrylamides in food products done in Brazil, it was found that the concentration of acrylamide in the samples ranged from <20 to 2528 mg kg^{-1}, with a considerable difference between individual foodstuffs within the same class. However acrylamide in maize-related products determined using LC-MS/MS showed levels below LOQ (Arisseto et al. 2007). In Colombia, bakery products made from corn flour and investigated using chromatography with mass spectrometry (GC/MS) for acrylamides showed a lower acrylamide content (< 75 $\mu g\ kg^{-1}$) in comparison with similar bakery products made of wheat flour (Pacetti et al. 2015). It was also observed that the acrylamide content of white corn flour (WCF) extrudates studied using liquid chromatography–tandem mass spectrometry (LC-MS/MS) (Masatcioglu et al. 2014) was below the limit of detection (LOD) value and this was attributed to the fact that the levels of glucose/fructose present in the sample could not aid its formation. However quantification of acrylamide has been reported in deep-fried potatoes samples collected and analyzed in Nairobi, Kenya (Ogolla et al. 2015). This is consistent with the reports published by other authors (Boroushaki et al. 2010; Hariri et al. 2015). French fries and other potato-based products usually contain a high level of acrylamide compared to roasted maize which is below LOQ. This is due to the fact that potatoes have a higher content of reducing sugars and asparagine that aids the Maillard reaction (Hariri et al. 2015).

5 CONCLUSIONS AND RECOMMENDATIONS

The main factors responsible for the formation of acrylamide in cereal products and starchy foods are reducing sugars (mainly glucose and fructose) and free asparagines (amino acids), respectively. In this current study glucose and fructose were monitored by polarimetry, however, the amount of acrylamide in raw and roasted maize sold in some Kenyan markets was low and could not be quantified. Different maize varieties need to be studied, taking into account seasonal and regional variability for high-temperature processes such as roasting, frying, and baking. There is therefore a need for further research on a wider range of foodstuffs and a larger number of samples, including other carbohydrate-rich foods, such as sweet potatoes, cassava, banana products, and other home-cooked foods. Sample preparation included water extraction under reflux conditions prior to defatting with n-hexane, derivatization with $KBrO_3$ and KBr, and liquid–liquid with hexane and ethyl acetate (4:1); further optimization of these processes should be performed in order to satisfy the preparation of a large number of samples and applied to different types of starchy foods for the determination of acrylamide content.

The relationship between the loss of content of reducing sugars and asparagine and the role they play in acrylamides formation in roasted maize should be explored further. It has been found that the use of inorganic fertilizers has an effect on acrylamide formation. A case study showed that decreasing nitrogen fertilization caused increases in the reducing sugar concentration from 60% up to 100% in potato tubers for all varieties reported. There was a high correlation between the reducing sugar content and the generation of acrylamide during frying. This resulted in a parallel increase in the acrylamide concentration of the French fries (De Wilde et al. 2006; Stockmann et al. 2018).

It is expected that these results will contribute to data accumulation for worldwide health risk assessment and be helpful in establishing approaches to lower acrylamide formation during cooking processes.

CONFLICT OF INTEREST

The authors declare that they have no potential conflict of interest in relation to the study in this paper.

ACKNOWLEDGMENT

Africa Center of Excellence II in Phytochemicals, Textiles and Renewable Energy (ACE II PTRE), Moi University, Eldoret, Kenya and Egerton University Kenya

REFERENCES

Ahn JS, Castle L, Clarke DB, Lloyd AS, Philo MR and Speck DR. (2002). Verification of the findings of acrylamide in heated foods. *Food Additives & Contaminants,* **19**: 1116–1124.

Alam Shahenvaz, Ahmad Razi, Pranaw Kumar, Mishra Prashant and Khare Sunil Kumar. (2018). Asparaginase conjugated magnetic nanoparticles used for reducing acrylamide formation in food model system. *Bioresource technology,* **269**: 121–126.

Arisseto Adriana Pavesi, Toledo Maria Cecilia, Govaert Yasmine, Loco Joris Van, Fraselle Stéphanie, Weverbergh Eric and Degroodt Jean Marie. (2007). Determination of acrylamide levels in selected foods in Brazil. *Food additives and contaminants,* **24**: 236–241.

Authority European Food Safety. (2012). Update on acrylamide levels in food from monitoring years 2007 to 2010. *EFSA Journal,* **10**: 2938.

Belkova Beverly, Hradecky Jaromir, Hurkova Kamila, Forstova Veronika, Vaclavik Lukas and Hajslova Jana. (2018). Impact of vacuum frying on quality of potato crisps and frying oil. *Food chemistry,* **241**: 51–59.

Boroushaki Mohammad Taher, Nikkhah Elham, Kazemi Abdollah, Oskooei Mojtaba and Raters Marion. (2010). Determination of acrylamide level in popular Iranian brands of potato and corn products. *Food and Chemical Toxicology,* **48**: 2581–2584.

De Wilde Tineke, De Meulenaer Bruno, Mestdagh Frédéric, Govaert Yasmine, Vandeburie Stephan, Ooghe Wilfried, Fraselle Stéphanie, Demeulemeester Kürt, Van Peteghem Carlos and Calus André. (2006). Influence of fertilization on acrylamide formation during frying of potatoes harvested in 2003. *Journal of agricultural and food chemistry,* **54**: 404–408.

Elmore J Stephen, Koutsidis Georgios, Dodson Andrew T, Mottram Donald S and Wedzicha Bronek L. (2005). Measurement of acrylamide and its precursors in potato, wheat, and rye model systems. *Journal of agricultural and food chemistry,* **53**: 1286–1293.

Fu Zhengjie, Yoo Michelle JY, Zhou Weibiao, Zhang Lei, Chen Yutao and Lu Jun. (2018). Effect of (-)-epigallocatechin gallate (EGCG) extracted from green tea in reducing the formation of acrylamide during the bread baking process. *Food chemistry,* **242**: 162–168.

Geng Zhiming, Wang Peng and Liu Aiming. (2011). Determination of acrylamide in starch-based foods by HPLC with pre-column ultraviolet derivatization. *Journal of chromatographic science,* **49**: 818–824.

Ghiasvand Ali Reza and Hajipour Somayeh. (2016). Direct determination of acrylamide in potato chips by using headspace solid-phase microextraction coupled with gas chromatography-flame ionization detection. *Talanta,* **146**: 417–422.

Gökmen Vural and Palazoğlu Tunç Koray. (2008). Acrylamide formation in foods during thermal processing with a focus on frying. *Food and Bioprocess Technology,* **1**: 35–42.

Gökmen Vural and Şenyuva Hamide Z. (2007). Acrylamide formation is prevented by divalent cations during the Maillard reaction. *Food chemistry,* **103**: 196–203.

Hariri Essa, Abboud Martine I, Demirdjian Sally, Korfali Samira, Mroueh Mohamad and Taleb Robin I. (2015). Carcinogenic and neurotoxic risks of acrylamide and heavy metals from potato and corn chips consumed by the Lebanese population. *Journal of Food Composition and Analysis,* **42**: 91–97.

Krishnakumar T and Visvanathan R. (2014). Acrylamide in food products: a review. *Journal of Food Processing and Technology,* **5**.

Li Dong, Chen Yaqi, Zhang Yu, Lu Baiyi, Jin Cheng, Wu Xiaoqin and Zhang Ying. (2012). Study on mitigation of acrylamide formation in cookies by 5 antioxidants. *Journal of food science,* **77**: C1144–C1149.

Lund Marianne N and Ray Colin A. (2017). Control of Maillard reactions in foods: Strategies and chemical mechanisms. *Journal of agricultural and food chemistry,* **65**: 4537–4552.

Manière Isabelle, Godard Thierry, Doerge Daniel R, Churchwell Mona I, Guffroy Magali, Laurentie Michel and Poul Jean-Michel. (2005). DNA damage and DNA adduct formation in rat tissues following oral administration of acrylamide. *Mutation Research/Genetic Toxicology and Environmental Mutagenesis,* **580**: 119–129.

Masatcioglu Mustafa Tugrul, Gokmen Vural, Ng Perry KW and Koksel Hamit. (2014). Effects of formulation, extrusion cooking conditions, and CO2 injection on the formation of acrylamide in corn extrudates. *Journal of the Science of Food and Agriculture,* **94**: 2562–2568.

Mesias Marta, Delgado-Andrade Cristina, Holgado Francisca and Morales Francisco J. (2018). Acrylamide content in French fries prepared in households: A pilot study in Spanish homes. *Food chemistry,* **260**: 44–52.

Mottram Donald S, Wedzicha Bronislaw L and Dodson Andrew T. (2002). Acrylamide is formed in the Maillard reaction. *Nature,* **419**: 448–449.

Normandin Louise, Bouchard Michèle, Ayotte Pierre, Blanchet Carole, Becalski Adam, Bonvalot Yvette, Phaneuf Denise, Lapointe Caroline, Gagné Michelle and Courteau Marilène. (2013). Dietary exposure to acrylamide in adolescents from a Canadian urban center. *Food and Chemical Toxicology,* **57**: 75–83.

Ogolla Jackline A, Abong George O, Okoth Michael W, Kabira Jackson N, Imungi Jasper K and Karanja Paul N. (2015). Levels of acrylamide in commercial potato crisps sold in Nairobi county, Kenya. *Journal of Food and Nutrition Research,* **3**: 495–501.

Ou Shiyi, Shi Jianjun, Huang Caihuan, Zhang Guangwen, Teng Jiuwei, Jiang Yue and Yang Baoru. (2010). Effect of antioxidants on elimination and formation of acrylamide in model reaction systems. *Journal of Hazardous Materials,* **182**: 863–868.

Pacetti Deborah, Gil Elizabeth, Frega Natale G, Álvarez Lina, Dueñas Pilar, Garzón Angélica and Lucci Paolo. (2015). Acrylamide levels in selected Colombian foods. *Food Additives & Contaminants: Part B,* **8**: 99–105.

Panpae Kornvalai, Jaturonrusmee Wasna, Mingvanish Withawat, Nuntiwattanawong Chantana, Chunwiset Surapon, Santudrob Kittisak and Triphanpitak Siriphan. (2008). Minimization of sucrose losses in sugar industry by pH and temperature optimization. *Malaysian Journal of Analytical Sciences,* **12**: 513–519.

Pedreschi Franco, Mariotti María Salomé and Granby Kit. (2014). Current issues in dietary acrylamide: formation, mitigation and risk assessment. *Journal of the Science of Food and Agriculture,* **94**: 9–20.

Stockmann Falko, Weber Ernst Albrecht, Schreiter Pat, Merkt Nikolaus, Claupein Wilhelm and Graeff-Hönninger Simone. (2018). Impact of Nitrogen and Sulfur Supply on the Potential of Acrylamide Formation in Organically and Conventionally Grown Winter Wheat. *Agronomy,* **8**: 284.

Swedish NFA. (2002). Information about acrylamide in food. *Swedish National Food Administration,* **4**: 24.

Vinci Raquel Medeiros, Mestdagh Frédéric and De Meulenaer Bruno. (2012). Acrylamide formation in fried potato products–Present and future, a critical review on mitigation strategies. *Food chemistry,* **133**: 1138–1154.

Wilson KM, Rimm EB, Thompson KM and Mucci LA. (2006). Dietary acrylamide and cancer risk in humans: a review. *Journal für Verbraucherschutz und Lebensmittelsicherheit,* **1**: 19–27.

Wyka Joanna, Tajner-Czopek Agnieszka, Broniecka Anna, Piotrowska Ewa, Bronkowska Monika and Biernat Jadwiga. (2015). Estimation of dietary exposure to acrylamide of Polish teenagers from an urban environment. *Food and Chemical Toxicology,* **75**: 151–155.

Zeng Xiaohui, Cheng Ka-Wing, Jiang Yue, Lin Zhi-Xiu, Shi Jian-Jun, Ou Shi-Yi, Chen Feng and Wang Mingfu. (2009). Inhibition of acrylamide formation by vitamins in model reactions and fried potato strips. *Food chemistry,* **116**: 34–39.

Zhang Yu, Dong Yi, Ren Yiping and Zhang Ying. (2006). Rapid determination of acrylamide contaminant in conventional fried foods by gas chromatography with electron capture detector. *Journal of Chromatography A,* **1116**: 209–216.

Zhang Yu, Ren Yiping and Zhang Ying. (2009). New research developments on acrylamide: analytical chemistry, formation mechanism, and mitigation recipes. *Chemical reviews,* **109**: 4375–4397.

Advances in Phytochemistry, Textile and Renewable Energy Research for Industrial Growth – Nzila et al. (Eds)

Industrial engineering and operation management in the ready-made garments industry

O. Bongomin*, J.I. Mwasiagi & E.O. Nganyi
Department of Manufacturing, Industrial and Textile engineering, Moi University, Eldoret, Kenya

I. Nibikora
Department of Polymer, Industrial and Textile engineering, Busitema University, Tororo, Uganda

ABSTRACT: Today's competitive advantage of the ready-made garments (RMG) industry depends on the ability to improve the efficiency and effectiveness of resource utilization through proper adoption of industrial engineering techniques. RMG industries have historically adopted fewer technological and process advancements. This is especially true for less developed regions like the East African Community (EAC), although significant amounts of textile and apparel products are produced in these regions. In most RMG industries, industrial engineering techniques have not been given enough attention, even though they need to compete globally and survive in this extremely competitive and dynamic business environment. Presently, only very few garment industries have comprehended the functions of the industrial engineering department. One of the base reasons for this shortage is that the garment industries suffer much from substantial inadequacy of information and literature on the practical application of industrial engineering techniques in garment manufacturing. In this paper, the application of industrial engineering tools: ABC classification, process mapping, time study, and brainstorming were demonstrated in a garment manufacturing factory. The empirical data obtained were utilized to determine the Standard Minute Value (SMV) and prepare an operation bulletin for trousers. The results from the present study are very useful to the garment industry for setting up a realistic production target and for measuring the production capability of a trouser assembly line as well as improving its efficiency.

1 INTRODUCTION

The ready-made garment (RMG) industry is not only one of the oldest, largest, labor-intensive, low-skilled, low-value, and most global industries but also the typical "beginner" industry for countries engaged in export-orientated industrialization (Abtew et al. 2019; Hamja et al. 2019). In 2016, the East African Community (EAC) pledged to phase out imports of second-hand clothing within three years to promote the development of the domestic garment sector (Calabrese, Balchin, & Mendez-parra 2017; Wolff 2020). But that can be hardly achieved without the proper implementation of industrial engineering (IE) and operation management (OM) functions in this sector.

The implementation of the IE and OM functions in the RMG industry are very crucial for sustainability in the business and need proper monitoring of success (Islam, Islam, & Gupta 2017). The OM aims at addressing problems related to low levels of sales and low turnover, over-inventory, and high manufacturing costs in textile-clothing companies in order to improve productivity and competitiveness (Maralcan & Ilhan 2017). It mainly strategizes at promoting the supply chain integration, adequate demand forecasting methods, lean manufacturing principles, implementation of information technologies, and production planning techniques for the long, medium, and short term (Cano & Zuluaga-mazo 2019). On the other hand, IE focuses on reducing the production time, which automatically reduces inventory cost to a minimum (Jana & Tiwari 2018; Khatun 2013). In RMG industry, the IE and OM functions are under IE department (Jana & Tiwari 2018).

In previous studies, IE has been applied in RMG to harness improvements in productivity (Khatun 2013) and increases in the agility of garment manufacturing (Khan, Islam, Elahi, Sharif, & Mollik 2019). Baset and Rahman (2016) and Hossain, Khan, Islam, Fattah, and Bipul (2018) applied IE techniques in the garments sector for reducing the cost of Standard Minute Value (SMV) and lead time to improve productivity by an implementation of proper line balancing. Howard, Essuman, and Asare (2019) applied IE as the strategies for determining the production cost and pricing of garments. Khan (2013) used lean manufacturing to achieve higher productivity in the apparel industry. Lean manufacturing has been increasingly used in the RMG industry to increase productivity for reducing costs and lead time (Hamja et al. 2019; International Labour Organization 2017; Khan 2016). Furthermore, Bashar and Hasin (2019) studied the impact of Just-in-time (JIT) production on organizational performance in the apparel industry.

*Corresponding author

DOI 10.1201/9781003221968-10

Tout ensemble, the key tasks of IE include product analysis through determining the optimum method of construction and establishing operation bulletin, production planning for effective and balanced flow of product, operator performance monitoring systems by hourly production monitoring and skill matrix, justifying all changes based on analysis of work content, continuous improvement, taking cost-saving opportunities, monitoring operator performances and taking action to improve performances, and eliminating causes of underperformance (Islam et al. 2017; Rahman & Amin 2016). In addition, the implementation of 5S and six sigma are important tasks of IE in RMG (Khan 2016).

The concept and functions of IE are indeed quite clear as they are the key to improve work nature and methods in RMG industry (Chandurkar, Kakde, & Bhadane 2015). Surprisingly, very few among the RMG industry in EAC have comprehended the IE functions and put in place the IE department. This is quite daunting, yet they need productivity improvement and sustainable competitiveness. But how can this be possible without a proper IE department.

The present case was built upon the empirical study conducted at Southern Range Nyanza Limited (NYTIL), Jinja, Uganda. NYTIL company is a vertically integrated textile industry with apparel manufacturing as the most critical department. The garment production manager is faced with the problem of being unable to determine a realistic production target, capacity setting, and workers' incentives due to the unavailability of SMV or standard allowed minute (SAM) and operation bulletin for garment assembly line. As the result, the garment department operates with low efficiency and is unable to meet the daily production target. SMV is the time required by an average skilled operator, working at a normal pace, to perform a specified task using a prescribed method (Elnaggar 2019). SMV is an important parameter used for garment assembly line balancing that is necessary for efficiency and productivity improvements (Gebrehiwet & Odhuno 2017; Yemane, Gebremicheal, Meraha, & Hailemicheal 2020). SMV is not only used in the apparel industry but also in the leather industry for product assembly lines, such as for bags and shoes (Moktadir 2017). Therefore, the present study aimed at determining SMV for a complex garment assembly line which was then used for developing the operation bulletin. In essence, the study demonstrated the important of IE and OM for determining a realistic operation bulletin, which is emblematic for productivity planning and improvement.

2 LITERATURE REVIEW

2.1 *RMG manufacturing challenges and opportunities*

Garment manufacturing, also known as apparel manufacturing, is labor intensive, which has led to the shifting of many apparel manufacturing facilities from developed countries to developing countries because of the cheap labor force. Although there is cheap labor in developing countries, garment industries are facing the greatest challenges, such as short production life cycles, high volatility, low predictability, high level of impulse, and quick market response (Rajkishore & Padhye 2015). In order to survive, the garment industries in developing countries are betting big on reducing the cost of production by sourcing cheaper raw materials and minimizing delivery cost rather than labor productivity because of the availability of cheap labor.

Global and local competition is still a major challenge amongst apparel manufactures. Therefore, one can only survive in the market if all unnecessary costs are reduced, the range of production is expanded, and consumers are considered individually. However, the local apparel manufacturers are gradually reducing the production and focusing on performing only the entrepreneurial functions involved in apparel manufacturing such as buying raw materials, designing clothes and accessories, preparing samples, arranging for the production, and distribution and marketing of the finished products (Rajkishore & Padhye 2015).

Rapid technological changes and customer expectations have also imposed a great challenge to apparel manufacturers, especially in developing countries. Therefore, there is high demand from the manufacturers to improve the quality of fashion products constantly, and thus to survive in the market (Karthik, Ganesan, & Gopalakrishnan 2017). In addition, the manufacturers are required to adjust their production system in order to meet market demand, so that they have to set a flexible production model that is capable of quick and easy adjustment to modern requirements (Babu 2012).

Another technological challenge facing apparel manufacturers in developing countries is the differences that exist in the process of making clothes of different fashions, which in one way or the other requires a different organization of technological processes (Colovic 2012). Therefore, this calls for the most economical ways of working and time required to perform work operations, change management, capacity, and planning. Further, it is necessary to implement new solutions in manufacturing, information systems, management techniques, and design (Colovic 2011).

The challenge facing the apparel industries in developing countries is the indispensability of the scientific approach and engineering applications for apparel manufacturing. This implies that the apparel manufacturers will find it very difficult to meet the cost of production unless and until manufacturing is done with a scientific approach, such as the implementation of a simulation model for line balancing and assembly line design, lean production, etc. (Babu 2012; Bongomin, Mwasiagi, Nganyi, & Nibikora 2020a).

Generally, the apparel industries across the whole world, especially in developing countries, will not achieve any pleasing results for the management unless they strive for necessary improvements that will lead to productivity growth, more rational usage of all-natural resources, and cost reduction. In most cases these companies do not see the necessity for changes in management, capacity, and planning, which are negatively impacting many apparel industries today (Karthik et al. 2017).

The disruption in textile and apparel industry triggered by digital transformation or disruptive technologies of industry 4.0 is enormous (Bongomin, Gilibrays Ocen, Oyondi Nganyi, Musinguzi, & Omara 2020). It is very likely that the RMG industry will undergo profound changes over the next few years. Smart Clothes, i.e., clothing characterized not only by its traditional protective and representational functions, but also by technological and digital features, have evolved as a promising opportunity for the RMG industry, and are known as fashion 4.0 or apparel 4.0 (Behr 2018; Bertola & Teunissen 2018).

2.2 *Garment manufacturing systems*

An apparel or garment production system is an integration of materials handling, production processes, personnel, and equipment that direct workflow and generate finished products (Babu 2012). There are three types of apparel production system that are widely adopted in garment industry: (1) group or modular production system (Sudarshan & Rao 2014); (2) progressive bundle production system; and (3) unit production system. In a modular production system, operations are done in contained and manageable work cells that include a number of specialized resources, such as an empowered work team, equipment, and work to be executed. This production system has achieved the success of flexibility. However, very high initial capital costs and investment in training are still the major limitations to its adaptation to most apparel industries.

The progressive bundle production system, normally referred to as the conventional production system, is still the most commonly installed production system to date in terms of garment production systems because of its cost effectiveness on high-tech machines. The operation in this system involves moving bundles of cut pieces manually to feed the line. Whereby, the operator inside the line drags the bundles by him/herself from the table and transfers the bundle to the next operator after completing his/her task. The major problem with the progressive bundle system is the tendency of accumulating a very large inventory which imposes the extra cost of controlling and handling inventory. In order to overcome the limitation of material handling in the progressive bundle system, a new system called the unit production system was developed. In this system, an overhead transporter is used to move the garment from one workstation to another. This system basically improves material handling in the assembly line. The success of the unit production system is that it improves the production lead times, productivity, and space utilization. However, this production system is extremely expensive. In general, the trade-off of these production systems depends on the production volume, product categories, and the cost effectiveness of high-tech machines (Karthik et al. 2017).

2.3 *Industrial engineering tools*

2.3.1 *ABC classification*

Traditionally, ABC analysis has been used to classify various inventory items into three categories: A, B, and C, based on the criterion of dollar volume. In the current globalized hyper-responsive business environment, a single criterion is no longer adequate to guide the management of inventories, and therefore multiple criteria have to be considered (Sibanda & Pretorius 2011). Other criteria that can be considered for ABC analysis include lead time, item criticality, durability, scarcity, reparability, stockability, commonality, substitutability, the number of suppliers, mode and cost of transportation, the likelihood of obsolescence or spoilage, and batch quantities imposed by suppliers. Consequently, ABC analysis has been adopted amongst researches to make decision on the selection of products, machines, production lines, etc. For instance, Pinho and Leal (2007) used ABC analysis to prioritize a production system for their study based on productivity per day criterion. Therefore, in the current situation, ABC analysis tool was also adopted to prioritize the product model and assembly line to be used in this study.

2.3.2 *Process mapping*

Process mapping is an exercise of identifying all the steps and decisions in a process in a diagrammatic form, with a view to continually improve that process. In literature, two commonly used types of process mapping are process flowchart (outline process map) and deployment charts. The former is useful for capturing the initial detail of the process. For instance, Kursun and Kalaoglu (2009), Kitaw et al. (2010), Bahadır (2011), and Yemane, Haque, and Haque (2017) used a process flowchart as a conceptual model in their simulation study with the aim of analyzing and understanding the current state of the studied system. The latter not only provides a basic overview but also shows who does what along with the interactions between people and departments. This one has been used as a stand-alone method amongst studies for process improvement. Uddin (2015) improved the production process using value stream mapping as a stand-alone method. Since then, the present study adopted process mapping as a tool for conceptual modeling; the process flowchart method was best suited for this study.

2.3.3 *Fishbone diagram*

Cause-and-effect diagram (fishbone diagram) is another method that has been widely used in studies (Barton 2004). It is an analysis tool that provides a systematic way of looking at effects (performance measures) and the causes (factors or independent variables) that create or contribute to those effects (Hekmatpanah 2011). One of the underlying benefits of this method is that it has nearly unlimited application in research, manufacturing, marketing, office operations, and so forth. One of its strongest assets is the participation and contribution of everyone involved in the brainstorming process (Hekmatpanah 2011). The ability of a cause-and-effect diagram to clearly identify and categorize factors that affect the performance of the system is one of the major reasons for its adoption in this study.

2.3.4 *Time study*

The definition of time study was first coined in the early 20th century in industrial engineering, referring to a quantitative data collection method where an external observer captured detailed data on the duration and movements required to accomplish a specific task, coupled with an analysis focused on improving efficiency (Lopetegui et al. 2014). The time study basically involves timing and observing the motion of the work associated with building the product. The collection processing times data are absolute requirements for improving the assembly operations in the facility (Ortiz 2006). The advantages of the time study method over other work measurement techniques include (Babu 2012) (1) it helps in developing a rational plan; (2) it helps in improving productivity; (3) it helps in balancing assembly lines; (4) it provides the time data for process design; and (5) it helps in determining operator skill levels. Nevertheless, conducting a time study is time-consuming and very tiresome, especially when the system has many elements to be measured.

However, time study has been the most commonly used as it determines accurate time standards, and it is economical for repetitive type work. A vast amount of research has been done using the time study method. For instance, Senthilraja, Aravindan, and Sathesh (2018) applied the time study technique for improving the operators' productivity in the rubber industry. While Khatun (2014) studied the effect of the time and motion study on productivity in the garment sector. The author postulated that the target productivity can be achieved by time study. Time study has fewer limitations than other work measurement methods, including activity sampling, predetermined time standards (PTS), and structured estimating (Babu 2012). The basic time study equipment consists of a stop watch, study sheet, and time study board (Ortiz 2006). The steps for conducting a time study have been reported in the literature (Babu 2012; Russell & Taylor 2011).

The number of timing cycles for specific activity basically depends on the end use of the time study data. For instance, if the time study data is to be used in probability distribution analysis, then a greater number of timing cycles or measurements give better results. For each work element/task, the processing time can be recorded 10 times (Kitaw, Matebu, & Tadesse 2010), or 15 times (Sudarshan & Rao 2014), 20 times (Kursun & Kalaoglu 2009), or more; the higher the number of measurements, the better the results. There are two common methods of measuring time with a stopwatch: fly back and continuous method (Puvanasvaran et al. 2013; Starovoytova 2017).

2.3.5 *Brainstorming*

Brainstorming or brainwriting is one of the most common techniques used to generate ideas from an individual or group of people. In most cases, it has been applied in both educational, industrial, commercial, and political fields (Al-khatib 2012). In the previous studies, brainstorming has been combined with a fishbone diagram (cause-and-effect analysis tool) for analyzing the current state of production systems. For instance, Barton (2004) used brainstorming and fishbone diagram to analyze and identify factors that affect the throughput of the production process. Many studies have shown the applicability of brainstorming as problem-solving techniques. Al-khatib (2012) confirmed the effectiveness of using brainstorming as a problem-solving tool.

2.3.6 *Observation*

Observation has been used for analyzing the current state of the system amongst studies. However, it has been used alongside interviews to capture more data on the current state of the system (Gebrehiwet & Odhuno 2017). Observation is a very important tool when conducting process mapping and time study. For instance, in garment assembly line, two major areas that were observed on the sewing machines are machine working (positioning, sewing, and dispose) and machine not working (waiting for repair, waiting for suppliers, personal need for workers and idle) (Babu 2012). The present study combined observation with process mapping for conceptual modeling of the garment assembly line.

3 METHODOLOGY

3.1 *Current state analysis*

Southern Range Nyanza Limited (NYTIL) garment manufacturing facility produces both knitwear and woven-wear garment. However, other garment manufacturing facilities in Uganda and Kenya produce either knitwear or woven-wear garment. The present study was focused on the woven-wear garment production department because of its complexity. The woven-wear garment manufacturing system consists of nine production sections or stages as shown in Figure 1. The ABC classification method was employed to prioritize the product style and the assembly line for the study. In this case, a trouser assembly line was given A- priority because of its complexity. Therefore, only the trouser assembly line section for production of a digital camouflage trouser model or style was selected for this study. This was based on the fact that the trouser assembly line entails a huge number of resources and high uncertainty. Therefore, improving its efficiency can lead to the overall improvement in productivity

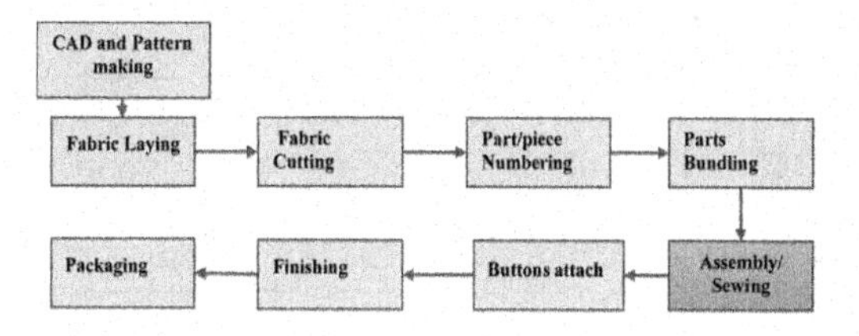

Figure 1. Woven wear garment manufacturing process.

of the garment manufacturing facility. In addition, the SMV is applicable for a garment assembly line section rather than other sections. In this study, brainstorming and fishbone diagram were used to analyze the garment assembly line.

3.2 *Time study*

A continuous stopwatch time study method was used to measure the operation time for each task involved in the trouser assembly process (Kayar 2017; Puvanasvaran, Mei, & Alagendran 2013). The observations were made, and the timing and recording were done at specific machines and helpers working points, as shown in Table 1.

The observed time was measured in seconds and then converted to standard time units (minutes). In the statistical method, the actual number of time measurements required for each task was determine by Eqn. 1 (Immawan & Kurniawan 2016). The actual number of time measurements was calculated for each task involved in the trouser sewing line.

$$N_m = \frac{\frac{z\alpha}{a}\sqrt{N\left(\sum (X_i)^2\right) - \left(\sum (X_i)^2\right)^2}}{\left(\sum X_i\right)} \quad (1)$$

where;
N_m is actual number of time measurements that should be taken.
N represents the number of preliminary time measurements that have been completed.
X_i is task time or observed time measured at the ith observation.
$z\,\alpha = 1.96$ for $\alpha = 5\%$ error.
$a =$ level of accuracy at 95%.

3.3 *Operation bulletin development*

The operation bulletin was developed following the steps outlined in Figure 2 (Babu 2012). The SMV or SAM was determined after obtaining the observed times (Bongomin et al. 2020b). The line target, target for an operation, and manpower requirement were calculated using Eqn 2, Eqn 3, and Eqn 4, respectively.

$$LT = \frac{WT \times PE \times PO}{Total\ SMV} \quad (2)$$

$$TO = \frac{WT \times PE \times ON}{SMV} \quad (3)$$

$$ME = \frac{LT}{TO} \quad (4)$$

where WT is working time, PE is the performance efficiency, PO is planned operators, ON is operator number, LT is line target, and TO is target for an operator.

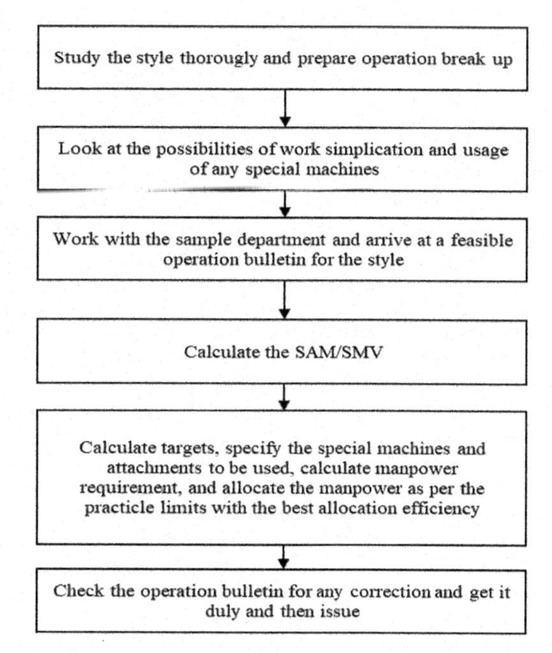

Figure 2. Operation bulletin construction approach.

Table 1. Component task classification and when to record the time.

S/N	Job	Component tasks and when to record the time		
1.	Sewing machine work	Taking the work piece and placing under needle	Sewing by machine	Placing on holding table
	When to record the time	Start of needle movement	When needle stops	When hand is taken from the workpiece
2.	Ironing	Taking and placing the workpiece	Ironing/pressing	Placing on holding table
	When to record the time	When iron is picked up to position	When iron is returned	When hand is taken from the workpiece

Table 2. Operation bulletin for trouser.

OP	Task description	Resource	OT (mins)	SMV/SAM	Target (pieces/day)	MPC	MPR
1	Buttonhole on flybox	Buttonhole machine	0.309	0.315	1142	0.46	1
2	Front rise overlocks	Overlock machine	0.243	0.283	1270	0.41	1
3	Knee patch attach	Single needle lockstitch	0.885	0.929	388	1.35	2*
4	Side pocket flatlock	Flatlock machine	0.255	0.352	1023	0.51	1
5	Side pocket overlock	Overlock machine	0.154	0.198	1816	0.29	1
6	Right flybox overlock	Overlock machine	0.259	0.334	1078	0.49	1
7	Side pocket attach	Single needle lockstitch	0.550	0.577	624	0.84	1
8	Side pocket topstitches	Single needle lockstitch	0.658	0.691	521	1.01	1
9	Right flybox attach	Single needle lockstitch	0.668	0.701	513	1.02	1
10	Fly attach	Single needle lockstitch	0.663	0.695	518	1.01	1
11	Front prep bundling	Helper	0.448	0.373	965	0.54	1
12	Back marking	Helper	0.227	0.214	1681	0.31	1
13	Back patch pressing	Iron press	1.659	1.356	266	1.98	2
14	Back patch attach	Single needle lockstitch	0.455	0.534	674	0.78	1
15	Hip pocket cutting	Automatic wallet machine	0.170	0.247	1460	0.36	1
16	Hip pocket overlocks	Overlock machine	0.800	0.927	388	1.35	2*
17	Hip flap folding	Helper	0.229	0.216	1666	0.32	1
18	Buttonhole on hip flap	Button hole machine	0.184	0.187	1923	0.27	1
19	Hip flap runstitch	Single needle lockstitch	0.316	0.331	1087	0.48	1
20	Hip flap turning	Turning machine	0.255	0.240	1499	0.35	1
21	Hip flap topstitches	Single needle lockstitch	0.402	0.421	854	0.61	1
22	Hip pocket finish	Single needle lockstitch	0.695	0.730	493	1.06	1
23	Hip flap attach	Single needle lockstitch	0.554	0.582	619	0.85	1
24	Back prep bundling	Helper	0.381	0.359	1001	0.52	1
25	F&B matching	Helper	0.458	0.432	834	0.63	1
26	Side seam overlock	Overlock machine	1.196	1.531	235	2.23	2
27	Side seam topstitches	Feed of arm machine	0.771	1.091	330	1.59	2
28	Knee pocket marking	Helper	0.384	0.362	995	0.53	1
29	Knee pocket folding	Helper	0.249	0.235	1532	0.34	1
30	Knee pocket hemming 1	Single needle lockstitch	1.108	1.163	310	1.70	2
31	Knee pocket tacking	Single needle lockstitch	0.532	0.558	645	0.81	1
32	Knee pocket overlock	Overlock machine	0.204	0.238	1514	0.35	1
33	Knee pocket hemming 2	Single needle lockstitch	1.210	1.270	284	1.85	2
34	Knee pocket ironing	Iron press	1.867	1.652	218	2.41	2
35	Knee pocket attach	Single needle lockstitch	1.192	1.251	288	1.82	2
36	Knee flap folding	Helper	0.427	0.403	894	0.59	1
37	Buttonhole on knee flap	Button hole machine	0.198	0.202	1783	0.29	1
38	Knee flap runstitch	Single needle lockstitch	0.211	0.221	1625	0.32	1
39	Knee flap turning	Turning machine	0.370	0.349	1033	0.51	1
40	Knee flap topstitch	Single needle lockstitch	0.373	0.392	919	0.57	1
41	Knee flap attach	Single needle lockstitch	1.639	1.721	209	2.51	3*
42	Bar tacking	Bartack machine	1.493	1.560	231	2.28	2
43	Back rise overlocks	Overlock machine	0.514	0.658	547	0.96	1
44	Back rise topstitches	Double needle lockstitch	0.408	0.576	625	0.84	1
45	Big loop part matching	Helper	0.081	0.076	4711	0.11	1
46	Big loop runstitch	Single needle lockstitch	0.248	0.260	1386	0.38	1
47	Big loop turning	Turning machine	0.160	0.151	2385	0.22	1
48	Big loop topstitches	Single needle lockstitch	0.222	0.233	1545	0.34	1
49	Big loop button hole	Button hole machine	0.093	0.095	3795	0.14	1
50	small loop runstitch	Loop stitch machine	0.141	0.149	2412	0.22	1
51	S&B loop, W.B attach	Single needle lockstitch	1.964	1.819	198	2.65	3
52	Waist band topstitch	Single needle lockstitch	1.076	1.129	319	1.65	2
53	Waist band closing	Single needle lockstitch	1.405	1.474	244	2.15	2
54	Small loop tacking	Single needle lockstitch	1.605	1.685	214	2.46	3*
55	Inseam overlock	Overlock machine	0.473	0.548	657	0.80	1
56	Trouser turning	Helper	0.376	0.355	1015	0.52	1
57	Inseam topstitch	Feed of arm machine	0.483	0.671	536	0.98	1
58	Adjustable prep	Helper	0.151	0.142	2535	0.21	1
59	1st adjustable attach	Single needle lockstitch	1.193	1.252	287	1.83	2
60	Button hole on bottom	Button hole machine	0.513	0.523	689	0.76	1
61	Adjustable hemming	Single needle lockstitch	0.156	0.164	2198	0.24	1
62	2nd adjustable attach	Single needle lockstitch	0.655	0.688	524	1.00	1
63	Bottom Rope attach	Helper	0.937	0.884	407	1.29	2*
64	Bottom hemming	Single needle lockstitch	0.870	0.913	394	1.33	2*
65	Final bar tacking	Bartack machine	0.856	0.895	402	1.31	2*

S-Small, B-Big, W.B-Waistband, OT-Observed time, WS-Workstation, OP-Operation, MPR-Manpower requirement, Mins-Minutes, MPC-Manpower Calculated.

4 RESULTS AND DISCUSSION

4.1 *SMV for trouser*

The SMV for each task on trouser assembly was calculated according to Bongomin, Mwasiagi, Nganyi, and Nibikora (2020b). The SMV for the 65 tasks on trouser assembly was determined by summation of the individual SMV. The result obtained shows an SMV of 41.763. The SMV achieved in this study is high because of the bottleneck on the trouser assembly line (Gebrehiwet & Odhuno 2017). However, the SMV can be reduced by the addition of manpower or resources (Mohibullah et al. 2019) for the operations such as knee patch attach, back patch pressing, side pocket topstitches, right flybox attach, fly attach, back patch attaches, hip pocket overlock, etc. These are the tasks whose SMV are relatively higher than the average SMV of 0.642 for the 65 tasks. In addition, SMV can be reduced by proper balancing of the trouser assembly line. The operator performance rating and machine allowances assignment have a significant effect on the SMV determination. Therefore, proper observation on operators' performance is requisite for achieving practically realistic SMV.

4.2 *Operation bulletin*

The operation bulletin was developed based on the following line specifications: the total SMV (41.763), planned efficiency (75%), total machine SMV (37.712), helper SMV (4.051), and minutes per day (480). The trouser assembly line with 65 operations and 61 planned operators was considered. The target and manpower calculations were done for each operation and are presented in Table 2. From the calculated manpower, the required manpower numbers were determined.

$$line\ target = \frac{480 \times 0.75 \times 61}{41.763} = 525\ pieces\ per\ day$$

$$Target\ for\ operation\ 1 = \frac{480 \times 0.75 \times 1}{0.312} = 1142\ pieces\ per\ day$$

$$Manpower\ requirement\ for\ operation\ 1 = \frac{525}{1142} = 0.46$$

The manpower requirement with (*) for an operation means that the exact number can further be determined after observing the level of work in progress (WIP) and idle time at the workstation. Most likely they represent the bottleneck workstations. The bottleneck workstations are the ones whose capacity is less than the demand placed on it and less than the capacities of all other resources. In order to determine the correct manpower requirement, prior knowledge of bottleneck workstations is of paramount importance. Therefore, this might result in an increase of manpower in the case of high WIP and a reduction of the manpower required in the case of high idle time in the workstation. To this end, for the line to perform to the expectations: two ironing operations require four ironers, 51 machine operations need 63 operators, and 12 helper operations require 13 helpers.

5 CONCLUSION

The present paper has demonstrated the function of IE for developing an operation bulletin that can be used for production planning. The accuracy and precision during the time study is very essential for obtaining a realistic SMV that is practically feasible. Therefore, digital technology for a time study method should be explored. Further study can take into consideration the non-value-added or non-productive operations such as separation of bundles, cutting of threads, and transfer of bundles by operator to the next operator. In addition, further study on line balancing is needed to reduce the number of workstations or cycle time which will improve the productivity and minimize the resource cost.

REFERENCES

Abtew, M. A., Kumari, A., Babu, A., & Hong, Y. (2019). Statistical Analysis of Standard Allowed Minute on Sewing Efficiency in Apparel Industry. *Autex Research Journal*, 1–7. https://doi.org/10.2478/aut-2019-0045

Al-khatib, B. A. (2012). The Effect of Using Brainstorming Strategy in Developing Creative Problem Solving Skills among Female Students in Princess Alia University College Department of Psychology and Special Education. *American International Journal of Contemporary Research*, *2*(10), 29–38.

Babu, V. R. (2012). *Industrial engineering in apparel production*. New Delhi, India: Woodhead Publishing India Pvt. Ltd. https://doi.org/10.1533/9780857095541

Bahadır, S. K. (2011). Assembly Line Balancing in Garment Production by Simulation. In W. Grzechca (Ed.), *Assembly Line - Theory and Practice* (pp. 67–82). Rijeka, Croatia: InTech.

Barton, R. R. (2004). Designing Simulation Experiments . In R. G. Ingalls, M. D. Rossetti, J. S. Smith, & B. A. Peters (Eds.), *Proceedings of the 2004 Winter Simulation Conference* (pp. 73–79). University Park, USA. https://doi.org/10.1109/WSC.2004.1371304

Baset, M. A., & Rahman, M. (2016). Application of Industrial Engineering in Garments Industry for Increasing Productivity of Sewing Line. *International Journal of Current Engineering and Technology*, *6*(3), 1038–1041.

Bashar, A., & Hasin, A. A. (2019). Impact of JIT Production on Organizational Performance in the Apparel Industry in Bangladesh. In *MSIE 2019, May 24–26* (pp. 184–189). Phuket, Thailand: Association for Computing Machinery. https://doi.org/10.1145/3335550.3335578

Behr, O. (2018). Fashion 4.0 – Digital Innovation in the Fashion Industry. *Journal of Technology and Innovation Management*, *2*(1), 1–9.

Bertola, P., & Teunissen, J. (2018). Fashion 4.0. Innovating fashion industry through digital transformation.

Research Journal of Textile and Apparel, 22(4), 352–369. https://doi.org/10.1108/RJTA-03-2018-0023
Bongomin, O., Gilibrays Ocen, G., Oyondi Nganyi, E., Musinguzi, A., & Omara, T. (2020). Exponential Disruptive Technologies and the Required Skills of Industry 4.0. *Journal of Engineering, 2020*, 1–17. https://doi.org/10.1155/2020/4280156
Bongomin, O., Mwasiagi, J. I., Nganyi, E. O., & Nibikora, I. (2020a). A complex garment assembly line balancing using simulation-based optimization. *Engineering Reports*, 1–23. https://doi.org/10.1002/eng2.12258
Bongomin, O., Mwasiagi, J. I., Nganyi, E. O., & Nibikora, I. (2020b). Improvement of garment assembly line efficiency using line balancing technique. *Engineering Reports, 2*(4), 1–18. https://doi.org/10.1002/eng2.12157
Calabrese, L., Balchin, N., & Mendez-parra, M. (2017). *The phase-out of second-hand clothing imports: what impact for Tanzania?*
Cano, J. A., & Zuluaga-mazo, A. (2019). Operations Management Strategies for the Textile-Clothing Sector in Colombia. In *Education Excellence and Innovation Management through Vision 2020* (pp. 7620–7625).
Chandurkar, P., Kakde, M., & Bhadane, A. (2015). Improve the Productivity with help of Industrial Engineering Techniques. *International Journal on Textile Engineering and Processes, 1*(4), 35–41.
Colovic, G. (2011). *Management of Technology Systems in Garment Industry*. New Delhi, India: Woodhead Publishing India Pvt. Ltd.
Colovic, G. (2012). *Strategic Management in the Garment Industry. Woodhead Publishing India Pvt Ltd.* https://doi.org/10.1533/9780857095855
Elnaggar, G. (2019). Effect of Operator Skill Level on Assembly Line Balancing in Apparel Manufacturing: A Multi-Objective Simulation Optimization Approach. In *Proceedings of the International Conference on Industrial Engineering and Operations Management, November 26-28* (pp. 308–315). Riyadh, Saudi Arabia: IEOM Society International.
Gebrehiwet, T. B., & Odhuno, A. M. (2017). Improving the Productivity of the Sewing Section through Line Balancing Techniques: A Case Study of Almeda Garment Factory. *International Journal of Sciences: Basic and Applied Research (IJSBAR), 36*(1), 318–328.
Hamja, A., Maalouf, M., & Hasle, P. (2019). The effect of lean on occupational health and safety and productivity in the garment industry – a literature review. *Production & Manufacturing Research, 7*(1), 316–334. https://doi.org/10.1080/21693277.2019.1620652
Hekmatpanah, M. (2011). The application of cause and effect diagram in the oil industry in Iran: The case of four liter oil canning process of Sepahan Oil Company. *African Journal of Business Management, 5*(26), 10900–10907. https://doi.org/10.5897/AJBM11.1517
Hossain, M. M., Khan, S. A., Islam, S. S., Fattah, J., & Bipul, M. R. (2018). SMV and Lead Time for T-Shirt Manufacturing. *American Journal of Engineering Research, 7*(5), 21–25.
Howard, P. M. A., Essuman, M. A., & Asare, T. O. (2019). Strategies for Determining the Production Cost and Pricing of Garments in Ghana: A Study of the Fashion Industries. *International Journal of Business and Social Science, 10*(3), 75–87. https://doi.org/10.30845/ijbss.v10n3p7
Immawan, T., & Kurniawan, R. (2016). Analysis of Line Balance Sound Board Glue Production on Assembly Grand Piano Process: Case Study PT Yamaha Indonesia Idle time. In *Advances in Ergonomics of Manufacturing* (pp. 349–359). https://doi.org/10.1007/978-3-319-41697-7
International Labour Organization. (2017). *Lean Manufacturing Techniques For Ready Made Garments Industry*. Geneva, Switzerland.
Islam, A., Islam, S. U., & Gupta, J. Das. (2017). Industrial Engineering and Operations Management Functions in the Apparel Manufacturing Industry. In *Proceedings of the International Conference on Industrial Engineering and Operations Management* (pp. 886–897).
Jana, P., & Tiwari, M. (2018). *Industrial Engineering in Apparel Manufacturing: A practitioner's guide*. US: Apparel Resources Pvt. Ltd.
Karthik, T., Ganesan, P., & Gopalakrishnan, D. (2017). *Apparel Manufacturing Technology*. Boca Raton, USA: Taylor & Francis.
Kayar, M. (2017). Study on the Importance of Employee Performance Assessment and Lost Productive Time Rate Determination in Garment Assembly Lines. *FIBRES & TEXTILES in Eastern Europe, 5*(125), 119–126. https://doi.org/10.5604/01.3001.0010.4638
Khan, A. M. (2013). Application of Lean Manufacturing to Higher Productivity in the Apparel Industry in Bangladesh. *International Journal of Scientific & Engineering Research, 4*(2), 1–10.
Khan, S. A., Islam, T., Elahi, S., Sharif, N., & Mollik, M. (2019). An attempt to increase agility of garment industry. *Journal of Textile Engineering & Fashion Technology Research, 5*(3), 154–161.
Khan, W. M. (2016). Implementation of Modern Garment Planning Tools & Techniques in Garment Industry of Bangladesh. *International Journal of Engineering and Advanced Technology Studies, 4*(3), 31–44.
Khatun, M. M. (2013). Application of Industrial Engineering Technique for Better Productivity in Garments Production. *International Journal of Science, Environment and Technology, 2*(6), 1361–1369.
Khatun, M. M. (2014). Effect of Time and Motion Study on Productivity in Garment Sector. *International Journal of Scientific & Engineering Research, 5*(5), 825–833.
Kitaw, D., Matebu, A., & Tadesse, S. (2010). Assembly Line Balancing Using Simulation Technique in a Garment Manufacturing Firm. *Journal of EEA, 27*, 69–80.
Kursun, S., & Kalaoglu, F. (2009). Simulation of production line balancing in apparel manufacturing. *Fibres and Textiles in Eastern Europe, 17*(4(75)), 68–71.
Lopetegui, M., Yen, P. Y., Lai, A., Jeffries, J., Embi, P., & Payne, P. (2014). Time motion studies in healthcare: What are we talking about *Journal of Biomedical Informatics, 49*, 292–299. https://doi.org/10.1016/j.jbi.2014.02.017
Maralcan, A., & Ilhan, I. (2017). Operations management tools to be applied for textile. *IOP Conference Series: Materials Science and Engineering, 254*(202005), 1–6.
Mohibullah, A. T. M., Takebira, U. M., Mahir, S. Q., Sultana, A., Rana, M., & Hossain, A. K. M. S. (2019). High-tech machine in RMG Industry: reducing SMV, lead time and boosting up the productivity. *Journal of Textile Engineering & Fashion Technology, 5*(1), 29–34. https://doi.org/10.15406/jteft.2019.05.00177
Moktadir, A. (2017). Productivity Improvement by Work Study Technique: A Case on Leather Products Industry of Bangladesh. *Industrial Engineering & Management, 6*(1).
Montevechi, J. A. B., Pinho, A. F. de, Leal, F., & Marins, F. A. S. (2007). Application of Design of Experiments on the Simulation of a process in an Automotive Industry. In *Proceedings of the 2007 Winter Simulation Conference* (pp. 1601–1609). IEEE.

Ortiz, C. A. (2006). *Kaizen Assembly: Designing, Constructing, and Managing a Lean Assembly Line. Assembly Automation* (Vol. 27). Boca Raton: Taylor & Francis. https://doi.org/10.1108/aa.2007.03327aae.001

Puvanasvaran, A. P., Mei, C. Z., & Alagendran, V. A. (2013). Overall Equipment Efficiency Improvement Using Time Study in an Aerospace Industry. *Procedia Engineering, 68*, 271–277. https://doi.org/10.1016/j.proeng.2013.12.179

Rahman, H., & Amin, A. (2016). An Empirical Analysis of the Effective Factors of the Production Efficiency in the Garments Sector of Bangladesh. *European Journal of Advances in Engineering and Technology, 3*(3), 30–36.

Rajkishore, N., & Padhye, R. (2015). *Garment Manufacturing Technology*. Cambridge,UK: Elsevier Ltd.

Russell, R. S., & Taylor, B. W. (2011). *Operations Management: Creating Value Along the Supply Chain* (7th ed.). US: John Wiley and Sons, Inc.

Senthilraja, V., Aravindan, P., & Sathesh, kumar A. (2018). Man Power Productivity Improvement through Operator Engagement Time Study. In *1st International Conference on Recent Research in Engineering and Technology 2018* (Vol. 4, pp. 1052–1065). Tamil Nadu, India: Exporations on Engineering Letters.

Sibanda, W., & Pretorius, P. (2011). Application of Two-level Fractional Factorial Design to Determine and Optimize the Effect of Demographic Characteristics on HIV Prevalence using the 2006 South African Annual Antenatal HIV and Syphilis Seroprevalence data. *International Journal of Computer Applications, 35*(12), 15–20.

Starovoytova, D. (2017). Time study of Rotary Screen Printing Operation. *Industrial Engineering Letters, 07*(04), 24–35.

Sudarshan, B., & Rao, D. N. (2014). Productivity improvement through modular line in Garment industries. In *5th International & 26th All India Manufacturing Technology, Design and Research Conference (AIMTDR)* (pp. 1–6). Assam, India.

Uddin, M. M. (2015). *Productivity Improvement of Cutting, Sewing and Finishing Sections of a Garment Factory Through Value Stream Mapping – a Case Study*. Master Thesis, Dhaka: Bangladesh University of Engineering and Technology,Department of industrial and production engineering.

Wolff, E. A. (2020). The global politics of African industrial policy: the case of the used clothing ban in Kenya, Uganda and Rwanda. *Review of International Political Economy*, 1–24.

Yemane, A., Gebremicheal, G., Meraha, T., & Hailemicheal, M. (2020). Productivity Improvement through Line Balancing by Using Simulation Modeling (Case study Almeda Garment Factory). *Journal of Optimization in Industrial Engineering, 13*(1), 153–165.

Yemane, A., Haque, S., & Malfanti, I. (2017). Optimal Layout Design by Line Balancing Using Simulation Modeling. In *Proceedings of the International Conference on Industrial Engineering and Operations Management* (pp. 228–245). Bogotá, Colombia: IEOM society international.

Advances in Phytochemistry, Textile and Renewable Energy Research for Industrial Growth – Nzila et al. (Eds)

Industrial output in Uganda: Does electricity consumption matter?

B. Akankunda & S.K. Nkundabanyanga
Department of Accounting, Faculty of Commerce, Makerere University Business School, Kampala, Uganda

M.S. Adaramola
Faculty of Environmental Sciences and Natural Resource Management, Norwegian University of Life Sciences, Ås, Norway

A. Angelsen
School of Economies and Business, Norwegian University of Life Sciences, Ås, Norway

ABSTRACT: Existing studies indicate that access to electricity is associated with gainful economic benefits. The main goal of the study was to examine the effect of electricity consumption on industrial output while holding other factors constant using data sets from the Uganda Investment Authority, Uganda Bureau of Statistics, Electricity Regulatory Authority, and World Bank enterprise report on workers' education, labour employed, and electricity consumed by the industries (agro based). Using the vector error correction model, results of the study indicated that that education, electricity consumption, and labour have a long-term causality on industrial output. Based on the research questions, the effect of electricity consumption on industrial output, the effect of educated workers on industrial output, and the labour employed influence industrial output. Furthermore, we find evidence that workers' education and the labour employed differentials correspond directly with industrial output differentials. Regulators must make efforts to increase an affordable and reliable supply of industrial electricity as a facilitating condition through subsidised electricity tariffs. Moreover, results suggest that industrial-led electricity consumption policies that can enhance industrial growth should be implemented. Likewise, the results also suggest efforts to encourage agro-based industries to employ more well-educated workers to increase the industrial output.

Keywords: Electricity consumption, Industrial output, capital, education, and labour employed.

1 INTRODUCTION

1.1 *Background of the study*

Industrial development has been an engine of any economy that desires to move from a non-industrial state to an industrial state (Daubaraitė 2015; Li et al. 2016; Mountjoy 2017). To achieve this, certain factors must be put in place to trigger and achieve this target (Izadmehr 2018; Shepherd 2015). Among these factors are availability and reliability of energy resources to meet the demand of the industry, which must be efficiently utilized by the industrial sector (Alley 2016; Ugwoke et al. 2016). The study objective is to examine the effect of electricity access on sustainable industrial output. Uganda is regarded as an agriculture-based economy and a food basket in the Eastern African region, given its ability to produce a variety of foods and in large quantities (Ssozi 2018). The agricultural output comprises food and cash crops, livestock, forestry, and fishing sub-sectors. In addition to being a good source of national income, agro-processing industries provide skilled and unskilled employment opportunities for many Ugandans to support their livelihood.

According to an Electricity Regulatory Authority tariff review report (2019), the Industrial Customer category includes Medium, Large, and Extra-Large Industrial Customers. Energy Sales to Industrial Customers reduced by 7% in 2019 compared to 11% in 2018. The decreasing rate in the Energy Consumption of Industrial Customers in 2019 was majorly attributed to the low industrial growth. The average consumption per Industrial Customer recorded a low growth rate in average consumption in 2019 as compared to 2018 among Industrial Customers (ERA 2019). In addition, according to Mawejje (2016), 80% of these industries rely heavily on electricity for their operations. Agricultural energy demand can be divided into direct and indirect energy needs. The direct energy needs include the energy required for land preparation, cultivation, irrigation, harvesting, post-harvest processing, food production, storage, and the transport of agricultural inputs and outputs (Dogan 2016), while indirect energy demand include the energy embedded in farm buildings and processing factories,

 DOI 10.1201/9781003221968-11

machinery, equipment, transport, food retailing, cooking and waste disposal. Hence, sustainable energy solutions for agriculture and food value chains are a central structural element to any support strategy for agro-based industries. The agro-processing industry transforms products originating from agriculture into both food and non-food commodities (Arranz-Piera 2017; Rovas & Zabaniotou 2015; Toyin et al. 2017). Upstream industries are engaged in the initial processing of products, with examples including rice and flour milling, leather tanning, cotton ginning, oil pressing, saw milling, and fish canning (Kumar et al. 2016). Furthermore, an energy input is required in food processing, as well as in packaging, distribution, and storage (Mawejje & Mawejje 2016).

Agriculture in Uganda, which is predominantly rain-fed, is increasingly adversely affected by the climate change and variability manifested in erratic rain patterns, prolonged dry spells, and floods. As a result, farm-level productivity is far below the attainable potential for most crops (Fermont & Benson 2011). Despite efforts, available evidence indicates that the industrial sector is experiencing slow growth and one of the factors responsible for this to a considerable extent is the poor industrial energy consumption. This is supported by the fact that the industrial growth especially for the Agricultural sector's value addition stands at 1.4, Human Development Index 0.484, small area equipped for irrigation (ha) 14,000 (2007-2017) (Food and Agriculture Policy Decision Analysis FAPDA 2017; Mawejje & Mawejje 2016). Access to clean energy for productive use is crucial if Uganda is to exploit its agricultural potential.

Besides, the majority of agro-processing industries indicated that electricity is still considered as the biggest obstacle (World bank 2019). The survey indicated that they continue to experience power outages: the percentage of firms reporting electricity as the most important obstacle for their day-to-day operations was 23%. This power interruption has made firms continue to experience loss in output, low value addition, high operating costs (total annual electricity costs), as well as failure to access information, limited access to production inputs (skilled labour, adequate capital (equipment and machinery), effective energy resources that are reliable and affordable (Izadmehr et al. 2018), and all these continue to hamper their industrial output (World Bank 2017). Nevertheless, studies on industrial output effect of electricity access are limited.

Nonetheless, Uganda has an installed capacity of 1252.3 MW, of which 1,246.5 MW supplies the main grid and 5.9 MW is off the main grid (Electricity Regulatory Authority 2019). Access to electricity in 2019 at the national level in Uganda is very low at 26.7% (Uganda Bureau of Statistics 2019). Uganda currently has one of the lowest electricity demand peaks: 700 MW peak electricity demand against installed generation capacity of over 1252.3 MW, with the highest average electricity end user tariffs being 669.5 sh/KWh (consumer customers), 599.2 sh/KWh (medium industrial consumers), 365.7 sh/KWh (large industries consumers), and 304.7 sh/kWh (extra-large consumer industries) (Electricity Regulatory Authority 2019; World Bank 2017).

Subsequently, the Government of Uganda has focused on increasing access to energy in rural areas by constructing various hydropower plants and extending and improving transmission lines to improve agriculture modernization. However, agricultural production continues to be constrained by the lack of irrigation systems (FAO 2017; Wanyama et al. 2017). To this end, GoU allocated UGX 5 billion (US$ 1.92 million) in 2013 to the Ministry of Water and Environment to rehabilitate irrigation schemes and provide new irrigation and water harvesting technologies to increase the water supply by 10.1 million cubic meters by 2017, however all these continue to fail.

According to the World Bank enterprise survey (2016), 90% of these firms give gifts to get electricity connections. Similarly, the duration of a typical outage stands at over 7 hours. Losses due to power outages also stand at 6% of annual sales (World bank enterprise survey 2013), and the percentage of firms reporting electricity as the most important obstacle for their day-to-day operations was 23% in 2012. Likewise, electricity is still the most commonly chosen top obstacle in firms' operations. The production of goods and services requires energy as an input, which is called a factor of production. Energy sources vary in their effectiveness as a factor of production, depending on their energy characteristics. In this study, electricity is an augmenting input in the production of goods. Although agro-based industries have tried to appreciate the benefits of electrification, their performance has remained low. However less empirical studies have been carried out to this effect. Thus, this study's aim was to examine how electricity access can stimulate industrial output.

1.2 *Previous work*

The demand for electricity by industry has been the subject of empirical studies over many years (Lin & Wang 2019). Previous studies on the causal link between electricity consumption and industrial output holding other factors constant have revealed differing views (Carlsson et al. 2020; Mawejje & Mawejje 2016). The results of the existing studies point toward four main types of causal relationships (hypotheses), which have been summarized by Mawejje and Mawejje (2016) as follows. First is the growth hypothesis where causality is one-way: from electricity consumption to output growth (Bekun & Agboola 2019; Ozcan & Ozturk 2019). Second is the conservation hypothesis in which causality runs from output growth to electricity consumption (Balcilar et al. 2019; Dey 2019; Odhiambo 2010). Third, the feedback hypothesis proposes a two-way causality between electricity consumption and output growth (Gorus & Aydin 2019; Ndlovu & Inglesi-Lotz 2020; Rahman 2020). Fourth, the neutrality hypothesis is related to no causality between electricity consumption and output growth (Aydin 2019; Dogan et al. 2016; Kahouli 2018).

1.3 *Research questions*

This study's aim was to examine the extent to how electricity access influences industrial output in Uganda, while addressing the following research questions:

1) What is the effect of electricity consumption on industrial output?
2) What is the effect of education on industrial output?
3) How does labour employed influence industrial output.

To answer the above research questions, the study used a Cointegrated Vector Autoregression (CVAR) as an estimation technique. Secondary data (time series data) on the inputs (electricity consumption, labour, education, and industrial output) were extracted from the industries for a period of 10 years (2009–2019). Additionally, we propose relevant policy recommendations.

1.4 *Theoretical framework*

In analyses of the role played by electricity in production process, growth models (exogenous growth models such as the Solow's growth model, the Ramsey-Cass-Koopman's model, the Diamond overlapping generation model, as well as the endogenous growth models) have been extended. Besides capital, labour, and knowledge (technology), electricity has been explicitly modelled as an argument for the economic output (Alley et al. 2016; Akekere & Yousuo 2013; Danish et al. 2015; Maxim 2014; Palit & Bandyopadhyay 2016; Yuan, Kang et al. 2008). This research has implications for economic theory. The output elasticity of electricity is nearly triple its share of inputs in production. The marginal revenue product of electricity is nearly triple the marginal revenue products of labour and capital inputs at equilibrium. Electricity consumption, akin to R&D, has a much larger role in industrial output than postulated in production theory. Differences in the output elasticities between labour and the labour employed functions raise additional debates about the adequacy of the human capital theory in explaining industrial output.

2 REVIEW OF LITERATURE

Most studies on the effects of electricity as a component of energy on output and economic growth frequently assume that energy enters the production function independently or in a Hicks-neutral fashion (see Akekere & Yousou 2013; Samuel & Lionel 2013; Stern 2004; Yuan et al. 2008). Limited attention has been paid to which energy may enter the production functions. Given the fact that most industrial machines depend on electricity to perform, the capital-augmenting factor of energy performance in a production model appears to offer a promising way of appraising the role of electricity on industrial output. Studies on industrial electricity access have attracted a lot of information, however studies are limited with deviating evidence, and there has not been a consensus over the effects of electricity on industrial output.

In Pakistan, Qazi et al. (2012) used the Johansen cointegration approach based on VAR to conduct a study on the relationship between disaggregate energy consumption and industrial output. The study covered the period 1972–2010. There were three results obtained from the analysis. The results showed a positive long-term relationship between disaggregate energy consumption and industrial output. On the other hand, evidence of a unidirectional causality was observed running from electricity consumption to industrial output.

In another study, Husain and Lean (2015) used a demand function to investigate the relationship between electricity consumption, output, and price in Malaysia using time series data for the period 1978–2011. In the long term, electricity consumption, output, and price were found to be cointegrated. Evidence of a positive relationship was found between electricity consumption and manufactured output.

In the same vein, Alley (2016) investigated the effect of electricity supply, industrialization, and economic growth using evidence from Nigeria. The results showed that electricity supply had a significant positive effect on industrial output. This result contrasts with that of the OLS. Likewise, Olarinde and Omojolaibi (2014) examined electricity consumption, institutions, and economic growth in Nigeria for the period 1980–2011 using causality using the Autoregressive Distributed Lag (ARDL) model and WALD test approach, and found a positive direct relationship between institutions, electricity consumption, and economic growth.

In another study, Sun and Anwar (2015) in Singapore used a trivariate vector autoregressive framework and found a positive relationship between electricity consumption and industrial production using monthly data from 1983 to 2014. Mawejje and Mawejje (2016) in Uganda found a long-term causality running from electricity consumption to industry at the sectoral level (macro level); a unidirectional short-term causality running from the services sector to electricity consumption; and neutrality in the agricultural sector using a vector error correction techniques.

Ettah , (2017) also found that Nigeria's national output is significantly and consistently improved by an available and sustained electricity supply, using the vector error correction mechanism (VECM). Similarly, Alby, Dethier, and Straub (2013) found that in developing countries with a high frequency of power outages, electricity-intensive sectors have a low proportion of small firms since only large firms are able to invest in generators to mitigate the effects of outages.

On the other hand, Akekere and Yousou (2013) found that electricity supply negatively affects industrial output in Nigeria. Nwajinka et al. (2013) also found that that other things being equal electricity supply has no significant impact on industrial productivity in Nigeria.

Whereas empirical findings on the nexus differ across countries, the differences may be due to the heterogeneity of infrastructural and other characteristics between the countries. The literature documenting the impact of electricity on economic growth from single-country studies has shown divergent findings. This study will focus on the effect of electricity access on industrial output at a micro level (agro processing industries), while holding other factors constant by investigating the effect of electricity consumption, capital, labour, education, and other firm-specific characteristics like size, location, managerial experience, and government intervention.

2.1 *The relevant policies for the problem in question*

The Government of Uganda has over the past eight years embarked on a Power sub-Sector Reform Programme, which has resulted in the implementation of significant structural changes within the sector to promote industrial growth. The Uganda National Development Plan (NDP 2010/11-15) highlights the need to invest in priority areas, among them facilitating availability and access to critical production inputs especially in agriculture and industry. The National Agriculture Policy (NAP) approved in 2013 also is in place to achieve food and nutrition security through coordinated sustainable agricultural productivity and value addition. The Energy Policy for Uganda (2002), which exists to meet the energy needs and rural electrification policy, was also formulated to increase electricity access with a target to increase access to electricity from 7% to 26%.

2.2 *Theoretical model*

The theory of production shows the relation between input changes and output changes. It also shows the maximum amount of output that can be obtained by an industry from a fixed quantity of resources. The production function is expressed as:

$$\text{Industrial output } Q = f(K, L, E) \quad (1)$$

where Q is industrial output, K represents capital, L represents labour, and E stands for electricity.

Under standard assumptions underlying an explicitly specified concave production function, inputs have a positive but a decreasing impact on output. Thus, electricity is expected to have independent positive effects on output. Though the effects of individual inputs on output may be examined, output is not (often) produced by a single input; therefore, a combination of inputs is necessary for optimal production (Romer 2012).

While microeconomic theories, such as human capital and signalling (Mincer 1974; Weiss 1995), suggests several avenues through which education can affect productivity, little consensus exists among economists on how education is related to productivity. However, empirical research has been limited because few data sets contain information on both workers' output and their education.

This study thus estimates a capital-augmenting model where electricity enters the model multiplicatively with capital to reflect the role of electricity-operated capital in the production process of agro-processing industries. Uganda should thus increase electricity supply to stimulate industrial output and enhance economic growth.

The industrial output is produced by the output growth and value added. That is:

$$Y_i = f(Y_i^t, Y_i^n) \quad (2)$$

where, Y_i = industrial output for each industry i; I, Y_i^n = total value added, n is the number of industries, and t represents the period.

The industrial output is produced by capital and labour. However, the capital equipment in the industrial sector runs on electricity: this study then assumes electricity enters the model in a capital-augmenting factor.

$$Y_i = (K^{\gamma} E^{\beta} L^{\alpha} e_{j=1}^{n} \sum mizi + e_i \quad (3)$$

where K^{γ} = capital employed in the industrial sector, proxied by gross fixed capital and assets owned; E^{β} = electricity supply (consumption); L^{α} = labour employed in the industrial sector, proxied by labour force participation; $e_{j=1}^{n} \sum mizi$ = other specific factors (location, nature of establishment and government intervention); and e_i = error term.

The differential of the natural logarithm of Equation (3) yields the elasticity Equation (4):

$$InY_i = \gamma InK_i + \beta InE_i + \alpha InL_i +_{j=1}^{n} \sum mizi + e_i \quad (4)$$

where γ, β, α are the coefficients; $e_{j=1}^{n} \sum mizi$ is a set of other firm-specific characteristics (location, rural/urban; government intervention; and managerial experience, education). Differentiating Equation 3 to get elasticity yields:

$$dY_i \frac{dY_i}{Y_i} = \gamma \frac{dK_i}{K_i} + \beta \frac{dE_i}{E_i} + \alpha \frac{dL_i}{L_i} \quad (5)$$

Implying that γ = 1% change in K leads to γ % increase in Y, β = 1% change in E leads to β % increase in Y, and α = 1% change in L leads to α % increase in Y:

$$InY_i = \gamma InK_i + \beta InE_i + \alpha InL_i + \beta_i \quad (6)$$

Equation (6) contains endogenous explanatory variables, which are electricity consumption, capital, labour, and education, with $\gamma, \beta,$ andα as the coefficients.

3 METHODS

3.1 *Methodology and the data*

The study used time series data from the World Bank enterprise surveys (2016) and Uganda Bureau of statistics (2018). The study used a population of

645 agro-processing industries; the sample for agro-processing industries was selected using stratified random sampling, which is the preferred sampling when the whole population is considered (as in this the study). Three levels of stratification were used in this survey: type of agriculture production, establishment size, and region (rural/urban). For the Uganda's Enterprise Survey, size stratification was defined using number employees into small enterprise (5 to 19), medium enterprise (20 to 99), and large enterprise (>99). In Uganda, many agro-based industries are recognized and grouped under different classifications such as livestock, horticulture, cereals, grains, vegetables, and cash crops agro-industries.

3.2 *Statistical analysis*

The study examined how electricity access influences industrial output in Uganda; electricity access, labour, and capital were the independent variables, and industrial output was the outcome variable. The data on the study variables were extracted and managed in excel; they were later exported to Eviews 7.0 and Stata 15.0 (StataCorp, College Station, TX, USA) for analysis. In order to avoid spurious results, the study variables were tested for normality using the Jarque-Bera statistic. The descriptive statistics were summarized by mean, standard deviations (SD), and minimum and maximum values. Due to the time series nature of our data, we tested for stationarity using the Augmented Dickey Fuller (ADF) test for unit root. This study tested for cointegration to find out whether the study variables have a long-term relationship. The null hypothesis was there is no cointegration among the variables, while the alternative hypothesis is that the variables are cointegrated. We used a cointegrated vector autoregression (CVAR) as an estimation technique. According to Gujarati and Porter (2009), the term autoregressive is due to the appearance of the lagged value of the dependent variable on the right-hand side, and the term vector is since one is dealing with a vector of two or more variables. The VAR model is one of the most successful and easy to use for the analysis of multivariate time series. Johansen's (1995) cointegration technique based on VAR is employed to determine the long-term relationship between industrial output and its explanatory variables using the 3SLS regression technique.

Table 1. Descriptive statistics and testing for normality.

VARIABLES	N	Mean	Max	Min	SD	Kurt	Jarque-Bera	*P*-value
LN (INDOUT)	44	8.1	8.576	7.435	0.367	2.039	3.975	0.137
LN (ELECCSP)	44	19.6	19.996	19.258	0.202	2.062	1.841	0.398
LN (EDUC)	44	14.1	14.263	13.825	0.118	2.174	1.965	0.374
LN (LBR)	44	7.2	7.316	7.018	0.092	1.667	4.276	0.118

4 RESULTS AND FINDINGS

This study examined the indirect role of electricity consumption, labour, and capital on industrial output in Uganda. The model assumed that the productivity of capital, labour, and education is augmented by electricity consumption, which provides the energy for the modern production technology. The direction of causality in the nexus was traced along the type of electricity used, and the effect of labour, education, and capital were estimated using Stata version 15 and E-views. The results in Table 1 show that variables are normally distributed given that the *P*-value is greater than 0.05 (Benjamin et al. 2018; Ioannidis 2018; Kennedy-Shaffer 2019). The results show that electricity consumption and education were stationary at the first level, while industrial output and labour were stationary at the first difference. Detailed analysis is presented in Table 2.

Table 2. Testing for stationarity of the variables (DFGLS unit root test).

	DFGLS Test at level Trend and intercept				DFGLS test at first difference Trend and intercept			
Variables	ADF Stat.	p-value	Decision	ADF Status	p-Stat.	value	Decision	Status
LN (INDOUT)	−1.79	0.081	Accept	Not stationary	−6.62	<0.001	Reject	Stationary
LN (ELECCSP)	−3.54	0.001	Reject	Stationary				
LN (EDUC)	−3.04	0.004	Reject	Stationary				
Ln (LBR)	−1.53	0.134	Accept	Not stationary	−5.18	<0.001	Reject	Stationary

Ho: The null hypothesis is that the series have a unit root (Not stationary)
Ha: The alternative hypothesis is that the series have no unit root (Stationary)
Reject the null if *P*-value is less than 0.05

Table 3. Lag length selection criteria.

Lag	LogL	LR	FPE	AIC	SC	HQ
0	251.4832	NA	1.82e-11	−13.37747	−13.2033	−13.31607
1	384.3432	229.8119	3.31e-14	−19.69423	−18.8235	−19.38724
2	423.3976	59.10935*	9.84e-15	−20.94041	−19.37303*	−20.38784*
3	438.0904	19.06093	1.15e-14	−20.86975	−18.6058	−20.07159
4	461.8432	25.67875	8.99e-15*	−21.28882	−18.3282	−20.24507
5	482.5036	17.86842	9.56e-15	−21.54074	−17.8835	−20.25139
6	503.3171	13.50064	1.28e-14	−21.80092*	−17.4471	−20.26599

*indicates lag order selected by the criterion
LR: sequential modified LR test statistic
(each test at 5% level)
FPE: Final prediction error
AIC: Akaike information criterion
SC: Schwarz information criterion
HQ: Hannan-Quinn information criterion

4.1 *Lag length selection*

The validity of the empirical results depends on the careful specification and the appropriateness of the choice of the cointegrating rank specifications of the underlying vector autogressive (VAR) model.

The testing of the lag structure was based on the maximum likelihood function and is supported for our data by the Akaike Information Criteria (AIC), Schwartz Information Criteria (SIC), Hannan–Quinn Information Criteria (HQIC), and Final Prediction Error (FPE) lag reduction tests. SC and HQ indicated an optical lag structure of 2 as per Table 3.

4.2 *Testing for cointegration*

After determining the lag structures of the data generating structures, we tested for cointegration to find out whether the study variables were cointegrated or whether the study variables have a long-term relationship. The null hypothesis was there is no cointegration among the variables, while the alternative hypothesis was that the variables are cointegrated. To test for this, we adopted a lag length of 2 and then used the Johansen (1988) procedure to test the existence of long-term equilibrium relations using the trace statistic test for cointegration.

The results in Table 4 show that the trace statistic is greater than the critical value (86.78 > 63.88); this means that the researcher rejected the null hypothesis that there is no cointegration among variables at 1% level of significance. The result further shows that there is one cointegrating equation. Therefore, there is one error term and the variables have a long-term relationship. Since the variables are cointegrated, the researcher ran a Vector Error Correction model (VECM). Table 4 shows the results of the analysis.

4.3 *Testing for Granger causality between the variables*

When variables are cointegrated, at least one or all the error correction terms should be negative and statistically significant in the short-run model showing convergence of the variables in the long term. In line with most of the literature in econometrics, one variable is said to Granger cause the other if it helps to make a more accurate prediction of the other variable than had we only used the past of the latter as predictor. Granger causality between two variables cannot be interpreted as a real causal relationship but merely shows that one variable can help to predict the other one better. An important point to note here is that even though the error correction term in the industrial output model is significant, it does not signify long-term convergence. Therefore, we cannot conclude that electricity consumption Granger causes industrial output. In addition, the F statistics for the joint significance of independent variables do not provide sufficient evidence to support the existence of short-term Granger causality running in either direction. We ran the Granger causality tests between the variables. Table 5 shows the results of the Granger causality test. The variables that were used are:

- A. Industrial output
- B. Electricity consumption
- C. Labour
- D. Education

Table 4. Results of unrestricted cointegration rank test (trace).

Hypothesized No. of CEs	Trace statistic	Critical value	*P*-value
None**	86.78	63.88	<0.001
At most 1	38.13	42.92	0.139
At most 2	15.36	25.87	0.545

CEs: Cointegrating equations
**Denotes rejection of the hypothesis at the 0.01 level

4.4 *Vector Error Correction model (VEC)*

To test whether there is a long-term or short-term relationship between industrial output and

Table 5. Granger causality test of the study variables.

Null Hypothesis:	Obs	F-Statistic	Prob.	Type of Decision	Causality
B does not Granger Cause A	41	1.56	0.22	Not Reject	No causality
A does not Granger Cause B	41	1.61	0.21	Not Reject	No causality
D does not Granger Cause A	41	1.31	0.28	Not Reject	No causality
A does not Granger Cause D	41	0.96	0.39	Not Reject	No causality
C does not Granger Cause A	41	0.66	0.52	Not Reject	No causality
A does not Granger Cause C	41	9.77	0.00	Reject	Uni-directional causality
D does not Granger Cause B	42	3.57	0.04	Reject	Bi-directional causality
B does not Granger Cause D	42	6.32	0.00	Reject	Bi-directional causality
Cdoes not Granger Cause B	41	1.03	0.37	Not Reject	No causality
B does not Granger Cause C	41	0.05	0.95	Not Reject	No causality
C does not Granger Cause D	41	2.83	0.07	Not Reject	No causality
D does not Granger Cause C	41	0.70	0.50	Not Reject	No causality

Reject the null hypothesis if the *P*-values is less than 0.05 level.

Table 6. Vector error correction model.

Independent variables	Coefficient	t-Statistic	Standard error
Electricity consumption	0.582	2.38	0.245
Education	−0.205	−0.6	0.341
Labour	2.301	5.11	0.45
Constant	−0.018	−1.055	0.017
ECT_1	−0.626**	−3.247	0.193
R-squared	0.489		
Adj. R-squared	0.415		
Akaike AIC	−2.746		
Schwarz SC	−2.495		
F-statistic	6.686**		
Log likelihood			
Durbin-Watson	62.85		
statistic	1.98		

**Significance at 0.01

electricity consumption, education, and labour we ran a VEC model. The Johansen and max trace statistics have shown the existence of at least one cointegrating relationship between the study variables and thus we proceeded to estimate this using the vector error correction framework.

Table 6 shows the results from the vector correction model; the results of the Durbin Watson test for the null hypothesis show that there is no serial correlation of the residuals, since the values of the DW-statistics are close to 2 in both models, thus we fail to reject the null hypothesis and conclude that the residuals are not serially correlated. The coefficient of the ECT_1 in model one was negative and significant at 5% level of significance. This means that there is a one period lags residual of the cointegrating equation. This implies that education, electricity consumption and labour have a long-term causality on industrial output. The coefficient of ECT_1 of −0.626 implies a deviation from the long-term growth rate in industrial output is corrected by 62.6% by the following year.

5 DISCUSSION OF THE RESULTS

The results of the study indicated that that education, electricity consumption, and labour have a long-term causality on industrial output. Based on the research questions, the effect of electricity consumption on industrial output, the effect of educated workers on industrial output, and how the labour is employed influence industrial output. The study's findings support earlier theories and studies that have consistently shown that education, electricity consumption, and labour have a long-term causality on industrial output (Antonioli et al. 2011; Asumadu-Sarkodie & Owusu 2017; Klein, Spady & Weiss 1991; Vandenberghe 2018).

5.1 *Does electricity consumption matter?*

The results indicated a positive and significant relationship electricity consumption and industrial output. Energy is a necessary condition for industrial and economic survival. Presently, energy still holds a decisive significance for economic activity because economic growth is determined by the energy resource of the country (Velasquez & Pichler 2010).

The findings of this study corroborate well with those of Sanchis (2007), who stated that electricity as an industry is responsible for a great deal of output using vector error correction model (VECM) and VAR analysis. Ciarreta et al. (2010) used panel data from 1970 to 2007 to analyse the causality relationship between electricity consumption, real GDP, and energy price. They revealed the long-term equilibrium relationship between variables. The causal relationship running from electricity consumption to GDP was revealed. Also, they found a bidirectional relationship between electricity consumption and growth in the short-term and long-term.

Similarly, this study collaborates well with findings by Abosedra et al. (2009) using a direction of causality between electricity consumption and industrial growth for Lebanon, using monthly data covering the period 1995 to 2005. The outcome of the study substantiates the absence of a long-term equilibrium relationship between electricity consumption and industrial growth and the existence of a unidirectional causality without feedback running from electricity consumption to industrial growth.

Velasquez and Pichler (2010) also agree that a sufficient and an affordable supply of electricity has a decisive significance for industrial productivity and economic growth. Since a country's economic growth is a composite of economic activities of enterprises, the less cost they must tolerate, the better a country's chance at harnessing their input towards greater levels of gross domestic product and growth. Okpara (2011) also supports this argument that industrial productivity can contribute immensely towards economic growth and poverty reduction. Rud (2012a, 2012b) investigate the effects of electricity provision on firms in India. This finding is also supported by Rud (2012a) where an increase in rural electrification in Indian States starting in the mid-1960s led to an increase in aggregate manufacturing output in the affected states.

According to Fisher-Vanden, Mansur, and Wang (2014), resource availability and input factor reliability are important for firm productivity and are especially problematic in developing countries like Nigeria. Olarinde and Omojolaibi (2014) examined electricity consumption, institutions, and economic growth in Nigeria for the period 1980–2011. They tested for causality using the ARDL and WALD test approach, and found a positive direct relationship between institutions, electricity consumption, and economic growth.

Furthermore, the study results also correspond well with Mawejje and Mawejje (2016), who confirmed a long-run unidirectional causality running from electricity consumption to GDP At the macro-level as well as a long-run causality running from electricity consumption to industry; a unidirectional short-run causality running from services sector to electricity consumption; and neutrality in the agricultural sector at a sectoral level.

These results suggest that current efforts to improve the supply and affordability of the electricity are a facilitating condition for industrial output. Moreover, results suggest industrial-led electricity consumption policies that can enhance industrial growth.

5.2 *Does workers' education and labour employed matter?*

Specifically, we find evidence that workers who are graduates are more productive than those with secondary school education; workers with secondary school education are more productive than those with primary school education; and workers with primary school education are more productive than those with no formal education. Furthermore, we find evidence that workers' education and the labour employed differentials correspond directly with industrial output differentials. Several studies provide empirical evidence in support of this viewpoint. Syverson (2011) find a positive relationship between the amount of education completed and labour employed. Earlier studies (see Solow 1956) argued that changes in national income are determined by changes in a country's stock of physical and human capital. More recently, the new growth theories, such as those formulated by Romer (1993) and Lucas (1988), also agree with this finding by focusing on the importance of idea gaps and learning externalities in explaining why some countries are richer than others. In another study, Vandenberghe (2018) also agrees that educated workers relate to increased industrial efficiency gains.

Cahuc and Zylberberg (2014) also argue that low-educated workers are too costly relative to their added value. As a result, firms are willing to substitute low-educated workers by capital, to outsource part of their activities to cheap-labour countries and (especially in the case of excess labour supply) to hire more educated workers as their productivity to wage cost ratio is more favourable. Okumu and Buyinza (2018) also confirmed that education has a positive and significant effect on output among firms in Uganda. Mitana, Muwagga, Giacomazzi, Saint Kizito & Ariapa, (2019) also found that industries that employ well-educated workers have increased productivity. Bartelsman, Dobbelaere & Peters, (2015) also confirmed that the quantile return to educated labour corresponds to the marginal change in productivity due to a marginal change in the share of that type of workers being conditional on being in a firm belonging to the quantile of the overall outcome distribution (i.e., the outcome being labour productivity in their case).

Levinsohn and Petrin (2003) showed a robust upward-sloping profile between education and productivity; they also systematically highlighted that educational credentials have a stronger impact on productivity than on wage costs. Firms' profitability (i.e., productivity–wage gap) is indeed found to rise when lower educated workers are substituted by higher educated ones (and vice versa). Estimates thus support the existence of a 'wage-compression effect', i.e., a situation in which the distribution of wage costs is more compressed than workers' education–productivity profile (Vandenberghe & Lebedinski 2014). On the other hand, earnings differentials between workers with different levels of education do not reflect genuine industrial productivity differentials. This would explain why workers earn such large returns from investing in education, yet, at the same time, positive changes in a nation's stock of human capital have only a small impact on aggregate productivity.

6 CONCLUSION AND FURTHER RESEARCH

This paper examined the effect of electricity consumption, capital, labour, and education on industrial output in Uganda. The study results found strong evidence that electricity consumption, labour employed, and industrial output are positively correlated. To examine the study objectives, we incorporate several input variables into an augmented Cobb–Douglas production function. The results indicate that education, electricity consumption, and labour have a long-term causality on industrial output using the vector error correction model (VECM). Specifically, we find evidence that electricity is a necessary condition for an industrial and economic survival. When we estimate the augmented Cobb–Douglas production function with all variables (electricity consumption, educational, and labour employed), there is evidence that well educated workers are more productive. To investigate whether labour employed differentials reflect increased output differentials, we present a model which simultaneously estimates the labour productivity function and production function for workers and the firms where they are employed.

These results suggest that industries that consume electricity have higher output than their counterparts, likewise those that have well-educated workers with productive labour employed produce more because they contribute more to the firm's output. The key positive results that emerged from the findings are the important roles of electricity consumption, education, and the nature of the labour employed in agro-based industries. On average, the relative factor differentials, i.e., electricity consumed, education, labour employed, and industrial output differentials, between different industries are equivalent.

These results suggest that current efforts to improve the supply and affordability of the electricity are a facilitating condition for industrial output. Moreover, results suggest industrial-led electricity consumption policies can enhance industrial growth. Likewise, the results also suggest efforts to encourage agro-based industries to employ more well-educated workers to increase the industrial output. The need to understand how industrial electricity consumption can be improved calls for further studies that can employ qualitative approaches to examine the cost of power and the causes of power fluctuations in Uganda.

REFERENCES

Abokyi, E. A.-K.-A. 2018. *Consumption of Electricity and Industrial Growth in the Case of Ghana.* Journal of Energy.

Alley, I. E. 2016. *Electricity supply, industrialization and economic growth: evidence from Nigeria.* International Journal of Energy Sector Management, 10(4), 511–525.

Arranz-Piera, P. K. 2017. Electricity generation prospects from clustered smallholder and irrigated rice farms in Ghana. *Energy, 121*, 246–255.

Asumadu-Sarkodie, S. &. 2017. *The causal effect of carbon dioxide emissions, electricity consumption, economic growth, and industrialization in Sierra Leone.* Energy Sources, Part B: Economics, Planning, and Policy, 12(1), 32–39.

Asumadu-Sarkodie, S., & Owusu, P. A. 2017. *Recent evidence of the relationship between carbon dioxide emissions, energy use, GDP, and population in Ghana: a linear regression approach.* Energy Sources, Part B: Economics, Planning, and Policy, 12(6), 495-503.

Bartelsman, E., Dobbelaere, S., & Peters, B. 2015.*Allocation of human capital and innovation at the frontier: firm-level evidence on Germany and the Netherland's.* Industrial and Corporate Change, 24(5), 875–949.

Daubaraitė, U. &. 2015. *Creative industries impact on national economy in regard to sub-sectors.* Procedia-Social and Behavioral Sciences, 129–134.

Dogan, E. S. 2016. *Exploring the relationship between agricultural electricity consumption and output:* New evidence from Turkish regional data. Energy Policy, 95, 370–37.

Ettah Bassey Essien, G. E.-A. 2017. *Public Supply Of Electricity And Economic Growth In Nigeria.* Nternational Journal Of Social Sciences, 53.

Fermont, A. &. 2011. *Estimating yield of food crops grown by smallholder farmers.* International Food Policy Research Institute, Washington DC, 1–68.

Fox, J., & Smeets, V. 2011. Does input quality drive measured differences in firm productivity? *International Economic Review, 52*(4), 961–989.

Ghali, K. H. 1999. *Financial Development and Economic Growth: The Tunisian Experience. Review of Development Economics., 3(3)*, 310–322.

Harlan, T. 2018. *Rural utility to low-carbon industry: Small hydropower and the industrialization of renewable energy in China. Geoforum*, 59–69.

Izadmehr, M. D. 2018. *Determining influence of different factors on production optimization by developing production scenarios.* Journal of Petroleum Exploration and Production Technology, 8(2), 505–520.

Jebli, M. B. 2017. *The role of renewable energy and agriculture in reducing CO2 emissions: Evidence for North Africa countries. Ecological Indicators, 74*, 295–301.

Klein, R., Spady, R., & Weiss, A. 1991. *Factors affecting the output and quit propensities of production workers.* The Review of Economic Studies, 58(5), 929–953.

Kumar, R. D. (2016).*Agro-processing industries in Haryana: Status, problems and prospects.* Economic Affairs, 61(4), 707.

Li, G., Fang, C., Wang, S., & Sun, S. 2016. *The effect of economic growth, urbanization, and industrialization on fine particulate matter (PM2. 5) concentrations in China.* Environmental science & technology, 50(21), 11452–11459.

Liedholm, C. E. 2013. *Small enterprises and economic development: the dynamics of micro and small enterprises.* Routledge.

Lin, B. &. 2016. *Why is electricity consumption inconsistent with economic growth in China?* Energy Policy, 88, 310–316.

Lucas, R., 1988, *On the mechanics of economic development,* Journal of Monetary Economics 22, 3–42.

Mawejje, J. &. 2016. *Electricity consumption and sectoral output in Uganda: an empirical investigation.* Journal of Economic Structures, 5(1), 21.

Mitana, J. M. V., Muwagga, A. M., Giacomazzi, M., Saint Kizito, O., & Ariapa, M. 2019.*Assessing educational outcomes in the 21st century in Uganda: a focus on soft skills.* Journal of Emerging Trends in Educational Research and Policy Studies, 10(1), 62–70.

Mountjoy, A. B. 2017. *Industrialization and Underdeveloped Countries.* Routledge.
Nwajinka, C. C. 2013. *National Electric Energy Supply and Industrial Productivity in Nigeria from 1970 to 2010.* Journal of Economics and Sustainable Development, Vol.4, No.14, 2013, 1–129.
Okumu, I. M., & Bbaale, E. 2019. *Technical and vocational education and training in Uganda: A critical analysis.* Development Policy Review, 37(6), 735–749.
Okumu, I. M., & Buyinza, F. 2018. *Labour productivity among small-and medium-scale enterprises in Uganda: the role of innovation.* Journal of Innovation and Entrepreneurship, 7(1), 13.
Ouedraogo, N. S. 2017. *Africa energy future: Alternative scenarios and their implications for sustainable development strategies.* Energy Policy, 106, 457–471.
Rafał, K. (2014). Electricity Consumption and Economic Growth: Evidence from Poland. *Journal of International Studies, 7, No 1, 2014*, pp. 46–57.
Romer, P. M., & Griliches, Z. 1993. *Implementing a national technology strategy with self-organizing industry investment boards. Brookings Papers on Economic Activity.* Microeconomics, 1993(2), 345–399.
Rovas, D. &. 2015. *Exergy analysis of a small gasification-ICE integrated system for CHP production fueled with Mediterranean agro-food processing wastes: The SMARt-CHP.* Renewable energy, 83, 510–517.
Salahuddin, M. &. 2015. *Internet usage, electricity consumption and economic growth in Australia: A time series evidence.* Telematics and Informatics, 32(4), 862–878.
Salahuddin, M. &. 2016. Information and Communication Technology, electricity consumption and economic growth in OECD countries: A panel data analysis. *International Journal of Electrical Power & Energy Systems, 76*, 185–193.
Sanchez, L. F. 2016. *Drivers of industrial and non-industrial greenhouse gas emissions.* Ecological Economics, 17–24.
Shepherd, R. W. 2015. *Theory of cost and production functions.* Princeton University Press.
Solow, R., 1956, A Contribution to the Empirics of Economic growth. Quarterly Journal of Economics, 70, 65–94.
Ssozi, R. 2018. *The implications of the world trade organization (WTO) agreement on agriculture on trade and agricultural liberalization on food security in Uganda.* Doctoral dissertation, Kampala International University.
Syverson, C. 2011. *What determines productivity?* Journal of Economic Literature, 49(2), 326–365.
Toyin, M. E. 2017. *Agro-processing Output and Agricultural Sector Employment: Evidence from South Africa.* Acta Universitatis Danubius Œconomica, 13(2).
Ugwoke, T. I. 2016. *Electricity consumption and industrial production in Nigeria.* Journal of Policy and Development Studies, 289(3519), 1–12.
Vandenberghe, V. 2018. *The Contribution of Educated Workers to Firms' Efficiency Gains: The Key Role of Proximity to the 'Local' Frontier.* De Economist, 166(3), 259–283.
Vandenberghe, V., & Lebedinski, L. 2014. *Assessing education's contribution to productivity using firm-level evidence.* International Journal of Manpower, 35(8), 1116–1139.
World Bank. 2013. *Uganda Enterprise survey.* Kampala: World Bank Group.
World Bank. 2017. *State of Electricity Access Report (SEAR) 2017.* Kampala: World Bank Group.
Xin-gang, Z. &. 2018. Technological progress and industrial performance: A case study of solar photovoltaic industry. *Renewable and Sustainable Energy Reviews, 81*, 929–936.
Yousuo, A. J. 2013. National Electric Energy Supply and Industrial Productivity in Nigeria from 1970 to 2010. *Journal of Economics and Sustainable Development,* 4(14).

Advances in Phytochemistry, Textile and Renewable Energy Research for Industrial Growth – Nzila et al. (Eds)

Antibacterial efficacy of the aqueous and ethanolic extracted dyes from *Datura stramonium*, *Racinus communis*, and *Galinsoga parviflora* plant leaves

A. Musinguzi, J.I. Mwasiagi & C. Nzila
Moi University, Eldoret, Kenya

I. Nibikora
Busitema University, Tororo, Uganda

ABSTRACT: Traditionally, *Datura stramonium*, *Racinus communis*, and *Galinsoga perviflora* plant leaves are well-known for their healing powers against various human diseases and infections when applied orally and externally. They possess several phytochemicals which are key for their antibacterial and antifungal attributes. In this study, aqueous and ethanolic extraction methods were used to extract the dyes, which were investigated for their antibacterial activity using the disc diffusion method and minimum inhibition concentrations (MIC) by serial dilutions of the original plant dye extracts. Results showed a bigger zone of inhibition (8–18 mm) against both *Staphylococcus aureus* (ATCC 25923) and *Pseudomonas Aeruginosa* (ATCC 27853) bacterial strains with the aqueous method as compared to the ethanolic method (6–13 mm). The MIC was at 3 and 30 mg/mL for aqueous and ethanolic extracted dyes, respectively, whereas the controls (water and ethanol) were found to not influence the inhibition outcome, thus confirming the presence of antibacterial properties in all selected plants with aqueous being the best extraction method.

Keywords: Antibacterial activity, *pseudomonas aeruginosa*, *Staphylococcus aureus,* Aqueous and ethanolic extraction, medicinal plants

1 INTRODUCTION

In the last decade, the transmission of healthcare-associated infections and infectious diseases within the healthcare surroundings became the second leading cause of death worldwide (Hasara et al. 2019). Annually, it is estimated that 4 million people get infected in Europe and 7 million people in the USA (Adlhart et al. 2018). In most developing countries, surgical site-related infections include Gram-positive cocci like *Staphylococcus aureus*, *Staphylococcus epidermidis* among others, and Gram-negative bacilli such as *Escherichia coli* and *Pseudomonas aeruginosa*, which are all well-known for burns and wound infections, are on the rise (Pallavali et al. 2017). Recently, it has been proved that human pathogenic microorganisms have a tendency to develop resistance to various antibiotics, thus posing a therapeutic threat (Yavuz et al. 2017). In a bid to overcome such hindrances, antibiotic resistance inhibitors, mostly from the plant kingdom, have been used since they are well-known to contain a variety of phytochemicals and other compounds that are used to protect them from pathogens (Sen et al. 2012; (Thakare et al. 2016). They also have pharmacological properties and are economically viable compared to synthetic drugs (Atef, Shanab, Negm, & Abbas 2019). For a long time, various medicinal plants have been used as traditional treatments for different human diseases and infections (Marwat et al. 2017; Sayyed & Shah 2014; (Zameer & Yaqoob 2017). Therefore, to confirm this ethnotherapeutic claim, dyes extracted from *Datura stramonium, Racinus communis*, and *Galinsoga parviflora* plant leaves by aqueous and ethanolic methods were evaluated for antibacterial efficacy.

2 LITERATURE REVIEW

On Earth, the plant kingdom is estimated to contain around 250,000-500,000 different species of which a small portion of around 10% can be of use in the food sector (Shuaib et al. 2013). There are higher possibilities that a bigger portion of more than 10% could exist of plant species with medicinal properties (Malpani 2013). The ethnobotanical survey done by Nambejja, Tugume, Nyakoojo & Kamatenesi-mugisha, (2019) and Kamatenesi-Mugisha and Oryem-origa (2007) clearly shows that there are many different medicinal plant species specifically in Uganda which are used

DOI 10.1201/9781003221968-12

by people to treat several diseases and infections and others are used to induce child birth.

2.1 *Datura stramonium plant species*

Thorn apple or Jimsonweed are other names of *Datura stramonium* plant, which is mostly grown in temperate and subtropical regions. It belongs to the family Solanaceae and its leaves are irregularly undulate, soft, and toothed. The flowers are trumpet-shaped and white to creamy in colour (Ahad et al. 2012). The plant takes a year to grow from 0.3 to 1.5 m tall and is proved to have medicinal compounds such as tropane alkaloids which also find applications in defence industries (Iranbakhsh, Ebadi, & Bayat 2010). The plant has many other biologically active compounds such as atropine, scopolamine, tannin, saponin, glycosides, phenol, sterols, lignins, fats, carbohydrates, and proteins. All these listed phytochemicals are normally found in different parts of the plant (leaves, flowers, seeds, and roots) in varying quantities and concentrations although studies have shown that leaves have the highest concentration of these active principles (Gutarowska, Machnowski, & Kowzowicz 2013). Based on the extraction method used to obtain plant dye extracts, their bioactive ingredients may vary thus leading to variations in the antibacterial properties of the extract. A review on the pharmacological and toxicological aspects of *Datura stramonium* done by Gaire and Subedi (2013) clearly shows that ethanolic extracts demonstrated the highest inhibitory activity compared to aqueous and methanolic methods. Also, phytochemicals like saponins, steroids, alkaloids, and flavonoids among others are normally common in ethanolic and crude aqueous plant extracts which are responsible for the antibacterial activity (Sayyed & Shah 2014).

2.2 *Ricinus communis plant species*

Ricinus communis, the castor oil plant, is a species of flowering plant in the Euphorbiaceae family. It is a soft wooden small tree, widely spread all through the tropics and temperate regions of the world. Its leaves have long petiole and palm-like lobed blades, the flowers are categorised as male or female depending on their arrangement at the top of the axis in panicles form. The fruits of the plant are in three-chambered format, there is a globose capsule with soft spines such that after the capsules mature, they split open into three cavities, thus expelling the seeds (Marwat et al. 2017). The phytochemistry of the plant shows the presence of the amino acids, fatty acids, flavonoids, phenolic compounds, phytosterol, terpenoids, alkaloids, saponin, tannins, insecticidals, ovicidals, and cytotoxic activity (Jena & Gupta 2012). The plant also exhibits a number of pharmacological activities like anticonceptive activity, antidiabetic activity, antifertility effects, anti-inflammatory activity, and antimicrobial activity, thus demonstrating the required antibacterial properties against different bacterial strains. However, the degree of bacterial activity exhibited by these plant extracts is dependent on the solvent used during the extraction process. This was well explored by Jena and Gupta (2012), whereby hexane and methanol extracts showed the highest antimicrobial activity as opposed to other solvents like acetone, petroleum ether, and water. A related study by Marwat et al. (2017) further confirmed that the antibacterial activity shown by plant extracts primarily depends on the type of the solvent used. In this case, alcohol and water were used in varying concentrations.

2.3 *Galinsoga parviflora plant species*

The plant grows well on fertile soils that are sunny or shady and uncultivated. It normally grows up to 0.6 m high and belongs to the Asteraceae family. The flowers produced are small headed with yellow florist discs (Damalas 2008). The plant's flowering time is normally from May to September and it bears fruits with hairy achenes. The squeezed liquid/juice from the plant is traditionally well-known for treating wounds, whereas its roots give an excellent remedy against beetle bites. The phytochemistry of the plant shows the presence of flavonoids, aromatic esters, diterpenoids, caffeic acid derivatives, steroids, and phenolic acid derivatives among others (Samar Ali et al. 2017). The plant also exhibits a number of pharmacological activities like antibacterial, antifungal, antioxidant, nematicidal, anti-inflammatory, cytotoxic, urease, α-glucosidase, lipoxygenase, hepatoprotective, and hypoglycemic among others. All these bioactive ingredients can be obtained from different plant parts with the help of various solvents. A review done by Zameer and Yaqoob (2017) shows that hexane extracts of *Galinsoga parviflora* demonstrated better antibacterial activity against selected bacterial strains compared to methanol and water extracts.

2.4 *Medicinal dye extraction methods*

The extraction process can be interpreted in a number of different ways based on how it is done and the intended end use of the extracts among others. According to Colvin (2018), extraction is the separation of the medicinally active component from its parent source using selective solvents through suitable standard procedures. The extraction of natural dyes for antimicrobial activity from plants can be done in various ways depending on the character, the therapeutic value, and the stability of the drug. Examples of some of the methods are maceration, infusion, percolation and decoction, Soxhlet extraction or hot continuous extraction, ultrasound-assisted extraction (UAE) or sonication extraction, accelerated solvent extraction (ASE), and supercritical fluid extraction (SFE) (Jansirani, Saradha, Salomideborani, & Selvapriyadharshini 2014; Murugan & Parimelazhagan 2013; Nn 2015). Prior to the extraction process, it is important to know a number of possible basic parameters that can influence the quality of an extract mostly for the case of plants. These include the part of the plant used, the extraction solvent, and the extraction procedure, among others (Prashant Tiwari et al. 2011).

2.5 *Maceration process*

This technique has been developed mainly to be applied in the extraction of dyes from medicinal plants. Its operation covers the soaking of grounded plant materials in a solvent (i.e., ethanol, water, methanol, etc.) which is left to stand for a period of time considering temperature and agitation as key parameters in the process. Then the extract is filtered and evaporated in the rotavapour to obtain a dry mass of the extract (Colvin 2018). The most important role of using this extraction procedure for crude drugs is to obtain the therapeutically desirable portion and eliminate the inert material by treatment with a selective solvent commonly known as the Menstruum. The maceration method is wellknown as a traditional and conventional extraction process thus being easy and more convenient as compared to other methods (Colvin 2018).

2.6 *Infusion, percolation, and decoction processes*

The same principles are followed as for the case of maceration though with slight modifications. Infusion involves macerating the solids for a short time with cold or boiling water (Prashant Tiwari et al. 2011). In percolation the dried powdered samples are usually packed in the percolator, boiling water is added and they are macerated for 2 hours. Thereafter, the process continues at a moderate rate (e.g. 6 drops/min) until the extraction is complete before evaporation to get a concentrated extract. Decoction is only suitable for extracting heat-stable compounds from hard plants materials (e.g. roots and barks) and usually results in more oil-soluble compounds compared to maceration and infusion (Nn 2015a).

2.7 *Antimicrobial efficacy and Minimum Inhibition Concentration (MIC) of medicinal plant extracts*

Plants and their products have been in use in folk medicine for a long time and are linked with traditional medicine (Ajayi & Ojelere 2014). Since then, different plant species have been reported to have phytochemical and pharmacological properties with antimicrobial activities against a number of different microorganisms (Alapati 2015; Gaire & Subedi 2013; McArthur, Tuckfield & Baker-Austin 2012; Sayyed & Shah 2014; Zameer & Yaqoob 2017). Considering the thousands of different herbal plant species in existence worldwide, studies done by Das, Tiwari & Shrivastava (2010) clearly show that a small portion of them have been assessed for both phytochemicals and pharmacological properties and yet there is still a very big portion of plants not investigated yet, hence there is an urgent need for efficient, simpler, and cost-effective methods to evaluate the efficacy of herbal plant extracts as well as their MICs. Sakha, Hora, Shrestha, Acharya, and Dhakal (2019) evaluated the antimicrobial activity of ethanolic extract from medicinal plant parts (like seeds, buds, and leaves) against various pathogenic bacteria using disc diffusion method. Then MICs of the extracts were investigated using micro broth dilution. The results exhibited good antimicrobial activity against bacterial strains though the extracts from leaves of different plants showed better inhibition as compared to extracts from other plant parts (seeds and bud). The MICs were found to be 12.5–25 mg/mL. Similarly, Batra (2012) used the agar well-diffusion method and serial dilution technique (96-well microliter plates) to evaluate the antimicrobial activity and MIC, respectively, of different solvent extracts from *Melia azedarach l* plant. The results clearly showed that all the solvents used resulted in extracts having significant activity against the bacterial strains, although ethanolic extracts demonstrated a better inhibition zone compared to others (methanol, ethanol, petroleum ether, and water). For the case of MIC, petroleum ether and aqueous extracts demonstrated the lowest activity against the bacterial strains compared to the rest. There are other recent innovative techniques which are used to evaluate the antimicrobial efficacy of different plant extracts and their MIC as discussed by different researchers (Balouiri, Sadiki, & Ibnsouda 2016; Elisha, Botha, Mcgaw, & Eloff 2017; Eloff 2019; Ohikhena, Wintola & Afolayan 2017; Zeeshan A. Khan 2019). These techniques are not commonly used since they require specified equipment and more assessment to achieve reproducible and standardised results. Also, it was noted that these techniques can be influenced by many factors like inoculum size, selection of positive controls among others.

3 MATERIALS AND METHODS

3.1 *Collection of medicinal plant leaves*

On the basis of their traditional uses, phytochemicals and pharmacological properties reported in the literature, fresh and mature *Datura stramonium* (DS), *Ricinus communis* (RC), and *Galinsonga parviflora* (GP) plant leaves were considered for this study. They were collected from the wild in Biharwe, Mbarara District, Uganda, washed thoroughly, and allowed to dry under the shed.

3.2 *Source of microorganisms*

Staphylococcus aureus (ATCC 25923) and *Pseudomonas aeruginosa* (ATCC 27853) bacterial strains were obtained from Microbiology lab, Busitema University, Mbale campus, Uganda. They were recovered from the storage media following the manufacturer's standard operating procedures and identified as per Atef et al. (2019).

3.3 *Preparation of medicinal plant extracts*

The extraction process was done according to Jansirani et al. (2014) with modifications. Briefly, the dried mature plant leaves of the selected plants were ground into moderately course powder with the help of a pestle and mortar, weighed using a digital balance, and subjected to aqueous and ethanolic extraction by a maceration process keeping the material to liquor ratio (MLR) of 15:150 w/v and 15:150 w/v, respectively. After the powder was soaked in different solvents (distilled water and 99% ethanol) for 24 hours at 50°C with occasional stirring, the extract was filtered through Whatman No.1 filter paper to obtain a clear first filtrate. The second filtrate was obtained by soaking the used grounded leaves in both solvents (water and ethanol) while keeping the same temperature and MLRs for 24 hours. Both filtrates were mixed and concentrated in a rotavapour in controlled conditions to 300 mg/mL and stored in an airtight container at 4°C for further studies.

3.4 *Preparation of inoculum*

Each bacterial strain was sub-cultured in Mueller Hinton agar slats at 37°C overnight. The bacterial isolated colonies were picked and dissolved in the 1 mL of the sterile normal saline and their turbidity adjusted to the equivalent of 0.5 McFarland standard solution (1×10^8 CFU/mL).

3.5 *Antibacterial activity of the plant extracts*

The plant extracts were tested for their antibacterial activity by using Disc diffusion method following the Kirby-Bauer Test Protocol (Hudzicki 2009) that is in line with recommended standards of Clinical and Laboratory Standards Institute (CLSI) (Jean B. Patel et al. 2015).

3.5.1 *Disc diffusion method*

The Mueller Hinton agar plates were prepared by pouring 20 mL of molten media on to sterile Petri dishes at 56°C and they were left to solidify. Thereafter, they were labelled and overnight incubated for sterility. Using sterile cotton swabs, 0.1% suspension of each bacterial strain was spread evenly over the Mueller Hinton agar plates and six depressions were created in the agar medium at a spacing of 2 cm apart with non-toxic pipette tips and filled with 0.1 mL (300 mg/mL) of plant extract in each well. Standard agar plates with Ciproflaxin (30 μg) and Amikacin (30 μg) were considered as positive control for the bacteria while distilled water and 99% ethanol was used as the negative control. All the plates were left to stand for 45 minutes for the diffusion of the extracts into the medium. Then they were covered and incubated aerobically at 37°C overnight. All the tests were performed in triplicate. The antibacterial activity of the extracts was then evaluated basing on the inhibition zones measured using an inhibition zone ruler that was placed edge to edge across the zone of inhibition over the plate disc and the mean diameters recorded in millimetres.

3.5.2 *Minimum Inhibition Concentration (MIC) determination of the aqueous and ethanolic extracts*

Minimum Inhibitory Concentration is the lowest concentration that can inhibit any growth of bacteria on the prepared culture plates (Batra 2012), It was determined for the aqueous and ethanolic extracts using serial dilutions as described by Fg et al. (2016) with slight modifications. Serial dilutions of the aqueous and ethanolic extracts of *Datura stramonium* (DS), *Ricinus communis* (RC), and *Galinsonga parviflora* (GP) were prepared at various concentrations (30, 3, 0.3 and 0.03 mg/mL) of extracts and put in 5 cm test tubes. Using non-toxic pipette tips, depressions were created in overnight incubated Mueller Hinton agar plates. Then 0.1 mL of different concentrations (30, 3, 0.3, and 0.03 mg/mL) of each extract were added to each depression on the Mueller Hinton agar plates and incubated overnight at 37°C. The lowest concentration that inhibited the growth of microorganisms was considered to be the minimum inhibitory concentration.

4 RESULTS

4.1 *Antibacterial efficacy of extracts from selected medicinal plants*

4.1.1 *Disc diffusion assay*

The results in Table 1 and Figures 1 and 2 show that the aqueous and ethanolic extracts of selected medicinal plants exhibited varied antibacterial activity. Based on the mean diameters of the inhibition zones, *Racinus Communis* aqueous dye extracts exhibited higher antibacterial activity against *Staphylococcus aureus* and *Pseudomonas* spp. bacterial strains compared to ethanolic dye extracts. Then *Galinsoga parviflora* aqueous extracts also showed better antibacterial activity as opposed to that demonstrated by ethanolic extracts. The trend was not different for *Datura stramonium* where its aqueous dye extracts demonstrated higher activity in comparison to ethanolic extracts. It was also clearly shown that for all selected plants, their aqueous and ethanolic dye extracts are more susceptible to Gram-positive bacteria (*Staphylococcus aureus*) than to Gram-negative bacteria (*Pseudomonas* spp).

4.1.2 *Antibacterial activity of the controls*

The control experiment demonstrated that there was no influence in the recorded inhibition zones of both aqueous and ethanolic extracts of different plants against used Gram-positive and Gram-negative bacterial strains, as clearly shown in Figure 3 and Table 2.

Table 1. Antibacterial efficacy of the extracts from selected medicinal plants.

			Mean diameter of zone of inhibition (mm)			
			Aqueous extract		Ethanolic extract	
Plant species	Plant family	Part used	*Staphylococcus aureus*	*Pseudomonas aeruginosa*	*Staphylococcus aureus*	*Pseudomonas aeruginosa*
Galinsoga parviflora (GP)	Asteraceae	Leaves	14	10	11	7
Racinus communis (RC)	Spurge	Leaves	18	14	13	8
Datura stramonium (DS)	Solanaceae	Leaves	12	8	10	6

Antibacterial activity: <6 mm: Weak; 7–12 mm: Moderate; >12 mm: Strong

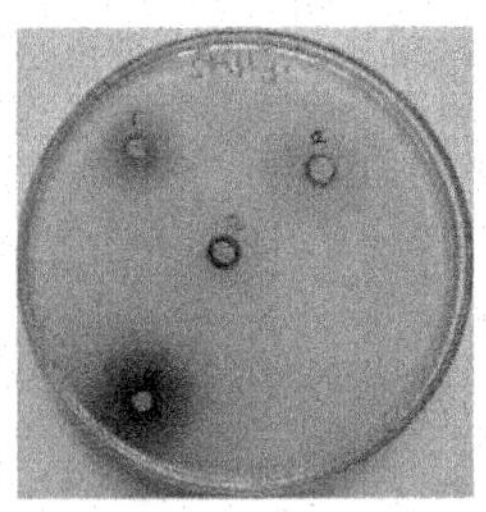

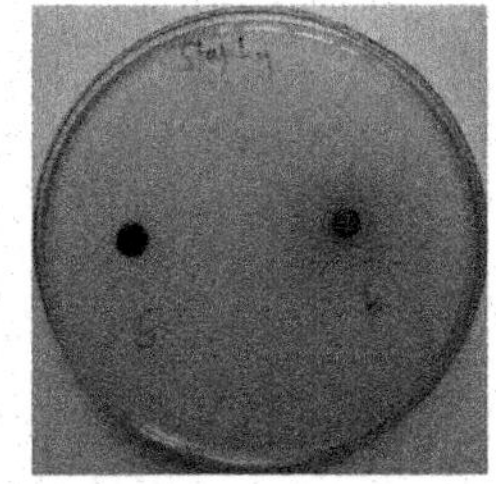

Key: 1 – Aqueous extract DS, 3 – Ethanolic extract DS, 2 - Ethanolic extract RC, 4 – Aqueous extract RC, 5 – Ethanolic extract GP, 6 – Aqueous extract GP.

Figure 1. Shows the antibacterial activity of Aqueous and Ethanolic extracts from selected plants against *Staphylococcus aureus* bacterial strain.

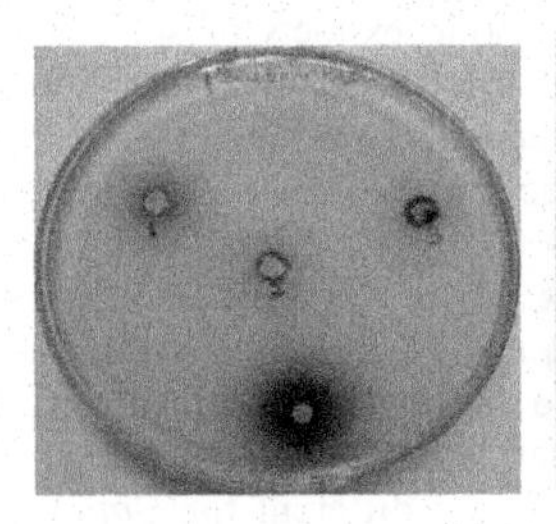

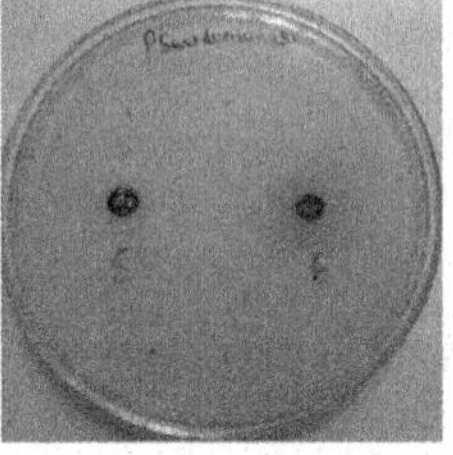

Key: 1 – Aqueous extract DS, 3 – Ethanolic extract DS, 2 – Ethanolic extract RC, 4 – Aqueous extract RC, 5 – Ethanolic extract GP, 6 – Aqueous extract GP.

Figure 2. Shows the antibacterial activity of Aqueous and Ethanolic extracts from selected plants against *Pseudomonas Aeruginosa* bacterial strain.

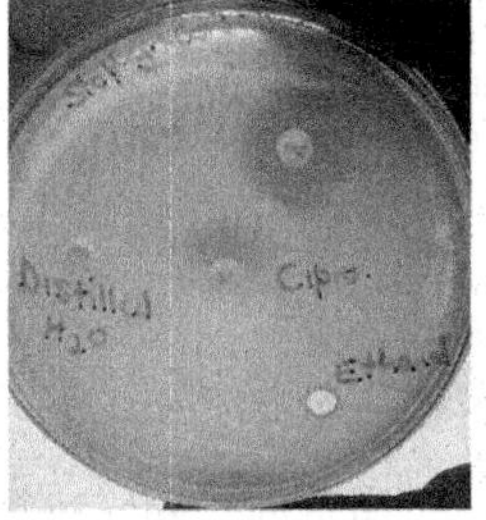

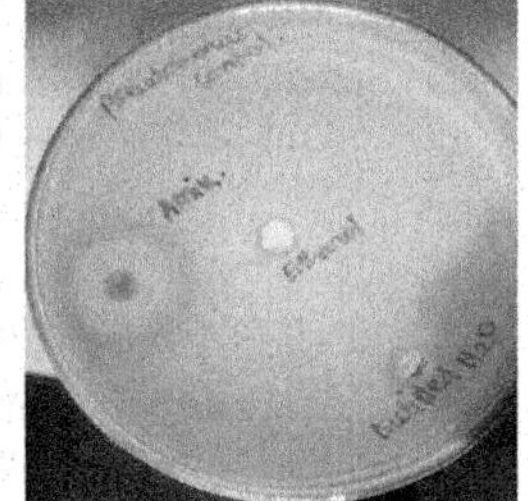

Figure 3. Antibacterial Activity of pure solvents and selected standard antibiotics.

Table 2. Antibacterial activity of controls.

	Zone of inhibition (mm)	
Controls	*Staphylococcus aureus*	*Pseudomonas aeruginosa*
Ciproflaxin (30 μg)	22	20
Distilled water (100 μL)	–	–
Ethanol (100 μL)	–	–

Antibacterial activity: Staphylococcus aureus; ≥22 mm, Pseudomonas Aeruginosa; ≥15 mm.

4.2 *Minimum Inhibition Concentration (MIC) of the aqueous and ethanolic herbal extracts from selected plants against both Staphylococcus aureus and Pseudomonas aeruginosa bacterial strains*

The MIC values obtained from plant extracts demonstrated antibacterial activity which ranged from 3 to 30 mg/mL (Table 3). Aqueous and ethanolic extracts of *Datura stramonium* gave MIC values of 30 mg/mL and 30 mg/mL, respectively, against both *Staphylococcus* and *Pseudomonas* bacterial strains. Ethanolic extracts of *Galinsoga parviflora* gave MIC values of 30 mg/mL against both *Staphylococcus* and *Pseudomonas* bacterial strains and none was observed for aqueous extracts. Then, aqueous and ethanolic extracts of *Racinus communis* gave MIC values of 3 mg/mL and 30 mg/mL, respectively, against both bacterial strains. Generally, aqueous extracts were observed to have the most effective antimicrobial activity compared to ethanolic extracts.

Table 3. Minimal Inhibition Concentration (MIC) of aqueous and ethanolic extracts from selected plants against *Staphylococcus aureus* and *Pseudomonas aeruginosa* bacterial strains.

Plant specie	Bacteria specie	Extract type	Minimum Inhibition Concentration (mg/mL)		
			30	3	0.3
Datura Stramonium	*Staphylococcus*	Aqueous	+	–	–
	Pseudomonas	Aqueous	+	–	–
Galinsoga Parviflora	*Staphylococcus*	Aqueous	–	–	–
	Pseudomonas	Aqueous	–	–	–
Racinus Communis	*Staphylococcus*	Aqueous	+	+	–
	Pseudomonas	Aqueous	+	+	–
Datura Stramonium	*Staphylococcus*	Ethanolic	+	–	–
	Pseudomonas	Ethanolic	+	–	–
Galinsoga Parviflora	*Staphylococcus*	Ethanolic	+	–	–
	Pseudomonas	Ethanolic	+	–	–
Racinus Communis	*Staphylococcus*	Ethanolic	+	–	–
	Pseudomonas	Ethanolic	+	–	–

'–' No bacterial activity, '+' Bacterial activity present

5 DISCUSSIONS

5.1 *Antibacterial efficacy of aqueous and ethanolic plant extracts*

With the increasing search for suitable antimicrobial agents that can replace the existing synthetic ones, natural resources like plants have been the major focus since they have been proven to have phytochemical properties that are capable of controlling the growth of microorganisms without having any toxic effects and they are eco-friendly (Akanmu, Bulama, Balogun, & Musa 2019; Fg et al. 2016; Mansour, Saif, & Al-fakih 2017). In the present study, the antibacterial activity of different plant extracts was evaluated against Gram-positive and Gram-negative bacterial strains. It was indicated (Table 1) that all extracts inhibited bacterial growth, although with variation. This may be attributed to the varying phytochemicals that could have been contained within the extracts that are capable of imparting antibacterial activities via various mechanisms. The findings done by Mansour et al. (2017) on phytochemical screening of plant extracts revealed that there are mainly five phytochemicals (flavonoids, tannins, alkaloids, glycosides, and terpenoids) that are found naturally in most plants and are well-known for their biological activities, which include bactericidal, fungicidal action among others, thus conferring the antibacterial activity of the analysed aqueous and ethanolic plant extracts. Also, more studies have confirmed that in addition to the five main phytochemicals responsible for bacterial activity against different bacterial strains, there are others like phenols, lignins, saponins, and sterols that are mainly found in plant leaves (Ali et al. 2017; Sayyed & Shah 2014). All these compounds have been found to have also pharmacological properties like antidiabetic activity, anti-inflammatory activity, antimicrobial activity, among others (Marwat, Khan, & Baloch 2017).

The current study further revealed that aqueous extracts of all plants gave improved resistance, thus a bigger zone of inhibition, ranging from 8–18 mm against both *Staphylococcus aureus* and *Pseudomonas aeruginosa* bacterial strains, compared to ethanolic extracts with 6–13 mm. This result differs from what is well-known from various studies (Udochukwu et al. 2015; Yu et al. 2014) where organic extracts showed a greater or same activity than aqueous extracts. This could be attributed to the chemical composition of plants and variations in the ability of the solvent to dissolve the grounded plant leaves, thus influencing the phytochemicals extracted. In related findings, Oluwajobi et al. (2019) evaluated the antibacterial and antifungal activities of aqueous and methanol leaf extracts from *Psidium guajava, Vernonia amygdalina*, and *Azadiracta indica* against different microbial isolates. The obtained range of zone of inhibition for aqueous extracts against both bacteria and fungi was greater than for methanol extracts. This was so due to the presence of all the analysed phytochemicals including tannins which were found in large quantities as opposed to methanol extracts where glycoside and anthraquinone were found to be missing. Furthermore, Kadi et al. (2011) used the disc diffusion test to evaluate the antibacterial activity of ethanolic and aqueous extracts from *Punica granatum L.* bark against different bacterial strains. The results showed aqueous macerate extracts to have higher antibacterial activity than ethanolic macerate extracts. This is agreement with the findings of Atef et al. (2019) where water extracts of *M. oleifera* and *M. recutita* plants demonstrated better activity against the selected sensitive isolates compared to ethanol extracts from both plants.

The study also demonstrated that there was higher activity against *Staphylococcus aureus* (Gram-positive) bacteria for all plant extracts (aqueous and ethanolic) compared to *Pseudomonas aeruginosa*

(Gram-negative) bacteria. This variation could have been attributed to differences in the morphological nature of these microorganisms. Normally, Gram-negative bacteria contain an outer phospholipid membrane with structural lipopolysaccharide parts, thus making the cell wall impermeable to several antimicrobial agents, whereas Gram-positive bacteria contain an outer peptidoglycan membrane which is more permeable, and thus more susceptible to many antimicrobial agents. This is consistent with Hiroshi Nikaido and Marti Vaara's (1985) findings on the molecular basis of bacterial outer membrane permeability and further explained by Hasara et al. (2019) in their study on medicinal plants against some human pathogenic bacteria. Also, Chanda and Baravalia, (2010) evaluated the antioxidant and antimicrobial potentiality of different medicinal plants against various skin diseases which are a result of different bacteria and fungi. The results obtained showed that indeed the antibacterial activities of *C. longa* and *C. amada* were higher on Gram-positive bacteria (*S. aureus* and *B. subtilis*) compared to Gram-negative bacteria.

On the control experiment, it was confirmed that the solvents (water and ethanol) used in extraction process had no any effect on the bacterial activity against the selected bacterial strains, compared to Ciproflaxin and Amikacin standard antibiotics which had antimicrobial influence on the tested microorganisms. These findings were in agreement with previous research done by Yavuz et al. (2017) where dimethyl sulfoxide was found to have no antimicrobial effect, compared with ceftriaxone and gentamicin which affected the bacterial strains used.

To accurately compare the effectiveness of different plant extraction methods concerning their respective inhibition zone diameters obtained from disc diffusion tests, MIC was employed and the values obtained demonstrated slight variation in antibacterial activity from all the selected plants. This variation could have been a result of the different solvents used in the extraction processes, and the chemical and volatile nature of different plant constituents among other factors. In the study by Mostafa et al. (2018), who used the disc diffusion method to evaluate the efficiency of the most effective plant extracts that had shown better antibacterial activity against food poisoning pathogens, the results also showed variation in MIC values of the different plant extracts considered. Furthermore, the present study demonstrated that *Galinsoga parviflora* ethanolic extract had activity (MIC) at 30 mg/mL against the tested bacterial strains whereas aqueous extracts didn't have any activity. This negative result may not necessarily mean that there are no bioactive constituents or the plant is inactive. The active agents may have been available in small quantities and thus unable to show any activity with the extract concentration employed (Chanda & Baravalia 2010). On the other hand, *Racinus communis* aqueous extract showed activity (MIC) at 3 mg/mL against tested bacterial strains which is the lowest compared to that obtained with ethanolic extracts, thus demonstrating the highest bacteriostatic activity. This dispersion could have been attributed to the presence of various phytochemicals, like phytate, cyanogenic glycosides, among others, which are capable of dissolving completely in water and are slightly or sometimes insoluble in ethanol (Udochukwu et al. 2015). In addition to that, this result is in agreement with previous studies (Akanmu et al. 2019; Fg et al. 2016).

6 CONCLUSIONS AND RECOMMENDATIONS

This study has shown that, all the extracts (aqueous and ethanolic) of *Datura stramonium, Racinus communis,* and *Galinsoga parviflora* medicinal plant leaves demonstrated bacteriostatic activity against the selected bacterial strains even though aqueous extracts proved to be better. High susceptibility of the plant extracts to Gram-positive bacteria was confirmed and based on the concentration used, their antibacterial activities were significantly lower than those of ciproflaxin and amikacin standard antibiotics. It has also proven that *Racinus communis* leaf extract was more potent against bacterial strains than other plants tested. These plants are known to possess various phytochemicals and pharmacological properties that contribute to their potential activity against different pathogens. This also justifies the persistent use of these herbs in traditional medicine. Further attention and research should be on using a mixture of water and ethanol at varying ratios during the extraction process, identifying the phytochemical constituents obtained, and evaluating their antimicrobial properties against Gram-positive and Gram-negative bacterial strains.

ACKNOWLEDGMENT

My acknowledgment goes to Busitema University in collaboration with the African Centre of Excellence II in Phytochemicals, Textile and Renewable Energy (ACE II PTRE) at Moi University for financial support. On the same note, I am thankful to Dr. Iramoit Jacob and his team who guided me through several experiments in the Microbiology labs, Busitema University-Mbale Campus.

REFERENCES

Adlhart, C., Verran, J., Azevedo, N. F., Olmez, H., Minna, M., & Gouveia, I., …Crijns, F. (2018). Surface Modifications For Antimicrobial Effects in the Healthcare Setting: A Critical Overview. *Journal of Hospital Infection*, (1), 1–29. https://doi.org/10.1016/j.jhin.2018.01.018

Ahad, H. A., U, A. B., Nagesh, K., D, S. K., & K, B. M. (2012). Fabrication of Glimepiride Datura Stramonium Leaves Mucilage and Polyvinyl Pyrolidone Sustained Release Matrix tablets: In Vitro Evaluation. *Journal of Science and Technology*, *8*(I), 63–72.

Ajayi, I. A., & Ojelere, O. (2014). Evaluation of the Antimicrobial Properties of the Ethanolic Extracts of some Medicinal Plant Seeds from South-West Nigeria. *Journal of Pharmacy and Biological Sciences*, *9*(4), 80–85.

Akanmu, A. O., Bulama, Y. A., Balogun, S. T., & Musa, S. (2019a). Antibacterial activities of aqueous and methanol leaf extracts of Solanum incanum Linn (Solanaceae) against multi-drug resistant bacterial isolates. *African Journal of Microbiology Research*, *13*(4), 70–76. https://doi.org/10.5897/AJMR2018.8969

Akanmu, A. O., Bulama, Y. A., Balogun, S. T., & Musa, S. (2019b). Antibacterial activities of aqueous and methanol leaf extracts of Solanum incanum Linn (Solanaceae) against multi-drug resistant bacterial isolates. *African Journal of Biotechnology*, *13*(4), 70–76. https://doi.org/10.5897/AJMR2018.8969

Alapati, P. and K. sulthana S. (2015). Phytochemical Screening of 20 Plant Sources for Textiles Finishing. *International Journal of Advanced Research*, *3*(10), 1391–1398.

Antara Sen and Amlabatra. (2012). Evaluation of Antimicrobial Activity of Different solvent extracts of Medicinal Plant: Melia Azedarach. L. *International Journal of Current Pharmaceutical Research*, *4*(2), 1–7.

Atef, N. M., Shanab, S. M., Negm, S. I., & Abbas, Y. A. (2019). Evaluation of antimicrobial activity of some plant extracts against antibiotic susceptible and resistant bacterial strains causing wound infection. *Bulletin of the National Research Centre*, *9*, 1–11.

Balouiri, M., Sadiki, M., & Ibnsouda, S. K. (2016). Methods for in vitro evaluating antimicrobial activity: A review. *Journal of Pharmaceutical Analysis*, *6*(2), 71–79. https://doi.org/10.1016/j.jpha.2015.11.005

Batra, A. S. and A. (2012). Evaluation of Antimicrobial Activity of Different solvent extracts of Medicinal Plant: Melia Azedarach L. *International Journal of Current Pharmaceutical Research*, *4*(2).

Chanda, S., & Baravalia, Y. (2010). Screening of some plant extracts against some skin diseases caused by oxidative stress and microorganisms. *African Journal of Biotechnology*, *9*(21), 1–9.

Colvin, D. M. (2018). A Review on Comparison of the Extraction Methods Used in Licorice Root: Their Principle , Strength and Limitation. *Medicinal & Aromatic Plants*, *7*(6), 1–4. https://doi.org/10.4172/2167-0412.1000323

Damalas, C. A. (2008). Distribution , biology , and agricultural importance of Galinsoga parviflora (Asteraceae). *Weed Biology and Management*, *153*(8), 147–153. https://doi.org/10.1111/j.1445-6664.2008.00290.x

Das, K., Tiwari, R. K. S., & Shrivastava, D. K. (2010). Techniques for evaluation of medicinal plant products as antimicrobial agent: Current methods and future trends. *Journal of Medicinal Plants Research*, *4*(2), 104–111. https://doi.org/10.5897/JMPR09.030

Elisha, I. L., Botha, F. S., Mcgaw, L. J., & Eloff, J. N. (2017). The antibacterial activity of extracts of nine plant species with good activity against Escherichia coli against five other bacteria and cytotoxicity of extracts. *BMC Complementary and Alternative Medicine*, *17*(133), 1–10. https://doi.org/10.1186/s12906-017-1645-z

Eloff, J. N. (2019). Avoiding pitfalls in determining antimicrobial activity of plant extracts and publishing the results. *BMC Complementary and Alternative Medicine*, *3*, 1–8.

Fg, M., Dw, N., Mw, A., Mp, N., Pk, G., Enm, N., & Jjn, N. (2016). Antimicrobial Activity of Aqueous Extracts of Maytemus putterlickoides , Senna spectabilis and Olinia usambarensis on Selected Diarrhea-Causing Bacteria. *Journal of Bacteriology and Parasitology*, *7*(2), 1–6. https://doi.org/10.4172/2155-9597.1000270

Gaire, B. P., & Subedi, L. (2013). A review on the pharmacological and toxicological aspects of Datura stramonium L. *Journal Integrative Medicine*, (3), 1–8. https://doi.org/10.3736/jintegrmed2013016

Gutarowska, B., Machnowski, W., & Kowzowicz, Ł. (2013). Antimicrobial activity of textiles with selected dyes and finishing agents used in the textile industry. *Fibers and Polymers*, *14*(3), 415–422. https://doi.org/10.1007/s12221-013-0415-x

Hasara, M., Zoysa, N. De, Rathnayake, H., Hewawasam, R. P., Mudiyanselage, W., Gaya, D., & Wijayaratne, B. (2019). Determination of In Vitro Antimicrobial Activity of Five Sri Lankan Medicinal Plants against Selected Human Pathogenic Bacteria. *International Journal of Microbiology*, *2019*(5), 1–9.

Hiroshi Nikaido and Marti Vaara. (1985). Molecular Basis of Bacterial Outer Membrane Permeability. *Microbiological Reviews*, *49*(1), 1–32.

Hudzicki, J. (2009). *Kirby-Bauer Disk Diffusion Susceptibility Test Protocol*.

Iranbakhsh, A., Ebadi, M., & Bayat, M. (2010). The Inhibitory Effects of Plant Methanolic Extract of Datura stramonium L. and Leaf Explant Callus Against Bacteria and Fungi. *Global Veterinaria*, *4*(2), 149–155.

Iyanuloluwa, O., Adamu, K. Y., & Audu, J. A. (2019). Antibacterial and Antifungal activities of aqueous leaves extract of some medicinal plants. *GSC Biological and Pharmaceutical Sciences*, *09*(01), 62–69.

Jansirani, D., Saradha, R., Salomideborani, N., & Selvapriyadharshini, J. (2014). Comparative evaluation of various extraction methods of curcuminoid from curcuma longa. In *Journal of chemical and Pharmaceutical sciences* (pp. 286–288).

Jean B. Patel, Franklin R. Cockerill, G. M. E. J. A. H. (2015). CLSI (Clinical and Laboratory Standards Institute); Methods for Dilution Antimicrobial Susceptibility Tests for Bacteria That Grow Aerobically; Approved Standard—10th Edition. National Committee for Clinical Laboratory Standards, Wayne, USA, 2015, 88., (January).

Jena, J., & Gupta, A. K. (2012). Ricinus communis linn: A phytopharmacological review. *International Journal of Pharmacy and Pharmaceutical Science*, *4*(4), 1–6.

Kadi, H., Moussaoui, A., Benmehdi, H., Lazouni, H. A., Benayahia, A., & Nahal, N. (2011). Antibacterial activity of ethanolic and aqueous extracts of Punica granatum L. bark. *Journal of Applied Pharmaceutical Science*, *01*(10), 180–182.

Kamatenesi-mugisha, M., & Oryem-origa, H. (2007). Medicinal plants used to induce labour during childbirth in western Uganda. *Journal of Ethnopharmacology*, *109*(May 2006), 1–9. https://doi.org/10.1016/j.jep.2006.06.011

Malpani, S. R. (2013). Antibacterial Treatment on Cotton Fabric from Neem Oil , Aloe vera & Tulsi. *International Journal of Advanced Research in Science and Engineering*, *8354*(2), 35–43.

Mansour, M., Saif, S., & Al-fakih, A. A. (2017). Antibacterial activity of selected plant (Aqueous and methanolic) extracts against some pathogenic bacteria. *Journal of Pharmacognosy and Phytochemistry*, *6*(6), 1929–1935.

Marwat, S. K., Khan, E. A., & Baloch, M. S. (2017). Ricinus cmmunis: Ethnomedicinal uses and pharmacological activities. *Pakistan Journal of Pharmaceutical Sciences*, (9), 1–14.

McArthur, J. V., Tuckfield, R. C., & Baker-Austin, C. (2012). Antimicrobial textiles. In *Handbook of Experimental Pharmacology* (Vol. 211, pp. 135–152). Elsevier Ltd. https://doi.org/10.1007/978-3-642-28951-4-9

Mostafa, A. A., Al-askar, A. A., Almaary, K. S., Dawoud, T. M., Sholkamy, E. N., & Bakri, M. M. (2018). Antimicrobial activity of some plant extracts against bacterial strains causing food poisoning diseases. *Saudi Journal of Biological Sciences*, *25*(2), 361–366. https://doi.org/10.1016/j.sjbs.2017.02.004

Murugan, R., & Parimelazhagan, T. (2013). Comparative evaluation of different extraction methods for antioxidant and anti-inflammatory properties from Osbeckia parvifolia Arn. – An in vitro approach. *Journal of King Saud University – Science*, (9), 1–9. https://doi.org/10.1016/j.jksus.2013.09.006

Nambejja, C., Tugume, P., Nyakoojo, C., & Kamatenesi-mugisha, M. (2019). Medicinal plant species used in the treatment of skin diseases in Katabi Sub- County , Wakiso District , Uganda. *Ethnobotany Research and Applications*, (7), 1–18. https://doi.org/10.32859/era.18.20.1-17

Nn, A. (2015a). A Review on the Extraction Methods Use in Medicinal Plants , Principle , Strength and Limitation. *Medical and Healthcare Textiles*, *4*(3), 3–8. https://doi.org/10.4172/2167-0412.1000196

Nn, A. (2015b). Review on the Extraction Methods Use in Medicinal Plants, Principle, Strength and Limitation. *Medicinal & Aromatic Plants*, *4*(3), 3–8. https://doi.org/10.4172/2167-0412.1000196

Ohikhena, F. U., Wintola, O. A., & Afolayan, A. J. (2017). Evaluation of the Antibacterial and Antifungal Properties of Phragmanthera capitata (Sprengel) Balle (Loranthaceae), a Mistletoe Growing on Rubber Tree, Using the Dilution Techniques. *Scientific World Journal*, *2017*, 1–9.

Pallavali, R. R., Degati, V. L., Lomada, D., Reddy, C., Raghava, V., & Durbaka, P. (2017). Isolation and in vitro evaluation of bacteriophages against MDR-bacterial isolates from septic wound infections. *Plos One*, (7), 1–16.

Prashant Tiwari, Bimlesh Kumar, Mandeep Kaur, Gurpreet Kaur, H. K. (2011). Phytochemical screening and Extraction: A Review. *International Pharmaceutica Sciencia*, *1*(1), 1–9.

Sakha, H., Hora, R., Shrestha, S., Acharya, S., & Dhakal, Dinesh, Srijana Thapaliya, K. P. (2019). Antimicrobial Activity of Ethanolic Extract of Medicinal Plants against Human Pathogenic Bacteria. *TUJM*, *5*(1), 1–6.

Samar Ali, Sara Zameer, M. Y. (2017). Ethnobotanical , phytochemical and pharmacological properties of Galinsoga parviflora: a review. *Tropical Journal of Pharmaceutical Research*, *3*(6), 274–279.

Sayyed, A., & Shah, M. (2014). Phytochemistry , pharmacological and traditional uses of Datura stramonium L. *Journal of Pharmacognosy and Phytochemistry*, *2*(5), 123–125.

Shuaib, M., Ali, A., Ali, M., Panda, B. P., & Ahmad, M. I. (2013). Antibacterial activity of resin rich plant extracts. *Journal of Pharmacy and Bioallied Sciences*, *5*(4), 265–270. https://doi.org/10.4103/0975-7406.120073

Thakare, P. V, Sharma, R. R., & Ghanwate, N. A. (2016). Antimicrobial Activity and Phytochemical Screening of Methanol Extract of Chlorophytum kolhapurens and Chlorophytum baruchii. *International Journal of Pharmacognosy and Phytochemical Research*, *8*(5), 794–799.

Udochukwu, U., Omeje, F. I., Uloma, I. S., & Oseiwe, F. D. (2015). Phytochemical Analysis of Vernonia amygdalina and Ocimum gratissimum Extracts and their Antibacterial Activity on some Drug Resistant Bacteria. *American Journal of Research Communication*, *3*(5), 225–235.

Yavuz, C., Kiliç, D. D., Ayar, A., & Yildirim, T. (2017). Antibacterial Effects of Methanol Extracts of Some Plant Species Belonging to Lamiaceae Family. *International Journal of Secondary Metabolite*, *4*(2), 429–433. https://doi.org/10.21448/ijsm.376691

Yu, H., Wang, Y., Wang, X., Sun, S., & Peng Shuai. (2014). Antibacterial Activity of Aqueous and Ethanolic Extracts of Portulaca Oleracea L and Taraxacum Mongolicum Against Pathogenic Bacteria of Cow Mastitis. *Intern J Appl Res Vet Med*, *12*(3), 210–213.

Zameer, S., & Yaqoob, M. (2017). Ethnobotanical , phytochemical and pharmacological properties of Galinsoga parviflora (Asteraceae): A review. *Tropical Journal of Pharmaceutical Research*, *16*, 3023–3033.

Zeeshan A. Khan, M. F. S. and S. P. (2019). Current and Emerging Methods of Antibiotic and Susceptibility Testing. *Diagnostics*, (5), 1–17.

Advances in Phytochemistry, Textile and Renewable Energy Research for Industrial Growth – Nzila et al. (Eds)

The use of solar evacuated tubes as an alternative method of drying

Jepkosgei Joan*, Isaiah Muchilwa & Jerry Ochola
School of Engineering, Moi University, Eldoret, Kenya

Owino George Omollo
School of Engineering, Egerton University, Nakuru, Kenya

David Tuigong
School of Engineering, Moi University, Eldoret, Kenya

ABSTRACT: Drying is one of the traditional methods of food preservation. It is performed in various ways which include sun drying (natural sun) and forced convection (diesel heaters and solar evacuated tubes). The sun drying method takes a long time and can be affected by seasons, thereby increasing the risk of spoilage of the food being dried. This may be avoided by using more efficient methods of drying that enable quick drying and can be used in all seasons. The aim of this study was to design and fabricate a heating system using solar evacuated tubes to produce heat that could dry farm products. Fabrication was done using waste aluminium sheets collected from Rivatex East Africa Limited. Tests were then carried out to determine the amount of heat required for drying, the arrangement of the solar evacuated tubes housing, and the quantity of air necessary for drying. Changes in temperature and relative humidity were also measured to determine the effectiveness of the dryer. And the arrangement of the solar tubes was parallel. The solar heat collector was able to heat the air from a temperature of 27.1 to 56.7°C, and relative humidity dropped from 52.4% to 36.5%. Results show that the designed solar dryer can be used as an alternative for natural sun drying.

Keywords: Drying, Solar Evacuated tubes, Solar Energy

1 INTRODUCTION

Maize drying is a process of removing moisture from the grain to a definite value of operation, therefore allowing ease of processing, storage of the product, and reduced the growth of microorganisms. Maize drying includes heat and mass transfer with varying rate processes that include physical or chemical transformation (Cengel, 2000). The typical maize dryers are also used to dry farm produce such as coffee, corn, vegetables, milk products, and commercial fruits (Wang, 2005). When the grains are harvested, they have a high moisture content of 25%–40% (Radhika, 2011) that must be reduced for ease of storage or processing. Poor drying of the maize grain will lead to the growth of mycotoxins, a low-molecular-weight natural product which is produced by filamentous fungi as a secondary metabolite under suitable temperature and humid conditions. Mycotoxins are poisonous chemical compounds, which when ingested by vertebrates cause sickness or even death (Korir, 2012). To ensure proper drying of high production capacity, an appropriate temperature should be used; if high temperature is used for drying it will affect the quality of the grains, and hence will affect consumers such as wet milling industries.

1.1 *Drying technique*

Drying techniques vary among farmers from sun drying which is a natural method to the artificial methods using forced hot air through the material. The various ways of selecting the method of drying depend on the cost and the production scale. Sun drying is the most common method that is practiced by farmers because it uses the sun as a source of energy which is abundant, less costly, simple, and environmentally friendly (Akinola, 2006). Around the world there is encouragement of the use of renewable energy due to the depletion of fossil energy. Its use by farmers with advanced technology will result in an increase in farm productivity (Weawsak, 2006).

Artificial forced hot air dryers are commonly used to speed up the process of drying. The product is conditioned in a chamber (Ahmet., 2013). Maize stored on a horizontal grid is dried by heat from a fire that has been set below. The disadvantage of this method is that the grain changes colour and produces an odour due to the effect of smoke from the fire. To overcome this problem a hot air chamber/heat exchanger unit and a chimney were developed, such

*Corresponding author

DOI 10.1201/9781003221968-13

as shallow layer dryers (John, 2017), continuous flow dryers (Kasiviswanathan., 2016), and fluidized dryers (Mujumdar., 2000). The mechanized drying method is advantageous in the sense that it speeds up the time required for drying, less labour is required, and a better quality of the product is produced. Its major challenges are the higher operation costs due to the use of electricity and the substantial amount of fuel (Ajay, 2009).

Solar dryers are specialized devices that control the process of drying and prevent the products from the damage caused by insects, rain, and dust (Bala., 2002), (Senadeera., 2007). Solar energy is absorbed and is converted into heat energy which then heats the air and dries the product. They have been beneficial in the sense that they generate high temperatures, lower relative humidity, lower product moisture content, reduce the spoilage of product during drying processes, use less space, take less time, and are relatively inexpensive compared to other artificial drying methods (Mohanraj, 2008).

1.2 *Components of the solar dryer*

a) Drying chamber: The drying chamber is the section where the drying process will take place, it is the section where the product to be dried is placed. (Folaranmi, 2008).
b) Air flow system: The flow of air into the drying system can be either be natural or forced (convection). For a natural system, the air is blown into the system by wind or hot air moves up and cold air moves down into the drying chamber. For forced convection, the air is blown in with the help of a fan. The fan can be powered by a generator or the utility electricity. When using the fan, the drying time is reduced and the product quality is maintained, hence optimizing the drying process (Mathew, 2001). The fan works either to create a negative pressure in the system which will prevent hot air leakage during drying to positive pressure which prevents cold air and dust entering into the system.
c) Solar collector: The solar collector is used to absorb the solar energy and converts it into heat, hence it is used for thermal application. It is used to absorb shorter wavelengths of sunlight of 0.3–2 mm and prevents heating wavelengths of 2–10 mm from being lost into the atmosphere using a greenhouse effect. Some of the types of solar collector are flat plate, heated pipes, and solar evacuated tubes (Norton, 2006), (Kalogirou., 2004).

The flat plate collectors have commonly been used for several solar experiments. Their operation parameters are mass flow rate of the fluid, inlet, outlet and ambient temperature, solar radiation, air speed, glass cover, and environment condition (Akpinar, 2010). Their performance depends on the design parameters which include type and thickness of the glazing, number and type of coating on the collectors plate, evacuated space between the collector and the inner glass, insulation type, and convection movement of the air in the system (Alghoul, 2005).

Solar evacuated tubes have been used for many years in Germany, Canada, China, and the UK. There are various types of evacuated tube but the most commonly used have a double glass tube, because of their reliability, excellent performance, and ease to manufacture (Zulovich, 2013). An evacuated tube has two glass tubes, i.e., the outer tube and the inner tube. The outer tube is made of very strong transparent borosilicate glass that can resist impact from hail and is 38 mm in diameter. The inner tube is made of borosilicate glass but is coated with a special selective coating that is excellent in absorbing solar energy and has a minimal reflection property (Kalogirou., 2004). The air is evacuated from the space between the two layers of glass to form a vacuum that will eliminate loss of heat through conduction and convection. At the bottom is a layer of barium that is used to absorb CO, CO_2, N_2, O_2, H_2O, and H_2 during operation and storage to ensure that the vacuum is maintained; also the barium shows the status of the vacuum in that when the vacuum ceases the silver-coloured barium layer will turn white, as shown in Figure 1.

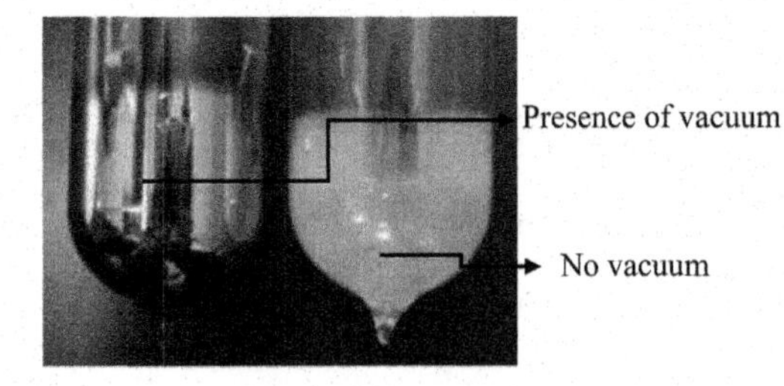

Figure 1. Solar evacuated tube, with vacuum and with no vacuum (Green Spec, 2018).

1.3 *The desirable features of a solar collector include the following:*

(a) Transparent cover: It traps heat from thermal radiation, this ensures fewer radiation and convection losses into the atmosphere. It protects the absorber from damage during hostile weather conditions. Covers are commonly made of low iron glasses such as fibreglass, flexiglass, thin plastic films, and reinforced polyester and ultraviolet-resistant plastic sheeting. The low iron glasses have a high transmission and low reflection of sunshine and are thin thereby increasing its efficiency (Joshua, 2008).
(b) Insulation: The insulation property prevents the thermal energy loss which minimizes the overall heat loss of the system when placed below the absorber plates. It must withstand the stagnating temperature, not be damaged by moisture or insects, and should be fire resistant (Joshua, 2008). Insulators are made from mineral wool, Styrofoam, fibreglass, urethanes, and selective grades of CFC-free polyurethane foam (PUF).

They must be kept dry or they may lose all their insulating qualities. Since the air that is flowing into the system contains moisture, the insulator should have a desiccant to absorb the moisture (Alghoul, 2005).

The benefits of using solar dryers include the following: dried grains are tasty, nutritious (high in carbohydrates and low in fat), the nutritional value and flavour in food is only minimally affected by drying, dried grains are easy to store (minimizing space of storage) and easy to prepare (processing), and the grains remain clean since the grains are protected from rain and pollution from dust

Some of the limitations of the solar dryer are inadequate infrastructure in Kenya, growing market difficulties by intensifying competition in the worldwide agricultural market, as well as the need for an improvement in the population income and supply situation. The major limitation for the automated solar powered dryer is the use of a fan. The fan should be inexpensive, durable, and produce high flow rates at a high pressure while having a low power consumption, in order to keep the price of the solar crop dryer down and at them same time ensure an efficient drying process.

2 MATERIAL AND METHODS

The solar system consisted of seven pieces of 0.74 m long open-ended evacuated glass tubes to heat the air, seven 0.5 m length aluminium to allow flow of air within the system, and wooden bed frames to hold the evacuated tubes in position. A blower was used to blow air into the solar evacuated tubes, as shown in Figure 2.

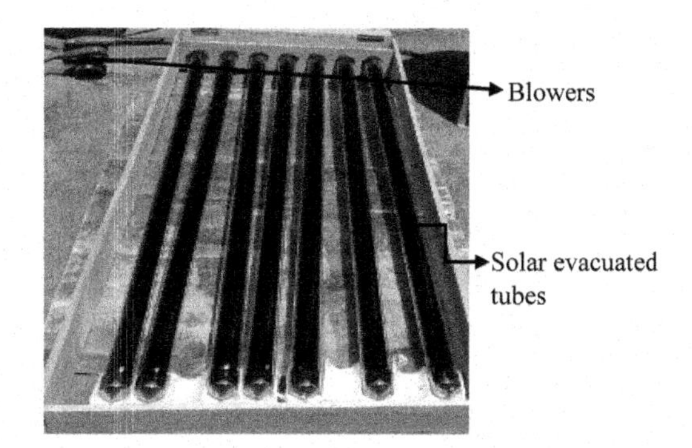

Figure 2. Arrangement of solar evacuated tubes.

The design and arrangement of the solar evacuated tube system consist of the following parts:

- Solar evacuated tubes
- Manifold chamber

2.1 *Solar evacuated tube*

The solar evacuated tubes are arranged parallel to each other in this system, as shown in Figure 2. Each evacuated tube consists of two concentric glass tubes made of extremely strong borosilicate glass. It consists of two glass tubes, and in-between the glass tubes there is a vacuum. The outer tube is transparent which allows rays of light to pass through with minimal reflection. The inner tube is coated with a special selective coating (Al-N/Al) which absorbs the solar radiation superbly with negligible reflection properties. The length of the evacuated tube is 1.8 m and the outer and inner tube diameters are 0.057 m and 0.047 m, respectively.

2.2 *Manifold chamber*

The manifold chamber consists of a square chamber made of wood which is a poor conductor of heat, and its measurements are 0.74 by 0.10 by 0.14 m, and a circular pipe of diameter 0.196 m made of aluminium pipe. The circular pipe is centrally passed through the square chamber and is closed at one end. Its surface contains seven holes in which chromium pipes are attached, which direct air into the solar evacuated tubes. Seven holes are also made on the square chamber where the solar evacuated tubes are attached and the closed ends are supported by a frame, as shown in Figure 3.

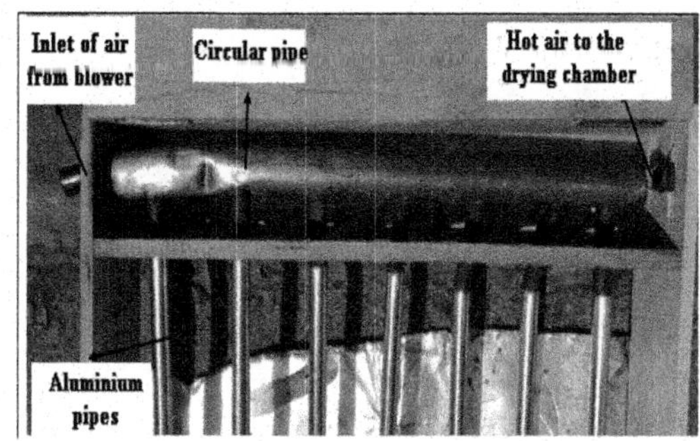

Figure 3. Manifold with the aluminium pipes.

3 RESULTS AND DISCUSSION

The system was tested for two days to show the variation that may occur on different days depending on the UV intensity.

Figure 4 shows the relationship between temperature and relative humidity against time; t1 and h1 are the temperature and relative humidity of the ambient air respectively. The ambient temperature t1 is lower with an average of 21.7°C, and relative humidity has a high average of 64.7%. As the air passes through the solar system the temperature of air t2 is increased to an average temperature of 46.7°C, and relative humidity h2 decreases to 45.3%. This is because as the temperature increases, moisture in the air is converted to vapour gas, hence decreasing the amount of water in the air.

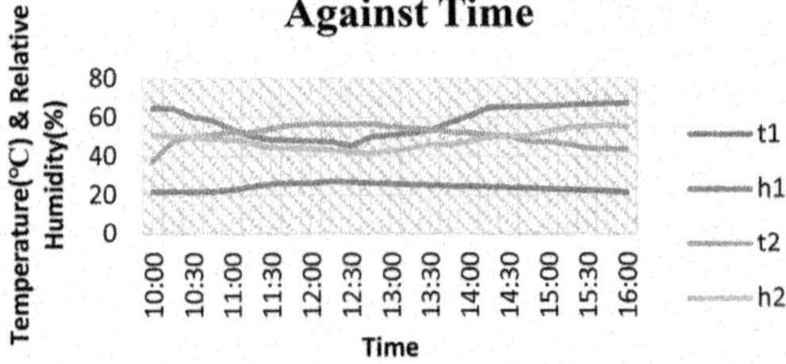

Figure 4. Relationship between average temperature & relative humidity against time.

3.1 *Relative between temperature and relative humidity against time*

Figures 5 and 6 show the relationship temperature and relative humidity against time. Figure 5 shows the change in the temperature and relative humidity of the ambient air, where t1 and h1 are the temperature and relative humidity of the ambient air. At 1,000 hrs t1 was 21.7°C which increased to 37.5°C at 1245 hrs, while the relative humidity decreased from 64.7% to 50.8%.

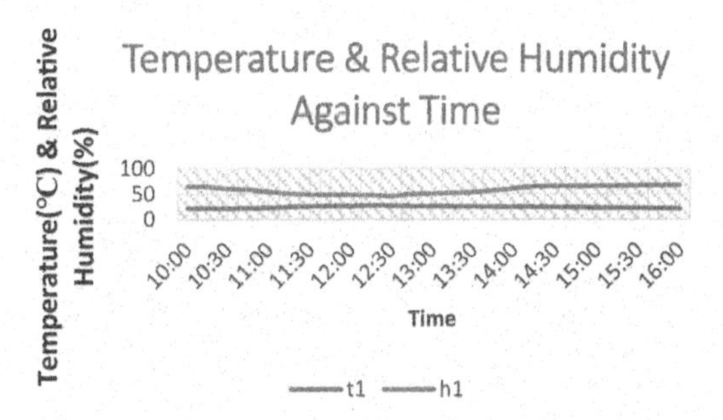

Figure 5. Relationship between temperature & relative humidity against time (ambient air).

Figure 5 shows the change in the temperature and relative humidity of the air from the solar evacuated tubes, where t1 and h1 are the temperature and relative humidity of the air out of the solar system, respectively. At 1,000 hrs t1 was 26.8°C which increased to 56.3°C at 1245 hrs, while the relative humidity decreased from 45.3% to 42.1%.

3.2 *Temperature against time*

Figure 7 shows the variation of temperature during drying before loading of maize on the cob, represented by t1 and t2, where t1 is the ambient temperature, and t2 is the temperature at the inlet of the dryer. The temperature increases with time up to 1245 hours, after which it starts decreasing.

The temperature of the ambient air is increased from 26.8 to 56.3°C at 1,300 hours; this is because the sun is at its peak intensity.

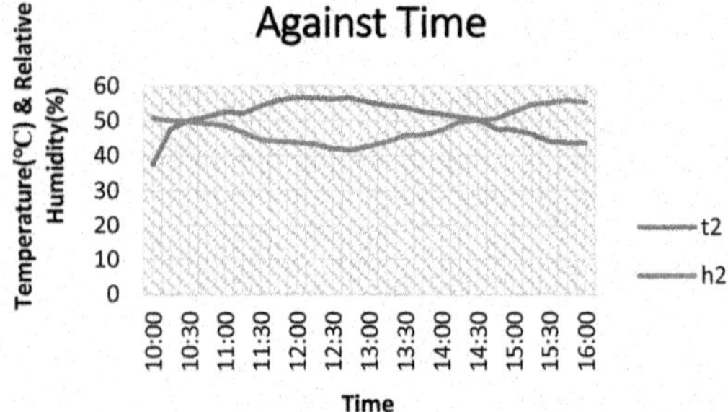

Figure 6. Relationship between temperature and relative humidity against time (Air in the Solar Evacuated Tube).

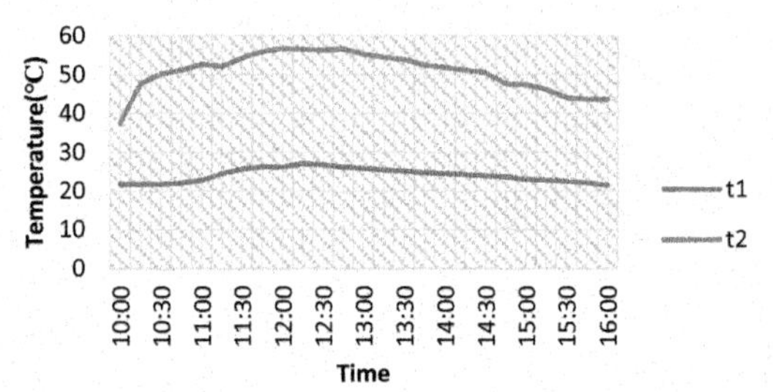

Figure 7. Relationship between temperature and time.

3.3 *Relative humidity against time*

The relative humidity of the drying system was measured when the solar system was operating, giving results of h1 and h2, where h1 is the ambient relative humidity, and h2 is the inlet relative humidity of air entering the drying chamber, as shown in Figure 8.

Relative humidity (representing the general moisture from the environment) decreased with time, h1 was 64.7% to 50.8% at midday, because the air was continually heated. Later the humidity started increasing, due to the lowering of the sun's intensity.

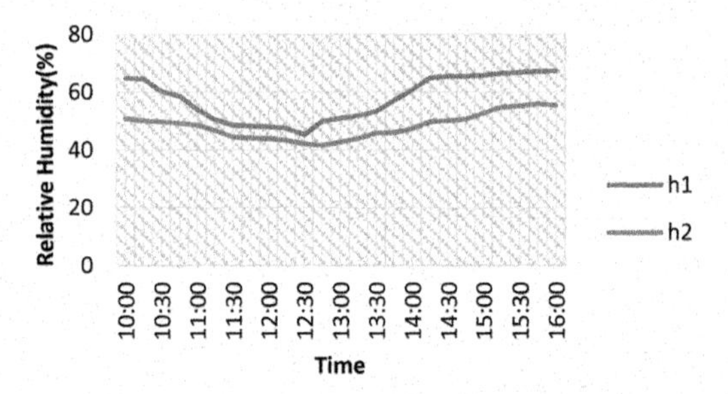

Figure 8. Relationship between relative humidity and time.

4 CONCLUSION

A solar collector system was fabricated, consisting of seven solar evacuated tubes whose absorbing area was 2.21 m^2 for each tube, a frame to hold the tubes in position made of wood, and a blower with an average speed of 2.4 m/s. The system could heat the air from a temperature of 21.7 to 26.8°C, decrease relative humidity from 64.7% to 50.8% at 1,000 hrs and raise temperature from 26.8 to 56.3°C, drop relative humidity from 45.3% to 42.1% at 1245 hrs. The found temperature is within the optimum range of heat needed to dry maize grain, i.e., 30–60°C (Vidya, 2013). This system will encourage the early harvest of maize and the reduction of growth of fungi, such as mycotoxins, which are harmful when consumed by humans and animals.

REFERENCES

Ahmet., D. U. (2013). Modeling of a Hot Air Drying Process by Using Artificial Neutral Network Method . *Termothehnica*, 51–57.

Ajay, C. S. (2009). Design of solar dryer with Turboventilatir and fireplace. Deopur, India: suman foundation.

Akinola, A. a. (2006). Exergetic nalysis of a Mixed-Mode Solar Dryer. *J. Engin. Appl. Sci 1*, 205–210.

Akpinar, E. K. (2010). Energy and exergy analysis of a new flat-plate solar air heater having different obstacles on absorber plates. *Applied Energy*, 3438–3450.

Alghoul, M. S. (2005). Review of material for solar thermal collectors. *Physics Department, University Putra* , 199–206.

Amyot, J. (1983). Social and Economic Aspects of Dryer Use for Paddy and Other Agricultural Produce in Thailand. *Chulalongkorn University Social Research Institute and International Development Research Center*.

Bala., B. (2002). Solar drying of pineapples using solar tunnel drier.

Cengel, Y. (2000). *Heat and Mass Tranfer an Engneering Approach*. McGraw-Hill.

Folaranmi, J. (2008). Design, Construction and testing of a siple Maize Dryer. *Leornado Electronic Journal Of Practices and technology*, 122–130.

Hussein, H. B. (2001). Toxicity, metabolism, and impact of mycotoxins on humans and animals. *Toxicology 167*, 101–134.

John, J. (2017, June 14). *University of Georgia Extension*. Retrieved from University of Georgia Extension: http://extension.uga.edu/publications/detail.html?number=B873&title=Grain%20and%20Soybean%20Drying%20on%20Georgia%20Farms

Joshua, F. (2008). Design, construction and testing of simple maize dryer. *Federal University of Technology, Leonardo Electronic Journal of Practices*, 122–130.

Kalogirou., S. a. (2004). Solar thermal collectors and applications. *Progress in Energy and Combustion Science*, 231–295.

Kasiviswanathan., M. S. (2016). Grain Drying Systems. *research gate*.

Korir, K. B. (2012). Mycological Quality of maize flour from Aflatoxin "Hot" Zone Eastern Province-Kenta. *African journal of healt science*, 143–146.

Lanyasunya, T. W. (2005). The risk of mycotoxins contamination of dairy feed and milk on smallholder dairy farms in Kenya. *Pak. J. Nutr. 4*, 162–169.

Mohanraj, M. (2008). Drying of Copra in a forced convection solar drier.

Mujumdar., A. S. (2000). Fluidized bed drying. In &. S. S. Mujumdar, *Developments in drying. Vol 1: food dehydration* (pp. 59–111). Bankngkok: kasetsart university press.

Norton, B. (2006). Anatomy of a solar collector: Developments in materials, components and efficiency improvements in solar thermal collector systems. *Refocus*, 32–5.

Rachaputi, N. W. (2002). Management practices to minimise pre-harvest aflatoxin contamination in Australian groundnuts. *Austr. J. Exp. Agric. 42*, 595–605.

Radhika, G. B. (2011). Mathematical Model on Thin Layer of Drying of Finger Millet (Eluesine coracana). *Adnvance Journal of Food Science Technology*, 127–131.

Senadeera., W. I. (2007). Performance Evaluation of an affordable solar dryer for drying of crops.

Wagacha, J. M. (2008). Mycotoxin problem in Africa: current status, implications to food safety and health and possible management strategies. *Int. J. Food Microbiol. 124*, 1–12.

Wang, Y. X. (2005). Drying characteristics and drying quality of carrot using a two-stage microwave process. *Journal of Food Engineering*, 505–511.

Weawsak, J. C. (2006). a mathematical modeling study of hot air drying for some agricultural products. *Trmmasat international journal of science and Techonology, 11(1)*, 14–20.

Zulovich, J. (2013). *Active solar collectors for farm*. Retrieved from http://muextension.missouri.edu/

Section 2: Insights in progressive textiles, transformative industrialization & phytochemistry

Advances in Phytochemistry, Textile and Renewable Energy Research for Industrial Growth – Nzila et al. (Eds)

Effects of alkaline and microwave surface modification on *Calotropis procera* bast fibres for development of fibre-reinforced polylactic acid composite

E.K. Langat & D.G. Njuguna
Department of Manufacturing, Industrial and Textile Engineering, School of Engineering, Moi University, Eldoret, Kenya
Africa Centre of Excellence II in Phytochemicals, Textiles and Renewable Energy (ACE II PTRE), Moi University, Eldoret, Kenya

ABSTRACT: Natural fibres have been used as reinforcement in a polymer matrix. One of the major challenges is incompatibility with matrices. In this study, *Calotropis procer*a plant bast fibres were extracted from stems using the decortication method. First, the samples were treated in varied alkaline solution concentrations of sodium hydroxide for different durations of treatment time. Second, the samples were subjected to microwave treatment at various power levels for different duration of treatment time. Fibres with optimum tensile properties were used with polylactic acid (PLA) matrix to fabricate biocomposite materials. Both alkaline-treated and microwave-treated samples showed a significant increase in tensile properties and decrease in fibre linear density when compared with untreated samples. Alkaline treatment 5 w/v% solution for 1 hour and microwave irradiation with 231Watts for 4 minutes treatment period were optimum for fibres modification. Tensile strength and tensile modulus of PLA/alkaline-treated-CPBF and PLA/microwave-treated-CPBF improved by 34.27% and 20.46% and by 21.22% and 15.54%, respectively.

1 BACKGROUND

1.1 *Materials used for composites*

In recent years, there has been a great awareness towards preserving our natural resources and the environment. With increased economic development and energy consumption, alternative methods and materials have been developed to replace petroleum-based products by bio-based materials. Moreover, the emphasis has been to reduce the use of petroleum-based products which are a key contributor of greenhouse gas emissions due to the production of high levels of carbon dioxide during extraction and processing. The other emphasized issue is the utilization of new renewable resources (Huda & Yang, 2008; Mohanty, Misra, & Drzal, 2002). Most of the polymer materials are made up of non-renewable and non-degradable synthetic plastics and their waste occupies the top position in landfills (Ariadurai, 2013). Plastics cause severe environmental and health damage (Bashir, 2013). Therefore there is a need to reduce plastic use and disposal in Africa and increase the use of more eco-friendly materials (Fuqua, Huo, & Ulven, 2012; Puglia, Biagiotti, & Kenny, 2005; Rayne, 2008) through further developing alternative methods and exploring new natural materials to replace or reduce petroleum-based products by bio-based materials.

1.2 *Biocomposites*

Biocomposites are renewable, recyclable, biodegradable and environmentfriendly resources which can be used for daily life applications in the construction, automobile, and biomedical sectors (Li, Panigrahi, & Tabil, 2009). Natural fibre-reinforced biocomposites using biodegradable polymers have been claimed as the most environmentfriendly bioproducts (dos Santos Rosa & Lenz, 2013). Studies have been conducted on the manufacture of biocomposites with the use of biodegradable polymers. Polymers such as polylactic acid (PLA) and polycaprolactone (PCL) are used to manufacture biocomposites depending on their desired properties and the end-use of the product (Gunatillake & Adhikari, 2003). The realization of a biodegradable composite is essentially possible by combining a biodegradable matrix with natural fibre reinforcement. Natural fibre-based PLA composites present the opportunity of achieving several desired properties. The factors considered in characterization are fibre volume/weight fraction, stacking sequence of the fibre layers, methods of processing,

DOI 10.1201/9781003221968-14

treatment of fibres, and the effect of environmental conditions (Sticklen, Kamel, Hawley, & Adegbite, 1992).

The reinforcement of the matrix is also an important aspect in composites. Textile fibres from various sources have gained a great deal of usage as reinforcement and this has led to the exploration of new plants that can produce fibres for composites. *Calotropis* is a perennial shrub found chiefly in China, India, Malaysia, as well as in most of Asia and large parts of Africa and South America (Yuanhui et al., 2018). The genus *Calotropis* (*Asclepiadaceae*) comprises two species: C. *gigantea* and C. *procera*, The plant grows mainly in arid regions such as Tharaka Nithi, Baringo, Makueni, Turkana and Kajiado counties in Kenya (Muriira, Muchugi, Yu, Xu, & Liu, 2018; Yuanhui et al., 2018). Two kinds of fibres can be obtained from the *Calotropis* plant: fruit and stem/bark (Ramasamy, Obi Reddy, & Varada Rajulu, 2018) and they can be used for various purposes (Chen et al. 2013; Zheng et al. 2016). The feasibility study done by Ramasamy et al. (2018) reported that both its fruit and bast fibres have potential as reinforcement in composites. Hydroxyl groups, hemicellulose, pectin, waxes and lignin affect the composite strength matrix adhesion and thermal stability of fibres during hot pressing processes.

There are various methods used in the manufacture of composites and different researchers have investigated and used them for developing a number of composite materials. A compression moulding method has been found to be the better option to obtain a composite based on natural fibres in terms of mechanical properties (Satyanarayana, Arizaga, & Wypych, 2009). This method also uses small amounts of energy, so the green concept is reinforced. Nevertheless, the process needs to be optimised to address issues such as delamination and production of composites with high fibre plies (Rubio-López, Olmedo, Díaz-Álvarez, & Santiuste, 2015). Also, the presence of low thermally stable fibre components such as waxes, hemicellulose and pectin can induce thermal degradation of the fibre at higher loadings (Elsabbagh, Steuernagel, & Ring, 2017; Santos, Mauler, & Nachtigall, 2009). Alkaline and microwave treatment are some of the chemical treatments of natural fibres to enhance their workability. Both alkaline treatment and microwave irradiation treatment have a potential to modify fibre surface through disruption of hydrogen bonding in the network structure, thereby increasing surface roughness (Imoisili, Tonye, Victor, & Elvis, 2018). They also remove some of the lignin, wax and oils covering the fibre cell wall external surface, depolymerizing cellulose and exposing the short length crystallites. This can help in reducing fibre damage by heat during hot processing by increasing the thermal stability of the fibres and improve fibre–matrix adhesion (Mohanty, Misra, & Hinrichsen, 2000). The thermal degradation of natural fibres is an important concern for the processing of Natural Fibre Reinforced Composites (Gassan & Bledzki, 2001) and therefore the treatment of fibres has the potential to improve the thermal properties of fibres (Kalia, Thakur, Celli, Kiechel, & Schauer, 2013). The influence of heating temperature, from 175 to 200°C, was analysed through comparison tests (Ochi, 2008). The specimens were manufactured from 10361D PLA matrix reinforced with basket weave flax woven plies, and the applied pressure was 32 MPa. Heating temperatures over 200°C produced fibre damage due to overheating. On the other hand, the matrix was not completely melted for heating temperatures below 175°C for short processing times. Thus, the temperature must be high enough for matrix melting, but not so high as to cause fibres damage. In this research, temperature of 200°C was used for the fabrication of PLA sheets, the pressure was kept constant. Ochi (Ochi, 2008) also found that natural fibres can be degraded at temperatures higher than 180°C, but if this temperature is applied for a short time period, fibres do not get damaged. In Rubio-López, et. Al., (2015) work, preheating time was set to 2 minutes and heating under pressure time to 3 minutes to compress the PLA sheets and fibres to form a composite structure." A temperature of 200°C was used and was kept constant to ensure PLA melting, its homogenous distribution in the laminate, and there was a slow application of pressure to avoid fibre misalignment.

In this study, *Calotropis procera* bast fibres were extracted manually and treated using two methods: alkaline solution and microwave irradiation in order to examine the applicability and sustainability of these two methods. Microwave irradiation was used because it affects the material thermally whereby microwaves heat the material by interacting with the molecules of material via the electromagnetic field produced by microwave energy. It also affects the material non-thermally, whereby it interacts with the polar molecules and ions in the materials, causing physical, chemical, and biological reactions (Thostenson & Chou, 1999). Through this treatment, the properties of the fibres and fibre–matrix adhesion could be improved. Biocomposite materials were also fabricated using POLYLACTIC acid as matrix and *Calotropis procera* bast fibres as reinforcement. The heating temperature was stated at 185°C to manufacture 4032D PLAbased biocomposite reinforced with *Calotropis procera* bast fibres. According to Ochi (Ochi, 2008), a pressure range between 8 and 32 MPa can be used to produce biocomposites. Tensile strength increases with manufacturing pressure until 8 MPa, then there is a plateau until a manufacturing pressure of 32 MPa, at which tensile strength decreases with pressure causing fibre breakage, while pressure under 8 MPa means a lack of cohesion in the biocomposite.

This research focuses on the exploitation of natural abundant resources for the benefit of humanity and environment. The specific objectives of the current work are as follows:

i. To extract and characterize *Calotropis procera* bast fibres.
ii. To treat *Calotropis procera* bast fibres with alkali solution and microwaves to modify the surface.

iii. To develop a biodegradable composite structure from polylactic acid and *Calotropis procera* bast fibres and characterize the tensile strength and water absorption properties.

2 MATERIALS AND METHODS

2.1 *Fibre samples and chemicals used*

The plant material, *Calotropis procera* stems, were collected by cutting and removing leaves from the wild-growing *Calotropis procera* shrubs near Marigat in Baringo County, Kenya. The treatment reagents: sodium hydroxide (NaOH), acetic acid and distilled water were obtained from Indo Kenya enterprises. Polylactic acid (PLA), Ingeo™ biopolymer 4032D were obtained in the form of pellets from Nature Works LLC and used as the matrix. The polymer has a density of 1.24 g/cm^3, its MFI is 3.9 g/10 min at 190°C and load is 2.16 kg.

2.2 *Extraction and preparation of calotropis procera bast fibres*

Figure 1. *Calotropis* bast fibres extraction process: (a) *Calotropis procera* plant; (b) a bundle of stems cut from the plant; (c) peeling off the bark; (d) fibres and the outer cort; (e) decorticated and dried fibres.

The bast fibres were extracted manually by the decortication method and sun-dried for two days to remove moisture. After drying, any extraneous matter that may still be adhering to the fibres was removed and the fibres were stored well in polythene bags. The steps used in the extraction of the bast fibres are further elaborated in Figure 1.

2.3 *Fibres surface treatment*

Fibre surface modification in this work was done by soaking the *Calotropis procera* bast fibres in alkali solution for various durations at room temperate and microwave treatment of *Calotropis procera* bast fibres was done by exposing the fibres to microwave irradiation at various powers for different lengths of time. A summary is given in Table 1.

Table 1. Alkali solution and microwave irradiation fibre surface treatment

Notation	Treatment	After treatment
First treatment (NaOH)	Soaking in (NaOH) 1% for 1 hour	The bast fibres were rinsed thoroughly with distilled water acidified with acetic acid (1% w/v) to neutralise the pH Check the rinse solution pH value. Digital pH meter SIN-PH-100 (Hangzhou automation technology co. ltd, P.R. China) was used. The alkali treated *Calotropis* bast fibres were then dried in an oven at 85°C.
	Soaking in (NaOH) 1% for 2 hours	
	Soaking in (NaOH) 1% for 3 hours	
	Soaking in (NaOH) 3% for 1 hour	
	Soaking in (NaOH) 3% for 2 hours	
	Soaking in (NaOH) 3% for 3 hours	
	Soaking in (NaOH) 5% for 1 hour	
	Soaking in (NaOH) 5% for 2 hours	
	Soaking in (NaOH) 5% for 3 hours	
Second treatment (Microwave irradiation), Using a microwave oven with an output power of 700 watts with varied percentage of power. At medium position, the power is 55 percent of 700 which is 385watts.	Exposure to 119 watts for 2 minutes	
	Exposure to 119 watts for 4 minutes	
	Exposure to 119 watts for 8 minutes	
	Exposure to 231 watts for 2 minutes	
	Exposure to 231 watts for 4 minutes	
	Exposure to 231 watts for 8 minutes	
	Exposure to 385 watts for 2 minutes	
	Exposure to 385 watts for 4 minutes	
	Exposure to 385 watts for 8 minutes	

2.4 *Characterization of Calotropis procera bast fibres*

The alkaline-treated, microwave-treated and untreated *Calotropis procera* fibres were characterised by determining their linear density and tensile properties. Linear density for both treated and untreated *Calotropis* fibres was determined as per ASTM D1577-2001 which is a standard test method for linear density of fibres. Four bundles of fibres from treated and four bundles from untreated *Calotropis* fibres were picked randomly and then cut to a specific length. Each of the four samples of treated and four samples of untreated *Calotropis* fibres were weighed. From the number of fibres in each bundle and the weights of the bundles, the weight of each fibre was determined. The linear density in Tex was determined by dividing the fibre weights in grams by its length in kilometres. The tensile strength of both treated and untreated *Calotropis* fibres were determined as per ASTM D3822M-2001, done at ambient conditions using a universal tensile testing machine (UTM-TH2730, Rycobel, Belgium). At a gauge length of 25 mm and speed of 5 mm/min, Tensile strength was determined for the four test specimen bundles of treated and the four untreated *Calotropis* fibres. An average of 30 test replicates for each were conducted. The tensile strength of treated and untreated fibres were determined in terms of breaking tenacity (cN/Tex) using the breaking force (cN) and the linear density (Tex) of the respective fibres. The optimum conditions obtained in this experiment were used to treat fibres used for the composite fabrication.

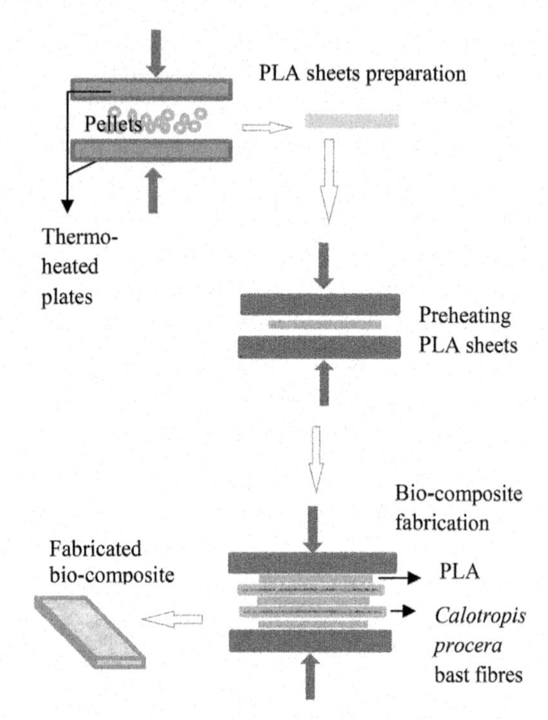

Figure 2. Figure schematic illustration of the compression molding composite fabrication process

2.5 *Fabrication of polylactic acid/Calotropis bast fibres (PLA/CBF) composite structure*

The polylactic acid/*Calotropis procera* bast fibres (PLA/CPBF) composites were fabricated by using *Calotropis* bast fibres as reinforcement of PLA matrix by applying a compression moulding method as described by Wu et al. (2012) with slight modification. A mould measuring 310 × 310 × 30 mm with a lid die was fabricated using a polished iron metal sheet. Aluminium foil was used to cover the inner surfaces to avoid the composites sticking onto the mould surface. The *Calotropis* bast fibres were cut into approximately 1.5cmlong lengths before being used for the fabrication of composite panels with dimensions 300 mm wide and 300 mm long. The compression moulding composite fabrication process is schematically shown in Figure 2.

To obtain fine PLA sheets, the pellets were placed between two thermo-heated plates and hot pressed at 200°C for 25 minutes where all the pellets melted. A load of 900 kN was used to apply 2 MPa pressure. A 1.5mmthick uniform PLA sheet was obtained. The matrix sheets were then stacked alternately with a layer of *Calotropis procera* bast fibres spread manually to form a fibre mat where each fibre mat weighed 10g. Normally, the stacked plies are placed between the thermo-heated plates and after a pre-heating time, a hot compression moulding is used to form a composite structure. But in this study, each PLA sheet was pre-heated first at 185°C for 12 minutes after which the fibres were laid and cold compressed for 4 minutes with a weight to give 10 MPa. Every subsequent layer of fibres was added after compressing the first layer of fibres and the PLA sheets. This was repeated for every additional layer of PLA sheets and fibres for a thicker structure. A calibrated digital analytical balance, Mettler PM200 (Marshall Scientific, Hampton, NH, USA) was used in all the analytical weighing. Composites with one, two and three layers of fibres were fabricated. The fabricated polylactic acid/*Calotropis* bast fibres (PLA/CPBF) composites using untreated fibres were designated as PLA/ut-CPBF while those fabricated using alkaline treated fibres and microwavetreated fibres were designated PLA/al-CPBF and PLA/mw-CPBF respectively. The fibres were treated using the optimum values of fibre alkali treatment and microwave fibre treatment obtained in this study.

2.6 *Evaluation of composite tensile properties and fractography studies*

The samples of treated and untreated PLA/*Calotropis procera* bast fibre composites were conditioned at temperature (23 ± 2°C) and relative humidity (65%) for 48 hours before testing. Tensile properties were measured according to the ASTM D638-2014 standard using a Universal Testing machine (UTM-TH2730, Rycobel, Belgium) with a maximum load cell of 5 kN at a loading rate of 2 mm/min. the Surface morphology of treated and untreated PLA/*Calotropis procera* bast fibre composites were examined using an MSX-500Di

Scopeman Digital Microscope (Herter Instruments, Barcelona, Spain). Five specimens were prepared and analysed.

2.7 *Evaluation of water absorption*

The water absorption test was done according to ASTM D 570–98 standard. The test specimens were in the form of discs 50 mm in diameter and 6.23 mm in thickness. The specimens were oven dried at 105°C for 1 hour and then placed in a container of distilled water maintained at a temperature of 23 ± 1°C, to rest on edge and entirely immersed. At the end of 0, 12 and 24 hours, the specimens were removed from the water one at a time, all surface water wiped off with a dry cloth, and weighed to the nearest 0.001g immediately. The percentage increase in weight during immersion, was calculated to the nearest 0.01%. In the current study, the specimens were prepared and analysed at least in triplicate.

3 RESULTS AND DISCUSSION

3.1 *Properties of the treated and untreated calotropis bast fibres*

The average linear density for untreated *Calotropis procera* bast fibres was found to be 18.08 ± 6.42 Tex while those of alkalitreated *Calotropis procera* bast fibres in 1% w/v, 3% w/v and 5% w/v sodium hydroxide solution for 1 hour, 2 hours and 3 hours exhibited a reduction in linear densities. The higher mean linear density was recorded for the fibres treated at 1% w/v for 1 hour (15.21 ± 3.22 Tex) while the lowest was the fibres treated in 5% w/v solution for 3 hours (11.91 ± 4.51 Tex). The linear densities for the *Calotropis procera* fibres irradiation were found to be higher than those of alkalinetreated fibres with the higher value obtained from those exposed to radiation of 119 watts power for 2 minutes (17.26 ± 5.31 Tex) while the lowest was recorded for microwavetreated fibres at 385 watts for 2 minutes (13.78 ± 4.29 Tex). It was observed that fibres treated by microwave irradiation at 385 watts for 8 minutes were entirely damaged and were discarded for they could not be analysed. These variations and differences could be ascribed to the exposure of the inner central lumen of the fibre which may affect the fibre and reduction in fibre diameter due to the loss of weight due to the removal of carbonaceous materials after treatment (Imoisili et al., 2018; Raghu & Goud, 2019)

The tenacity (in cN/Tex) of the treated and untreated *Calotropis procera* bast fibres were also determined and the mean tenacity of the untreated *Calotropis procera* bast fibres was found to be 37.88 cN/Tex. The highest mean tenacity of 72.06 cN/Tex was recorded for *Calotropis procera* bast fibres treated in 5% w/v alkali solution for 1 hour while the lowest mean tenacity of 28.79 cN/Tex was recorded for *Calotropis procera* bast fibres treated in 5% w/v alkali solution for 3 hours. On the other hand, high mean tenacity of 59.68 cN/Tex was observed for *Calotropis procera* bast fibres treated in microwave at 231 watts power for 4 minutes and the lowest mean tenacity of 25.33 cN/Tex for *Calotropis procera* bast fibres treated in microwave at 385 watts power for 2 minutes. The high energy radiation and long period of alkaline exposure led to the heavy damage to the fibre surface thus resulting in the decrease in mechanical properties. (Imoisili et al., 2018; Valadez-Gonzalez, Cervantes-Uc, Olayo, & Herrera-Franco, 1999). It was therefore noted that the optimum alkaline treatment and microwave treatment were 5% w/v alkali solution for 1 hour and microwave exposure of 231 watts power for 4 minutes. The micrographs of the fibres at different magnifications given in Figure 3, Figure 4, Figure 5, and Figure 6 show that the diameter of the treated fibre is smaller compared to the untreated fibre and this confirms that there were reductions in the diameter of the treated *Calotropis* bast fibres. The damage of fibres at high microwave irradiation is also noted in Figure 7 and Figure 8

Figure 3. Micrograph of *Calotropis procera* bast fibres untreated taken at ×17.

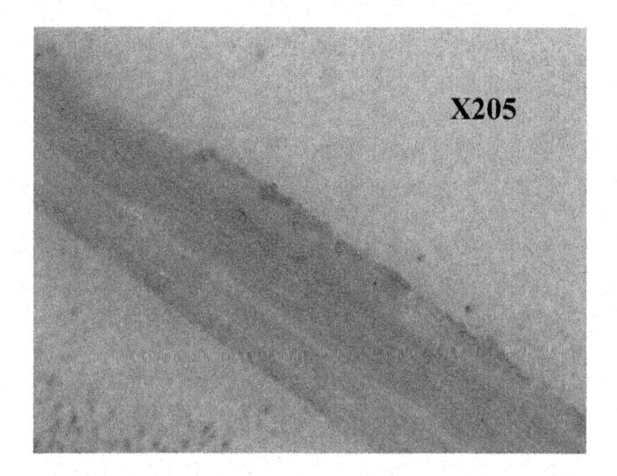

Figure 4. Micrographs of *Calotropis procera* bast fibres untreated taken at ×205.

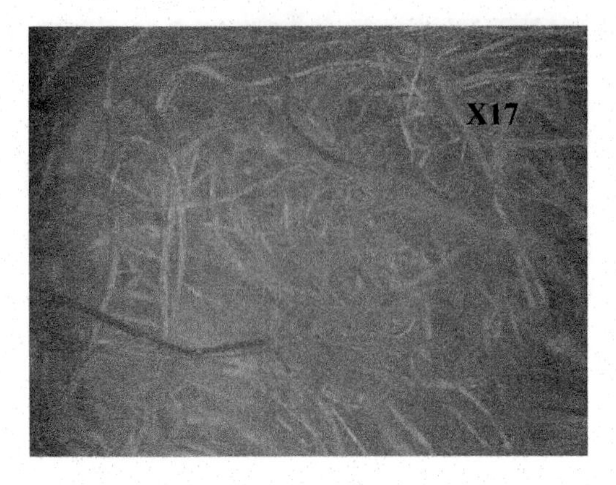

Figure 5. Micrographs of *Calotropis procera* bast fibres alkaline-treated taken at ×1.

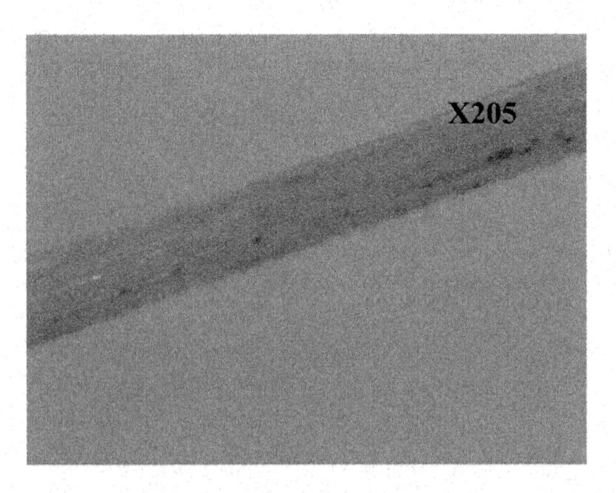

Figure 6. Micrographs of *Calotropis procera* bast fibres alkaline-treated taken at ×205.

Figure 7. Micrographs of *Calotropis procera* bast fibres microwave-treated at 385 watts for 2 minutes taken at ×17.

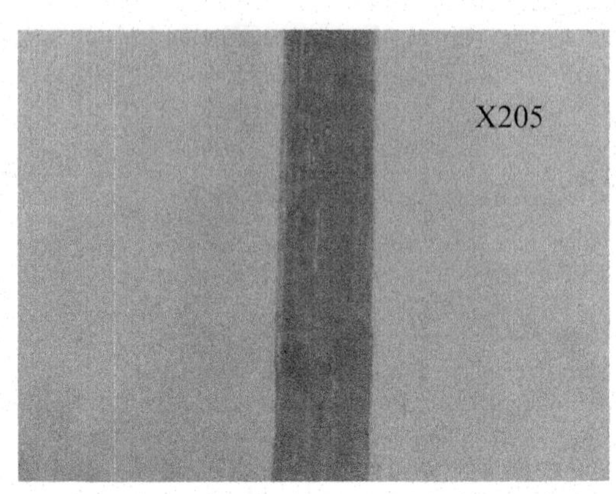

Figure 8. Micrographs of *Calotropis procera* bast fibres microwave-treated at 385 watts for 2 minutes taken at 205.

Figure 9. Micrographs of *Calotropis procera* bast fibres Microwave-treated at 231 watts power for 4 minutes taken at 17.

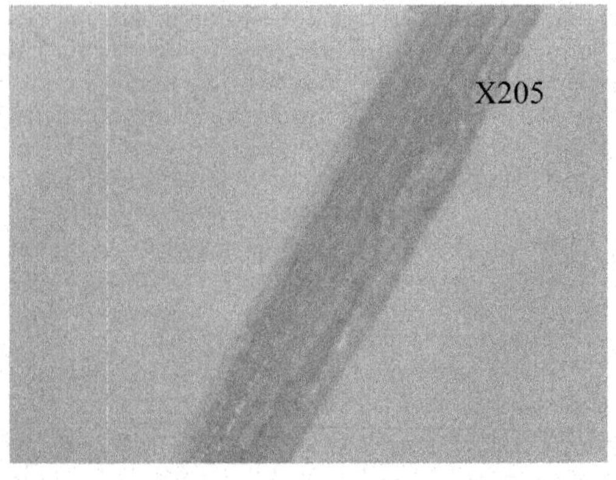

Figure 10. Micrographs of *Calotropis procera* bast fibres microwave-treated at 231 watts power for 4 minutes taken at 205.

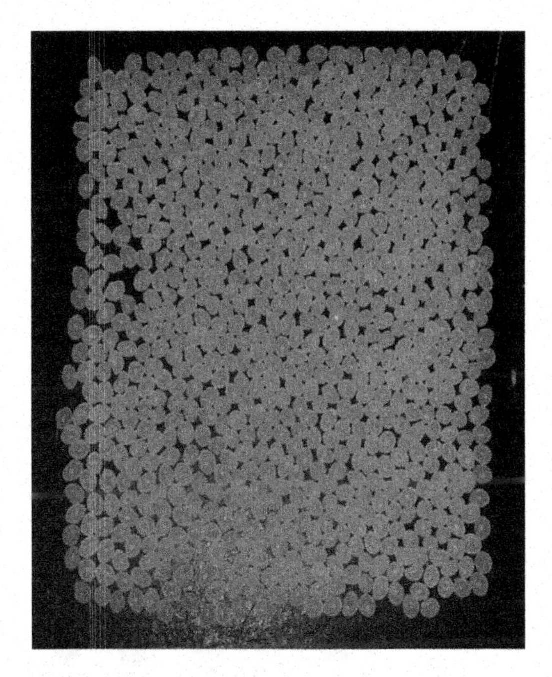

Figure 11. Polylactic acid (PLA) pellets.

Figure 12. PLA/*Calotropis procera* bast fibres composites.

3.2 *Effect of alkali and microwave irradiation treatment on tensile strengths of the PLA composites*

The optimum alkaline and microwave treatments of the extracted *Calotropis procera* fibres (5% w/v alkali solution for 1 hour and microwave exposure of 231 watts power for 4 minutes) obtained during fibre analysis in this study were used for fabrication of the PLA/*Calotropis procera* bast fibres composites. Figure 11 shows the PLA pellets and Figure 12 shows a fabricated composite. In composites prepared from untreated *Calotropis procera* bast fibres (PLA/ut-CPBF), the composite tensile strength increased with an increase in the number of fibre layers with that of three layers of fibres exhibiting better strength (Figure 13). Consequently, this was also observed in the composites with the treated fibres. The tensile strength of PLA/al-CPBF and PLA/mw-CPBF improved by 34.27% and 20.46% respectively.

High-resolution fracture behaviour images after tensile testing of the composites as illustrated in Figure 14, Figure 15, and Figure 16 show that there were fibre pullouts from the matrix in untreated composites signifying that there was poor adhesion between the fibres and the matrix.

Treatments clean the impurities and roughen the fibres' surface, therefore improving mechanical interlocking between the fibre and the matrix and bonding the hydrophilic cellulosic fibre and the hydrophobic plastic with functional molecules which react with both the cellulose OH and the polymer functional groups. For the untreated fibres, there is poor fibre–matrix adhesion. The pull-out of fibres during tensile strength testing is related to poor fibre–resin adhesion. The electrostatic, chemical bonding as well as mechanical interlocking mechanisms are the main contributors to interface bonding in fibre-reinforced composites (Harris, 1999) although different types of bonding could occur and act synergistically (Thakur, Thakur, & Kessler, 2017)

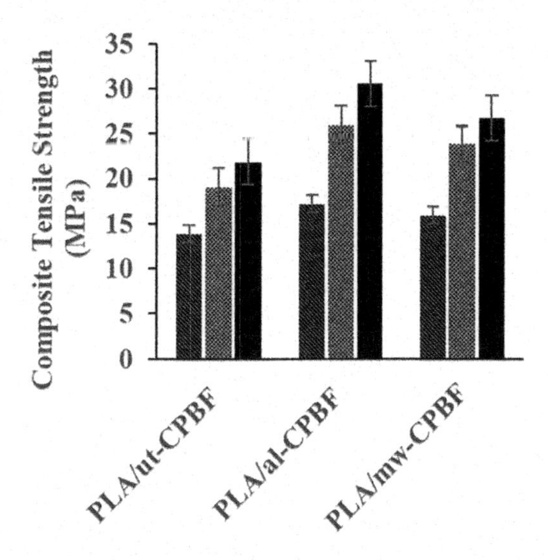

Figure 13. Tensile strength (MPa) of the fabricated composites.

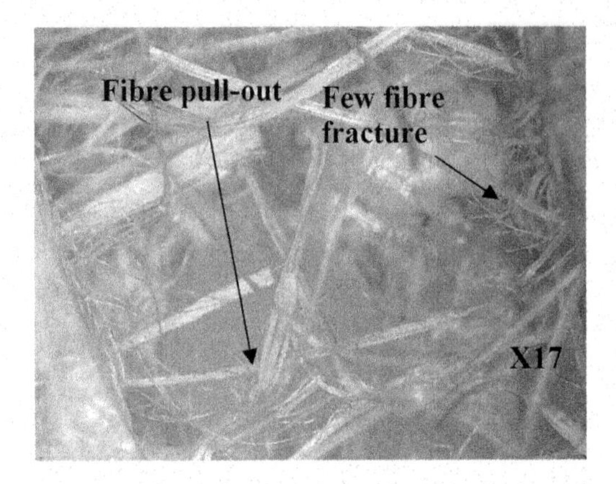

Figure 14. Micrographs of PLA/untreated *Calotropis procera* bast fibre composites showing the fracture surfaces.

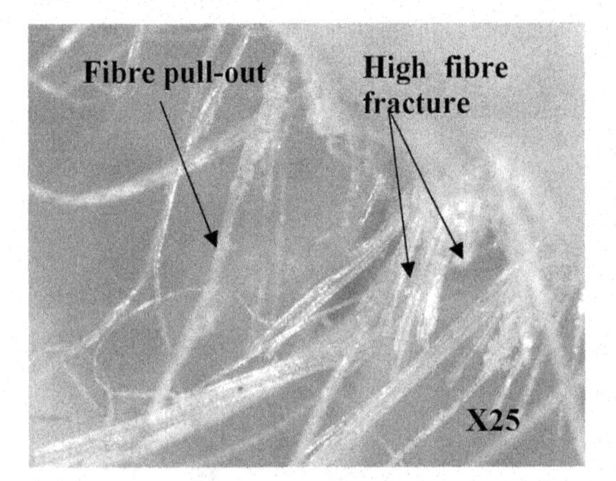

Figure 15. Micrographs of PLA/alkali-treated *Calotropis procera* bast fibre composites showing the fracture surfaces.

The minimal pull-outs of the fibres in the treated fibre composites than in untreated fibre composites indicated that there was better bonding of fibre and matrix in treated fibre composites. The fabrication process used in this study shows that the delamination problem as observed by Rubi-López et al. (Rubio-López et al., 2015) is greatly reduced which is one of the major issues relating to compression moulding with layers of fibres and PLA sheets.

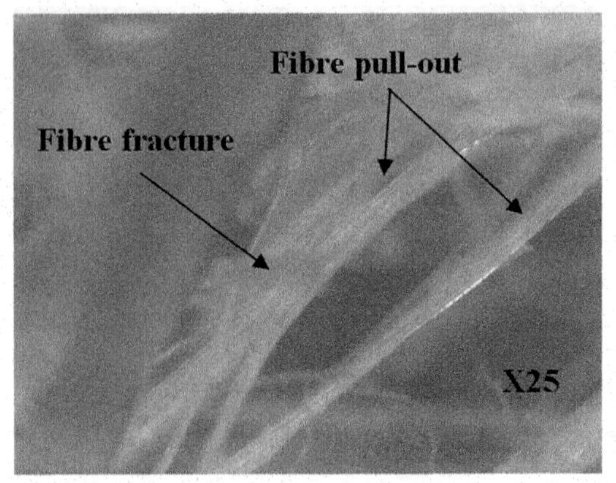

Figure 16. Micrographs of PLA/microwave-treated *Calotropis procera* bast fibre composites showing the fracture surfaces.

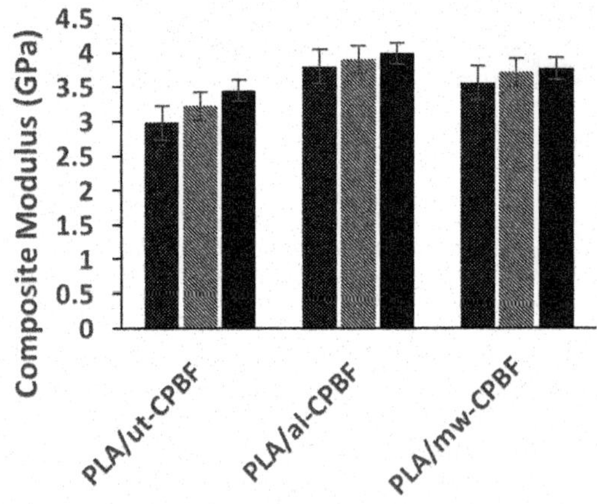

Figure 17. Effects of alkali and microwave irradiation treatments on tensile modulus of the PLA composites.

3.3 *Effects of alkali and microwave irradiation treatment on tensile modulus of the PLA composites*

The effect of alkali and microwave treatments on the tensile modulus of treated PLA/*Calotropis procera* bast fibres composites are given in Figure 17. It is notable from the results that the tensile modulus increased by 21.22% and 15.54% for the alkalitreated fibre composites and microwavetreated fibre composites respectively.

The mechanical properties such as tensile modulus are improved when treated with microwave irradiation because natural fibres contain polar molecules, for instance water, lignin, pectin and hemicellulose which efficiently absorb microwaves making the molecules volatilize adequately quickly (Zhang, Sun, Liang, Lin, & Xiao, 2017). Alkaline treatment also removes these molecules exposing hydroxyl groups thus causing an increase in surface roughness of fibres (Rizal et al., 2019).

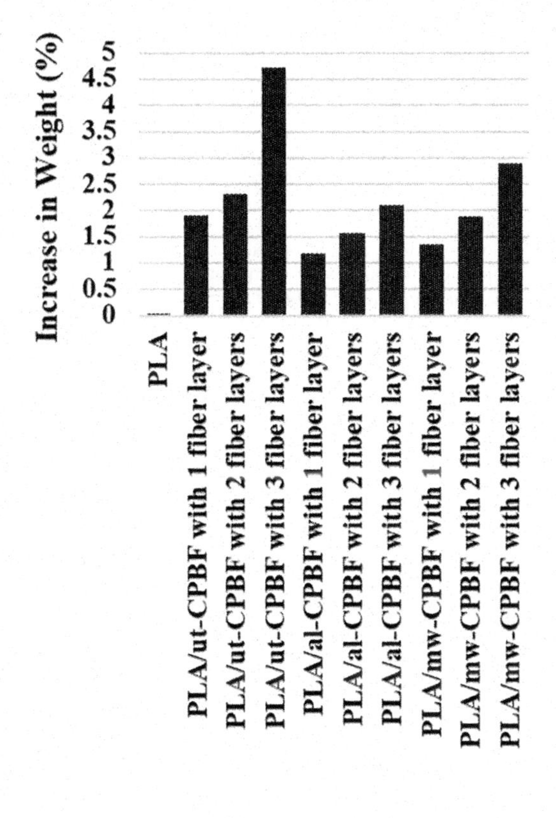

Figure 18. Effects of alkali and microwave irradiation treatment on water absorption of the PLA composites.

3.4 *Effects of alkali and microwave irradiation treatment on water absorption of the PLA composites*

In Figure 18, the water absorption is illustrated and it is evident from the results that the composites with untreated *Calotropis procera* bast fibres absorbed more water than the alkali and microwave-treated fibre composites. It is also noted that the composites with three layers of fibres had higher water absorption percentages, the highest was 4.721% recorded for the untreated fibre composites with three layers of fibres. This can be attributed to the hydrophilicity of *Calotropis procera* fibre as reported by Raghu and Goud (Raghu & Goud, 2020) who also reported a reduction in mechanical properties of the specimen on exposure to moisture. This may be possibly due to degradation of bonding at the fibre–matrix interface.

4 CONCLUSIONS AND RECOMMENDATIONS

4.1 *Conclusions*

The *Calotropis procera* bast fibres can be extracted manually using the decortication method and the results of the characterized fibres show that fibres with good properties that can be used as reinforcement in composites can be obtained. The treatment of *Calotropis procera* bast fibres with alkali solution and microwaves to modify the surface can improve the tensile properties of the fibres as well as the composites. Mild treatment of the fibre surface will prevent fibre damage. But on the contrary, higher concentration and long period of alkali solution treatment as well as higher microwave energy exposure or long period of exposure cause great damage to the fibres. It can therefore be concluded that the alkaline treatment at 5 w/v% alkaline solution for 1 hour and microwave irradiation with 231 watts of power for a 4-minute treatment period were best for *Calotropis procera* bast fibres modification. In this work it was demonstrated that an environment-friendly biocomposite structure from polylactic acid and *Calotropis procera* bast fibres can be developed. The results of the characterized composite structure showed good tensile strength and water absorption properties. In addition, this study has demonstrated that the fabrication process used in this study shows that the delamination problem, which is one of the major issues relating to compression moulding of layers of fibres and PLA sheets, is greatly reduced. This gives an opportunity for the developed composite material to be best applied in various fields such as electronic and food packaging materials, and non-structural packaging.

4.2 *Recommendations*

Further development of bast fibre extraction using mechanical methods such as designing extraction machine and optimization of the process is recommended. Microwave irradiation can be used as one of the environmentfriendly methods of modifying the *Calotropis procera* bast fibres to reduce the hydrophilic nature of the fibres so as to improve fibre–matrix adhesion. Further development of combined treatment methods such as using microwave and environmentfriendly chemicals or reagents or enzymes for fibre surface modification is recommended. Other benefits of alkali and microwave fibre surface treatment such as improved thermal stability to reduce fibre damage by heat during hot compression moulding fabrication process of PLA composite can also be investigated. Although the composites produced could be used in various applications such as food packaging materials, non-structural, further investigation needs to be carried out to evaluate other properties such as thermal properties and other mechanical properties to enhance their efficient practical applications.

ACKNOWLEDGMENT

The author is grateful to Africa Centre of Excellence II in Phytochemicals, Textiles and Renewable Energy (ACE II PTRE), Moi University, Uasin Gishu County,

Eldoret, Kenya for financial support and Moi University Textile Department, for the laboratory services and analytical success of this research.

REFERENCES

Ariadurai, S. (2013). *Bio-Composites: Current Status and Future Trends.*

Bashir, N. H. (2013). Plastic problem in Africa. *Japanese Journal of Veterinary Research, 61*(Supplement), S1–S11.

dos Santos Rosa, D., & Lenz, D. M. (2013). Biocomposites: Influence of matrix nature and additives on the properties and biodegradation behaviour. *Biodegradation–Engineering and Technology; InTech: Rijeka, Croatia*, 433–475.

Elsabbagh, A., Steuernagel, L., & Ring, J. (2017). Natural Fibre/PA6 composites with flame retardance properties: Extrusion and characterisation. *Composites Part B: Engineering, 108*, 325–333.

Fuqua, M. A., Huo, S., & Ulven, C. A. (2012). Natural fiber reinforced composites. *Polymer Reviews, 52*(3), 259–320.

Gassan, J., & Bledzki, A. K. (2001). Thermal degradation of flax and jute fibers. *Journal of Applied Polymer Science, 82*(6), 1417–1422.

Gunatillake, P. A., & Adhikari, R. (2003). Biodegradable synthetic polymers for tissue engineering. *Eur Cell Mater, 5*(1), 1–16.

Harris, B. (1999). Engineering composite materials.

Huda, S., & Yang, Y. (2008). Chemically extracted cornhusk fibers as reinforcement in light-weight poly(propylene) composites. *Macromol Mater Eng, 293*, 235–243.

Imoisili, P. E., Tonye, D. I., Victor, P. A., & Elvis, O. A. (2018). Effect of High-frequency Microwave Radiation on the Mechanical Properties of Plantain (Musa paradisiaca) Fibre/Epoxy Biocomposite. *Journal of Physical Science, 29*(3).

Kalia, S., Thakur, K., Celli, A., Kiechel, M. A., & Schauer, C. L. (2013). Surface modification of plant fibers using environment friendly methods for their application in polymer composites, textile industry and antimicrobial activities: A review. *Journal of Environmental Chemical Engineering, 1*(3), 97–112.

Li, X., Panigrahi, S., & Tabil, L. (2009). A study on flax fiber-reinforced polyethylene biocomposites. *Applied Engineering in Agriculture, 25*(4), 525–531.

Mohanty, A., Misra, M., & Drzal, L. (2002). Sustainable biocomposites from renewable resources: opportunities and challenges in the green material world. *J Polym Environ 1, 10*, 9–26.

Mohanty, A., Misra, M. a., & Hinrichsen, G. (2000). Biofibres, biodegradable polymers and biocomposites: An overview. *Macromolecular materials and Engineering, 276*(1), 1–24.

Muriira, N. G., Muchugi, A., Yu, A., Xu, J., & Liu, A. (2018). Genetic Diversity Analysis Reveals Genetic Differentiation and Strong Population Structure in Calotropis Plants. *Scientific Reports, 8*(1), 7832. doi: 10.1038/s41598-018-26275-x

Ochi, S. (2008). Mechanical properties of kenaf fibers and kenaf/PLA composites. *Mechanics of Materials, 40*, 446–452. doi: 10.1016/j.mechmat.2007.10.006

Puglia, D., Biagiotti, J., & Kenny, J. (2005). A review on natural fibre-based composites—Part II: Application of natural reinforcements in composite materials for automotive industry. *Journal of Natural Fibers, 1*(3), 23–65.

Raghu, M., & Goud, G. (2019). *Tribological properties of calotropis procera natural fiber reinforced hybrid epoxy composites.* Paper presented at the Applied Mechanics and Materials.

Raghu, M., & Goud, G. (2020). Effect of Water Absorption on Mechanical Properties of Calotropis Procera Fiber Reinforced Polymer Composites. *Journal of Applied Agricultural Science and Technology, 4*(1), 3–11.

Ramasamy, R., Obi Reddy, K., & Varada Rajulu, A. (2018). Extraction and characterization of calotropis gigantea bast fibers as novel reinforcement for composites materials. *Journal of Natural Fibers, 15*(4), 527–538.

Rayne, S. (2008). The need for reducing plastic shopping bag use and disposal in Africa. *African Journal of Environmental Science and Technology, 2*(3).

Rizal, S., Nakai, Y., Shiozawa, D., Khalil, H., Huzni, S., & Thalib, S. (2019). Evaluation of interfacial fracture toughness and interfacial shear strength of Typha spp. fiber/polymer composite by double shear test method. *Materials, 12*(14), 2225.

Rubio-López, A., Olmedo, A., Díaz-Álvarez, A., & Santiuste, C. (2015). Manufacture of compression moulded PLA based biocomposites: A parametric study. *Composite Structures, 131*, 995–1000.

Santos, E. F., Mauler, R. S., & Nachtigall, S. M. (2009). Effectiveness of maleated-and silanized-PP for coir fiber-filled composites. *Journal of Reinforced Plastics and composites, 28*(17), 2119–2129.

Satyanarayana, K. G., Arizaga, G. G., & Wypych, F. (2009). Biodegradable composites based on lignocellulosic fibers—An overview. *Progress in polymer science, 34*(9), 982–1021.

Sticklen, J., Kamel, A., Hawley, M., & Adegbite, V. (1992). Fabricating composite materials-a comprehensive problem-solving architecture based on generic tasks. *IEEE Expert, 7*(2), 43–53.

Thakur, V. K., Thakur, M. K., & Kessler, M. R. (2017). *Handbook of composites from renewable materials, biodegradable materials* (Vol. 5): John Wiley & Sons.

Thostenson, E., & Chou, T.-W. (1999). Microwave processing: fundamentals and applications. *Composites Part A: Applied Science and Manufacturing, 30*(9), 1055–1071.

Valadez-Gonzalez, A., Cervantes-Uc, J., Olayo, R., & Herrera-Franco, P. (1999). Effect of fiber surface treatment on the fiber–matrix bond strength of natural fiber reinforced composites. *Composites Part B: Engineering, 30*(3), 309–320.

Yuanhui, Q., Fang, X., Longdi, C., Ruiyun, Z., Lifang, L., Wenhong, F., …Jie, L. (2018). Evaluation on a Promising Natural Cellulose FiberCalotropis Gigantea Fiber. *Trends Textile Eng Fashion Technol., 2*(4). doi: 10.31031/TTEFT.2018.02.000543

Zhang, L., Sun, Z., Liang, D., Lin, J., & Xiao, W. (2017). Preparation and Performance Evaluation of PLA/Coir Fibre Biocomposites. 2017, 12(4), 14.

Advances in Phytochemistry, Textile and Renewable Energy Research for Industrial Growth – Nzila et al. (Eds)

Thermal properties of sisal/cattail fiber reinforced polyester composites

S.M. Mbeche*
Department of Manufacturing, Industrial and Textile Engineering, Moi University, Eldoret, Kenya
Africa Centre of Excellence II in Phytochemicals, Textiles and Renewable Energy (ACE II PTRE), Moi University, Eldoret, Kenya
Department of Project Management, E&M Technology House Limited, Nairobi, Kenya

P.M. Wambua & D.G. Njuguna
Department of Manufacturing, Industrial and Textile Engineering, Moi University, Eldoret, Kenya

ABSTRACT: Natural fibres have been the subject of intensive research in the recent past due to their eco-friendly and renewable nature, in addition to other attractive advantages. The current study evaluated the thermal properties of sisal/cattail fibre-reinforced polyester composites. The composites were fabricated by a hand lay-up technique at varying hybrid fibre weight fractions (5 to 25 wt.%) while maintaining a constant fibre blend ratio of 50/50. Composites were also prepared at constant fibre weight fraction of 20% while varying the fibre blend ratio between 0 and 100%. The thermal conductivity of the hybrid composites decreased by 2.93%, 23.4%, 4.31%, and 20.39% for 5–10, 10–15, 15–20, and 20–25wt.% hybrid fibre loadings, respectively. At a constant hybrid fibre weight fraction of 15%, thermal conductivity of the hybrid composites increased as the percentage of sisal fibres in the hybrid increased from 0%–100%. Thus the hybrid composites may be suitable for non-structural applications as ceiling boards, walls, room partitioning, door panels, and electronic and food packaging. Further studies should investigate the wettability, biodegradability, and flammability of the hybrid composites.

Keywords: *Agave sisalana*, hand lay-up technique, *Typha angustifolia*, thermal conductivity.

1 INTRODUCTION

Natural fibres have been a subject of intensive research in the recent past due to their eco-friendly and renewable nature, in addition to other attractive advantages such as lower density, high specific strength and stiffness, low toxicity, low cost of production, biodegradability, better thermal and insulating properties (Bongomin, Ocen, Nganyi, Musinguzi, & Omara 2020; Gupta, Akash, Rao, & Kumar 2016; Khanam, Reddy, Raghu, John, & Naidu 2007; Mbeche & Omara 2020; Nasir, Gupta, Hossen Beg, Chua, & Asim 2014; Venkateshwaran, ElayaPerumal, Alavudeen, & Thiruchitrambalam 2011). They present lower energy requirements for processing and are readily available in comparison to reinforcing fibres such as glass, aramid, and carbon. Thus, natural fibre-reinforced composites are cheap and attractive for their sustainability and find increasing commercial applications (Bongomin et al. 2020).

As a continuation of our study (Mbeche, Wambua, & Githinji 2020), we evaluated the effects of varying sisal/cattail fibre content on the thermal conductivity of sisal/cattail polyester commingled composites.

*Corresponding author

2 MATERIALS AND METHODS

Sisal fibres were supplied by Lomolo Sisal Estate Limited, Baringo county, Kenya. Green mature cattail leaves were obtained from *Typha angustifolia* from a swamp in the vicinity of Moi University, Eldoret, Kenya (Figure 1). Unsaturated polyester resin (UPR) and methyl ethyl ketone peroxide (MEKP) were supplied by Henkel Chemicals Limited, Nairobi, Kenya.

2.1 *Experimental procedures*

Cattail leaves were separated from the stalk grouping at the leaf base followed by the mechanical decortication process to extract the fibres. Fibres were subsequently dried at 80°C in an oven to constant weight to eliminate excess moisture that would otherwise result in poor fibre–matrix adhesion. Sisal fibres as supplied were cleaned with warm distilled water to remove chlorophyll, leaf juices, adhesive solids, and soluble impurities, after which they were dried (Mbeche & Omara 2020; Mbeche et al. 2020).

Sisal/cattail fibre-reinforced polyester composites were prepared by a simple hand (wet) lay-up technique at 5, 10, 15, 20, and 25% fibre weight fraction with sisal/cattail ratio kept at 50/50 and then at 0/100, 25/75,

DOI 10.1201/9781003221968-15

Figure 1. Natural fibres used in the study: (a) Cattail plant, (b) Sisal fibres, (c) Cattail fibre extraction process.

75/25, and 100/0 sisal to cattail fibre ratios at 15% fibre weight fraction, as described by Mbeche et al. (2020).

Thermal conductivity tests were done using a thermal conductivity apparatus (Model P5687, 199 Cussons Technology, UK) while surface morphology of cattail fibres was investigated using an MSX-500Di Scopeman Digital Microscope (Herter Instruments, Barcelona, Spain), as described by Mbeche et al. (2020).

All reagents used were of analytical grade. All equipment used was calibrated prior to use. Quality control was ensured through analysis of all samples in triplicate.

3 STATISTICAL ANALYSIS

Data were presented as means of triplicates with standard deviations attached. One-way analysis of variance was done followed by the Turkey test to identify any significant differences between the means. All analyses were performed at a 95% confidence interval using Sigma Plot statistical software (v14.0, Systat Software Inc., USA) (Omara et al. 2019).

3.1 *Limitations of the study*

Cattail fibres were extracted manually in this study. This manual extraction was done using a sharp knife (Figure 1c) and therefore could impact the thermal conductivity properties of the fibres negatively and that of the resultant hybrid composite. This is because aerenchyma tissues that influence the thermal properties of the fibre can be destroyed during the fibre extraction process.

4 RESULTS

The results of thermal evaluation of the hybrid composites are given in Table 1, Figures 2 and 3. Surface analysis of cattail fibres is given in Figure 4.

Table 1. Thermal conductivity of reinforced composites at 15 wt.% fibre weight fraction and varying sisal/cattail fibre ratios.

Sisal/cattail fibre ratio (%)	Thermal conductivity, λ (W/mK)
0/100	0.309 ± 0.099
25/75	0.385 ± 0.046
50/50	0.534 ± 0.086
75/25	0.558 ± 0.128
100/0	0.666 ± 0.046

Values are presented as means ± standard deviations

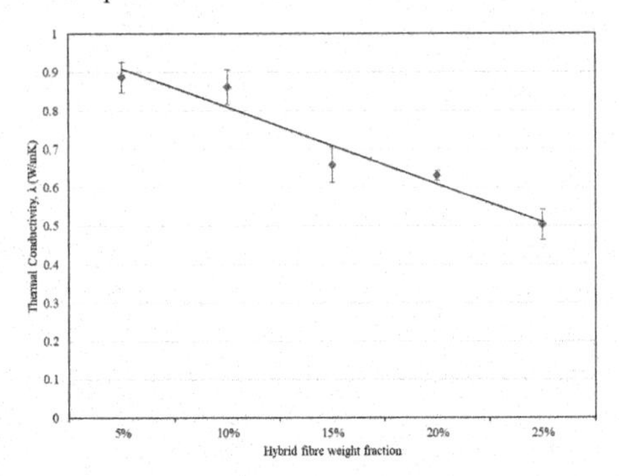

Figure 2. Thermal conductivity of polyester composites at 50/50 sisal/cattail fibre ratio and varying fibre weight fraction.

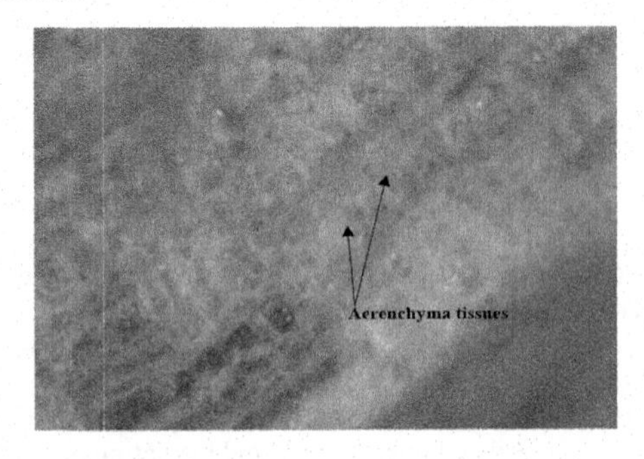

Figure 3. Micrograph (×270) showing aerenchyma tissues in cattail fibres.

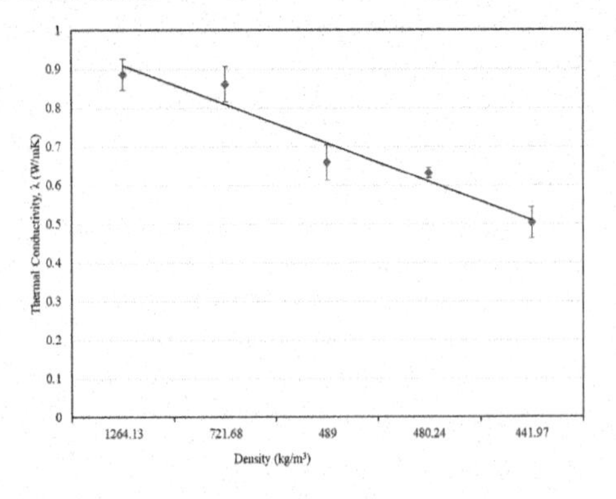

Figure 4. Effect of composite density on the thermal conductivity of the composites at 50/50 sisal/cattail fibre content and 15 wt.% in the hybrid.

5 DISCUSSION OF RESULTS

The effect of varying the percentage of sisal/cattail fibres in the polyester hybrid composites at a constant hybrid fibre weight fraction of 15 wt.% on the thermal conductivity of the composites is given in Table 1. Thermal conductivity of the hybrid composites increased with increase in the percentage of sisal fibres in the hybrid from 0 to 100%. This behaviour could be due to the fact that as sisal fibre content in the hybrid increased (0 to 100%) and the cattail fibre decreased (100 to 0%). The former increased thermal conductivity as they have a relatively better thermal conductivity compared to cattail fibres with good insulation properties (Colbers et al. 2017). A low mean thermal conductivity of 0.309 ± 0.09 W/mK was reported at 0/100 sisal/cattail fibre blend while the highest mean thermal conductivity of the composites (0.666 ± 0.046 W/mK) was reported at 100/0 sisal/cattail fibre blend. Statistical analysis indicated that there was a significant difference ($p = 0.023$) in the mean thermal conductivities of the hybrid composites. The results of the study were similar to previous reports. Alsina et al. (2005) reported thermal conductivities between 0.19 and 0.237 W/mK and 0.19 and 0.22 W/mK for jute/cotton hybrid polyester and ramie/cotton hybrid polyester composites, respectively. Further, Ramanaiah et al. (2011) concluded that the thermal conductivity of cattail-reinforced polyester composites were 0.32–0.385 W/mK at a fibre volume fraction of 0.15–0.32. A comparable thermal conductivity value of 0.163 W/mK at 85% clay was reported by Dieye, Sambou, Faye, Thiam, and Adj et al. (2017) while investigating the effect of the binder (clay) weight on the thermal conductivity of cattail fibre-reinforced composites.

The effects of varying the hybrid fibre weight fraction while the percentage of sisal/cattail fibres in the hybrid was kept at 50/50 is illustrated in Figure 2. Thermal conductivity of the composites decreased as the hybrid fibre loading increased from 5 to 25 wt.% with a minimum mean thermal conductivity of 0.5024 ± 0.04 W/mK. The thermal conductivity of the hybrid composites dropped by 2.93%, 23.4%, 4.31%, and 20.39% for 5–10, 10–15, 15–20, and 20–25 wt.% hybrid fibre loadings, respectively. Significant differences ($p = 0.001$) between the means of various composites at different hybrid fibre weight loadings were recorded. This could be due to the presence of cattail fibres in the hybrid as well as their increment in the composite as hybrid fibre weight fraction was increased from 5 to 25 wt.%. Cattail fibres generally have lower thermal conductivity due to the presence of aerenchyma tissues (Figure 3) (Colbers et al. 2017). Similar trends have been reported by previous studies. Ramanaiah et al. (2013), studying the effect of fibre weight fraction on thermal conductivity of fish tail palm tree fibre-reinforced polyester composites, reported a decrease in thermal conductivity with increase in fibre content from 0.1 to 0.4. Similarly, Ramanaiah et al. (2011) inferred that thermal conductivity of *Typha angustifolia*-reinforced composites decreased with increase in fibre loading. Thermal insulation properties reported in this paper (at 25 wt.%) were found to be better than those reported by Colbers et al. (2017) for cattail fibres.

The effect of composite density on thermal conductivity of sisal/cattail polyester hybrid composites at 50/50 sisal/cattail fibre content in the hybrid are shown in Figure 4. Thermal conductivity was directly proportional to the hybrid composite density where at high composite density, high thermal conductivity values were reported. This may be due to the fact that, as the density of the hybrid composite decreases, available voids between the fibres in the composite increases. It is these air-filled voids that result in lower thermal conductivity of the hybrid composites (Luamkanchanaphan, Chotikaprakhan, & Jarusombati 2012). The same behaviour was reported by Sair et al. (2018) investigating the effect of density on thermal behaviour of hemp/polyurethane composites.

6 CONCLUSION

At constant fibre blend ratio of 1:1 and 15 wt.% hybrid fibre weight fraction, composite density was reported to be directly proportional to its thermal conductivity and an optimal thermal insulation property was reported at a fibre weight fraction of 25 wt.%. Further, low thermal conductivity was reported as the ratio of cattail fibres increased from 0–100% in the hybrid at a constant hybrid fibre weight fraction of 15wt%. The composites produced, can be used in non-structural applications such as ceiling board and electronic and food packaging among many other applications where heat conservation is required.

Further studies on the physical properties for the resultant composite, i.e., water absorption, flammability, and biodegradability tests should be done. Additionally, further research should be done to examine the effect of using a woven blend of sisal/cattail fibres on thermal conductivity properties of the resultant composite.

REFERENCES

Alsina, O. L. S., De Carvalho, L. H., Filho, F. G. R., & Almeida, J. R. M. (2005). Thermal properties of hybrid lignocellulosic fabric-reinforced polyester matrix composites *Polymer Testing*, *24*, 81–85.

Bongomin, O., Ocen, G. G., Nganyi, E. O., Musinguzi, A., & Omara, T. (2020). Exponential Disruptive Technologies and the Required Skills of Industry 4.0. *Journal of Engineering*, *2020*, 1–17.

Colbers, B., Cornelis, S., Geraets, E., Gutiérrez-Valdés, N., Tran, L. M., Moreno-Giménez, E., & Ramírez-Gaona, M. (2017). A feasibility study on the usage of cattail (Typha spp.) for the production of insulation materials and bio-adhesives *Wageningen University and Research Centre*, 1–71.

Dieye, Y., Sambou, V., Faye, M., Thiam, A., & Adj, M. (2017). Thermo-mechanical characterization of a building material based on Typha australis *Journal of Building Engineering*, *9*, 142–146.

Gupta, N., Akash, Rao, K., & Kumar, S. (2016). Fabrication and evaluation of mechanical properties of alkaline treated sisal/hemp fiber reinforced hybrid composite. *IOP Conference Series: Materials Science and Engineering, 149*.

Khanam, N. P., Reddy, M. M., Raghu, K., John, K., & Naidu, V. S. (2007). Tensile, flexure and compressive properties of sisal–silk hybrid composites *Journal of Reinforced Plastic Composites*, *26*, 1065–1070.

Luamkanchanaphan, T., Chotikaprakhan, S., & Jarusombati, S. (2012). A study of physical, mechanical and thermal properties for thermal insulation from narrow-leaved cattail fibers *APCBEE Procedia*, *1*, 46–52.

Mbeche, S. M., & Omara, T. (2020 a). Effects of alkali treatment on the mechanical and thermal properties of sisal/cattail polyester commingled composites. *PeerJ Material Science Chemistry*, *2*, e5. doi: 10.7717/peerj-matsci.5

Mbeche, S. M., Wambua, P. M., & Githinji, D. N. (2020b). Mechanical properties of sisal-cattail hybrid reinforced polyester composites *Advances in Materials Science and Engineering*, *2020*, 1–9. doi: 10.1155/2020/6290480

Nasir, M., Gupta, A., Hossen Beg, M. D., Chua, G. K., & Asim, M. (2014). Laccase application in medium density fibreboard to prepare a bio-composite. *RSC Adv*, *4*, 1–6.

Omara, T., Karungi, S., Kalukusu, R., Nakabuye, B. V., Kagoya, S., & Musau, B. (2019). Mercuric pollution of surface water, superficial sediments, Nile tilapia (Oreochromis nilotica Linnaeus 1758 [Cichlidae]) and yams (Dioscorea alata) in auriferous areas of Namukombe stream, Syanyonja, Busia, Uganda. *PeerJ*, *7*, e7919.

Ramanaiah, K., Prasad, A. V. R., & Reddy, K. H. C. (2013). Mechanical and Thermo-Physical Properties of Fish Tail Palm Tree Natural Fiber–Reinforced Polyester Composites. *Int. J. Polym. Anal. Charact.*, *18*, 126–136.

Ramanaiah, K., RatnaPrasad, A. V., & Reddy, H. C. K. (2011). Mechanical properties and thermal conductivity of typha angustifolia natural fiber-reinforced polyester composites. *International Journal of Polymer Analysis and Characterization*, *16*, 496–503.

Advances in Phytochemistry, Textile and Renewable Energy Research for Industrial Growth – Nzila et al. (Eds)

Dyeing of cotton fabric with natural dye from *Flavoparmelia caperata*

Linet Jelagat* & Jackson Cherutoi
School of Sciences & Aerospace Studies, Department of Chemistry & Biochemistry, Moi University, Eldoret, Kenya

Achisa C. Mecha
School of Engineering, Department of Chemical & Process Engineering, Moi University, Eldoret, Kenya

ABSTRACT: Natural dyes have a natural origin, while synthetic dyes are from organic molecules and are made up of chemical compounds that may be harmful to human health due to their toxic nature. Textile industries have been using synthetic dyes in colouring fabrics for a long period of time. Recently, natural dyes have become of interest due to the health hazards associated with the synthetic dyes. The degradation of synthetic dyes produces a lot of by-products that are associated with health hazards, whereas such hazardous compounds have so far not been found in by-products of natural dye degradation. *Flavoparmelia caperata* is a lichen species that has been evaluated for its anti-fungal, anti-bacterial, anti-inflammatory, and cytotoxic effects. This study focuses on the application of dye from the lichen extract on a cellulosic fabric with and without a mordant. The dye was extracted using boiling water method (BWM) and ammonia fermentation method (AFM) where both resulted in a brown coloured extract. Phytochemical screening confirmed the presence of flavonoids, fixed oils, anthraquinones, terpenoids, tannins, steroids and alkaloids. The dye was applied on a cotton fabric and assessed for colour fastness properties to wash, light, and rub. The dyed fabrics were brown with extract from BWM and light purple from AFM without a mordant, on mordanting the purple colour darkened. The extract from AFM mordanted fabrics showed best results upon colour fastness tests to rubbing and washing of 3–5 while extracts from BWM displayed best results on light of 5 against a standard grey scale of 1–5, which is above the acceptable levels of 3. Vinegar was used as a mordant. *Flavoparmelia caperata* is therefore a good source of phytochemicals and it has potential for dyeing cotton fabrics.

Keywords: Natural dye, colour fastness, *Flavoparmelia caperata*

1 INTRODUCTION

The major environmental pollution today is from industries, with textile industries being the largest contributors of liquid effluent due to the large amount of water used in textile processing (Malik & Grohmann 2012). Although some other industries like paper and pulp mills, distilleries and tanneries also produce wastewater, textile industries are the largest producers of aqueous wastes and dye effluents are discharged from the dyeing process. It is estimation that globally 280,000 tons of textile dyes are released as textile industrial effluent yearly. Apart from being beautiful, dyes have disadvantages because they make living things in water uncomfortable when discharged as effluent into the environment. Effluents from industries containing synthetic dyes limit light penetrating into water bodies, and hence affect the activities of life. They also reduce the amount of dissolved oxygen by forming a thin layer on the water's surface affecting aquatic fauna.

Traditionally, the only source of colours was of natural origin but the extraction costs and the high rate of instability of natural dyes initiated the development and use of artificial dyes by Sir William Henry Perkin in 1856 (Kalra, Conlan, & Goel 2020). The use of synthetic dyes from organic sources and aniline continued in many sectors including textiles, pharmaceuticals, and cosmetics until negative anxiety against these dyes was reported by the University of Southampton in 2007 with a connection between some synthetic dyes and hyperactivity in children. These include colourants such as Sunset yellow FCF (E110), Quinoline yellow (E104), Carmoisine (E122), Allura red (E129), Tartrazine (E102), and Ponceau 4R (E124), commonly known as the Southampton six. Due to the influence of dyes on human health, other bodies such as the European Union (EU), World Health Organizatio (WHO), and United States Food and Drug Adminstration (US-FDA) warned against the use of synthetic colourants and this initiated research on natural dyes which are environmentally friendly.

Dyes are chemical compounds that are attracted to the substrate in a more or less permanent state giving

*Corresponding author

DOI 10.1201/9781003221968-16

colours or molecules which have the ability to absorb or reflect the visible part of light at a specific wavelength, imparting to the human eye a sense of colour. Dyeing is a process of imparting beauty to the fabric by applying many colours and their shades in the textile (Daberao, Kolte, & Turukmane 2016). Natural dyes are colourants, which are derived from plants, invertebrates, minerals, and organic sources such as fungi and lichens. These dyes are used for colouring textiles, food, and other materials. They are environmentally friendly, biodegradable, and their by-products can be used as fertilisers.

The textile industry is the largest consumer of dyestuffs. During the process of colouring, most of the synthetic dye does not adhere to the fabric, and hence is lost in the waste stream, thus effluents from the textile industry carry a number of dyes and also other additives which are added during the colouring process (An et al. 2017). These are difficult to get rid of, thus they are transported to sewers and rivers. They are highly water soluble and may also undergo degradation to form other by-products that are highly toxic and carcinogenic in nature. It is only in the textile industry that the largest amount of aqueous wastes and dye effluents are released from the dyeing process with both strong and persistent colour and highly biological oxygen demand (BOD), both of which are not environmentally friendly and are unacceptable (Singh & Arora 2011).

According to Mozumder and Majumder (2016), natural dyes gained usage due to environmental awareness,. These natural dyes are better in terms of biodegradability, good for ultraviolet protection, antibacterial activity, and have the best compatibility with the environment. Most dyes from natural sources are not substantive and they require the use of mordants to fix the dye properly on the fabric. The colour fastness of natural dyes is different based on the method used and the application used. Hence this study will use the best possibilities and methods of dyeing cotton fabric. The colour fastness property depends on the selection of the dye type to the type of the fabric to be applied (Mozumder & Majumder 2016). Thus during application, the choice and method yields different results in terms of colour fastness.

Lichens are organisms which are pleasant to the eye. They are symbiotic organisms composed of algae and fungi, which together form an independent and unique physiological unit (Shaheen, Iqbal, & Hussain 2019). They grow in terrestrial habitats like on trees, rocks, and soil. Their slow growth and harsh living conditions are due to the fact that they produce secondary metabolites, which are the main source of natural dyes. The first dye producing lichen was from a species of *Rocella* which produced a purple dye colour through a fermentation method of extraction. Lichen dyes have despides and despidones that are composed of two or three phenolic groups and are formed through the acetate–polymelonate pathway by the phycobiont partner. These are the sources of dyes from lichens which are mainly used to dye natural fabrics. As they grow on tree trunks, branches, and twigs of trees and rocks, unfortunately many people have not identified or used them due to lack of information on their importance, which is why they are unnoticed and unexplored.

Natural dyes are classified according to their source, i.e., (a) from plants: flowers, bark, leaves, and seeds; (b) from insects; (c) from minerals; and (d) from fungi. These natural colours are due to chromophores. Dyes are also classified as reactive dyes, vat dyes, azo dyes, direct dyes, and alkaline dyes (Benkhaya, Harfi, & Harfi 2018). These authors classified natural dyes according to their classes, application, and also principles, thus natural dyes are classified according to the chemical structure or according to their application on the textiles. Another classification of natural dyes is into three classes: (a) natural dyes from plants, e.g., indigoid; (b) those obtained from animals, e.g., cochineal; and (c) those obtained from minerals, e.g., ochre. Jihad (2014) stated that these dyes are safe due to their non-toxic, non-carcinogenic, and biodegradable nature and are grouped into three types: (a) substantive dyes which link directly with the fabric during dyeing process—this type of dye does not require mordants to fix the dye into the fabric as they contain their mordants in the dye structure, e.g. extracts of tea, walnut, and onion; (b) traditional dyes which consist of natural dyes that require mordants to form a link between the dye and the fabric; and (c) according to the dye chemical composition and hue.

The methods of extraction of natural dyes are found through trials of different methods. Mphande and Pogrebnoi (2014) used three extraction methods: Soxhlet, aqueous, and cold soxhlet extraction methods for the extraction of natural dyes. They found out that cold ethanol and hot soxhlet extraction methods gave the best results.

Ain, Sakinah, and Zularisam (2016) extracted anthroquinones from the roots of *Marinda citrifolia* using a solvent extraction method, which the authors described as the best and most convenient method as it is widely used for the extraction of natural dyes. The method is referred to as a safe and easy method of natural dye extraction compared to other methods. It has two phases: liquid (solvents) and solid (plant matrix), and two extraction processes whereby there is transfer of solute from plant material to the solvent by diffusion and osmotic pressure. The type of the solvent, pH, extraction time, and techniques of extraction are the factors that determine the quantity of the analyte.

According to Shaheen et al. (2019) the following extraction methods were used in the extraction of dyes from lichens: cow urine method (CUM), boiling water method (BWM), ammonia fermentation method (AFM), and dimethyl sulfoxide extraction method (DEM). It was reported that the three extraction methods gave different colours upon dyeing of silk fabric, with AFM being the best method of extracting natural dyes from lichens as it gave the brightest colours.

Mohamed, Ahmad, Abd Kadir, Ismail, & Wan Ahmad (2015) extracted natural dyes from two lichen species: *parmotrema praesorediosum* and *Heterodermia leucomeles*, using the boiling water method (BWM) and ammonia fermentation method (AFM). They used the extracted dye to dye silk fabrics, and found that both the two extraction methods gave good results.

Cotton fabrics were applied with natural dye using pre-mordanting, simultaneous mordanting, and post-mordanting methods of mordanting. Using sodium sulfate mordant, various colour shades were obtained from the fabric due to different mordanting methods. Post-mordanting gave out the best results overall (Coriaria, Species, Janani, Hillary, & Phillips 2014).

Janani (2015) dyed cotton fabric using natural dye from *Vernonia amygdalina* using pre-mordanting, simultaneous mordanting, and post-mordanting methods, with post-mordanting again registering the best results.

PrabhavathI, Devi, and Anitha (2015) used natural dyes from the bark of eucalyptus tree to dye cotton fabrics and also to find out the colour fastness of the dyed fabric with alum, stannous chloride, vinegar, and ferrous sulfate as mordants. Among the dyed cotton fabrics, post-mordanted fabrics with vinegar as a mordant produced the best shades.

Grover and Patni (2011) performed an experiment on the effects of dyeing with mordants and colour fastness of cotton natural dyed fabric. A dark brown yellow dye which had no effect on human skin was obtained, which showed that the use of combination of different mordants at different concentrations produces different shades of colours.

Sharma and Grover (2011) dyed cotton fabric using natural dye from walnut using three mordanting methods: pre-mordanting, simultaneous mordanting, and post-mordanting, and using alum, chrome, copper sulfate, and ferrous sulfate as the mordants, Post-mordanted cotton gave good results in terms of colour fastness.

Kumar and Prabha (2018) used natural dye from different plants to dye cotton and wool fabrics and using different mordants like cooking salt, vinegar, and potassium dichromate. Cooking salt and vinegar showed good results using the pre-mordanting method.

2 MATERIALS AND METHODS

The following chemicals, reagents, and equipment were used in this study: concentrated sulfuric acid, distilled water from Moi University Textile Engineering Laboratory, ammonia solution, acetic acid, Wagner's reagent, chloroform, and ferric acid purchased from Merck Ltd. All the reagents and chemicals used in this study were of analytical grade. The following equipment were used during the study: analytical balance (Citizen Ltd), spectrophotometer, Crock rubbing and colour fastness tester, MBTL sun for colour fastness against sunlight and Wash master for washing fastness tester from Paramount Company.

2.1 *Sample collection*

Flavoparmelia caperata was collected from Tunochun forest in Baringo county, Kenya with GPS coordinates of latitude 0.4667 and longitude 35.9667 during the month of October 2019. Purposive sampling was used for sampling the area and simple stratified sampling was used to come up with the sample. The sampling area was divided into strata and from each stratum, the sample was picked and then from the groups samples were picked randomly to represent the laboratory sample.

2.2 *Extraction methods*

The sample was washed off the substratum using distilled water and air dried at room temperature for 4 weeks in a chemistry laboratory at Moi University. The dried sample was ground into powder form and used for the extraction of dyes.

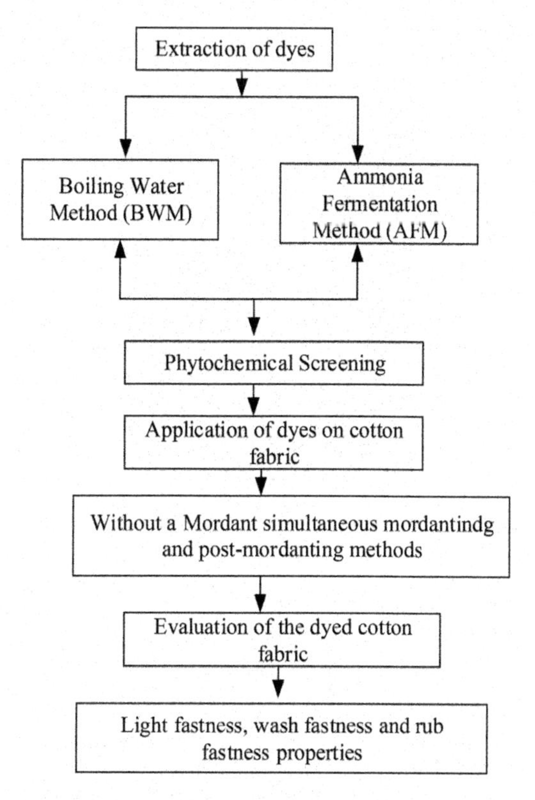

Figure 1. Simple illustration of methodology employed.

2.2.1 *Boiling water extraction method (BWM)*

Powdered sample of 50 g was put in 500 mL distilled water and allowed to stand for 30 minutes. The mixture was heated until it boiled for 30 minutes, the content was allowed to cool and filtered using No .1 Whatman filter paper. The filtrate was brown in colour.

Figure 2. Extract from the boiling water method.

2.2.2 *Ammonia fermentation method (AFM)*

Powdered sample of 50 g was added to dilute ammonium hydroxide solution, the content was shaken thoroughly and left for 30 days to ferment in a closed jar at room temperature. The content was filtered using No.1 Whatman filter paper and the filtrate obtained was dark brown.

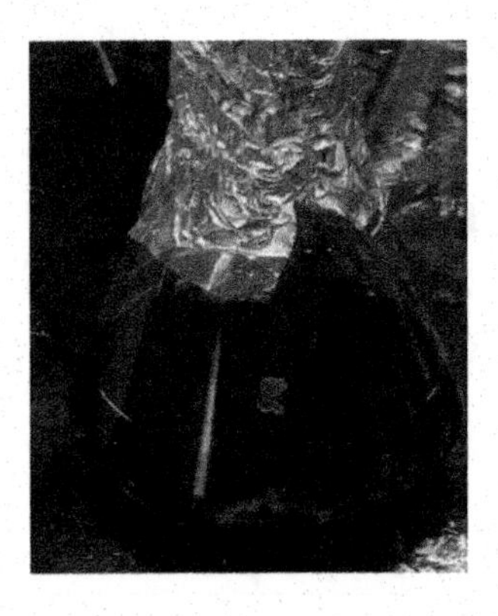

Figure 3. Extract from the ammonia fermentation method.

2.3 *Phytochemical screening*

The filtered extract was used for screening of phytochemicals using standard procedures.

2.3.1 *Test for anthraquinones*

A sample of 10 mL of the extract was boiled with 20 mL of sulfuric acid. The solution was shaken with 10 mL chloroform and the chloroform layer was pipetted into another test tube, 1 mL of dilute ammonia was added, the resulting solution was observed for colour changes.

2.3.2 *Test for flavonoids*

About 5 mL of dilute ammonia was added to 10 mL of the extract, concentrated sulfuric acid (1 mL) was added, a yellow colouration that disappeared on standing indicates the presence of flavonoids.

Extract of 10 mL was heated with 10 mL ethyl acetate over a steam bath for 3 minutes. The mixture was filtered and 4 mL of the filtrate was shaken with 1mL dilute ammonia solution, a yellow colouration indicates the presence of flavonoids.

2.3.3 *Test for saponins*

An extract of 10 mL was added 10 mL distilled water in a test tube, the solution was shaken vigorously and observed for a stable persistent growth.

2.3.4 *Test for tannins*

An extract of 10 mL was boiled with 10 mL of distilled water in a test tube and then filtered. Two drops of ferric chloride were added and observed for colour changes.

2.3.5 *Test for alkaloids*

Wagner's test: to 10 mL of the extract, 5 mL of Wagner's reagent (iodine in potassium iodide) was added.

2.3.6 *Test for terpenoids*

To the extract of 10 mL, 2 mL of dhloroform was added and 1mL of concentrated sulfuric acid was added to the mixture.

2.3.7 *Test for fixed oils*

An extract of 10 mL was heated for 3 minutes, the heated extract was filtered using No. 1 Whatman filter paper.

2.3.8 *Test for steroids*

An extract of 10 mL was added 10 mL acetic acid, 10 mL chloroform followed by 5 mL concentrated sulfuric acid.

2.4 *Application of the dye on a cellulosic fabric*

2.4.1 *Sample 1*

The fabrics were dyed without a mordant in hot dyeing conditions, the fabrics were dyed with 250 mL of the extract at a temperature between 60 and 80°C for 30 minutes and left to cool.

2.4.2 *Sample 2*

The fabrics were treated with both dye extract and vinegar as a mordant simultaneously using 250 mL of the extract and 10 mL of the mordant at a temperature of 60–80°C for 30 minutes.

2.4.3 *Sample 3*

The fabrics were treated with the dye extract and the vinegar mordant through the post-mordanting method. The dyed fabrics were treated with 10 mL of the mordant, the dyeing was carried out for 30 minutes at 60–80°C and temperature was maintained with the help of a thermometer.

After dyeing, the dyed fabrics were removed, left to cool, washed gently in distilled water to remove dye particles loosely adhered to the fabric surface, air dried, stretched, and stored for analysis.

3 RESULTS AND DISCUSSION

3.1 *Phytochemical analysis of the Lichen extract*

Phytochemical analysis was carried out for various classes of phytochemicals. Table 1 shows the results for the phytochemical tests.

Brown and purple shades were obtained from the dye extract, vinegar mordant played an important role in imparting colours to the fabrics, as shown by the fabrics' colour being darkened as compared to the unmordanted fabrics (Foisal & Nahar 2016); Wanyama, Kiremire, & Murumu 2014). Mordants have the affinity for colourants and the cellulosic fabrics give rise to firm complexes with the colourants in the fabric matrix. Brown colour shades indicates a major role played by flavonoids and tannins moieties during the colouring process; anthraquinones are responsible for different colour shades with the interaction with flavonoids. Saponins are good in solubilisation of dyes, their presence in the extract indicates the dye solubilises. Fixed oils help during dyeing of fabrics since it makes the fabric swell and increases the diameter of the fabric leading to the high penetration of dyes into the fabric. Terpenoids serve as a source of flavours, fragrance, and odour.

Table 1. Phytochemical analysis results.

Phytochemical	Observation	Results AFM	Results BWM
Anthraquinones	Colour change from Brown to colourless	+	+
Flavonoids	Presence of yellow colour	+	+
Saponins	Stable persistent growth	+	+
Tannins	Brown green colour present	+	+
Alkaloids	Red brown precipitate present	+	–
Terpenoids	A red colouration on the interface of the two solutions	+	+
Fixed oils	Oil smearing on the filter paper	+	+
Steroids	Blue green solution	+	+

+ presence of the phytochemical tested
– absence of the phytochemical tested

3.2 *Colour fastness properties*

Colour fastness of the dyed cotton fabrics ranged from good to excellent in the range of 3–5 against a standard grey scale of 1–5.

The colour fastness to washing, light, and rubbing were evaluated for the dyed cotton fabric with the dye extract from *Flavoparmelia caperata*. The rubbing fastness was determined in terms of wet and dry rubbing, where wet rubbing gave results of 4/5 in all dyeing methods except in fabrics dyed without the use of a mordant (3/5), while dry rubbing was 5 in all the dyed fabrics. Light fastness of the dyed fabrics dyed using extract from AFM gave grades of 4/5 while those from BWM gave excellent grades of 5 (Mohamed et al. 2015). This is an indication that the shades change in colour upon exposure to light due to photo-oxidation of chromophores (colour producing structure) and the use of a mordant has no effect on the colour fastness properties.

Wash fastness of the dyed samples were analysed as per the Paramount Company Washmaster for washing fastness tester using ISO and AATCC standards. Light fastness was analysed as per the Paramount Company MBTL, for colour fastness against sunlight using ISO and AATCC standards. Rubbing fastness was analysed as per the Paramount Company –Crock Rubbing colour fastness tester using ISO and AATCC standards in conjunction with the standard procedures of Texanlab Laboratories Pvt, Ltd. Colour Fastness Testing series.

Table 2. Colour fastness scale.

Method of Mordanting	Mode of extraction	Light fastness	Wash fastness	Rub fastness	Wet rub	Dry rub
Simultaneous	BWM	5	4	4	5	
Mordanting	AFM	4	4	4	5	
Post	BWM	5	5	4	5	
Mordanting	AFM	4	4		5	5
Unmordanted	BWM	4	5	3	5	
(Control)	AFM	3	5	4	5	

Colour development and dye absorption capability of the cotton fabric were evaluated in terms of CIE lab coordinates and k/s (colour strength) values, thus the colour fastness ranged from very good to excellent in the range of 3 and 5 against a grey scale of 1–5 for all the tested fastness.

The good wash fastness of the cellulosic fabric is due to (i) the presence of tannin moieties in the dye extract - since they improve the cellulosic fabric fastness by introducing extra hydroxyl and carboxyl functional groups for coordination, (ii) a stronger mordant–fabric–dye complex inside the fabric matrix (Wanyama et al. 2014), (iii) large particles of water insoluble dyes, and (iv) the covalent bonding of the fabric and the dye colourants achieved during the dyeing process. Meanwhile, the good results of light fastness might be due to the chemical structure of the dye. This is because the chemical structure is related directly to photochemical exposure, hence dyes which have a larger chemical structure have the ability to resist higher exposure to light (Arthur & Csiro, n.d.).

3.3 *Statistical analysis*

In the present study, relative colour strength (%), colour fastness to rub, wash, and light, were affected by the method of extraction and the use of a mordant, thus the data were analysed statistically using t-test one way analysis of variance (ANOVA), with significant values of 0.05.

4 CONCLUSIONS

Globally, serious efforts and awareness are in place to favour the shift towards the use of natural resources, especially as a source of dyes since they are friendly to the environment. This research showed Flavoparmelia caperata lichen species can be used as a source of natural dyes. The use of a vinegar mordant gave rise to different shades using different mordanting methods to achieve reasonable fastness properties.

The presence of anthraquinones, flavonoids, steroids, tannins, saponins, terpenoids, fixed oils, and alkaloids in the *Flavoparmelia caperata* extract indicates that it can be a source of natural dyes with good colour fastness properties to light, rub, and wash.

5 LIMITATIONS OF THE STUDY

The following limitations were established in this research work:

- Crude extract was used to dye cotton fabric in this study.
- Only 100% cotton fabric was dyed in this work.
- Only beaker dyeing was used in this work which was slow as compared to continuous dyeing process.
- Vinegar mordant was used, no other fixing agent was used in this project.

6 RECOMMENDATION

Further research needs to be done to determine the chromophoric structures of the colouring components present in the crude extract that are responsible for the characteristic brown and purple colour development on the cellulosic fabric.

ACKNOWLEDGMENT

The authors would like to acknowledge the Africa Centre of Excellence in Phytochemicals, Textile and Renewable Energy (ACEII – PTRE) for funding this work.

REFERENCES

Ain, J. N., Sakinah, A. M. M., & Zularisam, A. W. (2016). Effects of Different Extraction Conditions on The Production of Anthraquinone. *Australian Journal of Basic and Applied Sciences*, *10*(17), 128–135.

An, S., Zhao, L. P., Shen, L. J., Wang, S., Zhang, K., Qi, Y., Zheng, J., Zhang, X. J., Zhu, X. Y., Bao, R., Yang, L., Lu, Y. X., She, Z. G., & Tang, Y. Da. (2017). USP18 protects against hepatic steatosis and insulin resistance through its deubiquitinating activity. *Hepatology*, *66*(6), 1866–1884. https://doi.org/10.1002/hep.29375

Arthur, M., & Csiro, F. (n.d.). *Colour fastness Contemporary wool dyeing and finishing.* https://www.woolwise.com/wp-content/uploads/2017/05/04.1-Colour-Fastness-Presentation.pdf

Benkhaya, S., Harfi, S. El, & Harfi, A. El. (2018). *Classifications, properties and applications of textile dyes: A review. January 2017.*

Coriaria, A., Species, P., Janani, L., Hillary, L., & Phillips, K. (2014). *Mordanting Methods for Dyeing Cotton Fabrics with. 4*(10), 1–5.

Daberao, A. M., Kolte, P. P., & Turukmane, R. N. (2016). Cotton Dying with Natural Dye. *International Journal of Research and Scientific Innovation, March 2018*, 2321–2705.

Foisal, A., & Nahar, N. (2016). *Institutional Engineering and Technology (IET) Green Global Foundation ©. April.*

Grover, N., & Patni, V. (2011). Extraction and application of natural dye preparations from the floral parts of Woodfordia fruticosa (Linn.) Kurz. *Indian Journal of Natural Products and Resources*, *2*(4), 403–408.

Janani, L. (2015). *Dye for the Future: Natural Dye from Morinda Lucida Plant for Cotton and Silk Fabrics. 2*(October), 601–606.

Jihad, R. (2014). Dyeing of Silk Using Natural Dyes Extracted From Local Plants. *International Journal of Scientific & Engineering Research*, *5*(11), 809–818.

Kalra, R., Conlan, X. A., & Goel, M. (2020). Fungi as a Potential Source of Pigments: Harnessing Filamentous Fungi. *Frontiers in Chemistry*, *8*(May), 1–23. https://doi.org/10.3389/fchem.2020.00369

Kumar, V., & Prabha, R. (2018). Extraction and analysis of natural dye. *J. Nat. Prod. Plant Resour*, *8*(2), 32–38.

Malik, A., & Grohmann, E. (2012). Environmental protection strategies for sustainable development. *Environmental Protection Strategies for Sustainable Development, January*, 1–605. https://doi.org/10.1007/978-94-007-1591-2

Mohamed, N.A., Ahmad, M.R., Abd Kadir, M.I., Ismail, A., & Wan Ahmad, W.Y. (2015). Dyeing of Silk Fabric with Extracted Dyes from Lichens. *Advanced Materials Research*, *1134*(December), 165–170. https: //doi.org/10.4028/www.scientific.net/amr.1134.165

Mozumder, S., & Majumder, S. (2016). *Comparison between Cotton and Silk Fabric Dyed With Turmeric Natural Colorant. March.*

Mphande, B. C., & Pogrebnoi, A. (2014). *Impact of extraction methods upon light absorbance of natural organic dyes for dye sensitized solar cells application. 3*(3), 38–45. https://doi.org/10.11648/j.jenr.20140303.13

Prabhavathi, R., Devi, A. S., & Anitha, D. (2015). Improving the colour fastness of the selected natural dyes on cotton. *Asian Journal of Home Science*, *10*(1), 240–244. https://doi.org/10.15740/has/ajhs/10.1/240-244

Shaheen, S., Iqbal, Z., & Hussain, M. (2019). *First report of dye yielding potential and compounds of lichens: a cultural heritage of himalayan communities, pakistan. 51*(1), 341–360. https://doi.org/10.30848/PJB2019

Sharma, A., & Grover, E. (2011). Colour fastness of walnut dye on cotton. *Indian Journal of Natural Products and Resources*, *2*(2), 164–169.

Singh, K., & Arora, S. (2011). Removal of synthetic textile dyes from wastewaters: A critical review on present treatment technologies. *Critical Reviews in Environmental Science and Technology*, *41*(9), 807–878. https://doi.org/10.1080/10643380903218376

Wanyama, P. A. G., Kiremire, B. T., & Murumu, J. E. S. (2014). *Extraction, characterization and application of natural dyes from selected plants in Uganda for dyeing of cotton fabrics*. *8*(April), 185–195. https://doi.org/10.5897/AJPS12.065

Advances in Phytochemistry, Textile and Renewable Energy Research for Industrial Growth – Nzila et al. (Eds)

Biosynthesis of zinc oxide nanoparticles as a potential adsorbent for degrading organochlorines

C.O. Ondijo
Moi University, Eldoret, Kenya

O. K'owino
Masinde Muliro University of Science and Technology, Kakamega, Kenya

F.O. Kengara
Bomet University College, Bomet, Kenya

ABSTRACT: Water pollution due to organic contaminants has been a serious issue in developing countries because of the acute toxicities and carcinogenic nature of these pollutants. Among various water treatment methods, adsorption is purported to be one of the best because it is cheap and easy to prepare and use. Initially, activated carbon was being used in water treatment to remove contaminants but it proved to be expensive and also it did not degrade these contaminants after adsorption. For this reason, zinc oxide nanoparticles are synthesized for recommended use in the degradation of pesticides. The zinc oxide nanoparticles were synthesized using *Cissus quadrangularis* plant leaf extract. The surface analysis of the synthesized nanoparticle was analyzed using a particle analyzer and X-ray diffraction (XRD) crystallography. The synthesized nanoparticles were found to have a mean diameter of 14.83 nm and the XRD pattern revealed the formation of ZnO nanoparticles showing crystallinity. The synthesized ZnO nanoparticle showed a characteristic peak at a wavelength of 368 nm for electron excitation. This simple and cost-effective phytochemical approach for the formation of ZnO nanoparticles has a promising application in biosensing, photocatalysis, electronics, and photonics.

Keywords: Water pollution, Organic contaminants, nanoparticle, Carcinogenic, adsorption

1 INTRODUCTION

Nanoparticles or ultrafine particles are particles of matter that are between 1 and 100 nanometers (nm) in diameter. The term at times is used for larger particles, up to 500 nm, or fibers and tubes that are less than 100 nm in only two directions (Trojanowski & Fthenakis 2019). The various methods used to synthesize nanoparticles include coprecipitation, hydrothermal synthesis, inert gas condensation, ion sputtering scattering, microemulsion, microwave, pulse laser ablation, sol–gel, sonochemical, spark discharge, template synthesis, and biological synthesis (Rane, Kanny, Abitha, & Thomas 2018). Among these methods it is noted that "Green synthesis" (biological synthesis) of nanoparticles makes use of environment-friendly, non-toxic, and safe reagents (Mirzaei & Darroudi 2017). Nanoparticles synthesized using biological techniques or green technology have diverse natures, with greater stability and appropriate dimensions since they are synthesized using a one-step procedure (Parveen, Banse, & Ledwani 2016).

The principal parameters of nanoparticles which make them unique in chemical reactions and reactivity are their shape, size, and the morphological structure (Ealias & Saravanakumar 2017). Nanoparticles can be present as an aerosol (mostly solid or liquid phase in air), a suspension (mostly solid in liquids), or an emulsion (two liquid phases). However, in the presence of chemical agents (surfactants), the surface and interfacial properties may be modified (Saxena, Goswami, Dhodapkar, Nihalani, & Mandal 2019). Indirectly nanoparticles can stabilize against coagulation or aggregation by conserving particle charge and by modifying the outmost layer of the particle (Hong 2019). Nanoparticles have been in use in a wide range of technologies, including adhesion, lubrication, stabilization, and controlled flocculation of colloidal dispersions (Raghu, Parkunan, & Kumar 2020).

Nanoparticles such as colloidal gold may intrude into complex folded biological molecules which are 1 nm in size to both plants and animals which are normally evidenced through immunolabeling and related surface functionalization techniques to target nanoparticles to biomolecules as markers for high-resolution transmission electron microscopy and optical imaging systems (Nativo, Prior, & Brust 2008).

DOI 10.1201/9781003221968-17

Green routes are used for the synthesis of zinc oxide nanoparticles because of the small number of chemicals that are used, which then produces the least amount of pollutants, and they are energy efficient and cost-effective. A number of natural products such as plants, fungi, algae, bacteria, and viruses can be used to synthesize the zinc oxide nanoparticles (Naveed Ul Haq et al. 2017).

In the past zinc oxide nanoparticles have been in gas sensors, biosensors, cosmetics, storage, optical devices, window materials for display, solar cells, and drug delivery, due to their unique properties (Pascariu & Homocianu 2019). Zinc oxide nanoparticles have diameters of less than 100 nanometers. Their large surface area relative to their size results in higher catalytic activity (Siripireddy & Mandal 2017).

Nanomaterials can function in the phytoremediation system through directly removing pollutants, promoting plant growth and increasing pollutant bioavailability to ease the remediation process. This means the development of nanotechnology can provide an effective way of cleaning the environment with less cost and harmful end products.

Zinc oxide (ZnO) nanoparticles have a wide bandgap semiconductor with an energy gap of 3.37 eV at room temperature which then allows the transfer of electrons in the presence of light to initiate the phytoremediation reactions. The pathway in Figure 1 shows iron nanoparticles being used in environmental cleanup.

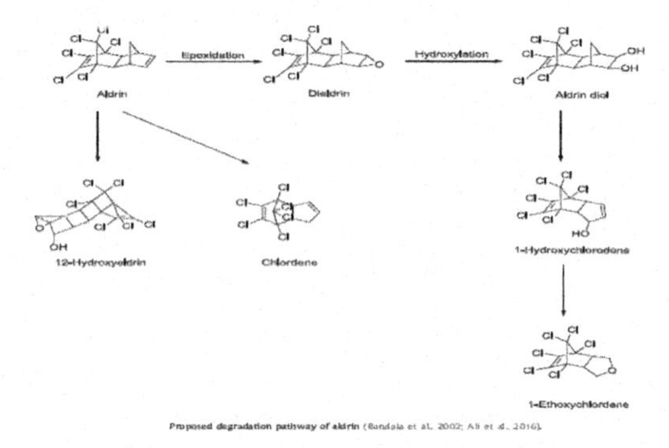

Figure 1. Proposed pathway for degradation of Aldrin.

2 LITERATURE REVIEW

Nanoparticle have proved to be very useful in the modern field of science because of their unique properties that have been used in catalysis, medicine, and electronics (Sharma et al. 2019). Nanoparticles have been used primarily in photocatalysis, which involves the use of light to initiate chemical reaction, such as degrading methyl blue to a harmless end product. Nanoparticles were used in this chemical reaction because it had a mobile valence electron between the conduction band and the valence band that is activated in the presence of light to initiate a reduction reaction to methyl blue (Boruah, Samantaray, Madras, Modak, & Bose 2020). This property of nanoparticles has made them a very important new technology for the development of adsorbents to remediate environmental pollutants to less harmful products. Specifically, nanoparticles have been used in remediating trace amounts of dyes, heavy metals, and pesticides from contaminated environments, i.e., soil and water (Mehta et al. 2019).

Nanoparticles can be prepared primarily by two methods: chemical methods and green synthesis (Sanjukta et al. 2016). Green synthesis is thought to be the best method since it uses less toxic reagents in the synthesis of nanoparticles, hence resulting in the release of less toxic by-products into the environment. In the case of green synthesis, a plant which contains either in its leaves, bark, or any other part a natural product component, which has the capability of enhancing binding of the nanoparticles, is selected for the synthesis of the nanoparticle (Saratale et al. 2018).

A nanoparticle which is friendly to the environment should be selected for phytoremediation so that it ensures that the end products of the adsorption process are not harmful to the environment or introduce pollutants to the environment (Kuppusamy, Palanisami, Megharaj, Venkateswarlu, & Naidu 2016).

Cissus quadrangularis plant was chose in the synthesis of the nanoparticles because it contains natural products that are essential in the binding of the synthesized nanoparticles (Velammal, Devi, & Amaladhas 2016). These compounds include flavonoids, glycosides, tannins, phenolics, triterpenoids, saponins, and glycosides (Dhanasekaran 2020).

ZnO nanoparticles have been employed in the photodegradation of dyes and pesticides to render these pollutants to less harmful products (Ong, Ng, & Mohammad 2018). For these reasons ZnO nanoparticles were synthesized as a potential adsorbent for the degradation of organochlorines which are persistent in the environment.

2.1 *Research design*

The research was carried out in Kenya. The research design was done at Moi university in the chemistry laboratory.

2.2 *Apparatus and reagents*

99% pure zinc nitrate, 99% pure methanol, and filter paper grade 1 were both obtained from Sigma Aldrich Chemicals Co., USA.

Cissus quandangularis plants were collected from Kisumu County 0°06′00.2″S 34°46′59.5″E.

2.3 *Synthesis of zinc oxide nanoparticles*

The cleaned, healthy plant materials were cut into small sections. They were dried under shade for 3 weeks. The dried material was ground into fine powder in an electric grinder. The powder obtained was then stored in desiccators to await extraction. Extraction was carried out using 1 g of each sample of coarsely powdered plant material with 25 mL of methanol solvent (HPLC) and kept for 48 hrs with slight shaking. All the extraction was performed at room temperature. All the extracts were filtered through Whatman No.1 paper to get filtrate as extracts which was dried to concentrate the samples (Prabhavathi, Prasad, & Jayaramu 2016). 50 mL of *Cissus quadrangualis* leaf extract was measured into a conical flask, which was then boiled at 60–80°C by a stirrer heater. 5g of zinc nitrate was added to the solution when the temperature reached 60°C. The mixture was heated until the suspension turned deep yellow in color. The paste was then collected in a ceramic crucible where it was transferred into an air furnace followed by heating at 400°C for 2 hrs. A light white colored solid of zinc oxide was collected and then powdered to form zinc oxide nanoparticles (Fazlzadeh et al. 2017).

2.4 *Characterization of zinc oxide nanoparticles*

The UV–Vis reflectance spectra (U–Vis DRS) measurements were carried out with UV140404B in the wavelength range of 200–850 nm in reflectance mode. The crystalline structure of the samples was analyzed by using PANalytical X'PERT PRO model X-Ray diffractometer, with the instrument operating at a voltage of 50 kV and a current of 30 mA.

3 RESULTS AND DISCUSSION

The following were the data from the zinc oxide nanoparticles after characterization using particle analyzer, scanning electron microscopy, and UV–Vis spectroscopy.

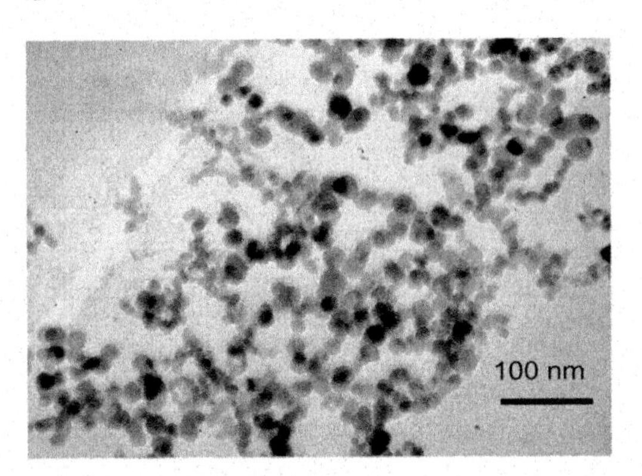

Figure 2. SEM image of agglomerated ZnO nanoparticles.

From the above data the synthesized nanoparticle using the*Cissus quandangularis* plant extract had a mean radius of 14.83 nm from Figure 3, which makes the process of synthesis viable for the synthesis of nanoparticles because the dimension of the synthesized particles were in the range of 1–100 nm, i.e., the required range for nanoparticles (Sun & Xia 2003).

Optical properties of the as-prepared ZnO nanostructure sample were revealed by UV–Vis spectroscopy at room temperature, as shown in Figure 4.

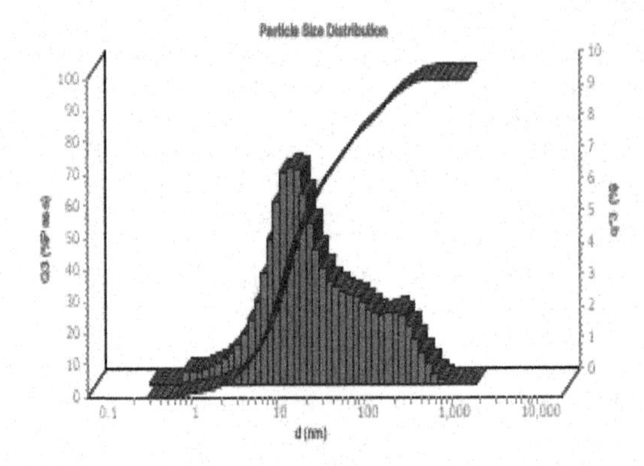

Figure 3. Particle analyzer analysis of ZnO nanoparticles.

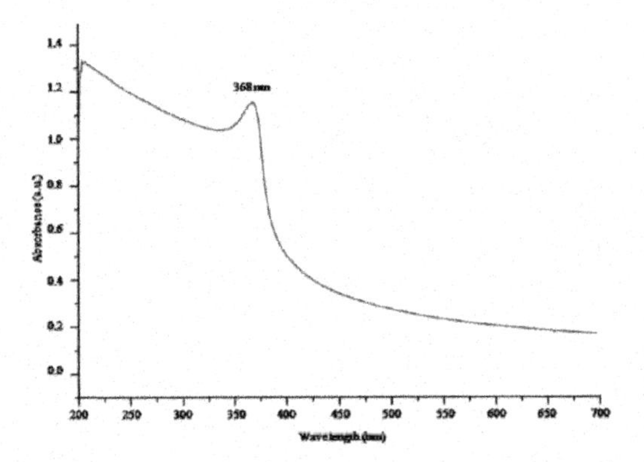

Figure 4. UV–Vis absorption spectrum of as-prepared ZnO NPs, treated *Cissus quandangularis* using optimal 0.5 mL volume in the proposed incubation cum precipitation method.

It can be seen from Figure 4 that there was intensive absorption in the ultraviolet band of about 200–400 nm. The absorption wavelength at about 368 nm of ZnO suggested the excitonic character at room temperature (Sun & Xia 2003).

The ZnO NPs embedded in *Cissus quandangularis* matrix with little agglomeration had sizes of about 5 nm throughout the carbon-coated copper grid and average particle size and shape in the range of 5–40 nm. The SEM image revealed that the particles are spherical and granular nanosized in nature, as shown in Figure 2.

4 CONCLUSION

My findings could be targeted for promising potential applications including biosensing devices and nanoelectronics because of the pollution-free and eco-friendly approach. This green synthesis approach shows that the environmentally benign and renewable latex of *Cissus quandangularis* can be used as an effective stabilizing and reducing agent for the synthesis of zinc oxide nanoparticles. Zinc oxide nanoparticles synthesized by this approach are quite stable and no visible changes are observed even after a month. Synthesis of zinc oxide nanoparticles using *Cissus quandangularis* is an alternative to chemical synthesis. We anticipate that the smaller particles are mostly stabilized by alkaloids and proteins. Further experiments on the systematic mode of mechanism of size-selective synthesis of zinc oxide nanoparticles using this very useful *Cissus quandangularis* should be done. Moreover, further research should be done on the applicability of the synthesized nanoparticles in remediation.

5 RECOMMENDATION

A study should be done to study the photoluminescence and luminescence properties of ZnO NPs synthesized from *Cissus quandangularis* plant extract.

ACKNOWLEDGMENT

I am thankful and grateful to Dr. Kengara, Dr. K'owino, Head, Department of Chemistry for providing all necessary facilities to carry out the present work. Furthermore, I amgrateful to ACEII-PTRE-Moi University for the financial support.

REFERENCES

Bandala, E. R., Gelover, S., Leal, M. T., Arancibia-Bulnes, C., Jimenez, A., & Estrada, C. A. (2002). Solar photocatalytic degradation of Aldrin. *Catalysis Today*, *76*(2–4), 189–199.

Boruah, B., Samantaray, P. K., Madras, G., Modak, J. M., & Bose, S. (2020). Sustainable photocatalytic water remediation via dual active strongly coupled AgBiO3 on PVDF/PBSA membranes. *Chemical Engineering Journal*, 124777.

Dhanasekaran, S. (2020). Phytochemical characteristics of aerial part of Cissus quadrangularis (L) and its in-vitro inhibitory activity against leukemic cells and antioxidant properties. *Saudi Journal of Biological Sciences*.

Ealias, A. M., & Saravanakumar, M. P. (2017, November). A review on the classification, characterisation, synthesis of nanoparticles and their application. In *IOP Conf. Ser. Mater. Sci. Eng* (Vol. 263, p. 032019).

Hong, N. H. (2019). Introduction to nanomaterials: basic properties, synthesis, and characterization. In *Nano-Sized Multifunctional Materials* (pp. 1–19). Elsevier.

Kuppusamy, S., Palanisami, T., Megharaj, M., Venkateswarlu, K., & Naidu, R. (2016). Ex-situ remediation technologies for environmental pollutants: a critical perspective. In *Reviews of Environmental Contamination and Toxicology Volume 236* (pp. 117–192). Springer, Cham.

Mehta, A., Mishra, A., Basu, S., Shetti, N. P., Reddy, K. R., Saleh, T. A., & Aminabhavi, T. M. (2019). Band gap tuning and surface modification of carbon dots for sustainable environmental remediation and photocatalytic hydrogen production–A review. *Journal of environmental management*, *250*, 109486.

Mirzaei, H., & Darroudi, M. (2017). Zinc oxide nanoparticles: Biological synthesis and biomedical applications. *Ceramics International*, *43*(1), 907–914.

Nativo, P., Prior, I. A., & Brust, M. (2008). Uptake and intracellular fate of surface-modified gold nanoparticles. *ACS nano*, *2*(8), 1639–1644.

Naveed Ul Haq, A., Nadhman, A., Ullah, I., Mustafa, G., Yasinzai, M., & Khan, I. (2017). Synthesis approaches of zinc oxide nanoparticles: the dilemma of ecotoxicity. *Journal of Nanomaterials*, *2017*.

Ong, C. B., Ng, L. Y., & Mohammad, A. W. (2018). A review of ZnO nanoparticles as solar photocatalysts: synthesis, mechanisms and applications. *Renewable and Sustainable Energy Reviews*, *81*, 536–551.

Parveen, K., Banse, V., & Ledwani, L. (2016, April). Green synthesis of nanoparticles: their advantages and disadvantages. In *AIP conference proceedings* (Vol. 1724, No. 1, p. 020048). AIP Publishing LLC.

Pascariu, P., & Homocianu, M. (2019). ZnO-based ceramic nanofibers: Preparation, properties and applications. *Ceramics International*, *45*(9), 11158–11173.

Raghu, H. V., Parkunan, T., & Kumar, N. (2020). Application of Nanobiosensors for Food Safety Monitoring. In *Environmental Nanotechnology Volume 4* (pp. 93–129). Springer, Cham.

Rane, A. V., Kanny, K., Abitha, V. K., & Thomas, S. (2018). Methods for synthesis of nanoparticles and fabrication of nanocomposites. In *Synthesis of inorganic nanomaterials* (pp. 121–139). Woodhead Publishing.

Riley, J. K., Matyjaszewski, K., & Tilton, R. D. (2018). Friction and adhesion control between adsorbed layers of polyelectrolyte brush-grafted nanoparticles via pH-triggered bridging interactions. *Journal of colloid and interface science*, *526*, 114–123.

Sanjukta, R. K., Samir, D., Puro, K., Ghataak, S., Shakuntal, L., & Sen, A. (2016). Green synthesis of silver Nanoparticles using plant. *Int J Nanomed Nanosurg*, *2*(2).

Saratale, R. G., Saratale, G. D., Shin, H. S., Jacob, J. M., Pugazhendhi, A., Bhaisare, M., & Kumar, G. (2018). New insights on the green synthesis of metallic nanoparticles using plant and waste biomaterials: current knowledge, their agricultural and environmental applications. *Environmental Science and Pollution Research*, *25*(11), 10164–10183.

Saxena, N., Goswami, A., Dhodapkar, P. K., Nihalani, M. C., & Mandal, A. (2019). Bio-based surfactant for enhanced oil recovery: Interfacial properties, emulsification and rock-fluid interactions. *Journal of Petroleum Science and Engineering*, *176*, 299–311.

Sharma, G., Kumar, A., Sharma, S., Naushad, M., Dwivedi, R. P., ALOthman, Z. A., & Mola, G. T. (2019). Novel development of nanoparticles to bimetallic nanoparticles and their composites: a review. *Journal of King Saud University-Science*, *31*(2), 257–269.

Siripireddy, B., & Mandal, B. K. (2017). Facile green synthesis of zinc oxide nanoparticles by Eucalyptus globulus and their photocatalytic and antioxidant activity. *Advanced Powder Technology*, *28*(3), 785–797.

Sun, Y., & Xia, Y. (2003). Gold and silver nanoparticles: a class of chromophores with colors tunable in the range from 400 to 750 nm. *Analyst*, *128*(6), 686–691.

Trojanowski, R., & Fthenakis, V. (2019). Nanoparticle emissions from residential wood combustion: A critical literature review, characterization, and recommendations. *Renewable and Sustainable Energy Reviews*, *103*, 515–528.

Velammal, S. P., Devi, T. A., & Amaladhas, T. P. (2016). Antioxidant, antimicrobial and cytotoxic activities of silver and gold nanoparticles synthesized using Plumbago zeylanica bark. *Journal of Nanostructure in Chemistry*, *6*(3), 247–260.

Advances in Phytochemistry, Textile and Renewable Energy Research for Industrial Growth – Nzila et al. (Eds)

Dyeing characteristics of different solvent extracts of *Euclea divinorum* on cotton fabric

S. Manyim & A.K. Kiprop
Department of Chemistry and Biochemistry, Moi University
Africa Center of Excellence in Phytochemicals, Textile and Renewable Energy, Moi University

J.I. Mwasiagi
Department of Manufacturing, Industrial and Textile Engineering, Moi University

A.C. Mecha
Department of Chemical and Process Engineering, Moi University

ABSTRACT: Most of the dyes used in the textile industry are of synthetic origin. Recently the use of natural dyes in dyeing has regained interest due to environmental hazards associated with synthetic dyes. Therefore there is need to introduce more natural dyes in order to satisfy the increasing demand. *Euclea divinorum* has been in use as a source of traditional medicine for toothache, chest pain, constipation, cancer, pneumonia, and snake bite. For many years the root and stem of *E. divinorum* has been used traditionally by different communities to color the mouth and lips and as a dye but its potential as a source of natural dye for textile dyeing has not been exploited. This study investigated the effect of different solvent extracts on the dyeing characteristics of dyestuffs from *Euclea divinorum* plant on cotton fabric. Light, wash, and rub fastness of the dyed cotton were tested using fad-o-meter, launder-o-meter, and crock-o-meter, respectively. Color coordinates, reflectance, and color strength were determined using the reflectance spectrophotometer. The aqueous and methanolic extracts showed the highest color strengths followed by ethyl acetate extract. The wash, light, and rub fastness values for aqueous and methanolic extracts were between 4 and 5 which is above the acceptable levels of 3. Quantitative phytochemical analysis indicated that methanolic extract had the highest content of phenols, tannins, and flavonoids compared to the other solvent extracts.

1 INTRODUCTION

Dyes from natural sources have started to gain popularity in the recent years due to the environmental hazards associated with synthetic dyes (Adeel et al. 2018). Synthetic dyes are mutagenic, toxic and carcinogenic (Khatri, Nidheesh, Anantha Singh & Suresh Kumar 2018; Rawat, Mishra, & Sharma 2016; Sharma, Dangi, & Shukla 2018). On the other hand, natural dyes are biodegradable and compatible with nature hence have found applications in various industries such as textile, food, cosmetics and drug industries as a source of color. Natural dyes are obtained from plants, animals, insects, and minerals (Prabhu & Bhute 2012). Extraction methods of natural dyes from their sources include aqueous, solvent/alcoholic, acid/base, fermentation, and enzymatic (Saxena & Raja 2014). Recent advanced methods of extraction are ultrasonic extraction (Ben Ticha et al. 2017) and microwave-assisted extraction (Sinha, Chowdhury, Saha, & Datta 2013).

A natural dye material is not a single chemical unit but forms part of the plant matrix that is made up of various materials which are not dyes (Baliarsingh, Panda, Jena, Das, & Das 2012). Methods of extracting natural dyes differ because the nature and solubility of these dyes vary depending on the source (Bechtold, Mahmudali, & Mussak 2007; (Souissi, Guesmi & Moussa 2018). The lack of standard methods of extraction that can be applied to various sources of natural dyes makes it a complex process (Gupta 2019; Vankar 2016). The process of determination of the suitable method for extraction of a dye from a particular source is a crucial step since it plays a significant role in the final concentration, color strength, and color fastness of the natural dye extract on the fabric (Velmurugan et al. 2017). Color measurements on a dyed fabric are given as the relative color strength and fastness properties. Color strength refers to the relative concentration of the dye absorbed by the fabric while the color fastness to wash, rub, and light is the ability of the dye to remain fixed on the fabric after subjecting the dyed fabric to respective fastness tests. Color fastness rating is between 1 (poor) and 5 (excellent) fastness (Souissi et al. 2018).

Euclea divinorum Hierns (Ebenaceae) is an evergreen shrub that is mainly found along escarpments

DOI 10.1201/9781003221968-18

and in rocky areas (Feyissa, Asres, & Engidaworkm 2013). *E. divinorum* plant has medicinal values and has been used traditionally by the Kalenjin community as an anti-venom and as a purgative drug (Kigen et al. 2016). *E. divinorum* has also been used locally as a source of natural dye. The roots and twigs of the plant were used as toothbrushes, mouth disinfectant, and to color the lips reddish brown (Maroyi 2011).

Phytochemicals in a plant extract are responsible for the particular color produced by the natural dye. Phenols, flavonoids, and tannins are the most significant classes of compounds in the dyeing process because they account for the ability of a dye to fix to the fabric (color fastness) and the shade formed on the fabric (Mongkholrattanasit et al. 2013). In addition, plants rich in tannins have been used as bio-mordants (Amin et al. 2020; Erdem İşmal et al. 2014; Prabhu & Teli 2014) and exhibit the potential to be used as alternatives to the toxic metallic mordants that are currently in use (İşmal & Yıldırım 2019). Qualitative phytochemical screening studies of *E. divinorum* has shown that it contains flavonoids, alkaloids, phenols, tannins, and terpenoids (Mwonjoria, Ngeranwa, Githinji, & Wanyonyi 2018). However quantitative phytochemical analysis of *E. divinorum* is still lacking.

The objective of this study was to determine the suitable solvent for extraction of natural dye from *E. divinorum* plant as well as the dyeing characteristics on cotton fabric. Quantitative phytochemical analysis of *E. divinorum* was also evaluated.

2 MATERIALS AND METHODS

2.1 *Materials*

The solvents (methanol, ethyl acetate, hexane and dichloromethane) used for extraction were of analytical grade. Folin–Ciocalteu, tannic acid quercetin and gallic acid were used as the analytical standards. All chemicals used were purchased from Pyrex East Africa Ltd. The cotton fabric (GSM 97.1) was obtained from Rivatex East Efrica Ltd (REAL).

2.2 *Methods*

2.2.1 *Collection and preparation of plant materials*

The root bark of *Euclea divinorum* plant was collected from Chemase escarpment in Nandi County. The collected plant materials were washed with tap water to remove dust and other particles then air-dried in the shade. The dry barks were ground to powder form in an electric grinder and weighed using a weighing scale.

2.2.2 *Dye extraction*

Dye extraction was done using sequential maceration of the ground samples with organic solvents (hexane, dichloromethane, ethyl acetate and methanol) and direct aqueous extraction. The organic solvents were used in order of their increasing polarity. Maceration for each solvent was done for 24 hours followed by filtration using Whatman No. 1 filter paper and the solvents were evaporated using a rotary evaporator to obtain a dry solid mass whose weight was determined and percentage yield calculated. The aqueous extract was filtered and directly used in the dyeing process.

2.2.3 *Quantitative phytochemical analysis*

Dichloromethane, ethyl acetate and methanolic extracts were subjected to quantitative evaluation of total tannins, phenols and flavonoids. All the spectrophotometric assays were measured by UV–Vis spectrophotometer (Model No: DU'720PC, Beckman Coulter). Measurements were done in triplicate and averaged (Baliarsingh et al 2012). MS Excel software was used to draw the standard curves and calculate the correlation coefficient (R^2).

2.2.3.1 Determination of total tannins

Tannin content was estimated using the standard procedure described by Petchidurai (Petchidurai et al. 2019). One milligram of the sample was dissolved in 1 mL of distilled water. An aliquot of 1 mL of the sample was mixed with 0.5 mL of 10% Folin–Ciocalteau's reagent and incubated for 3 minutes. 15% Na_2CO_3(1 mL) and distilled water (8 mL) were added and incubated in the dark for 30 minutes at room temperature. UV–Visible spectrophotometric analysis of absorbance was done at 725 nm in triplicate. A stock solution was prepared using tannic acid standard (1 g) in a 1000 mL volumetric flask. The stock solution was used to prepare tannic acid solutions in increasing concentrations (20, 40, 75, 100 and 125 mg/L). 0.5 mL of each concentration was measured using a micropipette and treated like the sample, and then absorbance was measured. A tannic acid calibration curve was plotted and the concentration of total tannins in the sample was expressed as mg tannic acid equivalence/g (mg TAE/g) of the dry sample extract.

2.2.3.2 Determination of total phenols

Total phenols was determined by the Folin–Ciocalteu method using the standard procedure (Shi et al. 2019). An aliquot of 0.5 mL of the sample was mixed with 10% Folin–Ciocalteu's reagent (2.5 mL) and 15% Na_2CO_3 (2.5 mL), and then it was incubated in the dark for 20 minutes at room temperature. UV–Vis analysis of absorbance was done at 725 nm in triplicate. A stock solution was prepared using gallic acid standard (1 g) in a 1,000 mL volumetric flask. The stock solution was used to prepare gallic acid solutions in increasing concentrations (20, 40, 75, 100 and 125 mg/L). 0.5mL of each concentration was measured using a micropipette and treated like the sample and then absorbance was measured. The gallic acid calibration curve was plotted and the concentration of total phenols in the sample was expressed as mg gallic acid equivalence/g (mg GAE/g) of the dry sample extract.

2.2.3.3 Determination of total flavonoids

Total flavonoids was evaluated by aluminium chloride calorimetric method using a standard procedure (Bahukhandi, Sekar, Barola, Bisht & Mehta 2019). The sample extract (1 mL) was mixed with methanol

(3 mL), 10% aluminum chloride (0.2 mL), 1M potassium acetate (0.2 mL) and 5.6 mL of distilled. Quercetin (1 g) was dissolved in distilled water in a 1 L volumetric to make a stock solution of 1000 mg/1L. From the stock solution of quercetin increasing concentrations of 10, 20, 40, 80 and 100mg/L were prepared and subjected to similar treatment as the sample extracts. All the samples were incubated at room temperature for 30 minutes and then the absorbance was measured at 420 nm in triplicate. The quercetin calibration curve was plotted and the concentration of the total flavonoids in the sample extract was expressed as mg of quercetin equivalent/g (mg QE/g) of the dry sample extract.

2.2.4 *Dyeing process*

Wetting of the cotton fabric was done using 5 g/L of non-ionic detergent for 30 minutes to enhance surface wettability. A concentration of 20 % on weight of the fabric (owf) and a material to liquor (M: L) of 1:40 was used during the preparation of the dye-bath using the different dye extracts.

The dyeing process was carried out as described by Yusuf et al. (2012). The dyeing time was 1 hour at a temperature of 80°C in open conical flasks with manual agitation of the dye bath. All dyeing processes were carried out in triplicate. Temperature was regulated using a water bath.

The dyed cotton fabrics were removed from the dye bath and immersed in a beaker containing 10 gL^{-1} of sodium chloride solution for 20 minutes (Savvidis et al. 2013). The saturated brine was used in the post-treatment process as a dye fixing agent. The dyed samples were washed with cold water to remove the unfixed dyestuff. The dyed samples were subjected to soaping with 2 gL^{-1} soap solution followed by washing with tap water and then dried at room temperature.

2.2.5 *Evaluation of the dyed cotton fabric*

The dyed cotton fabrics were subjected to color fastness tests to wash, light and rub.

2.2.5.1 Color fastness to washing

The wash fastness was measured with the SDL ATLAS M228 Rotawash Launder-o-meter, according to standard test ISO 105-C02:1989. The wash fastness values were determined using gray scale for assessing the staining of white standard material attached and gray scale for assessing change of color of the dyed samples. The color fastness rating is normally between 1 and 5, where 1 means very good fastness, 2 is poor, 3 is fairly good, 4 is good and 5 is excellent.

2.2.5.2 Color fastness to light

Light fastness was measured using a SDL ATLAS M237 light fastness tester fitted with 500 W mercury–tungsten lamp and a timer that had been preset. The testing was carried out according to the ISO 105 A02:1993 standard. The samples were then assessed for any fading using the gray scale.

2.2.5.3 Color fastness to rubbing

Rub fastness was measured using a SDL ATLAS M238AA Crock-meter according to the ISO 105-X12:2000 standard. The rubbed samples were evaluated for staining on white adjacent cotton using the gray scale.

2.2.5.4 Determination of color coordinates and color strength

Color coordinates of dyed samples which included L*, a*, b*C* and H° were measured by spectrophotometer X-rite SP60X using D65 source of light and 10° standard observer where the measurement was carried out in the range of 390–710 nm wavelength. All measurements were done in triplicate and an average was obtained. The R values obtained were used to calculate the relative color strength K/S as per the Kubelka–Munk equation (Hossen & Imran 2017). L*, a*, b*, R, C* and H° represent lightness, redness-greenness of color, yellowness–blueness of color, reflectance, saturation of color, and hue angle, respectively.

3 RESULTS AND DISCUSSION

3.1 *Preparation of plant materials and dye extraction*

Evaporation of the solvent led to formation of solid extracts for hexane, dichloromethane and ethyl acetate while methanolic extract formed a paste. Hexane extract was yellow in color, dichloromethane extract was light green, ethyl acetate extract was dark green and methanolic and aqueous extracts were brown in color. The results and the steps of obtaining the root extract of *Euclea divinorum* plant are as shown in Figure 1.

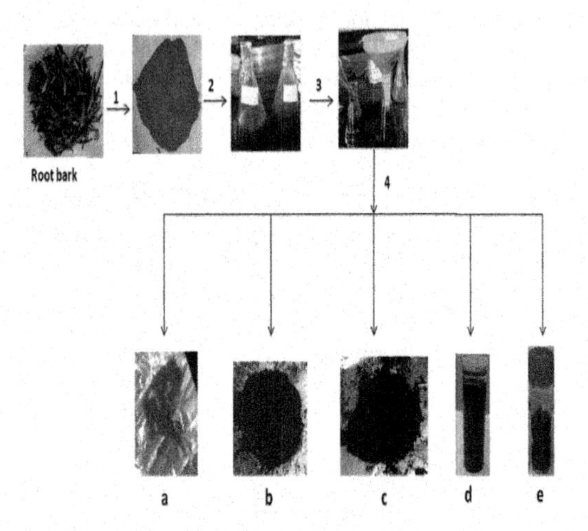

Figure 1. Dye extraction from the root bark of *E. divinorum*: (1) grinding, (2) maceration for 24 hours, (3) filtration, (4) solvent evaporation. Percentage yields for each solvent extracts were: (a) hexane 1.37%, (b) dichloromethane 2.52% (c) ethyl acetate 4.13%, (d) methanolic 18.81% and (e) aqueous 26.12%.

3.2 *Quantitative phytochemical analysis*

3.2.1 *Total tannins*

The linear regression equation for the calibration curve of tannic acid was $y = 0.007x + 0.0632$ with $R^2 = 0.9914$ (Figure 2) indicating a good linear relationship within the range of detection. As a result the equation was used to estimate total tannins in the different solvent extracts. The highest tannin content was observed in the methanolic extract (115.114 mg TAE/g) (Table 1). This can be attributed to the fact that tannins are polar and hence would preferably dissolve in more polar solvents than less polar solvents (Elgailani & Ishak 2016) The presence of tannins in the dye extract is significant because it enhances the textile finishing characteristics such as antioxidant, antimicrobial and antifungal properties (Fraga-Corral et al. 2020).

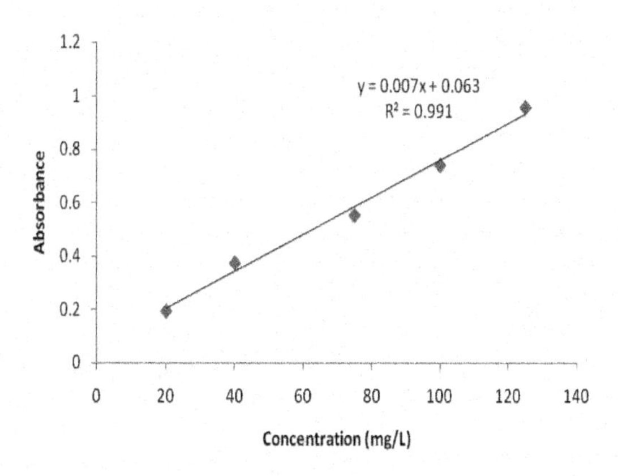

Figure 2. Calibration curve for tannic acid.

3.2.2 *Total phenols*

The linear regression equation for the calibration curve of gallic acid was $y = 0.0085x + 0.0242$ with $R^2 = 0.9938$ (Figure 3), indicating good linear relationship within the range of detection. As a result the equation was used to estimate total phenols in the different solvent extracts. Methanolic extract (123.741 mg GAE/g) showed the highest phenol followed by ethyl acetate (82.094 mg GAE/g) and dichloromethane (69.741 mg GAE/g) in that order. Comparable ranges to what was observed have been reported in other plants in the same family as *E. divinorum* (Ebeneacea) (Mekonnen, Atlabachew, & Kassie 2018). Moreover, the results were in line with the observations made on qualitative phytochemical screening of root extracts of *E. divinorum* (Nyambe, Hans, Beukes, Morris, & KandawaSchulz 2018).

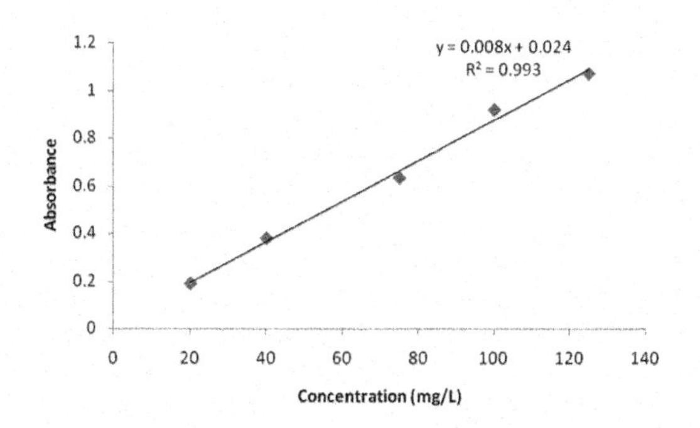

Figure 3. Calibration curve gallic acid.

3.2.3 *Total flavonoids*

The linear regression equation for the calibration curve of Quercetin standard was found to be $y=0.0048x + 0.0321$ with $R^2 = 0.9961$ (Figure 4) indicating good linear relationship within the range of detection. As a result the equation was used to determine total flavonoids in the different solvent extracts (Table 1). The flavonoids content across the three solvents was found to be lower compared to the other phytochemicals analyzed which was in agreement with qualitative phytochemical screening observations (Mwonjoria et al. 2018). The flavonoids content in dichloromethane (0.189 mg QE/g) was negligible which was similar to the findings of Al-Fatimi (2019) where flavonoids in dichloromethane extract were not detectable but moderate amounts were observed in methanolic extract.

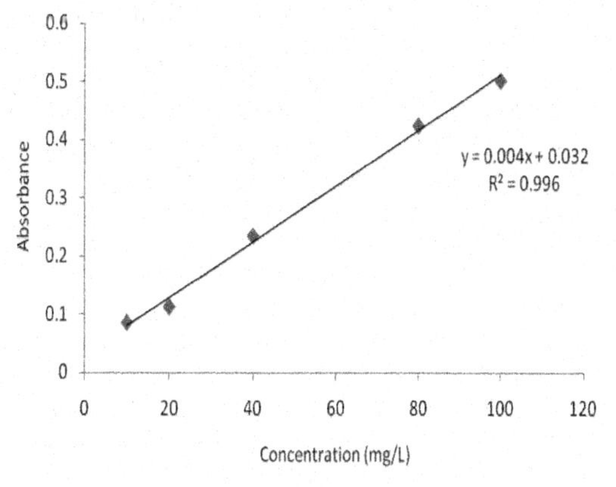

Figure 4. Calibration curve for quercetin.

Table 1. Estimated phytochemical content of the different solvent extracts of *E. divinorum*

Solvent extracts	Total tannins (mg TAE g^{-1})	Total phenols (mg GAE g^{-1})	Total flavonoids (mg QE g^{-1})
Methanol	115.114	123.741	4.979
Ethyl acetate	99.400	82.094	1.438
Dichloromethane	66.740	69.741	0.189

3.3 *Evaluation of the dyed fabric*

It was noted that there was variation in the shades of the dyed samples using the different solvent extracts (Table 2). The variation in the color shades of cotton fabric dyed with different solvent extract can be attributed to the difference in the phytochemical composition of each extract according to their polarity (Guinot, Gargadennec, Valette, Fruchier & Andary 2008)

Table 2. The different shades of the dyed samples using different solvent extracts

Undyed	Hexane	Dichloromethane
Ethyl acetate	**Methanol**	**Water**

3.3.1 *Color fastness*

The color fastness properties for washing, rubbing and light for the dyed cotton fabrics were as presented in Table 3.

Table 3. Color fastness properties of the different solvent extracts of *E. divinorum* plant

Solvent extract	Wash fastness		Rub fastness		Light fastness
	C.C	C.S	Dry	Wet	
Water	4–5	5	5	5	5
Methanol	4–5	5	5	5	5
Ethyl acetate	4	5	5	5	4–5
DCM	3–4	5	5	5	4
Hexane	2	5	5	5	3

3.3.1.1 Color fastness to washing

From Table 3 it is noted that the methanolic and aqueous dye extracts of *E. divinorum* plant showed good wash fastness properties among the five solvent extracts with a value between 4 and 5 which is approaching excellent. On the other hand, hexane and dichloromethane dye extracts showed the lowest color fastness values and do not meet the standard requirements for the color fastness test to washing (Souissi et al. 2018b). All the white cotton fabric attached to dyed samples during wash fastness testing did not show any staining and hence had an excellent value of 5. This indicates that the dyed fabric resists the color staining and color change during the washing process (Yang, Guan, Chen, & Tang 2018). Despite the little loss of dye by the dyed samples, the white cotton fabrics were not stained which is due to the fact that the conditions for wash fastness testing were not favorable for the white fabric to absorb the dye from the solution.

3.3.1.2 Color fastness to light

With regards to color fastness to light, it was observed that the fabric dyed with methanolic extract had the highest light fastness value followed by aqueous, ethyl acetate, dichloromethane and hexane (Table 3). Color fading due to exposure of the dyed samples to UV light was generally between fairly good and excellent which is quite acceptable in textile dyeing processes (Yusuf et al. 2015).

3.3.2 *Color fastness to rubbing*

Considering color fastness to washing, both dry and wet rubbing fastnesses for all the samples were excellent. The white samples sewed to the dyed samples during rub fastness testing did not show any staining at all, hence a rating of 5 (Table 3). Consequently these fastness results meet the requirement for color fastness which should be a rating of 3 and above (Pisitsak, Tungsombatvisit, & Singhanu 2018)

3.3.3 *Color strength and CIE L*a*b**

CIE-L*a*b* coordinates, L* (lightness/darkness), a* (redness or greenness), b* (yellowness to blueness) and R (percentage reflectance) of the dyed cotton samples determined were as shown in Table 4. K/S (relative color strength) on fabric was determined using percentage reflectance values using the Kubelka–Munk equation.

Table 4. Color measurements of the various dyed cotton samples

Dye Extract	L*	a*	b*	C*	H°	K/S
Methanol	63.52	+8.10	+14.73	16.1	60.3	0.592
Water	62.73	+7.92	+15.56	17.5	63.0	0.533
Ethyl acetate	72.78	+10.59	+10.83	15.1	45.6	0.330
DCM	70.45	+10.94	+8.46	17.6	51.7	0.214
Hexane	75.39	+12.18	+5.95	13.6	26.0	0.137

L* coordinates give the color range of the dyed samples in terms of lightness and darkness where 100 = white and 0 = black. The sample dyed with aqueous extract was the darkest followed by the methanolic extract. Regarding the relative color strength on the dyed fabric there was variation as the extraction solvent changed as reported by Hasan et al. (2015). The cotton fabric dyed with methanolic and aqueous extracts showed the highest relative color strength value (Figure 5). This indicates that the concentration of the dye substance was higher in these solvent extracts since there is always a direct relation between the relative color strength and the concentration of the dye absorbed by the fabric (Hossen & Imran 2017).

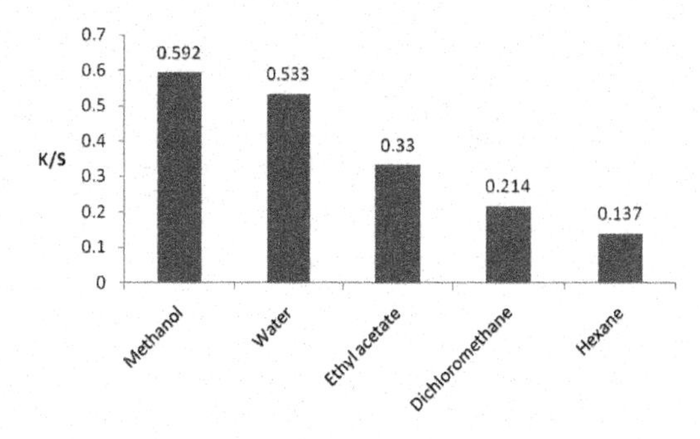

Figure 5. The relative color strength of *E. divinorum* extract with different solvents.

An increase in K/S ratio implies an increase in the strength of the color from the dye (Pisitsak et al. 2018). Trends in relative color strength on fabric were similar to those observed in the fastness testing results where methanolic and aqueous were the best and hexane the worst. This is ascribed to the fact that methanolic and aqueous extracts contain polar chemical compounds whose functional groups are mainly hydroxyls and carbonyls that easily attach to the cellulose in cotton fabric (Bukhari et al. 2017). As a result dye uptake and fixation is boosted hence there is increased color strength and fastness respectively.

4 CONCLUSION

Five different solvents extracted natural dyes from the *Euclea divinorum* plant. Quantitative phytochemical analysis indicated that methanolic extract had the highest content of phenols, tannins and flavonoids compared to the other solvent extracts. All the solvents extracted a dye but the methanolic and aqueous dye extracts had the best dyeing properties in terms of color fastness which was between 4 and 5. Certainly these two solvents are appropriate for the extraction of the dye from *E. divinorum* plant. Standard dye extraction conditions in terms of temperature, time, pH and material to liquor ratio are not available. Optimization of these dye extraction conditions using methanol or aqueous solvents will be of great significance in enhancing the dyeing properties of this *E. divinorum* dye extract.

ACKNOWLEDGMENT

This research was supported by Africa Centre of Excellence in Phytochemicals, Textile and Renewable Energy (ACE-PTRE) of which we are highly appreciative. Appreciation also goes to the School of Sciences and Aerospace Studies, Department of Chemistry and Biochemistry of Moi University and Rivatex East Africa Ltd (REAL).

5 CONFLICT OF INTEREST

The authors declare that there is no conflict of interest regarding the publication of this paper.

REFERENCES

Adeel, S., Hussaan, M., Rehman, F., Habib, N., Salman, M., Naz, S., Amin, N., & Akhtar, N. (2018). Microwave-assisted sustainable dyeing of wool fabric using cochineal-based carminic acid as natural colorant. *Journal of Natural Fibers*, 1–9.

Al-Fatimi, M. (2019). Antifungal Activity of *Euclea divinorum* Root and Study of its Ethnobotany and Phytopharmacology. *Processes*, *7*, 680.

Bahukhandi, A., Sekar, K. C., Barola, A., Bisht, M., & Mehta, P. (2019). Total Phenolic Content and Antioxidant Activity of *Meconopsis aculeata* Royle: A High Value Medicinal Herb of Himalaya. *Proceedings of the National Academy of Sciences, India Section B: Biological Sciences*, *89*(4), 1327–1334.

Baliarsingh, S., Panda, A. K., Jena, J., Das, T., & Das, N. B. (2012). Exploring sustainable technique on natural dye extraction from native plants for textile: Identification of colourants, colourimetric analysis of dyed yarns and their antimicrobial evaluation. *Journal of Cleaner Production*, *37*, 257–264.

Bechtold, T., Mahmudali, A., & Mussak, R. (2007). Natural dyes for textile dyeing: A comparison of methods to assess the quality of Canadian golden rod plant material. *Dyes and Pigments*, *75*(2), 287–293.

Ben Ticha, M., Meksi, N., Attia, H. E., Haddar, W., Guesmi, A., Ben Jannet, H., & Mhenni, M. F. (2017). Ultrasonic extraction of *Parthenocissus quinquefolia* colorants: Extract identification by HPLC-MS analysis and cleaner application on the phytodyeing of natural fibres. *Dyes and Pigments*, *141*, 103–111.

Bukhari, M. N., Shahid-ul-Islam, Shabbir, M., Rather, L. J., Shahid, M., Singh, U., Khan, M. A., & Mohammad, F. (2017). Dyeing studies and fastness properties of brown naphtoquinone colorant extracted from Juglans regia L on natural protein fiber using different metal salt mordants. *Textiles and Clothing Sustainability*, *3*(1).

Elgailani, I. E. H., & Ishak. (2016). Methods for Extraction and Charaterization of Tannins from Some Acacia Species of Sudan. *ResearchGate*, *7*(1), 43–49.

Erdem İşmal, Ö., Yıldırım, L., & Özdoğan, E. (2014). Use of almond shell extracts plus biomordants as effective textile dye. *Journal of Cleaner Production*, *70*, 61–67.

Feyissa, T., Asres, K., & Engidawork, E. (2013). Renoprotective effects of the crude extract and solvent fractions of the leaves of *Euclea divinorum* Hierns against gentamicin-induced nephrotoxicity in rats. *Journal of Ethnopharmacology*, *145*(3), 758–766.

Fraga-Corral, M., García-Oliveira, P., Pereira, A. G., Lourenço-Lopes, C., Jimenez-Lopez, C., Prieto, M. A., & Simal-Gandara, J. (2020). Technological Application of Tannin-Based Extracts. *Molecules*, *25*(3), 614.

Gargoubi, S., Tolouei, R., Chevallier, P., Levesque, L., Ladhari, N., Boudokhane, C., & Mantovani, D. (2016). Enhancing the functionality of cotton fabric by physical and chemical pre-treatments: A comparative study. *Carbohydrate Polymers*, *147*, 28–36.

Guinot, P., Gargadennec, A., Valette, G., Fruchier, A., & Andary, C. (2008). Primary flavonoids in marigold dye:

Extraction, structure and involvement in the dyeing process. *Phytochemical Analysis*, *19*(1), 46–51.
Gupta, V. K. (2019). Fundamentals of Natural Dyes and Its Application on Textile Fabrics. *Chemistry and Technology of Natural and Synthetic Dyes and Pigments*.
Hasan, Md. M., Abu Nayem, K., Anwarul Azim, A. Y. M., & Ghosh, N. C. (2015). Application of Purified Lawsone as Natural Dye on Cotton and Silk Fabric. *Journal of Textiles*, *2015*, 1–7.
İşmal, Ö. E., & Yıldırım, L. (2019). Metal mordants and biomordants. In *The Impact and Prospects of Green Chemistry for Textile Technology* (pp. 57–82).
Khatri, J., Nidheesh, P. V., Anantha Singh, T. S., & Suresh Kumar, M. (2018). Advanced oxidation processes based on zero-valent aluminium for treating textile wastewater. *Chemical Engineering Journal*, *348*, 67–73. https://doi.org/10.1016/j.cej.2018.04.074
Kigen, G., Maritim, A., Some, F., Kibosia, J., Rono, H., Chepkwony, S., Kipkore, W., & Wanjoh, B. (2016). Ethnopharmacological survey of the medicinal plants used in Tindiret, Nandi County, Kenya. *African Journal of Traditional, Complementary and Alternative Medicines*, *13*(3), 156.
Maroyi, A. (2011). The Gathering and Consumption of Wild Edible Plants in Nhema Communal Area, Midlands Province, Zimbabwe. *Ecology of Food and Nutrition*, *50*(6), 506–525.
Mekonnen, A., Atlabachew, M., & Kassie, B. (2018). Investigation of antioxidant and antimicrobial activities of Euclea schimperi leaf extracts. *Chemical and Biological Technologies in Agriculture*, *5*(1), 16.
Mohsin, M., Farooq, A., Ashraf, U., Ashraf, M. A., Abbas, N., & Sarwar, N. (2016). Performance Enhancement of Natural Dyes Extracted from Acacia Bark Using Eco-Friendly Cross-Linker for Cotton. *Journal of Natural Fibers*, *13*(3), 374–381.
Mongkholrattanasit, R., Klaichoi, C., Rungruangkitkrai, N., Punrattanasin, N., Sriharuksa, K., & Nakpathom, M. (2013). *Dyeing Studies with Eucalyptus, Quercetin, Rutin, and Tannin: A Research on Effect of Ferrous Sulfate Mordant* [Research Article]. Journal of Textiles; Hindawi.
Mwonjoria, J. K. M., Ngeranwa, J. J. N., Githinji, C. G., & Wanyonyi, A. W. (2018). Antinociceptive effects of dichloromethane extract of *Euclea divinorum* Lin. *Journal of Pharmacognosy and Phytochemistry*, *7*(6), 1104–1107.
Nyambe, M., Hans, R., Beukes, M., MORRIS, J., & KANDAWA-SCHULZ, M. (2018). Phytochemical and antibacterial analysis of indigenous chewing sticks, *Diospyros lyciodes* and *Euclea divinorum* of Namibia. *Biofarmasi Journal of Natural Product Biochemistry*, *16*, 29–43.
Petchidurai, G., Nagoth, J. A., John, M. S., Sahayaraj, K., Murugesan, N., & Pucciarelli, S. (2019). Standardization and quantification of total tannins, condensed tannin and soluble phlorotannins extracted from thirty-two drifted coastal macroalgae using high performance liquid chromatography. *Bioresource Technology Reports*, *7*, 100273.
Pisitsak, P., Tungsombatvisit, N., & Singhanu, K. (2018). Utilization of waste protein from Antarctic krill oil production and natural dye to impart durable UV-properties to cotton textiles. *Journal of Cleaner Production*, *174*, 1215–1223.
Prabhu, K. H., & Teli, M. D. (2014). Eco-dyeing using Tamarindus indica L. seed coat tannin as a natural mordant for textiles with antibacterial activity. *Journal of Saudi Chemical Society*, *18*(6), 864–872.
Rawat, D., Mishra, V., & Sharma, R. S. (2016). Detoxification of azo dyes in the context of environmental processes. *Chemosphere*, *155*, 591–605.
Savvidis, G., Zarkogianni, M., Karanikas, E., Lazaridis, N., Nikolaidis, N., & Tsatsaroni, E. (2013). Digital and conventional printing and dyeing with the natural dye annatto: Optimisation and standardisation processes to meet future demands. *Coloration Technology*, *129*(1), 55–63.
Saxena, S., & Raja, A. S. M. (2014). Natural Dyes: Sources, Chemistry, Application and Sustainability Issues. In S. S. Muthu (Ed.), *Roadmap to Sustainable Textiles and Clothing* (pp. 37–80). Springer Singapore.
Sharma, B., Dangi, A. K., & Shukla, P. (2018). Contemporary enzyme based technologies for bioremediation: A review. *Journal of Environmental Management*, *210*, 10–22.
Shi, P., Du, W., Wang, Y., Teng, X., Chen, X., & Ye, L. (2019). Total phenolic, flavonoid content, and antioxidant activity of bulbs, leaves, and flowers made from *Eleutherine bulbosa* (Mill.) Urb. *Food Science & Nutrition*, *7*(1), 148–154.
Sinha, K., Chowdhury, S., Saha, P. D., & Datta, S. (2013). Modeling of microwave-assisted extraction of natural dye from seeds of *Bixa orellana* (Annatto) using response surface methodology (RSM) and artificial neural network (ANN). *Industrial Crops and Products*, *41*, 165–171.
Souissi, M., Guesmi, A., & Moussa, A. (2018a). Valorization of natural dye extracted from date palm pits (*Phoenix dactylifera*) for dyeing of cotton fabric. Part 1: Optimization of extraction process using Taguchi design. *Journal of Cleaner Production*, *202*, 1045–1055.
Souissi, M., Guesmi, A., & Moussa, A. (2018b). Valorization of natural dye extracted from date palm pits (*Phoenix dactylifera*) for dyeing of cotton fabric. Part 2: Optimization of dyeing process and improvement of colorfastness with biological mordants. *Journal of Cleaner Production*, *204*, 1143–1153.
Vankar, D. P. S. (2016). *Handbook on Natural Dyes for Industrial Applications (Extraction of Dyestuff from Flowers, Leaves, and Vegetables) 2nd Revised Edition:* Niir Project Consultancy Services.
Velmurugan, P., Kim, J.-I., Kim, K., Park, J.-H., Lee, K.-J., Chang, W.-S., Park, Y.-J., Cho, M., & Oh, B.-T. (2017). Extraction of natural colorant from purple sweet potato and dyeing of fabrics with silver nanoparticles for augmented antibacterial activity against skin pathogens. *Journal of Photochemistry and Photobiology B: Biology*, *173*, 571–579.
Yang, T.-T., Guan, J.-P., Chen, G., & Tang, R.-C. (2018). Instrumental characterization and functional assessment of the two-color silk fabric coated by the extract from *Dioscorea cirrhosa* tuber and mordanted by iron salt-containing mud. *Industrial Crops and Products*, *111*, 117–125.
Yusuf, M., Khan, S. A., Shabbir, M., & Mohammad, F. (2016). Developing a Shade Range on Wool by Madder (*Rubia cordifolia*) Root Extract with Gallnut (*Quercus infectoria*) as Biomordant. *Journal of Natural Fibers*, 1–11.
Yusuf, M., Shahid, M., Khan, M. I., Khan, S. A., Khan, M. A., & Mohammad, F. (2015). Dyeing studies with henna and madder: A research on effect of tin (II) chloride mordant. Journal of Saudi Chemical Society, *19*(1), 64–72.

Advances in Phytochemistry, Textile and Renewable Energy Research for Industrial Growth – Nzila et al. (Eds)

Identification of phenolic compounds in *Prosopis juliflora* by liquid chromatography tandem mass spectrometry

S. Chepkwony & A. Kiprop
Department of Chemistry and Biochemistry, Moi University, Eldoret, Kenya

S. Dumarçay, H. Chapuis, P. Gerardin & C. Gerardin-Charbonnier
Inra, LERMaB, Faculté des Sciences et Technologies, Université de Lorraine, Vandœuvre-Lès-Nancy cedex, France

ABSTRACT: A substantive study on the identification of phenolic compounds present in *Prosopis juliflora* was undertaken. *P. juliflora* is known to be highly adapted to dry lands because of its deep taproots which tolerate dry and waterlogged soils. However, because of its invasive nature, it has spread to undesirable areas like sea shores, arable lands and roads becoming a great menace. Towards its valorization, this study aimed at identifying the various phenolic compounds present in it. Plant samples were collected from three different geographical areas in Kenya; Baringo, Garissa and Turkana Counties. Its acetonic extracts were thereafter analyzed using a reversed-phase liquid chromatography coupled to electrospray ionization – tandem mass spectrometry. Identification of the compounds present was achieved by comparison of experimental retention times and UV-MS spectra to bibliographic data and standard compounds. Twenty chemical compounds were tentatively identified out of which fifteen were found either as flavonoid aglycones or some of their glycosylated forms. For the first time, a new compound believed to be a B-type proanthocyanidin of mesquitol was identified based on the MS/MS diagnostic ions that resulted from retro-Diels-alder reaction, quinone methide cleavage and heterocyclic ring fission. The results illustrate a rich array of phenolic compounds present in *P. juliflora.*

1 INTRODUCTION

Prosopis juliflora, an exogenous plant that is normally known as 'mathenge' in Kenya is believed to have been first introduced in Kenya in 1973. The reason for this was to rehabilitate the quarries near the coastal regions (Jama & Zeila 2005; Mwangi & Swallow 2008; Oduor & Githiomi 2013). It was later introduced to Baringo County in the early 1980's by the Fuel wood Afforestation Extension Project (FAEP) with the aim of reducing soil erosion and combating desertification (Dubow 2011; Lenachuru 2003;. It is traditionally used as a remedy to various diseases like dysentery, measles, cold, gout, sore throat, diarrhea, excrescences, dermatological ailments, flu, indigestion, and eye problems like cataracts, inflammation and wound healing (Khandelwal et al. 2015). Studies have also shown that *P. juliflora* has high calorific value making its wood a good source of fuel wood (Oduor & Githiomi 2013). Its aqueous fraction has been found to show antiseptic, astringent and antibacterial properties (Khandelwal, Sharma, & Agarwal 2015; Thakur, Singh, Singh, & Mani 2014). It is also known to present anti-termite and antifungal properties (Sirmah 2009).

Chilume (2016) posted a feature entitled 'When a blessing becomes a curse – A case study of invasive *Prosopis juliflora*' in the University of Auckland blog. She talks about the drastic changes over the years that have been caused by the wide spread of *P. juliflora* after its introduction in Botswana. She highlights the fact that there is a public outcry from the locals living in the invaded areas who prefer complete eradication of the plant (Chilume 2016). This cry is shared by many in several other countries like Kenya, Eritrea and India (Bokrezion 2008; Mwangi & Swallow 2008; ;atnaik, Abbasi, & Abbasi 2017). The growing dislike towards *P. juliflora* has emerged due to its invasive property that has seen it rapidly spreading into areas where it was not initially intended to grow like forests, rangelands and croplands causing great havoc by affecting both the natural vegetation and also causing a decrease in land value towards livestock grazing (Golubov. Mandujano, & Eguiarte 2001). There is therefore a need of reducing this negative economic impact of *P. juliflora* by devising ways of valorizing it. This study therefore aimed at identifying the various phenolic compounds present in *P. juliflora* towards its valorization.

There is an increasing interest in the identification of novel phenolic compounds due to their versatile health benefits like antioxidant, anti-carcinogenic, anti-inflammatory, anti-allergic, antimicrobial, antihypertensive, anti-arthritic, cardio protective and antimicrobial activities (Bhuyan & Basu 2017; Kaurinovic & Vastag 2018; Seeram, Lee, & Scheuller 2006). In wood, the phenolic extractives protect the lingocellulose from fungal and microbial attack and are found

DOI 10.1201/9781003221968-19

mostly in the heartwood and the bark (Pecha & Perez 2015). Previous studies on the compounds present in *P. juliflora* revealed the presence of some alkaloids; juliprosopine, juliflorine, prosoflorine and juliprosine (Dos Santos et al. 2013; Henciya et al. 2017). The few flavonoids in *P. juliflora* that have been reported so far include mesquitol, catechin, quercetin, luteolin, 4'-*O*-methylgallocatechin and kaempferol (Chepkwony et al. 2020; Khandelwal, Sharma, & Agarwal 2016; Sirmah et al. 2009).

2 EXPERIMENTAL

2.1 *Reagents and chemicals*

All solvents (water, formic acid, acetonitrile and methanol) were HPLC grade apart from acetone and dichloromethane which were of analytical grade and were all purchased from Merck Company, Germany.

2.2 *Sample preparation and extraction*

The *P. juliflora* stem samples were collected from three different areas in Kenya; Baringo (0°28′0″ N, 35°58′0″E), Turkana (03°09′ N, 35°21′ E) and Garissa Counties (0°27′09″ S, 39°38′45″ E). The samples consisted of small trees aged less than 4 years and big trees aged more than 8 years. The stems were further divided into five parts which included the bark, sapwood, heartwood, knot wood and the pith. These samples were air dried, then crushed and ground by a Fritsch pulverisette 9 laboratory vibrating cup mill at 1,200 rpm and finally sieved through a 115-mesh sieve.

Serial extractions of the different samples were done using the Accelerated Solvent Extractor Dionex (ASE) using dichloromethane and acetone solvents in order of increasing polarity. An ASE Dionex extraction method was developed according to the method used by Sirmah et al. (2009) with some modifications. The extractions were performed on 8 g of sample powder at 100°C using a 34 mL cell size. Each extraction entailed three static cycles of 5 minutes each. The samples were later concentrated to dryness by a rotary evaporator.

2.3 *LC-ESI-MS/MS analysis*

Liquid chromatography-mass spectrometry analyses of samples were carried out using a Shimadzu (Noisiel, France) LC-20A ultra-HPLC (UHPLC) system equipped with an auto sampler and interfaced to a PDA UV detector SPD-20A, followed by an LC-MS 8030 triple-quadruple mass spectrometer. The separation was carried out at a flow rate of 0.4 mL/min on a Luna C18 analytical column (inner diameter, 150 mm by 3 mm; Phenomenex, Le Pecq, France) using a 10 minutes gradient as follows: starting from 2 % of acetonitrile which contained 0.1 % formic acid in water also containing 0.1 % formic acid solution, acetonitrile proportion was increased linearly to 20 % in 3 min then to 80 % in 6.25 min. Initial conditions were then reached in 0.25 min. The injection volume was 1 μL. UV-visible spectra were recorded between 190 and 800 nm.

Negative and positive electrospray mass spectrometric analyses were performed at a unit resolution between 100 and 2,000 m/z at a scan speed of 15,000 U/s. The desolvation line and heat-block temperatures were 250°C and 400°C, respectively. Nitrogen was used as a nebulizing (3 liters/min) and drying (15 liters/min) gas. The ion spray voltage was +/- 4,500V. Approximately 10 mg of each of the acetonic extracts were dissolved in 1000 μl of methanol before aspiration into the LC-MS/MS.

3 RESULTS AND DISCUSSION

3.1 *General*

The identification of the phenolic compounds present in *P. juliflora* by the LC-ESI-MS/MS was done in the negative ion mode which described the corresponding deprotonated pseudomolecular ions. This is because it has been demonstrated that the negative ion mode has higher sensitivity and selectivity compared to the positive ion mode detection which generates a higher background signal (Cuyckens & Claeys 2004; Gu et al. 2003; Sun, & Miller 2002). Another reason is because of the free phenol groups present in the compounds identified since acids deprotonate easily in the negative ion mode while they form adducts with cations in the sample or mobile phase in the positive mode (Swatsitang, Swatsitang, Robards, & Jardine 2000).

The analysis of the acetonic extracts revealed that the major class of phenolic compounds found in *P. julifora* was the flavonoids which either occurred as aglycones and / or their glycosylated forms. According to Sherwood and Bonello (2013), the sugar unit of phenolic glycoside serves to improve solubility for storage in cell organs. Reference standards were used to substantiate the identification of peaks whenever available otherwise, identification of the compounds was obtained by comparing their molecular ions obtained by the theoretical molecular weights from literature.

The LC chromatograms from the different regions were studied and analyzed as shown in Figure 1.

The selected chromatograms are examples of some of the many LC chromatograms obtained and show the peaks corresponding to the different compounds that were tentatively identified. Some peaks were found in higher percentages in one geographic region compared to the others while some peaks were found only in some regions and were absent in some.

3.2 *Phenolic glycosides*

Peak 1, was tentatively identified as coniferin (abietin). This is a glycosylated phenolic compound that usually serves as an intermediate in cell wall lignification. It is believed to aid in the transport of monolignols (Wang, Chantreau, Chantreau, & Hawkins 2013). This peak

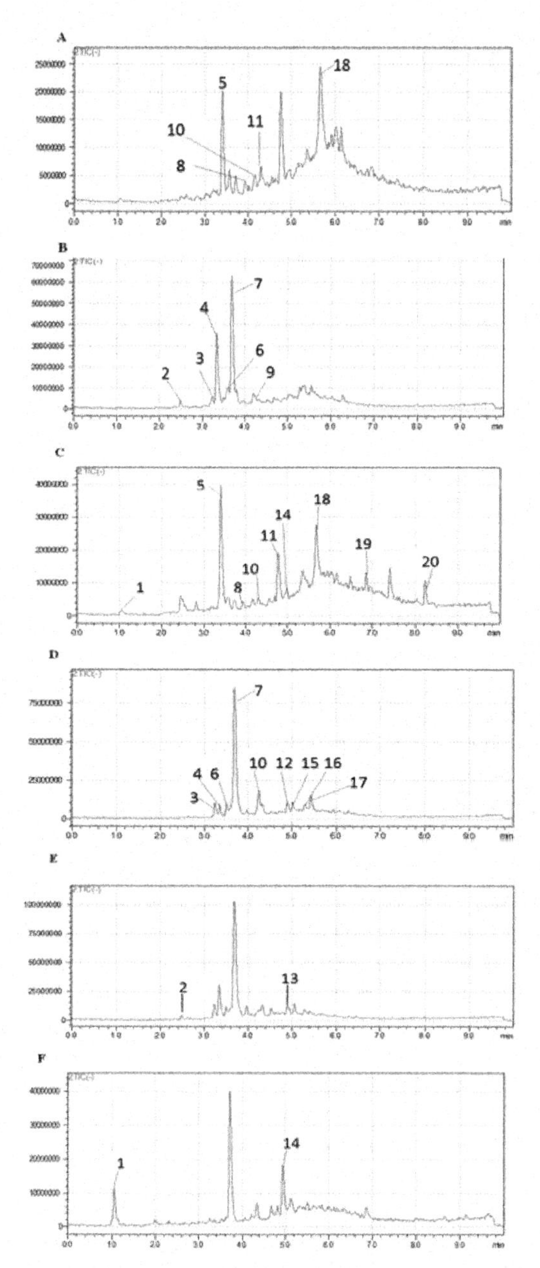

Figure 1. Various LC chromatograms of the acetonic extracts from *P. juliflora* samples of A) the bark of the big trees from Turkana County, B) the heartwood of the small trees of Turkana County, C) the bark of the small trees of Turkana County, D) the heartwood of the big trees of Turkana County, E) the heartwood of the small trees of Baringo County, F) the sapwood of the big trees of Garissa County.

was observed in the sapwoods and barks from all the three geographic regions, with it dominating more in the sapwoods than the barks. It was more abundant in the sapwood of Garissa and Baringo counties (13–25 %). Turkana County showed little amounts of coniferin in their sapwood (5 %). This compound was observed at a retention time of 1.06 min; UV 284 nm and MS,

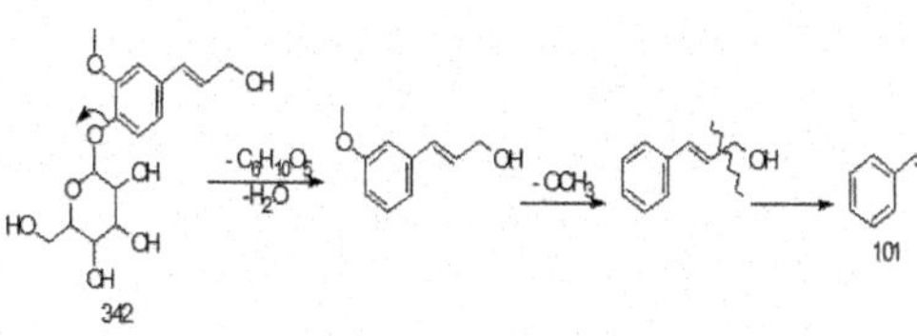

Scheme 1. Fragmentation pattern of coniferin (abietin)

341 [M-H]$^-$. A product ion scan of the [M-H]$^-$ion of this peak showed one major fragment ion at m/z 101. The fragmentation started with the loss of the sugar and water moieties which was followed by the loss of a methoxy and COH moieties yielding an ion with the m/z 101. Its fragmentation pathway is shown in scheme 3.1.

3.3 *Flavonoids*

Flavonoid aglycones are usually identified by fragmentations that require cleavage of the two C–C bonds at the C-ring. The diagnostic ion fragments of this cleavage can be justified by the retro-Diels-Alder (RDA) reactions (Cuyckens & Claeys 2004). They generally have characteristic absorption wavelength with two maxima either around 300-350 nm for bands from A-ring and 240-285 nm for bands from B-ring (Awouafack, Awouafack Tane, & Morita 2017).

Peak 2, was tentatively identified as gallocatechin with UV maxima at 280 nm and m/z 305 [M-H]$^-$; MS/MS [M-H]$^-$, 167,139,111,109. It was observed in small quantities and was found only in the small trees from all the three regions (1–4%). It is interesting to note that this compound was not found in the bark and sapwood but was only present in the knot wood, heartwood and pith samples. Its MS fragmentation pattern started from a retro-Diels Alder reaction (RDA) which is typical of flavan-3-ols (Rockenbach et al. 2012). This led to the cleavage of the C–C bond at the C-ring resulting in product ions of m/z 167. Product ions of m/z 139 and m/z 111 were also observed. These were due to the loss of a CO (28 Da) moiety from m/z 167 ion and two CO moieties for the later subsequent daughter ion.

Peak 4, 5 and 7, which were identified as catechin, 4′-*O*-methylgallocatechin and mesquitol respectively, had earlier been identified, quantified and reported in our previous article Chepkwony et al. (2020). Mesquitol and 4′-*O*-methylgallocatechin were among the most abundant compounds that were found in the *P. juliflora* samples.

Peak 13 (UV, 281 nm; MS, 315 [M-H]$^-$; MS/MS [M-H]$^-$; 300, 271, 243, 163, 151, 107) was a rare peak found only in the knot wood and heartwood samples from Baringo County at around 4–7%. It was not observed in samples from Garissa and Turkana Counties. This peak was tentatively identified as isorhamnetin with a [M-H]$^-$ peak of 315 Da. The reason why it was present only in samples from Baringo County and absent in the samples from the other regions has not yet been established. *Isorhamnetin* is usually known as

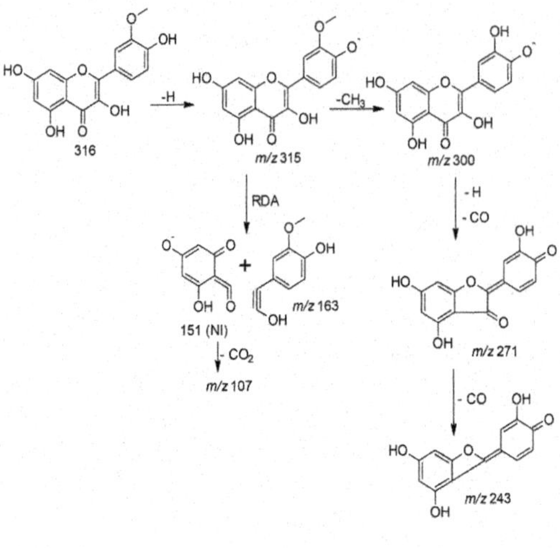

Scheme 2. Fragmentation pattern of isorhamnetin

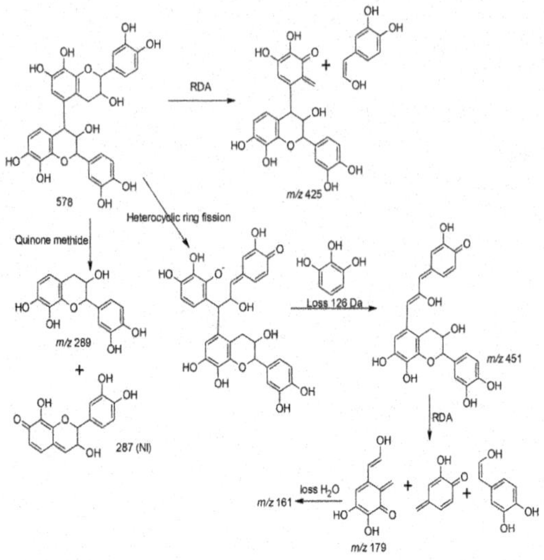

Scheme 3. Fragmentation pattern of the novel B-type proanthocyanidin of mesquitol

the methylated metabolite of quercetin which was earlier identified as present in *P. juliflora* by Khandelwal et al. (2016). Its fragmentation pattern agrees to what was suggested by Chen et al. (2015) and is elaborated in Scheme 3.2.

Peak 15 was identified as Taxifolin. It had UV, 285 nm; MS, 303 [M-H]; MS/MS $[M-H]^-$, 257, 150, 109. Its fragmentation pattern also started with a RDA yielding a product ion with m/z150. The fragment with an m/z 257 observed was obtained by the loss of both a CO (28 Da) and H_2O (18 Da) moieties from the $[M-H]^-$ ion of m/z 303. This compound was confirmed by the use of its standard which was observed to have the same retention time, same λ_{max} and same peaks.

Peak 17 was tentatively identified as Kaempferol. It had a UV maxima of 264 and 312 nm; MS, 285 $[M-H]^-$; MS/MS $[M-H]^-$, 256, 239, 227, 133, 109. The m/z 256 was attributed to loss of –COH (29 Da) moiety from the molecular ion; fragment m/z 239 was attributed to loss of OH (17 Da) moiety from m/z 256 while m/z 227 was attributed to loss of two COH (58 Da) moieties from m/z 285. Fragment m/z 133 and 109 were both fragment ions obtained from a retro-Diels-Alder cleavage of the molecular ion. This compound had also been earlier identified by Khandelwal et al. (2016) as present in *P. juliflora.*

3.4 *Proanthocyanidins*

Proanthocyanidins are oligomeric flavonoids which are responsible for the colors of fruit and flowers of many plants (Seeram et al. 2006). They are used to prevent rancidity caused by oxidation of unsaturated fats and are known for their chemopreventive biological effects (Rue, Rue, & Breemen 2017).

For the first time, a new compound believed to be a B-type proanthocyanidin of mesquitol (Peak 6) was identified. This compound was present in the knot wood, heartwood and pith of all regions. The amounts present were between 2–7%. The 7% was observed in the samples of both the big and small trees from Turkana County. The identification was based on the MS/MS diagnostic ions that resulted from retro-Diels-Alder reaction, heterocyclic ring fission and quinone methide cleavage where the cleavage occurs at the bond connecting the two flavanol molecules forming a molecule with a quinone in its structure. These fragmentation pathways are distinctive of proanthocyanidins (Hellström, Sinkkonen, Karonen, & Mattila 2007; Rue et al. 2017;. The peak was identified as a B-type proanthocyanidin as the $[M-H]^-$ ion observed was at m/z 577 whereas it would have been at m/z 575 if it was an A- type proanthocyanidin. This peak had a maximum absorption wavelength at 277 nm with MS, 577 $[M-H]^-$; MS/MS $[M-H]^-$, 425, 301, 289, 179, 161. This is elaborated in the fragmentation pattern shown in scheme 3.3.

Peak 3, was identified as catechin proanthocyanidin. It was also identified as a B-type proanthocyanidin. It was eluted much earlier at a retention time of 3.23 min than the mesquitol proanthocyanidin eluted at 3.49 min. This compound showed a UV maxima at 277 nm; MS, 577 [M-H]; MS/MS $[M-H]^-$, 451, 425, 289, 126. The m/z 425 was due to the retro-Diels-Alder reaction, the m/z 289 was due to the quinone methide reaction, the m/z 451 was obtained from a loss of the m/z 126 compound from the m/z 577 that had been obtained after a heterocyclic ring fission of the molecular ion. Its fragmentation pattern was similar to that of the mesquitol proanthocyanidin and it agreed to the one that had been proposed by Demarque, Crotti, Vessecchi, Lopesa, and Lopes (2015).

3.5 *Glycosylated flavonoids*

These are the most abundant phenolic compounds found in plants. Flavonoids usually occur in their glycosylated forms to enhance their water solubility increasing their bioavailability. It also makes the flavonoids less reactive and allows prevention of cytoplasmic damage (Cuyckens & Claeys, 2004; Slámová, Kapešová & Kapešová 2018;.

Peak 8, 10 and 11, were identified as glycosylated compounds of apigenin with peak 8 containing a dihexose moiety, peak 10 had a pentose and an hexose moiety while peak 11 had one hexose moiety. For peak 8, the retention time was 3.93 min; UV, 271, 331 nm; MS, 593 [M-H]; MS/MS [M-H]$^-$, 383, 353, 335. This compoud was identified as 6,8-di C-glucosyl apigenin (vicenin 2). Peak 10 had a retention time 4.30 min; UV, 271, 335 nm; MS, 563 [M-H]; MS/MS [M-H]$^-$, 383, 353, 335. This compoud was identified as 6-C-glucosyl-8-C-pentosyl apigenin or 6-C-pentosyl-8-C-glucosyl apigenin. The exact location of the sugar moieties could not be established but were either at position 6 or 8 of the A-ring. Peak 11 had a retention time of 4.77 min; UV, 269, 336 nm; MS, 431 [M-H]; MS/MS [M-H]$^-$, 311, 283, 163, 119, and was identified as apigenin glucoside.

For the three peaks (8, 10, 11), the MS fragmentation pattern started firstly with the fragmentation of the sugar moieties. The sugar moeities (hexose and pentose) of these compounds did not fragment from the main aglycone forming two sugar moieties of -162 Da and -132 Da respectively as it normally does for most glycosylated compounds. This shows that these glycosylated apigenin compounds had *C*-glycosidic bonds which are more stable than the *O*-glycosidic bonds. This is explained by Lenachuru (2017), who describes this as a characteristic fragmentation pattern of di *C*-glycosyl flavones. Fragmentation of the *C*-glycosylated flavonoids usually commence with the elimination of fragments from the sugar ring (Bonta 2017; Kumar 2017;. The glycosyl groups of C-glycosides (just like in this study) are usually connected to the C6 and / or C8 (Feng, Hao, & Li 2016).

For peak 8 and 10, the sugar moieties fragmented resulting in two product ions of *m/z* 353 and *m/z* 383. The two peaks observed an ion at *m/z* 335 which was attributed to the loss of H_2O moiety of 18 Da from the *m/z* ion 353. For peak 11, only the *m/z* 353 from fragmentation of the sugar moiety is observed which further fragments into products ions of *m/z* 283 and 163. The peak at *m/z* 119 is attributed to the loss of a CO_2neutral molecule (44 Da) from the *m/z* 163 ion. These fragmentation patterns have been deccribed in scheme 3.4. The fragmentation pathways agree to a study done by Ferreres, Silva, Andrade, Seabra, & Ferreira (2003).

Peak 14 was tentatively identified as isorhamnetin-3-rutinoside. Although Garissa and Turkana Counties did not contain the isorhamnetin aglycone, a glycosylated isorhamnetin was found in samples from Garissa County in high amounts of 14 %. It had a retention time of 4.94 min; UV 329 nm and a [M-H]$^-$ of 623, MS/MS [M-H]$^-$, 461, 315, 300, 271. Unlike the *C*-glycosylated flavonoids, where the sugar moieties were linked to a carbon atom at the A-ring, the two

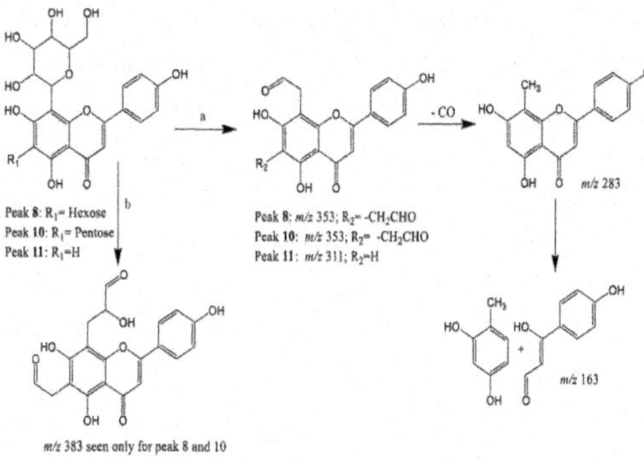

Scheme 4. Fragmentation patterns of the various glycosylated apigenin

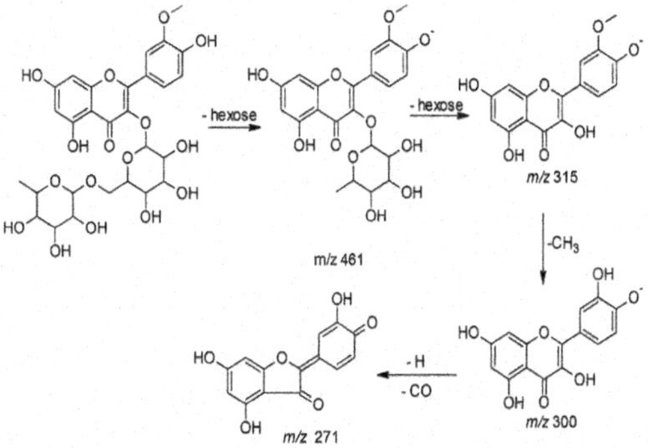

Scheme 5. Fragmentation pattern of isorhamnetin-3-rutinoside

sugar moieties for the isorhamnetin-3-rutinoside had an *O*-glycosidic bond and were bonded through the aliphatic hydroxyl group at the C-3 position. Its fragmentation started by the loss of the hexose sugar moiety to obtain product ion of *m/z* 461. This was followed by the loss of the remaining hexose moiety yielding a product ion of *m/z* 315. Loss of a methyl group (15 Da) yielded subsequent ion of *m/z* 300. Further loss of hydrogen and CO moieties gave the *m/z* 271 ion. This fragmentation pathways are shown in scheme 3.5. The isorhamnetin-3-rutinoside had earlier been reported by Prabha, Dahms, and Dahms (2014) to be present in *P. juliflora.*

Peak 18 was attributed to mesquitol diglucoside which was observed to be abundant in the bark of *P. juliflora* with ranges of 21–29%. This compound had earlier been identified, quantified and reported in our previous article Chepkwony et al. (2020).

3.6 *Unidentified compounds (9, 12, 16, 19, 20)*

Some of the peaks observed were for unknown compounds. These compounds have been included as they showed significant fraction in the *P. juliflora* extracts.

Peak 9, was observed with a retention time of 4.24 min; MS 287 [M-H]$^-$; MS/MS [M-H]$^-$, 242, 185, 153, 125, 123. Although this compound is unknown it was

Table 1. Compounds present in the *Prosopis juliflora* by the use of LC-ESI-MS/MS.

Peak number	Retention time t_R (min)	Wavelength λ_{max} (nm)	Molecular weight (M)	$[M-H]^-$	MS/MS Peaks MS^2 ions	Tentative Identification
1	1.06	284	342	341	101	Abietin (coniferin)
2	2.40	280	306	305	167;139;111109	Gallocatechin
3	3.23	277	578	577	577;451;425288;126	Catechin proanthocyanidin
4	3.32	278	290	289	151;137;109	Catechin
5	3.40	278	320	319	166;151;137109	4′-*O*-Methylgallocatechin
6	3.49	277	578	577	425;301;289179;161	Mesquitol proanthocyanidin
7	3.70	278	290	289	151;137;109	Mesquitol
8	3.93	271331	594	593	383;353;335	6,8-di *C*-glucosyl apigenin (vicenin 2)
9	4.24	–	288	287	242;185;153125;123	Unknown
10	4.30	271335	564	563	383;353;335	Glucosyl, pentosyl apigenin
11	4.77	269336	432	431	311;283;163119	Apigenin glucoside
12	4.90	–	302	301	245;227;199149;132;121109	Unknown
13	4.92	281	316	315	315;300;271243;163;151 107	Isorhamnetin
14	4.94	329	624	623	461;315;300271	Isorhamnetin-3-rutinoside
15	5.02	285	304	303	257;150;109	Taxifolin
16	5.31	281	330	329	144	Unknown
17	5.43	264312	286	285	256;239;227133;117;109	Kaempferol
18	5.67	274278	614	613	461;181;153	Mesquitol diglucoside
19	6.86	–	306	305	305;159;147145;132;129	Unknown
20	8.25	–	432	431	431;413;269255;243;169	Unknown

found only in the big trees from Turkana County. Peak 12, 16 and 19 were also unknown and were found only in the knot wood, heartwood and the pith from all three geographic regions. Peak 20, was observed with a retention time of 8.25 min. This compound was found present only in the bark and sapwood samples from the small trees of Turkana County and was absent in Baringo and Garissa Counties. More studies should be done in its identification, the role it plays in the tree and why it is found only in Turkana County.

A summary of the 20 peaks that were tentatively identified have been tabulated in Table 1 below according to their corresponding peak numbers.

4 CONCLUSION

A LC-ESI-MS/MS analytical method was employed to separate and characterize phenolic compounds present in *P. juliflora*. Phenolic compounds are of importance due to their potential health benefits like antioxidant, anti-inflammatory, anti-carcinogenic, antimicrobial, antihypertensive and cardioprotective. In *P. juliflora*, 20 peaks were detected out of which 15 were tentatively identified either as flavonoid aglycones or some of their glycosylated forms. Among these compounds identified, a new B-type proanthocyanidin of mesquitol has been reported here for the very first time. Quantification of some of the identified compounds has also been done in the negative ion mode. These results illustrate a rich array of phenolic compounds present in *P. juliflora*. The study recommends further study on the properties of these identified compounds in *Prosopis juliflora*.

ACKNOWLEDGEMENT

Financial support of the study was received from the Pamoja PHC project Curien Hubert program; the French Embassy in Kenya, the French National Research Agency (ANR) and the Africa Centre of Excellence in Phytochemicals, Textile and Renewable energy Moi University.

REFERENCES

Awouafack, M. D., Tane, P., & Morita, H. 2017. Isolation and Structure Characterization of Flavonoids. *Flavonoids-From Biosynthesis to Human Health.*

Bhuyan, D. J., & Basu, A. 2017. Phenolic compounds : potential health benefits and toxicity. In Q. V. Vuong (Ed.), *Utilisation of Bioactive Compounds from Agricultural and Food Waste* (pp. 27-59).

Bokrezion, H. 2008. The ecological and socio-economic role of *Prosopis juliflora* in Eritrea. *Academic Dissertation, Johannes Gutenberg-Universität Mainz, Germany, (PhD report).*

Bonta, R. K. 2017. Application of HPLC and ESI-MS techniques in the analysis of phenolic acids and flavonoids from green leafy vegetables (GLVs). *Journal of Pharmaceutical Analysis.* pp. 349–364.

Chen, Y., Yu, H., Wu, H., Pan, Y., Wang, K., Jin, Y. & Zhang, C. 2015. Characterization and Quantification by LC-MS/MS of the Chemical Components of the Heating Products of the Flavonoids Extract in Pollen Typhae for Transformation Rule Exploration. *Molecules* 2015, *20*, 18352–18366; DOI: 10.3390/molecules201018352.

Chepkwony, S. C., Dumarçay, S., Chapuis, H., Kiprop, A., Gerardin, P., & GerardinCharbonnier, C. 2020. Geographic and intraspecific variability of mesquitol amounts in *Prosopis juliflora* trees from Kenya. *European Journal Of Wood And Wood Products.*

Chilume, T. 2016. When a blessing becomes a curse – A case study of invasive *Prosopis juliflora*. [Blog post]. Retrieved from https://aucklandecology.com

Cuyckens, F. & Claeys, M. 2004. Mass spectrometry in the structural analysis of flavonoids. Journal of mass spectrometry. Vol 39 (1). pp. 1–15. doi.org/10.1002/jms.585.

Demarque, D. P., Crotti, A. E. M., Vessecchi, R., Lopesa, J. L. C. & Lopes, N. P. 2015. Fragmentation reactions using electrospray ionization mass spectrometry: an important tool for the structural elucidation and characterization of synthetic and natural products. *Natural Product Reports,* 33, 432. DOI: 10.1039/c5np00073d.

Dos Santos, E. T., Pereira, M. L. A., da Silva, C. F. P., Souza-Neta, L. C., Geris, R., Martins, D., & Figueiredo, M. P. 2013. Antibacterial activity of the alkaloid-enriched extract from *Prosopis juliflora* pods and its influence on in vitro ruminal digestion. *International journal of molecular sciences*, *14*(4), 8496–8516.

Dubow, A.Z. 2011. Mapping and managing the spread of *Prosopis Juliflora* in Garissa County, Kenya. Masters desertification. Kenyatta University.

Feng, W., Hao, Z. & Li, M. 2016. Isolation and Structure Identification of Flavonoids. *IntechOpen* DOI: org/10.5772/67810

Ferreres, F., Silva, B. M., Andrade, P. B., Seabra, R. M. & Ferreira, M. A. 2003. Approach to the Study of *C*-Glycosyl Flavones by Ion Trap HPLC-PAD-ESI/MS/MS: Application to Seeds of Quince (*Cydonia oblonga*) *Phytochem. Anal.* 14, 352–359. DOI: 10.1002.pca.727.

Golubov, J., Mandujano, M. & Eguiarte, L. (2001). The paradox of mesquites (*Prosopis spp.*): invading species or biodiversity enhancers? *Boleti'N De La Sociedad Bota' Nica De Me'Xico*, 2001, 69, 23.

Gu, L., Kelm, M.A., Hammerstone, J.F., Zhang, Z., Beecher, G., Holden, J., Haytowitz, D. & Prior, R.L. 2003. Liquid chromatographic/electrospray ionization mass spectrometric studies of proanthocyanidins in foods. *J. Mass spectrom.* 38: 1272–1280. DOI: 10.1002/jms.541.

Hellström, J., Sinkkonen, J., Karonen, M. & Mattila, P. 2007. Isolation and structure elucidation of procyanidin oligomers from saskatoon berries (*Amelanchier alnifolia). J Agric Food Che*m. Vol 55:157–164.

Henciya, S., Seturaman, P., James, A. R., Tsai, Y., Nikam, R., Wu, Y. C., Dahms, H. U. & Chang, F. R. 2017. Biopharmaceutical potentials of *Prosopis spp.*(*Mimosaceae, Leguminosa*). *Journal of food and drug analysis.* 25. pp. 187–196.

Jama, B. & Zeila, A. 2005. Agroforestry in the drylands of eastern Africa: a call to action. ICRAF Working Paper – no. 1. Nairobi: World Agroforestry Centre.

Kaurinovic, B. & Vastag, D. 2018. Flavonoids and phenolic acids as potential natural antioxidants. In *Antioxidants*. IntechOpen. DOI: 10.5772/intechopen.83731.

Khandelwal, P., Sharma, R. A., & Agarwal, M. 2015. Pharmacology and Therapeutic Application of *Prosopis juliflora*: A Review. Journal of Plant Sciences. Vol. 3 (4). pp. 234–240. DOI: 10.11648/j.jps.20150304.20.

Khandelwal, P., Sharma, R. A., & Agarwal, M. 2016. Isolation and identification of flavonoids from *Prosopis juliflora*. *Mintage Journal of Pharmaceutical and Medical Sciences*, 1–3.

Kumar, B. R. 2017. Application of HPLC and ESI-MS techniques in the analysis of phenolic acids and flavonoids from green leafy vegetables (GLVs). *Journal of pharmaceutical analysis*, *7*(6), 349–364.

Lenachuru, C. 2003. "Impacts of Prosopis species in Baringo District." In: S.K. Choge and B.N. Chikamai (eds) Proceedings of Workshop on Integrated Management of Prosopis Species in Kenya. Jointly published by the Global Environmental Facility (GEF), the Kenya Forestry Research Institute (KEFRI), and the Kenya Forest Department (FD), Nairobi.

Mwangi, E., & Swallow, B. 2008. *Prosopis juliflora* invasion and rural livelihoods in the Lake Baringo area of Kenya. *Conservation and Society*, *6*(2), 130.

Oduor, N. M. & Githiomi, J. K. 2013. Fuel-wood energy properties of *Prosopis juliflora* and *Prosopis pallida* grown in Baringo District, Kenya. African Journal of Agricultural Research. Vol. 8(21), pp. 2476–2481, DOI: 10.5897/AJAR08.221.

Patnaik, P., Abbasi, T. & Abbasi, S. A. 2017. *Prosopis* (*Prosopis juliflora*): blessing and bane. Tropical Ecology 58(3): 455–483.

Pecha, B. & Perez, M. G. 2015. Bioenergy. Biomass to biofuels. pp. 413–442. Doi: 10.1016/13978-0-12-407909-0.00026-2.

Prabha, D. S., Dahms, H. U. & Malliga, P. 2014. Pharmacological potentials of phenolic compounds from *Prosopis spp.*-a review. *Journal of Coastal Life Medicine.* 2(11): 918–924.

Rockenbach, I. I., Jungfer, E., Ritter, C., Santiago-Schübel, B., Thiele, B., Fett, R., & Galensa, R. 2012. Characterization of flavan-3-ols in seeds of grape pomace by CE, HPLC-DAD-MSn and LC-ESI-FTICR-MS. *Food research international*, *48*(2), 848–855.

Rue, E. A., Rush, M. D. & Breemen, R. B. 2017. Procyanidins: a comprehensive review encompassing structure elucidation via mass spectrometry. Phytochem. Vol 17(1): 1–16. Doi: 10.1007/s11101-017-9507-3.

Seeram, N. S., Lee, R., Scheuller, H. S. & Heber, D. 2006. Identification of phenolic compounds in strawberries by liquid chromatography electrospray ionization mass spectroscopy. Food Chemistry. Vol: 97:1–11.

Sherwood, P. & Bonello, P. 2013. Austrian pine phenolics are likely contributors to systemic induced resistance against *Diplodia pinea*. *Tree Physiol.* 33 (8). 845–854. doi.org/10.1093/treephys/tpt063.

Sirmah, P.K. 2009. Valorisation du *Prosopis juliflora* comme alternative à la diminution des ressources forestières au Kenya. Academic dissertation, Université de Lorraine, (PhD report).

Slámová, K., Kapešová, J. & Valentová, K. 2018. "Sweet Flavonoids": Glycosidase-Catalyzed Modifications. Int J Mol Sci. 19(7): 2126. doi: 10.3390/ijms19072126

Sun, W. & Miller, J. M. 2002. Tandem mass spectrometry of the B-type procyanidins in wine and B-type dehydrodicatechins in an autoxidation mixture of (+)-catechin and (−)-epicatechin. *J. Mass Spectrom.* 38: 438–446.

Swatsitang, P., Tucker, G., Robards, K. & Jardine, D. 2000. Isolation and identification of phenolic compounds in *Citris sinensis*. *Analytical chemistry. Acta.* 417, 231.

Thakur, R., Singh, R., Saxena, P., & Mani, A. 2014. Evaluation of antibacterial activity of Prosopis juliflora (SW.) DC. leaves. *African Journal of Traditional, Complementary and Alternative Medicines*, *11*(3), 182–188.

Wang, Y., Chantreau, M., Sibout, R., & Hawkins, S. 2013. Plant cell wall lignification and monolignol metabolism. *Frontiers in plant science*, *4*, 220.

Advances in Phytochemistry, Textile and Renewable Energy Research for Industrial Growth – Nzila et al. (Eds)

Phytochemical screening, total phenolic and flavonoid content of *Senna didymobotrya*

B.O. Sadia* & J.K. Cherutoi
School of Sciences and Aerospace Studies, Department of Chemistry & Biochemistry, Moi University, Eldoret, Kenya

C.M. Achisa
School of Engineering, Department of Chemical and Process Engineering, Moi University, Eldoret, Kenya

ABSTRACT: *Senna didymobotrya* has been used in Kenya by the Kipsigis community to control malaria as well as diarrhoea. The Pokot prepare charcoal from the stem for milk preservation. Research has not been done to investigate the effect of different extraction solvents on yield, total phenolic and flavonoid content of *Senna didymobotrya* plant roots. The aim of this study was to compare root extract yield of diethyl-ether, methanol, and aqueous solvents; phytochemical screening; and total phenolic and flavonoid content of *Senna didymobotrya* plant roots. Extraction was done by the Soxhlet method. Phytochemical screening was done using Harborne's (1973) method with a slight modification. Total flavonoid content was determined by aluminium chloride colourimetric assay at 420 nm. Total phenolic content was determined by Folin–Ciocalteu at 760 nm using UV-Vis spectrophotometry. Extraction yield of diethyl ether, methanol, and distilled water were 3.72 g (7.44%), 4.97 g (9.94%), and 9.09 g (18.18%), respectively, showing a significant difference ($p < 0.05$) in the yields obtained using the different solvents. Phytochemical screening was positive for phenols, tannins, saponins, gladiac glycosides, anthraquinones, alkaloids, and flavonoids. Total flavonoid content was found to be 48.3 ± 1.5 (QEmg/g) and total phenol content was calculated as 34.5 ± 0.1 (GAEmg/g). Distilled water can be utilized as the best extraction solvent. Senna has a high amount of flavonoid and phenolic content. The limitation of this research is that it only tested root extracts and not leaves, flowers, or seeds. More studies need to be done to isolate the different compounds identified.

Keywords: Phytochemicals, Flavonoid, Phenolic, *Senna didymobotrya*

1 INTRODUCTION

Medicinal plants are widely distributed throughout the world and have been used to promote human health (Ngulde, Sandabe, Tijjani, Barkindo, & Hussaini 2013), and in tackling excess mortality and morbidity among marginalized and poor populations. Currently, although accessibility to modern healthcare has become faster and easier, there are still populations that prefer to promote their health by using fresh medicinal plants. The most known are the traditional Chinese medicine and the Ayurveda of India (Johari & Khong 2019).

Most African countries are poor, with highly underdeveloped healthcare systems. A large percentage of the African population is living below the poverty line and cannot afford expensive conventional medicines. Moreover, the belief among certain African communities that certain diseases can only be managed using traditional medicines has greatly contributed to the continued use of herbal medicine. The World Health Organization (WHO) estimates that one-third of the world's population has no regular access to essential modern medicines and that about half the population in some parts of Africa, Asia, and Latin America faces these shortages (WHO 2012). Herbal medicine has therefore provided an alternative method for disease treatment and management.

Natural products have played an important role in modern drug development due to their structural diversity. It is estimated that about 25% of all modern medicines are directly or indirectly derived from higher plants (WHO 2012; Ganga, Rao, & Pavani 2012). Plant biomolecules serve as drug entities as well as chemical models for the design and synthesis of therapeutics for communicable and non-communicable diseases (Veeresham 2012). Important drugs such as paclitaxel (Mirjalili, Farzaneh, Bonfill, Rezadoost, & Ghassempour 2012), camptothecin (Kusari et al. 2009), morphine (Powers, Erickson, & Swortwood 2017), aspirin, cocaine (Brachet, Rudaz, Mateus, Christen & Veuthey 2001), codeine (Fakhari, Nojavan, Ebrahimi, & Evenhuis 2010), digitoxin (Kohls, Scholz-Botttcher, Teske, & Rullkotter 2015),

*Corresponding author

DOI 10.1201/9781003221968-20

quinine, artemisinin (Misra, Mehta, Mehta, & Mehta 2014), and silymarin (Saleh et al. 2015), among others, have been isolated from plants (Sofowora 2008).

The medicinal value of plants lies in some chemical substances that produce definite physiological actions on the human body. The most important of these compounds are alkaloids, flavonoids, tannins, and phenolic compounds. Biological and pharmacological activities of phytochemical compounds depend on factors such as the ecological factors, age of the plant, species, and method of extraction. Thus, each plant has different chemical composition, toxicity, and bioactivity (Jeruto, Arama, Anyango, & Maroa 2017). The type and amount of phytochemical compounds present usually differ from one part of the plant to another. This explains why in traditional medicine, different parts of the same plant may be used in treating different diseases (Ngulde et al. 2013).

Senna didymobotrya belongs to the genus *Senna* and family Fabaceae. It is a hairy, aromatic shrub 5–9 m tall. The plant flowers in each inflorescence are arranged sequentially in raceme of bright yellow petals. The fruit is a flat brown legume pod. Different ethnic communities in Kenya have various names for the species *didymobotrya*, for example, in Meru, it is called Murao/Kirao (Gakuubi & Wanzala 2012), Senetwet in Nandi and Kipsigis (Jeruto, Lukhoba, Ouma, Mutai, & Otieno 2008), Owinu/Obino in Luo, and Ithaa/Muthaa in Kamba (Wagate et al. 2012). The plant is used for the treatment of fungal and bacterial infections, hypertension, haemorrhoids, sickle cell anaemia, a range of diseases affecting women such as inflammation of fallopian tubes, fibroids, and backache, to stimulate lactation, and to induce uterine contraction and abortion (Nyamwamu et al. 2015).

2 LITERATURE REVIEW

Herbal medicines contain active ingredients that are present in complex mixtures formulated as crude fractions of plants or combinations of plants. Herbal drugs are widely accepted as an alternative treatment for primary health care needs in both developing and developed populations. However, herbal medicines have a range of limitations including lack of evidence of safety, efficacy, standardization, varying production practices and absence of regulatory standards and implementation protocols (Chawla et al. 2013). The quality issue of herbal drugs can be ensured by conducting some important tests such as micro- and macroscopic investigation, moisture content, exclusion of foreign organic matter, extractive values, ash value, qualitative and quantitative chemical tests, chromatographic characterization, toxicological test, phytochemical evaluation, and microbial tests (Chawla et al. 2013; Sahil, Sudeep, & Akanksha 2011; Yadav, Mahour, & Kumar 2011).

Quality control of the medicinal plants starts right at the source of the plant material. The phytochemical composition of the plant material and the resulting quality can vary due to several factors including a number of environmental factors such as geographical location, soil quality, temperature, and rainfall; taxonomy, the time of collection, method of collection, cultivation, harvesting, drying and storage conditions, preparation and processing methods can also affect composition. Contamination by microbes, chemical agents such as pesticides, and heavy metals, as well as by insects and animals during any of these stages can also lead to a poor quality of the finished products (Sahil et al. 2011).

Senna didymobotrya is a potential medicinal plant and the medicinal values have been explored well in many parts of the world by traditional practitioners (Nagappan 2012). In Kenya the Kipsigis community has been using this plant to control malaria as well as diarrhoea (Korir, Mutai, Kiiyukia, & Bii 2012). The Pokot peel the bark, dry the stem, and burn it into charcoal that they use to improve digestibility and palatability and to preserve milk (Tabuti 2007). In addition, it has been used to treat the skin conditions of humans and livestock infections as well (Njoroge & Bussmann 2007). It is also used in the treatment of animal diseases such as the removal of ticks (Njoroge & Bussmann 2006). The plant leaves and roots are also used as fish poison (Nyamwamu et al. 2015; Thangiah & Ngule 2013).

In East and Central Africa, Senna root extract is applied in the treatment of malaria, jaundice, intestinal worm, and ringworm (Nagappan 2012). The plant is also used for the treatment of fungal and bacterial infections, hypertension, haemorrhoids, sickle cell anaemia, a range of diseases affecting women such as inflammation of fallopian tubes, fibroids, and backache. Root decoction from this plant is used to manage general poison due to its emetic and purgative effect (Tabuti 2007).

Previous studies have isolated anthraquinones such as chrysophanol, physcion, emodin, tarosachrysone, aloe-emodin, fallacinol, rhein, and parientinic acid; flavonoids such as quercetin, ombuin, apigenin, kaemferol, A pigenin-5,7,4-trimethyl ether; and flavonoid glycosides such as isoguercitrin and kaempferol-3-rhamnoside from Senna (Alemayehu, Abegaz, Snatzke, & Duddeck 1989; Mahadevan, Upendra, Subburaju, Elango, & Suresh 2002;. In the past, researchers have used only one solvent in the *S. didymobotrya* extraction, for example, methanol (Jeruto et al. 2017); solvent ratios using methanol:water 9:1 (Thangiah et al. 2013), dichloromethane:methanol 1:1 (Alemayehu, Tadesse, Mammo, Kibret, & Endale 2015); and sequential extraction using hexane, ethylacetate, dichloromethane, and methanol (Mining et al. 2014; Nyamwamu et al. 2015;. Research has not been done to investigate the effect of different extraction solvents on yield, total phenolic and flavonoid content of *Senna didymobotrya* plant roots. This study will compare root extract yield of diethyl-ether, methanol, and aqueous solvents; phytochemical screening; and total phenolic and flavonoid content of *Senna didymobotrya* plant roots.

3 MATERIALS AND METHODS

3.1 *Chemicals and reagents*

Folin Ciocalteu's Phenol reagent, acetic acid, Wagners reagent, iron(III) chloride, hydrochloric acid, sodium carbonate, ammonia solution, gallic acid, quercitin, methanol, and diethyl ether were purchased from Merck Ltd. All the chemicals and reagents were of analytical grade and were used without further purification. Reagents used for standard preparations are quercetin and gallic acid were purchased from Sigma-Aldrich.

3.2 *Equipment*

Sohxlet assembly, Rotary Evaporator (R-200, Buchi Larbortechnik), Analytical Balance CITIZEN scale CX-220 (CITIZEN Private Ltd), 20~325 Mesh Grinder Vertical hammer mill pulverizer, Beckman Coulter DU 720 UV–Visible spectrophotometer (700 series (6584) R), Calibrated micropipettes used to accurate measurement and transfer.

3.3 *Sample collection*

Senna didymobtrya roots were collected from natural geographical landscapes of West Uyoma sub-location, Siaya County, Kenya. They were identified and authenticated at the Department of Biological sciences, Moi University where voucher specimen (SD 2018/03) were preserved for future reference.

3.4 *Preparation of Senna didymobtrya roots extract*

The collected roots were washed several times with distilled water to remove dust. They were dried at room temperature under shade for 3 weeks, chopped into small pieces (approximately one centimetre) and pulverized using a laboratory mill.

The extraction was carried out according to the method described by Kigondu, Rukunga, Keriko, Tonui, and Gathirwa (2009) with slight modifications. Using an electric analytical beam balance, for each extraction solvent 50 g of powdered dried plant material was weighed, placed in Soxhlet apparatus, and separately extracted with 250 mL of diethyl ether, methanol, and distilled water solvents for 48 hours. The extracted sub samples were transferred in a conical flask, and then placed in a rotary vacuum evaporator in a water bath at 40°C to recover the solvent and concentrate the crude extract. The semi-solid extracts were placed in sterile beakers and were placed in a desiccator containing anhydrous sodium sulfate for 24 hours for complete evaporation of the solvent. The total yield of the solid crude extracts were weighed and put in a tightly screwed capped glass containers and stored in the refrigerator at 4°C prior to use in the phytochemical screening and determination of total phenol and flavonoid content.

Yield of the extract obtained was calculated as follows:

$$\text{Extractive yield value} = \frac{\text{Weight of concentrated extract}}{\text{Weight of plant dried powder}} \times 100$$

3.5 *Qualitative phytochemical analysis*

The phytochemical analysis of the extracts for identification of bioactive chemical constituents was done using standard procedures by Trease and Evans (1989) and Sofowara (1983).

Tannins

A sample of 0.5 g was put in a test tube and 20 mL of distilled water was added and heated to boiling. The mixture was then filtered and 0.1% of $FeCl_3$ was added to the filtrate and observations made. A brownish green colour or a blue black colouration indicate the presence of tannins.

Saponins

The crude solvent extract was mixed with 5 mL of water and vigorously shaken. The formation of a stable form indicates the presence of saponins.

Flavonoids

About 1 g of the plant extract was mixed with a few fragments of magnesium ribbon (0.5 g) and a few drops of concentrated hydrochloric acid were added. Development of a pink or magenta red colour after 3 minutes indicates the presence of flavonoids.

Terpenoids

The solvent extracts of the plant material were taken in a clean test tube, 2 mL of chloroform was added, the test tube was vigorously shaken, and then evaporated to dryness. To this, 2 mL of concentrated sulfuric acid was added and heated for about 2 minutes. A grayish colour indicates the presence of terpenoids.

Glycosides

Keller–Kilani test

The solvent plant material extract was mixed with 2 mL of glacial acetic acid containing 1–2 drops of 2% solution of $FeCl_3$, the mixture was then poured into a test tube containing 2 mL of concentrated sulfuric acid. A brown ring at the interface of the two solutions indicates the presence of cardiac glycoside.

Alkaloids

The crude extract was mixed with 1% of HCl in a test tube. The test tube was then heated gently and a few drops of Mayers and Wagners reagents were added by the side of the test tube. A resulting precipitate confirms the presence of alkaloids.

Steroids

Liebermann's Burchard reaction

About 2 g of the solvent extract was put in a test tube and 10 mL of chloroform was added and filtered. The filtrate, 2 mL, was mixed with 2 mL of a mixture of acetic acid and concentrated sulfuric acid. A blue green ring indicates the presence of steroids.

Phenols

The plant extract was put in a test tube and treated with a few drops of 2% $FeCl_3$. A blue green or black colouration indicates the presence of phenols.

Anthraquinones

Borntreger's test

A sample of 5 g of the extract was put in a test tube and 10 mL of benzene was added. The mixture was shaken and filtered. 5 mL of ammonia solution was added to the filtrate and the mixture shaken. The presence of a violet colour in the ammonia phase (lower phase) indicates the presence of anthraquinones.

3.6 *Determination of total flavonoid content*

Total flavonoid content (TFC) was determined by aluminium chloride colourimetric assay adapted from Sandip et al. (2014) with a slight modification. Quercetin was used as the standard and the flavonoid content of the extract was expressed as mg of quercetin equivalent per gram of dried extract. 10 mg of quercetin was dissolved in 10 mL of 80% methanol (v/v) and filtered. This was labelled as stock solution. Then serial dilutions were performed: Five 50 mL volumetric flask were cleaned and 10, 20, 30, 40, and 50 μL of the stock solution added to the flask and made to the mark and the final concentrations were 0.2, 0.4, 0.6, 0.8, and 1.0 ppm, respectively. The blank consisted of all the reagents, except for the extract or quercetin in standard solution which was substituted with 1 mL of methanol. With a pipette 3 mL of plant extract or standard of different concentration solution was transferred into a test tube. Then 1 mL of 2% $AlCl_3$, 80% methanol solution and 1 mL of 1 M sodium acetate solution was added to 3 mL of the extract or standard. The test tube of the mixture was then allowed to stand at room temperature for 1 hour. The extract mixture was diluted to 50 mL to give absorbance below 1.0 in the UV–Vis spectrophotometer. The absorbance of the solution was measured at 446 nm using a UV–Vis spectrophotometer against a blank. The total content of flavonoid compounds in plant extracts in quercetin equivalents was calculated by the following equation

$$C = DF\left(\frac{c \times V}{m}\right) \tag{1}$$

Where C is the total content of flavonoid compounds in mg/g plant extract, in Gallic equivalents (GAE); DF is the dilution factor; c is the concentration of quercetin established from the calibration curve in mg/mL; V is the volume of the extract in mL; and m is the weight of pure plant extract in g.

3.7 *Determination of total phenolic content*

Total phenolic content (TPC) of methanol extract of *Senna didymobotrya* was determined by the method involving Folin–Ciocalteau reagent as oxidizing agent and gallic acid as standard. Gallic acid 10 mg was dissolved in 10 mL distilled water, this is 1 mg/mL concentration. It was labelled stock solution. Then serial dilutions were performed: Six 50 mL volumetric flask were cleaned and 20, 40, 60, 80, 100, and 120 μL of the stock solution added to the flask and made to the mark to give the final concentrations as 0.4, 0.8, 1.2, 1.6, 2.0, and 2.4 ppm respectively. The blank consisted of 1 mL Folin-Ciocalteau reagent, 3 mL distilled water, and 1 mL sodium carbonate solution. With a pipette 1 mL of the plant extract or standard of different concentration solution was transferred into a test tube. About 1 mL of Folin–Ciocalteau reagent (diluted 10 times) was added into the test tube. Finally, 1 mL of sodium carbonate (7.5%) solution was added into the test tube. The test tubes were left standing for 1 hour at 25 °C to complete the reaction. The test tube with extract solution was diluted to 20 mL to get absorbance below 1.0. The absorbance of the solutions was measured at 760 nm using UV–Vis spectrophotometry against a blank. The total content of phenolic compounds in plant extracts in gallic acid equivalents was calculated by equation (1) above.

4 RESULTS AND DISCUSSION

4.1 *Extraction of compounds of Senna didymobotrya roots*

Fifty grams of *S. didymobotrya* root powder yielded 3.72 g (7.44%), 4.97 g (9.94%), and 9.09 g (18.18%) of *S. didymobotrya* crude extracts when extracted with diethyl ether, methanol, and distilled water respectively. There was a significant difference ($p < 0.05$) in

Table 1. Results of phytochemical screening of aqueous extract of *Senna didymobotrya*.

No	Phytochemical	Observation	Test result
1	Phenols	Bluish black colour present	Positive
2	Tannins	Brown colour	Positive
3	Saponins	Formation of stable form	Positive
4	Gladiac glycosides	A brown ring at the interface of the two solutions	Positive
5	Anthraquinones	Violet colour observed in the ammonia phase (lower phase)	Positive
6	Alkaloids	Formation of precipitate	Positive
7	Flavonoids	Magenta red colour observed	Positive
8	Terpenoids	Greyish colour not observed	negative
9	Steroids	Blue ring not observed	negative

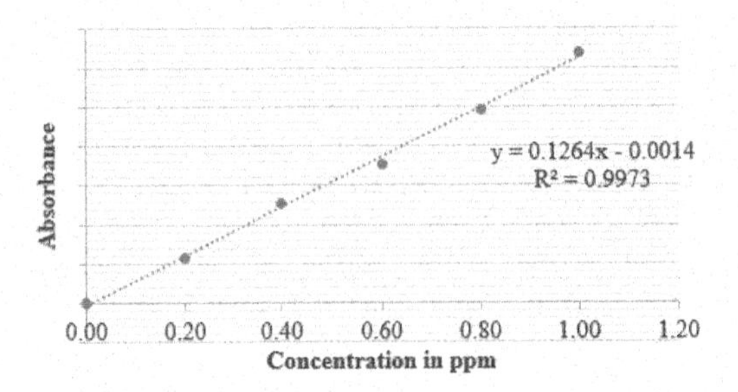

Figure 1. Calibration curve of quercetin.

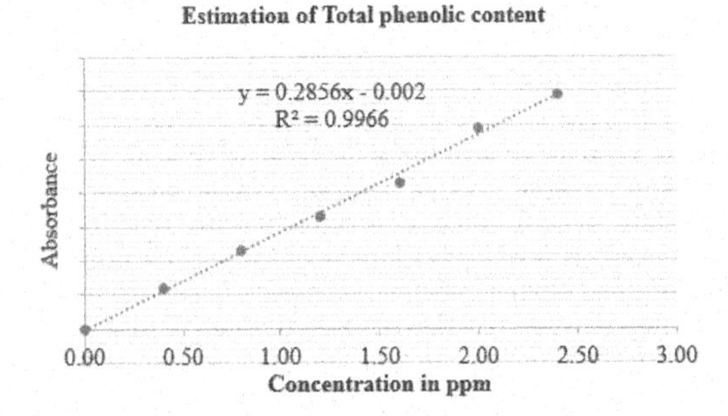

Figure 2. Calibration curve for gallic acid.

the yields obtained using the different solvents. Extraction with water gave the highest yield, followed by methanol, and diethylether gave the lowest yield.

4.2 *Phytochemical screening of Senna didymobotrya root extract*

Phytochemical screening, as shown in Table 1, tested positive for phenols, tannins, saponins, gladiac glycosides, anthraquinones, alkaloids, and flavonoids; terpenoids and steroids tested negative. The phytochemicals that tested positive are polar and this could be the reason why the yield was highest when water is used as extraction solvent since water is the most polar of the three solvents.

The presence of different classes of phytochemicals such as phenols, tannins, saponins, gladiac glycosides, anthraquinones, alkaloids, and flavonoids may be the reason why this plant has been used to treat a wide range of diseases traditionally.

4.3 *Total flavonoid and phenol content determination*

The total flavonoids content was calculated as quercetin equivalent (mg QE/g dry weight) using the equation based on the calibration curve of quercetin in Figure 1: $y = 0.1264x - 0.0014$, $R^2 = 0.9973$, where

Table 2. Preparation of calibration curve of quercetin.

Concentration of quercetin in (ppm)	Absorbance
0.00	0.000
0.20	0.023
0.40	0.051
0.60	0.071
0.80	0.098
1.00	0.128

Table 3. Calculation procedure for total flavonoid.

Concentration of gallic acid (ppm)	Absorbance
0.00	0.000
0.40	0.120
0.80	0.230
1.20	0.330
1.60	0.429
2.00	0.588
2.40	0.688

x is the quercetin equivalent (QE) and y is the absorbance.

Total phenolic compound content (TPC) was estimated as garlic acid equivalent (GAE), that is mg GAE per g dry weight, using the equation obtained from the calibration curve of garlic acid in Figure 2: $y = 0.2856 \times -0.002$, $R^2 = 0.9966$, where x is GAE and y is the absorbance.

Table 3 shows that the TFC is 48. 3354 ± 1.4775 mg QE/g dry weight and Table 5 shows that TPC is 34.4771 $\pm$ 0.08734 mg GAE/g dry weight.

5 CONCLUSIONS AND RECOMMENDATIONS

Phytochemical screening showed positive for phenols, alkaloids, anthraquinones, flavonoids, saponins, tannins, gladiac glycosides in the roots of *Senna didymobotrya.* Distilled water gave the highest yield of root extract of *S. didymobotrya* and can be utilized as the best extraction solvent. *Senna* has a high amount of flavonoids and phenols. The limitation of this research is that it only tested root extracts and not leaves, flowers, or seeds. More studies need to be done to isolate the different compounds identified.

CONFLICT OF INTEREST

The authors declare no conflict of interest.

Table 4. Preparation of calibration curve of garlic acid.

Sample solution µg/mL	Weight of dry extract in gram	Absorbance	GAE Concentration (mg/mL)	TPC as GAE $C = DF\left(\frac{c \times V}{m}\right)$	Mean SEM
1000	0.001	0.492	1.7297	34.5938	
1000	0.001	0.489	1.7191	34.382	34.48 ± 0.09
1000	0.001	0.490	1.7226	34.452	

Table 5. Calculation procedure for total phenolic.

Sample solution µg/mL	Weight of dry extract in gram	Absorbance	QE Concentration (mg/mL)	TFC as QE $C = DF\left(\frac{c \times V}{m}\right)$	Mean SEM
3000	0.001	0.038	0.3117	46.755	
3000	0.001	0.041	0.3354	50.31	48.33 ± 1.48
3000	0.001	0.039	0.3196	47.94	

ACKNOWLEDGMENT

The authors are thankful to Africa Centre of Excellence in Phytochemicals, Textile, and Renewable Energy (ACE II-PTRE) for the financial support.

REFERENCES

Alemayehu, G., Abegaz, B., Snatzke, G., & Duddeck, H. (1989). Quinones of *Senna didymobotrya* a. *Bulletin of the Chemical Society of Ethopia, 3(1).*

Alemayehu, I., Tadesse, S., Mammo, F., Kibret, B., & Endale, M. (2015). Phytochemical analysis of the roots of *Senna didymobotrya. Journal of Medicinal Plant Research, 9(34).*

Anpin Raja, R.D., Jeeva S., Prakash, J.W., Johnson, M. & Irudayaraj, V. (2011). Antibacterial activity of selected ethnomedicinal plants from South India. *Asian Pac J Trop Med*; 4: 375–378.

Brachet, A., Rudaz, S., Mateus, L., Christen, P., &Veuthey, J. (2001). Optimization of accelerated solvent extraction of cocaine and benzoylecgonine from coca leaves. *Journal of Separation Sciences, 24*; 865–873.

Chawla, R., Thakur P., Chowdhry A., Jaiswal S., Sharma A., Goel R., Sharma J., Priyadarshi S.S., Kumar V., Sharma R. K., & Arora R. (2013). Evidence based herbal drug standardization approach in coping with challenges of holistic management of diabetes: A dreadful lifestyle disorder of 21st century. *Journal of Diabetes and metabolic disorders*, 12, 33–35.

Fakhari, A. R., Nojavan, S., Ebrahimi, S. N., &Evenhuis, C. J. (2010). Optimized ultrasound-assisted extraction procedure for the analysis of opium alkaloids in papaver plants by cyclodextrin-modified capillary electrophoresis. *Journal of Separation Science*, 33, 2153–2159.

Gakuubi, M. M. & Wanzala, W. (2012). A survey of plants and plant products traditionally used in livestock health management in Buuri district, Meru County, Kenya. *Journal of Ethnobiology and Ethnomedicine, 8*(39).

Ganga, R. B, Rao, V. Y,& Pavani, V. S. P (2012). Quantitative and qualitative phytochemical screening and invitro antioxidant and antimicrobial activities of Elephantopus Scarber Linn. Recent Research in Science and Technology. Vol. 4 (4). pp 15–20.

Jeruto, P., Arama, P. F., Anyango, B., &Maroa, G. (2017). Phytochemical screening and antibacterial investigations of crude methanol extracts of Senna didymobotrya (Fresen.) H. S. Irwin & Barneby. *Journal of Applied Biosciences.* 114: 11357–11367.

Jeruto, P., Lukhoba, C., Ouma G., Mutai C., & Otieno D., (2008). An Ethnobotanical study of medicinal plants. *Journal of Ethnopharmacology.*

Johari, M. A., & Khong, H. Y. (2019). Total Phenolic Content and Antioxidant and Antibacterial Activities of *Pereskia bleo. Advances in Pharmacological Sciences,* 7428593.

Kigondu, E. V. M., Rukunga, G. M., Keriko, J. M., Tonui, W. K. & Gathirwa, J. W. (2009). Anti – parasitic activity and cytotocity of selected medicinal plants from Kenya. J. of Ethnopharm, 123: 504–509.

Kohls, S., Scholz-Botttcher, B. M., Teske, J., & Rullkotter, J. (2015). Isolation and quantification of six cardiac glycosides from the seeds of Thevetia peruviana provide a basis for toxicological survey. *Indian Journal of Chemistry, 54B*, 1502–1510.

Korir, R.K., Mutai, C., Kiiyukia, C. & C. Bii, C. (2012). Antimicrobial Activity and Safety of two Medicinal Plants Traditionally used in Bomet District of Kenya. *Research Journal of Medicinal Plants*; 6: 370–382.

Kusari, S. (2009). An endophytic fungus from Camtotheca acuminate that produces campothecin and analogues. *Journal of Natural Products, 72(1)*; 2–7.

Mahadevan, N., Upendra, B., Subburaju, T., Elango, K., & Suresh, B. (2002). Purgative and anti-inflamatory activities of *Cassia didymobotrya*, fresen. *Ancient Science of Life, 22(1)*, 9.

Mining J., Lagat Z. O., Akenga T., Tarus P., Imbuga M., & Tsamo M. K. (2014). Bioactive metabolites of Senna didymobotrya used as biopesticide against Acanthoscelides obtectus in Bungoma, Kenya. *Journal of Applied Pharmaceutical Science, 4*(9), 56–60. doi:10.7324/JAPS.2014.40910

Mirjalili, M. H., Farzaneh M., Bonfill M, Rezadoost H., & Ghassempour, A. (2012). Isolation and characterization of Stemphylium sedicola SBU-16 as a new endophytic taxol-producing fungus from Taxus baccata grown in Iran. *Federation of European Microbiological Societies Lett 328*, 122–129.

Misra, H., Mehta, D., Mehta, B. K., & Jain D. C. (2014). Extraction of Artemisinin, an Active Antimalarial Phytopharmaceutical from Dried Leaves of *Artemisia annua* L., Using Microwaves and a Validated HPTLC-Visible Method for Its Quantitative Determination. *Chromatography Research International, 2014*.

Nagappan, R. (2012). Evaluation of aqueous and ethanol extract of bioactive medicinal plant,Cassia didymobotrya (Fresenius) Irwin & Barneby against immature stages of filarial vector, Culex quinquefasciatus Say (Diptera: Culicidae). *Asian Pacific Journal of Tropical Biomedicine*; 707–711.

Ngulde, S. I., Sandabe, U. K., Tijjani M. B, Barkindo, A. A., & Hussaini, I. M. (2013). Phytochemical constituents, antimicrobial screening and acute toxicity studies of the ethanol extract of *Carissa edulis* Vahl. root bark in rats and mice.*American Journal of Research Communication, 1(9)*.

Njoroge, G. N. & Bussmann, R. W. (2006). Diversity and utilization of antimalarial ethnophytotherapeutic remedies among the Kikuyus (Central Kenya). *Journal of Ethnobiology and Ethnomedicine, 2*(8).

Powers, D., Erickson, S., & Swortwood, M. J. (2017). Quantification of Morphine, Codeine, and Thebaine in Home-Brewed Poppy Seed Tea by LC-MS/MS. *Journal of Forensic Sciences, 2017*.

Sahil, K., Sudeep, B., & Akanksha, M. (2011). Standardization of medicinal plant materials. *International Journal of Research in Ayurveda & Pharmacy*, 2(4), 1100–1109. ISSN 2229-3566.

Saleh, I. A., Vinatoru, M., Mason, T. J., Abdel-Azim, N. S., Aboutabl, E. A., & Hammouda F. M. (2015). Ultrasonic-Assisted Extraction and Conventional Extraction of Silymarin from *Silybum marianum* seeds; A Comparison. *Research Journal of Pharmaceutical, Biological and Chemical Sciences*, 6(2). ISSN: 0975-8585.

Sofowara, A. (1993). Medicinal plants and traditional medicine in Africa. *Spectrum books ltd, Ibadan Nigeria*, 191–289.

Sofowora, E. A. (2008). *Medicinal Plants and Traditional Medicine in Africa*. 3rd edn Spectrum books Ltd, Ibadan, Nigeria. 117–139, 195–248.

Tabuti, J.R.S. (2007). Senna didymobotrya (Fresen.) H.S. Irwin & Barneby. In: Schmelzer, G.H. and Gurib-Fakim, A. (Editors). Prota 11(1): Medicinal plants/Plantes médicinales 1. PROTA,Wageningen, Netherlands.

Thangiah, A. S., & Ngule, M. C., (2013). Phytopharmacological analysis of methanolic-aqua extract (fractions) of *Senna didymobotrya* roots. *International Journal of Bioasssays, 2(11)*, 1473–1479.

Trease, G.E. & Evans, W.C. (1989). Pharmacognosy, 11th end, Brailliere tindall, London, pp. 45–50.

Veeresham, C. (2012). Natural produts derived from plants as a source of drugs. *Journal of Advanced Pharmaceutical Technology & Research*, 3(4); 200–201.

Wagate, C. G., Mbaria, M. J., Gakuya, D.W., Nanyingi, M. O., Kareru, P. G., Njuguna, A., Gitau, N., Macharia, J. K. & Njonge, F. (2012). Screening of some Kenyan medicinal plants for antibacterial activity. *Phytotherapy Research, 24*.

WHO (2012). WHO Global Atlas of Traditional, Complementary and Alternative Medicine. World Health Organization, Geneva. 1 and 2.

Yadav, P., Mahour, K., & Kumar, A. (2011). Standardization and evaluation of herbal drug formulations. *Journal of Advanced Laboratory research in Biology*, 2(4).

Advances in Phytochemistry, Textile and Renewable Energy Research for Industrial Growth – Nzila et al. (Eds)

Structural and electronic properties of light atom doped 2D MoS_2: Quantum mechanical study

Kibet Too Philemon*
Department of Physics, University of Nairobi, Nairobi, Kenya

Kiptiemoi Korir Kiprono
Department of Physics, Moi University, Eldoret, Kenya

Musembi J. Robinson
Department of Physics, University of Nairobi, Nairobi, Kenya

Jackson K. Cherutoi
Department of Chemistry and Biochemistry, Moi University, Eldoret, Kenya

ABSTRACT: 2D MoS_2 has been identified as a potential material for optoelectronic, energy, and environmental applications. However, in pristine form its performance in the identified areas is lackluster and requires further bandgap optimizations. It has been shown that non-metallic light atoms when introduced in 2D MoS_2 can modify its electronic properties. In this work, we use density functional theory to study the effects of light atom dopants on structural and electronic properties of 2D MoS_2. It is found that O, Cl, P, and Se dopants can be introduced in 2D MoS_2 on S substitutional site under non-equilibrium growth conditions. It is noted that O and Cl substitutional dopants on S site induce bandgap narrowing while P and Se induce bandgap broadening of 2D MoS_2. While co-dopants Cl-P, O-P, and O-Se induce bandgap reduction while Cl-Se broaden the bandgap. Therefore, with proper dopant dosage, bandgap and electron carrier concentrations can be effectively moderated to suit various applications.

Keywords: Molybdenum disulfide, Density functional theory, Doping, Semiconductors

1 INTRODUCTION

Electron carrier manipulation is essential for the design and development of a new generation of solid state devices based on 2D materials. The substitutional doping of the non-metal element in such systems with an atom species with lower or higher valences modifies the electron-carrier concentration of the 2D host material. Such modifications allow for the creation of an atom-thick hetero-structure with potential applications in the photovoltaic and electronic industries.

Molybdenum disulfide (MoS_2), silicene, and graphene are some of the 2D materials that have recently attracted intense interest due to their desirable electronic and optical properties that are ideal for the development of nanoscale devices [1–5]. MoS_2 has been studied extensively due to its semiconductor character that allows for potential application in nano-photonics, photo-voltaics, photo-catalysis, and nano-electronics [6, 7]. Indeed, field effect transistors based on 2D MoS_2 have demonstrated performance comparable to those of silicon films and graphene ribbons [8]. Recent studies have shown that phototransistors based on bilayer and monolayer MoS_2 have limited sensitivity to various light spectrum such as green, yellow, and yellow [8, 9], and the addition of impurities tends to improve photo-response [10, 11]. The thermodynamically stable 2D MoS_2 is a semiconductor with a direct bandgap of 1.8 eV [12], which can be moderated with the introduction of defects [14]. Indeed, it has been shown that non-metallic elements such as carbon, hydrogen, boron, nitrogen, and fluorine can be used to substitute sulfur atoms in MoS_2, which has an overall modification of the bandgap [15]; information derived from such studies can help guide the experimental work in the fabrication of superior devices. Desirable electronic and optical properties observed in transition metal dichalcogenides can be attributed to intrinsic defects formed during the synthesis process [16]. Recently, techniques for extrinsic doping of MoS_2 have been developed, which showed that the desirable character can be induced in host

*Corresponding author

DOI 10.1201/9781003221968-21

materials via such approaches [17]. However, comprehensive studies that consider substitutional doping of 2D MoS_2 with non-metallic elements as a potential route towards realization of superior properties are limited, yet such studies may provide critical insights that may enable further optimization.

In this work, we perform *ab initio* density functional theory (DFT) calculations for selected sulfur substitutional dopants denoted by X (where X=Cl, O, Se, and P). To ensure negligible structural distortion of 2D MoS_2, the selected dopants should have an atomic radius comparable to that of the sulfur atom but with more/less valence electrons that can modify the electronic properties of the host material. Thereafter, the preferred doping site was determined and then the effects of dopants concentration on structural and electronic properties explored. Whereas the *ab initio* DFT calculations have previously focused on transition metal substitutional doping in 2D MoS_2 [18], studies that evaluate non-metallic dopants for tuning the electronic properties are lacking. A comprehensive understanding of atomic scale doping structure and corresponding change in properties can provide pathways for the synthesis of doped 2D MoS_2 for various applications. Moreover, these findings can be extended to other semiconductors, whose pure electronic structures are similar to those of 2D-MoS_2.

2 COMPUTATIONAL DETAILS

All calculations reported in this work were performed within the framework of DFT [19], as implemented in Quantum espresso suite [20]. The generalized gradient approximation (GGA) to the exchange and correlation was employed in the form proposed by Perdew–Burke–Ernzerhof (PBE) [21]. The core electrons were replaced with ultra-soft pseudo-potential as per Vanderbilt's formalism [22] and the charge density in the system was expanded on a plane wave basis set with energy cutoff of 70 Ry (700 Ry).

A well converged Monkhorst-Pack grid of $4\times4\times1$ was used for the integration of the Brillouin zone [23]. In all cases the geometrical optimizations were performed using the conjugate gradient algorithm until all forces were lower than 0.01 eV/Å, and the total energy for ionic minimization was considered converged when the total energy changes were lower than 10^{-8} eV between consecutive self-consistent steps.

The simulation of X-doped systems for the different dopant concentrations considered in this work used similar convergent parameters to those utilized in pristine 2D MoS_2. Super-cell calculations were performed using an optimized cell with a vacuum of 16 Å along the perpendicular direction, which ensured no interaction between the periodic images of the monolayer. Grimme-term was incorporated in all the calculations to ensure that the Van der Waals interaction is described correctly [24].

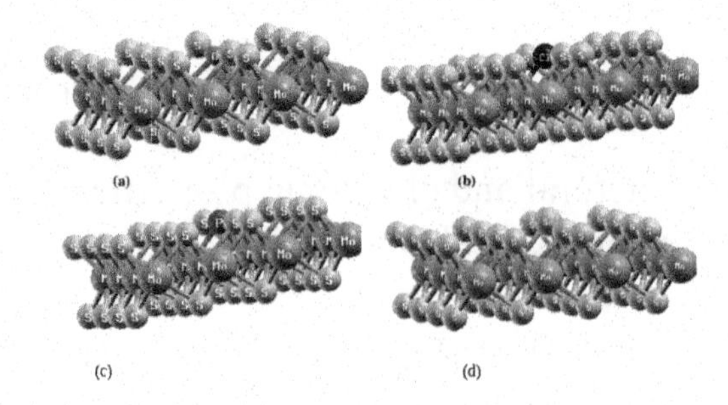

Figure 1. Side-view representation of doped 2D-MoS_2 (a) O, (b) Cl, (c) P, and (d) Se dopant site.

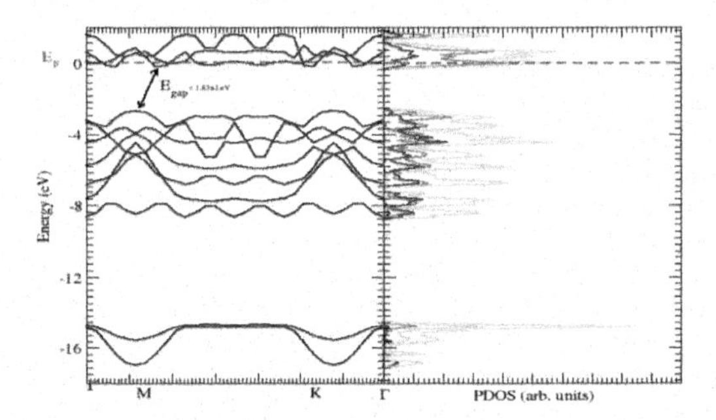

Figure 2. Band-structure and PDOS of Pristine 2D-MoS_2.

3 RESULTS AND DISCUSSION

3.1 *Structural properties*

Our calculations showed that the 2D MoS_2 doped with either chlorine (Cl), phosphorus (P) ,and selenium (Se) tends to have minimal surface distortion and maintains its planar surface as shown in Figure 1(b, c), with Mo–Cl and Mo-P bond lengths being 2.4267 Å and 2.4178 Å, respectively, which is an increase of 2.70% and 2.33% for Mo–Cl and Mo-P, respectively, as compared to S–Mo in pristine 2D-MoS_2. In the case of Se doping, the Se atom protrudes out of the planar surface as in Figure 1(d), with Mo-Se bond-length of 2.4393, which represents an increase of 3.23% as compared to S–Mo in pristine 2D-MoS_2. This is attributed to the fact that Se has a larger atomic radii as compared to the S atom and this tends to induce significant restructuring within the neighborhood of the dopant.

While for O doping, the O atom induces a deep depression on the surface, as shown in Figure 1(a) due to reduced Mo-O bond-length by up to 5.19% compared to S–Mo in pristine 2D-MoS_2. This results can be attributed to the smaller atomic radii of O compared to S and it tends to relax inwardly on the surface.

3.2 *Electronic properties*

The electronic properties of 2D-MoS_2 can be correlated to the free electrons, thus doping with appropriate dopant offers means to achieve the desired electronic properties. To have a realistic description of X-doped

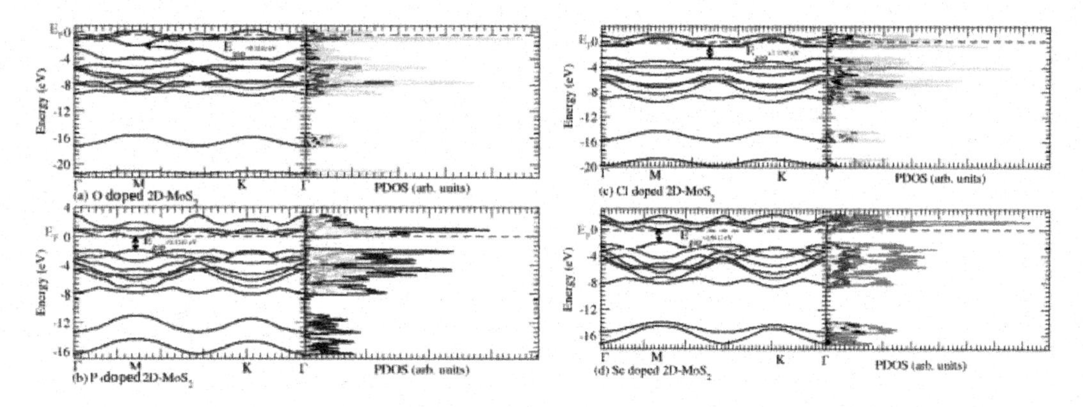

Figure 3. Band-structure and PDOS of doped 2D MoS_2.

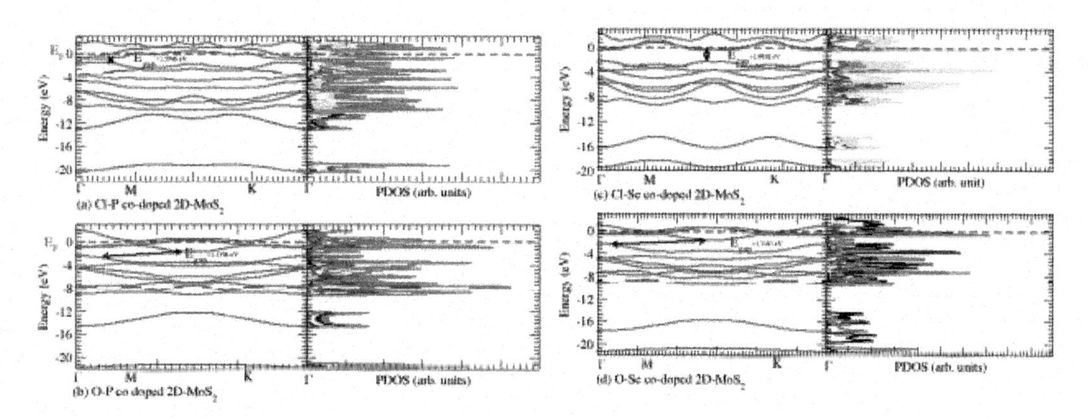

Figure 4. Band-structure and PDOS of co-doped 2D MoS_2.

2D-MoS_2, we consider a scenario where the dopant is at a preferred site, i.e., with lowest formation energy. Here, we report the effects of X substitutional dopant on electronic properties of 2D-MoS_2. The choice of the X dopant is dictated by the need for such elements to have atomic sizes comparable to that of **S** but with more or less valence electrons needed to modify the electronic properties of the pristine material. The effects of dopants considered in this work are summarized in Table 1. The band-structure of the same modification is presented in Figure 3, for O-, Cl-, P-, and Se-doped systems. The calculated bandgap energy for pristine 2D MoS_2 was determined to be consistent with the previous works [12, 24].

It is observed that O and Cl dopants reduce the bandgap by introducing new states on top of the valence band maximum (VBM) and below the conduction band minimum (CBM), which reduces the bandgap by 87% and 61%, respectively, as shown in Figure 3(a, c). While P and Se dopants induce bandgap broadening, which can be attributed to the downward shift of VBM and the upward shift of CBM, as shown in Figure 3(b, d).

To explore other potential bandgap tuning opportunities, we consider co-doping 2D MoS_2 with elements that reduce and increase the bandgap, using the combination below: Cl-P, Cl-Se, O-P, and O-Se. Using the same dopant concentration, it is noted that Cl-P, O-P, and O-Se tend to reduce the bandgap by 11%, 32%, and 5%, respectively, compared to the pristine system, as shown in Table I and Figure 4 (a, b, and d). Thus the systems co-doped with the above combination of dopants (Cl-P, O-P, O-Se) are anticipated to have higher electron carrier concentration and improved electron conductivity compared to a pristine system.

On the other hand, Cl-Se induces bandgap broadening of up to 8% compared to pristine 2D MoS_2, thus a significant reduction in electron carrier is anticipated, and the system is projected to have lower conductivity compared to a pristine system, as shown in Table I and Figure 4 (c).

In addition, we assessed the effect of dopant concentration on the bandgap, and it was observed that for both P and Se, the bandgap energies increase with an increase in dopant concentrations, as shown in Figure 5. While for Cl and O the bandgap energies

Table 1. Effects of dopants on the bandgap of 2D MoS_2 at 4.17% dopant concentration.

Dopant atom(s)	Bandgap energy (eV)
O	0.2282
Cl	1.1192
P	2.5242
Se	2.9412
Cl+P	1.5946
Cl+Se	1.9958
O+P	1.2396
O+Se	1.7287

decrease with an increase in dopant concentrations. At higher (above 31%) dopant concentration, it is observed that the bandgap becomes insensitive to any further increase for O, Se, and P, while Cl shows a continuous decrease in bandgap for dopant concentration of up to 37%.

3.3 *Formation energies*

The formation energy is essential for understanding the required condition for the introduction of the dopant in the host material. The formation energies of O doping, Cl doping, P doping, and Se doping are calculated using equations 1–4,

$$E_{form.}(O) = E(Mo_{16}S_{32-n_1}O_{n_1}) \\ -(E(Mo_{16}S_{32}) + n_1\mu_1 - n_O\mu_O) \quad (1)$$

$$E_{form.}(Cl) = E(Mo_{16}S_{32-n_1}Cl_{n_1}) \\ -(E(Mo_{16}S_{32}) + n_2\mu_2 - n_O\mu_O) \quad (2)$$

$$E_{form.}(P) = E(Mo_{16}S_{32-n_1}P_{n_1}) \\ -(E(Mo_{16}S_{32}) + n_3\mu_3 - n_O\mu_O) \quad (3)$$

$$E_{form.}(SE) = E(Mo_{16}S_{32-n_1}Se_{n_1}) \\ -(E(Mo_{16}S_{32}) + n_4\mu_4 - n_O\mu_O) \quad (4)$$

where $E(Mo_{16}S_{32-n_1}O_{n_1})$, $E(Mo_{16}S_{32-n_1}Cl_{n_1})$, $E(Mo_{16}S_{32-n_1}P_{n_1})$, $E(Mo_{16}S_{32-n_1}Se_{n_1})$, and $E(Mo_{16}S_{32})$ are the total energies of the O-doped 2D-MoS_2, Cl-doped 2D-MoS_2, P-doped 2D MoS_2, Se-doped 2D-MoS_2, and pure 2D-MoS_2, respectively. While $\mu_S, \mu_O, \mu_{Cl}, \mu_P$, and μ_{Se} are chemical potentials of the S, O, Cl, P, and Se elements, respectively, and n_S, n_O, n_{Cl}, n_P, and n_{Se} denote the numbers of the S, O, Cl, P, and Se atoms, respectively.

It is determined that the formation energies of the dopants considered are positive, implying that they can be introduced into the host material under non-equilibrium condition, in particular, O, Cl, P, and Se had a formation energy of 0.52 eV, 0.59eV, 0.78 eV, and 0.69 eV, respectively.

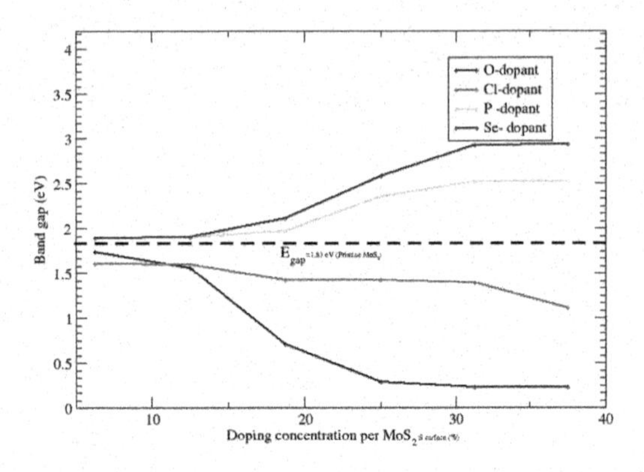

Figure 5. Bandgap vs. dopant concentration for O-, Cl-, P-, and Se-doped 2D MoS_2. Dopant concentration ranging between 0% and 37% is considered and compared to that of a pristine system at 1.84 eV, see dotted line.

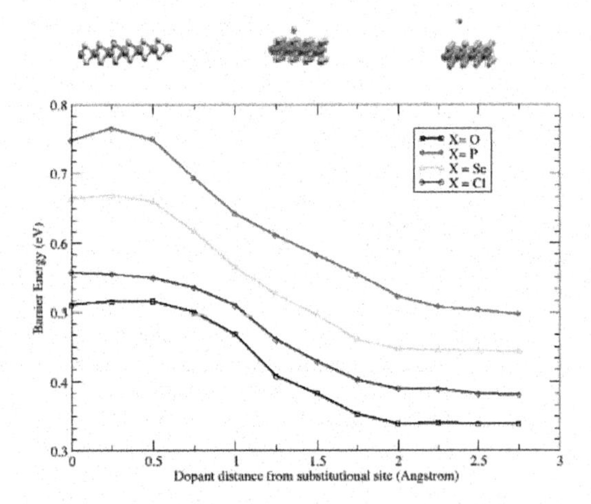

Figure 6. Barrier energy as a function of dopant distance from the substitutional site for O, P, Se, and Cl dopants.

To simulate the introduction of dopant into the host material, we consider a dopant at 3 Å from the surface, then using the nudged elastic band approach (NEB), we determine the energy required to move the dopant to the substitutional site on the 2D MoS_2 and we plot barrier energy as a function of distance from the substitutional site, as shown in Figure 6.

Our results show that O and Cl have the least barrier energy and would require lower activation energy to be incorporated into the host material, while Se and P have higher barrier energy, thus higher activation energy is required compared with O and Cl dopants. Further, O doping is anticipated to be easily achieved due to the lower barrier energy, which is in agreement with previous work [25].

4 CONCLUSIONS

In summary, we have performed *ab initio* DFT calculations to explore the effects of light atom dopants on structural and electronic properties of 2D-MoS_2 that may guide synthesis and utilization in various areas. It is found that the dopants considered in this study can be achieved under non-equilibrium conditions with O and P having the lowest and highest formation energy, respectively. It is noted that O and Cl dopants induce bandgap narrowing and at higher concentration the system becomes insensitive to O, while it responds to Cl up to 31%.

On the other hand, P and Se dopants induce bandgap broadening of 2D MoS_2 and at a higher concentration above 31% the system becomes insensitive. It was further observed that co-dopants Cl-P, O-P, and O-Se induce bandgap narrowing while Cl-Se broadens the bandgap and reduces the electron carrier concentration. Our findings may assist in the development of robust synthesis routes and the optimization of 2D MoS_2 for energy, optoelectronics, and environmental applications.

ACKNOWLEDGMENTS

The authors acknowledge ACEII-PTRE of Moi University for funding and CHPC-Cape Town for the provision of High Performance Computing Resources and support.

REFERENCES

[1] Wang, Q. H., Kalantar-Zadeh, K., Kis, A., Coleman, J. N., & Strano, M. S. (2012). Electronics and optoelectronics of two-dimensional transition metal dichalcogenides. *Nature nanotechnology*, *7*(11), 699–712.

[2] Lalmi, B., Oughaddou, H., Enriquez, H., Kara, A., Vizzini, S., Ealet, B., & Aufray, B. (2010). Epitaxial growth of a silicene sheet. *Applied Physics Letters*, *97*(22), 223109.

[3] Geim, A. K., & Novoselov, K. S. (2010). The rise of graphene. In *Nanoscience and technology: a collection of reviews from nature journals* (pp. 11–19).

[4] Korir, K. K., & Philemon, K. T. (2020). Tailoring Single Walled Carbon Nanotube for Improved CO2 Gas Applications: Insights from Ab initio Simulations. *Materialia*, 100694.

[5] Philemon, K. T., & Korir, K. K. (2020). Carbon dioxide gas sensing, capture, and storage potential of calcium oxide surface and single walled carbon nanotube: insights from ab initio simulation. *Journal of Physics: Condensed Matter*, *32*(24), 245901.

[6] Krasnozhon, D., Lembke, D., Nyffeler, C., Leblebici, Y., & Kis, A. (2014). MoS2 transistors operating at gigahertz frequencies. *Nano letters*, *14*(10), 5905-5911.

[7] Liao, J., Sa, B., Zhou, J., Ahuja, R., & Sun, Z. (2014). Design of high-efficiency visible-light photocatalysts for water splitting:MoS_2/AlN(GaN) heterostructures. *The Journal of Physical Chemistry C118*(31), 17594–17599.

[8] Lopez-Sanchez, O., Lembke, D., Kayci, M., Radenovic, A., & Kis, A. (2013). Ultrasensitive photodetectors based on monolayer MoS_2. *Nature nanotechnology*, *8*(7), 497–501.

[9] Kaasbjerg, K., Thygesen, K. S., & Jacobsen, K. W. (2012). Phonon-limited mobility in n-type single-layer MoS_2 from first principles. *Physical Review B*, *85*(11), 115317.

[10] Lee, H. S., Min, S. W., Chang, Y. G., Park, M. K., Nam, T., Kim, H., & Im, S. (2012). MoS_2 nanosheet phototransistors with thickness-modulated optical energy gap. *Nano letters*, *12*(7), 3695–3700.

[11] Wang, H., Yu, L., Lee, Y. H., Shi, Y., Hsu, A., Chin, M. L., & Palacios, T. (2012). Integrated circuits based on bilayer MoS2 transistors. *Nano letters*, *12*(9), 4674–4680.

[12] Mak, K. F., Lee, C., Hone, J., Shan, J., & Heinz, T. F. (2010). Atomically thin MoS_2: a new direct-gap semiconductor. *Physical review letters* *105*(13), 136805.

[13] Scalise, E., Houssa, M., Pourtois, G., Afanas'ev, V., & Stesmans, A. (2012). Strain-induced semiconductor to metal transition in the two-dimensional honeycomb structure of MoS_2. *Nano Research*, *5*(1), 43–48.

[14] Ramasubramaniam, A., Naveh, D., & Towe, E. (2011). Tunable bandgaps in bilayer transition-metal dichalcogenides. *Physical Review B*, *84*(20), 205325.

[15] Ma, Y., Dai, Y., Guo, M., Niu, C., Lu, J., & Huang, B. (2011). Electronic and magnetic properties of perfect, vacancy-doped, and nonmetal adsorbed $MoSe_2$, $MoTe_2$ and WS_2 monolayers. *Physical Chemistry Chemical Physics*, *13*(34), 15546–15553.

[16] McDonnell, S., Addou, R., Buie, C., Wallace, R. M., & Hinkle, C. L. (2014). Defect-dominated doping and contact resistance in MoS_2 *ACS nano*, *8*(3), 2880–2888.

[17] Gao, J., Kim, Y. D., Liang, L., Idrobo, J. C., Chow, P., Tan, J., & Koratkar, N. (2016). Transition-metal substitution doping in synthetic atomically thin semiconductors. *Advanced Materials* *28*(44), 9735–9743.

[18] Fu, Y., Long, M., Gao, A., Wang, Y., Pan, C., Liu, X., & Miao, F. (2017). Intrinsic p-type W-based transition metal dichalcogenide by substitutional Ta-doping*Applied Physics Letters* *111*(4), 043502.

[19] Kohn, W., & Sham, L. J. (1965). Self-consistent equations including exchange and correlation effects. *Physical review* *140*(4A), A1133.

[20] Giannozzi, P., Baroni, S., Bonini, N., Calandra, M., Car, R., Cavazzoni, C., & Wentzcovitch, R. M. (2009). QUANTUM ESPRESSO: a modular and open-source software project for quantum simulations of materials. *Journal of physics: Condensed matter*, *21*(39), 395502.

[21] Ernzerhof, M., & Scuseria, G. E. (1999). Assessment of the Perdew–Burke–Ernzerhof exchange-correlation functional. *The Journal of chemical physics*, *110*(11), 5029–5036.

[22] Vanderbilt, D. (1990). Soft self-consistent pseudopotentials in a generalized eigenvalue formalism. *Physical review B41*(11), 7892.

[23] Monkhorst, H. J., & Pack, J. D. (1976). Special points for Brillouin-zone integrations. *Physical review B13*(12), 5188.

[24] Ataca, C., Sahin, H., Akturk, E., & Ciraci, S. (2011). Mechanical and electronic properties of MoS2 nanoribbons and their defects. *The Journal of Physical Chemistry C115*(10), 3934–3941.

[25] Nolan, M. (2011). Charge compensation and Ce3+ formation in trivalent doping of the CeO2 (110) surface: the key role of dopant ionic radius. *The Journal of Physical Chemistry C115*(14), 6671–6681.

Advances in Phytochemistry, Textile and Renewable Energy Research for Industrial Growth – Nzila et al. (Eds)

Chemical composition and insecticidal activity of *Pinus caribaea* Morelet var. hondurensis needles against *Sitophilus zeamais* Motschulsky and *Callosobruchus maculatus* Fabricius

John Mary Kirima
Department of Chemistry, Faculty of Science, Kyambogo University, Kampala, Uganda
Environmental Quality Management and Consultancy Department, Nkerebwe New Hope, Kampala, Uganda

Timothy Omara*
Africa Center of Excellence II in Phytochemicals, Textiles and Renewable Energy (ACE II PTRE), Moi University, Eldoret, Kenya
Department of Chemistry and Biochemistry, School of Sciences and Aerospace Studies, Moi University, Eldoret, Kenya
Department of Quality Control and Quality Assurance, Product Development Directory, AgroWays Uganda Limited, Jinja, Uganda

ABSTRACT: Plant allelochemicals from essential oils have recently received considerable attention in pharmaceutical, cosmetic and agricultural sectors due to their biodegradability and low toxicity. This study analyzed the chemical composition and bioinsecticidal activity of essential oils of *Pinus caribaea* Morelet var. hondurensis needles. Thirty-nine (39) organic compounds were identified using gas chromatography/mass chromatography and gas chromatography, and the most abundant components were limonene (38.6%), α-pinene (27.6%), borneol (6.7%) and myrcene (3.5%). The chemical composition of the needles was dominated by monoterpene hydrocarbons (77.2%) followed by oxygenated monoterpenes (12.0%), sesquiterpene hydrocarbons (4.7%) and then lastly oxygenated sesquiterpenes (1.7%). In fumigant toxicity, 100% mortality was recorded at 10 μL/ml for bean weevils after 2 hours of exposure whereas the same concentration caused 100% mortality of maize weevils after 5 hours of exposure. In repellency bioassay using aliquots of acetonic essential oils, 100% repellence was recorded in bean weevils after 60 minutes of exposure while the same concentration (8 μL/ml) gave 100% repellence activity in maize weevils after 150 minutes. The essential oils showed higher insecticidal activity against bean weevils than maize weevils. Based on the results of this study, pine needles could be a suitable source of green insecticides for control of maize and bean weevils in stored food products.

1 INTRODUCTION

Pests have been the worst enemy of man since time immemorial. Maize weevils (*Sitophilus zeamais* Motschulsky, 1855) and bean weevils (*Callosobruchus maculatus* Fabricius) are some of the most destructive pests of stored food crops worldwide (Liao et al., 2016). They cause extensive quantitative loss, deterioration in food quality, seed viability, promote secondary pest and fungal infestations that further initiate the development of mycotoxins (Omara et al., 2020). Currently, their control relies heavily on synthetic insecticides which present problems such as environmental pollution, pest resurgence (resistance) and effects on non-target organisms in addition to direct toxicity to the users (Jagadeesan, Nayak, Pavic, Chandra, & Collins, 2015; Thompson & Reddy, 2016). Thus, the current efforts are directed towards the development of novel biofumigants for the management of stored product pests that are readily available, eco-friendly, effective and do not compromise food quality (Gerwick & Sparks, 2014). A feasible alternative has been to use plant products such as essential oils, ashes and extracts (Gaire, Scharf, & Gondhalekar, 2019; Omara et al., 2018).

Plant products are known for their properties of low residue formation, high selectivity, and difficulty to generate cross-resistance by insects (Isman, 2016). This is mainly attributable to their complex constituents and different modes of action against pests (Athanassiou, Hasan, Phillips, Aikins, & Throne, 2015). Over 2,000 plants have been reported to possess

*Corresponding author

DOI 10.1201/9781003221968-22

insecticidal activity with the renown bioinsecticides being nicotine, pyrethrum, rotenone, neem and plant essential oils (Philogene, Regnault-Roger, & Vincent, 2005). Essential oils have been shown to control stored product pests by fumigant and contact toxicity or as repellents and this maybe by affecting growth rate and oviposition (Chen, Akinkurolere, & Zhang, 2011). Thus, essential oils from aromatic plants such as *Pinus caribaea* could provide environmentally friendly alternatives to the currently used synthetic pesticides (Seixas et al., 2018).

Pinus caribaea Morelet (Pinaceae) is a resinous woody tree with evergreen needle-like leaves and grows up to 30 m tall, usually free from branches to a considerable height (Lee & Lee, 1991). It has many varieties native to Central America and the Caribbean (Chowdhury, Bhuiyan, & Nandi, 2008). The plant bark is grey to reddish-brown. Seeds are usually mottled grey or light brown, narrowly ovoid and about 6 mm long, with a well-developed usually persistent wing (Stanley & Ross, 1989). On the essential oils of *P. caribaea* needles, there are a few reports on its phytochemical composition (Barnola & Cedeño, 2000; Barnola, Cedeño, & Hasegawa, 1997; Chowdhury et al., 2008; Moronkola et al., 2009; Sonibare & Olakunle, 2008). Thus, the current study investigated the chemical constituents of essential oils in fresh needles *P. caribaea* growing in Buikwe district of Uganda, East Africa and evaluated their potential as insecticides for control of maize and bean weevils.

2 MATERIALS AND METHODS

2.1 *Collection, extraction and phytochemical analysis of Pinus caribaea needles*

Fresh *P. caribaea* needles (3 kg) were collected from Ferdsult Pine Forest, Luwombo village, Buikwe district of Uganda from identified cultivated plants. Needles were collected from various parts of the crowns to overcome plant plasticity phenomena (Bradshaw, 1965; Roussis, Petrakis, Ortiz, & Mazomenos, 1995). They were identified as *P. caribaea* Morelet var. hondurensis by a botanist at Makerere University Herbarium, Kampala (Uganda) where a voucher sample (No. JM-001) was deposited.

Aliquots (500 g) of crushed needles were hydrodistilled using a modified Clevenger-type apparatus for 3 hrs. The oil was dried using anhydrous sodium sulphate and stored in amber bottles away from UV light at 4°. Gas chromatography/mass spectrometry analysis was carried out with an Agilent 5975 GC/MSD system. One mL of the oil was withdrawn by an autosampling system. Innowax FSC column (60 m × 0.25 mm, 0.25 mm) was used with helium as the carrier gas (0.8 mL/min). GC oven temperature was kept at 60° for 10 minutes, programmed to 220° at a rate of 4°/min, kept constant at 220° for 10 minutes, and then programmed to 240° at a rate of 1°/min. Split ratio was adjusted at 40:1. The injector temperature was set to 250°. Mass spectra were recorded at 70 eV. GC analysis was carried out using an Agilent 6890N GC system. The flame ionization detector temperature was set at 300°. To obtain the same elution order with GC/MS, simultaneous autoinjection was done on a duplicate of the same column applying the same conditions described before. Relative percentage amounts of the separated compounds were calculated from FID chromatograms. Constituents were identified by comparison of their retention indices with those reported in literature (Akın, Saraçoğlu, Demirci, Başer, & Küçüködük, 2012; Baser, Demirci, Kurkcuoglu, Satin, & Tumen, 2009; Baser, Demirei, Özek, Akalin, & Özhatay, 2002a; Baser et al., 2002b; Demirci et al., 2017; Kaya, Benirci, & Baser, 2007; Moronkola et al., 2009; Saglam, Gozler, Kivcak, Demirci, & Baser, 2001; Tabanca, Demirci, Ozek, Tumen, & Baser, 2001). Matching against libraries such as Adams Library and those of reference compounds from NIST web book were done.

2.2 *Insecticidal assay*

Maize weevils (*S. zeamais*) and bean weevils (*C. maculatus*) were obtained from Banda stores, Kyambogo, Kampala (Uganda) and the colonies were mass reared on whole maize grains and beans, respectively in glass containers at optimal growth conditions (ambient temperature of 25–28°C, 60–70% relative humidity) in the dark at Kyambogo University Chemistry laboratory for three days prior to the bioassays.

Fumigant toxicity assay

Fumigant toxicity was tested using the method described by Liu and Ho (1999). Range-finding studies were run to determine the appropriate testing concentrations. A serial dilution of the essential oil (four concentrations) was prepared in acetone. Different concentrations of 2, 4, 6, 8 and 10 μL and were prepared by diluting each quantity in 1 ml acetone. They were then applied on filter papers separately and air dried for 10 minutes. Each dried filter paper was placed at the bottom of a glass jar. Counted 10 adult maize or bean weevils were placed in muslin clothes each with 12 g of whole maize grain or bean seeds. The clothes were tightly closed with rubber bands and hung at the center of the jars. The control was performed using a filter paper treated with acetone alone. The experiments stood for five days. Every day, the clothes were removed from the jars and the number of dead insects were counted. Each experiment was replicated three times and averaged.

The percentage mortality (PM) was calculated using Equation 1 (Asawalam, Emosairue, & Hassanali, 2006; Omara et al., 2018).

$$PM = \frac{N_D}{N_T} \times 100 \tag{1}$$

Where N_D and N_T are the number of dead and total number of test insects per jar, respectively.

Repellent activity

The repellent effects of the acetonic essential oils against maize and bean weevils were evaluated using the modified area preference method (Omara et al., 2018). The test area consisted of a filter paper cut into two halves. Petri dishes (9 cm in diameter) were used to confine the weevils during the experiment. The crude essential oil was diluted in 1 ml of acetone to four concentrations (2μL, 4μL, 6μL and 8μL). Acetone was used as the control. The filter paper was cut into half and 500 μL of each concentration was applied separately to half of the filter paper as uniformly as possible with a micropipette. The other half (control) was treated with 500 μL of acetone. Both halves were left for 5 minutes for acetone to evaporate. The treated and control halves were attached together using adhesive tapes and placed in petri dishes.

Counted 10 adult weevils were released at the center of each filter paper and the petri dishes covered and kept in an incubator at 27 ± 2 C and observations made on the number of insects present on both the treated and untreated halves and then recorded every after 30 minutes for 2.5 hours of exposure. The experiment was repeated thrice using both maize and bean weevils. The number of insects present on the control and treated areas of the filter paper were recorded every after 30 minutes for 24 hours of exposure.

Percentage repellency (PR) of the essential oils was calculated using **Equation 1** (Abdelghany, Awadalla, Abdel-Baky, El-Syrafi, & Fields, 2010; Omara et al., 2018).

$$PR = \frac{N_c - N_t}{N_c + N_c} x100 \quad (2)$$

From which N_c is number of insects on control experiment and N_t is number of insects on treated areas of the filter paper.

2.3 *Statistical analysis*

Data obtained from each dose-response bioassay were subjected to probit analysis (Finney, 1971) in which probit-transformed mortality was regressed against $\log_{10}$ transformed doses. Median lethal concentration (LC_{50}) was determined from the linear regression equation using the intercept and slope generated. Analyses were performed using Minitab statistical software (version 17, Minitab Inc., USA) and Microsoft Excel 365 (Microsoft Corporation, USA) at 95% confidence interval.

3 RESULTS AND DISCUSSION

The essential oil was a pale-yellow liquid with a strong aroma. The mean yield of the oils was 0.33% (v/w). This is comparable to 0.02-1.0% reported for fresh needles of *P. caribaea* by Sonibare and Olakunle (2008), Chowdhury et al. (2008) and Moronkola et al. (2009). A total of 39 compounds were identified and quantified, representing 95.6% of the essential oils

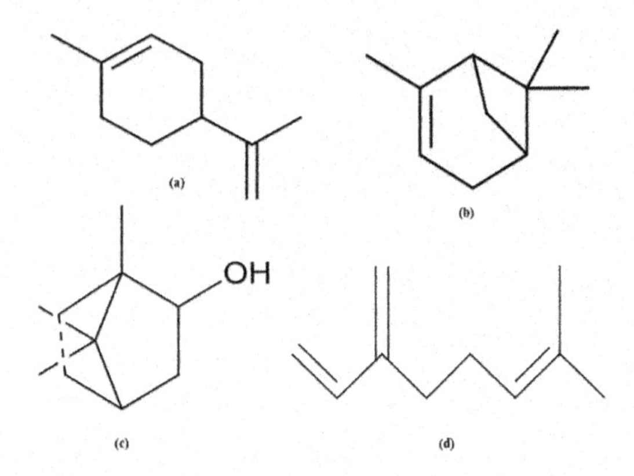

Figure 1. Structure of the major compounds identified in *P. caribaea* needle oils: (a) Limonene, (b) α-Pinene, (c) Borneol, and (d) Myrcene.

(Table 1). Structures of the most abundant components are shown in Figure 1.

The major components of the oils were limonene (38.6%), α-pinene (27.6%), borneol (6.7%) and myrcene (3.5%). Chowdhury et al. (2008) reported similar results in which limonene was not detected in essential oils of *P. caribaea* resins but was the dominant component in the fresh needles (48.84%), dried needles (31.58%) and inflorescence (32.14%) followed by caryophyllene (23.82% and 14.45%), germacrene D (8.40% and 2.59%) in fresh and dried needles, respectively. In an earlier study in Nigeria (Sonibare & Olakunle, 2008), the major constituents of essential oils from *P. caribaea* air-dried needles were β-phellandrene (67.9%), β-caryophyllene (10.2%) and α-pinene (5.4%). In another investigation, Moronkola et al. (2009) reported that the major components of oils from fresh needles of *P. caribaea* in Nigeria were limonene (42%), β-phellandrene (24.4%) and β-caryophyllene (7.6%). The differences in the chemical composition of the essential oils recorded in this study and those reported by preceding authors could be because the climatic conditions and soil properties of Uganda (East Africa) is different from that of Bangladesh (South Asia) and Nigeria (West Africa). Previous studies support that the chemical composition of essential oils and oleoresins of pines exhibit qualitative and quantitative variations both between and within the same species (Barnola, Cedeño & Hasegawa, 1997; Coppen, James, Robinson, & Subansenee, 1998; Dob, Berramdane, & Chelghoum, 2007; Ekundayo, 1978; Ghosn, Saliba, & Talhouk, 2006; Hmamouchi, Hmamouchi, & Zouhdi, 2005; Jantan, 2002; Kirima, Okuta, & Omara, 2020; Macchioni et al., 2003; Moronkola et al., 2009; Pagula & Baeckström, 2006; Venskutonis, Vyskupaityte, & Plausinaitis, 2000).

As reported for essential oils from other *Pinus* species (Ekundayo, 1978, 1988; Macchioni et al., 2003; Stevanoic et al., 2004), the chemical composition of *P. caribaea* needles in this study was dominated by monoterpene hydrocarbons (77.2%) followed

Table 1. Phytochemicals identified in essential oils of *P. caribaea* fresh needles from Uganda.

Peak	Constituent	RI [a]	Area (%)
1	**α− Pinene**	**1032**	**27.6 ± 0.17**
2	α− Thujene	1035	0.7 ± 0.00
3	Camphene	1076	1.6 ± 0.20
4	Hexanal	1093	Trace
5	β− Pinene	1118	0.6 ± 0.10
6	Sabinene	1132	0.9 ± 0.00
7	**Myrcene**	**1174**	**3.5 ± 0.20**
8	α− Phellandrene	1176	2.1 ± 0.17
9	α− Terpinene	1188	0.6 ± 0.10
10	**Limonene**	**1203**	**38.6 ± 0.10**
11	1,8-Cineole	1213	0.2 ± 0.00
12	β− phellandrene	1218	1.0 ± 0.10
13	γ− Terpinene	1255	Trace
14	*Trans*-linalool oxide	1450	0.5 ± 0.10
15	α− Copaene	1497	0.3 ± 0.00
16	Camphor	1532	0.3 ± 0.10
17	β− Cubebene	1547	Trace
18	Linalool	1553	1.1 ± 0.17
19	Aristolene	1589	0.9 ± 0.17
20	Bornyl acetate	1590	1.1 ± 0.00
21	β− Caryophyllene	1612	1.2 ± 0.10
22	Citronellyl acetate	1668	0.4 ± 0.00
23	α− Humulene	1687	0.2 ± 0.10
24	γ− Muurolene	1704	0.3 ± 0.00
25	α− Terpinyl acetate	1709	Trace
26	**Borneol**	**1719**	**6.7 ± 0.00**
27	Germacrene D	1726	0.8 ± 0.10
28	α− Muurolene	1740	0.3 ± 0.00
29	β− Selinene	1742	0.3 ± 0.00
30	Carvone	1751	0.2 ± 0.00
31	δ− Cadinene	1773	0.4 ± 0.10
32	*Trans*-Carveol	1845	0.5 ± 0.00
33	*p*-cymen-8-ol	1864	0.1 ± 0.00
34	*Cis*-Carveol	1882	0.5 ± 0.17
35	Geranyl butyrate	1901	0.2 ± 0.10
36	Caryophyllene oxide	2008	0.2 ± 0.00
37	Ledol	2057	0.7 ± 0.00
38	Globulol	2098	0.8 ± 0.10
39	Guaiol	2103	0.2 ± 0.10
Monoterpene hydrocarbons			77.2%
Sesquiterpene hydrocarbons			4.7%
Oxygenated monoterpenes			12.0%
Oxygenated sesquiterpenes			1.7%
Total			**95.6%**

Trace = <0.01%. *RI = Retention index as determined on an Innowax FSC column. Peak area presented as mean ± standard deviation of triplicates.

by oxygenated monoterpenes (12.0%), sesquiterpene hydrocarbons (4.7%) and then oxygenated sesquiterpenes (1.7%). The major constituents of the monoterpene hydrocarbons were limonene (38.6%), α-pinene (27.6%), myrcene (3.5%) and α-phellandrene (2.1%). Sesquiterpene hydrocarbons was dominated by β-caryophyllene (1.2%), aristolene (0.9%) and germacrene D (0.8%). The dominant monoterpene hydrocarbon constituents in our study differed from those reported by Coppen, Gay, James, Robinson, & Mullin, (1993; 1998; 1988), Ekundayo (1978), Dagne et al. (1999), and Barnola and Cedeño (2000) for the same species in which α-pinene (63.2–87.1%) was the major constituent. The abundance of monoterpenes in the essential oils of *P. caribaea* seems to be in congruence with published reports where limonene andβ-phellandrene dominated (Barnola et al., 1997; Valterová, Sjödin, Vrkoc, & Norin, 1995) and β-phellandrene occurred in non-quantitatively larger amounts (Valterová, Sjödin, Vrkoc, & Vrkoc, 1995). However, β-myrcene, sabinene and other monoterpenoids that were prominent compounds in previous reports (Barnola & Cedeño, 2000; Barnola et al., 1997; Valterová et al., 1995) were detected in lower quantities in this study, corroborating a recent observation (Moronkola et al., 2009). Similarly, α-ocimene was not identified as one of the components, which is in good agreement with previous reports (Barnola et al., 1997; Valterová et al., 1995).

Such differences in the chemical composition of the essential oils may be attributed to factors such as part of the plant used, time of collection, plant disease, genetic factors (chemotype), soil and climatic conditions, and age of the plant (Barnola & Cedeño, 2000; Barnola et al., 1997; Barnola, Hasegawa, & Cedefm, 1994; Bradshaw, 1965; Coppen et al., 1988; Moronkola et al., 2009; Roussis et al., 1995; Sonibare & Olakunle, 2008; Valterová et al., 1995). For example, a study on *P. caribaea* (var. caribaea, var. bahamensis and var. hondurensis) xylem resins in different provenances of Zimbabwe (Coppen et al., 1988) reported that α-pinene (20.8–66.6%) and β-phellandrene (19.4–59.9%) predominated and jointly accounted for 80-90% of the total monoterpene hydrocarbons. Barnola, Hasegawa, & Cedefm, (1994) reported that seasonal changes between dry and rainy seasons may be associated with the caryophyllene content variation (in conjunction with that of α-pinene) in *Pinus caribaea* needles.

Insecticidal activity of pine needles

All concentrations of the essential oil used caused mortality in *S. zeamais* and *C. maculatus* during the fumigant toxicity bioassay as shown in Table 2. The table show percentage mortality which is dose and time dependent. The highest concentration of 10 μL/ml recorded 100% mortality of *C. maculatus* after 2 hours of exposure. On the other hand, subjecting *S. zeamais* to the same concentration resulted in 100% mortality after 5 hours of exposure. Thus, the essential oil of pine needles was more toxic to *C. maculatus* than *S. zeamais* in the fumigant toxicity assay. The median lethal concentrations (LC_{50}) calculated by Probit analysis were 6.3 μL/ml and 5.2 μL/ml for *S. zeamais* and *C. maculatus*, respectively.

Table 3 shows the dose and time dependent repellence activity of the essential oils of pine needles against *S. zeamais* and *C. maculatus*. A dose of 8 μL/ml against *C. maculatus* had 100% repellence effect after 60 minutes of exposure but against *S. zeamais* 100% repellence activity was observed after 150 minutes of exposure.

Table 2. Percentage mortality due to fumigant toxicity of *P. caribaea* needle essential oil against the storage pests.

Pest	Time (hours)	Concentration of essential oil (μL/ml)			
		1	2	5	10
Sitophilus zeamis	01	0	20	20	80
	02	10	30	30	80
	03	10	40	40	90
	04	20	50	50	90
	05	40	50	50	100
	06	40	60	80	100
	24	60	70	90	100
	48	60	80	100	100
C. maculatus	01	0	10	40	90
	02	10	20	60	100
	03	20	30	80	100
	04	20	30	90	100
	05	30	40	100	100
	06	30	50	100	100
	24	40	80	100	100
	48	50	100	100	100

Table 3. Percentage repellency of *P. caribaea* needle essential oil against the storage pests.

Pest	Time (minutes)	Concentration of oil (μL/ml)			
		2	4	6	8
S. zeamis	30	0	20	20	80
	60	10	30	30	80
	90	10	40	40	90
	120	20	50	50	90
	150	40	50	50	100
C. maculatus	30	0	20	20	20
	60	20	60	60	100
	90	60	60	100	100
	120	60	80	100	100
	150	100	100	100	100

Essential oils from plants are among the recent options for cheaper, safer and eco-friendly substitutes (or adjuvants) for commercially available synthetic insecticides (Regnault-Roger, Vincent, & Arnason, 2012; Stevenson, Isman, & Belmain, 2017). In the current study, it was shown that use of pine needle essential oils adequately controlled *S. zeamais* and *C. maculatus* on stored food products. Essential oils such as that from pine needles are complex natural mixtures of compounds at different concentrations with 2 or 3 major components **(Table 1)** that establishes their biological properties (Bakkali, Averbeck, Averbeck, & Idaomar, 2008). Nonetheless, compounds in essential oils are known to act synergistically against the physiology of many insects (Lucia, Licastro, Zerba, Gonzalez, & Masuh, 2009; Showler, Osbrink, Morris, & Wargovich, 2017). Among the essential oil components, monoterpenes have drawn the greatest attention for insecticidal activity against stored product pests (Abdelgaleil, Mohamed, Shawir, & Abou-Taleb, 2016; Rajendran & Sriranjini, 2008). Various monoterpenes like limonene, carene, linalool α-pinene, eucalyptol, camphene and α-terpinene identified in the pine needle essential oils in this study have been reported to have contact and fumigation toxicity against stored product pests (Omara et al., 2018; Papachristos, Karamanoli, Stamopoulos, & Spiroudi, 2004; Stamopoulos, Damos, & Karagianidou, 2007). These monoterpenes (typically volatile, and rather lipophilic compounds) were present in the pine needle essential oils in significant amounts and thus the toxicity of the essential oils may be attributed to the high total concentration of monoterpenes in the oil. Limonene was the main component of our essential oil. It has been previously reported to possess good repellent and toxic properties against several arthropods (Guo et al., 2016; Hieu, Kim, Kwon, & Ahn, 2010; Hollingsworth, 2005; Ibrahim, Kainulainen, & Aflatuni, 2001; Karr & Coates, 1988; Kassir, Mohsen, & Mehdi, 1989; Mursiti, Estari, da Febriana, Rosanti, & Ningsih, 2019; Showler, Harlien, & Perez de Léon, 2019) and it is an ingredient of more than 15 insecticide and repellent products (Hebeish, Fouda, Hamdy, El-Sawy, & Abdel-Mohdy, 2008).

The mortality observed in fumigant toxicity assay could be as a result of volatile constituents entering the cuticle of the insects or due to nerve impulse inhibition of acetylcholine impulse which leads to paralysis and death of the insects (Abdelgaleil et al., 2016; Keane & Ryan, 1999). The higher susceptibility of *C. maculatus* to the essential oils could be as a result of its softer cuticle that allows easier penetration of the essential oil. This is supported by the observation that the survival of adult weevils is known to partly depend on its exoskeleton or cuticle (Casem, 2016).

One limitation of the current study was that the fresh needles were taken during only one season of the year and may not fully reflect the composition of the essentials throughout the year. Further, the essential oils were applied on filter papers to investigate their insecticidal potential against the storage pests. Although this showed good insecticidal activity under laboratory conditions and may have potential applications at small farmer's level, this delivery system may suffer from draw backs inherent to the volatile nature of essential oils when used in larger storage facilities such as silos. As such, rapid biodegradation of these compounds due to their poor physicochemical stability will require some controlled-release system such as nanotechnological formulations to optimize the action of the active ingredients.

4 CONCLUSIONS AND RECOMMENDATIONS

The chemical composition of essential oils of *P. caribaea* fresh needles grown in Buikwe district of Uganda is dominated by monoterpenes followed by oxygenated monoterpenes, sesquiterpenes and oxygenated sesquiterpenes. The essential oils showed higher insecticidal activity against bean weevils than maize weevils. Further studies should investigate the variation of the chemical composition of the essential oils at different times of the year as well as isolate pure compounds from the oils and test their bioinsecticidal activity against the storage pests.

ACKNOWLEDGEMENTS

The authors are grateful to the Government of the Republic of Uganda for the scholarship awarded to JMK and the Africa Center of Excellence II in Phytochemicals, Textiles and Renewable Energy (ACE II PTRE), Moi University that made this communication possible. The support of Department of Chemistry, Kyambogo University, Kampala, Uganda is highly acknowledged for the analytical success of this research.

REFERENCES

Abdelgaleil, S. A., Mohamed, M. I., Shawir, M. S., & Abou-Taleb, H. K. 2016. Chemical composition, insecticidal and biochemical effects of essential oils of different plant species from Northern Egypt on the rice weevil, Sitophilus oryzae L. *Journal of Pest Science, 89*, 219–229.

Abdelghany, A. Y., Awadalla, S. S., Abdel-Baky, N. F., El-Syrafi, H. A., & Fields, P. G. 2010. Stored-product insects in botanical warehouses. *Journal of Stored Products Research, 46*, 93–97.

Akın, M. H., Saraçoğlu, T., Demirci, B., Başer, K. H. C., & Küçüködük, M. 2012. Chemical composition and antibacterial activity of essential oils from different parts of *Bupleurum rotundifolium* L. *Records of Natural Products, 6*, 316–320.

Asawalam, E. F., Emosairue, S. O., & Hassanali, A. 2006. Bioactivity of *Xylopia aetiopica* (Dunal) a rich essential oil constituents on maize weevil *Sitophilus zeamais* Motch. (Coleoptera: Curculionidae). *Electronic Journal of Environment, Agriculture and Food Chemistry, 5*, 195–204.

Athanassiou, C. G., Hasan, M. M., Phillips, T. W., Aikins, M. J., & Throne, J. E. 2015. Efficacy of Methyl Bromide for Control of Different Life Stages of Stored-Product Psocids. *Journal of Economic Entomology, 108*, 1422–1428.

Bakkali, F., Averbeck, S., Averbeck, D., & Idaomar, M. 2008. Biological effects of essential oils-a review. *Food and Chemical Toxicology, 46*, 446–475.

Barnola, L. F., & Cedeño, A. 2000. Inter-population differences in the essential oils of *Pinus caribaea* needles. *Biochemical Systematics & Ecology, 28*, 923–931.

Barnola, L. F., Cedeño, A., & Hasegawa, M. 1997. Intraindividual Variations of Volatile Terpene Contents in *Pinus caribaea* Needles and Its Possible Relationship to Atta laevigata Herbivory. *Biochemical Systematics & Ecology, 25*, 707–716.

Barnola, L. F., Hasegawa, M., & Cedefm, A. 1994. Mono- and sesquiterpene variation in *Pinus caribaea* needles and its relationship to Atta laevigata herbivory. *Biochemical Systematics & Ecology, 22*, 437–445.

Baser, K. H. C., Demirci, B., Kurkcuoglu, M., Satin, F., & Tumen, G. 2009. Comparative morphological and phytochemical charactertization of *Salvia cadmica* and *S. smyrnaea*. *Pakistan Journal of Botany, 41*, 1545–1555.

Baser, K. H. C., Demirei, B., Özek, T., Akalin, E., & Özhatay, N. 2002a. Micro-distilled volatile compounds from *Ferulago* species growing in Western Turkey. *Pharmaceutical Biology, 40*, 466–471.

Baser, K. H. C., Nuriddinov, H. R., Ozek, T., Demirci, A. B., Azcan, N., & Nigmatullaev, A. M. 2002b. Essential oil of *Arischrada korolkowii* from the Chatkal mountains of Uzbekistan. *Chemistry of Natural Compounds, 38*, 51–53.

Bradshaw, A. D. 1965. Evolutionary Significance of Phenotypic Plasticity in Plants. *Advances in Genetics, 13*, 115–155.

Casem, M. L. 2016. Cell systems. In: Case studies in Cell biology. Academic Press, Elsevier. p. 1–406.

Chen, H., Akinkurolere, R. O., & Zhang, H. 2011. Fumigant activity of plant essential oil from *Armoracia rusticana* (L.) on *Plodia interpunctella* (Lepidoptera: Pyralidae) and *Sitophilus zeamais* (Coleoptera: Curculionidae). *African Journal of Biotechnology, 10*, 1200–1205.

Chowdhury, J. U., Bhuiyan, M. N. I., & Nandi, N. C. 2008. Essential oil constituents of needles, dry needles, inflorescences and resins of *Pinus caribaea* Morelet growing in Bangladesh. *Bangladesh Journal of Botany, 37*, 211–212.

Coppen, J. J. W., Gay, C., James, D. J., Robinson, J. M., & Mullin, I. J. 1993. Xylem resin composition and chemotaxonomy of three varieties of *Pinus caribaea*. *Phytochemistry, 33*, 1103–1111.

Coppen, J. J. W., James, D. J., Robinson, J. M., & Subansenee, W. 1998. Variability in Xylem Resin Composition Amongst Natural Populations of Thai and Filipino *Pinus merkusii* de Vriese *Flavour and Fragrance Journal, 13*, 33–39.

Coppen, J. J. W., Ronbinson, J. M., & Mullin, I. J. 1988. Composition of xylem resin from five Mexican and Central American *Pinus* species growing in Zimbabwe. *Phytochemistry, 27*, 1731–1734.

Dagne, E., Bekele, T., Bisrat, D., Alemayehu, M., Warka, T., & Elokaokich, J. P. 1999. Essentail oils of Resins from three *Pinus* species growing in Ethiopia and Uganda *Ethiopia Journal of Sciences, 22*, 253–257.

Demirci, B., Yusufoglu, H. S., Tabanca, N., Temel, H. E., Bernier, U. R., Agramonte, N. M., .&Demirci, F. 2017. *Rhanterium epapposum* Oliv. essential oil: Chemical composition and antimicrobial, insect-repellent and anticholinesterase activities. *Saudi Pharmaceutical Journal, 25*, 703–708.

Dob, T., Berramdane, T., & Chelghoum, C. 2007. Essential oil composition of *Pinus halepensis* Mill. from three different regions of Algeria. *Journal Essential Oil Research, 19*, 40–43.

Ekundayo, O. 1978. Monoterpenes composition of the needles oils of *Pinus* species. *Journal of Chromatographic Science, 16*, 294–295.

Ekundayo, O. 1988. Volatile constituents of *Pinus* needle oils. *Flavour Fragrance, 3*, 1–11.

Finney, D. J. 1971. Probit Analysis, 3rd ed. Cambridge University Press, Cambridge, England. p. 333.

Gaire, S., Scharf, M. E., & Gondhalekar, A. D. 2019. Toxicity and neurophysiological impacts of plant essential oil components on bed bugs (Cimicidae: Hemiptera). *Scientific Reports, 9*, 3961.

Gerwick, B. C., & Sparks, T. C. 2014. Natural products for pest control: An analysis of their role, value and future. *Pest Management Science 70*, 1169–1185.

Ghosn, M. W., Saliba, N. A., & Talhouk, S. Y. 2006. Chemical composition of the needle-twig oils of *Pinus brutia* Ten. *Journal of Essential Oil Research, 18*, 445–447.

Guo, S., Zhang, W., Liang, J., You, C., Geng, Z., Wang, C., & Du, S. 2016. Contact and repellent activities of the essential oil from *Juniperus formosana* against two stored product pests. *Molecules, 21*, 504.

Hebeish, A., Fouda, M. M. G., Hamdy, I. A., El-Sawy, S. M., & Abdel-Mohdy, F. A. 2008. Preparation of durable insect repellent cotton fabric: limonene as insecticide *Carbohydrate Polymers, 74*, 268–273.

Hieu, T. T., Kim, S. I., Kwon, H. W., & Ahn, Y. J. 2010. Enhanced repellency of binary mixtures of *Zanthoxylum piperitum* pericarp steam distillate or *Zanthoxylum armatum* seed oil constituents and *Calophyllum inophyllum* nut oil and their aerosols to *Stomoxys calcitrans*. *Pest Management Science, 66*, 1191–1198.

Hmamouchi, M., Hmamouchi, J., & Zouhdi, M. 2005. Chemical and Antimicrobial Properties of Essential Oils of Five Moroccan Pinaceae. *In: The Antimicrobial/Biological Activity of Essential Oils. Eds. Lawrence B.M. p. Allured Publ. Corp.Carol Stream, IL*, 368–372.

Hollingsworth, R. G. 2005. Limonene, a citrus extract, for control of mealybugs and scale insects. *Journal of Economic Entomology, 98*, 772–779.

Ibrahim, M. A., Kainulainen, P., & Aflatuni, A. 2001. Insecticidal, repellent, antimicrobial activity and phytotoxicity of essential oils: With special reference to limonene and its suitability for control of insect pests *10*, 243–259.

Isman, M. B. 2016. Pesticides Based on Plant Essential Oils: Phytochemical and Practical Considerations. Medicinal and Aromatic Crops: Production, Phytochemistry, and Utilization: ACS Publications. p. 13–26.

Jagadeesan, R., Nayak, M. K., Pavic, H., Chandra, K., & Collins, P. J. 2015. Susceptibility to sulfuryl fluoride and lack of cross-resistance to phosphine in developmental stages of the red flour beetle, Tribolium castaneum (Coleoptera: Tenebrionidae). *Pest Management Science, 71*, 1379–1386.

Jantan, B. 2002. A Comparative Study of the Oleoresins of Three *Pinus* Species from Malaysian Pine Plantations. *Journal of Essential Oil Research, 14*, 327–332.

Karr, L. L., & Coates, J. R. (1988). Insecticidal properties of d-limonene. *Journal of Pesticide Science, 13*, 287–290.

Kassir, J. T., Mohsen, Z. H., & Mehdi, N. S. 1989. Toxic effects of limonene against *Culex quinquefasciatus* Say larvae and its interference with oviposition. *Anz. Schädlingskde, Pflanzenschutz, Umweltschutz., 62*, 19–21.

Kaya, A., Benirci, B., & Baser, K. H. C. 2007. Micromorphology of glandular trichomes of *Nepeta congesta* Fisch. Mey. var. congesta (Lamiaceae) and chemical analysis of the essential oils. *South African Journal of Botany, 73*, 29–34.

Keane, S., & Ryan, M. 1999. Purification, characterisation, and inhibition by monoterpenes of acetylcholinesterase from the waxmoth, *Galleria mellonella* (L.). *Insect Biochemistry and Molecular Biology, 29* 1097–1104.

Kirima, J. M., Okuta, M., & Omara, T. 2020. Chemical composition of essential oils from *Pinus caribaea* Morelet needles. *French-Ukrainian Journal of Chemistry, 08*, 142–148.

Lee, Y. S., & Lee, S. T. 1991. *Modern Systematic Botany. U. Song Publishing, Seoul. p2.*

Liao, M., Xiao, J.-J., Zhou, L.-J., Liu, Y., Wu X-W, & Hua, R.-M. 2016. Insecticidal Activity of *Melaleuca alternifolia* Essential Oil and RNA-Seq Analysis of Sitophilus zeamais Transcriptome in Response to Oil Fumigation. *PLoS ONE, 11*, e0167748.

Liu, Z. L., & Ho, S. H. 1999. Bioactivity of the essential oil extracted from *Evodia rutaecarpa* Hook f. et Thomas against the grain storage insects, *S. zeamais* Motsch. and *Tribolium castaneum* (Herbst) *Journal of Stored Product Research, 35*, 317–328.

Lucia, A., Licastro, S., Zerba, E., Gonzalez, A. P., & Masuh, H. 2009. Sensitivity of *Aedes aegypti* adults (Diptera: Culicidae) to the vapors of *Eucalyptus* Essential oils. *Bioresource Technology, 100*, 6083–6087.

Macchioni, F., Cioni, P. L., Flamini, G., Morelli, I., Maccioni, S., & Ansaldi, M. 2003. Chemical Composition of Essential Oils From Needles, Branches and Cones of *Pinus pinea, P. halepensis, P. pinaster* and *P. nigra* From Central Italy. *Flavour and Fragrance Journal, 18*, 139–143.

Moronkola, D. O., Ogunwande, I. A., Oyewole, I. O., H., C. B. K., Ozek, T., & Ozek, G. 2009. The Needle Oil of *Pinus caribaea* Morelet From Nigeria. *Journal of Essential Oil Research, 21*, 342–344.

Mursiti, S., Estari, N. A. L., da Febriana, Z., Rosanti, Y., & Ningsih, T. W. 2019. The Activity of d-Limonene from Sweet Orange Peel (*Citrus sinensis* L.) Extract as a Natural Insecticide Controller of Bedbugs (*Cimex cimicidae*). *Oriental Journal of Chemistry, 35*, 1420–1425.

Omara, T., Kateeba, K. F., Musau, B., Kigenyi, E., Adupa, E., & Kagoya, S. 2018. Bioinsecticidal Activity of Eucalyptol and 1R-Alpha-Pinene Rich Acetonic Oils of *Eucalyptus saligna* on *Sitophilus zeamais* Motschulsky, 1855 (Coleoptera: Curculionidae). *Journal of Health and Environmental Research, 4*, 153–160.

Omara, T., Nassazi, W., Omute, T., Awath, A., Laker, F., Kalukusu, R., & Adupa, E. 2020. Aflatoxins in Uganda: An Encyclopedic Review of the Etiology, Epidemiology, Detection, Quantification, Exposure Assessment, Reduction and Control. *International Journal of Microbiology, 2020*(4723612), 1–18.

Pagula, F. P., & Baeckström, P. P. 2006. Studies on Essential Oil-Bearing Plants From Mozambique: Part II. Volatile Leaf Oil of Needles of *Pinus elliottii* Engelm. and *Pinus taeda* L. *Journal of Essential Oil Research, 18*, 32–34.

Papachristos, D. P., Karamanoli, K. I., Stamopoulos, D. C., & Spiroudi, U. M. 2004. The relationship between the chemical composition of three essential oils and their insecticidal activity against *Acanthoscelides obtectus* (Say). *Pest Management Science, 60*, 514–520.

Philogene, B. J. R., Regnault-Roger, C., & Vincent, C. 2005. Biopesticides of Plant Origin. *Journal of Natural Products, 68*, 1138–1139.

Rajendran, S., & Sriranjini, V. 2008. Plant products as fumigants for stored-product insect control. *Journal of Stored Products Research, 44*, 126–135.

Regnault-Roger, C., Vincent, C., & Arnason, J. T. 2012. Essential oils in insect control: Low-risk products in a highstakes world. *Annual Reviews of Entomology, 57*, 405–424.

Roussis, V., Petrakis, P. V., Ortiz, A., & Mazomenos, B. E. 1995. Volatile constituents of needles of five *Pinus* species grown in Greece. *Phytochemistry, 39*, 357–361.

Saglam, H., Gozler, T., Kivcak, B., Demirci, B., & Baser, K. H. C. 2001. Volatile compounds from *Haplophyllum myrtifolium*. *Chemistry of Natural Compounds, 37*, 442–444.

Seixas, P. T. L., Demuner, A. J., Alvarenga, E. S., Barbosa, L. C. A., Marques, A., de Sá Farias, E., & Picanço, M. C. 2018. Bioactivity of essential oils from *Artemisia* against Diaphania. *Scientia Agricola, 75*, 519–525.

Showler, A. T., Harlien, J. L., & Perez de Léon, A. A. 2019. Effects of Laboratory Grade Limonene and a Commercial Limonene-Based Insecticide on *Haematobia irritans* irritans (Muscidae: Diptera): Deterrence, Mortality, and Reproduction. *Journal of Medical Entomology, 56*, 1064–1070.

Showler, A. T., Osbrink, W. L. A., Morris, J., & Wargovich, M. J. 2017. Effects of two commercial neem-based insecticides on lone star tick, *Amblyomma americanum* (L.) (Acari: Ixodidae): deterrence, mortality, and reproduction. *Biopesticides International, 13* 1–12.

Sonibare, O. O., & Olakunle, K. 2008. Chemical composition and antibacterial activity of the essential oil of Pinus caribaea from Nigeria *African Journal of Biotechnology, 7*, 2462–2464.

Stamopoulos, D. C., Damos, P., & Karagianidou, G. 2007. Bioactivity of five monoterpenoid vapours to *Tribolium confusum* (du Val) (Coleoptera: Tenebrionidae). *Journal of Stored Products Research, 43*, 571–577.

Stanley, T. D., & Ross, E. M. 1989. Flora of Southeastern Queensland. *Queensland Department of Primary Industries Publ., Misc. Pub. 81020, Queensland, Australia.*

Stevanoic, T. F., Garneau, F., Jean, F. I., Gagnon, H., Vilotic, D., Petrovic, S., & Pichette, A. 2004. The essential oil composition of *Pinus mugo* Terra from Serbia. *Flavour Fragrance, 20*, 96–97.

Stevenson, P. C., Isman, M. B., & Belmain, S. R. 2017. Pesticidal plants in Africa: a global vision of new biological control products from local uses. *Industrial Crops and Products, 110*, 2–9.

Tabanca, N., Demirci, F., Ozek, T., Tumen, G., & Baser, K. H. C. 2001. Composition and antimicrobial activity of the essential oil of *Origanum x dolichosiphon* P. H. Davis. *Chemistry of Natural Compounds, 37*(3), 238–241.

Thompson, B. M., & Reddy, G. V. P. 2016. Effect of temperature on two bio-insecticides for the control of confused flour beetle (Coleoptera: Tenebrionidae). *Flavour Entomology, 99*, 67–71.

Valterová, I., Sjödin, K., Vrkoc, J., & Norin, T. 1995. Contents and enantiomeric compositions of monoterpene hydrocarbons in xylem oleoresins from four *Pinus* species growing in Cuba. Comparison of trees unattacked and attacked by *Dioryctria horneana. Biochemical Systematics & Ecology, 23*, 1–15.

Venskutonis, P. R., Vyskupaityte, K., & Plausinaitis, R. 2000. Composition of Essential Oils of *Pinus sylvestris* L. From Different Locations of Lithuania. *Journal of Essential Oil Research, 12*, 559–565.

Advances in Phytochemistry, Textile and Renewable Energy Research for Industrial Growth – Nzila et al. (Eds)

Phytochemical screening, total phenolic content, total flavonoid content and GC-MS evaluation of crude acetonic extracts of *Prosopis juliflora*

M.P. Odero & A.K. Kiprop
Africa Centre of Excellence in Phytochemicals, Textile and Renewable Energy (ACE-PTRE II), Eldoret, Kenya
Department of Chemistry and Biochemistry, Moi University, Eldoret, Kenya

I.O. K' Owino
Department of Pure and Applied Chemistry, Masinde Muliro University of Science and Technology, Kakamega, Kenya

M. Arimi
Department of Chemical and Process Engineering, Moi University, Eldoret, Kenya

ABSTRACT: Commonly referred to as the Mathenge plant in Kenya, the plant *Prosopis juliflora* has been considered as a noxious weed within the boundaries of Kenya occupying ASAL lands, colonizing water bodies and affecting livestock by the blocking of their rumens, causing diarrhoea and in some cases leading to the loss of teeth which eventually leads to death. Despite these disadvantages, the plant has been identified in several nations as a plant of great medicinal value being used for the treatment of eye problems, digestive problems, lung problems, and sore throats among other ailments. These medicines have mainly been extracted from the pods, leaves and bark. Acetonic extracts of the heartwood have in recent research been shown to be high in flavonoid content with mesquitol being an isolated compound of interest. This is because mesquitol, in comparison to the existing antioxidants such as catechin, probucol and alpha-tocopherol has been shown to have better antioxidant and radical scavenging properties potentials hence it can be useful for managing diseases such as cancer and diabetes. With the heartwood being significantly different from the other parts of the plant, limited research has been done on the phytochemicals present, their total flavonoid contents, total phenolic content and potential medicinal uses. Phytochemical screening of acetonic extracts showed the presence of flavonoids, alkaloids, terpenoids, steroids, tannins and no saponins. The total flavonoid content was 33.7 ± 1.22 while the total phenolic content was 53.5 ± 2.6 GAE. GC-MS evaluation of the crude extracts identified five potential compounds namely 2(4H)-benzofuranone,5,6,7,7a-tetrahydro-4,4,7a-trimethyl; 2(4H)-benzofuranone 5,6,7,7a-tetra hydro-4,4,7a-trimethyl-(R); cyclopentane-carboxylic acid 3-(3-fluorophenylcarbamoyl)-1,2,2-trimethyl; 6,6-dimethyl-10-methylene-1-oxa-spiro[4.5] decane; and 2-pentanol, 2,4-dimethyl.

1 INTRODUCTION

The need for survival of plants in different environments over the evolution of time has led to the development of different mechanisms of survival, including the development of phytochemicals. Different plant extractives have different uses with some being medicinal or repellents, while others are poisonous (Azevedo et al., 2018). The *Prosopis juliflora* plant, also referred to as the honey mesquite plant, is a hardy and woody tree belonging to the genus *Prosopis*. In Kenya, it is commonly referred to as the mathenge plant and grows primarily in the arid and semi-arid regions of Kenya (Sathiya & Muthuchelian, 2011).

Due to its potential as a medicinal plant, and the unusually high amounts of flavonoids in its heartwood, *P. juliflora* has elicited a lot of curiosity within the scientific community. Whereas mesquitol, catechin, and 4′-O-methylgallocatechin have previously been isolated from the acetonic extracts of heartwood (Chepkwony et al., 2020; Sirmah, 2009), limited knowledge exists as to the phytochemicals present, their total flavonoid content, and their total phenolic content, which are major pointers to its medicinal potential.

In this study, we have undertaken phytochemical screening, measurement of total flavonoid content and total phenolic content, and GC-MS evaluation of the acetonic extracts of the heartwood of *P. juliflora*.

2 LITERATURE REVIEW

The *Prosopis juliflora* plant, locally referred to as "mathenge" is a noxious weed, having been listed among the top 100 most undesirable plant species. This is because it is known to occupy vast areas of lands

 DOI 10.1201/9781003221968-23

especially in ASAL regions in the process interfering with biodiversity, colonizing water bodies while affecting livestock by the blocking of their rumens, causing diarrhoea and in some cases leading to the loss of teeth and eventually death (Vijayakumar et al., 2016). In fact surveys done in regions where the trees dominate such as Baringo county in Kenya showed that up to 90% of the natives wanted the *Prosopis* plant to be totally eradicated from their habitat (Mwangi & Swallow, 2008).

Despite all the negative reports on the plant, attempts to valorise the plants have shown that the plant is full of valuable phytochemicals and has potent ability to be used as a source of food, medicine and lead compounds for drug development with flavonoids, phenolics, alkaloids, and terpenes considered as the most significant (Henciya, Seturaman, & Rathinam, 2016). Previous literature suggests that the plant has been used as a form of traditional medicine with the leaf used for treatment of cancerlike conditions, sore throat, diarrhoea, measles, eye infections and flu. The bark has been used for the treatment of wounds with research showing it has both antifungal and antibacterial properties (Prabha, Dahms, & Malliga, 2014).

Several studies have been done on the medicinal properties of the *Prosopis juliflora* plant with ethanolic extracts of the leaves showing positive antibacterial and anticancer activities (Sathiya & Muthuchelian, 2008, 2011). The extracts of the plant's bark have been shown to have antifungal and antioxidant activities. Previous work on the heartwood of the *P. juliflora* plant shows that it contains several compounds of interest such as mesquitol, catechin, and epi-catechin among other flavonoids. These have been shown to have antitermite, antifungal, and antioxidant activities as well as radical scavenging properties.

3 MATERIALS AND METHODS

3.1 *Sample collection*

The *P. juliflora* plant samples were collected from Marigat, Baringo county in Kenya (latitude 0°, 28′ 0.01″ N, longitude 35°, 57′ 0.01″E) and air dried.

3.2 *Extraction*

Extraction was done based on the method previously used by Odero et al. (Odero, Munyendo, & Munyendo, 2017) with some modifications. Briefly, the heartwood was separated first from other parts of the plant and ground into fine powder using a hammer mill. It was then dried under a shade until the achievement of a constant weight. Approximately 15 grams was then serially extracted in a Soxhlet machine with solvents of increasing polarity starting from hexane, dichloromethane, and then acetone. The extracts were then evaporated under vacuum conditions and the extracts put in pre-weighed bottles for future use.

3.3 *Chemicals and equipment*

All the chemicals used were of analytical grade procured by Moi University and supplied by Pyrex and Kobian Laboratories.

3.4 *Phytochemical screening*

Phytochemicals were screened using previously discussed methods (Anyalogbu, Anyalogbu, & Nwalozie, 2013; Lakshmibai & Amirtham, 2018; Sivanandham, 2016; Wangila, 2017). The following phytochemicals were tested: terpenoids, alkaloids, flavonoids, tannins, saponins and steroids.

3.4.1 *Test for alkaloids*

Two millilitres of the extracts were dissolved in 2 mL of Wagners reagent. The colour changes were observed with the appearance of a reddish-brown precipitate suggesting the sample is positive for alkaloids.

3.4.2 *Test for saponins*

This was done using the foam test. 2 mL of the extracts were diluted in 20 mL of distilled water in a test tube and shaken for 20 minutes. Formation of foam on top of the test tube showed the presence of saponins in the plant extracts.

3.4.3 *Test for flavonoids*

Two millilitres of the plant extract was put in a test tube and two drops of dilute NaOH added. The appearance of a yellow colour which becomes colourless after two drops of dilute sulfuric acid were added confirmed the presence of flavonoids in the plant extract.

3.4.4 *Test for steroids*

Two millilitres of the plant extract were put in a test tube with 2 mL of chloroform. One millilitre of sulfuric acid was added to it with a dark reddish colour confirming the presence of steroids.

3.4.5 *Test for tannins*

Two millilitres of the plant extract were dissolved in 45% of ethanol. The test tube was then boiled for five minutes and 1 mL of 15% ferric chloride solution was added. The appearance of a greenish to black colour confirmed the presence of tannins.

3.4.6 *Test for terpenoids*

To 5 mL of the plant extract was added 2 mL of chloroform and 3 mL of concentrated sulfuric acid. A reddish brown colour at the interface confirmed the presence of terpenoids.

3.4.7 *Total Phenolic Content (TPC)*

The total phenolic content in both *P. juliflora* and *A. succotrina* were determined using a previously discussed method (Singleton, Orthofer, & Lamuela-Ravent, 1999). Acetonic extracts of *P. juliflora* and fresh extraction of *A. succotrina* were prepared and read in a Beckman Coulter Model DU^{R}720 UV/VIS spectrophotometer. This was done by mixing 0.5 mL of the extracts with 2.5 mL of 10% Folin–Coicalteu reagent dissolved in distilled water with 2.5 mL of 7.5% $NaHCO_3$. The samples were prepared in triplicate. Blanks were also prepared with 50% methanol, 2.5 mL of 10% Folin–Coicalteu reagent dissolved in distilled water and 7.5% of $NaHCO_3$. The samples were then incubated for 45 minutes at a temperature of 45°C and the absorbance was measured at 770 nm.

Standard solutions of gallic acid were also used for the calibration line. Based on the measured absorbance, the total phenolic content was measured in gallic acid equivalent (mg of GAE/100g of extract). This was calculated using the formula below:

$$\text{T.P.C} = \text{Df}\frac{\text{CV}}{\text{M}} \tag{1}$$

where C is the concentration of gallic acid established from the calibration curve, mg/mL; V is the volume of the extract used; and M is the mass of the extract.

3.4.8 *Total flavonoid content*

The procedure for total flavonoid content was as described by Odero et al. (Odero et al., 2017), with some modifications. Approximately 0.25 grams of each of the dyes was taken and dissolved in 1.25 mL of distilled water 75 μL of 6% $NaNO_3$ was added and the solution was shaken together to mix. It was then incubated in a dark oven at room temperature for approximately 6 minutes.

A solution of 10% $AlCl_3$, was then added and the mixture was incubated further in a dark cabinet for 5 minutes. A calibration curve of the standard quercetin at concentrations of 0, 20, 40,60, 80 and 100 μg/mL diluted in methanol. The absorbance was measured at 510 nm λ max using Beckman Coulter Model DU^R 720 UV–Vis and the results were obtained in triplicate. The total flavonoid content was calculated using the formula:

$$\text{TFC} = \frac{\text{Absorbance of crude extracts x Mass of quercetin in Mg}}{\text{Absorbance of standards xmass of extracts in Mg}} \tag{2}$$

3.4.9 *GC-MS*

The crude extracts of both dyes were initially extracted using C-18 Solid Phase Extraction cartridges and filtered through 0.45 μm syringe filters. They were then transferred into auto-sampler vials for GC-MS Shimadzu QP 2010 Model for further analysis. The carrier gas used was ultrapure helium with the flow rate set at 1 mL/minute. A BP-X5 non-polar column, 30m; 0.25 mm ID; 0.25 μm film thickness, was used for separation.

The GC-MS machine was set and programmed as follows: temperature of 50°C (1 minute). This was subsequently increased at a rate of 5°C/min up to 250°C (9 minutes) with the total run-time being exactly 50 minutes. One microliter of the sample was injected into the GC at 200°C in split mode. This was in split ratios of 10:1 with the interface temperatures set at 250°C. The Electron Ionization (E.I) ion source was set at 200°C.

Mass analysis was done in full scan mode within the ranges 50–600 m/z and the detected peaks matched against the NIST libraries for possible identification. Both fragmentation patterns and retention index were used for matching.

Table 1. Phytochemical screening of acetonic extracts of the heartwood of *P. juliflora*.

Phytochemical components	Results
Terpenoids	+
Saponins	–
Tannins	+
Alkaloids	+
Flavonoids	+
Steroids	+

Key: +: present –: absent.

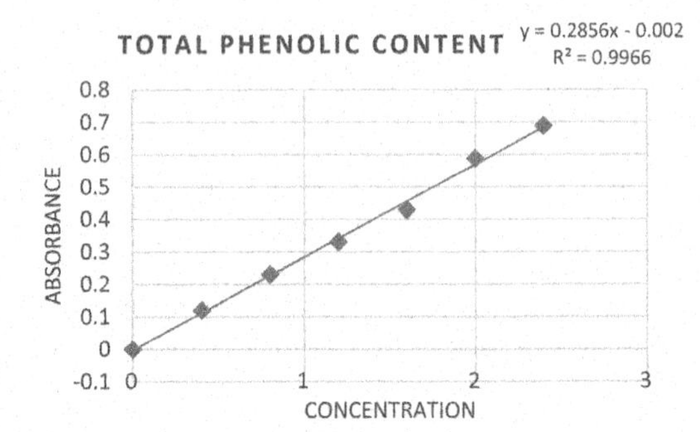

Figure 1. Graph of absorbance vs. concentration of GAE.

4 RESULTS AND DISCUSSION

4.1 *Phytochemical screening results*

Phytochemical screening is important because it shows whether the tested compounds are present in the sample. The results of the six compounds that were screened are given in Table 1.

Acetonic extracts of *P. juliflora* yielded positive results for terpenoids, tannins, alkaloids, flavonoids and steroids and negative results for saponins. These results are similar to those previously discussed by (Lakshmibai, Amirtham, & Radhika, 2015 who reported slightly similar results on the phytochemistry of the plant.

4.2 *Total phenolic content*

The natural antioxidant activity of the plant is determined by the phenolic composition in the extract. A calibration curve using gallic acid as a standard was drawn as shown in Figure 1, and used for the calculation of TPC.

As shown in the Figure 1, a good correlation coefficient ($R^2 = 0.9966$) was obtained from the standard curve. The TPC of the acetonic extracts were found to be 53.5 ± 2.6 gallic acid equivalent (GAE). The results are slightly lower than those of another species

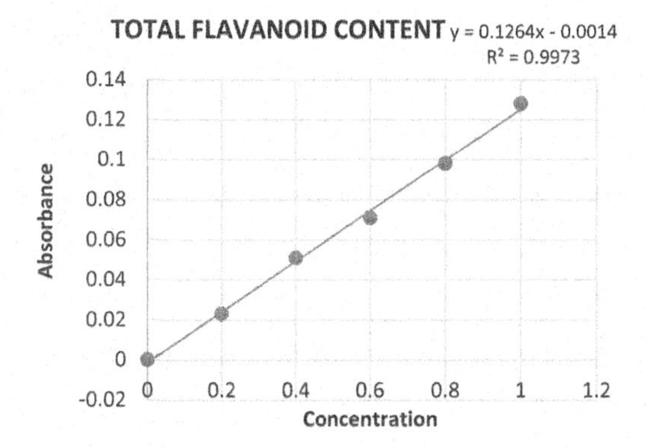

Figure 2. Graph of concentration of quercetin against absorbance.

of *Prosopis* i.e., *Prosopis cineraria* whose leaves were found to have a TPC of 530 mg/g GAE (Rathore et al., 2019). However, the results indicate what other studies confirm, that the *Prosopis juliflora* has potent antioxidant properties (Almaraz-Abarca et al., 2007).

4.3 *Total flavonoid content*

Figure 2 was used for the calculation of total flavonoid content. The correlation coefficient of the graph ($R^2 = 0.9973$) was obtained for the standard curve. The total flavonoid content described in mg of quercetin equivalent per gram of dry weight of the heartwood of *P. juliflora* for the acetonic extract, was found to be 33.7 ± 1.22.

4.4 *GC-MS identified compounds*

Considered a hybrid system that combines the superior GC separation technique and the sensitive mass spectrometer, the GC-MS has become a common identification instrument in natural products for oily and volatile compounds due to its sensitivity, simplicity and accuracy in measuring and identifying compounds (Al-rubaye, Hameed, & Kadhim, 2017). In this experiment, the total run time was 50 minutes. Several compounds were detected and their fragmentation patterns and retention indexes compared to NIST 2014, MS library. The GC chromatogram is shown in Figure 3.

Several compounds that are significant were eluted at retention times 5.182 and 25.180. The GC chromatogram is captured in Figure 3 and the subsequent mass spectrum captured by total ion chromatogram in Figure 4 for 2-pentanol, 2,4-dimethyl and Figure 5 respectively for 2(4H)-benzofuranone,5,6,7,7a-tetrahydro-4,4,7a-trimethyl; 2(4H)-benzfuranone,5,6,7,7 a-tetra hydro-4,4,7a-trimethyl-(R); cyclopentanecarboxylic acid, 3-(3-fluorophenylcarbamoyl)-1,2,2-trimethyl; and 6,6-dimethyl-10-methylene-1-oxaspiro[4.5] decane.

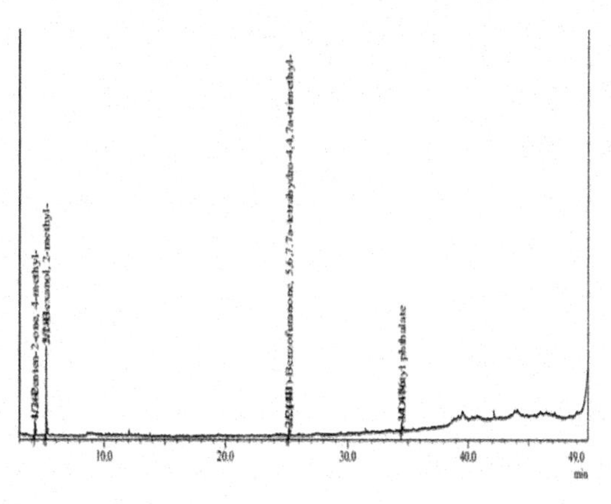
Figure 3. GC chromatogram of acetonic extracts of *Prosopis juliflora*.

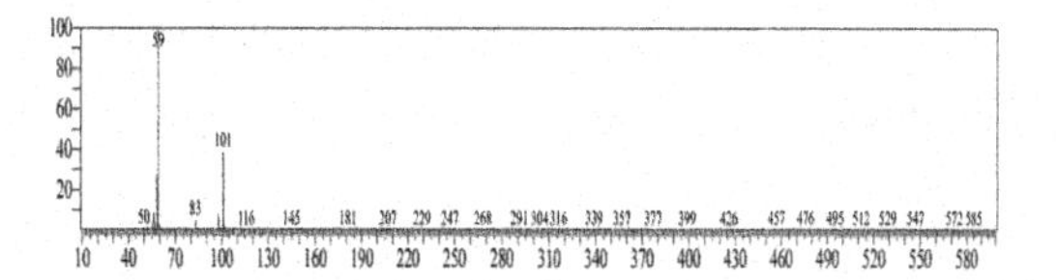
Figure 4. Total ion chromatogram for 2-pentanol, 2,4-dimethyl.

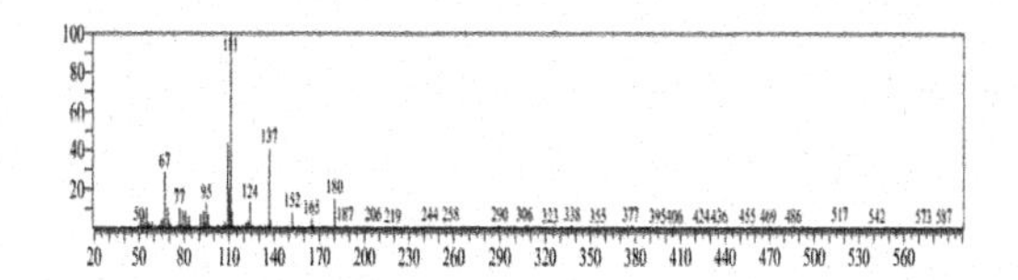
Figure 5. Total ion chromatogram for compounds a,b,c,d.

4.4.1 *Identified compounds of Prosopis juliflora from the NIST library*

A summary of GC retention time (RT), retention index (RI) and information obtained GC-MS showing molecular weights, chemical formulas, names and possible compounds based on compared NIST entries. The list of identified compounds is shown in Table 2 and the structures of each of the identified compounds is given in Figure 5.

The mass spectrometer analyses different compounds eluted at different times; this is based on their fragmentation patterns which are usually unique to certain compounds already in the library called fingerprints in the NIST library (Siwe et al., 2019). The GC-MS analysis yielded five potential natural products from the heartwood of the plant. Of these compounds, some have been previously isolated in other plants such as benzofuranones and their derivative previously isolated from *Asystasia mysorensis* using GC-MS (Maina, Madivoli, Ouma, Ogilo, & Kenya, 2019) and 6,6-dimethyl-10-methylene-1-oxaspiro[4.5] decane which has been isolated also using a GC-MS from the leaves of *Eremomastax speciose (Siwe et al., 2019).*

Table 2. The list of identified compounds in *Prosopis juliflora* extract.

Retention time	M.W	Chemical formula	Structure	Name
25.180	180	$C_{11}H_{16}O_2$	a	2(4H)-Benzofuranone, 5,6,7,7a-tetrahydro-4,4,7a-trimethyl
25.180	180	$C_{11}H_{16}O_2$	b	2(4H)-Benzofuranone, 5,6,7,7a-tetrahydro-4,4,7a-trimethyl-, (R)-
25.180	293	$C_{16}H_{20}FNO_3$	c	Cyclopentane carboxylic acid, 3-(3-fluorophenylcarbamoyl)-1,2,2-trimethyl
25.180	180	$C_{12}H_{20}O$	d	6,6-Dimethyl-10-methylene-1-oxa-spiro[4.5]decane
5.182	116	$C_7H_{16}O$	e	2-Pentanol, 2,4-dimethyl-

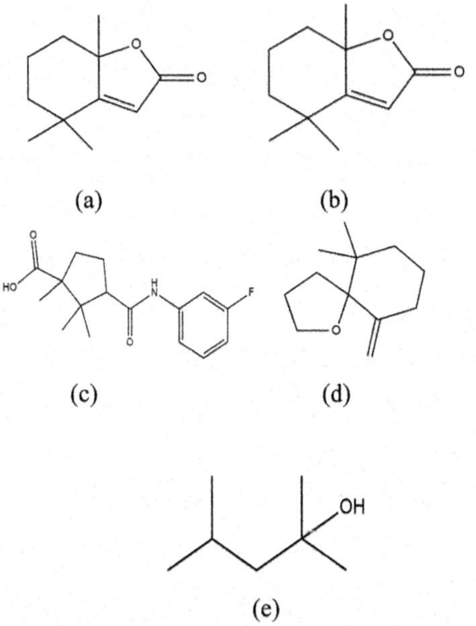

(a) (b)

(c) (d)

(e)

Figure 6. Structures of compounds identified in *Prosopis juliflora* extract.

5 CONCLUSION AND RECOMMENDATION

The results of this study showed the presence of important phytochemicals such as terpenoids, tannins, alkaloids, flavonoids and steroids which usually have medicinal properties. This supports the notion that the plant has potential medicinal properties. With the use of a GC-MS machine, the following compounds were also identified 2(4H)-benzofuranone, 5,6,7,7a-tetrahydro-4,4,7a-trimethyl; 2(4H)-Benzofuranone, 5,6,7,7a-tetrahydro-4,4,7a-trimethyl-,(R)-; cyclopentane carboxylic acid, 3-(3-fluorophenylcarbamoyl)-1,2,2-trimethyl; 6,6-Dimethyl-10-methylene-1-oxa-spiro[4.5] decane; and 2-Pentanol, 2,4-dimethyl-.

Further studies on the bioactivity of the acetonic and identified compounds from the heartwood of the plant should be studied further.

ACKNOWLEDGMENT

This study was made possible by a Postgraduate scholarship awarded to the first author by the Africa Centre of Excellence II in Phytochemicals, Textile and Renewable Energy (ACE II-PTRE). A World Bank funded project through the Inter-University Council of East Africa (IUCEA) and hosted by Moi University.

REFERENCES

Almaraz-Abarca, N., da Graça Campos, M., Avila- Reyes, J.A., Naranjo-Jimenez, N., Corral, J.H.& Gonzalez-Valdez, L.S., (2007). Antioxidant activity of polyphenolic extract of monofloral honeybee-collected pollen from mesquite (Prosopis juliflora, Leguminosae). *Journal of Food Composition and Analysis*, *20*(1), pp. 119–124.

Al-rubaye, A. F., Hameed, I. H., & Kadhim, M. J. (2017). A Review?: Uses of Gas Chromatography-Mass Spectrometry (GC-MS) Technique for Analysis of Bioactive Natural Compounds of Some Plants A Review?: *International Journal of Toxicological and Pharmacological Research, March*. https://doi.org/10.25258/ijtpr.v9i01.9042

Anyalogbu, E. A., Ezeji, E. U., & Nwalozie, C. J. (2013). Phytochemical Screening and Anti-malaria / Typhoid Fever Activities of Alstonia boonei (De Wild) Stem Bark Powder. *1989*, 1–3.

Azevedo, G., Damasceno, D. B., Souto, A. L., Bezerra, I., Roque, A. D. A., & Ferrari, M. (2018). *Prosopis juliflora?: Phytochemical*. 1–21.

Chepkwony, S. C., Dumarçay, S., Chapuis, H., Kiprop, A., Gerardin, P., & Charbonnier, C. G. (2020). Geographic and intraspecific variability of mesquitol amounts in Prosopis juliflora trees from Kenya. *European Journal of Wood and Wood Products*. https://doi.org/10.1007/s00107-020-01535-8

Henciya, S., Seturaman, P., & Rathinam, A. (2016). ScienceDirect Biopharmaceutical potentials of Prosopis spp . *Journal of Food and Drug Analysis*, *25*(1), 187–196. https://doi.org/10.1016/j.jfda.2016.11.001

Khandelwal, P., Sharma, R. A., & Agarwal, M. (2016). Phytochemical analyses of various parts of Prosopis juliflora. *Mintage Journal of Pharmaceutical & Medical Sciences*, *5*(1), 16–18. https://doi.org/10.11648/j.ijpc.20160201.12

Lakshmibai, R., & Amirtham, D. (2018). Phytochemical analysis and antioxidant activity of Prosopis Juliflora thorn extract. *5*(1), 21–27.

Lakshmibai, R., Amirtham, D., & Radhika, S. (2015). Preliminary Phytochemical Analysis and Antioxidant Activities of Prosopis Juliflora and Mimosa Pudica Leaves. 04(30), 5766–5770.

Maina, E. G., Madivoli, E. S., Ouma, J. A., Ogilo, J. K., & Kenya, J. M. (2019). Evaluation of nutritional value of Asystasia mysorensis and Sesamum angustifolia and their potential contribution to human health. *Food Science and Nutrition*, *7*(6), 2176–2185. https://doi.org/10.1002/fsn3.1064

Mwangi, E., & Swallow, B. (2008). Prosopis juliflora Invasion and Rural Livelihoods in the Lake Baringo Area of Kenya *JSTOR 6*(1), 130–140.
Odero, M. P., Munyendo, W. L., & Kiprop, A. K. (2017). Quantitative Analysis of the Flavonoid Mesquitol in the Medicinal Plant Prosopis juliflora with Seasonal Variations in Marigat, Baringo County-Kenya. *5*(6), 107–112. https://doi.org/10.11648/j.sjac.20170506.16
Prabha, S., Dahms, H.-U., & Malliga, P. (2014). P harmacological potentials of phenolic compounds from P rosopis spp. -a. *2*(11), 918–924. https://doi.org/10.12980/JCLM.2.2014J27
Preeti, K., Avatar, S. R., & Mala, A. (2015). Pharmacology and Therapeutic Application of Prosopis juliflora?: A Review. *3*(4), 234–240. https://doi.org/10.11648/j.jps.20150304.20
Rathore, A., Dadhich, R., Purohit, K., Sharma, S. K., Vaishnava, C. S., Joseph, B., & Khatri, A. (2019). Phytochemical screening and total phenolic and flavonoid content in leaves of Prosopis cineraria (L.) Druce. *7*(3), 1853–1855.
Sathiya, M., & Muthuchelian, K. (2008). Investigation of Phytochemical Profile and Antibacterial Potential of Ethanolic Leaf. *Ethnobotanical Leaflets 12:*, *12*, 1240–1245.
Sathiya, & Muthuchelian, K. (2011). Anti-tumor potential of total alkaloid extract of Prosopis juliflora DC. leaves against Molt-4 cells in vitro. *10*(44), 8881–8888. https://doi.org/10.5897/ AJB10.875
Singleton, V. L., Orthofer, R., & Lamuela-Ravent, R. M. (1999). Analysis of Total Phenols and Other Oxidation Substrates and Antioxidants by Means of Folin-Ciocalteu Reagent. *Methods in Enzymology*, *299*(1974), 152–178.
Sirmah, P. (2009). Natural Product Research?: Formerly Natural Product Letters Unusual amount of (-) - mesquitol from the heartwood of Prosopis juliflora. 37–41. https://doi.org/10.1080/14786410801940968
Sivanandham, V. (2016). *Phytochemical techniques – a review. December 2015.*
Siwe, G. T., Ernestine, N. Z., Amang, A. P., Mezui, C., Choudhary, I., & Tan, P. V. (2019). Comparative GC-MS analysis of two crude extracts from Eremomastax speciosa (Acanthaceae) leaves. *7*(1), 25–29.
Vijayakumar, L., Pavithra, S., Satheshkumar, M., Balasubramaniam, L., Prakaash, S., Sriram, K., Yaashikaa, P., & Nadu, T. (2016). A novel and cost effective method for textile dye degradation using sawdust of prosopis. *International Journal of Advanced Engineering Technology*, *VII*(II), 166–169.
Wangila, P. (2017). Pharmaceutical Analytical Chemistry?: Open Access Phytochemical Analysis and Antimicrobial Activities of Cyperus rotundus and Typha latifolia Reeds Plants from Lugari Region of Western Kenya. *3*(3). https://doi.org/10.4172/2471-2698.1000128

Advances in Phytochemistry, Textile and Renewable Energy Research for Industrial Growth – Nzila et al. (Eds)

Evaluation of shading on tea yield and phenolics in aerated and unaerated products

R.K. Korir & S.M. Kamunya
KALRO-Tea Research Institute, Kericho, Rift Valley, Kenya

R.C. Ramkat
Moi University, Eldoret, Rift Valley, Kenya
African Center of Excellence in Phytochemicals, Textiles and Renewable Energy (PTRE)

R.C. Muoki*
KALRO-Tea Research Institute, Kericho, Rift Valley, Kenya

ABSTRACT: Varying climatic conditions have pronounced effects on the tea plant. This study was to evaluate the effects of shade on yield and quality in aerated and unaerated products. Cultivar TRFK 6/8 was maintained under three shading regimes (30, 60, and 90%) and unshaded control in three seasons of a year. Yield components included shoot growth, shoot density, and monthly seasonal yields. Biochemical parameters included catechins, caffeine, and polyphenols. Yield components, catechin, and caffeine were negatively affected while polyphenols was positively affected by an increase in shading. Product diversification is possible in existing populations through the production of unaerated tea under moderate shading (30 and 60%) during the cold/wet season and high-quality aerated product during warm/hot seasons while maintaining optimum yields. There is a need to undertake genomic studies so as to provide insights into complex regulatory networks and the identification of genes relevant to biochemical changes due to shading in tea.

1 INTRODUCTION

The tea (*Camellia sinensis* (L.) O. Kuntze) plant plays a significant role to the economy of most developing countries cultivating the crop. Tea made from the leaves of the plant, processed into green or black tea, is drunk as a mild stimulant due to the caffeine content, since time in memorial (Tong et al., 2014). Overproduction of black cut, tear, and curl (BCTC) tea in Kenya has led to the stagnation/decline in the unit prices of processed tea despite increased costs of production, resulting in decreased returns for tea growers. As the world market for black tea continues to shrink, that for green tea and other forms of specialty teas is growing. For instance, in the year 2018, Japan earned 27.27 USD per kilogram made (kg.mt) for green tea compared to 2.93 USD per kg.mt for black CTC in Kenya. Japan also produces Matcha; the highest-quality green tea made from tencha which is grown under shade and contains a high amino acid content together with a low catechin content. Sweetness is associated with amino acids, particularly theanine, which has a taste that is described as "umami" or "brothy," while catechins and caffeine contribute to the astringency of the liquor (Liu et al., 2017).

Shading has been shown to effectively improve the quality of tea beverages by causing a reduction of the concentration of flavonoids, which are the main compounds that contribute to astringency, in the leaves (Wang et al. 2012). Further, tea plants in their original habitat grow under natural shade. Therefore, the use of shade trees in tea plantations was adopted to emulate the natural conditions (Eden, 1976). Shading is also a cultural practice used to mitigate the effects of adverse weather conditions. Shade trees in tea plantations are important in the reduction of temperatures, that ultimately reduces excessive transpiration and moisture losses. In addition, shading protects the plant against losses arising from hail damage and frost bite which is approximated at 2 million kilograms per year (Figure 1).

The present study was therefore designed to guide the tea industry to ameliorate the environmental conditions in the existing plantations. This would improve tea yields and quality for the processing of specialty and diversified products without necessarily uprooting and replanting, which is a very costly venture for the resource-constrained smallholder farmers.

*Corresponding author

DOI 10.1201/9781003221968-24

Figure 1. Shaded tea plants protected against severe hail damage.

2 MATERIALS AND METHODS

This study was superimposed on an existing tea plantation located at Kenya Agricultural and Livestock Research Organization – Tea Research Institute (KALRO-TRI), Kericho Centre (Latitude 0°22′S, longitude 35°21°E; elevation 2180m a.m.s.l). One cultivar; TRFK 6/8, a green tea cultivar used in the processing of high-quality black tea due to the high level of polyphenol content (27.07%) and moderate yields (4441 kh.mt/ha/yr) was used in this study. To test the effect of shading intensity, three different black shade nets were used (30, 60, and 90% shading regimes) and a negative control (unshaded regime) over three seasons in 2015–16 (September–December, January–March; and April–August, representing the warm/wet (WW), hot/dry (HD), and cool/wet (CW) seasons, respectively). The trial was laid in a randomized complete block design (RCBD) and replicated three times, each plot having 12 plants spaced at 0.91 m by 1.22 m (Figure 2). The shade nets were mounted on rectangular frames 1m above the bush table. Each treatment block had two guard rows all round. Prior to data collection, the trial was maintained for a period of 1 month for the plants to acclimatize to the shading effects. The environmental conditions (Table 1) were

Figure 2. Shaded experimental field trial.

recorded in a KALRO-TRI weather station situated 100m from the shade trial.

2.1 *Yield and physiological parameters*

Yield and physiological data collection was done in three seasons over a period of 1 year from September, 2015 to August, 2016. Shoot growth rate (SGR) (millimeters per day (mmd^{-1})) was determined as described by Nyabundi et al. (2016). The shoot extension (mm) was measured using a vernier caliper. In each season, three shoots were tagged in every plot and their shoot length measured at intervals of three days until the pluckable shoots had reached two leaves and a bud stage. Shoot growth rate was determined by the following formula;

$$SGR = \frac{FR\text{-}IR}{ND} \tag{1}$$

Where;
FR is final reading
IR is initial reading
ND is number of days
Shoot density was determined by counting the number of mature harvestable shoots (two leaves and a bud) captured within a 0.3 m × 0.3 m grid that was randomly thrown on to the plucking tables, as outlined by Odhiambo (1989). The mean of three randomly selected bushes at every plucking round was taken and extrapolated to shoot density per m^2.

2.2 *Sampling, extraction, and quantification of catechins, caffeine, and total polyphenols*

Sampling for biochemical analysis was done once per season in the same year and shoots were processed into two products, namely, Black Curl, Tear and Cut (BCTC) and Green Orthodox (GO). After grinding the shoots, 0.2 g was placed in a 10 mL falcon tube labeled (a) and 5 mL of pre-heated 70% methanol was added. The samples were then incubated at 70°C for 10 minutes in a water bath and vortexed at 0, 5, and 10min intervals. Thereafter, the samples were cooled to room temperature, centrifuged at 3,500 rpm for 10 min and the supernatant extract decanted into a clean dry 10 mL falcon tube labeled (b). To the original tube (a), methanol extraction was repeated, the supernatant decanted to the new tube (b), and the volume brought to 10 mL with cold 70% methanol. Catechins and caffeine were analyzed as per ISO 14502-1:2005(E) standard, whereas total polyphenol content was determined following ISO 14502-1: 2003 standard.

High-performance liquid chromatography (HPLC) was used to assay for the tea catechins and caffeine. One milliliter of the sample extract was transferred into a graduated tube and diluted to 5 mL with a stabilizing solution (10% v/v acetonitrile with 500 μg/ml EDTA and ascorbic acid). The solution was further filtered through a 0.45 μm membrane filter. A 20 μL aliquot

Table 1. Summary of meteorological observations during the time of the study.

	Rainfall (mm)	Daily values of seasonal mean temperatures (°C)	Mean wind run at 2 m above ground (km/day)	Mean daily sun shine hours (hrs)
Warm/wet	201.4	23.3	74.7	6.2
Hot/dry	98.8	26.0	80.6	7.8
Cold/wet	234.9	23.1	69.0	5.8

of this solution was injected into HPLC. Reverse-phase HPLC analysis was utilized.

A Shimadzu LC 20 AD HPLC system fitted with a SIL 20AC auto-sampler and a SPD-M20A photodiode array detector with a glass LC10 chromatography application with a Gemini 5 μm C6- Phenyl, 250 mm × 4.6 mm (Phenomenex, Torrance, CA, USA) separation column, fitted with a Phenomenex Security Guard column (4mm × 3.0mm) Phenyl cartridge was used. A gradient elution was carried out using the following solvent systems: mobile phase A (acetonitrile/acetic acid/EDTA/double distilled water- 9/2/0.2/88.8 v/v/v/v) and mobile phase B (acetonitrile/acetic acid/EDTA/double distilled water- 80/2/0.2/17.8 v/v/v/v).

The mobile phase binary gradient conditions were 100% solvent for 10 minutes coupled with a 15 minute linear gradient to 68% mobile phase A, and 32% mobile phase B then held at this composition for 10 minutes. The mobile phase flow rate was set at 1.0 mL/min with the column temperature maintained at 35±0.5°C. Peak detection was performed at 278 nm. The identification of individual catechins was carried out by comparing the retention times and unknown peaks with peaks obtained from the mixed known standards of (−)gallocatechin (GC), (−)epigallocatechin (EGC), (+)catechin (C), (−)epicatechin (EC), (−)epigallocatechin gallate (EGCg), (−)picatechin gallate (ECg), and (−)gallocatechin gallate (GCg) from Sigma Aldrich, UK under the same chromatographic conditions. Quantitation of the catechins was done using consensus individual catechin relative response factor (RRF) values with respect to caffeine.

The finally total catechin content was determined by the following formula:

% Total catechin (TC) = %GC + %EGC + %C +%EC + %EGCg + %GCg + %ECg

Caffeine content was quantified by the following formula:

$$\%\text{Caffeine} = \frac{A_{sample} - A_{intercept} \times RRF_{std} \times V \times d \times 100}{Slope_{Caffeine} \times m \times 1000 \times DM} \quad (2)$$

where:

A_{sample} is peak area of the individual component in the test sample.

$A_{intercept}$ is peak area at the point of interception on y-axis.

$Slope_{caffeine}$ is caffeine calibration line slope.

V is sample extraction volume.

D is dilution factor.

M is mass in grams of test sample.

DM is dry matter content of test sample.

2.3 *Data analysis*

All the determinations were carried out in triplicate and the data were subjected to analysis of variance, and the means separated by the least significant difference (LSD) test, using Genstat 15th Edition.

3 RESULTS AND DISCUSSION

3.1 *Shoot Growth Rate (SGR)*

The highest SGR (1.47 mm per day) was achieved during the WW season in 30% shading regime while the lowest (0.81 mm per day) was recorded in the CW season in unshaded regime (Table 2). Although no significant difference was observed between the shading regimes, significant variations occurred between seasons (Table 2). Warm and wet season had significantly higher SGR (1.42 mm per day) compared to the other seasons (Table 2). Apart from the unshaded regime, which was exposed to hail damage (Figure 2), shoot growth rate reduced with increase in shading. This occurrence was due to moderate protection of the plants from adverse weather conditions. Sheppard and Sheppard (1976) found that in *Cornus stolonifera,* a woody species, that plant growth was optimized with 25% shading, but reduced with more shading.

3.2 *Shoot density*

Highest shoot density (224 shoots/m^2) during the CW season as compared to under 90% shading regime which had 92 shoots/m^{2}during the HD season (Table 2). There was a significant difference for the factors and their interactions ($P \leq 0.05$). Generally, 30% shading had the highest number of shoots compared to all other shading regimes. Seasonal variations in shoot density showed a significantly higher density in CW (198 shoots/m^2) season while HD had the least density (106 shoots/m^2) (Table 2). This can be explained by the fact that moderate shading offers favorable micro climate for bud formation. Shoot density was higher during CW, indicating that humid conditions coupled with low temperatures are ideal conditions for budding in tea plants. Semchenko,Lepik, Götzenberger, and Zobel (2012) noted in their study on perennial herbaceous species that moderate shade had a strongly facilitative effect on plant growth.

3.3 *Green leaf yield*

Monthly mean seasonal yields ranged between 32 kg.mt/ha for unshaded regime during the HD season to 339 kg.mt/ha for 30% shaded regime during CW season (Table 2). Significant difference was observed within the shading regimes and seasons and between their interactions. Overall, the lowest yields (201 kg.mt/ha) were recorded in the unshaded regime compared to the significantly higher yields (260 kg.mt/ha) recorded under 30% shading regime (Table 2), whereas significantly higher seasonal yields were observed during the WW season (254 kg.mt/ha) as compared to the HD season (122 kg.mt/ha) that had the lowest yields. The significantly low yields recorded in the HD season for the unshaded regime of 32 kg.mt/ha must have been caused by low rainfall precipitation and photoinhibition caused by high temperatures coupled with long sunshine hours. Further, Barua (1969) in a study conducted in North East India reported that tea yields in plants grown under 35% light intensity were higher than in plants grown in full sun. Fu et al. (2015) also noted prolonged shading might lead to reduced tea leaf biomass.

3.4 *Total catechins*

The total catechin (TC) contents in tea products ranged from 1.21% in BCTC processed from 90% shade regime during the HD season to 22.24% in GO processed from unshaded regime during CW season (Table 2). An increase in shading intensity subsequently reduced the accumulation of catechins, with 90% shading regime giving significantly lower TC (10.81%) levels compared to the unshaded regime (12.45%) (Table 2). The CW season accumulated maximum catechins (13.01%) compared to HD (10.7%) season that recorded the lowest level with higher TC content being recorded in the GO product (19.82%) compared to BCTC (3.19%) (Table 2). All variables and interactions were significantly ($P \leq 0.05$) different. Song et al. (2019) concluded that the concentration of total catechins was higher in unshaded than in shaded leaves, and lower in early January than at other times, indicating that catechin accumulation is proportional and inversely proportional to humidity and temperature, respectively. Similar observations by Astill, Birch, Dacombe, Humphrey, and Martin (2001) explained that catechin content of the CTC-manufactured black teas is lower as a function of the greater leaf disruption, and enzymatic oxidation during withering and fermentation in CTC manufacture results in the conversion of catechins to theaflavins and thearubigins.

3.5 *Caffeine content*

As depicted in Table 2, caffeine content was significantly ($P \leq 0.05$) influenced by all factors and their interactions. Caffeine content decreased significantly with an increase in shade intensity to a minimum of 2.59% under 90% shading regime for BCTC product, with the highest content of caffeine being recorded in the HD season (3.64%) under unshaded regime in GO product (Table 2). Caffeine content for GO product during HD season under 30 and 60% shading regimes was 2.64% and 2.61%, respectively, which is below Yatakamidori (Japanese green tea cultivar) with 2.67% (Kerio, Wachira, Wanyoko, & Rotich, 2013).

Caffeine content was affected positively by high temperature and dry conditions (Table 1). This is in order with the findings of Lee et al. (2010) that caffeine content was higher in tea under high temperature and long-time sun exposure. Kirakosyan et al. (2004) also explained that tea cultivars had high levels of caffeine during the dry season due to the accumulation of secondary metabolites such as caffeine by plants as a form of defense mechanism. In order to respond and adapt to environmental stresses, caffeine acts through allelopathy (Kim & Sano, 2008) as repellant (Chou & Waller, 1980) and signaling molecules to activate plant defense responses (Nathanson, Owuor, Netondo, & Bore, 1984).

3.6 *Total polyphenols*

Total polyphenols content analysis revealed a varied difference between shading regimes, seasons, and tea products. The amount of TP ranged from 18.95% for BCTC product of 30% shaded regime during the HD season to 24.70% for GO product of 90% shaded regime. The densest shading regime (90%) had the highest level of total polyphenols (22.35%) compared to unshaded regime (21.61%) (Table 2), supporting the idea of Lee et al. (2013) that individual phenolic compounds resulted in increased levels of total phenolic compounds in the shaded tea plants. Also, Wang et al. (2012) noted that there was a marked increase in concentration of phenolic acids in shaded leaves in tea. The content of total polyphenol was also affected by seasonal variations, with CW season having the highest level (23.04%) followed by WW and finally HD, showing that TP content is elevated by high humidity and low temperatures (Table 2). This finding was also inferred from the results of Ghabru and Sud (2017), i.e., that the synthesis of TP was assisted by minimum temperature and evaporation. Cherotich et al. (2013) also explained that polyphenols content are lower in the dry season compared to the wet season in most tea cultivars, because water is one of the raw materials for photosynthesis and it has direct impacts on the organic synthesis of plants of both the primary and the secondary metabolites. Generally, GO product registered higher content of total polyphenols (23.45%) (Table 2). Relatively lower levels of polyphenols in black tea can be attributed to the conversion of tea polyphenols into

Table 2. Effects of different shade intensity and seasons on yield and biochemical contents in tea.

Shading regime	Season	Shoot growth rate	Shoot density/m^2	Monthly mean seasonal yields (kg.mt/ha)	Total catechin %		Caffeine %		Polyphenols %	
					BCTC	GO	BCTC	GO	BCTC	GO
0%	WW	1.46	128	295	3.56	19.63	2.96	2.98	19.97	23.03
	HD	0.84	106	32	3.26	19.38	3.38	3.64	19.23	22.70
	CW	0.81	183	275	6.66	22.24	3.13	3.26	20.50	24.23
30%	WW	1.47	120	274	3.47	19.32	2.80	2.93	22.80	23.37
	HD	1.01	126	165	1.67	18.68	3.18	3.64	18.90	22.87
	CW	1.11	224	339	4.15	21.97	2.76	2.64	21.33	23.87
60%	WW	1.43	104	230	2.99	19.38	2.83	2.79	22.03	23.23
	HD	1.12	100	146	1.67	18.31	3.18	3.47	18.93	22.97
	CW	0.98	203	322	3.81	21.18	2.76	2.61	21.90	24.27
90%	WW	1.32	94	214	2.58	19.01	2.69	2.69	20.27	23.60
	HD	1.17	92	143	1.21	18.02	3.01	2.86	19.03	23.00
	CW	0.86	182	253	3.29	20.77	2.59	2.61	23.50	24.70
CV%		17.6	3.3	9.8	0.1		0.3		3.1	
Overall means (0%)		1.04	139	201	12.45		3.23		21.61	
	(30%)	1.20	157	260	11.54		2.99		22.19	
	(60%)	1.18	136	233	11.22		2.94		22.22	
	(90%)	1.11	122	204	10.81		2.74		22.35	
Season	(WW)	1.42	111	254	11.24		2.83		22.29	
	(HD)	1.03	106	122	10.27		3.30		20.95	
	(CW)	0.94	198	297	13.01		2.79		23.04	
Product	(BCTC)				3.19		2.93		20.70	
	(GO)				19.82		3.01		23.49	
$P \leq 0.05$	(SR)	NS	4.4	21.4	0.009		0.007		0.46	
	(S)	0.17	3.8	18.5	0.007	0.006	0.40			
	(P)	NS	7.6	37.0	0.006	0.005	0.33			
	(SR.S)				0.013	0.012	0.80			
	(SR.P)				0.011	0.010	NS			
	(S.P)				0.010	0.008	0.56			
	(SR.S.P)				0.019	0.017	1.13			

WW, warm/wet; HD, hot/dry; CW, cold/wet; SR, shading regime; S, season; P, product; SR.S, shading regime*seasons; SR.P, shading regime *products; SR.S.P, shading regime* season*product; BCTC, black cut, tear and curl; GO, green orthodox and NS, not significant.

theaflavin and thearubigins during the fermentation process (Fernando & Soysa, 2015).

4 CONCLUSION AND RECOMMENDATION

Shoot growth rate, monthly mean seasonal yields, catechin content, and caffeine content were negatively affected by the increase of shading intensity in tea while total polyphenol content was positively affected by an increase in shading. The different trends depicted by total catechin content and total polyphenol contents could indicate that there are other phenolic compounds which were affected differently from the catechins used in this study. Product diversification can be adopted in existing tea populations in the production of unaerated tea under moderate shading (30% and 60%) during the cold/wet season and the production of high-quality aerated product during warm/hot seasons of the year while maintaining optimum yields. There is a need to undertake genomic study in future to understand the patterns of gene expression so as to provide insights into complex regulatory networks and the identification of genes relevant to biochemical changes due to shading in tea. The study was limited at the time of setting up the trial by the unavailability of a white shading net which could have acted as a positive control.

ACKNOWLEDGMENT

We acknowledge contribution of Kenya Agricultural and Livestock Research Organization-Tea Research Institute, National Research Fund and African Center of Excellence in Phytochemicals, Textiles and Renewable Energy for assisting in the study and publication of the findings.

REFERENCES

Astill, C., Birch, M. R., Dacombe, C., Humphrey, P. G., & Martin, P. T. (2001). Factors affecting the caffeine and polyphenol contents of black and green tea infusions. *Journal of agricultural and food chemistry, 49*(11):5340-5347. https://doi.10.1021/jf010759.

Barua, D. N. (1969). Light as a factor in metabolism of the tea plant (Camellia sinensis L.). In *Long Ashton Symp, 2d, Univ of Bristol.*

Cherotich, L., Kamunya, S. M., Alakonya, A., Msomba, S. W., Uwimana, M. A., Wanyoko, J. K., & Owuor, P. O. (2013). Variation in catechin composition of popularly cultivated tea clones in East Africa (Kenya). In *JKUAT ANNUAL SCIENTIFIC CONFERENCE PROCEEDINGS:* 121–141. http://hdl.handle.net/123456789/3405.

Chou, C. H., & Waller, G. R. (1980). Possible allelopathic constituents of Coffea arabica. *Journal of chemical ecology*, *6*(3):643–654. https://doi.org/10.1007/BF00987675

Eden, T. (1976). *Tea*. Longman Group Limited.

Fernando, C. D., & Soysa, P. (2015). Extraction Kinetics of phytochemicals and antioxidant activity during black tea (Camellia sinensis L.) brewing. *Nutrition journal*, *14*(1):74. https://doi.org/10.1186/s12937-015-0060-x.

Fu, X., Chen, Y., Mei, X., Katsuno, T., Kobayashi, E., Dong, F., Naoharu, W.., & Yang, Z. (2015). Regulation of formation of volatile compounds of tea (Camellia sinensis) leaves by single light wavelength. *Scientific reports*, 5: 16858. https://doi.org/10.1038/srep16858.

Ghabru, A., & Sud, R. G. (2017). Variations in phenolic constituents of green tea [Camellia sinensis (L) O Kuntze] due to changes in weather conditions. *J Pharmacogn Phytochem*: 1553–1557. E-ISSN: 2278-4136.

Hirai, M., Yoshikoshi, H., Kitano, M., Wakimizu, K., Sakaida, T., Yoshioka, T. & Maki, T. (2008, October). Production of value-added crop of green tea in summer under the shade screen net: Canopy microenvironments. In *International Workshop on Greenhouse Environmental Control and Crop Production in Semi-Arid Regions 797*: 411–417. https://doi.10.17660/ActaHortic.2008.797.59.

Kerio, L. C., Wachira, F. N., Wanyoko, J. K., & Rotich, M. K. (2013). Total polyphenols, catechin profiles and antioxidant activity of tea products from purple leaf coloured tea cultivars. *Food chemistry*, *136*(3–4): 1405 1413. https://doi.org/10.1016/j.foodchem.2012.09.066.

Kim, Y. S., & Sano, H. (2008). Pathogen resistance of transgenic tobacco plants producing caffeine. *Phytochemistry*, *69*(4):882–888. https://doi.org/10.1016/j. phytochem.2007.10.021.

Kirakosyan, A., Kaufman, P., Warber, S., Zick, S., Aaronson, K., Bolling, S., & Chul Chang, S. (2004). Applied environmental stresses to enhance the levels of polyphenolics in leaves of hawthorn plants. *Physiologia plantarum*, *121*(2):182–186. https://doi.org/10.1111/j.1399-3054.2004.00332.x

Lee, J. E., Lee, B. J., Chung, J. O., Hwang, J. A., Lee, S. J., Lee, C. H., & Hong, Y. S. (2010). Geographical and climatic dependencies of green tea (Camellia sinensis) metabolites: a 1H NMR-based metabolomics study. *Journal of agricultural and food chemistry*, *58*(19):10582–10589. https://doi.10.1021/jf102415m.

Lee, L. S., Choi, J. H., Son, N., Kim, S. H., Park, J. D., Jang, D. J., & Kim, H. J. (2013). Metabolomic analysis of the effect of shade treatment on the nutritional and sensory qualities of green tea. *Journal of agricultural and food chemistry*, *61*(2):332–338. https://doi.10.1021/jf304161y.

Liu, G. F., Han, Z. X., Feng, L., Gao, L. P., Gao, M. J., Gruber, M. Y., & Wei, S. (2017). Metabolic flux redirection and transcriptomic reprogramming in the albino tea cultivar 'Yu-Jin-Xiang'with an emphasis on catechin production. *Scientific reports*, *7*:45062. https://doi.org/10.1038/srep45062.

Nathanson, J. A. (1984). Caffeine and related methylxanthines: possible naturally occurring pesticides. *Science*, *226*(4671):184–187. https://doi.org/10.1126/science.6207592.

Nyabundi, K. W., Owuor, P. O., Netondo, G. W., & Bore, J. K. (2016). Genotype and environment interactions of yields and yield components of tea (Camellia sinensis) cultivars in Kenya. https://repository.maseno.ac.ke/handle/123456789/291.

Odhiambo, H. O. (1989). Nitrogen rates and plucking frequency on tea: The effect of plucking frequency and nitrogenous fertilizer rates on yield and yield components of tea in Kenya. *Tea-Tea Research Foundation (Kenya)*.

Semchenko, M., Lepik, M., Götzenberger, L., & Zobel, K. (2012). Positive effect of shade on plant growth: amelioration of stress or active regulation of growth rate? *Journal of ecology*, *100*(2), 459–466. https://doi.org/10.1111/j.1365-2745.2011.01936.x.

Sheppard, R., & Sheppard, R. I. (1976). Light intensity effects on redosier dogwood. Song, K. E., Jeon, S. H., Shim, D. B., Jun, W. J., Chung, J. W., & Shim, S. (2019). Strong Solar Irradiance Reduces Growth and Alters Catechins Concentration in Tea Plants over Winter. *Journal of Crop Science and Biotechnology*, *22*(5):475–480. https://doi.org/10.1007/s12892-019-0215-0.

Song, K. E., Jeon, S. H., Shim, D. B., Jun, W. J., Chung, J. W., & Shim, S. (2019). Strong Solar Irradiance Reduces Growth and Alters Catechins Concentration in Tea Plants over Winter. *Journal of Crop Science and Biotechnology*, *22*(5):475–480. https://doi.org/10.1007/s12892-019-0215-0.

Tong, X., Taylor, A. W., Giles, L., Wittert, G. A., & Shi, Z. (2014). Tea consumption is inversely related to 5-year blood pressure change among adults in Jiangsu, China: a cross-sectional study. *Nutrition journal*, *13*(1):98. https://doi.org/10.1186/1475-2891-13-98.

Wang, Y., Gao, L., Shan, Y., Liu, Y., Tian, Y., & Xia, T. (2012). Influence of shade on flavonoid biosynthesis in tea (Camellia sinensis (L.) O. Kuntze). *Scientia horticulturae*, *141*:7–16. https://doi.org/10.1016/j.scienta.2012.04.013

Advances in Phytochemistry, Textile and Renewable Energy Research for Industrial Growth – Nzila et al. (Eds)

Colorimetric study of natural dye from *Beta vulgaris* peels and pomace on cellulosic substrate

Vincent Rotich*
Department of Chemistry & Biochemistry, School of Sciences and Aerospace Studies, Moi University, Eldoret, Kenya

Phanice Wangila
Department of Chemistry & Biochemistry, School of Sciences and Aerospace Studies, Moi University, Eldoret, Kenya
Department of Physical Sciences, School of Science & Technology, University of Kabianga, Kabianga, Kenya

Jackson Cherutoi
Department of Chemistry & Biochemistry, School of Sciences and Aerospace Studies, Moi University, Eldoret, Kenya

ABSTRACT: Synthetic dyes are associated with carcinogenic, toxic, and allergic effects on humans and our environment. Natural dyes have attracted attention globally because of their non-hazardous nature. *Beta vulgaris* (Beetroot) plant wastes such as peels and pomaces are an unexploited resource. The present study involved solvent extraction of natural dye from *B. vulgaris* peels and pomace, and its application on cellulosic fabrics alongside natural mordants (alum and tannic acid) in comparison to metallic mordants (potassium dichromate, ferrous sulfate, and copper sulfate) to improve the colour fastness of the cotton substrate and establish colour strength equivalence (ceq) relating to synthetic Reactive Orange HER. In mordanting, the three methods (pre-mordanting, simultaneous, and post-mordanting) were employed. Response surface methodology and central composite design were used to optimize extraction and dyeing conditions, namely temperature, M:L ratio, time, and pH. The optimized extraction conditions were M:L ratio of 1:20 and time of 11 hours. This resulted in a moderate (40%) yield of natural dye from the plant, proving to be better than conventional methods. Optimized dyeing resulted in temperature 55′C, time 75 minutes, and pH 6. The CIE L*, a*, b*, C*, and h° values were studied by standard methods. The dyed fabrics exhibited very good to excellent colour fastness test (light, washing, rubbing, and perspiration fastness) in the range of 4–5 in gray scale. These findings reveal that *B. vulgaris* peels and pomace can be potential alternatives to synthetic dyes in the colouration of cotton fabrics.

1 INTRODUCTION

Up to the late 19th century people were using natural dyes for the colouring of textiles (Frose, Schmidtke, Sukmann, Sukmann, & Ehrmann, 2018; Yusuf et al., 2017). Europeans applied archaic dyeing technique involving sticking plants to fabric and rubbing crushed pigments onto clothes (Ado, Yahaya, Kwalli, & Abdulkadir, 2014). Synthetic dyes emerged when in 1856 a teenager William Perkin accidentally discovered a dye called mauve while trying to make quinine in his home lab. People then liked synthetic dyes due to the good repeatability of shade and brilliance in colour performance (Stewart, 2017). The introduction of synthetic dyes led to an almost complete displacement of natural dyes. Synthetic dyes have been produced from non-renewable and non-biodegradable petrochemicals. Approximately 30 million tons of dyes are being consumed globally in textile industries, with about 70,000 tons being released to the environment (Yusuf, Shabbir, & Mohammad, 2017). According to Business Week, humans who are allergic to textile auxiliary chemicals will rise by up to 60% by 2020 (Arora, Arora, & Gupta, 2017; Chaudhry et al., 2019).

The interest in the use of synthetic dyes has reversed and is in rapid decline due to the result of adverse environmental effects. This has compelled many countries to impose sanctions against the synthetic dyes as a result of the carcinogenicity, toxicity, and allergic reactions associated with them, such as toxic amines (Li et al., 2015). Their visible residues in effluents are also a menace to the ecosystem (Yamjala, Yamjala, & Ramisetti, 2016). In addition, costly methods are employed to eliminate them from the environment. More eco-friendly dyes are now being developed to replace synthetic dyes that are toxic and hazardous to health (Li et al., 2015). Researchers and industrialists are actively involved in this ecological revolution (Khan et al., 2014; Rather et al., 2016; Uddin, 2015; Yusuf, Mohammad, Shabbir, & Khan, 2016). For instance, the German Act of 1994 forbidding azo dyes and some eco-labelling standards for

*Corresponding author

DOI 10.1201/9781003221968-25

textiles such as Oeko-Tex Standards 100 led to technological innovation and the domestic development of environment-friendly dye substitutes (Almahy & band Ali, 2013).

Natural dyes are vastly superior to synthetic dyes in that they age well and develop a patina, soft, lustrous, and soothing shades to the human eye (Kusumawati, Santoso, Sianita, & Muslim, 2017; Sanjeeda et al., 2014). They are non-pollutants, non-toxic, non-carcinogenic, easy to handle, and biodegradable (Alsehri, Naushad, Ahamad, Alothman, & Aldalbahi, 2014; Ghoulia et al., 2012). They are obtained from renewable sources viz. plants, animals, and minerals (Yusuf et al., 2017). For example, in plants they are derived from roots, barks, leaves, fruits, and flowers. Natural dyes also contain bioactive compounds beneficial to our health. Many natural dyes have UV protective (Simpson, Simpson, & Aytug, 2015), antimicrobial (Wangatia, Wangatia, & Moyo, 2015), and deoxidizing (Baião et al., 2017) properties. The limitations of the study were that the textile industry has not readily considered the use of natural dyes in their processes as there are no standard shade cards, application procedures, set conditions, and parameters, thus they frequently require extraction and process optimization. The apparatus and equipment for extraction and analysis are limited in availability or expensive. Moreover, colour fastness performance ratings are inadequate for modern textile usage. Natural dyes require mordants to fix, modify, and hold colour onto the fabric. Mordants can be metallic or biomordants (Rather et al., 2016).

Beet (*Beta vulgaris* L. ssp. vulgaris) is a flowering and true biennial, hence rarely perennial plant (see Figure 1). The plant is well spread and widely cultivated in Europe, America, and throughout Asia (Chawla, Parle, Sharma, & Yadav, 2016).

Optimization of extraction and dyeing conditions is very important in order to obtain maximum colouring property from the extract as well as optimum dyeing results. Response surface methodology (RSM) is a common statistical approach for process optimization, modelling, and establishing the effects of the interaction of several factors concurrently. Central composite design (CCD) is a very useful process optimization model tool in RSM (Sun et al., 2010). The aim of this study was to investigate the effect of extraction pH, time, and material to liquor ratio on the extraction of natural dye from pomegranate using CCD of RSM in Minitab 17 statistical software. The effect of extraction time and M:L ratio has not been previously reported for the extraction of colour from *B. vulgaris* peels and pomace. Similarly, dyeing parameters (temperature, pH, and time) have not been studied in *B. vulgaris* peels and pomace. CCD developed 14 and 20 experiments for extraction and dyeing, respectively. MS Excel software was used for regression analysis by way of ANOVA for the levels of input variables that influence and optimize a response (Hamanthraj, Desai, & Bisht, 2014). The extraction yield and relative colour strength were placed as responses in regression for extraction and dyeing, respectively.

Figure 1. Beetroot.

Therefore, this research work mainly focused on the determination of the optimum crude dye yield from *B. vulgaris* and the chemistry it has on cellulosic substrates with and without mordant while applying optimized dyeing conditions. Colour strength equivalences to commercial Reactive Orange HER were determined by applying the method according to Thomas Bechtold et al. (Thomas Bechtold, et al., 2006). *B. vulgaris* is believed to be the main commercial source of betalains (Baião et al., 2017). The bioactive antioxidants and colour-giving compounds may be present in the waste part of beetroot such as peel (Singh, Ganesapillai, & Gnanasundaram, 2017), pomace (Kushwaha, Kumar, Vyas, & Kaur, 2018), and stalk (Maran and Priya, 2016). The red and orange pigments in *B. vulgaris* make it ideal to compare with Reactive Orange HER for colour equivalence. Therefore, this research seeks to establish the best extraction and dyeing conditions of *Beta vulgaris* by utilizing its wastes as a potential source of natural dye alongside the usage of mordants to improve the colour.

2 MATERIALS AND METHODS

2.1 *Chemicals, reagents, and materials*

All chemicals used in the experiments were of analytical grade and were used without further purification: methanol, formic acid, alum, hydrochloric acid, acetic acid (Loba Chemie); iron(II) sulfate, copper(II) sulfate pentahydrate, and potassium dichromate (Blulux); tannic acid, Gulbar salt, or sodium sulfate, sodium carbonate (Narcolab); Commercial Reactive Orange HER (Roop). *B. vulgaris* plant peels and pomace for extraction material were collected from hotels, restaurants, and fresh juice dealers around Eldoret town. Scoured and bleached cotton fabric was obtained from Rivatex East Africa Ltd, Eldoret. Purposeful and stratified random sampling methods were applied.

2.2 *Equipment*

Rotary vacuum evaporator (Hahnvapor Rotary Evaporator HS-2005S, Germany), DU 720 UV–Vis spectrophotometer (Beckman Coulter, USA), standard fabric sample cutter/swatch master, absorbency spray master, dyeing water bath, 5 L colour matching booth, motorized crock master, MBTL sun master (Paramount Instruments Pvt Ltd, India) and Perspirometer (SDL Atlas, USA), SP60 Spectrophotometer (X-Rite, USA).

2.3 *Selection of the extraction experimental variables and statistical modelling*

Two experimental variables, i.e., time of dye extraction and material to liquor ratio (M:L) were considered in extraction. The experimental variables and their levels are given in Table 1.

Table 1. The factor levels and variables.

Variables		Coded factor levels				
Symbol	Name	$-\alpha$	-1	0	$+1$	$+\alpha$
A	Time	5.17	6	8	10	10.8
B	M:L	1.89	5	10	20	23.1

2.4 *Extraction and concentration of betalain pigment*

The fresh peels and pomace of *Beta vulgaris* were sliced into small pieces, placed on a layer of wax paper and dried in an oven at 40°C for 8–10 hours until the samples were completely dry and crunchy. The dried plant materials were then ground using a pestle and mortar into very fine powder. Finely sieved *B. vulgaris* powder (20 g) were weighed and carefully placed in a 20 g Whatman cellulose extraction thimble. Methanol was measured into a 500 mL Schlenk flask of the Soxhlet apparatus and extraction was carried out at the boiling point of methanol (68° C). The M:L ratios (plant powder to solvent) and time of extraction were performed according to the experimental runs obtained from Central Composite Design. Due to the thimble size, the plant quantity was maintained at 20 g in all the experiments. The stability of betanin depends directly on its pH; the optimum pH being between 4 and 5 (Antigo, Rita de Cássia Bergamasco, & Grasiele Scaramal Madrona, 2017). Hence constant pH of 4 was maintained in this study by applying a few drops of formic acid. The dye extracts obtained from each of the experiments were filtered using Whatman No. 1 filter paper and concentrated by Vacuum Rotary Evaporator to obtain crude extracts.

2.5 *Spectrophotometry analysis and total betalains content calculation*

The spectrophotometric analysis for quantification of the main colour compounds was done by using DU 720 UV–Visible spectrophotometer. The absorbance values were recorded at 535 nm for betalain compound (Singh et al., 2017). Appropriate dilutions were done for each type of measurements with distilled water. The concentrations of the respective betalain compounds were calculated according to equation (1) as determined by Ravichadran and co-workers (2013).

$$\text{Total betalains content (mg/100 g)} = A \times DF \times MW \times 1\,000/€L \quad (1)$$

where A is the absorption value at 535 nm density; DF is the dilution volume; MW is molecular weight of betalain (550g/mol); € is the extinction coefficient for betalain 60000 L/mol; and L is the path length of the cuvette (Singh et al., 2017).

2.6 *Statistical modelling of extraction process*

A second-order linear polynomial equation was used to express the betalain content in each extraction as a function of independent variables which affect extraction.

$$\text{Response, } y = b_0 + (b_1F_1) + (b_2F_2) + (b_3F_3) + (b_4F_1F_1) + (b_5F_2F_2) + (b_6F_3F_3) + (b_7F_1F_2) + (b_8F_1F_3) + (b_9F_2F_3) \quad (2)$$

where y (response) represents the betalain content (unit: mg/g), while b and F are the coefficients and factor variables, respectively.

2.7 *Statistical modelling and application of dye on cotton fabric*

This is an experimental design with three factors, namely, temperature, pH, and dyeing time (Table 2). Equation (2) was applied where y relative colour strength is the response. The dyeing method reported by Hong (2018) was performed in a laboratory scale water bath dyeing machine at pre-optimized dyeing conditions, i.e., fabrics must be desized, scoured, bleached, and mercerized. Preparation of dye stock solution was done as described by Bukhari et al. (2017). The control measures were unmordanted fabrics and reactive dyeing (Reactive Orange HER). The dyed cotton fabrics were then washed with distilled water to get rid of excess dye molecules at the surface of cotton fabrics (Ding & Freeman, 2017).

Table 2. Experimental variables and their levels for dyeing.

Variables		Coded levels				
Symbol	Name	$-\alpha$	-1	0	$+1$	$+\alpha$
A	Temp (°C)	−2.155	20	55	90	112.16
B	Time (mins)	−1.515	30	75	120	148.49
C	pH	−2.4175	4	6.5	9	10.5825

2.8 *Mordanting*

Mordanting was then performed following optimized dyeing conditions using natural mordants and metallic mordants at varying concentrations (20–50% of weight of fabric (Rather et al., 2016) following the three methods of mordanting.

2.9 *Estimation of colour strength equivalence, (Ceq)*

Colour strength equivalence between the natural dye and Reactive Orange HER synthetic dye was estimated by using Equation (3) (Bechtold, et al., 2006).

$$\text{Ceq} = \frac{\text{Aextr}}{\varepsilon\ \text{RO84d}} \tag{3}$$

where Ceq is colour strength equivalence of Reactive Orange HER (g dm-3); Aextr is absorbance of dye plant extract; εRO84 is extinction coefficient of Reactive orange HER (7.94 $dm^3\ cm^{-1}\ g^{-1}$ at $\lambda = 491$ nm); and d is path length of cuvette (cm). The absorbance value for the natural dye (Aextr) was multiplied by the corresponding dilution factor (DF).

2.10 *Colour fastness properties tests*

The colour fastness to washing and light of pre-conditioned dyed cotton fabrics was determined according to American Association of Textile Chemists and Colourists (AATCC) Test Method 61 (2009) and AATCC Test Method 16 (2004), respectively. Colour fastness to crocking of dyed cotton fabrics (dry and wet rubbing) and perspiration (alkaline and acidic) were determined according to AATCC Test Method 8 (2001).

2.11 *Evaluation of CIELab colour coordinates*

The characteristic CIELab colour coordinates, that is (light/darkness (L*), tones (a* & b*), and ΔE*ab of the 10 cm^2 dyed samples were determined by SP60X spectrophotometer colour meter (Khan, Hussain, & Jiang, 2018). Both unmordanted and mordanted natural dyed cellulosic fabrics were automatically illuminated under D65 10 as described by Rather et al. (2016).

3 RESULTS AND DISCUSSION

3.1 *Response surface and optimization of extraction conditions*

The interaction of adsorbents (dyes and mordants) with the adsorbent materials (cotton fibre) is discussed in this study. Optimization is critical to determine the best adsorption design mechanism pathways and systems (Rather et al., 2016).

The plant powder quantity (20 g) and pH (4) value were applied constantly in all the runs. Fitting equation (2) betanine as a function of independent variable using time (X1) of extraction and M:L ratio (X2) response, y (yield/betanine) was obtained in a linear model as;

$$y = 193.15 + 13.66X_1 + 1.11X_2 \tag{4}$$

3.2 *Effect of time and M:L ratio on extraction yield and total betalain content*

From Equation (4) and Table 3 results, there is a direct correlation between extraction yield and the variables affecting extraction. The longer the time of extraction, the higher the extraction yield and quantity of betalain, and vice versa (Figure 2). Very long extraction

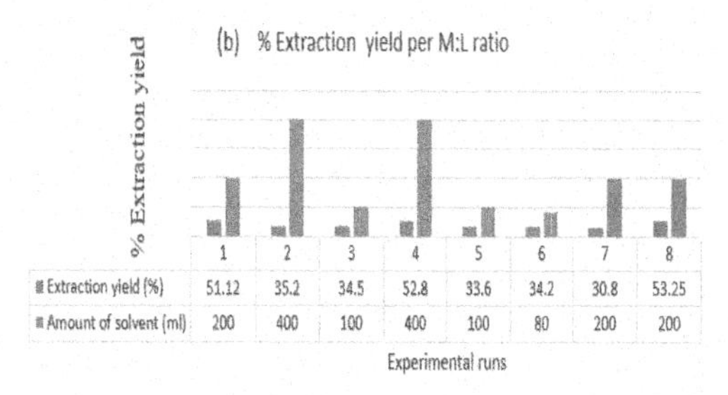

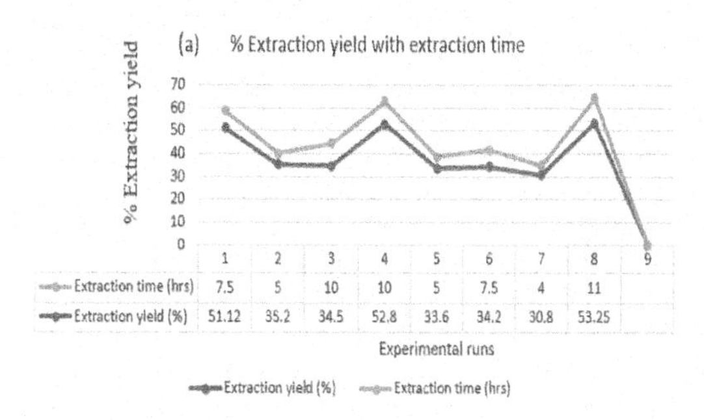

Figure 2. Effect of extraction time (a) and M: L ratio (b) on % extraction yield.

Table 3. Central composite design for betalain pigment quantification.

Runs	Amount of solvent (ml); (M:L Ratio)	Extraction time (hrs)	A max, 535–537 nm	Betalains (mg/100 g)	Yield (%)
1	200	7.5	0.82	375	51.12
2	400	5	0.68	311	35.2
3	100	10	0.65	297.9	34.5
4	400	10	0.79	362	52.8
5	100	5	0.66	302.5	33.6
6	80	7.5	0.78	357.5	34.2
7	200	4	0.67	307	30.8
8	200	11	0.81	371.3	53.25

Table 4. ANOVA for dye extraction.

Properties	df	SS	MS	F	Significance F	p-value
Intercept	2	6533.2761	3266.638	6.1545	0.01608	0.00022
A (Time, hrs)	11	5838.5010	530.7728			0.00641
B (M:L ratio)	13	12371.777				0.32811
Regression Statistics						
Multiple R	0.72669					
R^2	0.52808					
Adjusted R^2	0.44228					
Standard Error	23.0385					

time leads to degradation of betalain. Large quantities of solvent with less time limits the interaction of the solvent with plant molecules results in a low betalain yield. The content of total betalains averaged 335 mg/100 g on fresh weights. The result obtained was in accordance with Zakharova and Petrova (1997), who found that the total betalain content of red beet was 250–850 mg/100g on fresh weight.

By way of ANOVA (Table 4) the R2 value shows there was a moderate positive relationship among the variables, hence the regression model fits the data. That is 52% of variance in betalain content was attributed to the variables. The slope (b_0) was 193.15, showing that the expected change of response, y by one-unit change of variables is immense. The significant F value is less than 0.05. Therefore the relationship is significant. Time is very significant from the p-value ($p < 0.05$), while the M:L ratio (amount of plant material to solvent) has $p < 0.05$ demonstrating it to be a weak evidence or insignificant to extraction. The slope (b0) has a low p value (<0.05) meaning it is different from zero. This generated the optimized extraction conditions shown in Table 5.

Table 5. Optimized extraction conditions.

Parameter	Optimized conditions
Amount of solvent (M:L)	1:10
Extraction time (hrs)	11

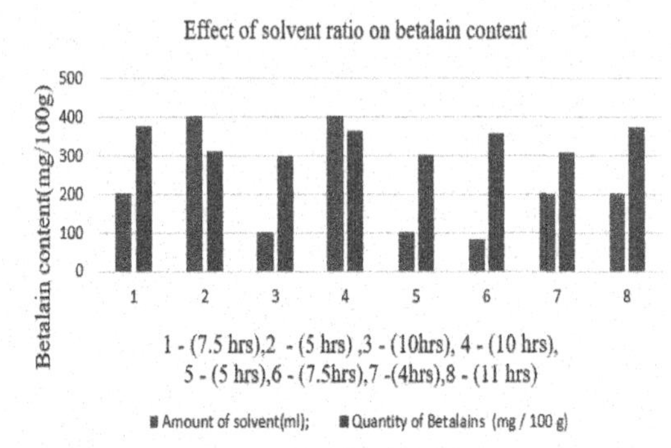

Figure 3. Comparison of time (hrs) and M:L ratio with quantity of betalains.

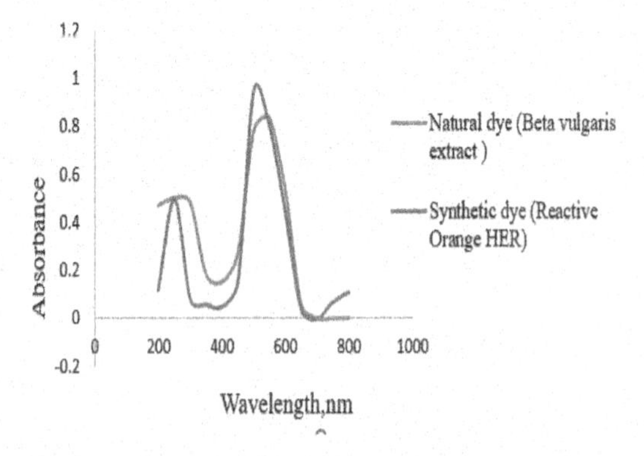

Figure 4. Absorption spectra of *B. vulgaris* and reactive orange HER.

3.3 *UV–Vis analysis*

UV–Visible spectral analysis was used to identify the chromophoric groups present in dye molecules that are responsible for enhanced chemical interactions (Bukhari et al., 2017; Rather et al., 2016a, b). The UV–Visible spectrum of *B. vulgaris* plant extract and that of Commercial Orange HER is presented in Figure 4 with λ max at = 492 nm and 534 nm respectively. The peak at 534 nm is attributed to the betanin and this finding is similar to that of Hernandez-Martinez, Hernandez, Vargas, and Rodríguez (2013). Absorbance peaks at 300 nm and 534 nm are characteristic for the red violet betalain group, betacyanin (Singh et al., 2017). Reactive Orange HER dye has a shoulder, indicating it is a non-homogeneous mixture. The absorption in the 300–800 nm region is attributed to various chromophores responsible for colour as well as conjugate systems (Saxena, Tiwari, & Pandey, 2012). Their close λ max demonstrate that their colour intensities are also very close.

3.4 *Optimization and statistical modelling of dyeing conditions of cotton fabrics*

With the use of un-coded values of parameters and relative colour strength as y (response), Equation 2 was fitted to form a regression equation as:

$$y = 5.845 + 0.00828X_1 + 0.01308X_2 - 0.00087X_3 \quad (5)$$

The regressed variables (temp, X_1; time, X_2; and pH, X_3) are demonstrated in Table 5. The p-values of the variables were greater than the common p-level of 0.05, which indicated that they were not statistically significant. There are other factors which also affected dyeing and colour strength such as dye concentration and mordanting. Therefore, the use of mordants to fix colour onto fabrics and modify colour was required according to Kulkarni, Gokhale, Bodake, and Pathade (2011) and Samantha et al. (2011). The adjusted R^2 values were well within the acceptable limits indicating a good fit model. The response surface plot represented in Figure 5 reveals a non-linear (zig-zag) relation between the response and the variables. The investigated variables have some effect on dyeing in terms of linear, quadratic, and cross terms leading to optimized dyeing conditions in Table 6. Increase in dye bath concentration leads to more dye transfer to the fabric, hence a higher apparent depth of colour (Rather et al., 2016).

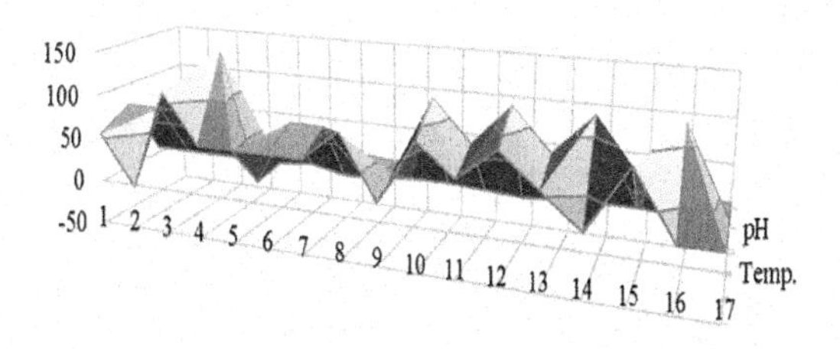

Figure 5. Response surface plot for dyeing process.

3.5 *Effect of mordants on relative colour strength and absorption*

Natural dyes need metal ions for adsorption of the dye by forming an insoluble composition precipitate on the surface of the fibres, and this gives the fabric a range of bright colours (Uddin, 2015). Mordants fix and modify colour onto fabrics as well as enhancing the colour fastness of dyed fabrics (Kulkarni et al., 2011; Samantha et al., 2011). *B. vulgaris* dye is water soluble and contains carbonyl and hydroxyl groups, hence it can interact with cotton fabric via hydrogen bonding (Sarhan, & Salem, 2018). Natural mordants of alum-tannic acid gave the best shades with post-mordanting (see Figure 6). Aluminium ions have a strong affinity for cellulose and readily serve as a bridge between multiple dye molecules and/or between the fibre and dye. It was observed that the colour strength increases gradually with a dyeing time of up to 75 minutes and decreased when the time increased further (Brahma, Islam, Shimo, & Shimo, 2019).

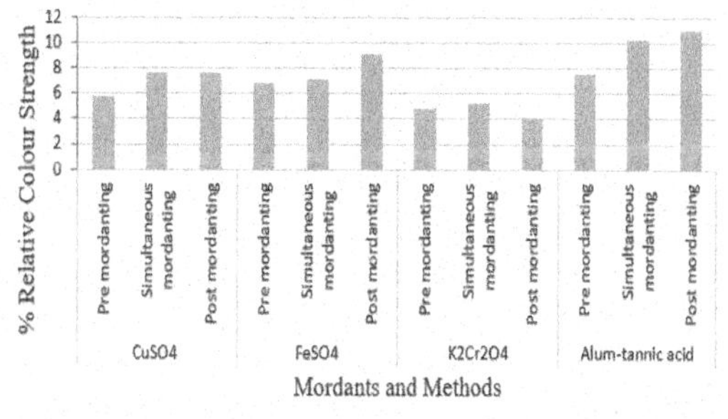

Figure 6. Effect of mordant concentration and methods on relative colour strength.

The post-mordanting method showed the best result, as shown by its chroma (C*) and relative colour strength percentages, especially by using alum and tannic acid as well as iron(II) sulfate, having 11% and 9.1%, respectively, as seen in Table 7. Dye molecules from solution diffuse into the fabric matrix, followed by adsorption of dye molecules at the surface of the fabric. The adsorbed dye molecules are then absorbed into the fabric matrix, followed by chemical fixation of dye molecules onto the mordanted fabric (Bukhari et al., 2017).

Table 7. Optimized dyeing conditions.

Parameter	Optimal values
Temperature (°C)	55
Time (minutes)	75
pH	6.5

Table 6. ANOVA for dyeing process.

Properties	df	SS	MS	F	Significance F	p-value
Intercept	3	5.74209	1.9140	0.4833	0.69962	0.01026
Temp.,^{0}C	13	51.4885	3.9607			0.60370
Time, hrs	16	57.2306				0.29970
pH						0.99684
Regression Statistics						
Multiple R	0.316753					
R^2	0.100333					
Adjusted R^2	−0.10728					
Standard Error	1.990139					

3.6 *Colour fastness properties tests*

See Table 8.

Table 8. Wash fastness rating to colour change and staining.

		Washing IS 105-C06:2010		Light Rubbing IS 105-D02:2000				IS 105-B02:2000		Perspiration IS 105-E04:2000			
				Dry		Wet				Acid		Alkaline	
Mordant	Method	C	S	C	S	C	S	C	S	C	S	C	S
$CuSO_4$	Pre mordanting	3/4	4	4	4	3/4	3/4	4	3	4	4	3/4	3/4
	Simultaneous	4	4	4	4	4	4	4	4	4	4	4	4
	Post mordanting	4	4/5	5	4/5	4	4/5	4	4/5	4	4/5	4	4/5
$FeSO_4$	Pre mordanting	4	4/5	4	4/5	4	4	4	3/4	2/3	4/	3	4/5
	Simultaneous	4/5	4	4/5	4	4	4	4/5	4	4/5	4	4	4
	Post mordanting	4	4/5	4	4/5	4	4	4	4/5	3/4	4/5	4	4/5
$K_2Cr_2O_4$	Pre mordanting	3	3/4	3	3/4	3	3/4	3	3/4	4	3/4	3/4	4
	Simultaneous	4	4	4	4	4	4	4	4	4	4	4	4
	Post mordanting	3	3	3	4	3/4	4	3	4	3	4	3	4
Alum-tannic acid	Pre mordanting	4	4	4	4	4	4	4	4	4	4	4	4
	Simultaneous	4	4	4	4/5	3/4	3/4	4/5	4	4	4/5	3/4	3
	Post mordanting	4/5	4	5	4/5	4	4	4/5	3/4	4	4/5	4	4
Without mordant		4	4/5	5	4	4	3/4	4	4	3	3/4	4	4/5

C = colour change, s = staining.

3.7 *Evaluation of mordants effect on (CIELab coordinates)*

Post-mordanted cotton fabrics demonstrated better fastness results among wash, light, rubbing, and perspiration fastness as seen in alum-tannic acid, iron(II) sulfate, and copper(II) sulfate (Table 9) with mean ratings of 4–5 in colour change and staining gray scale. Mordanting increases interaction between cotton functional groups and dye functional groups (hydroxyl and carbonyl groups), resulting in increased dye exhaustion. This is directly attributed to the increase in colour strength values of dyed and mordanted cotton samples compared with dyed unmordanted fabrics (Rather et al., 2016a, b). The colour change ratings in laundering of mordant natural dyes

Table 9. Average colour coordinates of specimen fabrics.

		CIELab cordinates						
Mordant/D ve	Method	L*	a*	b*	C*	h°	ΔE	Relative Colour strength %
Not Mordanted								3.4
$CuSO_4$	Pre mordanting	77.56	−0.19	8.52	8.52	−88.72	35.62	5.1
	S.mordanting	75.62	−0.03	10.0	10.04	−89.83	33.26	7.7
	Post mordanting	73.52	0.44	11.8 4	11.85	87.87	30.59	7.4
$FeSO_4$	Pre mordanting	76.70	1.94	8.85	9.06	77.64	34.35	5.5
	S. mordanting	75.60	0.42	9.48	9.49	87.46	33.36	6.7
	Post mordanting	71.13	0.50	11.8	11.89	87.59	28.41	9.1
$K_2Cr_2O_4$	Pre mordanting	78.86	−0.57	7.07	7.09	94.61	37.44	4.2
	S. mordanting	77.88	0.14	7.03	7.03	88.86	36.46	4.8
	Post mordanting	75.86	0.50	7.11	7.13	85.98	34.57	5.3
Alumtannic acid	Pre mordanting	77.09	2.74	9.34	9.73	73.65	34.40	5.3
	S. mordanting	73.91	5.48	7.53	9.31	53.95	32.07	7.0
R.Orange	Post mordanting	69.41	9.47	11.7	12.13	51.13	26.25	11.0
HER	Exhaust dyeing	66.47	22.01	–	22.64	−13.58	37.94	11.3

S = Simultaneous.

Mordant	Pre mordanting	Simultaneous mordanting	Post mordanting
Without mordant			
Copper Sulphate			
Ferrous Sulphate			
P. dichromate			
Alum-Tannic acid			
R. Orange HER			

Figure 7. Effect of mordanting on the production of shades of *B. vulgaris* extract.

on cotton varied depending on mordant (Francine, Jeannette, & Pierre, 2015), with ratings of 4–5 (slight-to-no change) to 2 (visible change) on the colour change gray scale, while staining ratings also ranged from 4–5 to 2 (Francine et al., 2015). In general, the high L* and low a* values mean brighter and less red colour shade. This is attributed to the concentration gradient of dye on fibre via adsorption (Ali, Islam, & Mohammad, 2016, Rather et al., 2016). These observations agreed with Sufian, Hannan, Rana, and Huq, (2016).

3.8 *Characteristics of dyed cellulosic fabrics*

Alum-tannic acid and copper sulfate of mordanted *B. vulgaris* dyed cotton fabrics were of deeper shade (Figure 7). Pre-mordanted dyed fabrics had the lightest shade among all mordanted dyed cotton fabrics except for dichromate, corroborating Yusuf et al.'s (2016) findings which showed that mordant/dye interaction reduces quantity of dye molecules diffusing into the cotton matrix. Similarly, the same interaction reduced the quantity of dye molecules diffused in and absorbed by the cotton matrix during the simultaneous mordanting–dyeing process. This observation agreed with findings made by Yusuf et al. (2016).

3.9 *Colour strength equivalence (Ceq)*

By using Equation 3 and dilution factor of 10 of absorbance (Aextr), Ceq of Reactive Orange 84 to Beta vulgaris was 0.168 at the highest absorbance of 1.33. The mass equivalent (Meq) values indicated that 1 kg *B. vulgaris* dye could yield approximately 3.36 g kg^{-1} equivalent of commercial Reactive Orange HER. This procedure and findings were according to the procedure of Kechi, Chavan, and Moeckel, (2013).

4 CONCLUSIONS AND RECOMMENDATIONS

From the UV–Vis analysis the total betalain (which is responsible for the red colour) content was estimated to range between 350–380 mg betalain and 100 g on fresh weight. The study established the optimized extraction conditions as being M:L ratio of 1:10 and time of 11 hours while that of dyeing temperature, time, and pH were 55 ^{o}C, 75 minutes, and pH 6, respectively. The post-mordanting method with copper sulfate and alum-tannic acid showed a higher relative colour strength. The plant dye extract showed very stable colour fastness (good to excellent) such as wash fastness, light fastness, rub fastness, and fastness to perspiration test. This could be because the covalent bond between the dye molecules and the carboxyl groups in the cotton fibres is strong (Singam et al., 2019). These observations also agreed with Geelani, Ara, Mir, Bhat, & Mishra, (2016) in dyeing and fastness properties of *Quercus robur* with natural mordants on natural fibres. More research needs to be done on the optimization of extractions and dyeing parameters of *B. vulgaris* using different techniques to maximize the full potential. Furthermore, research can also be carried out in the dyeing process such as double mordanting of substrates with different types of mordants and advanced techniques of analyzing colour measurements of *B. vulgaris* dyed substrates.

ACKNOWLEDGMENTS

The authors are thankful to the Africa Centre of Excellence in Phytochemicals, Textile and Renewable Energy (ACEII-PTRE) for the financial support.

REFERENCES

Ado A, Yahaya H, Kwalli A, & Abdulkadir R.S (2014). Dyeing ofTextiles with Eco-Friendly Natural Dyes: A Review. International Journal of Environmental Monitoring and Protection, 1(5), 76-81.

Ali, M. K., Islam, S., & Mohammad, F. (2016). Extraction of natural dye from walnut bark and its dyeing properties on wool yarn. Journal of Natural Fibers, 13, 458–469.

Almahy, HA, & band Ali, AA. (2013). Extraction of carotenoids as natural dyes from the Daucus carota Linn (carrot) using ultrasound in Kingdom of Saudi Arabia. Research Journal of Chemical Sciences, 3(1), 63–66.

Alsehri SM, Naushad M, Ahamad T, Alothman ZA, & Aldalbahi A (2014) Synthesis, characterization of curcumin based ecofriendly antimicrobial bio-adsorbent for the removal of phenol from aqueous medium. Chem Eng J 254:181–189.

Antigo J.L., Rita de Cássia Bergamasco R., & Grasiele Scaramal Madrona G.S. (2017). Effect of pH on the stability of red beet extract (Beta vulgaris l.) microcapsules produced by spray drying or freeze drying. Food Sci. Technol vol.38 .DOI /10.1590/1678-457x.34316 .

Arora, J., Agarwal, P., & Gupta, G. Rainbow of natural dyes on textiles using plants extracts: Sustainable and eco-friendly processes. Green and Sustainable Chemistry, (2017), 7(01), 35-47.

Baião D, Silva D, Mere Del Aguila E, Paschoalin V. Nutritional, Bioactive and Physicochemical Characteristics of Different Beetroot Formulations; 2017.

Bechtold T, Mussak R, Mahmud-Ali A, Ganglberger E & Geissler S. (2006). Extraction of natural dyes for textile dyeing from coloured plant wastes released from the food and beverage industry. Journal of the Science of food and Agriculture, 86, 233-242.

Brahma, S., Islam, M. R., Shimo, S. S., & Dina, R. B. Influence of Natural and Artificial Mordants on the Dyeing Performance of Cotton Knit Fabric with Natural Dyes. IOSR Journal of Polymer and Textile Engineering, (2019), 01-06.

Bukhari, M. N., Islam, S., Shabbir, M., Rather, L. J., Shahid, M., & Singh, U., (2017). Dyeing studies and fastness properties of brown naphtoquinone colorant extract from Juglans regia L. on natural protein fibre using different metal salt mordants. Textiles and Clothing Sustainability, 3(3), 1–9.

Chaudhry, Siddiqui, S. I., Fatima, B., Tara, N., Rathi, G., & Chaudhry, S. A. Recent advances in remediation of synthetic dyes from wastewaters using sustainable and low-cost adsorbents. The impact and prospects of green chemistry for textile technology. (2019), pp. 471-507).

Chawla, H., Parle, M., Sharma, K., & Yadav, M. (2016). Beetroot: A Health Promoting Functional Food. Nutraceuticals, 1, 0976-3872.

Ding, Y. I., & Freeman, H. S. (2017). Mordant dye application on cotton; optimization and combination with natural dyes. Coloration Technology, 133(5), 369–375.

Francine, U., Jeannette, U., & Pierre, R. J. Assessment of antibacterial activity of neem plant (Azadirachta indica) on Staphylococcus aureus and Escherichia coli. Journal of Med Plants, (2015), 3(4), 85-91.

Frose, A., Schmidtke, K., Sukmann, T., Junger, I. J., & Ehrmann, A. Application of natural dyes on diverse textile materials. International Journal for Light and Electron Optics, 181(2018), (2019), 215–219.

Geelani, S. M., Ara, S., Mir, N. A., Bhat, S. J. A., & Mishra, P. K. (2016). Dyeing and fastness properties of Quercus robur with natural mordants on natural fibre. Textile and Clothing Sustainability, 2(8), 1–10.

Ghouila, H., Meksi, N., Haddar, W., Mhenni, M., & Jannet, H. (2012). Extraction, identification and dyeing studies of Isosalipurposide, a natural chalcone dye from Acacia cyanophylla flowers on wool. Industrial Crops and Products, 35, 31-36.8 (paperback). pp 70.

Gokhale, S. V., & Lele, S. S. (2014). Betalain content and antioxidant activity of Beta vulgaris: Effect of hot air convective drying and storage. Journal of Food Processing and Preservation, 38, 585–590.

Hamanthraj K.P.M, Desai S.M, & Bisht S.S. (2014). Optimization of Extraction Parameters for Natural Dye form Pterocarpus Santalinus by Using Response Surface methodology. International Journal of Engineering Research and Applications, 4(9 (Version 3)), 100- 108.

Hernandez-Martinez A.R., Estevez M., Vargas S., & Rodríguez R. (2013). Stabilized Conversion Efficiency and Dye-Sensitized Solar Cells from Beta vulgaris Pigment .International journal of molecular sciences, 14, 4081-93, DOI 10.3390/ijms14024081.

Hong, K. H. (2018). Effects of tannin mordanting on coloring and functionalities of wool fabrics dyed with spent coffee grounds. Fashion and Textiles, 5, 33.

Kechi, A., Chavan, R., & Moeckel, R. (2013). Dye Yield, Color Strength and Dyeing Properties of Natural Dyes Extracted from Ethiopian Dye Plants. Textiles and Light Industrial Science and Technology, 2(3), 137-145.

Khan, A., Hussain M.T. & Jiang H. 2018. Dyeing of silk fabric with natural dye from camphor (Cinnamomum camphora) plant leaf extract. Coloration Technology. doi:10.1111/cote.12338.

Kulkarni SS, Gokhale AV, Bodake UM, & Pathade GR (2011) Cotton dyeing with natural dye extracted from pomegranate (Punica granatum) Peel. Univers J Environ Res Technol 1(2):135–139.

Kushwaha, R., Kumar, V., Vyas, G. & Kaur, J. (2018): Optimization of different variable for eco-friendly extraction of betalains and phytochemicals from beetroot pomace. Waste Biomass Valori, 9, 1485–1494.

Kusumawati, N., Santoso, A. B., Sianita, M. M., & Muslim, S. Extraction, Characterization and Application of Natural Dyes from the Fresh Mangosteen (Garcinia mangostana L.) Peel. International Journal on Advanced Science, Engineering and Information Technology, (2017), 7(3), 878-884.

Li, H. X., Xu, B., Tang, L., Zhang, J. H., & Mao, Z. G. Reductive decolorization of indigo carmine dye with Bacillus sp. MZS10. International Biodeterioration & Biodegradation, (2015), 103, 30-37.

Maran, J.P. & Priya, B. (2016): Multivariate statistical analysis and optimization of ultrasound-assisted extraction of natural pigments from waste red beet stalks. J. Food Sci. Tech., 53(1), 792–799.

Rather, L. J., Islam, S., Azam, M., Shabbir, M., Bukhari, M. N., Shahid, M., Khan, M. A., Haque, Q. M. R., & Mohammad, F. (2016). Antimicrobial and fluorescence finishing of woolen yarn with Terminalia arjuna natural dye as an ecofriendly substitute to synthetic Antibacterial agents. RSC Advances, 6, 39080–39094.

Rather, L. J., M. Shabbir, M. N. Bukhari, M. Shahid, M. A. Khan, & F. Mohammad. 2016. Ecological dyeing of woollen yarn with Adhatoda vasica natural dye in the presence of biomordants as an alternative copartner to metal mordants. Journal of Environmental Chemical Engineering 4:3041–49. doi:10.1016/j.jece.2016.06.019.

Ravichandran, K., Saw, N.M.T., Mohdaly, A.A.A., Gabr, A.M.M., & Smetanska, I. (2013): Impact of processing of red beet on betalain content and antioxidant activity. Food Res. Int., 50(2), 670–675.

Samanta A.K., & Konar A, (2011). Dyeing of Textiles with Natural Dyes. India.

Sarhan, T. M., & Salem, A. A. Turmeric dyeing and chitosan/titanium dioxide nanoparticle colloid finishing of cotton fabric. Indian Journal of Fibre & Textile Research, (2018), 43(4), 464-473.

Saxena H.O, Tiwari R, & Pandey A.K. (2012). Optimization of Extraction & Dyeing Conditions of Natural Dyes from Butea monosperma (Lam) Kuntze Flowers and Development of Various Shades. Environmental & We an International Journal of Science and Technology. 7, 29-35.

Simpson, J. T., Hunter, S. R., & Aytug, T. Superhydrophobic materials and coatings: a review. Reports on Progress in Physics, (2015), 78(8), 086501.

Singam, R. T., Marsi, N., Mamat, H., Mohd Rus, A. Z., Main, N. M., Huzaisham, N. A., & Mohd Fodzi, M. H. The Preparation and Characterization on Natural Dyes Based on Neem, Henna and Turmeric for Dyeing on Cotton with Superhydrophobic Coating. International Journal of Integrated Engineering, (2019), 11(7), 137-143.

Singh A, Ganesapillai M, & Gnanasundaram N: Optimization of extraction of betalain pigments from Beta vulgaris peels by microwave pretreatment. IOP Conf Ser Mater Sci Eng 2017; 263: 032004.

Stewart, C. W. Comparative Study of the Environmental Impacts of Standardized Dyeing Systems Using Natural and Synthetic Dyes on Knitted Cotton Fabric. Fiber and Polymer Science. (2017), 3(1), 76-84.

Sufian, A., Hannan, A., Rana, M., & Huq, M. Z. (2016). Comparative study of fastness properties and color absorbance criteria of conventional and avitera reactive dyeing on cotton knit fabric. European Scientific Journal, 12(15), 352–364.

Sun, Q., Xiao, W., Xi, D., Shi, J., Yan, X. and Zhou, Z. "Statistical optimization of biohydrogen production from sucrose by a co-culture of Clostridium acidisoli and Rhodobacter sphaeroides". International Journal of Hydrogen Energy. (2010). 35, 4076–4084.

Uddin, M. G. (2015). Extraction of eco-friendly natural dyes from mango leaves and their application on silk fabric. Textile& Clothing Sustainability, 1, 7.

Wangatia L.M, Tadesse K, & Moyo S. (2015). Mango Bark Mordant for Dyeing Cotton with Natural Dye: Fully eco-friendly Natural Dyeing. International Journal of Textile Science, 4(2), 36- 41.

Yamjala, K., Nainar, M. S., & Ramisetti, N. R. Methods for the analysis of azo dyes employed in food industry–a review. Food chemistry, (2016), 192, 813–824.

Yusuf, M., Mohammad, F., Shabbir, M., & Khan, M. A. (2016). Eco-dyeing of wool with Rubia cordifolia root extract: Assessmentof the effect of Acacia catechu as biomordant on color and fastness properties. Textile and Clothing Sustainability, 2(10), 1–9.

Yusuf, M., Shabbir, M., & Mohammad, F., (2017).Natural colorants: Historical, Processing and Sustainable Prospects. Nat. Prod. Bioprospect. 7(1), 123–145.

Zakharova, N.S. & T. A.Petrova 1997. Investigation of betalain and betalain oxide of leaf beet. Applied Biochem. And Microbiology, 33 (5):481–484.

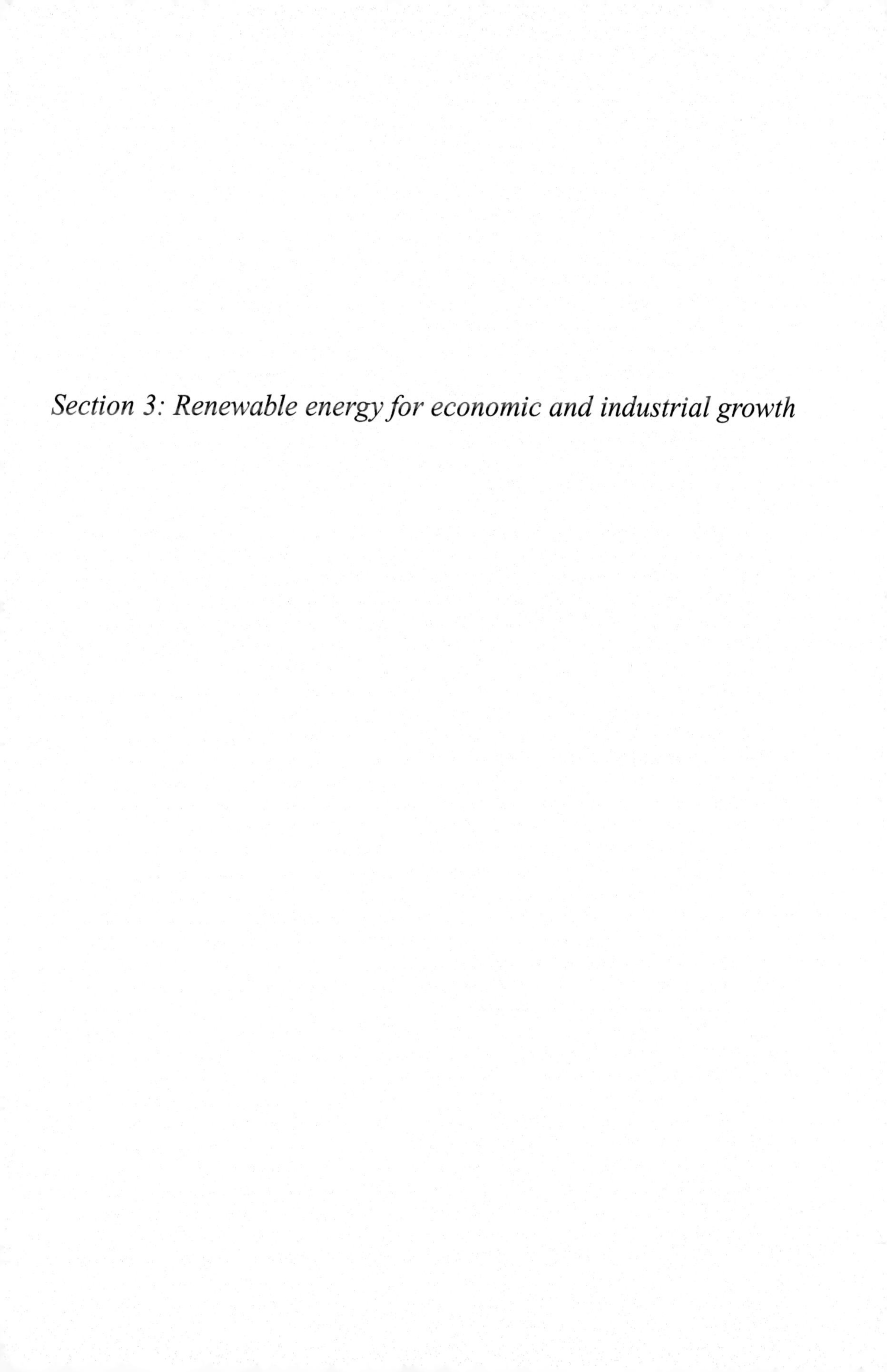

Section 3: Renewable energy for economic and industrial growth

Advances in Phytochemistry, Textile and Renewable Energy Research for Industrial Growth – Nzila et al. (Eds)

Role of modified Coulomb potential in determining stability of isobars

H.K. Cherop, K.M. Muguro & K.M. Khanna
Department of Physics, University of Eldoret, Eldoret, Kenya

ABSTRACT: The binding energy of an atomic nucleus is composed of a number of different forms of energy. The most important ones are the nuclear interaction energy and the Coulomb energy due to the Coulomb repulsion between the protons. The recently proposed modified Coulomb potential model has been used to calculate the most stable nuclei for a fixed mass number, for a few nuclei (rather isobars) with Z > 92. Calculations from this study show that the modified Coulomb potential model generates the most stable nuclei (Z_{STABLE}) for the isobars when $n > 21$. As the values of n increase, some occurrence of nuclear transformations that include beta decay and gamma decay, which the isobars undergo in order to gain stability are revealed. However, some unknown kind of radiations that require further investigations are predicted to be emitted when the value of n increases consecutively at the initial stages of nuclear decay.

1 INTRODUCTION

Theoretical and experimental research on the nucleus of an atom in nuclear theory has given rise to new information and data, which is vital in understanding the structure of the nucleus and the nucleon-nucleon interactions. In spite of the spectacular advances made in nuclear theory and experimental nuclear physics, it is still not known exactly as to how many protons and neutrons can constitute a bound atomic nucleus, especially in the region of periodic table when Z varies from Z > 92 to say Z = 120 or more.

It is now sufficiently confirmed experimentally and theoretically that the nucleus of an atom is composed of neutrons (N), protons (Z), and that its mass number (A), is given by A = N + Z. A proton being positively charged has slightly less mass compared to the mass of the neutron that has no charge. The magnetic moment of the proton is positive, but that of the neutron is negative. Inside the nucleus, the protons and neutrons are referred to as nucleons since the nuclear forces between the neutrons and protons are charge independent. In addition, there exist different types of nuclei. Some nuclei have proton number constant (Z constant) but different mass number (A), and such nuclei are called isotopes while the nuclei that have constant mass number but different atomic number are called isobars. Then there are nuclei with constant neutron number (N), but different A and Z. Such nuclei are called isotones. There is another set of nuclei, in which the proton number is equal to the neutron number ($Z = N$) in two or more nuclei. Such nuclei are called isomers or mirror nuclei, for instance, and in which has two neutrons and has two protons.

Quite a number of nuclear models that include liquid drop model, Bethe-Weizsäcker mass formula, collective model, evaporation model, Fermi gas model, shell model, individual particle model, nuclear pairing model, superfluid model among others, have been proposed from time to time to explain the properties of nuclei in different regions of mass number (Greiner & Maruhn 1996; Rowe & Wood 2010). However, none of these nuclear models can explain all the properties of nuclei. The basic parameters that come into play in the formulation of such nuclear models are the nuclear masses and the binding energy of the nuclei. As the nuclear size increases among the nuclei, the nuclear binding energies in the ground state also increase due to the effect of increased shells that are occupied by paired nucleons. The nucleons in this interaction experience several forces that are dominated by the Coulomb repulsion in the proton pairs. The Coulomb force being a long range type increases with increase in nuclear size, thus, Coulomb interaction is predicted to contribute greatly to the stabilization of the super heavy nuclei (SHN) that are likely to exist in the "island of stability" (Oganessian 2012). Therefore, investigation of some isobaric nuclei among the super heavy nuclei with Z > 92 is carried out in order to find out the role of the modified Coulomb potential in determining the stability of super heavy isobaric nuclei. The calculations of this study are essential in identifying the transuranic elements that are likely to reside in the island of stability. Consequently, providing some fundamental information in the synthesis of the super heavy nuclei.

2 LITERATURE REVIEW

Nuclei with different atomic number (Z) and constant mass number (A) are called isobars. The word

DOI 10.1201/9781003221968-26

isobar is derived from two Greek words “isos” that mean equal and the word “baros” that mean weight. A British chemist by the name Alfred Walter Steward suggested this category of nuclides in the year 1918 (Kauffman 1983; McArthur 1947;. Isobars just like isotopes have several applications in several fields, which include nuclear energy, nuclear medicine, and industrial applications. It is well established that the atomic number of elements in the periodic table determines the chemical properties of all the elements in the nuclear landscape (Epiotis & Henze 2003). Therefore, the atomic structures of all isobaric nuclei differ among themselves since their atomic numbers vary for a fixed mass number of nuclei leading to different chemical properties, and different applications in everyday life. For instance, among the isobars whose mass number is 60, a special mention is given to Cobalt, Nickel and Ferrous. Cobalt-60 is a synthetic radioactive nuclei, which produces gamma rays that are used in the treatment of malignant tissues in medical field as well as sterilization of medical equipment (Masefield et al. 2008). Similarly, Nickel-60 is a stable naturally occurring metal that can also be produced by fission reaction. The applications of Nickel-60 include stereotactic radiosurgery and radiotherapy (Akram et al. 2018; Guo 2018;. Another special isobar in this category is Ferrous-60. This is a rare isotope of iron which can only be created through massive supernova explosion and it is presumed to have been the primary source of planetary heating and chronometer in the early solar system (Kohman & Robison 1980; Wallner et al. 2016).

In production of nuclear power, some isobars have contributed immensely in the operations of the nuclear reactors. These isobars, namely Tellurium-135, Iodine-135 and Xenon-135 undergo beta minus decay and their decay chains lead to the Xenon poisoning in the nuclear reactors. Similarly, Neodymium-149 decays to Promethium-149 which subsequently decays to Samarium-149 that is responsible for Samarium poisoning (Stacey 2018). The presence of Samarium-149 and Xenon-135 in the nuclear reactors that are fueled by Uranium-235 plays a critical role in controlling the power levels generated. This is because Samarium-149 and Xenon-135 have a very large neutron capture cross-sections, thus, they are added to the control rods of the nuclear reactors in order to control the neutron flux hence regulating the fluctuations of the concentration of the radioactive nuclides (Cameron 2012; Jevremovic 2005).

Other properties of isobars are described by the isobaric analogue states (IAS), which are nuclei that have the same isospin and spin parity in a fixed mass number (A) (Xu et al. 2016). Isobaric analogue states among the medium-mass nuclei were discovered in 1961. This came as a surprise to many nuclear physicists due to the assumptions that the Coulomb interaction among the light nuclei was insignificant, and the isobaric spin could not carry any weight (Anderson 1962; Feshbach & Kerman 1967). Research has shown that, the measurement of the properties of isobaric analogue states such as Coulomb displacement energy and its width can reveal the halo structures of the neutron-dripline of nuclei (Takeuchi et al. 2001). Therefore, isospin is an important entity in the study of nucleons in nuclear spectroscopy as well as in recognizing the existence of elementary particles such as quarks, which were proposed by Murray Gell-Mann and his student George Zweig in 1964, and later discovered experimentally by SLAC-MIT team of researchers between 1968 and 1972 (Friedman 1991; Gell-mann & Fritzsch 2010).

From the above discussion, it is evident that nuclei especially isobars undergo nuclear decay in order to gain stability. Consequently, these nuclei release some energy in form of radiations that can be harnessed and applied in nuclear medicine, nuclear reactors and industrial applications. Nonetheless, the concept of nuclear stability, which is determined by the interaction of nucleons in the nucleus of an atom, has not been fully exploited since the nature of interactions between the nucleons and the limits of Coulomb stability (Oganessian 2012) are not known especially in the super heavy nuclei. Therefore, investigation of some isobaric nuclei among the super heavy nuclei with $Z > 92$ is carried out in this paper, in order to find out the role of the modified Coulomb potential in determining the stability of isobars as they undergo nuclear decay.

3 METHODS

The Coulomb potential according to the assumptions on the liquid drop model (Bjornholm & Lynn 1980) can be written as,

$$E_C = \frac{3}{5}\frac{Z_1 Z_2 e^2}{4\pi\varepsilon_0 r} \tag{1}$$

where, e is the electron charge, Z_1 and Z_2 are the proton nuclear charges, r is the distance between the charges and it is given by with being the nuclear radius parameter, A is the mass number and is the permittivity of free space.

Coulomb interaction is a long-range force whereas very small size protons are confined inside the nucleus whose size is also very small. Thus, the Coulomb potential inside the nucleus has to be modified in order to confine the Coulomb energy within the boundary of the nucleus and to make it more effective. The modification of Coulomb energy was based on the following assumptions: Firstly, the charge distribution inside the nucleus is assumed uniform. Secondly, for a nucleus of an atom with $N > Z$, the nuclear core is composed of equal number of protons and neutrons ($N = Z$) and the excess neutrons reside in the surface region. Finally, the nuclear core radius is given by and this radius provides the maximum nuclear charge radius resulting from the proton-proton repulsion. From the above assumptions and our own intuition, a multiplier exponential correction term (C_t) was proposed and Equation 1 was modified to obtain the modified Coulomb potential (Cherop et al. 2019), which is written as,

$$E_C(Mod) = \frac{3}{5}\frac{Z_1 Z_2 e^2}{4\pi\varepsilon_0 R_0} e^{\frac{R_0^n}{nR^n}} \tag{2}$$

The binding energy of an atomic nucleus is composed of a few different forms of energy. The most important ones are the nuclear interaction energy and the Coulomb energy due to the Coulomb repulsion between the protons. According to Weizsäcker semi-empirical mass formula (SEMF), the binding energy equation (Dai et al. 2017; Heyde 2004) is written as,

$$BE(A,Z) = a_1 A - a_2 A^{\frac{2}{3}} - a_3 \frac{Z(Z-1)}{A^{\frac{1}{3}}} - a_4 \frac{(A-2Z)^2}{A} \pm \delta(A) \tag{3}$$

In Equation 3, the first term is the volume term, the second term is the surface energy term, the third term is the Coulomb energy term, the fourth term is the asymmetry energy term and the fifth term is the pairing energy correction term. By substituting Equation 2 into Equation 3 we obtain,

$$BE(A,Z) = a_1 A - a_2 A^{\frac{2}{3}} - \frac{3}{5}\frac{Z^2 e^2}{4\pi\varepsilon_0 R_0} e^{\frac{R_0^n}{nR^n}} - a_4 \frac{(A-2Z)^2}{A} \pm \delta(A) \tag{4}$$

In terms of the mass defect, the binding energy of the nucleus can be written as;

$$BE(A,Z) = \{ZM_p + NM_n - M(A,Z)\}c^2 \tag{5}$$

where Z is the atomic number, $M_p = 1.00727650$u is the proton rest mass, N is the neutron number, $M_n = 1.0086650$u is the neutron rest mass, $M(A,Z)$ is the mass of an atom of mass number A and c is the velocity of light in vacuum. Equating Equation 4 to Equation 5 and re-arranging it yields,

$$\{M(A,Z)\}c^2 = \{ZM_p + (A-Z)M_n\}c^2 - a_1 A + a_2 A^{\frac{2}{3}} + \frac{3}{5}\frac{Z^2 e^2}{4\pi\varepsilon_0 R_0} e^{\frac{R_0^n}{nR^n}} + a_4 \frac{(A-2Z)^2}{A} \pm \delta(A) \tag{6}$$

To find the value of Z for which the nucleus for a given A is stable (Z_{STABLE}), we differentiate $M(A,Z)$ with respect to Z in Equation 6 and equate it to zero to get,

$$\left(\frac{\partial M(A,Z)}{\partial Z}\right)_{A=\text{Costant}} = \{M_p - M_n\}c^2 + \frac{3}{5}(2Z)\frac{e^2}{4\pi\varepsilon_0 R_0} e^{\frac{R_0^n}{nR^n}} + a_4 \frac{(A-2Z)}{A}(-4) = 0 \tag{7}$$

All the terms in Equation 7 are known, hence, it can be solved mathematically and written in terms of Z_{STABLE} as,

$$Z_{STABLE} = \frac{2a_4 A + 0.646695A}{4a_4 + a_3 A^{\frac{2}{3}} e^{\frac{R_0^n}{nR^n}}} \tag{8}$$

Therefore, Equation 8 is used to calculate the values of Z_{STABLE} for the isobars with $Z > 92$. For a given value of A, Equation 8 is used get the value of Z_{STABLE} that corresponds to the most stable isobar, and the values of n play an important role in determining the value of Z_{STABLE}.

The isobars selected for investigation are classified into two categories. The first category of nuclei is selected randomly, and it comprises of the super heavy elements that were synthesized in laboratories using particle accelerators. The properties of these elements are well known but their nucleon interactions are not exactly known. They include, $^{239}93$, $^{240}95$, $^{253}99$, $^{257}106$ and $^{266}109$. The second category of the super heavy nuclei are nuclei that have been predicted to exist using sophisticated theoretical models. These nuclei include $^{292}120$ which were predicted using relativistic models and some Skyrme interactions to have shell closures that are related to central density depression (Afanasjev & Frauendorf 2005; Afanasjev et al. 2018; Bender et al. 1999). Other nuclei were obtained from the studies on the biconcave disks and toroidal shapes of some nuclei using Skyrme-Hartree-Fock (SHF) calculations (Kosior et al. 2017), which have revealed that the nuclei $^{364}138$ yields the lowest energy in the toroidal solutions. The Gogny-Hartree-Fock-Bogoliubov (HFB) calculations have shown that, in the nuclei $^{416}164$ and $^{476}184$, their toroidal shapes represent the lowest in energy solutions at axial shape (Afanasjev et al. 2018; Warda 2007). Similarly, the calculations obtained from triaxial Relativistic Hartree-Bogoliubov (RHB) theory has also predicted for the existence of the nuclei $^{360}130$, $^{432}134$, $^{340}122$ and $^{392}134$, which have more pronounced triaxial deformations that tend to reduce the stability of the nuclei against spontaneous fission (Afanasjev et al. 2018).

Therefore, the crux of this work is to investigate the role of the modified Coulomb potential in determining the stability of the isobars of these nuclei and possibly to describe the nature of their nucleon interactions.

4 RESULTS AND DISCUSSIONS

The calculations of Z_{STABLE} using Equation (8) are tabulated in Tables 1–3.

The results of Table 1 predict that, the most stable isobar corresponds to $Z = 94$ which is Plutonium nucleus. However, the last stable known element in the periodic table today is Bismuth whose $Z = 83$, while all the nuclei with $Z > 83$ decompose through radioactive decay. Therefore, the values of Z_{STABLE} that are generated by the modified Coulomb model can be regarded as the longest-lived nuclei or stable nuclei against spontaneous fission.

Table 1. Calculations for Z_{STABLE} values for A = 239.

A = 239					
n	R_o^n	R^n	Ct	Z_{STABLE}	ELEMENT
1	7.42E-15	8.07E-15	2.50885	70	YTTERBIUM
2	5.51E-29	6.51E-29	1.52659	84	POLONIUM
3	4.09E-43	5.25E-43	1.29617	88	RADIUM
4	3.03E-57	4.24E-57	1.19597	90	THORIUM
5	2.25E-71	3.42E-71	1.14076	91	PROTACTINIUM
6	1.67E-85	2.76E-85	1.10621	91	
7	1.2E-99	2.2E-99	1.08284	92	URANIUM
8	9.2E-114	1.8E-113	1.06615	92	
9	6.8E-128	1.4E-127	1.05377	92	
10	5.1E-142	1.2E-141	1.04431	93	NEPTUNIUM
11	3.8E-156	9.4E-156	1.03692	93	
12	2.8E-170	7.6E-170	1.03104	93	
13	2.1E-184	6.1E-184	1.02629	93	
14	1.5E-198	4.9E-198	1.02242	93	
15	1.1E-212	4E-212	1.01921	93	
16	8.5E-227	3.2E-226	1.01655	93	
17	6.3E-241	2.6E-240	1.01431	93	
18	4.7E-255	2.1E-254	1.01242	93	
19	3.5E-269	1.7E-268	1.01081	93	
20	2.6E-283	1.4E-282	1.00944	93	
21	1.9E-297	1.1E-296	1.00827	93	
$n > 21$	$R_o^n < 1.9E\text{-}297$	$R^n < 1.1E\text{-}296$	1.00000	94	PLUTONIUM

Table 2. Calculated values of Z_{STABLE} for the isobars with A = 240, 253, 257 and 266.

n	Z_{STABLE} for A = 240	Z_{STABLE} for A = 253	Z_{STABLE} for A = 257	Z_{STABLE} for A = 266
1	70	73	73	75
2	84	87	88	90
3	88	92	93	95
4	90	94	95	97
5	91	95	96	99
6	92	96	97	100
7	92	96	97	100
8	92	97	98	101
9	93	97	98	101
10	93	97	98	101
11	93	97	98	101
12	93	97	99	102
13	93	98	99	102
14	93	98	99	102
15	93	98	99	102
16	94	98	99	102
17	94	98	99	102
18	94	98	99	102
19	94	98	99	102
20	94	98	99	102
21	94	98	99	102
$n > 21$	94	98	100	102

As the values of n increase from $n = 1$ to $n > 21$ in Table 1, the values of Z_{STABLE} that are generated occur in the series of $Z_{STABLE} = 70 \rightarrow 84 \rightarrow 88 \rightarrow 90 \rightarrow 91 \rightarrow 92 \rightarrow 93 \rightarrow 94$.

The analysis of these transformations show that, the Coulomb energy correction term displays some unknown form of decay transformation from Z = 70 to Z = 88, followed by subsequent beta minus decay whereby a neutron decays into a proton, electron and electron antineutrino. As n tends towards $n > 21$, the most stable isobar is obtained since no correction to the Coulomb law is required at large distance. Similar calculations as shown in Table 1, were carried out for Americium-240 (A = 240), Einsteinium-253 (A = 253), Seaborgium-257 (A = 257) and Meitnerium-266 (A = 266) using Equation 8. The results of the calculated values of Z_{STABLE} are shown in Table 2.

From the calculations in Table 2, it is found that, for A = 240, the most stable isobar corresponds to Z = 94, which is Plutonium nucleus. This isotope (Plutonium-240) has high rate of spontaneous fission and it can raise the neutron flux of samples in nuclear explosives (Şahin & Ligou 1980). For A = 253, the most stable isobar corresponds to Z = 98, which is Californium nucleus. Californium-253 has half-life of 17.81 days (Knauer & Martin 2012). For A = 257, the most stable isobar corresponds to Z = 100, which is Fermium nucleus. Fermium-257 is known to be the most stable isotope with half-live of 100.5 days (Wild et al. 1973). Similarly, for A = 266, the most stable isobar corresponds to Z = 102, which is Nobelium nucleus which has not yet been discovered experimentally.

The calculations of Z_{STABLE} for the stability of isobars among the hyper heavy nuclei are shown in Table 3. The elements in this table are theoretical nuclei

Table 3. Calculated values of Z_{STABLE} for A = 292, A = 340, A = 360, A = 364, A = 392, A = 416, A = 432 and A = 476.

n	Z_{STABLE} for A = 292	Z_{STABLE} for A = 340	Z_{STABLE} for A = 360	Z_{STABLE} for A = 364	Z_{STABLE} for A = 392	Z_{STABLE} for A = 416	Z_{STABLE} for A = 432	Z_{STABLE} for A = 476
1	80	90	94	94	100	102	108	111
2	97	111	116	116	124	128	134	141
3	103	117	122	123	131	136	142	151
4	105	120	125	126	135	140	146	155
5	107	122	127	128	137	142	148	158
6	108	123	128	129	138	143	149	159
7	108	123	129	130	139	144	150	160
8	109	124	130	131	139	145	151	161
9	109	124	130	131	140	145	151	162
10	109	124	130	131	140	146	151	162
11	110	125	131	132	140	146	152	163
12	110	125	131	132	140	146	152	163
13	110	125	131	132	140	147	152	163
14	110	125	131	132	140	147	152	163
15	110	125	131	132	141	147	152	164
16	110	125	131	132	141	147	152	164
17	110	125	131	132	141	147	152	164
18	110	125	131	132	141	147	152	164
19	110	125	131	132	141	147	152	164
20	110	125	131	132	141	147	152	164
21	110	125	131	133	141	147	152	164
$n > 21$	111	126	132	133	141	148	152	164

since the last known element currently in the periodic table of elements is Oganesson with atomic number Z = 118. However, atomic calculations suggest that the existence of nuclei on earth or in interstellar bodies may end at $Z \simeq 172$ (Fricke et al. 1971; Indelicato et al. 2011; Pyykkö 2011). This implies that, investigations are on course to discover the existence of such nuclei experimentally.

The results in Table 3 show that, the application of Equation 8 predicts the existence of the most stable or the longest-lived nuclei for the given mass numbers. These nuclei include $^{292}111$ which might fall under the category of Roentgenium isotopes. Other stable isobars predicted by the model include $^{340}126$, $^{360}132$, $^{364}133$, $^{392}141$, $^{416}148$, $^{432}152$ and $^{476}164$. The elements $^{340}126$ and $^{432}152$ have magic proton number and a semi-magic proton number respectively. Since occurrence of magic numbers corresponds to nuclei having extra stability, it is expected that these nuclei will be more stable than their neighbouring isotopes. The reduction in the correction term as n increases from $n = 1$ to $n > 21$, leads to generation of new values of the most stable nuclei for which the nucleus of a given A is stable.

Other methods of determining the stability of isobaric nuclei include the use of the mass parabolas. The mass parabolas predict accurately the stable isobaric nuclei among the light and intermediate mass nuclei. However, they are not accurate in determining the stability of super heavy isobaric nuclei. This is because the shapes of the mass parabolas change into exponential curves (Cherop 2020). Thus, the modified Coulomb potential enriches the mass parabolas in the sense that, it generates the stable values of all the super heavy isobaric nuclei accurately. Consequently, predicting the most stable isobaric nuclei that might exist in the island of stability.

5 CONCLUSIONS AND RECOMMENDATIONS

One of the properties of nuclei that determine the existence of elements in the nuclear landscape is the nuclear stability, which is determined by the neutron to proton ratio. As the number of protons increase in the nucleus of an atom, the number of neutrons also increases in order to maintain the stability of the nucleus. Any imbalance between the ratio of protons and neutrons may lead to imbalance in the nuclear forces causing nuclear instability. In this paper, the category of nuclei investigated fall under the group of super heavy nuclei, which are unstable. However, some mathematical models have predicted for the existence of stable super weights in the "island of stability". The type of nuclei that may exist in island of stability may include isobars, isotones and isotopes. In order to investigate the nature of isobaric nuclei that may exist in the island of stability, a modified Coulomb potential model having a multiplier exponential correction term has been used to calculate the values of Z_{STABLE} for which the nucleus for a given A is fixed. Our calculations reveal that, the modified Coulomb potential model generates the most stable nuclei (Z_{STABLE}) for a given isobar when $n > 21$. The model also reveals that as value of n increases from $n = 1$ to $n > 21$, some occurrence of nuclear transformations that the nucleus undergoes in order to gain stability are noted. These transformations include beta decay and gamma decay. However, some unknown kind of radiations are generated when the value of n increases consecutively from $n = 1$ to $n = 2$, 3, 4, or 5 for the hyperheavy isobars as shown in Tables 2 and 3. It is recommended that, some further investigations on the unknown form of radiations can be carried out in order to describe the kind of nucleon interactions that may take place as n increases at the initial stages of the nuclear transformations.

ACKNOWLEDGEMENT

We are indeed very grateful to the ministry of education, higher loans education board of kenya for supporting this work under scholarship award: HELB/45/003/VL.II/100 (2017/2018).

REFERENCES

Afanasjev, A. V., Agbemava, S. E., & Gyawali, A. (2018). Hyperheavy nuclei: Existence and stability. *Physics Letters B*, *782*,533–540. https://doi.org/10.1016/j.physletb.2018.05.070

Afanasjev, A. V., & Frauendorf, S. (2005). Central depression in nuclear density and its consequences for the shell structure of superheavy nuclei. *Physical Review C*, *71*(2), 1-5. https://doi.org/10.1103/physrevc.71.024308

Akram, M., Ullah, H. Z., Altaf, S., Iqbal, K., Altaf, S. M., Khan, M. A., & Buzdar, S. A. (2018). Radiation absorbed dose for cobalt-60 gamma source in phantoms for different materials. *The Journal of the Pakistan Medical Association*, *68*(2), 264–267.

Anderson, J. D., Wong, C., & McClure, J. W. (1962). Isobaric States in Nonmirror Nuclei. *Physical Review*, *126*(4), 2170–2173. https://doi.org/10.1103/physrev.126.2170

Bender, M., Rutz, K., Reinhard, P.-G., Maruhn, J. A., & Greiner, W. (1999). Shell structure of superheavy nuclei in self-consistent mean-field models. *Physical Review C*, *60*(3), 1–22. https://doi.org/10.1103/physrevc.60.034304

Bjørnholm, S., & Lynn, J. E. (1980). The double-humped fission barrier. *Reviews of Modern Physics*, *52*(2), 725–931. https://doi.org/10.1103/revmodphys.52.725

Cameron, I. R. (2012). *Nuclear fission reactors*. Springer Science & Business Media.

Cherop, K. H. (2020). The significance of the modified Coulomb energy model in the binding energy equation. *Scientific Israel-Technological Advantages*, *22*(2), 3–14.

Cherop, H. K., Muguro, K. M., & Khanna, K. M. (2019). The Role of the Modified Coulomb Energy in the Binding Energy Equation for Finite Nuclei. *Scientific Israel-Technological Advantages*, *21*(5–6), 82–89.

Dai, H., Wang, R., Huang, Y., & Chen, X. (2017). A novel nuclear dependence of nucleon–nucleon short-range correlations. *Physics Letters B*, *769*, 446–450. https://doi.org/10.1016/j.physletb.2017.04.015

Epiotis, N. D., & Henze, D. K. (2003). Periodic Table (Chemistry).

Feshbach, H., & Kerman, A. (1967). Isobar Analogue States. *Comments Nucl. Part. Phys.*, *1*, 69–74.

Fricke, B., Greiner, W., & Waber, J. T. (1971). The continuation of the periodic table up to Z = 172. The chemistry of superheavy elements. *Theoretica Chimica Acta*, *21*(3), 235–260. https://doi.org/10.1007/bf01172015

Friedman, J. I. (1991). Deep inelastic scattering: Comparisons with the quark model. *Reviews of Modern Physics*, *63*(3), 615–627. https://doi.org/10.1103/revmodphys.63.615

Gell-Mann, M., & Fritzsch, H. (2010). *Murray Gell-Mann: Selected Papers* (Vol. 40). World Scientific.

Greiner, W. & Maruhn, J. A. (1996). "Nuclear models," Berlin: Springer-Verlag, p.77.

Guo, F. (2018). 3-D treatment planning system—Leksell Gamma Knife treatment planning system. *Medical Dosimetry*, *43*(2), 177–183. https://doi.org/10.1016/j.meddos.2018.03.001

Heyde, K. (2004). *Basic ideas and concepts in nuclear physics: an introductory approach*. CRC Press.

Indelicato, P., Bieroñ, J., & Jönsson, P. (2011). Are MCDF calculations 101% correct in the super-heavy elements range? *Theoretical Chemistry Accounts*, *129*(3–5), 495–505. https://doi.org/10.1007/s00214-010-0887-3

Jevremovic, T. (2005). Nuclear Reactor Control: Methods of reactor control, Fission product poisoning and Reactivity coefficients. *Nuclear Principles in Engineering*, 397–424.

Kauffman, G. B. (1983). Alias J. J. Connington: the life and work of Alfred W. Stewart (1880-1947) chemist and novelist. *Journal of Chemical Education*, *60*(1), 38. https://doi.org/10.1021/ed060p38

Knauer, J., & Martin, R. (2012). Chemical Technology Division Oak Ridge National Laboratory. *Californium-252 Isotope for 21st Century Radiotherapy*, *29*, p. 10.

Kohman, T. P., & Robison, M. S. (1980). Iron-60 as a possible heat source and chronometer in the early Solar System. *Lunar and Planetary Science Conference* (Vol. 11, pp. 564–566).

Kosior, A., Staszczak, A., & Wong, C.-Y. (2017). Toroidal Nuclear Matter Distributions of Superheavy Nuclei from Constrained Skyrme–HFB Calculations. *Acta Physica Polonica B Proceedings Supplement*, *10*(1), 249. https://doi.org/10.5506/aphyspolbsupp.10.249

Masefield, J., Morrissey, B., Chitra, J., Bennett, N., Mathias, L., & Brinston, R. (2008). Cobalt-60 Use and Disposal: An Established Pathway. *International Meeting on Radiation Processing*.

McArthur, D. N. (1947). Prof. A. W. Stewart. *Nature*, *160*(4056), 116. https://doi.org/10.1038/160116a0

Oganessian, Y. (2012). Nuclei in the "Island of Stability" of Superheavy Elements. *Journal of Physics: Conference Series*, *337*, 012005. https://doi.org/10.1088/1742-6596/337/1/012005

Pyykkö, P. (2011). A suggested periodic table up to Z $\leq$ 172, based on Dirac–Fock calculations on atoms and ions. *Phys. Chem. Chem. Phys.*, *13*(1), 161–168. https://doi.org/10.1039/c0cp01575j

Rowe, D. J., & Wood, J. L. (2010). *Fundamentals of nuclear models: foundational models*. World Scientific Publishing Company.

Şahin, S., & Ligou, J. (1980). The Effect of the Spontaneous Fission of Plutonium-240 on the Energy Release in a Nuclear Explosive. *Nuclear Technology*, *50*(1), 88–94. https://doi.org/10.13182/nt80-a17072

Stacey, W. M. (2018). *Nuclear reactor physics*. John Wiley & Sons.

Takeuchi, S., Shimoura, S., Motobayashi, T., Akiyoshi, H., Ando, Y., Aoi, N., Fü, Z., Gomi, T., Higurashi, Y., Hirai, M., Iwasa, N., Iwasaki, H., Iwata, Y., Kobayashi,

H., Kurokawa, M., Liu, Z., Minemura, T., Ozawa, S., Sakurai, H., Ishihara, M. (2001). Isobaric analog state of 14Be. *Physics Letters B*, *515*(3–4), 255–260. https://doi.org/10.1016/s0370-2693(01)00890-5

Wallner, A., Feige, J., Kinoshita, N., Paul, M., Fifield, L. K., Golser, R., Honda, M., Linnemann, U., Matsuzaki, H., Merchel, S., Rugel, G., Tims, S. G., Steier, P., Yamagata, T., & Winkler, S. R. (2016). Recent near-Earth supernovae probed by global deposition of interstellar radioactive 60Fe. *Nature*, *532*(7597),69–72. https://doi.org/10.1038/nature17196

Warda, M. (2007). Toroidal structure of Super-Heavy Nuclei in the HFB Theory. *International Journal of Modern Physics E*, *16*(02), 452–458. https://doi.org/10.1142/s0218301307005880

Wild, J. F., Hulet, E. K., & Lougheed, R. W. (1973). Some nuclear properties of fermium-257. *Journal of Inorganic and Nuclear Chemistry*, *35*(2), 1063–1067. https://doi.org/10.1016/0022-1902(73)80176-9

Xu, X., Zhang, P., Shuai, P., Chen, R. J., Yan, X. L., Zhang, Y. H., Wang, M., Litvinov, Y. A., Xu, H. S., Bao, T., Chen, X. C., Chen, H., Fu, C. Y., Kubono, S., Lam, Y. H., Liu, D. W., Mao, R. S., Ma, X. W., Sun, M. Z., Xu, F. R. (2016). Identification of the Lowest T=2, $J\pi$=0+ Isobaric Analog State in Co52 and Its Impact on the Understanding of β-Decay Properties of Ni52. *Physical Review Letters*, *117*(18),0. https://doi.org/10.1103/physrevlett.117.182503

Advances in Phytochemistry, Textile and Renewable Energy Research for Industrial Growth – Nzila et al. (Eds)

Production of biogas from sized cotton yarn wastes

Maurice Twizerimana*
Department of Manufacturing, Industrial and Textile Engineering, Moi University, Eldoret, Kenya

Milton M. Arimi
Department of Chemical and Process Engineering, Moi University, Eldoret, Kenya

Eric Oyondi Nganyi
Department of Manufacturing, Industrial and Textile Engineering, Moi University, Eldoret, Kenya

Xumay Bura Hhaygwawu
Department of Mechanical, Production and Energy Engineering, Moi University, Eldoret, Kenya

ABSTRACT: Solid waste management is among the environmental challenges facing many industries in the world today. Biogas production is one of the most cost-efficient renewable energy technologies that use biodegradable wastes as feedstock. Furthermore, it is one of the methods for reducing greenhouse gas emission. Cotton Yarn Wastes (CYW) is among the biodegradable wastes that are commonly managed by dumping onto the open land or disposing in sanitary landfills where they undergo anaerobic decomposition. However, CYW could be used as substrate to generate energy in the form of biogas that can be utilized in other activities like powering textiles production. The aim of this study was to investigate the use of CYW as a substrate for biogas production using anaerobic batch reactor. The experiment was carried out in reactors of two-liter capacity. The CYW and inoculum were characterized before and after digestion. The CYW contained 93.18% total solids, 82.48% total volatile solids and 6.82% moisture content while the respective values for digested sludge were 21.61%, 23.61% and 78.38%, respectively. The carbon to nitrogen (C/N) ratio of inoculum was 20.5, which is in the suitable range to keep the anaerobic digestion in a stable condition. However, the CYW had high carbon content; resulting in a C/N ratio of 42.5. The effect of TS concentration at different ratios of CYW on biogas volume produced was investigated. The reactors loading was differentiated using a mixture with concentration corresponding to R_1 (1:1), R_2 (1:1.5), R_3 (1:2), R_4 (1:2.5), R_5 (1:3), R_6 (1:3.5), R_7 (1:4), R_8 (1:5), R_9 (1:6), and R_{10} (1:10) on TS content basis i.e. 50%, 40%, 33%, 28%, 25%, 22%, 20%, 18%, 14%, and 10% respectively. The total biogas yield was 6307 mL, 6519 mL, 6711 mL, 7178 mL, 4878 mL, 3868 mL, 3720 mL, 3306 mL, 1164 mL, and 932 mL respectively after 36 days. The results indicated that biogas production increased with increase in TS content. The ratio that provided 28% of TS content had the highest biogas yield. The average reduced TVS at the end of digestion was 88.49%. The results of this test indicated that CYW is a suitable substrate for AD due to its high biodegradability. Therefore, the reactors should run at 28% TS, for maximum biogas generation.

Keywords: batch reactor, total solids, textile waste, anaerobic digestion, moisture content.

1 INTRODUCTION

Since the beginning of the industrial revolution, the energy requirement for industries has gradually been increasing worldwide. Population growth and the promotion of living standards have always been one of the key drivers of increased energy demand and fiber consumption (Hasanzadeh, Mirmohamadsadeghi, & Karimi, 2018; Wang, 2010). Fossil fuel resources including crude oil and its derivatives, coal, and natural gas represent important world energy resources (Al-Hamamre et al., 2017). However, increasing world population along with reducing fossil fuel reserves have resulted in global interest to gradually change the energy source from fossils to renewable energy (Rajendran & Balasubramanian, 2011a). One of the environmental problems faced by the world is that of solid waste management especially biodegradable wastes such as textiles cotton yarn wastes (CYW). The problem of the final textile wastes industry has now assumed serious dimensions since it has

*Corresponding author

DOI 10.1201/9781003221968-27

no salability and pollutes the atmosphere (Sharma-Shivappa, 2008). Additionally, environmental pollution caused by dumping or landfilling of waste materials in the environment are among the most crucial issues the world is facing today (Deepanraj, Sivasubramanian, & Jayaraj, 2017). Normally, the burning, landfilling, and open dumping methods are used as treatment methods for organic wastes. Landfilling results in emissions of methane, carbon dioxide, and nitrous oxides, which contribute to the greenhouse effect (Wang, 2010). Currently, the management of textile wastes involves recycling them as second-hand textiles, filling materials in the textile industry, composting, landfilling, and burning (Hasanzadeh et al., 2018). Therefore, the annual global production of end life textile wastes is increasing, causing an increased interest in the impact of the disposed of wastes on the environment. However, textile wastes are an enriched source of energy and materials (Hasanzadeh et al., 2018). Textile wastes include wastes from streams of fiber, textile, and clothing manufacturing process, commercial service, and consumption (Hu et al., 2018).

Textile wastes mainly consist of cotton and viscose fibers. Reports from previous studies showed that cotton wastes have a significant potential to be used as a substrate for the production of different bioenergy such as biogas (Rasel et al., 2019). The environmental problems caused by organic wastes should be militated against. One effective way to avoid these problems is to use the wastes as a substrate for biogas production (Papacz, 2011). Biogas represents one of the most important renewable energy sources (Triolo, Pedersen, Qu, & Sommer, 2012). It is possible to mitigate the negative environmental effects of solid wastes by using them for the production of biogas. Moreover, the transformation of complex organic materials into biogas reduces the emission of greenhouse gases and can produce by-products like high-value fertilizer for growing crops (Jeihanipour, Aslanzadeh, Rajendran, & Balasubramanian, 2013; Treichel et al., 2019). Furthermore, about emissions, biogas production might be better for the environment than incineration of organic wastes. Methane from biogas has different applications. It may be utilized as a source of heat, steam, electricity and can be upgraded to vehicle fuels (Papacz, 2011). It may also be used as a household fuel for cooking and lighting or in fuel cells (Velmurugan, , Deepanraj, & Jayaraj, 2014). Putting all these advantages into consideration, biogas is one of the most environmentally friendly energy sources which could substitute fossil fuels (Manager et al., 2009). Biogas can be produced from a wide range of substrates such as industrial, municipal, wastewater, agricultural, and food wastes as well as plant residues (Phun et al., 2017). Biogas consists mainly of methane (40-75%), carbon dioxide (25-60%), and other impurities and the biogas composition exchanges depend on the type of the substrate (Andriani, Wresta, Atmaja, & Saepudin, 2014; Rajendran & Balasubramanian, 2011).

Anaerobic digestions (AD) are divided into three categories depending on the solid contents. The low solid reactors contain less than 10% total solids (TS) with a material to water ratio of 1:10 (Kleinheinz & Hernandez, 2016). The medium solid reactors contain 15-20% TS with a material to water ratio of 1:5-7. Finally, the high solid reactors have TS of 22-40% with a material to water ratio of 1:2.5-4.5 (MONNET, 2009). Generally, the organic dry matter content that is suitable substrates for AD is in a range of 70-95% of TS. The TS in cotton waste are in this range. The TS affect the survival growth and activities of microorganisms in anaerobic reactors (Budiyono, Syaichurrozi, & Sumardiono, 2014). Anaerobic batch reactors of solid wastes are more useful because they can perform quick digestion with simple and inexpensive equipment, and can help in assessing the rate of digestion easily (Khalid, Arshad, Anjum, Mahmood, & Dawson, 2011). Substrates with less than 60% of organic dry matter content are rarely considered valuable for AD (Vögeli, Lohri, Gallardo, Diener, & Zurbrügg, 2014). The performance of AD process is highly dependent on the characteristics of substrates as well as the activity of the microorganisms involved in different degradation steps (Horváth, Tabatabaei, Karimi, & Kumar, 2016). A substrate that provides carbohydrates, proteins, fats, cellulose and hemicelluloses is suitable for AD (Achinas, Achinas, & Euverink, 2017; Rajendran, 2015).

The optimization of the AD has been mainly concentrated on the operational parameters such as moisture content (MC), TS, VS, mixing, pH, C/N ratio, loading and retention times, temperature, feedstock composition, and pre-treatment methods (Ferguson, Villa, & Coulon, 2014). Facilities that are available run using mainly industrial wastes as feedstock. Nevertheless, the need for expanding AD to a range of new substrates has raised attention in key points which should be taken into consideration when new feedstocks are going to be used. The use of cotton wastes for the production of high-value compounds including biogas and industrial products provides a means to overcome disposal issues, reduce consumption of fossil fuels and mitigate adverse impacts on the environment (Sharma-Shivappa, 2008). From literature, there is very limited work that has used cotton waste as a substrate for biogas production (Ismail & Talib, 2016). Isci and Demirer (2007) studied the anaerobic treatability and methane generation potential of different cotton wastes in batch reactors. Results indicated that cotton wastes can be treated anaerobically and are a good source of biogas. Given its large potential for biogas production, cotton certainly merits more research attention for being used as a feedstock indigestion with manures. Ismail and Talib (2016) examined the potential of using recycled medical cotton industry waste as a source of biogas recovery. Further studying of biogas production from cotton waste by continuous system and regeneration of alkali solutions of CO_2 if large amounts are used is needed. Rasel et al. (2019) studied the cotton waste (spinning,

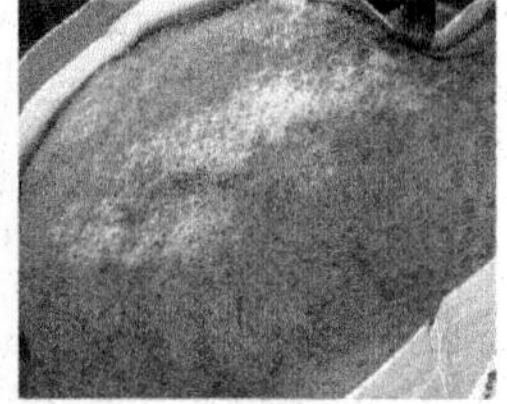

Figure 1. Cotton yarn waste sample.

knitting, and cutting waste) proper utilization via biogas production. However, to the best of our knowledge, no study focused on CYW to produce biogas. The objective of this research was therefore to investigate the potential of procuring biogas from CYW using an anaerobic batch digestion process.

2 MATERIALS AND METHODS

The CYW was the substrate used in this study. CYW were collected from Rivatex Eastern Africa Ltd, Kenya while fresh cow manure used as inoculum was collected from a farm at Moi University, Eldoret, Kenya. The CYW was cut into small pieces using a pair of scissors to facilitate biodegradability (Wang, 2010) and kept in the laboratory for one week (Figure 1). Measured 20% of the total volume of the working reactor was used as inoculum. The inoculum was kept in the refrigerator at 4° for two days and was used without any further treatment. The physicochemical properties of CYW and inoculum were characterized before digestion and the mixtures loaded were prepared according to those characterizations. The TS, VS, and MC were analysed according to standard Methods 2540 (ALPHA, 2012). Kheldahl method was used to determine the total nitrogen content. Total carbon analysis was determined using the Walkey-Black potassium dichromate method (Bakr & El-ashry, 2018; MYOVELA, 2018). The pH was analysed using PH-009 (I) A (pen-type pH meter). The experiment was carried out in batch type laboratory-scale reactors at Chemical and Process Engineering Laboratory, Moi University, Kenya between December 2019 and May 2020.

The reactors of 2 liters' total volume, 12 cm of diameter, and 25 cm height each made of aspiration plastic bottles were used for biogas production (Hasanzadeh et al., 2018). All the reactors with 50% working volume (1 kg) were run concurrently. These reactors were closed with suitable rubber plugs and some holes were dispersedly drilled in the center of the plug for water displacement and biogas collection. The flexible rubber piper and syringes were used to pass water in and out of the conical flask (1000 mL) for displaced water measurements. The reactor was sealed and then arranged for the entire setup. Fresh CYW were mixed with inoculum (needed only at initial state), fed to the reactors, and then the reactors were closed. The batch reactors were buried in a bucket filled with sawdust

Table 1. CYW to water ratio and the loaded materials.

Reactors	CYW to water	CYW in gram	Water in gram
R_1	1:1	427	400
R_2	1:1.5	342	480
R_3	1:2	285	533
R_4	1:2.5	244	571
R_5	1:3	214	600
R_6	1:3.5	190	622
R_7	1:4	171	640
R_8	1:5	142	667
R_9	1:6	122	686
R_{10}	1:10	84	727

Figure 2. Full biogas set up.

at depth of 30 cm to minimize temperature fluctuation during the day and night (Figure 2). The operating temperature was 30 ± 3°. The pH was maintained at 7.2 ± 0.4. The following equations were used to determine the TS, TVS, and MC. (Hasanzadeh et al., 2018). Both TS and MC content was calculated on a wet basis.

$$TS = \frac{W_3 - W_1}{W_2 - W_1} \times 100 \quad (1)$$

$$TVS = \frac{W_3 - W_4}{W_3 - W_1} \times 100 \quad (2)$$

$$\%MC = \frac{W_2 - W_3}{W_2} \times 100 \quad (3)$$

Where = Weight of crucible, = Weight of wet material and crucible, = Weight of dry material and crucible at 105° ovens, = Weight of material and crucible after ignition at 550°.

The volume of biogas collected in the conical flask was measured by the water displacement method daily for 15 days to 36 days. The operating parameters of the reactor were controlled to enhance microbial activity and thus increase the anaerobic degradation efficiency of the system. Evaluation of process parameters was done periodically to assess the efficiency of the anaerobic treatment. The CYW was mixed with water to maintain the TS content at 50%, 40%, 33%, 28%,

Table 2. Physicochemical characteristics of CYW and inoculum before digestion.

Characteristic items	CYW	Inoculum
pH	7.1	6.4
%MC	6.82	92.67
%TS	93.18	7.33
%TVS (% of TS)	82.48	88.64
%AC (% of TS)	17.52	11.36
Total Carbon (% TC)	35.7	32.25
Total Nitrogen (% TN)	0.84	1.57
C/N ratio	42.5	20.5

25%, 22%, 20%, 18%, 14% and 10% in the batch reactors. This was done to investigate the effect of %TS on biogas production. The experiment was run in triplicate. Each reactor was then filled with 200 g of fresh cow dung as inoculum to get the constant quantity of working volume.

3 RESULTS AND DISCUSSION

3.1 *Physicochemical characteristics of CYW and the inoculum*

The characteristics of feedstock are important in designing and operating anaerobic reactors. The initial characteristics of feedstock strongly affect the quality and quantity of biogas produced and as well as anaerobic stability (Phun et al., 2017).

The biodegradability of organic substrates is influenced by their physicochemical characteristics (Aslanzadeh, 2014). The characteristics of organic wastes determine the success of the AD process (e.g. high biogas production potential and degradability). The physicochemical characteristics of CYW and inoculum are given in Table 2.

The TS content of the feedstock was 93.18% and 7.33% for CYW and inoculum respectively. The TS content of loaded materials was between 10 and 50%; it can be categorized as low, medium, and high solid content (Anahita & Saad, 2019; MONNET, 2009). The potential of gas production from the substrate depends on the TVS loading of the reactor and the percentage of TVS reduction through digestion (Adebayo & Odedele, 2020). Therefore, a substrate with a high concentration of TVS is the best for AD. The TVS percentage of TS of CYW and inoculum were 82.48 and 88.64%, respectively (Table 2). This is in agreement with the suggested value (70-95%) for biogas production (Getahun & Gebrehiwot, 2014). This shows that the large fraction of CYW is biodegradable thus it can serve as a good feedstock for biogas production. The MC of CYW and inoculum were 6.82% and 92.67%, respectively. The low MC of CYW could have been due to its high solid content. Le et al. (2011) and Khalid, Arshad, Anjum, Mahmood, and Dawson (2013) reported that the substrate which contains 70-80% of MC is good for AD. Therefore, the MC of CYW is out of this range. However, the MC of loaded materials was in the range of 50-90% which was in the preferable range of AD.

The C/N ratio of the inoculum was 20.5 which was in the suitable range to keep the AD in a stable condition (Bambokela, Matheri, Belaid, Agbenyeku, & Muzenda, 2016). However, the CYW had a high C/N ratio of 42.5 and therefore this is not easily degradable carbon content. Although, this can lead to accumulations of fatty acids in the reactor, lower pH, and cause inhibition to methanogenic bacteria (Lee et al., 2015; Page et al., 2015). The optimum C/N ratios between 20:1 and 30:1 have been suggested as suitable (Matheri, Ndiweni, Belaid, Muzenda, & Hubert, 2017; Patinvoh, Mehrjerdi, Horváth, & Taherzadeh, 2016). In spite of that, C/N ratios of the feedstocks are often much lower or high than this (Habiba, Hassib, & Moktar, 2009; Patinvoh et al., 2016). A substrate with a high C/N ratio has poor buffering capacity while low C/N results in the accumulation of ammonia and increasing the pH which becomes toxic to methanogens (Anahita et al., 2019). Nevertheless, the inoculum (fresh cow dung) was suitable to increase the buffer capacity of the digestion process (Gu et al., 2014). However, there is possible variation in buffering capacity that may result in variation of substrate composition (Anahita et al., 2019). Additionally, Gu et al. (2014) suggested that a suitable inoculum can provide the nitrogen, micronutrients, and macronutrients for AD. Nevertheless, for solid wastes with a high C/N ratio, the ammonia inhibition effect can be compensated by dilution with water which lowers the concentration of potential inhibitors (Chen, Cheng, & Creamer, 2008).

According to Einarsson and Persson (2017), it makes sense to consider both the C/N ratio and TS content as can both contribute to biogas production. The results of physicochemical properties of effluents have been shown that there were decrease in the average of TS and TVS. After digestion, the TS, TVS, and MC were 21.61%, 23.61%, and 78.38%, respectively. The moisture content was high due to water addition before digestion and water that has increased during the hydrolysis process. Olanrewaju and Olubanjo (2019) explained that there was a reduction in the TS and TVS as biogas yield increased. In this study, the digestion sample diluted at 28% TS (28:72% ratio of solid to water ratio) exhibited relatively high biodegradation efficiency. Therefore, CYW exhibited characteristics indicating that it would be readily biodegradable, due to the high TS and TVS of the solid fraction. This indicated that the CYW used in this study had the potential for use as a substrate for biogas production.

3.2 *Biogas production*

One of the specific objectives of this research was to determine the performance of AD of CYW conducted at different loading ratios of TS concentration. For this reason, it was important to evaluate the process performance in terms of biogas composition as well as production of various loading rates of TS concentration. The experimental results of the daily biogas production of reactors R_1, R_2, R_3, R_4, R_5, R_6, R_7, R_8, R_9, and R_{10} are illustrated in Figure 3. The retention time for the reactors was between 15 to 36 days. This was in the range of 23 days reported by Isci and Demirer (2007) but lower than 9 days reported by Saravanan, Sendilvelan, Arul, and Raj (2009) where they produced biogas from cotton wastes in 23 and 45 days respectively. Each of the reactors had two peaks, while the value peaks and positions were different. The peaks of biogas production for the first days of digestion may be related to easily biodegradable substrates that were present in the CYW (high solid content, carbohydrates, proteins, and starch) (Parawira, Murto, Zvauya, & Mattiasson, 2004; Wei et al., 2019). It was also noted that, after the conversion of the easily biodegradable fraction, the system needed to start over the degradation of more complex compounds with a greater level of difficulty (Parawira et al., 2004; Xia et al., 2018); a fact that was evidenced by the lowering of the biogas production. Another reasonable explanation of these peaks of biogas could be the lack of oxygen at the beginning of the experiment which was caused by nitrogen flow in the reactor headspace (Parawira et al., 2004). The earliest production peaks in the reactors may be also associated with the capacity for adaptation to the AD process of the microorganisms already present in the inoculum.

Gu et al. (2014) reported that the rapid production of biogas in the early days was due to a large amount of organic matter available in the reactor. Furthermore, all the reactors displayed very similar trends in biogas production. Biogas production increased rapidly for the first days and then sharply declined in 7 days. Afterward, the production began increasing up to the highest production volume. Thus, the highest daily biogas production was observed between day 10 and day 16 for all reactors. The anaerobic digestion process of CYW was in three phases including fast digestion (1-7 days), steep descent digestion (8-16 days), and gradual descent (after 16 days). Figure 3 shows that biogas production was not ended at the same time. This could be because the carbon in the reactors was not equally degraded or converted to biogas. The highest daily biogas production for R_1, R_9, and R_{10} was 600 mL, 156 mL, and 147 mL on day 10, respectively. Similarly, the highest daily biogas production was also recorded on day 13 for R_2 and R8 with a daily amount of 330 mL and 245 mL respectively.

Finally, the highest biogas production for R_3, R_4, and R_5 was 395 mL, 617 mL, and 500 mL recorded on day 15 while R6 and R7 had 351 mL and 319 mL recorded on day 15, respectively. It was found that R_4 produced the highest biogas yield (7178 mL) after 36 days. R_{10} produced the lowest biogas volume of 932 mL during the entire experiment period. This was due to low substrate loaded in the reactor as the biogas production increases with an increase in substrate loading. Filer, Ding, and Chang (2019) showed that if the substrate loaded is too low, there is a possibility

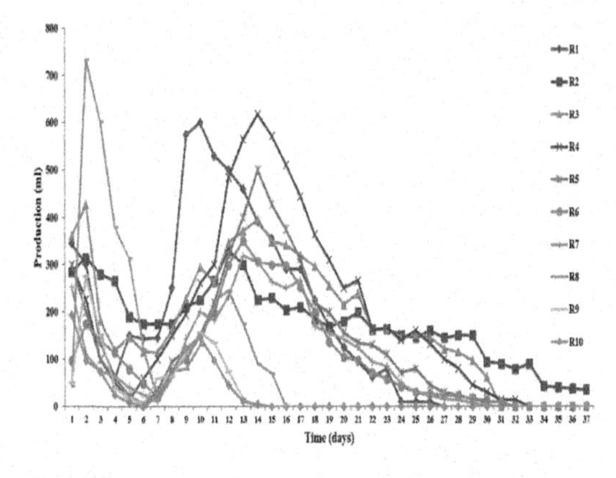

Figure 3. The cumulative daily biogas production.

of low quantities of biogas produced due to the low metabolic activity of the microorganisms. However, this has been contrasted by Creamer et al. (2008) who showed that the total biogas produced was not affected by the quantity of the substrate loaded but the quantity of TS digested. Therefore, the biogas production was related to the value of the %TS and MC presented in the reactors.

3.3 *The effect of total solids content on biogas produced*

Analysis of results showed that the volume of biogas produced is important for controlling and monitoring the process of AD. A good biogas production reflects the proper operation of the reactor. This study examined the effects of percentage %TS of CYW on biogas production to determine a suitable value of %TS for optimum biogas production. The effect of TS content on biogas production was studied by varying the TS from 10 to 50%. In relation to TS concentration, Hao et al. (2016) stated that the TS concentration is the most important factor in microbial community activity. The TS concentration in the substrate limits the loading capacity to an organic loading rate (OLR) to prevent system failure from overloading and accumulation of inhibitory compounds (Phun et al., 2017). The results showed that the biogas yield corresponded with the amount %TS concentration in the reactor (Figure 4). There was a gradual increase in biogas production with a corresponding increase in %TS. This showed that there was gradual acclimatization of microbial communities to the conditions in the reactor and probably new predominant microbial communities for high solid-state digestion were formed (Patinvoh et al., 2017). However, as the process continues, a time comes when any minimal increase in %TS concentration would no longer contribute to the increase in the volume of biogas produced (Figure 4). This is predicted to be due to the function of water in reactors since TS content will be directly corresponding to water content. According to Budiyono, Widiasa, and Johari (2014), water content is one of the very important factors that affects AD in solid wastes.

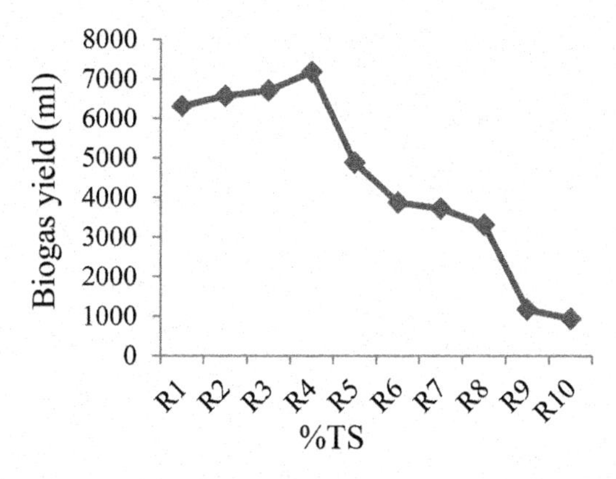

Figure 4. Effect of TS on biogas yield. R_1 to R_{10} represents 50, 40, 33, 28, 25, 22, 20, 16, 14, and 10% TS respectively.

They advanced two main reasons which included (1) water facilitates the movement and growth of bacteria and thus facilitating the dissolution and transport of nutrients and (2) water reduces the limitation of mass transfer of non-homogenous or particulate substrates. A similar trend was observed by Deepanraj, Sivasubramanian, and Jayaraj (2016), who studied the multi-response optimization of process parameters in biogas production from food wastes using Taguchi. In their study, they found that there was a decrease in biogas yield by volatile solid removal efficiency 1.12% and chemical oxygen demand removal efficiency by 12.85% when the %TS was increased from 7.5% to 10% due to poor microbial substrate contact with an increased amount of substrate in the reactor. Parawira et al. (2004) also reported that biogas yield from potato solid wastes increased as the TS increased from 10% to 40% and then decreased as TS was increased from 50% to 80%.

This is possible because when %TS increases, the amount of water decreases, thus reducing the level of microbial activity which then affects the amount of biogas produced. This is most evident at higher values of TS. Igoni, Abowei, Ayotamuno, and Eze (2008) showed that slurry of high TS concentration was more acidic than that of lower TS concentration, which is an additional reason why a higher value of TS concentration would not significantly lead to an increase in the volume of biogas produced. Finally, the most important finding of this research was that the best performance for biogas production was the reactor with 28% of TS.

4 CONCLUSION

The physicochemical characteristics of CYW showed that it has the potential for use as a substrate for biogas production. The CYW had an average TS content of 93.12% and TVS of 86.48% respectively which is appropriate for biogas production. However, the C/N ratio of the CYW was 42.5 which was far higher than expected for AD. Each of the reactors had two peaks which were related to the easily biodegradable substrate that presents into CYW, while the value peaks and positions are different. All the reactors displayed very similar trends in biogas production. The results obtained showed that the amount of biogas produced was related to the %TS in the rectors. There was a gradual increase in biogas production with a corresponding increase in %TS up to optimal value. The reactor (28% TS) showed the highest biogas volume yield (7178 mL), and also gave the highest daily biogas production (617 mL). The results obtained confirm that reactors should run at 28% TS for maximum biogas generation. The CYW has a high C/N ratio therefore further research is required on co-digestion of CYW with other substrates with a low C/N ratio to reduce the C/N ratio to the range of 20-30:1 to optimize biogas production.

ACKNOWLEDGEMENT

The authors acknowledge the sincere financial and moral support from the African Centre of Excellence in Phytochemicals, Textile and Renewable Energy (ACE II-PTRE), Moi University, Eldoret, Kenya which led to this communication.

CONFLICT OF INTEREST

Authors have declared that no competing interests exist.

REFERENCES

Achinas, S., Achinas, V., & Euverink, G. J. W. 2017. A Technological Overview of Biogas Production from Biowaste. Journal of Engineering, 3(3), 299–307.

Adebayo, G. B., & Odedele, O. S. 2020. Production and Characterization of Biogas from Domestic Waste by Anaerobic Digestion. International Journal of Environmental and Bioenergy, 15(1), 1–9.

Al-Hamamre, Z., Saidan, M., Hararah, M., Rawajfeh, K., Alkhasawneh, H. E., & Al-Shannag, M. 2017. Wastes and biomass materials as sustainable-renewable energy resources for Jordan. Renewable and Sustainable Energy Reviews, 67, 295–314.

Anahita Rabii & Saad Aldi, Y. D. and E. E. 2019. A Review on Anaerobic Co-Digestion with a Focus on the Microbial Populations and the Effect of Multi-Stage Digester onfuguration. Energies, 25.

Andriani, D., Wresta, A., Atmaja, T. D., & Saepudin, A. 2014. A review on optimization production and upgrading biogas through CO2removal using various techniques. Applied Biochemistry and Biotechnology, 172(4), 1909–1928.

Aslanzadeh, S. 2014. Pretreatment of cellulosic waste and high-rate biogas production. University of Borås.

Bakr, N., & El-ashry, S. M. 2018. Communications in Soil Science and Plant Analysis Organic matter determination in arid region soils?: loss-on-ignition versus wet oxidation. Communications in Soil Science and Plant Analysis, 00(00), 1–15.

Bambokela, E. J., Matheri, A. N., Belaid, M., Agbenyeku, E. E., & Muzenda, E. 2016. Impact of Substrate Composition in Biomethane Production under Thermophilic Conditions. International Conference on Advances in Science, Engineering, Technology and Natural Resources (ICASETNR-16) 44–48.

Budiyono, Widiasa, I. N., & Johari, S. 2014. Increasing Biogas Production Rate from Cattle Manure Using Rumen Fluid as Inoculums. International Journal of Science and Engineering, 6, 31–38.

Budiyono, Syaichurrozi, I., & Sumardiono, S. 2014. Effect of total solid content to biogas production rate from vinasse. International Journal of Engineering, 27(2), 177–184.

Chen, Y., Cheng, J. J., & Creamer, K. S. 2008. Inhibition of anaerobic digestion process?: A review. Elsevier, 99, 4044–4064.

Deepanraj, B., Sivasubramanian, V., & Jayaraj, S. 2016. Multi-response optimization of process parameters in biogas production from food waste using Taguchi – Grey relational analysis. Energy Conversion and Management, 10.

Deepanraj, B., Sivasubramanian, V., & Jayaraj, S. 2017. Effect of substrate pretreatment on biogas production through anaerobic digestion of food waste. International Journal of Hydrogen Energy, 42(42), 26522–26528.

Diane L. S., & Stephen. F. . T. 2009. Technology Investigation , Assessment , and Analysis. Task 1 Final Report GTI Project Number 20614. North American.

Einarsson, R., & Persson, U. M. 2017. Analyzing key constraints to biogas production from crop residues and manure in the EU — A spatially explicit model. Open Access, 1–23.

Ferguson, R. M. W., Villa, R., & Coulon, F. 2014. Bioengineering options and strategies for the optimization of anaerobic digestion processes. Environmental Technology (United Kingdom), 3(1), 1–14.

Filer, J., Ding, H. H., & Chang, S. 2019. Biochemical Methane Potential (BMP) Assay Method for Anaerobic Digestion Research. Water, 29.

Getahun, T., & Gebrehiwot, M. 2014. The potential of biogas production from municipal solid waste in a tropical climate. Environ Monit Assess, 4637–4646.

Gu, Y., Chen, X., Liu, Z., Zhou, X., & Zhang, Y. 2014. Bioresource Technology Effect of inoculum sources on the anaerobic digestion of rice straw. Elsevier Ltd, 158(2014), 149–155.

Habiba, L., Hassib, B., & Moktar, H. 2009. Bioresource Technology Improvement of activated sludge stabilisation and filterability during anaerobic digestion by fruit and vegetable waste addition. Bioresource Technology, 100(4), 1555–1560.

Hao, L., Bize, A., Conteau, D., Chapleur, O., Courtois, S., Quéméner, E. D.,& Mazeas, L. 2016. New insights into the key microbial phylotypes of anaerobic sludge digesters under different operational conditions. Water Research.

Hasanzadeh, E., Mirmohamadsadeghi, S., & Karimi, K. 2018. Enhancing energy production from waste textile by hydrolysis of synthetic parts. Fuel, 218,41–48.

Horváth, I. S., Tabatabaei, M., Karimi, K., & Kumar, R. 2016. Recent updates on biogas production - a review. Biofuel Research Journal, 10, 394–402.

Hu, Y., Du, C., Leu, S. Y., Jing, H., Li, X., & Lin, C. S. K. 2018. Valorisation of textile waste by fungal solid state fermentation: An example of circular waste-based biorefinery. Resources, Conservation and Recycling, 129(September 2017), 27–35.

Igoni, A. H., Abowei, M. F. N., Ayotamuno, M. J., & Eze, C. L. 2008. Effect of Total Solids Concentration of Municipal Solid Waste on the Biogas Produced in an Anaerobic Continuous Digester. The CIGR Ejournal, X, 1–11.

Isci, A., & Demirer, G. N. 2007. Biogas production potential from cotton wastes. Renewable Energy, 32(5), 750–757.

Ismail, Z. Z., & Talib, A. R. 2016. Recycled medical cotton industry waste as a source of biogas recovery. Journal of Cleaner Production, 112, 4413–4418.

Jeihanipour, A., Aslanzadeh, S., Rajendran, K., & Balasubramanian, G. 2013. High-rate biogas production from waste textiles using a two-stage process. Renewable Energy, 52, 128–135.

Khalid, A., Arshad, M., Anjum, M., Mahmood, T., & Dawson, L. 2011. The anaerobic digestion of solid organic waste. Waste Management, 31(8), 1737–1744.

Khalid, A., Arshad, M., Anjum, M., Mahmood, T., & Dawson, L. 2013. Table of Contents The anaerobic digestion of solid organic waste. Waste Management, 31(8), 1737–1744.

Kleinheinz, G., & Hernandez, J. 2016. Comparison of two laboratory methods for the determination of biomethane potential of organic feedstocks. Journal of Microbiological Methods, 130, 54–60.

Le, R., Chardin, C., Benbelkacem, H., Bollon, J., Bayard, R., Escudié, R., & Buffière, P. 2011. Influence of substrate concentration and moisture content on the specific methanogenic activity of dry mesophilic municipal solid waste digestate spiked with propionate. Bioresource Technology, 102(2), 822–827.

Lee, D. J., Lee, S. Y., Bae, J. S., Kang, J. G., Kim, K. H., Rhee, S. S., & Seo, D. C. 2015. Effect of Volatile Fatty Acid Concentration on Anaerobic Degradation Rate from Field Anaerobic Digestion Facilities Treating Food Waste Leachate in South Korea. Journal of Chemistry, 2015.

Matheri, A. N., Ndiweni, S. N., Belaid, M., Muzenda, E., & Hubert, R. 2017. Optimising biogas production from anaerobic co-digestion of chicken manure and organic fraction of municipal solid waste. Renewable and Sustainable Energy Reviews, 80, 756–764.

Rasel M., Israt Z., Sakib H. B., Kazi M., Hoque H., Mazadul H., & Alam M. M. 2019. View of Industrial Waste Management by Sustainable Way. EJERS, European Journal of Engineering Research and Science, 4(4), 4.

ALPHA, Methods 2540. 2012. Standard Methods for the Examination Water nad Wastewater.

Monnet, F. 2009. An Introduction to Anaerobic Digestion of Organic Wastes–Final Report. November 2003. AIP Conference Proceedings, 1169, 185–189.

Myovela, H. 2018. Anaerobic Digestion of Spineless Cacti (Opuntia Ficus-Indica (L.) Mill) Biomass in Tanzania?: The Effects of Aerobic Pre-Treatment.

Olanrewaju, O. O., & Olubanjo, O. O. 2019. Development of a Batch-Type Biogas Digester Using a Combination of Cow Dung , Swine Dung and Poultry Dropping. International Journal of Clean Coal and Energy, 15–31.

Page, L. H., Ni, J. Q., Zhang, H., Heber, A. J., Mosier, N. S., Liu, X., & Harrison, J. H. 2015. Reduction of volatile fatty acids and odor offensiveness by anaerobic digestion and solid separation of dairy manure during manure storage. Journal of Environmental Management, 152, 91–98.

Papacz, W. 2011. Biogas As Vehicle Fuel. Journal of Kones Powertrain and Transport, 18(1).

Parawira, W., Murto, M., Zvauya, R., & Mattiasson, B. 2004. Anaerobic batch digestion of solid potato waste alone and in combination with sugar beet leaves. Elsevier Ltd, 29, 1811–1823.

Patinvoh, R. J., Mehrjerdi, A. K., Horváth, I. S., & Taherzadeh, M. J. 2016. Dry fermentation of manure with straw in continuous plug flow reactor?: Reactor development and process stability at different loading rates. Bioresource Technology.

Patinvoh, R. J., Osadolor, O. A., Sárvári, I., & Taherzadeh, M. J. 2017. Bioresource Technology Cost effective dry anaerobic digestion in textile bioreactors?: Experimental and economic evaluation. Elsevier, 24: 549–559.

Phun, C., Bong, C., Lim, L. Y., Lee, C. T., Ho, C. S., Ho, W. S., & Ho, W. S. 2017. The characterisation and treatment of food waste for improvement of biogas production during anaerobic digestion – a review. Journal of Cleaner Production 16.

Sharma-Shivappa R. R., & Chen Y. 2008. Conversion of Cotton Wastes to Bioenergy and Value-Added Products. American Society of Agricultural and Biological Engineers, 51(6), 8.

Rajendran, K. 2015. Waste textiles to biogas. Lap Lambert Academic Publishing.

Rajendran, K., & Balasubramanian, G. 2011a&b. High rate biogas production from waste textiles. University of Borås.

Ojolo S. J., A. I. Bamgboye, B. S. Ogunsina, & S. A. O. 2008. Analytical Approach for Predicting Biogas Generation in a Municipal Solid Waste Anaerobic Digester. Journal of Environmental Health Science and Engineering, 5(3), 179–186.

Saravanan, C. G., Sendilvelan, S., Arul, S., & Raj, C. S. 2009. Bio Gas from Textile Cotton Waste - An Alternate Fuel for Diesel Engines. The Open Waste Management Journal, 2(1), 1–5.

Telliard, W. A. 2001. Method 1690 Ammonia-N in Water and Biosolids by Automated Colorimetry with Preliminary Distillation. U . S . Environmental Protection Agency Office of Water Office of Science and Technology Engineering and Analysis Division (4303). Environmental Protection.

Triolo, J. M., Pedersen, L., Qu, H., & Sommer, S. G. 2012. Bioresource Technology Biochemical methane potential and anaerobic biodegradability of non-herbaceous and herbaceous phytomass in biogas production. Bioresource Technology, 125, 226–232.

Velmurugan, S., Deepanraj, B., & Jayaraj, S. 2014. Biogas Generation through Anaerobic Digetsion Process – An Overview Biogas Generation through Anaerobic Digestion Process-. Research Journal of Chemistry and Environment.

Vögeli, Y., Lohri, C. R., Gallardo, A., Diener, S., & Zurbrügg, C. 2014. Anaerobic Digestion of Biowaste in Developing Countries. Dübendorf, Switzerland: Eawag – Swiss Federal Institute of Aquatic Science and Technology. Retrieved from www.sandec.ch

Wang, Y. 2010. Fiber and Textile Waste Utilization. Waste Biomass Valor, 135–143.

Wei, L., Qin, K., Ding, J., Xue, M., Yang, C., & Jiang, J. 2019. Optimization of the co-digestion of sewage sludge , maize straw and cow manure?: microbial responses and effect of fractional organic characteristics. Scientific Reports, 1–10.

Xia, T., Huang, H., Wu, G., Sun, E., Jin, X., & Tang, W. 2018. The characteristic changes of rice straw fibers in anaerobic digestion and its effect on rice straw-reinforced composites. Industrial Crops & Products, 121, 73–79.

Advances in Phytochemistry, Textile and Renewable Energy Research for Industrial Growth – Nzila et al. (Eds)

Optimization of liquid fuel from microwave pyrolysis of used tyres

R.K. Bett
School of Engineering, Department of Mechanical, Production and Energy Engineering, Moi University, Eldoret, Kenya
The Africa Centre of Excellence II in Phytochemicals, Textiles and Renewable Energy (ACE II PTRE), Moi University, Eldoret, Kenya

A. Kumar
School of Engineering, Department of Chemical & Process Engineering, Moi University, Eldoret, Kenya

Z. Siagi
School of Engineering, Department of Mechanical, Production and Energy Engineering, Moi University, Eldoret, Kenya

ABSTRACT: Used tyres pose a threat to the environment, especially in developing countries, since they are non-biodegradable. Liquid from used tyres can used as a source of fuel to replace petroleum diesel. Research has been done to prove that a pyrolysis technique can extract liquid fuel from used tyres. Microwave pyrolysis is an alternative valorization process which is efficient and environmentally friendly, and hence is effective in producing Tyre Pyrolysis Oil. Optimization was done and Central Composite Design was used to optimize the variables. The variables were microwave power, residence time, and particle size. The yield was correlated as a quadratic function of the reaction variables. Response surface and contour plots for the correlation were plotted to indicate the effects of operating variables and finally to identify the points of optimal yield. The highest yield of 38.4 (wt.%) corresponded to a microwave power of 50%, residence time of 17.5 minutes, and a particle size of 25 mm^2.

1 INTRODUCTION

The major source of tyres in Kenya is tyre imports. According to Dilewski (Gernod Dilewski, 2012) about 62% of all tyres in Kenya are imported. Twenty percent of the tyres in Kenya are those that are imported with cars while those produced in Kenya and the grey market account for 18%. The quantities of used tyres are expected to increase exponentially. This is due to the fact that the number of people and companies that buy new cars is increasing (Gernod Dilewski, 2012). In 2012, the total tons per annum of used tyres in Kenya were estimated at 51,000 tons/a and is expected to increase to about 133,000 tons/a in 2031n as shown in the Figure 1.

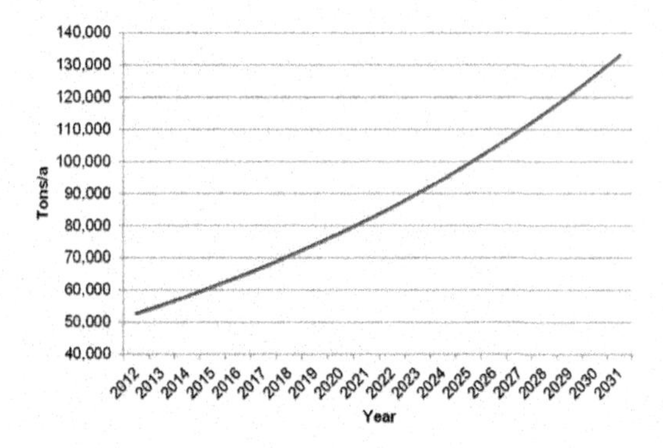

Figure 1. Forecasted quantities of waste tyres in Kenya (Gernod Dilewski, 2012).

According to Mamun et al. (2015) tyre pyrolysis oil has a calorific value greater than that of gasoline and petroleum diesel. This implies that it has the potential to be used as a substitute or a blend with petroleum diesel and can be used in internal combustion engines. Further research indicates that it can be blended with petroleum diesel in various proportions and be used in internal combustion engines (Murugan, Ramaswamy, & Govindan, 2008). Most tyres are made of isobutylene-isoprene copolymer rubber, styrene-butadiene copolymer rubber (SBR), and cis-polybutadiene rubber (CBR). Such rubbers have long-chained polymers bonded with sulfur atoms, therefore making the elastomer very stable and hence limiting degradability (Canon, Muñoz-Camelo, & Singh, 2018). Apart from the synthetic rubber, tyres contain carbon black plasticizers, antioxidants, lubricants, natural rubber, and some inorganic materials such as silica that makes their disposal very complicated. Therefore, due to the difficulty in disposal of used tyres, there are a lot of interests in waste-to-energy conversions using relevant technologies such

DOI 10.1201/9781003221968-28

as Thermochemical conversions. Pyrolysis is one such process that involves the thermal cracking of used tyre material at elevated temperatures in the absence of oxygen to produce liquid, solid, and gas products (Wikipedia, n.d.). Microwave pyrolysis is one of the latest technologies to replace the conventional heating. It is not only faster but also a cleaner process that leads to reduced energy consumption (Kumar, Chirchir, Namango, & Kiriamiti, 2016). The reaction variables in microwave heating include microwave power, reaction time, and particle size. The aim of the current study is to optimize the liquid fuel yield from microwave pyrolysis by varying microwave power, residence time and particle sizes.

1.1 *Problem statement*

The cost of fossil fuels has been escalating, thus rendering their use not only expensive but also unsustainable. Furthermore, the coal and petroleum deposits in the entire world will be exhausted due to continuous exploitation. According to some research, the natural gas, oil, and coal will be exhausted in 54, 53, and 110 years, respectively (Singh, 2015). This makes it necessary to come up with an alternative source of energy. Waste-to-energy conversion is one of the alternative sources of fuel. Used tyres pose a threat to the environment, especially in developing countries, since the current methods of disposal pollutes the environment. Liquid from used tyres has the potential to be used as a source of fuel to replace petroleum diesel (Mamun et al., 2015). It is not only a cheaper alternative but also sustainable due to the availability of used tyres and appropriate conversion technologies. However, despite this cheaper alternative of used tyres, there is a gap of inefficient production of tyre pyrolysis oil in the conversion technologies. This has resulted in lower production of liquid fuel from used tyres and thus reducing the interest in waste-to-energy technologies.

1.2 *Objectives*

The main objective of this study was to come up with optimal conditions for liquid fuel production from microwave pyrolysis of used tyres using experimental and statistical approaches.

1.3 *Limitations of the study*

a) Laboratory prototype may vary from large-scale production in terms of power consumption, reaction time, capital cost, and maintenance.
b) The particle size was limited to 25 mm^2 being the smallest particle size but even smaller particle sizes can be achieved. On the other hand, the sample size of the feedstock was limited to 100 g.

2 LITERATURE REVIEW

There is some research that has been done on microwave pyrolysis process. First, there is research that has been conducted to find out the calorific value of pyro-oil which is a product of microwave pyrolysis (Mamun et al., 2015). The main objective of the research was to find out the calorific value of the oil products and to compare it with that of fossil fuels. The higher calorific value of the pyro-oil was found to be 53 MJ/kg higher than that of petroleum diesel at 44.8 MJ/kg and close to that of methane at 55.5 MJ/kg.

Research to study the effects of microwave power on the microwave pyrolysis process was done by (Song et al., 2017). The tyre powder was treated under different power levels and expressed as specific microwave power (SMP) which is power per 1 g sample. The SMPs chosen were 9 W/g, 15W/g, and 24 W/g. The microwave oven used was rated with an output power of 900 W and SMPs of 9 W/g, 15 W/g, and 24 W/g corresponds to 270W (30%), 450W (50%), and 720W (80%), respectively. It was found that the highest yield of liquid fuel 45 (wt.%) was obtained at a SMP of 15 W/g (50%) while the highest gaseous yield of 18.5 (wt.%) was obtained at a SMP of 24 W/g (80%).

Another study was conducted in 2016 by Alex Lu Chia Yang on microwave pyrolysis of used tyres with and without activated carbon as a catalyst (Yang & Ani, 2016). In this experiment, the scrap tyre was heated at temperatures between 400 and 600°C to produce liquid fuel. The experiments were carried out with and without activated carbon as a microwave absorbent. The main objective was to study the effects of temperatures and activated carbon on the yields. Furthermore, the pyro-oil was characterized for the chemical composition, compound functional group, and the calorific value. It was found out that the optimal temperature for the pyrolysis process was 500 ^{o}C with the highest yield of pyrolytic oil with activated carbon as catalyst at 54.39 (wt.%). Without activated carbon, the highest liquid fuel yield was (28.63 wt.%). The high calorific value of the oil was in the range of 42–43 MJ/kg. It was also found out that the liquid fuel has a complex structure comprising long-chained hydrocarbons (Yang & Ani, 2016).

In 2018, a research on microwave pyrolysis of used tyres with carbonaceous susceptor for production of liquid fuel was done (Idris, Chong, & Ani, 2019). The main objective of the study was to study the effects of temperature on the liquid fuel yield and fuel properties. Activated carbon was used in this study to elevate the microwave temperatures and hence enhance the production of tyre pyrolysis oil. The results indicated that the optimum yield of liquid fuel was at 500°C with a yield of 38.12 (wt.%). The important chemical compounds that were found to be present in tyre pyrolysis oil are xylene (BTX), toluene, and benzene.

3 MATERIALS AND METHODS

3.1 *Feedstock preparation*

A used tyre (Triangle 1000 R 20 10.00X20 truck tyre) was used for the preparation of feedstock (tyre chips). It was shredded by knife to achieve the required size;

the unit of measure being the cross-sectional area. All sizes had a uniform thickness of 2 mm. The feedstock was then screened to remove impurities and dust that were on the surface of the tyre material. They were then dried in the air to remove all the moisture that could interfere with the pyrolysis process. The feedstock was then ready for pyrolysis.

Image 1. Feedstock Preparation.

3.2 *ANOVA and regression analysis*

A central composite design (CCD) with five levels and three factors was used for this analysis. The three variables particle size, microwave power, and residence time were optimized using the CCD. The central values, the step sizes and the range of variables were as follows: central microwave power of 50%, step of 10% and range of 40%–60%; central residence time of 17.5 minutes, step of 4.5 minutes and range of 13 minutes to 22 minutes; and central particle size of 112.5 mm^2, step of 52.5 mm^2, and range of 60–165 mm^2, as shown in Table 1 below.

The number of experimental data obtained at each number of factors is given by the formula, $N = 2^n + 2(n) + n_c$ where N is the number of runs, n is the number of factors and n_c is the number of replications at the centre points. A total of 17 experiments, including three replications at the centre point, were conducted. For a full factorial rotatable design, $\alpha = [2^k]^{1/4} = [2^3]^{1/4} = 1.682$ (Antony, 2014). $0_1, 0_2$ and 0_3 are the centre points while $-\alpha$ (1.682) and $+\alpha$ (1.682) are the axial points. To get the value for the axial point, the following equation is applied: Axial point = mean of both the upper and lower level $\pm\alpha$ (range between the upper and lower level divided by 2) (Anthony, 2014). Therefore, the axial point = X $\pm\alpha$ (range/2); $N = 2^n + 2(n) + n_c = 2^3 + 2(3) + 3 = 17$ runs. A full quadratic model for liquid fuel yield was tested and Design Expert 12–Trial Version was used for ANOVA and regression analysis.

3.3 *Experimental procedure*

An experimental investigation was done on the pyrolysis of a used tyre using a microwave oven (SAMSUNG GE0103MB1) with an output power of 900 W and 2450 MHz. Modification was done on the microwave so as to suit the pyrolysis process. A 50 mm hole was drilled in the ceiling of the microwave so as to hold the neck of the conical flask. In addition, the hole was meant to ensure that the mouth of the conical flask was outside of the microwave cavity. The mouth of the conical flask was then sealed using a wooden cork. The sprout of the conical flask was connected to the Liebig condenser. The 500 mL quartz glass roundbottomed flask was utilized as a reactor. 100g of used tyre was fed into the reactor. The residence time and microwave power were set and microwave was switched on. The feedstock was heated continuously in the absence of oxygen to produce fumes. Due to pressure buildup in the reactor, the fumes escape through the sprout of the conical flask which is connected to a Liebig condenser, as shown in Figure 2. The liquid fuel condenses and is collected in a sample bottle while the gaseous products escape. The solid residue remains in the roundbottomed flask.

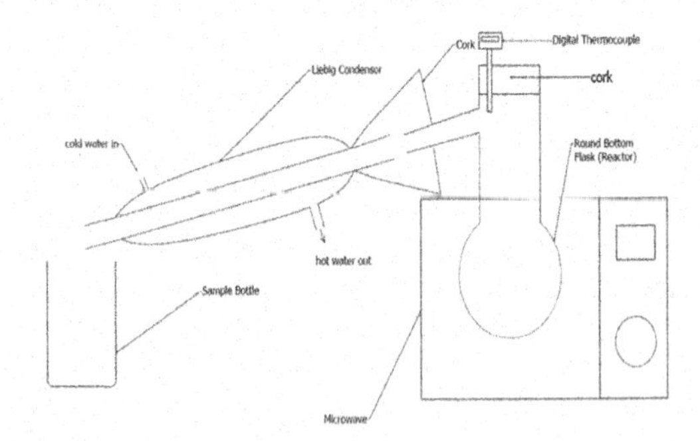

Figure 2. Microwave pyrolysis setup.

4 RESULTS AND DISCUSSIONS

Table 2 illustrates the CCD matrix with experimental and predicted yields after analysis using Design Expert.

where:

X1 is microwave power in %

X2 is residence time in minutes

X3 is particle size in mm^2

Y1 is Experimental yield in (wt.%)

Y2 is Quadratic model yield in (wt.%)

Table 3 gives the ANOVA for regression analysis for a full quadratic model.

Table 1. Independent variables and their levels in CCD.

Factors	Units	$-\alpha$	-1	0	$+1$	$+\alpha$
Microwave Power (X1)	%	30	40	50	60	70
Residence Time (X2)	Mins	10	13	17.5	22	25
Particle Size (X3)	mm^2	25	60	112.5	165	200

Table 2. CCD matrix with experimental and predicted yields.

Std	Run	X1	X2	X3	Y1	Y2	Residual
4	1	60.0	22.0	60.0	34.1	34.75	−0.6531
1	2	40.0	13.0	60.0	27.5	27.66	−0.1627
17	3	50.0	17.5	112.5	35.8	35.4	0.396
15	4	50.0	17.5	112.5	35.4	35.4	−0.004
10	5	70.0	17.5	112.5	28.5	28.54	−0.0448
9	6	30.0	17.5	112.5	28.0	27.3	0.6968
7	7	40.0	22.0	165.0	29.4	29.96	−0.5601
2	8	60.0	13.0	60.0	29.1	29	0.0991
6	9	60.0	13.0	165.0	28.1	27.55	0.5538
14	10	50.0	17.5	200.0	32.3	32.46	−0.1553
13	11	50.0	17.5	25.0	38.4	37.59	0.8073
12	12	50.0	25.0	112.5	33.2	31.92	1.28
11	13	50.0	10.0	112.5	23.3	23.93	−0.6279
5	14	40.0	13.0	165.0	26.4	26.21	0.192
3	15	40.0	22.0	60.0	33.6	34.61	−1.01
8	16	60.0	22.0	165.0	29.8	30.1	−0.2983
16	17	50.0	17.5	112.5	34.9	35.4	−0.504

Table 3. ANOVA for response surface quadratic model.

Source	Sum of Squares	df	Mean Square	F-value	p-value	
Model	253.36	9	28.15	33.71	< 0.0001	**Significant**
X1	1.86	1	1.86	2.23	0.1792	
X2	77.1	1	77.1	92.33	< 0.0001	
X3	31.86	1	31.86	38.15	0.0005	
X1.X2	0.72	1	0.72	0.8622	0.384	
X1.X3		1	0	0	1	
X2.X3	5.12	1	5.12	6.13	0.0425	
$(X1)^2$	78.84	1	78.84	94.41	< 0.0001	
$(X2)^2$	78.84	1	78.84	94.41	< 0.0001	
$(X3)^2$	0.2034	1	0.2034	0.2436	0.6367	
Residual	5.85	7	0.8351			
Lack of Fit	5.44	5	1.09	5.35	0.165	**Not significant**
Pure Error	0.4067	2	0.2033			
Cor Total	259.2	16				

As illustrated in Table 3, the Model F-value of 33.71 implies the model is significant. There is only a 0.01% chance that an F-value this large could occur due to noise. p-Values less than 0.0500 indicate model terms are significant. In this case X2, X3, X2.X3, $(X1)^2$, and $(X2)^2$ are significant model terms. Values greater than 0.1000 indicate the model terms are not significant. Although X1 was not significant, it cannot be dropped because it was part of model hierarchy. The lack-of-fit F-value of 5.35 implies the lack-of-fit is not significant relative to the pure error. There is a 16.50% chance that a lack-of-fit F-value this large could occur due to noise. The p-value for lack-of-fit was greater than 0.05 and therefore it was not significant.

As illustrated in Table 4, the "Predicted R^2" of 0.8372 is in reasonable agreement with the "Adjusted R^2" of 0.9485 i.e. the difference was less than 0.2. "Adeq Precision" measures the signaltonoise ratio. A ratio greater than 4 is desirable. In this particular case, the ratio of 19.497 indicates an adequate signal. The full quadratic model can therefore be used to predict the yield as a function of selected operation variables,

Table 4. Fit statistics.

PARAMETER	VALUE
Std. Dev.	0.9138
Mean	31.05
C.V. %	2.94
R^2	0.9774
Adjusted R^2	0.9485
Predicted R^2	0.8372
Adeq Precision	19.4967

The full quadratic model is given by:

$$\begin{aligned} \textit{Yield}, Y = {} & -72.00930 + 2.00002\,X1 + 5.85774 \\ & X2 + 0.042149\,X3 - 0.005657\,X1. \\ & X2 + (1.70384\textit{Exp} - 18)\,X1. \\ & X3 - 0.003448\,X2.X3 - 0.018700\,(X1)^2 \\ & - 0.132977\,(X2)^2 - 0.00005\,(X3)^2 \quad (1) \end{aligned}$$

Equation 1 can be further reduced by getting rid of terms that are not significant to give equation 2:

$$
\begin{aligned}
Yield, Y = & -72.00930 + 2.00002\, X1 + 5.85774\, X2 \\
& +0.042149\, X3 - 0.003448\, X2.X3 \\
& -0.018700\, (X1)^2 - 0.132977\, (X2)^2 \quad (2)
\end{aligned}
$$

Equation 2 was used to plot response surface and contours for optimization of liquid fuel yield

Figure 3 is a plot for yield as a function of microwave power and residence time. The optima lie close to a microwave power of 50% and residence time of 17.5 minutes.

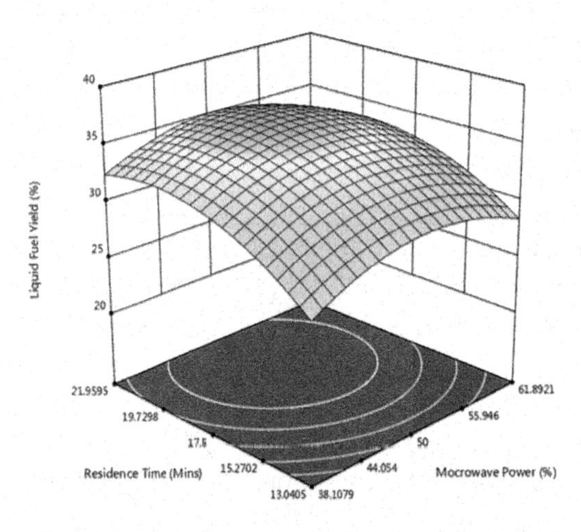

Figure 3. RSM plot for effect of microwave power and residence time on yield.

Figure 4 gives a plot for yield as a function of microwave power and particle size. The optima lie close to a microwave power of 50% and particle size of 25 mm^2.

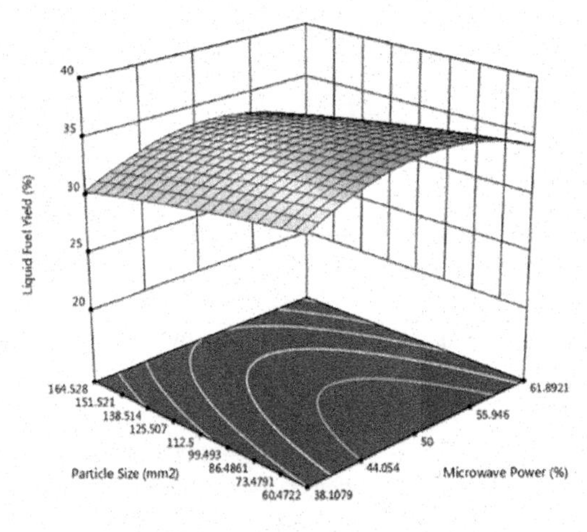

Figure 4. RSM plot for the effect of microwave power and particle size on yield.

Figure 5 gives a plot for yield as a function of particle size and residence time. The optima lie close to a particle size of 25 mm^2 and residence time of 17.5 minutes.

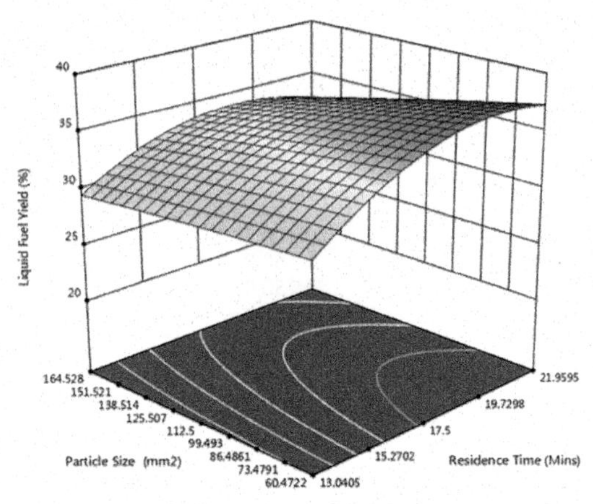

Figure 5. RSM plot for the effect of particle size and residence time on yield.

5 CONCLUSION AND RECOMMENDATIONS

In this study the optimization of liquid fuel yield was carried out using the microwave pyrolysis technique. The liquid fuel yield was correlated in a quadratic equation expressed as a function of reaction variables. ANOVA indicated that the correlations fitted the experimental data satisfactorily. Response surface and contour plots indicated that the highest yield of 38.4 (wt. %) corresponded to a microwave power of 50%, particle size of 25 mm^2, and a residence time of 17.5 minutes. In the published literature, a similar trend is observed for the effect of microwave power on the yield. When a microwave with an output power of 900 W was used to study the effect of microwave power on liquid fuel yield by Song et al. (2017), the optimum yield of tyre pyrolysis oil was at microwave power of 50%. This is because at 50% power level, all the oily products are cracked sufficiently because of sufficient temperatures. At lower power levels, the temperatures are lower and all complex components may not be cracked to form oily products. On the other hand, at higher power levels, temperatures will be elevated further and the oily products formed may further be cracked to form gaseous products, thus reducing the liquid products (Song et al., 2017). The particle size of 25 mm^2 was the smallest particle size used and gave the highest liquid fuel yield when interactions with other variables were held constant. Not much research has been done on the effects of particle size on microwave pyrolysis but the possible reason is increased surface area for the microwave heating, according to Hossain and Rahman (2015). The residence time of 17.5 minutes was the optimal time that allowed the pyrolysis process to be completed, beyond which there was no further increase in the liquid fuel yield. With other optimal conditions, it is the optimal time for a 100 g sample of used tyres to produce liquid fuel.

RECOMMENDATIONS FOR FURTHER RESEARCH

The liquid fuel produced is not refined despite the fact that it has a higher calorific value compared to petroleum diesel (Mamun et al., 2015). Research is still needed on refining the liquid fuel so as to be used directly in the internal combustion engines. Furthermore, more research is needed on the composition and possible use of gaseous products.

ACKNOWLEDGMENTS

The World Bank's Africa Centre of Excellence in Phytochemicals, Textile and Renewable Energy (ACEII –PTRE), Moi University is acknowledged for the full funding of the study. In addition, Moi University School of Engineering department of Mechanical, Production and Energy Engineering is much appreciated for the technical support and permission to use most of their facilities.

REFERENCES

Antony, J. (2014). Design of Experiments for Engineers and Scientists. (2nd ed.). *Elsevier*. https://doi.org/10.1016/C2012-0-03558-2

Canon, A. R., Muñoz-Camelo, Y., & Singh, P. (2018). Decomposition of Used Tyre Rubber by Pyrolysis: Enhancement of the Physical Properties of the Liquid Fraction Using a Hydrogen Stream. *Environments*, *5*(6). https://doi.org/10.3390/environments5060072

Gernod Dilewski. (2012). *Waste Tyre Management Kenya*. https://studylib.net/doc/8747300/content-of-the-presentation—kmi—kenya-motor-industry-.

Hossain, M. S., & Rahman, D. A. N. M. M. (2015). Production of Liquid Fuel from Pyrolysis of WasteTires. *International Journal of Scientific & Engineering Research*, *6*(11), 1224–1229. https://doi.org/10.14299/ijser.2015.11.013

Idris, R., Chong, C. T., & Ani, F. N. (2019). Microwave-induced pyrolysis of waste truck tyres with carbonaceous susceptor for the production of diesel-like fuel. *Journal of Energy Institute*, *92*(6), 1831–1841. https://doi.org/10.1016/j.joei.2018.11.009

Kumar, A., Chirchir, A., Namango, S. S., & Kiriamiti, H. K. (2016). Microwave Irradiated Transesterification of Croton Megalocarpus Oil – Process Optimization using Response Surface Methodology. *Proceedings of the 2016 Annual Conference on Sustainable Research and Innovation*, 177–184. https://www.researchgate.net/publication/311790689

Mamun, A. Al, Salam, B., Ani, F. N., Md, S., & Kabir, H. (2015). *Pyrolysis of Scrap Tyre by Microwave*. https://doi.org/10.13140/RG.2.2.32638.59208

Murugan, S., Ramaswamy, M. C., & Govindan, N. (2008). The Use of Tyre Pyrolysis Oil in Diesel Engines. *Waste Management*, *28*(12), 2743–2749. https://doi.org/10.1016/j.wasman.2008.03.007

Singh, S. (2015). *How long will fossil fuels last?* https://www.business-standard.com/article/punditry/how-long-will-fossil-fuels-last-115092201397_1.html

Song, Z., Yang, Y., Sun, J., Zhao, X., Wang, W., & Mao, Y. (2017). Effect of power level on the microwave pyrolysis of tire powder. *Energy*, *127*, 571–580. https://doi.org/10.1016/j.energy.2017.03.150

Wikipedia. (n.d.). *Pyrolysis*. Retrieved April 18, 2021, from https://en.wikipedia.org/wiki/Pyrolysis

Yang, A. L. C., & Ani, F. N. (2016). Controlled Microwave-Induced Pyrolysis of Waste Rubber Tires. *International Journal of Technology*, *7*(2). https://doi.org/10.14716/ijtech.v7i2.2973

Advances in Phytochemistry, Textile and Renewable Energy Research for Industrial Growth – Nzila et al. (Eds)

Renewable energy policy implementation sustainability in Uganda: Enablers and drawbacks

Bosco Amerit*
Department of Business Administration, Makerere University Business School, Kampala, Uganda

Kassim Alinda & Julius Opiso
Department of Accounting, Makerere University Business School, Kampala, Uganda

Muyiwa S. Adaramola
Faculty of Environmental Sciences and Natural Resource Management, Norwegian University of Life Science (NMBU), Ås, Norway

ABSTRACT: Due to the desire to increase the proportion of renewable energy in Uganda's energy mix, the Government of Uganda formulated the Renewable Energy Policy (REP) of 2007, with the view to enhance the use of modern renewable energy in the country from 4% in 2007 to 61% by 2017. Despite all efforts, after more than 10 years of implementation, the set target was not achieved. This paper sought to examine the policy implementation actions to date, and to develop an account of the deficiencies with regard to the governance architecture of the energy sector and how this served to enable or impede policy implementation. The study primarily entailed conducting integrative reviews and a critique of the sector plans retrieved from secondary data. Additional data and information was collected from government documents, national and international project reports, as well as utility companies. Additional information was collected through key informant interviews. It was found out that the major enablers of renewable energy policy implementation in Uganda included the existence of supportive legal and policy instruments, growth in local organizational capacity, and increased ongoing research efforts. The key drawbacks however included the high investment costs for renewable energy technologies, inadequate human capital and training, a weak regulatory framework, and poor enforcement, as well as uncoordinated institutional action. As such, specific measures to address the drawbacks have accordingly been suggested.

Keywords: Renewable energy policy (REP), Government of Uganda (G.O.U), Global Electricity Transfer Feed in Tariffs (GET FiTs), Renewable Energy Feed in Tariffs (REFiTs)

1 INTRODUCTION

1.1 *Background*

Matters of sustainable development of clean energy continue to rank highly in the world's development agenda, ranging from national to regional, up to the international sphere. The 2015 United Nations Sustainable Development Summit adopted the 2030 Agenda for Sustainable Development that sets out targets for the development of renewable energy (UN Report, 2015). Whereas the achievement of the 2030 agenda would significantly contribute to the realization of the SDG7, its effectiveness is largely dependent on the nature of national action taken to either boost the enabling factors or address the possible bottlenecks encountered in the course of policy implementation for renewable energy. According to the Ministry of Energy and Mineral Development (MEMD) of Uganda, renewable energy relates to resources whose sources are replenished continuously by natural processes. These have been categorized to include solar energy, hydropower, biomass, wind, and geothermal among others. Also, peat and wastes are considered as renewable sources of energy in Uganda's energy mix in the energy policy.

Based on the national vision 2040, the government of Uganda developed the energy policy (2002) with the aim of "meeting the energy needs of Uganda's population for social and economic development, in an environmentally sustainable manner." This policy laid the foundation for the development of Uganda's Renewable Energy Policy (REP) (2007) with the goal of increasing the use of modern renewable energy,

*Corresponding author

DOI 10.1201/9781003221968-29

through the improvement of the legal and institutional responsiveness to renewable energy investments. It also catered for the establishment of an appropriate financing and fiscal policy framework for investments in renewable energy technologies, promotion of research, international cooperation, technology transfer, and the realization of efficient and sustainable consumption of biomass energy. Whereas some success was registered in areas such as the hydro power subsector, research on efficient technologies for biomass energy use, wind turbine studies and tests, solar power generation, and geothermal studies and tests, these however have not been realized without challenges.

1.2 *Overview of energy resources and energy mix in Uganda*

Uganda is endowed with renewable and non-renewable energy resources. The renewable energy resources include biomass (firewood, charcoal and cogeneration), hydrological (water), solar energy, geothermal energy, and wind energy resources. Non-renewable energy resources, on the other hand, include crude oil, peat, and nuclear energy, as shown in Table 1. Indeed, even with the huge energy resource potential, access to modern and clean energy remains a critical concern in the country. The country's energy sector is still dominated by the use of biomass with fuel wood, charcoal, and agricultural residues contributing 88% to the national primary energy mix by mid-2019, while electricity (mainly from hydropower power plants) contributed just 2% and petroleum resources accounted for 10% (RNEP, 2019).

As a measure of increasing uptake of renewable energies, the Government of Uganda implemented a power subsector reform program aimed at providing adequate, reliable, and least cost power to meet the country's demand, promoting efficiency in the power sub-sector, and scaling up of rural and peri-urban access to maximize the impact on poverty reduction. This resulted in the implementation of significant structural changes within the policy and regulatory framework of energy sector.

In a bid to reinforce the development and application of renewable energy resources in the country, the Government of Uganda, established the Feed-in Tariff (FiT) structure as a policy tool designed to stimulate investment in the renewable energy sources. FiT usually entails promising producers of energy an above-market price for what they deliver to the grid. In Uganda, the FiT is technology based, covers all renewable energy resources, and is applicable to installed capacity ranging from 0.5–20 MW. Unlike previous phases (I–III), where FiT covered all feasible renewable technologies as long as they were considered priority technologies, it is worth noting that in the current Phase 4 (2019–2021), small hydropower plants and bagasse power generation are now considered priority renewable technologies too. Details of the approved FiT for priority and other technologies are shown in Tables 2 and 3 respectively.

Table 1. Energy resources potential in Uganda.

Energy resources	Potential	References
Renewable energy		
Hydropower (MW)	>4500 MW	NPA, 2013
Solar energy	Mean solar radiation of 5.1 kWh/m^2/day >5000 MW	ERA NPA, 2013
Geothermal energy (MW)	>1500 MW	NPA, 2013
Biomass (cogeneration) (MW)	>1700 MW	NPA, 2013
Wind energy	2 m/s to about 4 m/s @ less than 10 m height	ERA
Waste residues		
Crop residues (selected crops)	148.67 PJ/year	Okello et al., (2013)
Animal manures (selected animals)	65.23 PJ/year	
Forest residues	44 PJ/year	
Fossil fuels		
Crude oil	Reserves: 6.5 billion barrels; 2.2 billion is recoverable	Patey, 2015
Peat	6000 million m^3 (equivalent to 250 Mtoe) >800 MW	ERA NPA, 2013
Nuclear energy	>24000 MW	NPA, 2013

Table 2. Current approved FiT for priority technologies in Uganda (2019 – 2021).*

Technology	Installed capacity, IP (MW)	FiT (US$/kWh)
Hydro	$10 < IP \leq 20$	0.0751
	$5 < IP \leq 10$	0.0792 – 0.0751**
	$0.5 < IP \leq 5$	0.0792
Bagasse (co-generation)	$0.5 < IP \leq 20$	0.0793

*Source:https://www.era.or.ug/index.php/tariffs/generation-tariffs/feed-in-tariff

**Computed as a regressive allocation of costs with increases in plant's installed capacity

Table 3. Current approved FiT for other technologies in Uganda (2019–2021).*

Technology	Maximum Return on Equity (%)	FiT Ceiling (US$/kWh)
Biogas	13.5	0.115
Landfill Gas	13.5	0.066
Waste-to Energy/ Biomass	13.5	0.095
Wind Power	13.5	0.104
Solar PV	10.0	0.071

*Source: https://www.era.or.ug/index.php/tariffs/generation-tariffs/feed-in-tariff

1.3 *Objectives of the study*

Despite the implementation of the sector reforms, the country continues to experience significant power supply shortages, low rates of access to electricity, and high levels of power losses, all of which negatively impact on the country's economic growth. This study interrogated the nature of enablers and drawbacks associated with the status quo. Ultimately, it is hoped that the findings will serve to inform the ongoing review of the National Draft Energy Policy of 2019. With hindsight on the state of policy implementation, policy makers should be able to develop appropriate response strategies to address the existing impediments to renewable energy policy implementation in the country.

2 REVIEW OF LITERATURE

In order to deepen the analysis of the state of renewable energy policy implementation in Uganda, the study was guided by two concepts established under the theory of public policy; the concept of public policy implementation and the concept of public policy scenarios. The former is defined as an activity undertaken by the executor of the policy in the hope of obtaining an outcome in accordance with the goals or objectives of a policy itself (Edwards III, 1980). It describes the factors that influence public policy implementation. Such include communication, resources, disposition, and bureaucratic structures. The Public Policy Scenario concept was defined by Mowery and Rosenberg (1979) to mean a learning process that seeks to orients an organization to future dynamics that may justify the alteration of the status quo.

Quite relatedly, according to Grindle (1980), two issues influence the variability of success of policy implementation: the policy content and the context of implementation. The policy content covers a wide spectrum of issues ranging from the extent to which group interests are integrated within the policy, the achievable group benefits, the extent of association between desired change, and the policy targets. It also encompasses the correctness of the implementation-setting of the program, the detail of policy implementation specifications as well as the adequacy of resource allocation to the implementation effort. The policy context meanwhile concerns the implementation environment, which constitutes the nature of strategies that aid policy implementation, the power to implement, institutional dynamics, character of the regimes in power, as well as the level of target group compliance and responsiveness.

Malik et al. (2019) further examined the policies, enablers, and drawbacks to renewable energy deployment for promoting environmental sustainability in the Gulf corporation council. They found that the need for economic diversification to reduce reliance on a single resource enhanced the effort to implement renewable energy initiatives. Additionally, due to the diminished hydrocarbon reserves, loss of oil export revenue, climate change, and the associated mitigation pledges together with the abundant solar energy resource, diversifying energy resources to renewable energy was found to be inevitable. However, due to the apparent lack of combined policy framework for wide-scale renewable energy consumption the renewable energy resource development was curtailed.

According to Ahlborg and Hammar (2014), upon scrutinizing the state of enablers and drawbacks to rural electrification in Tanzania and Mozambique-grid extension, off-grid and renewable technologies, they found that political ambitions based on expected growth of demand was a key enabler to the growth of renewable energy; however, the bottom-up enablers such as local initiatives from industries or churches also contributed immensely to the success of policy implementation. The drawbacks were related to lack of access to human capital, hitches ranging from planning to donor dependency, to low rural market, less interest from private sector, and to technical matters. In addition, Khuong, McKenna, and Fichtner (2018) examined the enablers of renewable energy development in Southeast Asian countries and found that there was tremendous economic growth which had emerged from the impetus of renewable energy development. Furthermore, Mezher, Dawelbait, and Abbas (2012) studied renewable energy policy choices for Abu Dhabi. They found that climate change and fossil fuel depletion were the main enablers for the latest growth in renewable energy (RE) resources. However, despite these findings, the high cost of RE technologies had turned out to be the main hindrance to the diffusion of RE power generation.

2.1 *State of Uganda's energy sector*

Given the small nature of Uganda's economy, both the context and content issues are glaring; large-scale industrialization has been slow and as such there has been limited growth among large-scale consumers, which has consequently led to a mismatch in energy supply and demand (Meyer, Eberhard & Gratwick, 2018). Installed generation capacity increased to 1,252.4 MW with peak demand at 723.76 MW by the end of December 2019 (BMAU, 2020). Despite the mismatch, the local financing that would otherwise support the growth in demand in renewable energy is greatly impaired by the unfavorably high bank lending rates. The local capital market thus barely meets the financial requirements for long-term investments (Meyer, Eberhard, & Gratwick, 2018). Attempts by the Central Bank to reduce the commercial interest rates through the reduction of the Central Bank Rate to 10%, and now 7% during COVID-19, have consistently not attracted similar response from the commercial banks whose lending rates remain oscillating above 18%. Ultimately, besides limiting the growth in the consumer base, the funding for electricity capital development has largely been foreign sourced. Such funding came with quite unfair contractual obligations which resulted in the Government of

Uganda exorbitantly paying for excess capacity (also known as deemed energy) or idle capacity (Kojima and Trimble, 2016: 20, 45; OAG Report, 2016; Okoboi & Mawejje, Munyambonera, & Bategeka 2016). Consequently, the achievement of the overall target of the 2007 Renewable Energy Policy to diversify the energy supply sources and technologies in the country from 4% to 61% by the year 2017 unsurprisingly failed and still remains on paper.

3 MATERIALS AND METHODS

The study relied on secondary data sources and hence adopted an integrative literature review approach. Appropriate information and data were collected from different sources such as government documents and reports, scientific papers, and relevant project reports. The key variables assessed included the overall policy performance objectives and targets, inputs and outputs, and their degree of achievement of intermediate outcomes. Special focus was placed on establishing the nature of motivating success factors and the respective bottlenecks encountered. A combination of random and purposive sampling methods was deployed in selecting the target projects highlighted in the Uganda's Ministerial Policy Statements. Several progress reports of the government agencies also provided relevant information. However, priority was given to assessing targets that were physically verifiable and measurable.

Additionally, quite a significant amount of data was sourced from case studies and a review of renewable energy project reports, Ministry of Energy and Mineral Development reports, reports from Ministry of Finance and Economic Development, reports from the Parliamentary committee on natural resources, the World Bank's Private Participation in Infrastructure (PPI) database, Aid Data, and the China Africa Research Initiative (CARI) project database. In addition, reports from the following government agencies: Uganda National Renewable Energy and Energy Efficiency Alliance (UNREEEA), the Electricity Regulatory Authority (ERA), were reviewed. Moreover, reports from the following utilities: ESKOM Uganda, UETCL, UMEME, independent power producers, as well as relevant press reports and news articles were studied. Finally, we consulted with some of the key stakeholders within each of the relevant organizations, including the GET-FiT Secretariat.

The data collected was analyzed based on a combination of approaches; first the research team reviewed secondary data sources including but not limited to the Energy Policy (2002) and the Renewable Energy Policy (2007) targets. Subsequently, the research team conducted a few key informant interviews with selected project managers. A few call-backs were made to some key informants aimed at triangulating information that was critical in identifying the main enablers and drawbacks to renewable energy policy implementation.

4 RESULTS AND DISCUSSIONS

4.1 *Enablers for renewable energy policy implementation sustainability*

4.1.1 *Supportive legal and policy instruments*

The implementation of the renewable energy policy was guided by various legal and policy instruments, a majority of which derive authority from the 1995 Constitution of Uganda, under Article XI, regarding the promotion of energy policies to meet people's energy needs in an environment-friendly manner. Also among these is The Atomic Energy Act, 2008 and the Electricity Act 1999 which specifically set out the legal framework for reforms in the power sub-sector, the Rural Electrification Strategy and Plan as well as the regulatory framework for power generation from small renewable energy sources. Some of the salient features of this law included the liberalization of the electricity industry, the disbandment of the Uganda Electricity Board (UEB) (historically a vertically integrated monopoly) into three entities (generation, transmission, and distribution), the establishment of the Electricity Regulatory Authority (ERA) to regulate the sector, establishing the Rural Electrification Fund (REF), with the main objective of enhancing rural access to electricity, and the establishment of the Electricity Dispute Tribunal (EDT) that exercises powers to hear and determine electricity sector disputes.

4.1.2 *Growth in local organizational capacity*

The government of Uganda, in collaboration with international development partners occasionally worked out joint mechanisms, structures, and systems that supported the development of renewable energy in Uganda. This involved enlisting substantial participation of the private sector and Non-Governmental Organizations under Public Private Partnership arrangements with a view to enhancing investment in renewable and clean energy. The Global Electricity Transfer Feed in Tariff (GET-FiT) Programme being one such intervention that sought to promote the production of renewable energy from the Independent Power Producers (IPPS). The Energy Fund and Rural Electrification Fund (REF) were created to support, among other things, large-scale investment in renewable energy projects and increasing connectivity, respectively. Subsidies and incentives, such as energy rebates, long-term developed Standardized Power Purchase Agreements (PPAs), Renewable Energy Feed in Tariff (REFIT), as well as the implementation of the Clean Development Mechanisms all seek to support the enhancement of the renewable energy generation.

Additional capacity was provided under the Credit Support Instruments (CSIs) by Uganda Energy Credit Capitalization Company (UECCC), along with the possibility of the availing credit for renewable energy technologies (RETs) by commercial banks, including the soft loans provided by NGOs. Quality control and certification was undertaken by Uganda National Bureau of Standards (UNBS) and Uganda National

Alliance for Clean Cooking (UNACC). Also, following the introduction of the Chinese and Taiwanese silicon solar PV products to the country, the Government of Uganda launched the Rural Electrification Strategy and Plan (RESP) in 2001 but it is yet to align the energy policy to support prosumerism.

4.1.3 *Increased research effort*

The increased proliferation of research institutions simplified the monitoring efforts. There has been increased timely dissemination of the sector's home-grown solutions, challenges, and emerging issues. The output from the research institutions also served to guide policy formulation and review implementation strategies and plans. A cross section of these research institutions included the Global Green Growth Institute, which signed a 5-year working relationship with the government of Uganda, to foster green economic growth through the implementation of a planning framework that sought to support government efforts in expanding investments in renewable energy; the NRECA International which partnered with the Renewable Energy Agency (REA) to define the Uganda's electrification strategy through the Accelerated Rural Electrification Programme funded by the World Bank. Their aim was to develop a master electrification plan for one new electric service territory in Uganda. Today, the team is on course, laying the groundwork to produce master plans for all 13 territories of the country's electric service funded by the USAID/Power Africa. Likewise, an immense contribution was registered by the interventions of Energy4Impact, an organization that supports businesses serving off-grid communities with a range of services for business development services, and access to finance and project development for innovative models.

Efforts from research institutions such as Rocky Mountain Institute provide businesses, communities, institutions, and entrepreneurs with information that supports the accelerated adoption of market-based solutions towards a cost-effective shift from fossil fuels to efficient and renewable resources. These interventions supported the government of Uganda in the development and implementation of an integrated electrification strategy that sought to drive energy access and economic growth. Organizations such as Duke Nicholas Institute continuously engage in direct research on policy issues towards addressing the challenges around increased access to modern energy solutions among the disadvantaged communities. Notably, these efforts have led to the development of new disruptive tools, such as the means to evaluate electricity access through machine learning techniques that are applied to aerial imagery data. Additionally, some of these organizations enabled businesses, investors, development partners, and government to identify the appropriate impactful ways to support off-grid energy access. For instance, the Catalyst of Grid Advisor assisted the Rural Electrification Agency to develop an off-grid electrification strategy for Uganda. This entailed active engagement with the private sector actors and developers to coordinate the development of renewable energy mini-grids, including recommending stand-alone energy solutions as an integral part of a greater national electrification planning paradigm (ERA, 2017).

4.2 *Drawbacks to the renewable energy policy implementation sustainability*

The transition to a renewable-energy economy is indeed a collective and gradual but complicated long-term process that summons interventions from multiple actors for effective social change (Anderson et al., 2015). This covers a wide spectrum of technological, organizational, socio-economic, and political changes (Markard et al., 2012). Given the wide range of stakeholder involvement and resource requirement, government and the concerned implementing entities undoubtedly encountered a multiplicity of bottlenecks in a drive to achieve the transition to renewable energy consumption in Uganda. Briefly, some of the key drawbacks are reviewed below.

4.2.1 *Limited resources and inadequate capacity building*

Whereas efforts to extend the main power grid to rural areas have gradually taken shape in Uganda, its implementation calls for the deployment of a substantial amount of resources. The natural alternative would be to have solar-powered mini-grids as the ultimate answer to rural access to electricity; however, due to resource limitations, this has remained a challenge across the board. The otherwise small and remote communities that would be well-suited for solar mini-grids are not a "one size fits all" category and this further complicates the transition. Ultimately this works to reduce cross-subsidization of customers. Sadly, the costs of providing access are unbearably high due to the remoteness of the sites, dispersed populations, and difficulty of the terrain. Regrettably, the tariff for domestic consumers connected on the main grid is equally not motivating enough due to the relatively high domestic tariff currently operating at 0.15 €/0.19 USD per kWh (UMEME, 2014). This tariff is relatively higher than that for the industrial consumers because the supply at low voltage is associated with higher investment and systems operational costs than it is for the industrial consumers who are supplied at high voltage or medium voltage (Umeme, 2015).

The RET equipment requires a skilled workforce to operate and conduct maintenance which resonates with the works of Wilkins (2010). The need for specialized skills calls for routine training especially in the advent of sophistication in the architectures of the technologies used; first to conduct installation and to provide after sales extension services as well as to create awareness of the availability and operational features of RET. Technical support to support the diffusion of RETs, such as the solar PV systems among rural communities, calls for a large workforce with basic technical skills. Such artisans that ensure ready access to spare parts require specialized training. Generally, inadequacies in such auxiliary industry expertise results in increased cost of RE projects which further complicates the deployment of RET as supported by Murphy, Twaha, & Murphy (2014). The state of Uganda's affairs in relation to training facilities, operation, and maintenance of RETs is still deficient

4.2.2 *Prohibitive costs of investment and operation*

The high investment cost for renewable energy projects has generally been one of the limiting factors. This cascaded into the unaffordability of energy especially by rural communities even when provided by the independent power producers. The cross-cutting challenge has been the unavailability of micro-finance schemes for the renewable energy technologies customized for the rural areas. The available alternatives for such financing have been the banking sector whose lending costs have been prohibitively high (typically at rates as high as 25% per month). A good proportion of Uganda's population have no or limited access to institutional micro-finance services and thus rely largely on moneylenders, suppliers, family, and friends for short-term seasonal loans which may not sufficiently support such investments in energy project. Also, given the limited secure savings options available to the households at the local community level the populace has been unable to build a relevant asset base that would over time support investments in renewable energy initiatives.

Additionally, the marketing and maintenance structure for the renewable energy technology devices in rural areas have been hard to come by. Running rural outlets has not been considered profitable due to the high costs for the transportation and mobilization of the dispersed nature of the population, low incomes, and low demand that further complicate the status quo. A critical barrier to the development of RE technology rests on the high initial investment and installation costs of RE equipment. The also eroded the investor con?dence aggravated by the overall inadequacy of ?nancing tools. Further, given the increased costs of investment in RE equipment and devices some unscrupulous persons have taken advantage of this void to supply sub-standard and most times fake energy equipment such as the solar PV system components which fail shortly after they are installed or commissioned. It has therefore been very dif?cult for an average Ugandan/company to invest in RE technology systems. Additionally, before obtaining a license for the construction of a power generating facility, it has been a prerequisite to conduct feasibility study as well as environmental impact assessment (EIA). These require hiring consultants who may not be easily affordable to local developers on account of limited funding options.

4.2.3 *Bureaucracy and mis-coordination*

The inefficiencies associated with Uganda's public service, by defaults aligned the administrative systems to bureaucracy across the spectrum of decision-makers. Based on varying mandates, it has been difficult to find a common denominator among concerned actors. For instance, while the Electricity Regulation Authority (ERA) issues permits and licenses for the generation of power; the Ministry of Energy and Mineral Development (MEMD) and the Renewable Energy Agency (REA) issues support agreements and subsidies in relation to rural electrification schemes; the Directorate of Water Development sanctions the utilization of water resources; and the National Environmental Management Authority (NEMA) provides clearance on Environmental Impact Assessment; the Uganda Investment Authority (UIA) issues the investment licenses; and finally the local governments of the areas where the project is planned provide ground clearance to commence the initial operations, among others (MEMD, 2018). It is not strange to find that regulations are overlooked in certain areas for lack of a robust monitoring mechanism because of the level of manual configuration of the regulating agencies. There were therefore involuntary project delays occasioned by such manual operations of the concerned regulators which frustrate investment. Whereas the planning was gradually adapting to a long-term orientation, it had assumed a short run dimension due to the lack of data and information to support strategic decision-making based on informed long-run projections.

The private sector in Uganda has played an increasingly important role in the energy sector, especially since the enactment of 1999 Electricity Act. One of its critical deficiencies however was associated with government's failure to develop a comprehensive policy framework that would inspire investments in the nuclear and thermal energy. There is a need to revise the existing laws with an emphasis to gradually adopting smart grid technologies and to support the prosumer movement in the renewable energy sub-sector. International experience suggests that co-operation between the public and private sectors in the form of public–private sector partnerships (PPP) can be a powerful incentive for improving the quality and efficiency of public services, and means of public infrastructure financing. Sadly, the extensive and sophisticated level of corruption that apparently exists among the PPP has not only curtailed the efficiency of some of the energy projects but has rendered most partnerships less effective.

4.2.4 *High power tariffs and grid unreliability*

Due to low household connectivity capacity and the limited coverage of the national power grid, electricity access remains very low with only 6% of rural households have access to grid power as compared to 40% of urban households (MEMD, 2020). This has been partly attributed to the high prevailing power tariffs (National Development Plan III, 2020). Attempts to fast track the construction of new transmission lines to distribute power from new plants have been slow. A case in point was the Ayago dam in which government had to pay for deemed power after failure to construct a power evacuation grid before the completion of hydro power dam. Efforts to extend the network to the rural areas were slowed down by systemic corruption which ultimately interrupted the implementation effort, hence stalling the project (Parliament of Uganda, 2020). Attention needs to be paid to implementing community schemes as well as the provision of subsidies to independent power producers operating the mini grids (ERA, 2018)

The existing grid exhibits significant deficiencies and may not conform to the 4Ds (digitization, decentralization, decarbonization and democracy) modernization agenda any sooner. Most of the power challenges with the existing national grid are occasioned by reliance on old technology. The current grid still operates on conventional technology that is highly unreliable (ERA 2019) and can barely support increased grid load arising from increased connectivity and rural grid extension, albeit with over 16% losses (UMEME 2020). This technology insufficiently supports the current independent power producers (IPPs) considering that most IPPs are generally decentralized yet the grid was largely designed to support the needs of centralized systems (Twaha, 2016). There are no systems to monitor and enforce Decarbonization.

4.2.5 *Ineffectual quality control*

The inadequate capacity to superintend over the national technical standards and operational quality control created substantial drawbacks to the increase in renewable energy system. As such, concern for decarbonization which would otherwise be the foundation for the increasing the adoption of renewable energy remains a peripheral matter. This can be attributed to the limited awareness of the benefits of adopting renewable energy along with shortage of the required training meant to increase efficacy of RETs, in line with the works of by Murphy (2014). A number of RETs are described through the lenses of absence of minimum standards in relation to their performance, durability and reliability (Fashina, 2017) which subsequently upset the potential for large-scale commercialization (Raisch, 2016). Currently, there are limited standards and regulatory requirements within Uganda. Even where such exist, the effectiveness of the enforcement teams is quickly curtailed by the influence of powerful politicians with hidden interests through a process best described by Stigler (1975) as regulatory capture. Even certification of the installations is normally a tall order (Karekezi, 2012; Norton 2015). It is no wonder that there is a huge circulation of substandard/poor quality RE equipment such as solar components and systems into the country.

4.3 *Disruptions from the COVID-19 era*

Upon the advent of COVID-19, Uganda went under countrywide lockdown for over 4 months, further slowing down project implementation of the rural electrification programme. The travel ban disrupted the supplies, supervision, and normal operations of the electricity sub-sector and most contractors have since invoked the force majeure clauses established in their contracts upon failure to meet their contractual deadlines and obligations. The lockdown period stalled most operations and made it impossible to conduct Factory Acceptance Tests (FATs) on the imported key materials which included transformers and conductors acquired from China, India, and Turkey among others (MEMD, 2020). The 2020 annual monitoring reports for the Financial Year 2019/20 present evidence on the delay in the implementation of the rural electrification programme and transmission projects across the country. This was due to the difficulties in shipping the materials from overseas. The restriction also meant that the project field teams could not conduct monitoring and supervision of the work done, ultimately impacting the quality of the final works (BMAU, 2020)

4.3.1 *Reduced electricity demand and revenue collection due to increased costs*

Following the closure of several businesses, hotels and restaurants among others, there was a reduction in the demand for electricity from medium and large-scale consumers. The manufacturing sector was involuntarily forced to cut down production on account of limited market (BMAU, 2020). The data from ERA indicates that the manufacturing sector alone accounts for 78% of all electricity sales. This disrupts the government budget plans by placing extra financial pressure on the government to pay deemed power computed per the contractual terms to cater for the available generation capacity not utilized (MEMD, 2020). The reduction in revenue was further aggravated by aggravated by the government's directive to suspend disconnection of defaulters. Due to the travel restrictions, an increase in the cost of operation was inevitable among energy sector utility service providers. For example, the restrictions increased the risks and effectiveness of the field maintenance staff and the rectification of faults on the already challenged network amidst the requirement to practice social distancing. This rendered difficult the practice of shared means of transport as the number of travelers had been restricted and called for either the use of more vehicles or motorcycles, thus increasing the costs of operation.

4.3.2 *Reduction in the rate of new connections*

During the lockdown, the consumption at household level increased. However, as a measure of ensuring safety of staff and consumers, the utility companies opted to suspend the connection of new users on the national grid (MEMD, 2020), especially given the unreliability of the old technology. As such, a limited number of new household users and commercial enterprises were connected during the COVID-19 lockdown period (BMAU, 2020) down from the Financial Year 2019/2020 target connection of 300,000 households that had been earmarked for connection through the free connections policy. Unfortunately, by the end of the first half of the Financial Year 2019/20, only 30% of the target households were connected. There is a high chance that the targeted household connectivity for the Financial Year 2019/2020 will unlikely be met in a country where the access rate to electricity is just 23%.

5 AREAS OF FURTHER RESEARCH

The possible areas for further research should address the impact of GETFiT Program supported by donors [NORFUND, KfW, EU, DFID and the long run effect of inefficient utilization of loan capital investment on energy projects as well as the increased participation of DFIs and export credit organizations (KfW, WB, ATI, China Exim Bank, AFD, ADB, PTA Bank), large foreign investment funding requirements; Public vs. private interest. Also, there are various government projects in energy (ministerial energy statements). The Government of Uganda has invested in Power Factor Correction (PFC) equipment to reduce energy wastage with the hope that other industry operators would adapt accordingly. There is need to find out: (i) how far has the government pilot project gone to enhance efficiency of targeted industries; (ii) how far have the other neighborhood industries taken up the PFC equipment (technology diffusion); and (iii) what factors may be promoting or otherwise constraining the adoption of PFC equipment.

5.1 *Limitations of the research*

Some of the major challenges encountered during the study were the reluctance of some government agencies to share the requested data and information while others shied off from the key informant interviews. This however did not affect the study in either way as there were several options and sources of information.

6 CONCLUDING REMARKS

6.1 *Summary of findings*

The study examined the current state of the Renewable Energy Policy and discussed the enablers and drawbacks to its implementation in Uganda and suggests policy interventions to address the identified drawbacks.

Considering that Uganda is endowed with huge potential of renewable and non-renewable energy resources, the country could meet all the citizens' energy needs if these resources, especially renewable energy, were well developed. This paper identified three main factors, which are supportive legal and policy instruments, growth in local organizational capacity and increased research effort, as main motivations for the renewable energy implementation and development in Uganda.

However, we found that issues such as limited resources, prohibitive costs of investment and operation, bureaucracy and overlapping roles of the government agencies, inadequate human capacity and training, and a weak private sector, are some of the main obstacles to attainment of the REP targets and limit its success.

6.2 *Policy suggestions*

Based on the findings from this review paper, the following policies and actions are recommended:

The concerned authorities should work towards growing the demand for electricity by domestic customers through reduction of the electricity tariff so that power is made more affordable, thereby justifying consumer choice for electricity vis-a-vis the other dirty sources.

Higher Education Institutions need to consolidate their influence on policy formulation through enhanced capacity in the development of models and simulations, to capture all the kinds of risks in the industry which possess the potential to diminish the sustainability of policy implementation. The modeling and simulation should be able to inform policy actors on the nature of shocks that the industrial sector may guard against hence being able to take timely and appropriate preventive or remedial action.

As critical large-scale consumers of electricity, the Government should further incentivize local manufacturing in order to enhance their consumption. This could be done through deliberately promoting the assembling of essential electrical components like cables, conductors, transformers, and switch gear to reduce supply chain interruptions. The initiative of promoting the Buy Uganda Build Uganda (BUBU) should be extended to companies manufacturing renewable energy equipment.

Whereas the Rural Electrification Programme had registered good performance, and the number of individuals accessing the electricity grid had been increased, projects that were due for funding had not yet been commenced, partly on account of delays in procurement and failure to meet effectiveness conditions. The government therefore ought to earnestly review its procurement laws and make the process efficient but free of deficiencies.

ACKNOWLEDGMENT

Bosco Amerit, Kassim Alinda and Julius Opiso acknowledge the PhD scholarship support by Makerere University Business School (MUBS) Kampala Uganda/Norwegian University of Life Sciences (NMBU) Ås Norway Collaborative project titled 'Capacity Building in Education and Research for Economic Governance in Uganda', which is funded by the Norwegian Agency for Development Cooperation (Norad) through Norwegian Programme for Capacity Development in Higher Education and Research for Development (NORHED) program.

REFERENCES

Ahlborg, H., & Hammar, L. (2014). Drivers and barriers to rural electrification in Tanzania and Mozambique–Grid-extension, off-grid, and renewable energy technologies. Renewable Energy, 61, 117–124.

Akella, A.K., Saini, R.P. & Sharma, M.P. Social, *economic and environmental impacts of renewable energy systems*. Renew. Energy 2009, 34, 390–396.

Budget Monitoring and Accountability Unit: Semi Annual Monitoring Report FY 2019/20

Centre for Research in Energy and Energy Conservation document. Available at: www.creec.or.ug, http://www.infracoafrica.com/projects.

Chien, T. & Hu, J.L. Renewable energy: *An ef?cient mechanism to improve GDP. Energy Policy* 2008, 36, 3045–3052.

Cicale, N.J. The Clean Development Mechanism: Renewable Energy Infrastructure for China Electricity Regulation Authority report 2020 document. Available at http://dx.doi.org/10.1016/j.biombioe.2013.06.003.

Electricity Regulatory Authority document. Available at: https://www.era.or.ug/index.php?option=com_content&view=article&id=214:renewable-energy-opportunities&catid=100:renewable-energy-investment-guide

Electricity Regulatory Authority document. Available at: https://www.era.or.ug/index.php?option=com_content&view=article&id=214:renewable-energy-opportunities&catid=100:renewable-energy-investment-guide

Electricity Regulatory Authority document. Available at: https://www.era.or.ug/index.php?option=com_content& https://www.era.or.ug/index.php?option=com_content&view=article&id=214:renewable-energy-opportunities&catid=100:renewable-energy-investment-guide

Fashina, A.A.; Adama, K.K.; Oyewole, O.K.; Anye, V.C.; Asare, J.; Zebaze Kana, M.G. & ;oboyejo, W.O. *Surface Texture and Optical Properties of Crystalline Silicon Substrates*. J. Renew. Sustain. Energy 2015, 7, 063119-1–063119-11

Fashina, A.A.; Azeko, S.T.; Asare, J.; Ani, C.J.; Anye, V.C.; Rwenyagila, E.R.; Dandogbesi, B.; Oladele, O. &;yeris, M. *A Study on the Reliability and Performance of Solar Powered Street Lighting Systems*. Int. J. Sci. World 2017, 7, 110–118.

Hansen, U.E.; Pedersen, M.B. &;ygaard, I. *Review of Solar PV Market Development in East Africa;* UNEP Risø Centre, Technical University of Denmark: Kongens Lyngby, Denmark, 2014.

Huh, S. Y., Kwak, D., Lee, J., & Shin, J. (2014). Quantifying drivers' acceptance of renewable fuel standard: Results from a choice experiment in South Korea. Transportation Research Part D: Transport and Environment, 32, 320–333.

Kamese, G. *Renewable Energy Technologies in Uganda: The Potential for Geothermal Energy*; A Country Study Report under the AFREPREN/HBF Study; Heinrich Boell Foundation: Berlin, Germany, 2004.

Karekezi, S. *Disseminating renewable energy technologies in sub-Saharan Africa*. Annu. Rev. EnergyEnviron. 1994, 19, 387–421.

Karekezi, S. Renewables in Africa—*Meeting the energy needs of the poor. Energy Policy* 2002, 30, 1059–1069.

Khuong, P. M., McKenna, R., & Fichtner, W. (2019). Analyzing drivers of renewable energy development in Southeast Asia countries with correlation and decomposition methods. Journal of cleaner production, 213, 710–722.

Malik, K., Rahman, S. M., Khondaker, A. N., Abubakar, I. R., Aina, Y. A., & Hasan, M. A. (2019). Renewable energy utilization to promote sustainability in GCC countries: policies, drivers, and barriers. Environmental Science and Pollution Research, 26(20), 20798–20814.

Mawejje, J.; Munyambonera, E., &;ategeka, L. *Uganda's Electricity Sector Reforms and Institutional Restructuring; Economic Policy Research Centre*: Kampala, Uganda, 2012.

Mezher, T., Dawelbait, G., & Abbas, Z. (2012). Renewable energy policy options for Abu Dhabi: Drivers and barriers. Energy policy, 42, 315–328.

Ministry of Energy and Mineral Development document. Available at: https://www.energyandminerals.go.ug/site/assets/files/1081/draft_
revised_energy_policy_-_11_10_2019-1_1.pdf

Ministry of Energy and Mineral Development. Strategic Investment Plan 2014/15–2018/19; Ministry of Energy and Mineral Development: Kampala, Uganda, 2014.

Ministry of Energy and Minerals Development, Strategic Investment Plan 2014/15 – 2018/19, page 33.

Ministry of Finance Planning and Economic Development document. National Budget Framework Paper FY2019/20

Murphy, P.M., Twaha, S., & Murphy, I.S. *Analysis of the cost of reliable electricity: A new method for analyzing grid connected solar, diesel and hybrid distributed electricity systems considering an unreliable electric grid, with examples in Uganda*. Energy 2014, 66, 523–534. Review Report on Uganda's Readiness for the Implementation of Agenda 2030.

Oxford Energy document. Available at: https://www.oxfordenergy.org/wpcms/wp-content/uploads/2015/10/WPM-601.pdf

Pierson, P. *When effect becomes cause: Policy feedback and political change*. World Polit. 1993, 45, 595–628.

Raisch, V. *Financial assessment of mini-grids based on renewable energies in the context of the Ugandan energy market*, Energy Procedia 2016, 93, 174–182.

Szabo, S.; Bódis, K.; Huld, T.; & Moner-Girona, M. *Energy solutions in rural Africa: Mapping electrification costs of distributed solar and diesel generation versus grid extension*. Environ. Res. Lett. 2011, 6, 034002.

Twaha, S.; Ramli, M.A.; Murphy, P.M.; Mukhtiar, M.U.; & Nsamba, H.K. *Renewable based distributed generation in Uganda: Resource potential and status of exploitation*. Renew. Sustain. Energy Rev. 2016, 57, 786–798.

Uganda Electricity Transmission Company Limited. Document: System Performance Reports

UN General Assembly, *Transforming Our World: The 2030 Agenda for Sustainable Development* (A/RES/70/1).25 September 2015.

Wilkins, G. *Technology Transfer for Renewable Energy; Taylor & Francis: Abingdon, UK*, 2010. 135.

Zomers, A. *Remote access: Context, challenges, and obstacles in rural electri?cation*. IEEE Power Energy Mag. 2014, 12, 26–34. 138. Rotberg, I.R. (Ed.) China into Africa: Trade, Aid, and In?uence; Brookings Institution Press: Washington, DC, USA, 2008.

Advances in Phytochemistry, Textile and Renewable Energy Research for Industrial Growth – Nzila et al. (Eds)

A review of low-cost materials for biogas purification

Nyambane Doricah, Sombei Dorcas, Jepleting Anceita & Achisa C. Mecha
Department of Chemical and Process Engineering, Moi University, Eldoret, Kenya

ABSTRACT: The increasing demand for energy for both household and industrial use has necessitated the exploration of renewable energy sources such as biogas. Raw biogas contains 50–70% methane and impurities which comprise 30–40% carbon dioxide, 5% moisture, 0.5% hydrogen sulfide, and other trace compounds. In Kenya, biogas is used in households without purification. This poses challenges such as corrosion of equipment and low biogas calorific value. In this study, we assess the various low-cost adsorbents that have been recently employed for biogas purification. The materials evaluated include activated carbon from organic waste such as coconut shells, and iron oxide from lathe machine iron chips. The performance of these materials in biogas purification that has been reported in various studies is evaluated. The findings indicate that these materials have great potential in biogas purification and can be readily applied in small-scale systems. Such systems are expected to contribute significantly to increase the access to clean energy in rural areas.

Keywords: adsorption, biogas, energy, purification

1 INTRODUCTION

Biogas production is an important initiative aimed at increasing access to clean and safe energy with low environmental impact. The adoption of biogas as a source of energy is therefore a major way of contributing to the realization of Sustainable Development Goal 7: affordable and clean energy. Biogas is produced by anaerobic digestion of biomass such as plant and animal wastes. Raw biogas is composed of methane (50–70%), carbon (IV) oxide (30–40%), moisture (5%), hydrogen sulfide (0.5%), and other trace compounds. The main impurities in biogas are therefore CO_2, H_2S, and moisture (Vijay, Chandra, Subbarao, & Kapdi, 2006). Large-scale production of biogas in developed economies such as in Europe employ biogas purification technologies such as chemical absorption, pressure swing adsorption, membrane separation, and cryogenic separation among others (Niesner, Jecha, & Stehlík, 2013). These technologies are relatively expensive and suited for purification systems for these small-scale systems. Low-cost biogas upgrading systems that use adsorption techniques can be locally adapted for biogas purification at a small-scale level to address the challenge of corrosion of pipes by H_2S in biogas and to increase the heating value of biogas by removing CO_2 and moisture. This study therefore assesses potential candidates for adsorbents that can be used to develop low-cost biogas upgrading systems for small-scale use. The use of locally available materials that are less costly and easily available such as waste iron chips and activated carbon from organic waste is explored.

2 BIOGAS PURIFICATION TECHNIQUES

Biogas upgrading refers to the removal of contaminants in the raw biogas such as moisture, H_2S, and CO_2 to produce a bio-methane stream of a certain desired quality. The main technologies used are absorption, adsorption, membrane separation, and cryogenic separation. However, most of these target large-scale systems and little has been mentioned on technologies for small-scale applications (Awe, Zhao, Nzihou, Minh, & Lyczko, 2017).

Absorption can be achieved by physical or chemical means. Physical absorption using water scrubbing is commonly used and relies on the separation of CO_2 and H_2S from the biogas due to their increased solubility in water compared to methane. It uses pressurized water as an absorbent. Its advantages include the simplicity and high efficiency of methane recovery. The limitations are high investment costs, high operating costs due to high pressure pumping, possible clogging because of bacterial growth, foaming, low flexibility towards variation of gas input, a lot of consumption of water and energy as well as the need for gas drying (Angelidaki et al., 2018). Chemical methods of absorption work in the same principle as physical absorption but chemical reaction takes place between the solvent and the absorbed substances. This makes use of CO_2 reactive absorbents such as alkanol amines (mono ethanol amine) or di-methyl ethanol amine (DMEA), and alkali aqueous solutions such as potassium hydroxide, sodium hydroxide, ferric chloride, and ferric hydroxide, among others (Lasocki, Kołodziejczyk, & Matuszewska, 2015). The advantage

DOI 10.1201/9781003221968-30

of these methods is high bio-methane recovery (typically above 97%). The drawbacks associated with liquid solutions for CO_2 and H_2S removal are the high energy requirement, especially in regeneration of adsorbents, selectivity of chemicals used, negative environmental impact from waste liquids, and high corrosion rate (Awe et al., 2017).

Membrane separation relies on the principle of selective permeability of membranes allowing the separation of the biogas components. The technology is effective for the removal of CO_2, H_2S, and moisture from raw biogas. The advantages of membrane separation are the process is compact, light in weight (thin membranes used), has low energy and maintenance requirements, and easy to process. The drawbacks are high membrane costs and maintenance costs since commercial membranes are fragile (Ryckebosch, Drouillon, & Vervaeren, 2011).

Cryogenic separation exploits the fact that different gases liquefy under different temperature–pressure conditions. The difference in boiling points of biogas constituents can be exploited to separate other gases like CO_2, H_2S, N_2, O_2, and siloxanes from methane at high pressure above 80 bar and low temperatures of up to −160°C (Awe et al., 2017). This technique can purify raw biogas to produce a high-purity product with a methane concentration of 90%–99%. The limitations of this process are that it requires high capital and operating costs due to cooling and compression by a large amount of equipment and instruments such as compressors, turbines, heat exchangers, and distillation columns (Xiao, Avalos, Vinh, & Kaliaguine, 2015).

Adsorption is a relatively low-cost and effective process for biogas purification. Adsorption is deemed the most economical and feasible of all biogas purification techniques (Abdullah, Mat, Aziz, & Roslan, 2017). Adsorbent materials are able to selectively retain some compounds of a mixture by molecular size. Adsorption using processes such as pressure swing adsorption (PSA) separates the different gases from biogas based on their molecular characteristics and the affinity of the adsorbent material. This method is advantageous in that, high bio-methane recovery (95%–99%) is achieved and the gas can be directly delivered at high pressures. However, disadvantages associated with it are high investment and operational costs and an extensive process control is needed, hence it cannot be used for small-scale application. In fact, the most difficult aspect of PSA operation is controlling the high temperature and pressure, which has limited the application of this method on a wider scale (Noorain, Kindaichi, Ozaki, Aoi, & Ohashi, 2019).

Adsorption under ambient conditions using a variety of micro-porous materials, such as activated carbons, zeolites, and metal–organic frameworks have been considered to carry out CO_2 separation (Durán, Álvarez-Gutiérrez, Rubiera, & Pevida, 2018). The utilization of activated carbon offers advantages due to its high adsorption capacity at ambient conditions, low regeneration cost, long-term stability, and fast kinetics. The utilization of activated carbon for biogas upgrading is extensive; it can be used for the removal of CO_2 and H_2S from biogas (Vivo-Vilches et al., 2017). The production of carbon adsorbents from biomass feedstocks can involve physical or chemical activation to develop the porosity. Generally prior cleaning, washing, and drying of carbon-rich materials are required. There are two main steps for the preparation of activated carbon: the first step is the carbonization of carbonaceous raw material below 800°C under an inert atmosphere. The next step is activation of the carbonized product by a physical or chemical method (Mdoe, 2014). Activation chemicals include potassium iodide, zinc acetate, potassium hydroxide, potassium carbonate, and sodium hydroxide (Zulkefli et al., 2019).

Hydrogen sulfide can be removed using a catalyst of iron oxide in the form of oxidized steel wool or chips of iron cut from the lathe operation of any workshop. Iron chips are often disposed of from workshops as they are of no great value. These can therefore be utilized in biogas cleaning after being exposed in air. When raw biogas comes into contact with steel wool/chips, iron oxide gets converted to iron sulfide which forms elemental sulfur when exposed to air (Shah & Nagarseth, 2015). This review evaluates the performance obtained from various studies using activated carbon and iron oxide for the removal of moisture, H_2S, and CO_2 from raw biogas.

3 LOW-COST ADSORBENTS FOR BIOGAS PURIFICATION

3.1 *Activated carbon*

Agricultural wastes and by-products such as wood residues, fruit peels, hulls of rice, and coconut shells are potential sources of activated carbon for biogas purification (Durán et al., 2018). Coconut shells are appropriate for the preparation of activated carbon due to their high carbon content and low ash content. During the activation process, the spaces between the elementary crystallites become cleared of less organized, loosely bound carbonaceous material. The resulting channels through the graphitic regions, the spaces between the elementary crystallites, together with the tissues within and parallel to the graphitic planes constitute the porous structure, with large surface area (Marshall, Ahmedna, Rao, & Johns, 2000).

3.1.1 *Carbon(IV) oxide removal using activated carbon*

Activated carbon can adsorb CO_2 from raw biogas due to its high porosity. The changes in molecular dipoles of CO_2 through an asymmetric bond stretching motion results in some permanent polarity in CO_2 molecules at room temperature and pressure. The charged CO_2 molecules undergo adhesion and Van der Waals attraction towards the high surface area of activated carbon (Mamun, Karim, Rahman, Asiri, & Torii, 2016). The presence of water vapor in raw biogas affects the CO_2 capture process design. CO_2 adsorption capacity reduces to almost 54% under co-adsorption of water at high relative humidity conditions compared to dry conditions (Inés Durán, Rubiera, & Pevida, 2017). To improve the adsorption performance of activated carbon, the preparation conditions have to be optimized. For instance, the addition of coal tar pitch during the preparation process results in a high carbon yield, a superior mechanical resistance and a competitive adsorption performance (Plaza, Durán, Rubiera, & Pevida, 2015). Studies have shown that there is no evidence of adsorption of methane on activated carbon. Also, theoretically CO_2 is acidic and polar, hence it interacts with activated carbon. However, methane molecules are completely neutral and have a regular tetrahedral structure which favors a completely non-polar structure. Therefore, methane is an inert material toward the adsorption on activated carbon (Mamun & Torii, 2017).

3.1.2 *Performance of activated carbon in biogas purification*

Table 1 shows the typical performance of activated carbon in biogas purification.

Table 1. Removal of biogas impurities using activated carbon.

Material	Contaminant	Performance	Reference
Activated carbon (pine sawdust)	Carbon dioxide	Adsorption capacity 2.00–5.24 mmol CO_2/g of AC	(Durán et al., 2018)
Activated carbon	Carbon dioxide	Adsorption capacity of 2.55 mmol/g of AC	(Yang, Gong, & Chen, 2011)
Activated carbon	Carbon dioxide	Methane concentration increased from 62% to 91%	(Mamun et al., 2016)
Activated carbon and potash	Hydrogen sulfide	H_2S decreased from 0.57% to 0.01%	(Orhorhoro, Orhorhoro, & Atumah, 2018)
Activated carbon	Hydrogen sulfide	Adsorption capacity 7.3 mg H_2S/g of AC	(Zulkefli et al., 2019)
Impregnated activated carbon and potash	Carbon dioxide, H_2S, and water vapor	CO_2 reduced from 31.00% to 18.61%; H_2S decreased from 0.057% to 0.01%; Water vapor reduced from 0.93% to 0.01%	(Shah & Nagarseth, 2015)
Activated carbon	Carbon dioxide	2.1mmol/g of AC	(Gil et al., 2015)

Activated carbon shows high adsorption capacity of up to 5.24 mmol CO_2/g of AC. It is efficient in raising the bio-methane concentration by around 29% when used in biogas upgrading. This material is regarded as more effective in the removal of CO_2 from biogas than in biogas desulfurization. The high adsorption capacity of activated carbon makes it a very attractive low-cost adsorbent for biogas upgrading.

3.2 *Iron oxide*

Iron oxide can be made from exposing or reacting iron pellets to/with oxygen or exposing other ironic materials such as steel wool, iron rich soils, or iron chips to oxygen-rich air. Iron oxide is a good adsorbent because of its high surface area-to-volume ratio, its surface can be modified, it has excellent magnetic properties, great biocompatibility, ease of separation using external magnetic field, reusability, and it is comparatively low-cost. It can also coordinate with other elements due to variable oxidation states.

3.2.1 *Hydrogen sulfide removal using iron oxide*

Raw biogas contains hydrogen sulfide which must be eliminated due to its corrosiveness and toxicity. Associated effects are piping and equipment corrosion, and environment and user hazards are possible if hydrogen sulfide is still present when biogas is burned. Through chemisorption, H_2S reacts easily with Fe_2O_3 to form insoluble salts of Fe_2S_3 (Ryckebosch et al., 2011). The rusted iron chips will react with the hydrogen sulfide in raw biogas as indicated in the following equation.

$$Fe_2O_3 + 3H_2S \rightarrow Fe_2S_3 + 3H_2O \quad (1)$$

Iron oxides nanoparticles obtained from acid mine drainage are potential adsorbents for desulfurization, due to their low-cost and good adsorptive performance under ambient conditions (Awe et al., 2017). Hence they can be considered as a promising, cost-effective alternative for biogas desulfurization.

3.2.2 *Performance of iron oxide in biogas purification*

Table 2 shows the typical performance of iron(III) oxide in biogas desulfurization.

These studies illustrate the great potential of iron oxide for biogas desulfurization. It is observed that the efficiency of rusted iron chips (86.6%) is as good as that of commercial iron(III) oxide (90%). The low cost of the iron chips is an added advantage.

4 OTHER ASPECTS: MATERIAL OF CONSTRUCTION AND ADSORBENT REUSE

Fixed bed adsorption columns are employed for biogas purification. The column is packed with the adsorbents. The bed can be constructed using plastic materials which are less costly than metallic or glass materials. To increase the effectiveness and reduce

Table 2. Performance of iron oxide in biogas purification.

Material	Effectiveness	Reference
Rusted iron chips from lathe machine	86.6% H_2S removal efficiency	(Shah & Nagarseth, 2015)
Steel wool plus water and silica gel	Increase in bio-methane concentration from 68% to 90%	(Nallamothu, Teferra, & Rao, 2013)
Oxidized steel wool	95% H_2S removal efficiency	(Magomnang & Villanueva, 2015)
Iron (III) oxide	89% H_2S removal efficiency	(Mamun & Torii, 2017)
Iron (III) oxide	Over 90% removal efficiency	(Ryckebosch et al., 2011)
Nanostructured-iron oxide	Breakthrough capacity of 2.5 mg H_2S/g iron oxide	(Cristiano et al., 2020)

costs, it is important to consider the regeneration and reuse of the spent adsorbents. Regeneration can be achieved by heating the spent carbon in a high-temperature furnace (800^0C) under process conditions of steam, air, and inert atmosphere. The contaminants are vaporized, restoring the carbon's original pore structure, allowing for its reuse. Upon exhaustion of the adsorbent, it can be incinerated to produce energy. In cases where the spent adsorbent is deemed non-hazardous, it can be used for land-fill (Coppola & Papurello, 2018).

5 CONCLUSIONS AND RECOMMENDATIONS

Biogas production is increasing, especially in Africa, to meet the demand for clean and safe energy. However, the presence of impurities such as carbon dioxide, hydrogen sulfide, and moisture limit its use. Most of the purification methods available in the world today require sophisticated equipment and high skills to operate. This makes them unsuitable for developing countries. This study has demonstrated the potential of low-cost adsorbents employed under ambient conditions for biogas purification. The use of organic wastes such as coconut shells as precursors and waste iron chips from lathe machine for the production of activated carbon and iron oxide, respectively, presents an avenue for low-cost materials in biogas purification. Based on the promising results from previous studies, it is necessary to optimize the performance of these materials to increase their effectiveness and accelerate their uptake.

ACKNOWLEDGMENT

The authors acknowledge the Africa Centre of Excellence in Phytochemicals, Textile and Renewable Energy (ACE II-PTRE) at Moi University for funding this study.

REFERENCES

Abdullah, A. H., Mat, R., Aziz, A. S. A., & Roslan, F. (2017). Use of Kaolin as Adsorbent for Removal of Hydrogen Sulphide from Biogas. *Chemical Engineering Transactions, 56*, 763–768.

Angelidaki, I., Treu, L., Tsapekos, P., Luo, G., Campanaro, S., Wenzel, H., & Kougias, P. G. (2018). Biogas upgrading and utilization: Current status and perspectives. *Biotechnology Advances, 36*(2), 452–466. doi: https://doi.org/10.1016/j.biotechadv.2018.01.011

Awe, O. W., Zhao, Y., Nzihou, A., Minh, D. P., & Lyczko, N. (2017). A Review of Biogas Utilisation, Purification and Upgrading Technologies. *Waste and Biomass Valorization, 8*(2), 267–283. doi: 10.1007/s12649-016-9826-4

Coppola, G., & Papurello, D. (2018). Biogas Cleaning: Activated Carbon Regeneration for H_2S Removal. *Clean Technologies, 1*, 4. doi: 10.3390/cleantechnol1010004

Cristiano, D. M., de A. Mohedano, R., Nadaleti, W. C., de Castilhos Junior, A. B., Lourenço, V. A., Gonçalves, D. F. H., & Filho, P. B. (2020). H_2S adsorption on nanostructured iron oxide at room temperature for biogas purification: Application of renewable energy. *Renewable Energy, 154*, 151–160. doi: https://doi.org/10.1016/j.renene.2020.02.054

Durán, I., Álvarez-Gutiérrez, N., Rubiera, F., & Pevida, C. (2018). Biogas purification by means of adsorption on pine sawdust-based activated carbon: Impact of water vapor. *Chemical Engineering Journal, 353*, 197–207. doi: https://doi.org/10.1016/j.cej.2018.07.100

Durán, I., Rubiera, F., & Pevida, C. (2017). Separation of CO_2 in a Solid Waste Management Incineration Facility Using Activated Carbon Derived from Pine Sawdust. *Energies, 10*(6), 827.

Gil, M. V., Álvarez-Gutiérrez, N., Martínez, M., Rubiera, F., Pevida, C., & Morán, A. (2015). Carbon adsorbents for CO_2 capture from bio-hydrogen and biogas streams: Breakthrough adsorption study. *Chemical Engineering Journal, 269*, 148–158. doi: https://doi.org/10.1016/j.cej.2015.01.100

Lasocki, J., Kołodziejczyk, K., & Matuszewska, A. (2015). Laboratory-Scale Investigation of Biogas Treatment by Removal of Hydrogen Sulfide and Carbon Dioxide. *Polish Journal of Environmental Studies, 24*(3), 1427–1434. doi: 10.15244/pjoes/35283

Magomnang, A.-A. S. M., & Villanueva, E. P. (2015). Utilization of the uncoated steel wool for the removal of hydrogen sulfide from biogas. *International Journal of Mining, Metallurgy & Mechanical Engineering, 3*(3), 108–111.

Mamun, M. R. A., Karim, M. R., Rahman, M. M., Asiri, A. M., & Torii, S. (2016). Methane enrichment of biogas by carbon dioxide fixation with calcium hydroxide and activated carbon. *Journal of the Taiwan Institute of Chemical Engineers, 58*, 476–481. doi: https://doi.org/10.1016/j.jtice.2015.06.029

Mamun, M. R. A., & Torii, S. (2017). Enhancement of methane concentration by removing contaminants from biogas mixtures using combined method of absorption and adsorption. *International Journal of Chemical Engineering, 2017*, 1–7.

Marshall, W. E., Ahmedna, M., Rao, R., & Johns, M. M. (2000). Granular activated carbons from sugarcane bagasse: Production and uses. *International Sugar Journal, 102*, 147–151.

Mdoe, J. E. G. (2014). Agricultural Waste as Raw Materials for the Production of Activated Carbon: Can Tanzania Venture into this Business? *Huria: Journal of the Open University of Tanzania*, *16*, 1–15.

Nallamothu, R. B., Teferra, A., & Rao, B. V. A. (2013). Biogas purification, compression and bottling. *Global Journal of Engineering, Design and Technology*, *2*(6), 34–38.

Niesner, J., Jecha, D., & Stehlík, P. (2013). Biogas Upgrading Technologies: State of Art Review in European Region. *Chemical Engineering Transactions*, *35*, 517–522. doi: 10.3303/CET1335086

Noorain, R., Kindaichi, T., Ozaki, N., Aoi, Y., & Ohashi, A. (2019). Biogas purification performance of new water scrubber packed with sponge carriers. *Journal of Cleaner Production*, *214*, 103–111. doi: https://doi.org/10.1016/j.jclepro.2018.12.209

Orhorhoro, E. K., Orhorhoro, O. W., & Atumah, E. V. (2018). Performance Evaluation of Design AD System Biogas Purification Filter. *International Journal of Mathematical, Engineering and Management Sciences*, *3*(1), 17–27.

Plaza, M. G., Durán, I., Rubiera, F., & Pevida, C. (2015). CO_2 adsorbent pellets produced from pine sawdust: Effect of coal tar pitch addition. *Applied Energy*, *144*, 182–192. doi: https://doi.org/ 10.1016/j.apenergy.2014.12.090

Ryckebosch, E., Drouillon, M., & Vervaeren, H. (2011). Techniques for transformation of biogas to biomethane. *Biomass and Bioenergy*, *35*(5), 1633–1645. doi: https://doi.org/10.1016/j.biombioe.2011.02.033

Shah, D., & Nagarseth, H. (2015). Low Cost Biogas Purification System for Application Of Bio CNG As Fuel For Automobile Engines. *International Journal of Innovative Science, Engineering & Technology*, *2*(6), 308–312.

Vijay, V. K., Chandra, R., Subbarao, P. M. V., & Kapdi, S. (2006). *Biogas purification and bottling into CNG cylinders: producing Bio-CNG from biomass for rural automotive applications*.

Vivo-Vilches, J. F., Pérez-Cadenas, A. F., Maldonado-Hódar, F. J., Carrasco-Marín, F., Faria, R. P. V., Ribeiro, A. M., Rodrigues, A. E. (2017). Biogas upgrading by selective adsorption onto CO_2 activated carbon from wood pellets. *Journal of Environmental Chemical Engineering*, *5*(2), 1386–1393. doi: https://doi.org/10.1016/j.jece.2017.02.015

Yang, H., Gong, M., & Chen, Y. (2011). Preparation of activated carbons and their adsorption properties for greenhouse gases: CH_4 and CO_2. *Journal of Natural Gas Chemistry*, *20*(5), 460–464. doi: https://doi.org/10.1016/S1003-9953(10)60232-0

Zulkefli, N. N., Masdar, M. S., Isahak, W. N. R. W., Jahim, J. M., Rejab, S. A. M., & Lye, C. C. (2019). Removal of hydrogen sulfide from a biogas mimic by using impregnated activated carbon adsorbent. *PLoS ONE*, *14*(2), 1–25. doi: https://doi.org/10.1371/journal.pone.0211713

Advances in Phytochemistry, Textile and Renewable Energy Research for Industrial Growth – Nzila et al. (Eds)

Photovoltaic off-grid solar home system sizing using the charging current and total energy methods: A comparison of the two sizing methods

Sebastian Waita
Condensed Matter Research Group, Department of Physics, University of Nairobi, Nairobi, Kenya

ABSTRACT: Kenya receives some good sunshine owing to its location in the tropics. Solar energy, being environmentally friendly and an inexhaustible source of energy, has the potential to change people's lives for the better, especially in the rural communities in Kenya. Solar energy is converted to useful electrical energy using solar panels exposed to the sun's radiation. The solar panels and the other balance of system components are usually interconnected to provide this energy for powering loads. This connected system, popularly known as a Solar Home System (SHS) is creating a lot of impact in the rural communities in Kenya. Although the system's installation is key in the overall system performance, the system has to be well designed and sized as well for it to perform optimally. A number of commercially available sizing softwares are complex and way beyond the average person in terms of cost. There are two main sizing methods: the charging current method, herein abbreviated as CCM, and the total energy method, abbreviated as EOM, which could be easier and more accessible to solar designers and installers. In this paper, we have presented a comparison of the two sizing methods to assess if there is any significant difference between the two methods. On applying both methods to an example, it has been found that they give the same sizing details of the components. Furthermore, the EOM appears more appealing since it provides the total power rating of the system, the most commonly used and understood term in rating solar panels. The method is recommended for estimating the components for small solar PV systems. Large solar PV system sizing needs commercial sizing software.

Keywords: Solar Home Systems, Solar energy, communities, sizing, method, off-grid, charging current, total energy

1 INTRODUCTION

Since the Kenyan government enacted the energy act in 2006 through the sessional paper number 4 of 2004, and thereafter gazetted the renewable energy policy in 2012, there has been an upheaval of activities in the renewable energy sector in Kenya, especially on photovoltaic solar systems. The creation of the Energy Regulation Commission (ERC) (now renamed Energy and Petroleum Regulatory Authority (EPRA)) to regulate the energy sector created a "solar energy rush" to meet the EPRA requirements for licensing to deal with solar photovoltaic energy systems. As a result, a number of institutions are involved in the training of solar technicians mostly at the technician 2 (T2) level. The Department of Physics, University of Nairobi, through the solar academy has also been involved in training solar Photovoltaic technicians since 2012 (Justus Simiyu, et al., 2014).

An off-grid photovoltaic solar system (popularly called stand-alone solar system) has a number of key components: (1) the solar modules—they generate the required electricity on irradiation through the photoelectric effect; (2) the charge controller—it protects the battery bank from overcharging and over discharging; (3) the Battery bank—batteries store the energy for use at night or in the absence of sunlight during the day; (4) the inverter—it converts the direct current (DC) into alternating current (AC) (can be omitted if AC is not needed); (5) the load—the appliances powered by the solar system; and (6) the cables—they interconnect the various components (Michael Boxwell, 2019; Kefa V.O. Rabah, 2005).

A well-designed, sized, and installed solar system will serve the client satisfactorily. However, in cases where the client complains, it is about low performance and poor operational condition. Whereas some cases relate to component failures and are not within the installers' control, most of the above cases are due to failure to visit site, system design, sizing errors/oversights/mistakes, and incomplete understanding of solar systems (Kenya Renewable Energy Association KEREA, Report, 2009).

A professionally installed solar system has to go through a number of steps: site visit, system design, sizing and installation, testing, commissioning, and

DOI 10.1201/9781003221968-31

user training. These steps are presented in detail below. The designing, sizing, and the installation are very critical steps. When a system is wrongly designed, sized, or installed, it will not perform optimally. A number of sizing methods exist and have been tried and they work. However, most of them are expensive, complicated, and require great understanding of computer modeling (Majid Alabdul Salam et al., 2013), simulation (M. Chikh, Mahrane, & Bouachri, 2011; Mohit Jain, & Neha Tiwari, 2012; N.D. Kaushika, Nalin, & Nalin, 2005), and even programming (M. Sidrach-de-Cardona & Ll. Mora Lopez, 1998). In many developing countries, the sales person is both the installer and also the PV solar system sizer, although most have very limited knowledge in the area of solar photovoltaic (PV) sizing and installation. The tendency is therefore to use unjustified estimates which more often than not do not meet the clients' need. A simple straight-forward method is needed for the individual with basic education and who wants to do professional PV sizing, especially in the rural developing world.

In this work, we present and compare by way of an example, two simple sizing methods which can easily be adopted by the PV professional with basic education and limited computer knowledge and skill. We designate the first method as the charging current method (CCM) and the second method as the energy output method (EOM), but first we describe some key steps common to both methods.

2 METHODOLOGY

2.1 *Common steps in both methods*

2.1.1 *Visit the site of installation*

A site visit is always necessary before any work on system design and sizing begins. The site visit helps the installer to: have a real assessment of the solar resource in the area, create an impression of how and where the system will be installed, assess any impediments to the system working properly and optimally like nearby trees that can cause shading, type of roof thatching which determines how the panels will be mounted, area terrain which determines the means to the place, and assess the risk and hazards involved in the work etc.

2.1.2 *Determine the solar resource in Peak Sun Hours (PSH)*

The amount of irradiance in the area determines how much energy the modules can generate. The solar resource available in the area is therefore very necessary. We can get this from either the nearest meteorological station, or one can do actual site measurements if the right equipment is available—irradiance meter or pyranometer (but this may take some time)—or even get it from reliable internet resource maps. If the irradiance is given in W/m^2/day or kW/m^2/yr, then we have to calculate the value per day by dividing by the number of days in a year. In sizing calculation, we use PSH (how many hours we would have the same energy if the irradiance was the standard 1000 W/m^2 (1 kW/m^2)). The irradiance per day then needs to be divided by the standard value solar irradiance on the earth's surface, 1000 W/m^2 or 1 kW/m^2 to get PSH.

2.1.3 *Compute the Daily Energy Demand (DED)*

Next we need to determine the daily energy required by the user. We do this by talking to the user about their needs and future plans so we can factor that into the design and sizing. The items are entered in Table 1 as shown.

In Column 1, we enter the place/room in the house where the light or appliance will be located. Column 2 is for the number of appliances or lights per room that will be powered. In column 3 indicate the system voltage while the power rating of each appliance is entered in column 4. The estimated number of hours each appliance will be in use per day is for column 5. The total power for each appliance is obtained by multiplying column 2 and column 4 and the value is entered in column 6. Lastly the amount of energy needed for each appliance is obtained by multiplying the number in column 5 by that in column 6 and the product entered in column 7. The total power needed by all the appliances is the sum of the values in column 6 (we shall need this value for total power for the inverter sizing). The content of column 7 is also summed up giving the total Daily Energy Demand (DED) for the user in Whrs. Since system components are never ideal, we allow a margin of about 25% loss. Therefore we add 25% of the total in column 7 to get the final DED in Whrs.

Table 1. Details of the load demand and how to arrange the columns for calculation purposes.

Column 1	Column 2	Column 3	Column 4	Column 5	Column 6	Column 7
Room	No of Appliances	System Voltage (V)	Appliance Power Rating (W)	Daily usage (hrs) (Whrs)	Total Power (W)	Daily energy Demand
1						
2						
3						
4						

2.2 *Method 1: Charging Current Method (CCM)*

2.2.1 *Determine the number of solar modules needed:*

First, determine the system voltage, Vsyst. We can estimate this from the DED. Practice has established that for DED less than 1 kWhr (or a sum of module wattage of about 160 W), a 12 V system is required, 1–3 kWhrs, requires 24 V, and 3–4 kWhrs requires 48 V, etc. This can be used as a general guide but actual sizing must be done since other factors like cable sizing may make it necessary to adjust the system voltage.

2.2.2 *The next step is to estimate the charging current needed from the modules*

This is done by diving the final DED in Whrs with the system voltage, giving us the DED in Ahrs. The DED in Ahrs is then divided by the PSH to give the charging current required from the modules to meet the DED.

2.2.3 *Choosing the module(s)*

The information we have now can enable us to go to the market and choose a suitable module size or module sizes that has/have a maximum current rating equivalent or close to the charging current obtained in 2.2.2 above. If such a module is not available, then a combination of modules of the same rating whose total Imax is equivalent or close to the charging current is chosen. Note that some manufacturers use Pmax, I max, and Vmax while others use Pm, Im, and Vm to designate the maximum power point, maximum current, and maximum voltage, respectively. Here, we use the former designation. The module performance in the field will be less than the indicated. We assume a loss of about 20%, so our Imax will be (Imax) $\times 0.8$ to get the real field value, Ireal.

2.2.4 *Calculating the numbers of module strings*

The charging current is divided by the real maximum current, Ireal, to get the number of module parallel strings, Np, you will need in the system. Please note that this is not the number of modules but the number of parallel strings you will connect to give the calculated charging current.

2.2.5 *Calculating the numbers of modules per strings*

Next the number of series modules, Ns, connected for each parallel string is calculated. This is achieved by dividing the system voltage by 80% of the maximum power voltage (Vmax), i.e., system voltage/($0.8 \times$ Vmax).

2.2.6 *Calculating the total number of modules*

Lastly, we calculate the total number of modules, Nt, we need by multiplying the number of parallel strings, Np, by the number of series connected modules per string, Ns.

2.2.7 *Next is the determination of the battery bank capacity needed*

The battery bank capacity (Ahrs) (this is not the number of batteries) needed in a day (i.e., the total energy we need from the batteries in a day) is now determined from step 2.2.2 above. To calculate the battery bank capacity, we multiply the DED in Ahrs by the days of autonomy (the number of days we wish to have energy supply from the batteries when there is no or minimal sunshine) and divide this product by the depth of discharge (DOD) (how much energy you can safely use from the battery to avoid excess draining), given as a decimal. Some designers prefer a DOD of 50% (0.5) and others 20% (0.2). We use 20%. Take note that the lower the percentage of DOD, the larger will be the battery bank.

2.2.8 *Choosing the battery size*

Next choose the battery size (rating) depending on what is available in the market.

2.2.9 *Choosing the number of battery strings*

The number of battery parallel strings, Nbp is now determined by dividing the battery bank capacity (Ahrs) with the rating of battery chosen (Ahrs)

2.2.10 *Choosing the number of batteries in a string*

The number of batteries connected in series, Nbs, in each string is obtained by dividing the system voltage by the voltage of one single chosen battery.

2.2.11 *Calculating the total number of batteries*

The total number of batteries, Nbt needed is now calculated by multiplying the number of parallel strings by the number of series connected batteries in each parallel string.

2.2.12 *Sizing the charge controller*

The charge controller size is determined by the maximum current that can flow through it and the system voltage. The total short circuit current (Isc) that all the modules can generate is used. This is computed by multiplying the Isc of a single module by the number of module parallel strings. A margin of 25% is allowed so that the total current to the charge controller becomes the total current from all the modules multiplied by 1.25. A charge controller with a current rating higher than this calculated current is chosen. It is important that the charge controller is of the same voltage as the system voltage. If the user has plans of future expansion of the system by way of adding new appliances or another room, it should be factored into the controller sizing.

2.2.13 *Sizing the inverter*

The total power needed on a daily basis provides a basis for estimating the inverter size. So the sum of column 6 in Table 1 is used and a factor of 15% in inverter conversion losses is factored. So $0.15 \times$ the sum of column 6 is added to the sum of in column 6 to get the minimum power rating of the inverter. An inverter in the market that has such or near rating or higher is chosen. If the PV system is purely direct current (DC), then an inverter is not needed.

2.2.14 *Cable sizing*

Cable sizing must be done carefully to minimize voltage losses due to resistance. Cable size is usually given in terms of the cross sectional area of the cable in mm^2. The thicker the cable, the lower the resistance; and the higher the cost. Generally, the recommended maximum voltage drop is about 5% of the system voltage and is used as a guide for cable sizing. The cable cross-sectional area is related to the voltage drop, length, and current flowing through the cable by: $A = 2\rho LI/Vd$, where A is the conductor cross-sectional area (mm^2), ρ is the resistivity of the conductor material in $\Omega mm^2/m$ (for copper it is 0.0183), L is the conductor length one-way (no return) (m), I is the current flowing (A), and Vd is the voltage drop along the conductor ($0.05 \times$ system voltage) (Michael Boxwell, 2019). This calculation can also be used to calculate the maximum cable length to give a voltage drop within the limit of 5% for a particular cable size. If the cable size obtained by calculation is not in the market, then choose the next larger cable size.

2.3 *Method 2: The Energy Output Method (EOM)*

The energy output method calculates the energy output of the system components taking into account their inefficiencies. The starting point is the load, DED, then compute backwards towards the module as detailed below. This enables the calculation of the total power of all the modules we need for our solar system and then proceed on to size the other components. The method is demonstrated in Figure 1 below and the explanation beneath it.

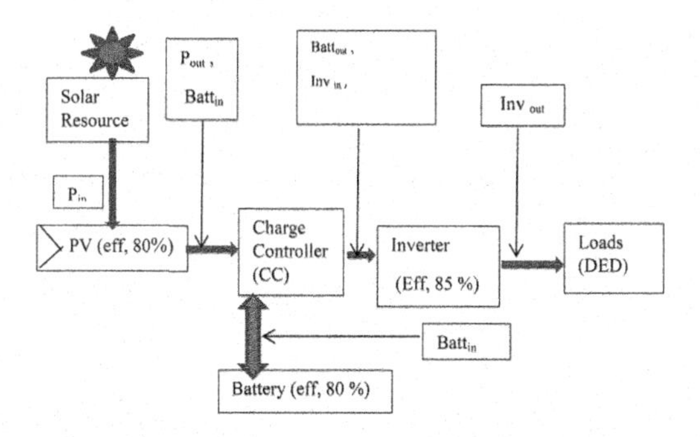

Figure 1. A block layout of the energy flow and component interconnection for a PV system using alternating current (AC).

2.3.1 *Calculation of the inverter output, Inv_{out}*

The energy that should be given out by the inverter is the energy that should go to our loads and so the DED is equal to our inverter output.

2.3.2 *Calculation of inverter input Inv_{in}/battery output, $Batt_{out}$*

The energy input to the inverter is what has been given out by the battery. So the inverter input is equal to the battery output. So the question is: how much energy is needed for the inverter input to get the output equivalent of DED? The question is answered by dividing the DED by the inverter efficiency, as a decimal 0.85, i.e., (DED/0.85).

2.3.3 *Determination of the battery input, $Batt_{in}$/module output, P_{out}*

The battery input is the same as the module output since the battery gets its energy from the module. Again the question is: how much energy is needed at the battery input to get the battery output we have obtained?. The question is answered by dividing the $Batt_{out}$ by the battery efficiency, as a decimal 0.80, i.e. ($Batt_{out}$/0.80).

2.3.4 *Determination of module input, Pin*

The Module input is computed by dividing the module output divided by the module efficiency i.e. Pin = (Pout/0.8).

2.3.5 *Lastly, calculate the module capacity, (Pw)*

Divide the module input with the product of solar resource (PSH) and module efficiency, 0.8, i.e. $Pw = (Pin/(PSH \times 0.8))$. A suitable module combination is then chosen from what is in the market.

2.3.6 *Number of module parallel strings*

The total system power in watts is divided by the rating of one module.

2.3.7 *Number of modules connected in series per string*

Divide the system voltage by the nominal voltage of one module.

2.3.8 *Total number of modules*

Multiply the numbers of parallel strings by the number connected in series in each parallel string.

2.3.9 *Sizing of charge controller*

Multiply Isc for one module and the number of modules strings. Then multiply this total by 1.25 (margin of 25%) to get final charge controller current rating. Lastly choose the charge controller with this current rating or higher and same system voltage.

2.3.10 *Battery bank capacity*

First multiply the battery output $Batt_{out}$, by 100 and divide by the DOD in %. Secondly, divide the resulting value by the system voltage to get the battery capacity in AHrs. Thirdly, choose a battery from what is in the market, and divide the battery capacity in AHrs by the rating in AHrs of one battery to get the number of parallel strings. Fourthly, calculate the number of series connected batteries per string by dividing the system voltage by the voltage of one battery. Fifthly, multiply the number of parallel strings by the number connected in series per string to get the total number of batteries needed. When days of autonomy are factored, it increases the number of batteries.

3 RESULTS AND DISCUSSION

The two methods were subjected to a similar client requirements and the component sizes calculated by applying the above methods. Consider a client whose DED is 904 Whrs and the total power consumed by all appliances is 168 W. The insolation at the place is 5.5 kWhrs/m^2.

3.1 *The CCM method*

On applying method 1 (CCM), we get an estimated a charging current of 16.95 A and if for example the modules available in the market are 80W, then from the module specifications, the current at maximum power point is 4.4 A. Thus, to generate the charging current of 16.95, four modules are needed. The computation on the battery bank capacity gives 1412.5 Ahrs and assuming 200 Ahrs as our choice of batteries, then seven of them are needed to create the battery bank capacity calculated. This high battery bank capacity takes into consideration 3 days of autonomy. Without the days of the autonomy, only about two batteries of similar size will be needed. The charge controller rating is calculated to be 26 A for a short current rating of the modules of 5.2 A. In the market, one may not easily get such a charge controller so the next higher rating available in the market, 30 A, is chosen. The voltage rating of the charge controller has to match the system voltage of the solar system, in this case 12V. Lastly, the inverter is rated depending on the devices that will be powered by the solar system assuming they are all on at the same time. The total power consumption in the example is 168 W so the next higher inverter rating in the market is obtained, for our example, a 200 W (VA) inverter.

3.2 *The EOM method*

The Energy method when applied to the example above assuming 80 W modules and 200 Ahrs batteries are the ones available gives the total power from the modules as 302 W. This means four modules are required. The computation gives a battery bank capacity of 1329.42 Ahrs, which is 6.65–7 batteries of 200 Ahrs, factoring in 3 days of autonomy. Both the charge controller rating as well as the inverter ratings are obtained as 30 A and 200 VA respectively.

3.3 *Comments on the two methods*

The two methods lead to the same results for each component as can be seen from above. The methods applied here are not new but the CCM is the most commonly used method (Michael Boxwell, 2019; Mark Hankins, 2010; Geoff Stapleton, Lalith Gunaratne, & Peter, 2002). In the CCM method, the technical specifications of the panel need to be known so as to make a final decision on the exact size and number of panels needed. This may pose a challenge to a number of installers and designers especially if they have no access to these details which need either searching in the internet or physically obtaining details in the shops. This is a drawback for this method of sizing. On the other hand, the EOM gives the total power needed from the panels to meet the daily demand, a value almost all dealers in solar PV easily identify with. This means once the computation is done, one does not need further details to make a final decision but can just call the seller to confirm what sizes are available and contact the client and advice on the combination of panels needed. This is easier and more straightforward and is advantageous compared to the CCM method.

In the real installation, care is taken to install solar modules in a location where shading should not occur at any time of the day. However, technically, modules have also bypass diodes installed to ensure that in case some cells are shaded, the current flow from the unshaded cells of the module bypass the shaded cells to minimize power loss, since the shaded cells act as loads and offer resistance. Again, once a solar PV system is installed, its entire lifetime depends on the individual components. For example solar modules have a factory warranty life span of 20–25 years, batteries can last for 1–7 years depending on the maintenance, and a charge controller can last for 10 years, while inverters can last for several years depending on quality, usage, and maintenance.

4 CONCLUSION

Two methods of sizing a SHS have been discussed in detail in this paper. It has been shown that the EOM is a better and more straightforward method than the CCM. The methods provide working estimates and that is why they are more suited for small-sized solar systems. More work still needs to be done to refine the methods, for example, by taking real field data and using it to size and compare with the initial sizing estimates. For large solar systems, commercial sizing softwares are recommended.

ACKNOWLEDGMENT

International Science Program, Sweden is thanked for support.

REFERENCES

Chikh, M., A. Mahrane, & F. Bouachri (2011), PVSST 1.0 sizing and simulation tool for PV systems, Energy Procedia 6, 75–84.

Geoff Stapleton, Lalith Gunaratne & Peter JM. Konings (2002), The Solar Entrepreneur's Handbook, Global Sustainable Energy Solutions

Justus Simiyu, Sebastian Waita, Robinson Musembi, Alex Ogacho & Bernard Aduda, (2014). Promotion of PV Uptake and Sector Growth in Kenya through Value Added Training in PV Sizing, Installation and Maintenance, Energy Procedia 57, 817–825.

Kaushika N.D., Nalin Gautam & Kshitiz Kaushik (2005), Simulation model for sizing of stand alone solar PV system with interconnected array, solar energy materials and solar cells, 85, 499–519.
Kefa V.O. Rabah, (2005), integrated solar energy systems for rural electrification in Kenya, Renewable Energy, Volume 30, Issue 1, 23–42.
Kenya Renewable Energy Association KEREA, Photovoltaic systems Field Inspection and Testing Report, 2009.
Majid Alabdul Salam, Ahmed Aziz, Ali H A Alwaeli, & Hussein A Kazem, (2013), Optimal sizing of photovoltaic systems using HOMER for Sohar, Oman, International Journal, of Renewable Energy Research, Vol.3, No.2, 301–307.
Mark Hankins (2010). Stand-alone Solar Electric Systems: The Earths can Expert Handbook for Planning, Design and Installation: 1st Edition, Routledge.
Michael Boxwell, (2019), Solar Electricity Handbook: A simple, practical guide to solar energy – designing and installing solar photovoltaic systems, Greenstream Publishing, 26–32; 97–147.
Mohit Jainl and Neha Tiwari (2014), Optimization and Simulation of Solar Photovoltaic cell using HOMER: A Case Study of a Residential Building, International Journal of Science and Research (IJSR), Volume 3 Issue 7, 1221–1223.
Sidrach-de-Cardona M. & Ll. Mora Lo' pez" (1998), A simple model for sizing stand alone photovoltaic systems, Solar Energy Materials and Solar Cells 55, 199—214.

Advances in Phytochemistry, Textile and Renewable Energy Research for Industrial Growth – Nzila et al. (Eds)

Economic analysis of a stand-alone residential home solar PV system in Western Kenya

W.K. Cheruiyot
Department of Physics, Faculty of Science, University of Eldoret, Eldoret, Kenya

ABSTRACT: Universal access to electricity is the current focus all over the world, especially in the developing countries where the majority of those without electricity live today. The SDG7 committed the world countries to work together and provide access to electricity to all by the year 2030. In response to this agenda, the Kenyan government has expanded the national grid supply across the country through the Rural Electrification Authority (REA), but the rate of connectivity in the rural regions is still very low despite the presence of the grid infrastructure. In addition, the government sought to increase access to electricity through solar energy in remote, low density and traditionally underserved counties through Kenya Off-Grid Solar Access Project (K-OSAP). However, integration of PV power generators into the energy mix requires the right approach to design and operational planning due to fluctuation of their outputs. Sizing of an off-grid PV system is necessary at the planning stage to make it cost effective with regard to load demand and upfront cost. This paper present an economic study design of solar energy potential in Western Kenya Region using 100 W installed stand-alone PV system in a residential home as a case study through life cycle cost analysis method. Results show that the investment will be recovered in 6.38 years and levelized cost of energy of 3.5/kWh is attainable.

Keywords: Solar energy, Off-grid PV system, residential energy demand, Life cycle cost analysis.

1 INTRODUCTION

Population growth and industrialization are the major drivers of rapidly increasing energy demand in developing countries where supply of energy is insufficient. It is important to note that the world is geared towards renewable sources of electricity away from fossil fuels, the use of photovoltaics (PV) is impressive for its sustainability as reported in popular literature works (1, 2). Regardless of the intermittency and seasonal variations of the solar resource, global capacity expected growth from 2019 to 2024 is 700-880GW indicating robust PV technology development (3). In the operational energy mix of Kenya, the year 2019 saw 11.6 billion kWh of electricity consumed of which geothermal, hydro, wind, thermal and solar contributed 45%, 27.5%, 13.5%, 11.2% and 0.8% respectively compared to 11.18 billion kWh consumed in the year 2018 where solar stood at 0.12% (4). This presents a clear indication of energy demand increase annually and a steady increase of solar generated electricity despite the vast availability of solar energy resources in the country. Off-grid electrification is currently the country's flagship project with estimates of over 700,000 connections as per the end of year 2018 that in return will lower the cost of energy. It is designed to increase energy access via stand-alone solar systems in 14 underserved Counties through the project duped the Kenya Off-Grid Solar Access Project (KOSAP) in conjunction with other funded programs and organizations that are in support of the last-mile distribution of solar solutions (5). The favourable features of solar energy have led to its rapid growth across governments (abundance, clean, well distributed and economically reliable for long term operation) with applications ranging from grid connected to stand-alone (Grid-Backup, Grid-Tied, Off-Grid and Solar direct) PV systems (6).

Majority of institutions in developing countries (Public and private organizations) and individual residential homes prefer or supplement grid electricity supply with independent sources commonly diesel generators (DG) or PV systems. The running and maintenance costs of a diesel generator deter many low income home owners from continued use of DG and as well due to their minimal daily energy demands that makes DG non-economical .This calls for a customizable, affordable and non-disruptive energy supply source. It has been proven over time that small to large electricity demands can be met sustainably by well sized PV energy systems and is as well viable technically and economically to use in the rural areas (7).

High investment costs of PV technology remains a major obstruction in many developing countries and to help lessen these, government initiatives to promote

DOI 10.1201/9781003221968-32

use of solar energy is encouraged. To make this a reality, outdoor evaluation of PV system financial benefits over a period of time at a specified location is necessary. On the other hand, PV and solar thermal systems are cheaper compared to DG over extended use as noted by (8) under reasonable site of solar insolation. Economic and energy analysis was carried by (9) in order to evaluate the application of the building integrated PV system. Economic examination by (10) and (11) of energy mix of DG and PV systems for a school in India and East Malaysia respectively based on life cycle cost (LCC) concluded that stand-alone PV system is suitable for use when energy demands are low. Recent studies considering economic aspects of stand-alone PV systems were investigated in the literature; (12–15).

Economic analysis of PV system has seen a number of studies done but further research are justifiable because many of the works did not validate with experimental results. In addition, updated analysis is required by customers of PV systems to know if their investment is still profitable with changing times. This present study is aimed at analyzing the economic viability and potential of stand-alone solar PV power through life cycle cost analysis method for a residential home in western Kenya.

2 DESIGN AND METHODOLOGY

2.1 *System configuration*

Configuration of a PV system under this study with complete components is as shown in Figure 1, wired together to supply electricity to the various load's (DC and AC) power rating presented in Table 1. The PV cell converts sunlight directly to electricity. The charge controller prevents battery from being overcharged or over discharged. The inverter converts DC voltage to AC voltage. The battery stores electrical energy for use during night and non-sun days.

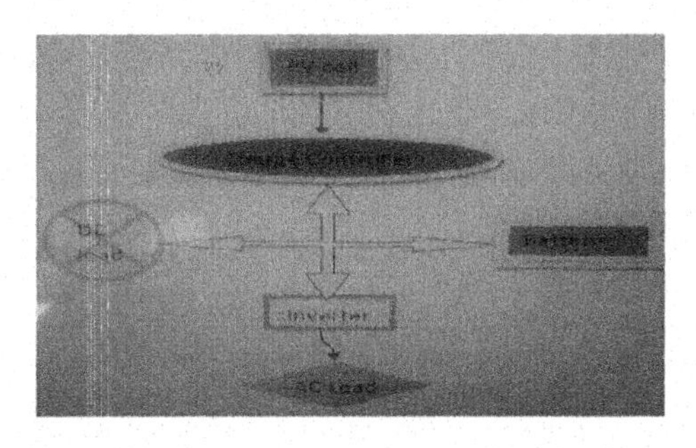

Figure 1. Configuration of a stand-alone PV system.

2.1.1 *Meteorological data*

The residential home PV system is located in Kenya, a location which has a geographical position of latitude and longitude of 0.42 N and 35.03 E respectively. The location receives an average solar insolation of 5.35 ($kWh/m^2/day$).

2.1.2 *Component characteristics*

Presented in this section are individual component properties of a residential PV system. Table 1 shows the specification of the installed PV module.

a. PV module

Table 1. PV module characteristics

Item	Specifications
Performance warranty (yrs)	25
Cell type	Poly silicon
Cell size (mm)	156 × 104 × 36
Open–Circuit Voltage, V_{oc} (V)	21.9
Short-Circuit Current, I_{SC} (A)	6.13
DC bus voltage (V)	24
Module Efficiency (%)	14.63
Max Power, P_{max} (kW)	0.1
Voltage at Maximum Power Point, V_{mpp} (V)	18.0
Current at Maximum Power Point, I_{mpp} (A)	5.56
Temperature Coefficients of, P_{max} (%/°C)	−0.45
Temperature Coefficients of, V_{oc} (%/°C)	−0.32
Temperature Coefficients of, I_{SC} (%/°C)	+0.04
Series Fuse Rating (AM)	10:00
Power Tolerance (%)	0%, +6
Cost (Ksh)	9000

b. Charge controller

Table 2. Charge controller characteristics

Battery voltage (V)	12–24
Maximum input voltage (V)	24
Maximum current output (A)	20
Efficiency (%)	95%
Safety factor (SF)	1.25
Cost (Ksh)	4500

c. Battery (Dry cell)

Table 3. Battery characteristics

Battery capacity	250 Ah
Voltage (V)	24
Depth of discharge (DOD)	0.8
Efficiency (%)	95
Life span (yrs.)	6
Cost (Ksh)	9500

d. Inverter

Table 4. Inverter characteristics

Type PWM	
Power requirement (kW)	0.08
AC voltage (V)	240
Efficiency (%)	90
Cost (Ksh)	4500

e. System Loads

This section presents residential home energy demands estimated from power rating summation of the appliances as shown in Table 1. It is observed from Table 1 that the power rating is approximately 0.4 kW but to cater for future appliances' upgrades we shall consider a value of 0.5 kW.

Table 5. Load power rating.

S/No	Appliance	Unit Power (W/Unit)	Quantity	Total power (kW)
1	PC (Laptop)	60	1	0.060
2	Television	45	1	0.045
3	Lighting Bulb	40	6	0.240
4	Phone charger	6	2	0.012
5	Electric shaver	20	1	0.020
6	Home sound system	12	1	0.012
			Total	0.389

f. Input data for design and economic analysis

This section presents additional input data for design and economic analysis as shown in Table 6 below.

Table 6. Additional input data

Item	Value
Average hours of operation	9
Continuous cloudy days	4
Temperature correction factor	0.95
Installation cost (Ksh)	10% of PV
O&M/year (Ksh)	2% of PV
PV surface area (m^2)	0.565
Tilt angle (*C*)	13.5
PV derating factor (%)	80@25 years
Interest rate (%)	7
Inflation rate (%)	5.47
Cost of electricity (Ksh)	22.5 per kWh

2.2 *Design and Economic analysis*

a. Sizing and Design analysis

For optimum utilization of an installed PV system, sizing process where each component ratings are determined are carried out in order to meet the residential home energy demand.

Daily home energy demand from the PV module, E_d is given by (16):

$$E_d = P_r H / \eta_{overall} \tag{1}$$

where P_r(kW) is the home power rating, H is the number of hours the PV system is in use per day and $\eta_{overall}$ is the overall efficiency expressed as the product of independent component efficiency as expressed below:

$$\eta_{overall} = \eta_{PV} \eta_{CON} \eta_B \eta_{INV} \tag{2}$$

where $\eta_{PV}$is the PV module efficiency, η_{CON} is the charge controller efficiency, η_B is the battery efficiency and η_{INV} is the inverter efficiency.

During maximum solar insolation (MSI) (kWm^{-2}), the PV peak power W_P (kW) is given as (17)

$$W_P = \eta_{PV} A_{PV} \times MSI \tag{3}$$

where A_{PV} is the PV module flat surface area.

The system total direct current I_{DC}(Ah) is as expressed below (16):

$$I_{DC} = W_P / V_{DC} \tag{4}$$

where V_{DC} (V) is the DC bus voltage.

The estimated battery storage capacity $B_{Capacity}$(kWh) is given by (18):

$$B_{Capacity} = N_C P_r H / DOD \times \eta_{CON} \eta_B \tag{5}$$

where N_C is the location's number of continuous cloudy days and DOD is the battery maximum depth of discharge.

Charge controller also known as voltage regulator sizing is obtained by determining the product of the module short circuit current I_{SC} and the safety factor SF which gives the rated current of the charge controller as shown below (18):

$$I_{Controller} = I_{SC} \times SF \tag{6}$$

For the inverter, its operational power requirement W_{INV} (kW) is given as:

$$W_{INV} = 1.25 \times P_{DF} \tag{7}$$

where P_{DF} is the residential home power demand.

The power output of the PV module at any given time is dependent directly on the amount of solar radiation striking the PV surface given as (19):

$$P_{PV} = P_R f_{PV} \left(G / G_{STC}\right) \tag{8}$$

where P_R is the rated capacity of the PV which is the output power under standard test condition (STC), f_{PV} is the PV derating factor (%), G is the solar radiation incident on the PV (kWm^{-2}) and G_{STC} is the incident solar radiation at STC.

b. Life cycle cost analysis

A life cycle cost analysis (LCC) technique is applied here to evaluate the residential PV system's costs. The initial PV system cost is normally relatively high while the rest of operational related costs are relatively low since there is no fuel cost. Total costs from PV component system acquisition, operating, maintenance to replacement over its lifetime are what life cycle cost analysis (LCC) entails normally expressed as the system present cost (17) as expressed:

$$LCC = \sum_{r=1}^{6} C_r \tag{9}$$

Where C_r is the individual component present worth further represented by r in equation below as:

$$r \in \langle 1, 2, 3, 4, 5, 6 \rangle \equiv [PV, B, CON, INV, INST, O\&M] \tag{10}$$

Where PV is the PV module cost, B represent battery cost, CON represent charge controller cost, INV represent the inverter cost, $INST$ represent installation cost and $O\&M$ represent operation and maintenance cost.

The interest rate i (%), inflation rate d(%), initial cost of the battery C_{BO} (Ksh), number of replacements j and battery life span n present worth of the battery and given as:

$$C_B = C_{BO} + \sum_{k=1}^{j} C_{BO}\left(1 + d/1 + i\right)^{kn} \tag{11}$$

For operation and maintenance costs $C_{O\&M}$, the present worth is expressed as (8):

$$C_{O\&M} = C_{O\&M/y}\left(\frac{1+i}{1+d}\right)\frac{1-\left(\frac{1+d}{1+i}\right)^{N}}{1-\left(\frac{1+d}{1+i}\right)} \tag{12}$$

Where $C_{O\&M/y}$ is the operation and maintenance cost per year and N is the life span of the PV module.

The total annualized cost is the sum of the annualized costs of each system component in terms of the present worth. To calculate the annualized LCC (ALCC) we use the following equation (20):

$$ALCC = LCC \times \frac{1-\left(\frac{1+d}{1+i}\right)}{\left(\frac{1+d}{1+i}\right)^{N}} \tag{13}$$

To calculate the system unit electrical cost UEC we use the following equation:

$$UEC = {}^{ALCC}/_{366E_d} \tag{14}$$

The number of years it takes to recover an investment's initial cost i.e. payback time (PBT) is calculated using the following equation:

$$PBT = {}^{LCC}/_{Q_{AP}} \times UEC_{KE} \tag{15}$$

Where UEC_{KE} is the cost of electricity supply in Kenya and Q_{AP} is the annual energy production from the PV system expressed as:

$$Q_{AP} = 8760E_d \tag{16}$$

c. Levelized cost of energy

This is a widely used technique that gives a more accurate energy cost calculation defined as the ratio of LCC of the PV system to the whole life cycle produced energy (LCE) as is expressed (21):

$$LCOE = {}^{LCC}/_{LCE} \tag{17}$$

The LCE is calculated as in equation below (20):

$$LCE = \sum_{i=0}^{n} AEP \times (1 - f_{PV})^{i}/_{(1-r)r} \tag{18}$$

Where AEP is the expected annual energy produced which is the estimated life span of the PV system. As the system time goes by, the output power yield will be degraded with a factor f_{PV} that is used here to get a better energy forecast.

3 RESULTS AND DISCUSSION

The PV module life span is 25 years and a total attainable revenue for the period is 182250 Ksh. It is further shown that the investment will be recovered in 6.38 years. Table 7 presents the outputs of economic and

Table 7. Economic and design data output.

Total units produced in one month: year	27 kWh:324 kWh
Revenue generated in one month	607.5 Ksh
Total units produced in 25 years	8100 kWh
Number of units consumed in one month	15 kWh
Excess units produced in one month	12 kWh
Total revenue generated in 25 years	182250 Ksh
Installation cost	900 Ksh
Operation and maintenance cost	180 Ksh
LCC	28580 Ksh
PBT	6.38 years
LCOE	3.5/kWh

design data results of the PV system under study for a residential home with energy demand specified earlier.

LCOE of 3.5/kWh is a clear indication that the cost of solar power is far much below the cost of electricity from the grid paid monthly as electricity bill. In addition, the loads in this residential home are not classified as critical loads meaning that they could be reduced at night hours. This implies that overall costs may reduce as a result of reducing the battery capacity. PBT variation is dependent on factors such as the type of solar cell, irradiation at the location, capacity of the system and degrading factor of the PV module

4 CONCLUSIONS AND RECOMMENDATIONS

The computation results obtained show that the PV system makes an economically efficient power source for the residential home under study. The investment will be recovered in 6.38 years a proof that off-grid PV systems should be encouraged for residential homes that their energy demands are low. Improvements on individual component conversion efficiencies will reduce the PBT of solar modules and eventually further improve the LCOE value. LCC major investment components are the PV module, battery, charge controller and inverter. Future work is recommended to put in consideration individual component tax relief in determination of LCC.

REFERENCES

[1] Liu, J., Xu, F., & Lin, S. (2017). Site selection of photovoltaic power plants in a value chain based on grey cumulative prospect theory for sustainability: A case study in Northwest China. *J. Clean. Prod.*, 148, 386–397.

[2] International Energy Agency. Snapshot of Global PV Markets. Available online: http://www.iea-pvps.org/

[3] Renewables 2019. IEA. Retrieved 28 May 2020

[4] Kenya National Bureau of Statistics; 2020. Economic survey report.

[5] Kiprop.E, Kenichi. M & Maundu. N. (2018). Can Kenya Supply Energy With 100% Renewable Sources? *International Scientific Journal of Environmental Science.* 141, 38–39.

[6] Brito, M., Gomes, N., Santos, T., & Tenedório, J. (2012). Photovoltaic potential in a lisbon suburb using lidar data. *Sol. Energy*, 86, 283–288.

[7] Headley, S. J. (2010). Solar-Diesel Hybrid Power System Optimization and Experimental Validation. Masters of Science, University Of Maryland, College Park. Solar Electricity, Wiley, London.

[8] Oliver, M. & Jackson, T. (2001). Energy and economic evaluation of building-integrated photovoltaics. *Energy*, 26, 431–439.

[9] Kolhe, M., Kolhe, S., & Joshi, J. (2002). Economic viability of stand-alone solar photovoltaic system in comparison with diesel-powered system for India. *Energy Econ*, 24, 155–165.

[10] Ajan, C.W., Ahmed, S.S., Ahmad, H.B., Taha, F., & Zin, A.A. (2003). On the policy of photovoltaic and diesel generation mix for an off-grid site: East Malaysian perspectives. *Sol. Energy*, 74, 453–467.

[11] Shaw-Williams, D., Susilawati, C., & Walker, G. (2018). Value of residential investment in photovoltaics and batteriesin networks: A techno-economic analysis. *Energies*, 11, 1022.

[12] Mahmud, M., Huda, N., Farjana, S., & Lang, C. (2018). Environmental impacts of solar-photovoltaic and solar-thermal systems with life-cycle assessment. *Energies*, 11, 2346.

[13] Shah, S., Valasai, G., Memon, A., Laghari, A., Jalbani, N., & Strait, J. (2018). Techno-economic analysis of solar PV electricity supply to rural areas of Balochistan, Pakistan. *Energies*, 11, 1777.

[14] SWERA. (2008). "Solar and Wind Energy Resource Assessment Project (SWERA), Kenya Country Report".

[15] Alamsyah T.M.I., Sopian K., & Shahrir A. (2003). In Techno-economics Analysis of a Photovoltaic System to Provide Electricity for a Household in Malaysia. Proceedings in International Symposium on Renewable Energy: *Environment Protection & Energy Solution for Sustainable Development*, 387–396.

[16] Mahmoud, M.M., & Ibrik, I.H. (2006). Techno-economic Feasibility of Energy Supply to Remote Villages in Palestine by PV-Systems, Diesel Generator and Electri Grid. *Renewable Sustainable Energy Rev.*, 10: 128–138.

[17] Assad, A. (2010). A Stand-Alone Photovoltaic System, Case Study: A Residence in Gaza. *J. of Applied Sciences in Environmental Sanitation*, 5 (1): 81–91

[18] Duffie J.A. & W.A. Bechaman (1991). Solar engineering of thermal processes. *John Wiley and Sons.*

[19] Abd El-Shafy, A.N. (2009). Design and Economic Analysis of a Stand-Alone PV System to Electrify a Remote Area Household in Egypt. *The Open Ren. Energy Journal*, 2: 33–37.

[20] Mulligan, C. J., Bilen, C., Zhou, X., Belcher, W. J., & Dastoor, P. C. (2015). Levelized cost of electricity for organic photovoltaics. *Solar energy materials and solar cells*, 133, 26–31.

[21] Myhr, A., Bjerkseter, C., Ågotnes, A., & Nygaard, T. A. (2014). Levelized cost of energy for offshore floating wind turbines in a life cycle perspective. *Renewable energy*, *66*, 714–728.

Advances in Phytochemistry, Textile and Renewable Energy Research for Industrial Growth – Nzila et al. (Eds)

Experimental investigation of thermal efficiency enhancement of improved biomass cookstoves

Waganesh Admase Wagaye*
Department of Mechanical, Production and Energy Engineering, Moi University, Eldoret, Kenya
Department of Mechanical and Production Engineering, Arba Minch University, Arba Minch, Ethiopia
Africa Center of Excellence II in Phytochemicals, Textiles and Renewable Energy (ACE II PTRE), Moi University, Eldoret, Kenya

Meseret Biazen Belete
Department of Mechanical, Production and Energy Engineering, Moi University, Eldoret, Kenya
Africa Center of Excellence II in Phytochemicals, Textiles and Renewable Energy (ACE II PTRE), Moi University, Eldoret, Kenya
Department of Mechanical Engineering, Haramaya Institute of Technology, Haramaya University, Harar, Ethiopia

ABSTRACT: Biomass is one of the best and widest areas that has attracted the world's attention as regards the production of energy from renewable sources. A large proportion of world energy sources is renewable energy. Improved biomass cookstoves are recently developed devices for domestic cooking utilizing biomass as fuel through gasification. In Ethiopia, biomass is the most widely used source of energy for cooking and heating applications. This study attempted to enhance the thermal efficiency of domestic cookstoves used in Ethiopia. The developed model was tested experimentally using the water boiling testing protocol (WBT). Results showed that the stove had a thermal efficiency of 45.17%, with specific fuel consumption of 11.53 g/L and 43–48 minutes of cooking time. The mathematical model was implemented and validated, and the experimental results showed that there was an improvement of the stove, as well as reduced fuel consumption and cooking time.

1 INTRODUCTION

Most of the world's population in low- and middle-income countries still relies on solid fuels (wood, animal dung, charcoal, crop wastes, and coal) which are burnt inefficiently (Quinn et al., 2018). Starting in the last decade, using solid biomass as a fuel for domestic cooking applications has become a common practice, especially for people who live in rural areas.

In most developing countries, more than 50% of the population depends on the traditional use of biomass (Bombaerts, Jenkins, , Sanusi, & Guoyu, 2019; Njiru & Letema, 2018; Sime, Tilahun, & Kebede, 2020). More specifically, in Sub-Saharan Africa, around 753 million people (i.e., 80% of the population) use biomass as an energy source (Adem et al., 2019). Harmful emissions from traditional biomass cookstoves causes around 4 million deaths per year globally (Samal, Mishra, Mukherjee, & Das, 2019). Ethiopia is one of those developing countries and more than 95% of the country's population relies on biomass feedstocks (Khatiwada, Purohit, & Purohit, 2019). This shows that most of the country's population does not have electricity. Therefore, biomass is the basic energy resource in the country (Bantelay, 2014). Mostly, they use three stone cookstoves to prepare their meal (Adem et al., 2019). Due to its simplicity, a three-stone open firing continues to be used for cooking and heating purposes (Jewitt, Atagher, & Clifford, 2020; Shiferaw, 2011). The use of inefficient combustion cookstoves expose individuals to indoor air pollution. Due to deficient ventilation, it has a negative health impact on the end-users, especially women and children (Bantelay, 2014; Jewitt et al., 2020). Globally, 3 billion people are exposed to household air pollution caused by solid fuel combustion (Pratiti, Vadala, Kalynych, & Sud, 2020). Using biomass as a fuel will not end, but the technology will be upgraded for the future (Sutar, Kohli, Ravi, & Ray, 2015). To reduce hazards caused by the burning of solid biomass using traditional inefficient cookstoves, cleaner and modern household cookstoves are being developed in advance to produce clean gases (Venkataraman, Sagar, Habib, Lam, & Smith, 2010).

*Corresponding author

DOI 10.1201/9781003221968-33

Improved cookstoves (ICS) can deliver "triple wins" by improving household health, local environments, and global climate (Pattanayak et al., 2019). Ethiopia will be expected to rely on biomass for the coming years as a primary source (Beyene & Koch, 2013; Jewitt et al., 2020). So far, the focus of development has been on the improvement of parts of the stove i.e., insulation, grate, and skirt. Current research has focused on minimizing emissions and enhancing energy efficiency (Adem et al., 2019; Manoj, Sachin, & Tyagi, 2013; Ochieng, Vardoulakis, & Tonne, 2017; Quansah et al., 2017). ICS are multipurpose with both internal and external benefits. Internally, they reduce the mortality rate due to air pollution, the concentration of smoke, reducing the demand for biomass fuel, money, and time saved in acquiring fuel, reducing the use of animal dung as a fuel instead of as a fertilizer. Externally improved biomass cookstoves are beneficial in a way of reducing greenhouse gas (GHG) emissions and deforestation (Barnes & Smith, 1993). Therefore, the overall finding of the study underlined the high importance of strengthening social groups to enhance the adoption of improved cookstoves. This paper was written to build the model of improved cookstoves with high thermal efficiency, less fuel consumption, and less time of cooking for household applications.

2 MATERIALS AND METHODS

The improved household cookstove was designed and built with different modifications using easily and locally available materials. Only clay soil and wood ash were used to build the entire stove. The model is based on the improvement of thermal efficiency using low thermal conductive materials. The modification includes insulating the cookstove by mixing clay soil with wood ash. Increasing firepower, reducing fuel consumption, and enhancing the combustion process by reducing the height of the combustion zone, as well as by adding secondary air through the developed secondary air holes as shown in Figure 1.

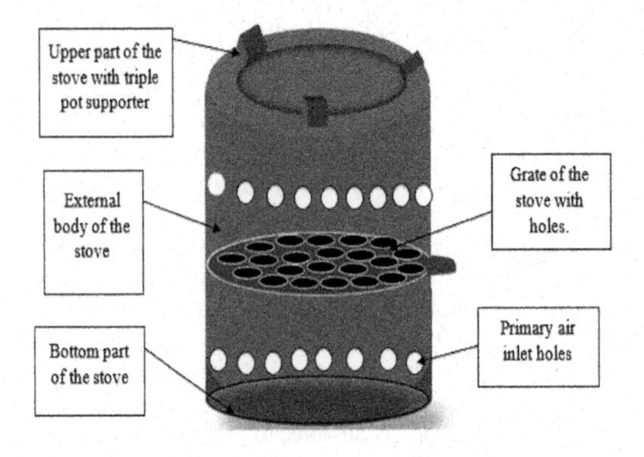

Figure 1. Schematic diagram of the introduced cookstove.

2.1 *Design of the desired cookstove*

The model of the desired improved household cookstove employed a simple design while utilizing input materials that are readily available for the rural community of Ethiopia.

The main part of the stove is clay soil and wood ash; their selection to be used for developing the entire stove was based on their low thermal conductivities (high insulation capacities), high workability, and availability. The new improved cookstove was designed by determining each design parameter. The construction of the improved cookstove model is shown in Figure 2.

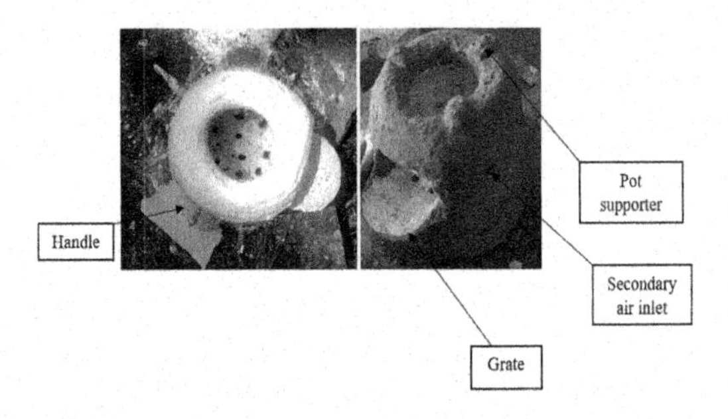

Figure 2. Stove under construction.

2.2 *Testing fuel characterization*

The experiment was conducted using sundried small pieces of eucalyptus tree wood as the feedstock. Eucalyptus is a common tree in Ethiopia, and it was preferred because of its availability. The physical and thermal characteristics of eucalyptus wood used for testing were determined by proximate analysis according to ASTM standard 19103.

2.3 *Energy demand*

This is the amount of energy required to cook the food or boil water. In Ethiopia, people cook stew such as beans, peas, and vegetables. Therefore, to compensate for this to aid in the design of the stove, rice was taken as the desired food. This is because the time taken to cook rice is almost the same as for stews (Adem & Ambie, 2017). The calorific value of rice was used to calculate the energy demand of the entire stove. The energy demand was calculated using Equation 1 (Adem & Ambie, 2017; Bantelay, 2014).

$$Q_n = \frac{M_f \times E_s}{t} \tag{1}$$

where Q_n is the energy needed (kJ/hr), E_s is the specific energy (kJ/Kg), M_f is the mass of food (kg), and t is the cooking time (hr).

A kilogram mass of the rice was used with the cooking time assumed to be 20 minutes. The specific energy of rice is 330.43 kJ/kg (Shiferaw, 2011). Substituting the above values in Equation 1 gave the energy needed (Qn) = 991.29 kJ/hr.

2.4 *Energy input/fuel consumption*

The amount of fuel expected from biomass fuel added to the combustion chamber was calculated using Equation 2.

$$\text{FCR} = \frac{Q_n}{HV_f \times \varepsilon_g} \quad (2)$$

where FCR is the fuel consumption rate (kg/hr), HV_f is the heating value of fuel, is the heat energy needed (kJ/hr), and ε_g is the efficiency of the gasifier (%).

The predicted efficiency of the stove was 40%. Thus, the calculated FCR by substitution in Equation 2 was found to be 0.13203 kg/hr.

2.5 *Reactor diameter*

This refers to the size of the reactor and is a function of the fuel consumed per unit time (FCR) to the specific gasification rate (SGR) of biomass material (Equation 3).

$$D = \sqrt{\frac{1.27 \times FCR}{SGR}} \quad (3)$$

where D is the diameter of a reactor in meters. SGR of the biomass fuel lies in the range of 50–210 kg/m^{2}hr.

Using assumed FCR of 2 kg/hr and SGR of 100 kg/m^{2}hr, the diameter (D) = 0.15937 m. This is equal to 159.37 mm ($\approx$160 mm). The diameter of the reactor was taken as 180 mm, with an allowance of 20 mm in order to accommodate the variation of cooking habits throughout Ethiopia (Adem & Ambie, 2017).

2.6 *Height of the reactor*

The reactor height (H) is the overall height of the reaction chamber and was the modified part from previously improved cookstoves. This dimension indicates the loading capacity of the reactor. It is calculated assuming density of feedstock (ρ_f) = 100 kg/m^3 employing Equation 4 (Adem & Ambie, 2017).

$$\text{H} = \frac{\text{SGR} \times \text{t}}{\rho_f} \quad (4)$$

By substitution, H = 200 mm (for SGR = 100 kg/m^{2}hr and t = 20 minutes). To make the stove manageable and compensate the increase in diameter (Adem & Ambie, 2017), the developed stove height was taken as 180 mm. This was also to reduce the time taken by fire to reach the bottom part of the stove. To increase the firepower, the height should be made shorter.

2.7 *Performance evaluation of the improved cookstove*

For this entire experiment, a revised version of the Water Boiling Test Protocol (Adem & Ambie, 2017) was used. This was conducted in a simulated kitchen to find the specific fuel consumption and thermal efficiency of the desired improved cookstove. The apparatus used included a thermometer with an immersible probe, analytical balance, a standard 5-liter cooking pot, and biomass feedstock (eucalyptus wood).

The thermal efficiency (η_{th}) of the stove was calculated as the ratio of the amount of heat gained by the water inside the pot and evaporated to the energy of the fuel used for heating. The complete mathematical relation is given in Equation 5 (Adem & Ambie, 2017).

$$\eta_{th} = \frac{m_{wi} \times C_{pw} \times (T_b - T_i) + m_e \times L}{m_f \times H_{vf}} \times 100 \quad (5)$$

where m_{wi} = the initial mass of water (kg), C_pw = specific heat capacity (kJ/kg/°C), T_i = initial temperature of the water (°C), T_b = final (boiling) temperature of the water (°C), m_e = mass of water evaporated (kg), L = latent heat of evaporation of water(kJ/kg), mf = mass of fuel, and H_{vf} = heating value of the fuel (kJ/kg).

The performance of the stove was expressed in terms of specific fuel consumption (SFC) which measured the amount of fuel required to cook the entire food (Equation 6) (Adem & Ambie, 2017).

$$SFC = \frac{Fuel\ used\ (Kg)}{Food\ cooked\ (Kg)} \quad (6)$$

2.8 *Statistical analysis*

The experiments were conducted in triplicate and the values obtained were averaged. Quantitative data were presented as means with errors as standard deviations attached. All statistical analyses and mathematical computations were performed using Minitab Statistical Software (Release 17, Minitab Inc., USA).

3 RESULTS AND DISCUSSION

3.1 *Physical and thermal properties of biomass used*

The characteristics of eucalyptus wood used are reported in Table 1. The results of the proximate

Table 1. Physical and thermal properties of eucalyptus wood used.

Characteristics*	Current study	Previous study (Adem & Ambie, 2017)
Size (mm)	250–350	150–200
Length (mm)	45–70	30–50
Dry density (kg/m^3)	395	480
Moisture content (% wb)	5.64	5.64
Volatile matter (% db)	80.81	80.81
Fixed carbon (% db)	13.02	13.02
Ash content (% db)	54	54
Calorific value (MJ/kg)	16.5	18.64

* wb: wet basis, db: dry basis

analysis of eucalyptus wood were comparable to those reported by previous authors (Adem & Ambie, 2017) in Ethiopia.

3.2 *Performance evaluation of the improved cookstove designed*

Table 2. Performance parameters and test results of the improved cookstove.

Parameter	Test 1	Test 2	Test 3	Mean*
Mass of fuel burnt (kg)	0.55	0.55	0.55	0.55 ± 0.00
Initial mass of water (kg)	5.00	5.00	5.00	5.00 ± 0.00
The initial temperature of water (°C)	20.90	21.21	20.80	20.97 ± 0.21
Final temperature of the water (°C)	97.83	98.56	98.87	98.42 ± 0.53
Mass of water evaporated (kg)	1.053	1.102	1.134	1.096 ± 0.041
Thermal efficiency (%)	43.97	45.29	46.25	45.17

*Means are presented as mean ± standard deviation of triplicates. Specific heat capacity of water = 4.187 kJ/kg/°C, latent heat of evaporation of water = 2,260 kJ/kg, and heating value of fuel = 16,500 kJ/kg

From the experimental results (Table 2), the average thermal efficiency of the developed stove is 45.17%. This shows that the stove is more efficient compared with the previously developed improved gasifier stoves in the country with 31% (Shiferaw, 2011), 17.2% (Bantelay, 2014), 26.5% (Panwar & Rathore, 2008) and 39.6% (Adem & Ambie, 2017) efficiency. Further, it had a lower average specific fuel consumption of 11.0 g/L, compared to 57.0 g/L reported previously (Adem & Ambie, 2017).

On the other hand, the cooking time was 43–48 minutes. Mostly, improved stoves are developed and shielded by stainless or mild steel. However, this makes it not simple to manufacture them. Additionally, those materials are not easily accessible in remote areas while others are unaffordable. Therefore, end-users cannot afford this price. The developed stove in this study excluded the use of metallic materials. Besides, to manufacture this stove, it does not require any advanced technology, modern instruments or any new skills. The biomass fuel is fed to the stove continuously to control the power of fire since the stove is not of a batch type. Therefore, this ICS differs from the other improved stoves as it is easier to control. The stove has three pot supports on the upper part of it, implying that the stove can be used for cooking using different pot sizes.

3.3 *Manufacturing cost of the stove*

The cost of materials that were used to develop the stove and labor cost to manufacture the ICS were found to be 9.167 USD or 275 Ethiopian birrs (Table 3).

Table 3. Manufacturing cost of the designed improved cookstove.

Item number	Description	Total cost (USD)
1	Clay	2.6667
2	Wood (ash)	1.5
3	Labor	5
	Total cost	9.167

4 CONCLUSIONS AND RECOMMENDATIONS

The study presented a clear view of the improvement in the efficiency of biomass cookstoves. The model can be used anywhere, and the fuel type is not fixed. The stove can use any type of solid biomass except rice husk and sawdust. The present study showed the thermal efficiency of improved cookstoves to be 45.17% and fuel consumed to boil 5 liters of water was 0.55 kg. Further studies should analyze the indoor air pollution (carbon dioxide and particulate matter concentration) from the biomass stove designed.

ACKNOWLEDGMENTS

The authors would like to acknowledge the World Bank and the Inter-University Council of East Africa (IUCEA) for the scholarship awarded to them through the Africa Center for Excellence II in Phytochemicals, Textiles and Renewable Energy (ACE II-PTRE) and Moi University, Kenya for organizing an opportunity to uplift the capacity of students in the research area through conferences.

REFERENCES

Adem, K. D., & Ambie, D. A. 2017. Performance and emission reduction potential of micro-gasifier improved through better design. *AIMS Energy*, *5*(1), 63–76.

Adem, K. D., Ambie, D. A., Arnavat, M. P., Henriksen, U. B., Ahrenfeldt, J., & Thomsen, T. P. (2019). First injera baking biomass gasifier stove to reduce indoor air pollution, and fuel use. *AIMS Energy*, *7*(2), 227–245.

Bantelay, D. T. 2014. Design, Manufacturing and Performance Evaluation of House Hold Gasifier Stove: A Case Study of Ethiopia. *American Journal of Energy Engineering*, *2*(4), 96.

Barnes, D. F., Openshaw, K., Smith, K. R., & Plas, R. V. D. 1993. The design and diffusion of improved cooking stoves. *The World Bank Research Observer*, *8*(2), 119–141.

Beyene, A. D., & Koch, S. F. 2013. Clean fuel-saving technology adoption in urban Ethiopia. *Energy economics, 36*, 605–613.

Bombaerts, G., Jenkins, K., Sanusi, Y. A., & Guoyu, W. 2020. *Energy justice across borders* (p. 305). Springer Nature.

Jewitt, S., Atagher, P., & Clifford, M. 2020. "We cannot stop cooking": Stove stacking, seasonality and the risky practices of household cookstove transitions in Nigeria. *Energy Research & Social Science, 61*, 101340.

Khatiwada, D., Purohit, P., & Ackom, E. K. 2019. Mapping Bioenergy Supply and Demand in Selected Least Developed Countries (LDCs): Exploratory Assessment of Modern Bioenergy's Contribution to SDG7. *Sustainability, 11*(24), 7091.

Manoj, K., Sachin, K., & Tyagi, S. K. 2013. Design, development and technological advancement in the biomass cookstoves: A review. *Renewable and Sustainable Energy Reviews, 26*, 265–285.

Njiru, C. W., & Letema, S. C. 2018. Energy poverty and its implication on standard of living in Kirinyaga, Kenya. *Journal of Energy*, 2018.

Ochieng, C., Vardoulakis, S., & Tonne, C. 2017. Household air pollution following replacement of traditional open fire with an improved rocket type cookstove. *Science of the Total Environment, 580*, 440–447.

Panwar, N. L., & Rathore, N. S. 2008. Design and performance evaluation of a 5 kW producer gas stove. *Biomass and Bioenergy, 32*(12), 1349–1352.

Pattanayak, S. K., Jeuland, M., Lewis, J. J., Usmani, F., Brooks, N., Bhojvaid, V., & Ramanathan, N. 2019. Experimental evidence on promotion of electric and improved biomass cookstoves. *Proceedings of the National Academy of Sciences, 116*(27), 13282–13287.

Pratiti, R., Vadala, D., Kalynych, Z., & Sud, P. 2020. Health effects of household air pollution related to biomass cook stoves in resource limited countries and its mitigation by improved cookstoves. *Environmental Research*, 109574.

Quansah, R., Semple, S., Ochieng, C. A., Juvekar, S., Armah, F. A., Luginaah, I., & Emina, J. 2017. Effectiveness of interventions to reduce household air pollution and/or improve health in homes using solid fuel in low-and-middle income countries: A systematic review and meta-analysis. *Environment international, 103*, 73–90.

Quinn, A. K., Bruce, N., Puzzolo, E., Dickinson, K., Sturke, R., Jack, D. W., & Rosenthal, J. P. 2018. An analysis of efforts to scale up clean household energy for cooking around the world. *Energy for Sustainable Development, 46*, 1–10.

Samal, C., Mishra, P. C., Mukherjee, S., & Das, D. 2019. Evolution of high performance and low emission biomass cookstoves-an overview. *AIP Conference Proceedings, 2200.*

Sime, G., Tilahun, G., & Kebede, M. 2020. Assessment of biomass energy use pattern and biogas technology domestication programme in Ethiopia. *African Journal of Science, Technology, Innovation and Development*, 1–11.

Sutar, K. B., Kohli, S., Ravi, M. R., & Ray, A. 2015. Biomass cookstoves: A review of technical aspects. *Renewable and Sustainable Energy Reviews, 41*, 1128–1166.

Venkataraman, C., Sagar, A. D., Habib, G., Lam, N., & Smith, K. R. 2010. The Indian National Initiative for Advanced Biomass Cookstoves: The benefits of clean combustion. *Energy for Sustainable Development, 14*(2), 63–72.

Yohannes, S. S. 2011. Design and performance evaluation of biomass gasifier stove. *An Unpublished Master Thesis prepared in Chemical Engineering Department, Institute of Technology, Addis Ababa University.*

Advances in Phytochemistry, Textile and Renewable Energy Research for Industrial Growth – Nzila et al. (Eds)

Mini-grids as the vehicle to rural development in Uganda: A review

R.G. Mugagga
Kenyatta University, Nairobi, Kenya

H.B.N. Chamdimba
Malawi University of Science and Technology, Limbe, Malawi

ABSTRACT: Mini-grids are regarded as the technology that shall equitably realize the last mile connectivity through a blinding pace. In this paper, the authors investigated the socio-economic impact of mini grid adoption on the rural communities which was undertaken through a review of literature on the existing large scale mini grids in Uganda. Simultaneously, the benefits, challenges and opportunities of mini-grid development in the country were also assessed. Among the fourteen mini grids currently operational in the country, five case studies based on three typical technology types i.e. solar, hydro and biomass were reviewed. The results showed increasing costs per kilowatt in the order of hydro, solar and biomass technologies corresponding to $0.11-0.18, $0.19-0.28 and $0.28 respectively. Secondly, that households in tiers 1-2 were the main beneficiaries as opposed to commercial customers and lastly that the prospects of stimulating productive uses of energy shall inevitably impact rural development in Uganda positively.

1 INTRODUCTION

Universal electricity access continues to be fronted with more concerted efforts more critically in this last decade of action for the UN Agenda 2030 (IEA IRENA UNSD WB WHO, 2019). Though this motive is faced with several impediments, several strides are underway ranging from several global, regional, and national plans (Bhatia & Angelou, 2015; Narayan et al., 2020). Rapid deployment technologies inclusive of mini grids (MGs) have been envisaged to expedite the last mile connectivity to the rural communities so as to reduce the electricity access gap (Weston et al., 2016). In Uganda, several institutional frameworks have been put in place being supplemented by nascent policy developments. Among these was the advent of the Rural Electrification Agency (REA) whose mandate is to widen the rural electricity access in more adequate and reliable approaches through off- and on-grid solutions (MEMD, 2002, 2019). The policy developments on the other hand continue creating favorable environments for the participation of the private sector in the deployment of emerging green technologies for distributed generation (DG) (MEMD, 2015b; Twaha et al., 2016).

In effort to increase rural electrification for socio-economic development, the country needs to invest more in MG projects (MEMD, 2015a). Increased adoption is being fueled by their numerous factors such as reducing technology costs, autonomy in nature, modularity, short lead times and increasing productive uses of energy (PUE) (Opiyo, 2019). MGs are regarded as intermediate power utilities that can serve the rural communities with sufficient energy demand through a blinding pace (Bahaj & James, 2019). Furthermore, they are essential solutions to rural electrification with preferences of higher value of power, reduced losses, reduced costs of transmission and distribution networks and micro generation within consumers' proximity (Marnay, Asano, Papathanassiou, & Strbac, 2008).

In the past, several authors have devoted several studies to the renewable energy (RE) resource potential (ERA, 2016; MEMD, 2015a; Twaha et al., 2016) as well as on the levels of deployment of nascent green technologies in Uganda (Fashina, Mundu, & Akiyode, 2018; Mugagga & Chamdimba, 2019). But few studies (Nygaard et al., 2018) have been conducted to reveal the practical implications of MG adoption to the society. Therefore this paper seeks to fill this research gap. The authors conducted a desk review methodology of existing literature on the available reported MGs upon which a further review of five of them was conducted. The selection was based on the diversity of technologies that have been employed inclusive of solar photovoltaics (PV), hydro and biomass which were analyzed to assess their independent penetration impacts. This shall aid in establishing the linkage between rural electrification and its related socio-economic impacts towards rural development on a local context hence provide the impetus for better policy and increased technological penetration.

This review has been organized as follows; first, is an analysis of the impact of rural electrification to rural

DOI 10.1201/9781003221968-34

development in section two. Secondly, the metrics on which the impacts of rural electrification could be based discussed in section three. Thirdly, a description of MGs with particular emphasis on the case studies in view are highlighted in section four while section five contains the challenges to the deployment of MGs in the country. Finally, the discussions, conclusions as well as recommendations from the review are covered in sections six, seven and eight respectively.

2 RURAL ELECTRIFICATION AND DEVELOPMENT

In contrast with its regional counterparts i.e. Kenya, Rwanda and Tanzania, it becomes evident that the country still falls behind in regard to attainment of the ultimate universal energy access ambition (Table 1). It is noteworthy that on-grid electrification has for long been the backbone of the electrification program throughout all these countries yet significant national electrification strides in Uganda have trailed (George et al., 2019; UBOS MEMD, 2014).

Table 1. Electricity access rates in East Africa.

Country	Population (millions) in 2016*	Access rate (%) in 2017		
		National	Urban	Rural
Kenya	44.2	64	81	58
Rwanda	11.3	34	85	24
Tanzania	47.7	33	65	17
Uganda	39.9	22	57	11

Source: (Bah et al., 2017)* (IEA IRENA UNSD WB WHO, 2019).

However, due to the difficulty posed in extending the grid to the hard to reach areas with regard to cost implications, grid unreliability, and increased transmission and distribution losses to mention, off- grid solutions like the MGs shall continue to step-up the low rural electrification rates in Uganda (MEMD, 2015a). The role of increased demand growth is substantial in driving this implementation (MEMD, 2013) through PUE so as to boost local development (ESMAP, 2019).

At the fore front of orchestrating this demand driven rural development is the REA established through the Energy Policy of 2002 (MEMD, 2002). Its fundamental role is in establishing and maintaining ties between the project beneficiaries and the project developers (government or private sector). These efforts are stimulated by proper planning, utilization and management of the RE resources so as to suit the planned developments in different service territories (ERA, 2016). Despite the past weaknesses and bottlenecks of the first Rural Electrification Strategy and Plan (RESP) (2001-10) that were meant to expedite rural electrification, corrective measures for the impediments of this former strategy were arrived at as outlined in the current RESP II (2013-22) (MEMD, 2013). This current plan has a goal of attaining 26% rural electricity access by 2022 through a centralized approach of 8,500 new MG connections (MEMD, 2013). However, in the most recent REA master plan (2018-28) that spanned 10 service territories, an opportunity to serve up to 62,000 households was identified considering the business case. In this plan, a potential MG site constituted of at least 50 households (UOMA, 2019). This therefore gives room for more than 1,200 potential sites for MG development for improved rural electrification.

Access to electricity has been argued by various researchers via two lines of thought that DG either acts as a catalyst (Kirubi, Jacobson, Jacobson, & Mills, 2009; Parhizi et al., 2015) to already established economic activities or as an enabler for economic transformation (Narayan et al., 2020; Romankiewicz, Marnay, Zhou, & Qu, 2014) in most of the underserved areas. However, this should be taken in context of the level of the underlying conditions. The Ugandan context has for long partaken of the both paradigms which are seemingly producing pragmatic results. It is worth noting that to reach the remaining underserved population, the call for increased private sector financing, stronger energy policies and comprehensive electrification programming shall drive a seamless transition (ESMAP, 2019). With this in place, rural employment through startups will inevitably improve livelihoods in disadvantaged communities (MEMD, 2013). With more than 70% of the country population residing in the rural areas, provision of MGs to this vast population shall be of essence in encompassing the concept of leaving no one behind. Additionally, benefits with regards to improved economics, environmental stability, technical performance and the ease of interconnection in the event of widening of the utility grid also comes in handy as and when the time comes (Williams, Jaramillo, Taneja, & Selim, 2015).

3 BENEFITS OF RURAL ELECTRIFICATION

Rural electrification broadly provides the following benefits i.e. social-economic, environmental, health benefits and reduced energy inequalities as further discussed below.

3.1 *Social-economic benefits*

Since most rural communities are reliant on agriculture and livestock farming, the likelihood of increased food preservation through adoption of various MG technologies stands of great importance as fighting malnutrition and food insecurity remains a priority (NPA, 2015). Reduction of food losses increases the shelf life of animal, diary and agricultural products. For instance, the Kitobo MG has supported the growth of the cold value chain on the Island which is credited for poverty reduction, increased food productivity and food security. Secondly, rural electrification also has the potential of creating jobs for the local people both directly and indirectly. In consideration of the former, MG infrastructure development and operation usually employs both skilled and unskilled labor. Indirectly, electricity in the rural areas has proved able to transform communities in terms of diversification of sources of livelihoods by presenting several business opportunities including agro- and non-agro-based enterprises through PUE (George et al., 2019; Lane, Hudson, Gous, & Kuteesa, 2018). Other community benefits due to electrification include the access to better educational services, less rural to urban migration in search for better livelihoods, economic diversification, cheaper and more reliable social services (Weston et al., 2016).

3.2 *Environmental benefits*

A great deal of attention has been focused on RE generation throughout the energy community due to their potential to reduce greenhouse gases (GHGs) in the power sector. Reduction of the carbon footprint accrued from promoting efficient and green power generation, is essential in the attainment of the country's Nationally Determined Contributions (NDC) aimed at 22% by 2030 which thus necessitate fast action (MEMD, 2019). The country continues to witness more levels of ascent in the energy ladder which provides an optimistic horizon towards the use of RE based MGs. In addition, solar technologies have an emission reduction potential of 1.5 million tonnes of CO_2 equivalent (MWE, 2015). The impact of off grid solar PV based generation on the reduction potential of GHG emissions in the short and long term was also studied through simulations of three futuristic scenarios of 2020, 2030 and 2040 by the authors (Zubi, Dufo-lópez, Pasaoglu, & Pardo, 2016). From their study, the long term provided an optimistic scenario for GHG reduction with focus to developing countries using the integrated Hybrid Optimization Generation Algorithms (iHOGA).

3.3 *Health benefits*

The study carried out by the Uganda Bureau of Statistics (UBOS MEMD, 2014) recorded that over 9.2% of the healthcare facilities in the country were un electrified yet electricity plays a vital role in improving access to health care services. In the vein of ensuring that no one is left behind, energy access to health facilities serving the more disadvantaged rural population necessitates concerted efforts. This has exceedingly been underscored by the current COVID-19 crisis. The utilization of MG technologies as accelerators to the last mile connectivity to community health care settings allows room for precisely meeting several prospects of reduced vaccine losses due to refrigeration, prolonged working hours inclusive of the ability to help in night child delivery, provision of energy for emergency life care and powering hospital equipment (Mohapatra et al., 2019). Off-grid rural electrification has the potential of improving the air quality from several detrimental anthropogenic emission activities such as the utilization of kerosene lamps for lighting to combat Chronic Respiratory Diseases (CRDs) predominant in children and women (IRENA, 2019). Furthermore, the switch to DG reduces health risks such as brain cancer exposed to the communities residing in the proximity of the high voltage transmission lines.

3.4 *Reduced energy inequality*

Through the Sustainable Development Goal (SDG) lens, SDG 10 seeks to among others reduce inequalities. In Uganda's context, this is amplified by the 2006 National Equal Opportunities policy goal. Just as it is in many other developing countries, the country gives priority to urban areas concerning access to basic services such as electricity. Though rather controversial, the time has come when energy investments are needed in more vulnerable localities where project proceeds are tougher to make. The current underlying conditions demonstrate that electricity access is marked by acute inequalities with 11.8% electrification rate in the northern region (i.e. mostly rural) in contrast to 64.9% in Kampala region (UBOS MEMD, 2014). Furthermore, households in rural areas are largely dependent on fuel wood characterized by severe health impacts and more time invested in drudgery than on productive activities such as education, agriculture and businesses. This exacerbates the plight faced by women whose fundamental societal and cultural role deters their economic welfare yet their entrepreneurial role is fundamental in low income countries (ESMAP, 2019). Therefore, provision of sustainable energy to rural households through MGs is paramount in addressing the urban-rural energy divide which shall positively impact on socio-economic development.

4 MINI-GRIDS

4.1 *Mini-grids defined*

A mini grid is basically a standalone localized cluster of electric power systems that entails all the stages of energy generation, distribution, storage, loads and load control that function semi-autonomously from the centralized utility grid meant to serve a few or numerous customers (ESMAP, 2019; Romankiewicz et al., 2014). They provide a building block for integrating various energy sources inclusive of wind, micro-hydro, solar as well as biomass to boost local energy resilience (Parhizi et al., 2015; Warneryd, Håkansson, & Karltorp, 2020). Used interchangeably with the term micro grids, these energy systems have been known as the most dynamic and rapidly changing global energy generation topology (Schnitzer, Lounsbury, Carvallo, Deshmukh, Apt, & Kammen, 2014). Though to avoid misperception, the term mini grid is used for the rest of this paper.

MGs can either be off or on-grid networks. In the event of a power outage, frequency drop, voltage sag or maintenance of the main grid, the latter networks can operate autonomously as they are coupled with a groundbreaking islanding feature (Romankiewicz et al., 2014). MGs are regarded as the better approach of delivering a fast and more economic approach of power generation to rural communities. They also serve as bridging technologies before grid arrival essentially through the smaller electrical distribution networks to compliment the utility network (Bahaj & James, 2019). New technologies with regard to payment schemes, load limits, load monitoring as well as remote control are further progressing MG assumption (UOMA, 2019).

MG sizes differ depending on the context of application as they are dependent on the organizational setup, financial model, local demand, customer base, resource potential, technology and their quality (Shrestha, Shrestha, Shrestha, Papadakis, & Maskey, 2020). The context of analysis discussed in this paper entails MGs that provide power to several customers and not only a single facility like a hospital, school or industry irrespective of the few kW or MW generated.

4.2 *Mini-grid as an option for driving Uganda's rural electrification efforts*

As of 2018, the country incurred 3.9 and 16.6% electricity losses due to transmission and distribution inefficiencies respectively (ERA, 2018). This definitely comes at an extra cost to power consumers. MG proximity to the loads reduces transmission and distribution losses, of which the latter poses as the major challenge of the Electricity Supply Industry (ESI) in the country (ERA, 2019). Correspondingly, MG are proposed with increased power efficiencies, resilience and economic operation with reduced end user cost (Parhizi et al., 2015). This thus makes them an outstanding option for the Ugandan rural communities. To the utilities, the cost of upgrading the grid to the rural areas with lower population densities is uneconomic and becomes extremely difficult (ERA, 2019). While the use of MG technologies provides an economically attractive and competitive option to fossil based power generation due to their (MG) technological improvements and their plummeting costs. For instance, in 2018, solar PV installation costs and the levelized cost of energy (LCOE) were reported as $1210 and $0.085 per kWh respectively (IRENA, 2019). The plummeting energy costs foster green growth, job creation, increased energy access, energy independence and security which will positively influence the country competitiveness and productivity.

4.3 *Tiers of energy access*

The level of electricity access has for the past been underestimated by governments where relative samples of the population were considered to make sweeping statements and decisions regarding energy access of the larger communities. Until recently, energy access was a "have" or "have-not" situation irrespective of the reliability, quantity or quality. It was through the efforts of the World bank Multi-Tier Framework (MTF) that unearthed electricity access as tiers of service ranging from Tier 0 to 5 (ESMAP, 2019). The MTF was conceptualized on the basis of seven multidimensional attributes of electricity service that are; the power quality, capacity, health and safety, affordability, legality, service hours and reliability. It also uncovered tier 0, a category where there is no meaningful electricity access (Bhatia & Angelou, 2015). A summary of the standards is presented (Table 2).

Concerted efforts to increase universal access to energy are leading a new wave from the binary assessment to the integration of a multi-tier system through the country's Sustainable Energy for All (SE4All) action agenda (MEMD, 2015b). This has been backed by recent policy interventions to ensure that newer projects follow suit. As a result, there is continued growth rate for the customers of tier 1 in Uganda (IEA IRENA UNSD WB WHO, 2019) majorly because they are of low cost and can be deployed rapidly almost everywhere (Bahaj et al., 2019).

4.4 *Mini-grid cases studies in Uganda*

Solar PV technology remains dominant in the continued acceleration of electricity access through MG

Table 2. Multi-tier standards for household electricity supply.

	Tier 0	Tier 1	Tier 2	Tier 3	Tier 4	Tier 5
Energy rating		3–<50 W	50–<200W	200–<800 W	800–<2000 W	>2 kW
Availability (per day)		>4 hrs	>4 hrs	>8 hrs	>16 hrs	>23 hrs
Reliability	Unscheduled outages				≤14 disruptions per week	≤3 disruptions per week
Power quality	Poor quality				Better power quality	
Affordability	Unaffordable			Cost of Standard consumption package of 365 kWh/year <5% of household income		
Legality	Not legal				Bill is paid to a legal entity	
Health and safety	Presence of past accidents				No past accidents and less perception of high future risks.	

Source: (Bhatia & Angelou, 2015).

development and is usually backed by the diesel gensets. The unit costs of RE power generation relatively competes favorably with that of the national grid whose cost reflections per kWh are \$0.2, \$0.18 and \$0.16 for the residential, commercial and large industrial customers respectively (ERA, 2018). The opportunity presented in the country of the use of blended tariffs for the MGs (Table 3) provides good prospects for advancing these developments.

It is worth noting however that the current MG tariffs are still lower than those from diesel generated electricity i.e. \$3 per day (Lane et al., 2018).

Now that MGs exist, a few cases have been discussed in brevity in order to highlight some key lessons that serve as useful benchmarks moving forward with a further analysis in section six discussed further on.

4.4.1 *Kyamugarura and Kanyegaramire*

These solar PV MGs were set up as part of the energy for Development (e4D) program in collaboration with the Uganda Rural Electrification Agency (REA) to act as an informative benchmark on how best to implement sustainable energy driven MGs to the target areas for socio-economic development (Bahaj & James, 2019; Nygaard et al., 2018). The setup of the two MGs is akin to each other with the power plant components accommodated in shipping containers. These MGs boast of customer bases inclusive of schools, businesses, health centers and households. The use of prepaid special card readers at the onset solved the problem of physical fees collection where upon paying for several units of energy, the card is then inserted in the consumer's meter to power up the their premise. This also provided additional security to the power supply. Both sized to deliver 28kWh/day, the maximum demand capacity was exceeded 30 months after commissioning due to the grid price equivalent tariff which usually led to power outages at night in order to suit the designed 70% depth of discharge of the battery (Bahaj & James, 2019).

4.4.2 *Kisiizi Hydro*

The Kisiizi MG is a fully owned subsidiary of Kisiizi Hospital that is governed by the Church of Uganda (KHPL, 2014). Generated power is used by the hospital and its conglomerates while the excess is sold to the surrounding school, polytechnic, trading center and homes. Initially, power generation was meant to cover only the now 60-bed hospital and its affiliated institutions with the surplus being sold to its wider Kisiizi community. However, the demand of power usage continued to soar even with the incorporation of the 294kVA Ossberger crossflow in 2009 meant to supplement the existing 60kVA Gilkes Turgo (KHPL, 2014). The number of electricity connections was 372 by 2013 but due to capacity constraints, this MG can no longer handle new customers. In the event of dry seasons and failures from the turbines, the system is backed by a 80kVA standby diesel generator (KHPL, 2018).

4.4.3 *Kabunyata*

Through the Anchor, Business and Community (ABC) model, this solar PV MG was constructed with the support of GIZ and Kirchner Solar Uganda Limited that is a subsidiary of Kirchner Solar. The MG has an installed capacity of 22.5 kW comprised of 48 high performance solar PV modules with a yearly generation of 22MWh (Nygaard et al., 2018). With the rapidly increasing demand for telecommunication services in rural Uganda, this provides great motivation for service providers to develop sustainable and green power base transceiver stations. This operational model was one of the pilot studies carried out by GIZ where electricity was supplied to both the local households and the telecom tower in Kabunyata village (Fraatz, 2013). This has accrued dual benefits to the beneficiaries where the telecom company (Airtel) has realized a 40% reduction in the costs of fuel utilized in powering the existing diesel genset while boosting the network coverage to the surrounding areas. Moreover, the local economy has also been promoted through increasing household incomes, powering ten Small Medium Enterprises (SMEs) and two institutions where consumption is metered through a prepaid system (Kurz, 2014).

4.4.4 *Kitobo MG*

Kitobo Island constitutes one of the 84 islands of Ssese archipelago where fishing is the predominant economic activity. Commissioned in 2016, this solar MG provides electricity to over 600 rural households through 880 PV panels each with nominal power of 260Wp which are backed up by an 80kVA diesel genset (Barelli et al., 2019). This solar – diesel hybrid powered system is a typical off-grid type and comprises of off-grid inverters whose frequency modulation is characteristic of its loads (ERA, 2019). Subsequently, productive energy use in the local fish value chain through flake ice making was revamped and as a result reduced the gap between supply of ice for refrigeration, increased product (fish) shelf life and profits of the rural fishermen. Collectively, this has catalyzed economic development through the creation of other employment opportunities as opposed to the previous use of a private diesel generator that was characterized with high diesel expenses where only 30 villagers were connected (Barelli et al., 2019). The provision of sustainable energy services to the community has ultimately promoted their livelihoods.

4.4.5 *Ssekanyonyi MG*

Developed by Pamoja Energy Limited under the initiative of using agricultural residues as sources of energy, the 32kW$_e$ project utilized downdraft gasification of maize cobs to produce syngas to run the generator. However, to date, this plant is not operational due to project failure in reaching commercial parity as the system was oversized whilst in the preliminary stages with consideration of the neighboring mill as an anchor load which was never actualized. Worse still, the use of un-chopped cobs also hindered the reactors from

Table 3. Mini grid descriptions.

Mini grid/ Project	Size & RE Technology	Electrical Storage	Capital Cost (*$10^6$$)	Tariff $/kWh	Connections	Developer	Location (District)	Year Launched
Kalangala	600kW Solar PV, 1000 kW Genset	240 Batteries	16	0.19	3100 HH 24 Comm.	Kalangala Infrastructure Services Ltd.	Bugala Island – Kalangala	2015
Kayanja	5 kW Solar PV	Battery	n/r	n/r	120	Remergy Energy	Kasese	2015
Kitobo	228.8 kW Solar PV & 80kVA Genset	520kWh battery	1.4	0.2675	600	Absolute Energy Africa Ltd.	Kalangala	2016
Kyamugarura & Kanyegaramire	13.5 kW Solar PV	24, 800Ah batteries	0.0675	0.28	> 500 100	E4D & REA	Kyenjojo	2015
Kabunyata	22.5 kW Solar PV	Battery	0.540	0.217 & 0.498*	120 & 10SMEs	Kirchner Solar	Luweero	2013
Kiboga	1 kW Solar	Battery	n/r	n/r	11 & 5 SMEs	CREEC	Kiboga	2011
Nyagak 1	Hydro -3.5MW & Genset - 8MW	Local Grid	3.782	0.18	14000 HH, 29 HC & 44 Sch.	Electro-Maxx	Zombo, West Nile	2012
Kisiizi	Hydro -300kW & genset - 80kVA	Local Grid	0.9	0.16	> 372	Kisiizi Power Co.	Rukungiri	2009
Tiribogo	$32kW_e$ Biomass	n/r	n/r	0.28	170	Pamoja Energy Limited	Muduuma Subcounty- Mpigi	2012
Ssekanyonyi	$32kW_e$ Biomass & Solar	Local grid	n/r	n/r	75	Pamoja Energy Limited	Mityana	
Bwindi Micro HPP	64kW Hydro	Local grid	n/r	n/r	62	Bwindi Community Hospital	Kanungu District	2014
Suam	40kW Hydro	Local grid	n/r	0.11-0.12	n/r	GIZ	Bukwo	2013
RMS Pico HPP	5kW Hydro	n/r	n/r	n/r	16	CREEC	Kasese	2013

Source: (ERA, 2018, 2019; Kurz, 2014; Nygaard et al., 2018).
* - households and SMEs respectively, HH – household, Sch. – Schools, Comm. – Commercial, n/r – not reported, CREEC – Centre for Research in Energy and Energy Conservation, HPP – Hydro Power Plant.

attaining higher temperatures which led to partial gasification thus reducing the overall efficiency of the gasifier. Ultimately, the consumers didn't benefit in the long run as the running costs couldn't allow for further generation (Owen & Ripken, 2017). It thus becomes a cautionary tale to involve several concerned actors during the planning, implementation and the operation of MGs (Warneryd et al., 2020).

5 CHALLENGES AND PRIORITIES TO MG DEVELOPMENT IN UGANDA

Rural electricity access is a complex issue and monumental challenge that calls for the need to address several factors collaboratively (Narayan et al., 2020). This is because the prospective sites vary on a case by case basis and there is no one size fits all procedure and solution. This poses several challenges to MG development however, consideration of several priorities beforehand provide necessary insights for the up scaling of MGs.

5.1 *Financial barriers*

Irrespective of the reducing technology costs, the expense of energy generation and supply with regard to operational, maintenance and management costs still remains high in developing countries (Bhatia & Angelou, 2015). This has stalled five MGs meant to be constructed by Absolute energy in Wakiso, Namayingo, Mukono and Kalangala Islands (ERA, 2018). This impediment amplifies the revenue risk leading to longer payback times yet private sector investment is reliant on grants, equity and debt financing (Mugagga & Chamdimba, 2019). Tackling the balance between the scale of anticipated risks and returns thus remains a key aspiration to private developers (Williams et al., 2015). Compounding the project risks is the difficulty in supporting developments in poor investment climates where the returns on investment would be more or less uncertain leading to large capital outlays (Gambino et al., 2019). The customers' willingness to pay (WTP) and ability to pay (ATP) also vary rather disproportionally within different regions of the country as most un-electrified areas are predominantly agrarian based which limits their income levels.

While corrective factors are critical in arriving at a commercially viable project as they help normalize the anticipated demand load profile, they also help maintain the balance of the tariff structure. Also, favorable political motivation helps secure private as well as public sector investments in addition to promoting subsidies (SEforALL BNEF, 2020; Shrestha et al., 2020). Reasonable tariffs and power subsidies are a reality in MG development as they stimulate and ensure project sustainability. Subsidies usually take the form of either capital or operating subsidies (Williams et al., 2015). The Ugandan government offers 50% capital subsidies as they tend to be more attractive for RE based generation due to long term sustainability (SEforALL BNEF, 2020).

5.2 *Institutional barriers*

Aside from the upfront costs, the role of efficient, robust and established regulatory frameworks are of great essence in expediting the stages of MG development (Lane et al., 2018). The existing frameworks in the country are characterized by lengthy bureaucratic stages which are not necessary. This red tape deters investment, increases transaction costs which prolongs project timelines leading to inherent in line ramifications (Chessin, Cook, Gessesse, & Solano-Peralta, 2017; Weston et al., 2016). Since MGs are regulated by ERA, those below 2MW are issued with certificates of exemption from annual licensing fees (ERA, 2019). However, for those projects above the 2 MW range, project developers have delineated the challenges of navigating through the existing frameworks; the lengthy durations within which the licensing permits are processed, unclear approval processes, with tariff adjustments varying on a case by case basis though on a positive note, the REA is either entirely responsible for distribution infrastructure development or reimbursement of 50% of the construction cost to the developer (Chessin et al., 2017; Nygaard et al., 2018). Guaranteed and clear long-term policies that are consistent over time avert damage, increase confidence to the projects in the pipeline and those under execution which is crucial in providing fair and equitable environments for MG development (Romankiewicz et al., 2014).

The regulatory environment with regard to MG development in Uganda is less mature compared to neighboring Kenya and Tanzania. This represents the biggest bottleneck in encouraging private investment in Uganda and this therefore explains the limited number of MG established so far (SEforALL BNEF, 2020). Benchmarking from government initiatives such as setting cost reflective tariffs – in Kenya, Rwanda and Tanzania, objectively defined projects like the Kenya Off-grid Solar Access Project (KOSAP) meant to increase modern energy access to 14 underserved counties – in Kenya, and clear regulations with respect to grid arrival as is the case in both Tanzania and Rwanda (Weston et al., 2016). This shall help address this eminent challenge

5.3 *Community acclimatization and involvement*

The primary energy sources utilized in MG power generation are usually prone to availability and reliability concerns due to the intermittent nature of the sun, wind and the hydro resources (Mohapatra et al., 2019). This is characteristic of most MGs in the country as most do not incorporate the hybrid (several RE resources) setup which would otherwise increase cost efficiency, competitiveness, flexibility and reliability (Parhizi, Lotfi, Khodaei, & Khodaei, 2015). Proper demand side management comes into perspective though it is still a difficult phenomenon to impose onto the rural populations to allow for modified electricity usage patterns in low generation seasons so as to at least sustain basic lighting services to the communities at night (Parhizi et al., 2015). Majorly, access to electricity services is usually characterized by lifestyle transformation due to improved standards of living. With increased energy affordability i.e. lower tariffs, the significance of increased customer satisfaction becomes apparent and comes with a lock-in effect that usually tends not to accommodate unscheduled power cuts (Zubi et al., 2016). This has often led to dissatisfaction and reduced customer confidence.

On the other hand, early involvement of the locals prior to the implementation of the proposed MG projects is essential with regards to acquisition of land, determination of the load profile, economic activities, past site history and freely sourcing of all the prerequisite data. This ushers good customer management that aids in future process implementation in regards to knowing; the level of incorporation of the local expertise, appropriate modes of payment, and their employability to mention (Kirubi et al., 2009).

5.4 *Technical barriers*

Project developers are always confronted with the concern of involvement of the community workmanship as the local reticulation personnel mainly because they are predestined to work hand in hand with the technical development team for technical project sustainability (Barelli et al., 2019). However, the skill set of the local workmanship has been lacking leading to installations that are well below standard (Mugagga & Chamdimba, 2019). Project technical sustainability has been partly hampered by the urge of the newly trained technical personnel shifting to more highly paying jobs in the urban dwellings. Also, in the event of major repairs, the rural MGs are faced with the problem of long lead times experienced in deliveries alongside the high repair and maintenance costs (Williams et al., 2015). These at times necessitate highly experienced technical expertise and consequently, downtimes become more pronounced which distorts customer confidence (Weston et al., 2016).

The Electricity Regulatory Authority (ERA) should participate in the enforcement of comprehensive technical standards that are paramount in ensuring proper installations, power and system quality for systematic MG deployment (Chessin et al., 2017). These standards ensure adequate training of the local personnel which subsequently contributes to project sustainability. It is imperative to adopt the DC- coupled MG for purposes of avoiding several power conversion sequences which in effect lower the system efficiency yet energy savings of up to 33% with Direct Current (DC) MG networks can be obtained (Opiyo, 2019).

5.5 *Inadequate load estimations*

Load forecasts are the domain of data acquisition for MG planning and operational purposes. Since MGs provide for higher tiers of energy access, it becomes problematic to fully anticipate the level of load growth. The severity of this challenge is more pronounced in areas where electricity access is an alien concept i.e. most rural areas and thereby necessitating the use of software tools such as HOMER to simulate the load profiles and growth (Mohapatra et al., 2019). Furthermore, the failure to establish contracts with the large anchor load customers during the initial stages of project development exposes developers to un-bankable load availability (Schnitzer et al., 2014). Also, with the catalytic nature of electricity, several startups of PUE tend to overshoot beyond the designed optimal capacity. This eventually leads to a mismatch in demand and supply (Bahaj et al., 2019).

The authors (Gambino et al., 2019), developed a methodology for studying the energy needs assessment for the effective design and deployment of MGs for rural electrification while mitigating uncertainties in electrical demand with specific concern to the green field communities. Proper MG sizing can be achieved by collecting the community and customer-based data through standard questionnaires and results studied with adequate corrective factors to generate the load profiles and determine the peak load (SEforALL BNEF, 2020). The utilization of linear growth profiles through the interpolation of the existing demand can somewhat be representative of the population growth, economic growth and typical consumption patterns. Forecast demand growth gives a realistic horizon to the anticipated payback times (Kirubi et al., 2009). Productive use monitoring using timers and relays helps to limit scrupulous customer connections. This ensures adequate load control and paves way for thorough load audits to be performed prior to new customer stepwise connections.

6 DISCUSSION

Of the fourteen MGs reviewed, those utilizing solar have gained ground in meeting the increasing energy demand with regard to DG mainly because of their plummeting technology costs. Solar MGs tend to have relatively higher energy costs than hydro i.e. \$0.19 – 0.28 /kWh and \$0.11 – 0.18/kWh respectively due to the energy storage and capital costs. However, the daytime power consumption tends to offset these energy storage costs though battery less systems are generally uncommon in Ugandan MGs. Private investments have proved to be more financially viable in areas with established bankable anchor loads which ultimately ensures the optimum and efficient utilization of generated power thus enabling the actualization of shorter payback periods. Anchor loads for the Kisiizi and Kabunyata MGs had a more certain power demand unlike the Ssekanyonyi MG. Key to note is the advantage of economies of scale played out in the Kalangala MG whose costs per kWh compete favorably with those of the hydro MGs. As witnessed in this context also, the MGs with PUE mainly predominant in the fishing, agro-industry sectors as well as the small scale business enterprises that operate during the day time hours have proved to provide more employment opportunities which continue to raise the socio-economic status of the beneficiaries. However, the hydro MGs are still heavily susceptible to the varying water levels which in turn affects power generation levels like the case of Kisiizi MG. Nevertheless, from this review, it has been noted that several MGs have adopted energy storage options to avert the challenge of intermittency but hybridization ought to be the ultimate solution.

Though the payback is dependent on numerous intricate factors, the technology used plays a big part. The Kitobo MG for instance, has higher tariffs than its counterparts due to the fact that vanadium flow batteries were employed which have a lifetime of close to 40 years as opposed to the lead acid and lithium ion batteries. It is by no doubt therefore that the project payback period is largely dependent on a case by case basis and usually spans between 4-10 years. Additionally, despite Alternating Current (AC) MGs often benefitting from grid arrival due to financial compensation, AC based MGs incorporate more components as opposed to DC MGs and are thus more costly. Lastly, the payback period of a particular project lessens with increased PUE which usually ensure maximum utility of generated power and thus maximize returns.

7 CONCLUSION

Indeed, MG development by taking advantage of the abundance of RE resources that are widely available in the country proves to be the least cost solution for increased energy access to the underserved rural communities by providing the opportunity to leap frog to faster socio-economic rural development. This is mainly because MGs are regarded as the better approach of delivering a fast and more economic approach of power generation as witnessed in this review. Other than the Kalangala and Kabunyata MGs which have had more than 50% impact on businesses comprised of commercial customers, the rest have enormously impacted households who mainly use energy for phone charging, lighting, and other small appliances in tiers 1-2. This therefore serves as a wakeup call for developers to look further into investing in selling services as well as stimulating PUE through joint ventures by funding locally led commercial activities. This shall necessitate the adoption of specific tailor-made business models for advancing profitable enterprises like refrigeration, ice making, and milling to mention which shall collectively favor rapid rural development.

8 RECOMMENDATIONS

To attain the 98% energy access spelt out in the country's SE4ALL action agenda by 2030, revamped efforts of the government in working with key stakeholders to develop strategic interventions shall remain pivotal. It's equally essential to keenly study and develop a MG policy that targets the existing policy gaps through the benchmarking on already established frameworks driving positive change. Furthermore, the ability to further incorporate the MTF in energy policy shall prove useful in providing the ability to track progress towards SDG 7; help in setting realistic goals for energy access; help formulate adequate policy interventions; establish the inter-linkages between energy access and other sectors and delineate the relevant supply and demand data. Correspondingly, studies to investigate the impact of the different tiers of electricity access on the level of rural development should be conducted so as to provide a useful benchmark for further related investment decisions for the wider national grid electrification programs. This is because the energy demand growth is eminent. Finally, increased initiatives should be put forth to advance the integration of hybrid renewables for better flexibility, reliability and reduced LCOE.

REFERENCES

Bah, E. M., Hanouz, M. D., Crotti, R., Hoffman, B., Marchat, J. M., & Verdier-Chouchane, A. (2017). *The Africa Competitiveness Report 2017.*

Bahaj, A., Blunden, L., Kanani, C., James, P., Kiva, I., Matthews, Z., Price, H., Essendi, H., Falkingham, J., & George, G. (2019). The Impact of an Electrical Minigrid on the Development of a Rural Community in Kenya. *Energies*, *12*(778), 1–21. https://doi.org/10.3390/en12050778

Bahaj, A., & James, P. (2019). Electrical Minigrids for Development: Lessons From the Field. *Proceedings of the IEEE*, 1–14. https://doi.org/10.1109/JPROC.2019.2924594

Barelli, L., Bidini, G., Cherubini, P., Micangeli, A., Pelosi, D., & Tacconelli, C. (2019). How Hybridization of Energy Storage Technologies Can Provide Additional Flexibility and Competitiveness to Microgrids in the Context of Developing Countries. *Energies*, *12*(3138), 1–22. https://doi.org/10.3390/en12163138

Bhatia, M., & Angelou, N. (2015). *Beyond Connections: Energy Access Redefined.*

Chessin, E., Cook, R., Gessesse, E., & Solano-Peralta, M. (2017). *Practical Guide to the Regulatory Treatment of Minigrids.*

ERA. (2016). *The Least Cost Generation Plan 2016 – 2025.*

ERA. (2018). *Annual Report FY 2017-18.*

ERA. (2019). *Sector Update Newsletter.*

ESMAP. (2019). *Mini Grids for Half a Billion People: Market Outlook and Handbook for Decision Makers. Executive Summary.*

Fashina, A., Mundu, M., & Akiyode, O. (2018). The Drivers and Barriers of Renewable Energy Applications and Development in Uganda: A Review. *Clean Technologies*, *1*, 9–39. https://doi.org/10.3390/cleantechnol1010003

Fraatz, J. (2013). *Mobile towers for rural electrification.* https://www.gsma.com/mobilefordevelopment/wp-content/uploads/2013/06/GIZ.pdf

Gambino, V., Del Citto, R., Cherubini, P., Tacconelli, C., Micangeli, A., & Giglioli, R. (2019). Methodology for the Energy Need Assessment to Effectively Design and Deploy Mini-Grids for Rural Electrification. *Energies*, *12*(574), 1–27. https://doi.org/10.3390/en12030574

George, A., Boxiong, S., Arowo, M., Ndolo, P., Chepsaigutt-Chebet, & Shimmon, J. (2019). Review of solar energy development in Kenya: Opportunities and challenges. *Renewable Energy Focus*, *29*, 123–140. https://doi.org/10.1016/j.ref.2019.03.007

IEA IRENA UNSD WB WHO. (2019). *Tracking SDG 7: The Energy Progress Report*. https://trackingsdg7.esmap.org/

IRENA. (2019). *Future of Solar Photovoltaic: Deployment, investment, technology, grid integration and socio-economic aspects (A Global Energy Transformation: paper)*.

KHPL. (2014). *Kisiizi Elecricity – Annual Report 2014*.

KHPL. (2018). *Power Company*. Kisiizi Hospital. http://www.kisiizihospital.org.ug/?page_id=89

Kirubi, C., Jacobson, A., Kammen, D. M., & Mills, A. (2009). Community-Based Electric Micro-Grids Can Contribute to Rural Development: Evidence from Kenya. *World Development*, *37*(7), 1208–1221. https://doi.org/10.1016/j.worlddev.2008.11.005

Kurz, K. (2014). *The ABC-Modell Anchor customers as core clients for mini-grids in emerging economies*.

Lane, J., Hudson, W., Gous, A., & Kuteesa, R. (2018). *Mini-Grid Market Opportunity Assessment: Uganda*.

Marnay, C., Asano, H., Papathanassiou, S., & Strbac, G. (2008). Policymaking for Microgrids: Economic and Regulatory Issuues of Microgrid Implementation. *IEEE Power & Energy Magazine*, 66–77. https://doi.org/10.1109/MPE.2008.918715

MEMD. (2002). *The Energy Policy for Uganda*.

MEMD. (2013). *Rural Electrification Strategy and Plan 2013–2022*.

MEMD. (2015a). *Scaling Up Renewable Energy Program Investment Plan*.

MEMD. (2015b). *Uganda's Sustainable Energy For All (SE4All) Initiative Action Agenda*.

MEMD. (2019). *Draft National Energy Policy*.

Mohapatra, D., Jaeger, J., & Wiemann, M. (2019). *Private Sector Driven Business Models for Clean Energy Mini-Grids: Lessons learnt from South and South-East-Asia*.

Mugagga, R. G., & Chamdimba, H. N. B. (2019). A Comprehensive Review on Status of Solar PV Growth in Uganda. *Journal of Energy Research and Reviews*, *3*(4), 1–14. https://doi.org/10.9734/JENRR/2019/v3i430113

MWE. (2015). *Uganda's Intended Nationally Determined Contributions*.

Narayan, N., Vega-garita, V., Qin, Z., Popovic-gerber, J., Bauer, P., & Zeman, M. (2020). The Long Road to Universal Electrification: A Critical Look at Present Pathways and Challenges. *Energies*, *13*(508), 1–20. https://doi.org/10.3390/en13030508

NPA. (2015). *National Development Plan II, 2015/16 – 2019/20*.

Nygaard, I., Bhamidipati, P. L., Andersen, A. E., Larsen, T. H., Cronin, T., & Davis, N. (2018). *Market for the integration of smaller wind turbines in mini- grids in Uganda*.

Opiyo, N. N. (2019). A comparison of DC- versus AC-based minigrids for cost-effective electrification of rural developing communities. *Energy Reports*, *5*, 398–408. https://doi.org/10.1016/j.egyr.2019.04.001

Owen, M., & Ripken, R. (2017). *Bioenergy for Sustainable Energy Access in Africa: Technology Country Case Study Report*.

Parhizi, S., Lotfi, H., Khodaei, A., & Bahramirad, S. (2015). State of the Art in Research on Microgrids: A Review. *IEEE Access*, *3*, 890–925. https://doi.org/10.1109/ACCESS.2015.2443119

Romankiewicz, J., Marnay, C., Zhou, N., & Qu, M. (2014). Lessons from international experience for China's microgrid demonstration program. *Energy Policy*, *67*, 198–208. https://doi.org/10.1016/j.enpol.2013.11.059

Schnitzer, D., Lounsbury, D. S., Carvallo, J. P., Deshmukh, R., Apt, J., & Kammen, D. M. (2014). *Microgrids for Rural Electrification: A critical review of best practices based on seven case studies*.

SEforALL BNEF. (2020). *State of the Global Mini-Grids Market Report 2020*.

Shrestha, P., Shrestha, A., Shrestha, N., Papadakis, A., & Maskey, K. (2020). Assessment on Scaling – Up of Mini – Grid Initiative: Case Study of Mini – Grid in Rural Nepal. *International Journal of Precision Engineering and Manufacturing-Green Technology*. https://doi.org/10.1007/s40684-020-00190-x

Twaha, S., Ramli, M. A. M., Murphy, P. M., Mukhtiar, M. U., & Nsamba, H. K. (2016). Renewable based distributed generation in Uganda: Resource potential and status of exploitation. *Renewable and Sustainable Energy Reviews*, *57*, 786–798. https://doi.org/10.1016/j.rser.2015.12.151

UBOS MEMD. (2014). *Uganda rural-urban electrification survey*.

UOMA. (2019). *Market Map of off-grid energy in Uganda*.

Warneryd, M., Håkansson, M., & Karltorp, K. (2020). Unpacking the complexity of community microgrids: A review of institutions' roles for development of microgrids. *Renewable and Sustainable Energy Reviews*, *121*(109690), 1–16. https://doi.org/10.1016/j.rser.2019.109690

Weston, P., Verma, S., Onyango, L., Bharadwaj, A., Peterschmidt, N., & Rohrer, M. (2016). *Green Mini-Grids in Sub-Saharan Africa:*

Williams, N. J., Jaramillo, P., Taneja, J., & Selim, T. (2015). Enabling private sector investment in microgrid-based rural electrification in developing countries: A review. *Renewable and Sustainable Energy Reviews*, *52*, 1268–1281. https://doi.org/10.1016/j.rser.2015.07.153

Zubi, G., Dufo-lópez, R., Pasaoglu, G., & Pardo, N. (2016). Techno-economic assessment of an off-grid PV system for developing regions to provide electricity for basic domestic needs: A 2020 – 2040 scenario. *Applied Energy*, *176*, 309–319. https://doi.org/10.1016/j.apenergy.2016.05.022

Advances in Phytochemistry, Textile and Renewable Energy Research for Industrial Growth – Nzila et al. (Eds)

Production of solketal, a fuel additive, through microwave heating and catalysis

Kenneth K. Shitemi
Eldoret, Kenya

Kirimi Kiriamiti
School of Engineering, Moi University, Eldoret, Kenya

ABSTRACT: Many efforts are being geared toward finding new sources of alternative energy to replace the currently used non-renewable fossil fuels. Among the renewable energy resources being looked into, biofuels are receiving intensive attention. The use of biofuels and fossil fuels lead to fuel gelling and chocking of nozzle as well as having an effect of corrosion on different parts of the engine. This can be mitigated by use of fuel additives. Fuel additives are materials that improve the cleanliness of different parts of the engine (such as fuel injector, intake valve, etc.), promote complete combustion of fuel, reduce fuel gelling and choking of nozzle as well as reduce corrosion impact on different parts of the engine. This in turn leads to improved engine performance, reduced emissions, and reduced fuel consumption. Fuel additives can also reduce particulate emissions of diesel fuel and increase oxygen concentration. Solketal, an oxygenate fuel additive, is one of the fuel additives used to improve engine and fuel performance. It is primarily produced from glycerol, a by-product obtained from the production of biodiesel. It helps to reduce the soot, reduce the particulate emission, and improve the cold flow properties of liquid.

Keywords: Solketal, catalyst, microwave heating

1 INTRODUCTION

The increasing awareness of the environmental effects of the by-product greenhouse gases resulting from the combustion of traditional fossil fuels coupled with their finiteness are fueling the race for research into the development of alternative fuels. However, because the global energy demand is significant, and growing by the day, advancements in establishing suitable and sustainable substitutes have been slow, but biomass is showing significant promise (Fatimah et al., 2019; Ilgen, Yerlikaya, & Akyurek, 2017; Priya, Selvakannan, Chary, Kantam, & Bhargava, 2017; Talebian-Kiakalaieh, Amin, Najaafi, & Tarighi, 2018; Vinicius Rossa, Gisel Chenard Díaz, Germildo Juvenal Muchave, Gomes Aranda, & Castellã Pergher, 2019). The conversion of bio-glycerol into glycerol constituent ethers and esters, through etherification and esterification, has received especially high interest because glycerol is a good platform for other chemicals, and oxygenated compounds in particular, according to a growing number of studies on the potential of value-added substances in industrial glycerol applications. As such, it continues to show high potential in the development of fuel additives, such as cyclic acetals and ketals with aldehydes and ketones respectively. Fuel additives are materials that help to improve the cleanliness of different engine parts (such as the fuel injectors and intake valves) by reducing the incidence of fuel turning to gel, and promote complete combustion, therefore, reducing the overall corrosion impact on engine parts. As a result, they help to improve overall engine efficiency and, reduce emissions and fuel consumption. There is also evidence that they also reduce particulate emissions and increase oxygen concentration in diesel fuels, and improve the thermal stability of jet fuels, therefore, reducing residue deposits in jet engines significantly (Jorge Sepúlveda et al., 2015; Mota, Silva, Nilton Rosenbach, Costa, & Silva, 2010; Talebian-Kiakalaieh et al., 2018; Vinicius Rossa et al., 2019). This paper, therefore, aims to evaluate the effectiveness of the different common solketal production methods and models (and microwave heating and catalyst action in particular) and highlight ways in which they can be improved by reviewing studies on solketal production published in the last 10 years.

2 MATERIALS AND METHODS

This is a retrospective study of the evolution of solketal production using microwave heating and catalysis and will use studies published within the last 10 years (from 2010–2020) to generate its information. The study will look at the different methods and processes different

DOI 10.1201/9781003221968-35

experiments used and analyze their effectiveness by looking at the product yields and the models' ability to recycle the catalysts used.

3 RESULTS AND DISCUSSIONS

The 1980s saw the introduction of reformulated gasoline (a composite of regular gasoline and oxygenated compounds) in the US in response to increasing public awareness and regulatory actions aimed at improving the quality of air in large urban areas. And, while methyl tert-butyl ether (MTBE) was the most popular additive from the onset, because of its cost effectiveness used by refineries to reduce production cost of Reformulated Gasoline (RFG) whose classification as a hazardous material by the International Agency of Research on Cancer (IARC) in 2000 resulted in it being phased out, and created opportunities for the development and penetration of renewable fuel substitutes, such as ethanol. However, despite ethanol experiencing a surge in popularity, investment and development since then, many studies have raised concerns about its practicality and sustainability because of its over-reliance on corn (for carbohydrate fermentation), and the infancy of its production from cellulosic materials (Mota et al., 2010; Talebian-Kiakalaieh et al., 2018). While glycerol's triol constitution with more than half its weight in oxygen atoms makes it a good base for the development of oxygenated fuel additives, its polar properties that make it insoluble in hydrocarbons coupled with its high boiling point make it unsuitable for blending with gasoline. However, its ketals and acetals (and those formed through reactions with acetone (illustrated in Scheme 1) and formaldehyde (illustrated in Scheme 2) in particular) form better combustion enhancers for gasoline (Mota et al., 2010).

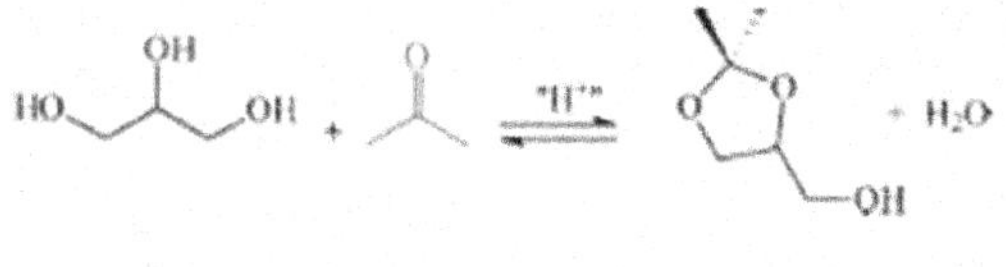

Figure 1. Scheme 1. Reaction of glycerol with acetone under heterogeneous acid catalysis (Mota et al., 2010).

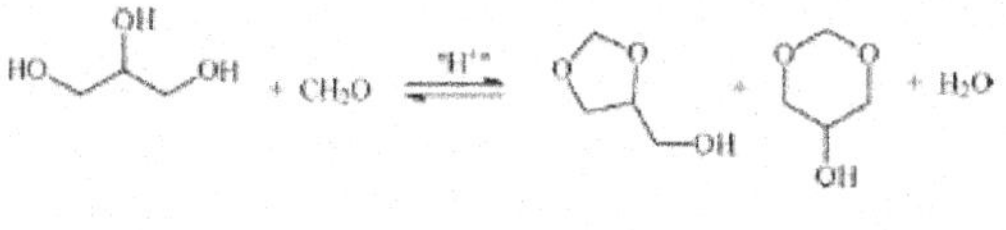

Figure 1. Scheme 2. Reaction of glycerol with formaldehyde solution under heterogeneous acid catalysis (Mota et al., 2010).

Ketalization processes between glycerol and acetone can also result in the production of 2,2-dimethyl-1, 3-dioxolane-4-methanol (also known as solketal) when acid is used to catalyze the reaction as illustrated in the scheme below (Talebian-Kiakalaieh et al., 2018).

3.1 *Glycochemistry*

The processes through which glycerol is processed into value added products, such as the acetalization process used to produce solketal, glycerol acetal and glycerol formal which are used as solvents, plasticizers, surfactants, disinfectants, and flavoring agents, is known as glycerochemistry (Vinicius Rossa et al., 2019). However, solketal production is so far the most economic and promising use of glycerol because of the versatility of its uses and applications across a variety of industries. In combustion, it is used as an oxygenate fuel additive and helps to reduce soot by increasing the octane number of fuels, therefore, increasing overall combustion (efficiency especially in gasoline), and reduces the incidence of fuel crystallization by improving the cold flow properties of fuels. In manufacturing, solketal's versatility as a solvent and a plasticizer make it an integral component in the production of polymers, and its versatility solubilizing and suspension agent make it an important component in the manufacture of drugs and in food processing. Solketal's sustainability on the environment despite its higher aquatox fish test LC50 level of 3162 ppm also increases its suitability for more uses and applications (Fatimah et al., 2019; Ilgen, Yerlikaya, & Akyurek, 2017; Luma Sh. Al-Saadi, Eze, & Harvey, 2019; Pandian Manjunathan, Sanjeev P. Maradur, A.B. Halgeri, & Shanbhag, 2014; Talebian-Kiakalaieh et al., 2018; Vinicius Rossa et al., 2019).

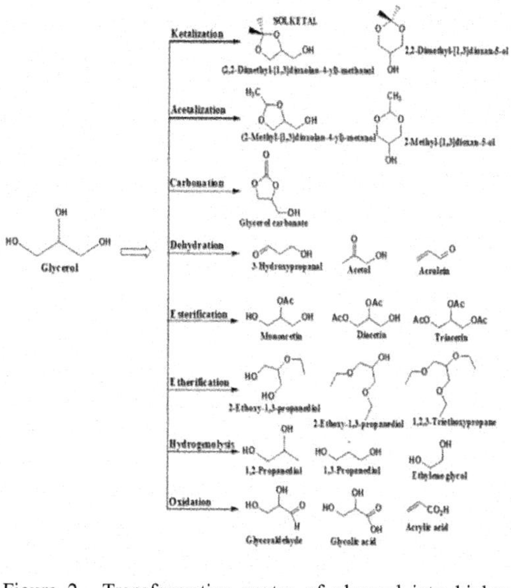

Figure 2. Transformation routes of glycerol into higher added value products (Vinicius Rossa et al., 2019).

In the ketalization process, the synthesis of glycerol and ketones produces two branched oxygenates, solketal (2,2-dimethyl- [1,3]-dioxan-4-yl methanol) and 2,2-dimethyl- [1,3] dioxane-5-ol, while substituting acetone for ketones increases the reaction's selectivity for solketal molecules, which have a five-membered

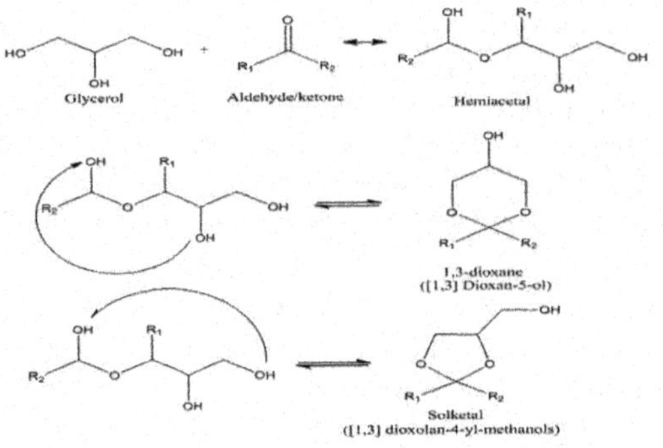

Figure 3. Reaction mechanism of glycerol with aldehydes/ketones (Talebian-Kiakalaieh et al., 2018).

ring. However, the reaction between glycerol and acetone in the presence of an acid catalyst yields solketal (2, 2-dimethyl-1, 3-dioxolane-4methanol or 1,2-isopropylideneglycerol) as a condensate (Nanda et al., 2016; Vinicius Rossa et al., 2019). Ketals can also be produced by substituting alcohol for glycerol in the same process. And, although there are records of ketalization dating back as early as late 19th Century when Fischer produced solketal using hydrogen chloride to catalyse the reaction between acetone and glycerol in a batch reactor, and a similar experiment between him and Pfahler in the early 20th century where they used anhydrous sodium sulfate instead of acetone, major breakthroughs in the sector came at the close of the century fueled by the increase in the availability of cheap glycerol from the growing biodiesel industry (Nanda et al., 2016). However, Renoll and Newmann recorded the highest solketal yield (87–90%) of the period, in 1948, in a lengthy and cumbersome process that used p-toluene sulfonic acid (pTSA) as the catalyst and petroleum ether as the reaction medium in a three-neck flask with reflux equipped with sealed mechanical stirrer, and employed low pressure distillation to separate the products (Nanda et al., 2016).

While there is a broad range of etherifying agents for the alkylation of glycerol that includes isobutylene, tertbutyl alcohol and C4 olefinic petrochemical fractions, using tertbutyl alcohol removes the need for solvents to dissolve glycerol. However, the resulting reaction produces water as a by-product (as shown in Figure 5) that might reduce the effectiveness of the process by deactivating the heterogenous catalysts used. With isobutylene, there is an increase in the possibility of the development of two phases in the reaction depending on the set conditions that may create mass transfer problems in the reactor (Jorge Sepúlveda et al., 2015).

Recently, there has been an increase in interest and investment in the development of alternative processes that use heterogenous acidic catalysts in glycerol esterification instead of the current mineral acids which are not environmentally sustainable. As such, there has been a surge in the development and use of

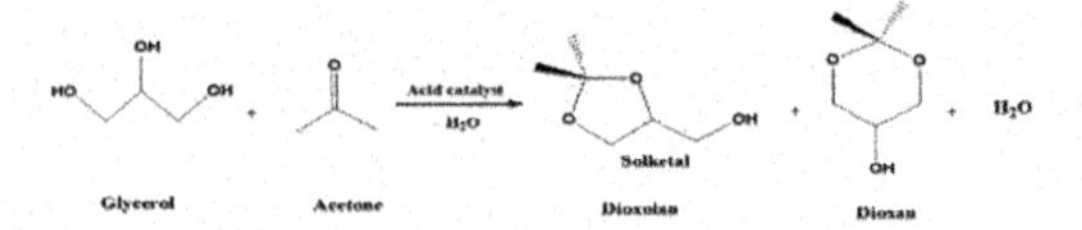

Figure 4. Solketal route production by ketalization of glycerol with acetone facilitated by major homogeneous and heterogeneous acid catalysts (Luma Sh. Al-Saadi et al., 2019; Vinicius Rossa et al., 2019).

new compatible materials containing immobilized sulfonic acid groups (such as sulfonic mesostructured silicas), zeolites (such as acidformed wide pore zeolites like H-Y and H-Beta), polyvinyl sulfonic resins (such as acidic ion-exchange resins like Amberlyst15 and 35) and niobic acids (such as sulfonated niobia and pillared clays), which behave as active and selective catalysts for esterification. The etherification of glycerol with tertbutyl alcohol (TBA) is a catalysis reaction that produces a mixture of mono-tert-butyl-glycerol (MTBG), di-tert-butyl-glycerol (DTBG) and tri-tert-butyl-glycerol (TTBG), and some unwanted by-products in some cases that result from polymerization reactions (Jorge Sepúlveda et al., 2015).

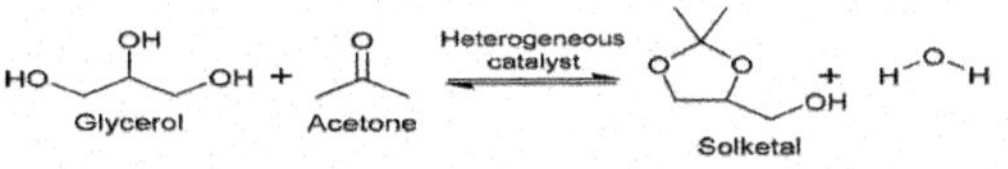

Figure 5. Synthesis scheme of glycerol to solketal.

Homogenous catalysts, such as H_2SO_4, HCl, HF, and p-toluene sulfonic acid, are also commonly used in etherification processes, and produce significantly higher glycerol conversion yields when used in ketalization reactions. However, the resulting large volumes of waste and toxic by-products make their use economically and environmentally impractical and unsustainable, especially because of the costs and complexities associated with their management. Other chlorides, such as tin chloride ($SnCl_2$), are also unsuitable in the long term because they are corrosive and reduce overall equipment reusability. On the other hand, heterogenous catalysts are more economic, practical and sustainable for use in ketalization processes because they can be reused, work without the need for solvents, have higher product selectivity and make it easier to separate catalysts and products. However, their solid compounds and constituents are uneconomic and unsustainable due to their low reusability because of their solubility in the polar solvents whose use they necessitate, and their low surface area which necessitates the use of high temperatures to activate and reduces the thermal stability of the reaction. As such, there is an immediate need to invest in the development of catalysts that are affordable, have a high efficiency, and are thermally stable if the development of glycerol acetalization is to be sustainable and economical in the long term (Fatimah et al., 2019; Ilgen et al., 2017; Nanda et al., 2016; Pandian Manjunathan et al., 2014; Priya et al., 2017).

Resin catalysts, and amberlyst in particular, have been known to increase the efficiency of the conversion of glycerol to solketal and promote the selectivity for its molecules, but their use is not practical for widescale production because their reactions have poor thermal stability that makes it impossible to recycle and reuse them. Hierarchical zeolites, on the other hand, which have higher thermal stability are so far the catalyst responsible for the highest recorded glycerol–solketal conversion at 72% efficiency with 72% selectivity for solketal molecules in a reaction using H-Beta (BEA framework) at a temperature of 60° C and stirring at 700 rpm with 5% of catalyst and glycerol:acetone molar ratio of 1:4 for H-BEA. Moreover, despite MFI zeolite being known to produce lower yields than those achieved using amberlyst at 80%, mainly because of the relatively narrow channel size that affects the transport of the reactant carried out and the shape selectivity, their selectivity rates are almost 100%. The basic mechanism of metal salt catalysis is a nucleophilic attack by the hydroxyl group of glycerol to the carbocation obtained from the protonation step, resulting in the formation of the intermediate, followed by a water elimination step. The carbocation is produced from the Lewis or Brønsted acid sites, which activate the ketone carbonyl group through a protonation step (i.e., Brønsted acids) or polarization (Fatimah et al., 2019; Ilgen et al., 2017; Nanda et al., 2016; Pandian Manjunathan et al., 2014; Priya et al., 2017).

To search for an effective heterogeneous catalyst for the ketalization process, the reaction in a stirred batch reactor over a series of silica-induced heteropolyacid catalysts, i.e., tungsto-phosphoric acid (PW), tungsto-silisic acid (SiW), molybdo-phosphoric acid (PMo), and molybdo-silisic acid (SiMo) has been studied. The reported catalytic activities for the catalysts are in the order of: SiMo<PMo<SiWS<PWS, mainly owing to the increase in acidity. The authors reported glycerol conversion of more than 97% with a very high selectivity of 99% towards solketal at the reaction conditions: 70 ° C, A/G of 12:1, catalyst (PW) loading of 0.2 g, and 2–3 h. The high yield of solketal in this work was attributed to the strong acidity of the catalyst that promoted the reaction kinetics and to the high A/G (12:1). Good catalytic stability was also observed, as the catalyst lost its activity by ~15% after four consecutive batch runs using the same catalyst (Nanda et al., 2016).

3.2 *Glycerol to solketal over resin catalysts*

According to Fatimah et al. (2019), Brønsted's acidity is the most important property of solid acid catalysts in the production of solketal from glycerol. A typical resin catalyst (i.e., amberlyst) catalyzed the reaction of glycerol with acetone to produce above 80% of the glycerol conversion. It has been reported that a resin, amberlyst-36, which was applied at different reaction temperatures from 25 to 70° C, was an excellent catalyst to convert glycerol with a conversion of 85% to 97% to solketal with a selectivity of 99%. The catalyst is also active at lower pressures with similar reaction parameters, either in pure glycerol or in an equimolar reactant. According to some references, the high conversion was influenced not only by the surface acidity but also by the resin structure. Moreover, the surface acidity was an important parameter that played a crucial role in improving the selectivity and the conversion in the production of solketal. All resins showed good selectivity to solketal (>80%), and the important catalytic parameter of the resin to conversion of glycerol is the acid capacity (oversulfonated resin). With the highest acid capacity (sulfonic acid), these catalyst materials can improve not only the selectivity to solketal production but also the conversion of raw glycerol to above 90%. A limitation of the catalyst activity is the presence of NaCl as a poison for the surface acidity, which is possibly due to the impurities in glycerol.

3.3 *Glycerol to solketal over mesoporous silica*

Hafnium and zirconium modified TUD-1 have been reported as being superior catalysts for the conversion of glycerol to solketal. These two catalysts (Hf-TUD-1 and Zr-TUD-1) are examples of active metal-modified mesoporous silica in which Hf and Zr are in the framework. Their activity was higher than FAU(USY) and Al(TUD-1). The highest conversion of glycerol to solketal was more than Ž0%. The catalytic activity was a function of:

(i) the number of acid sites,
(ii) the presence of mesopores,
(iii) the existence of a large surface area, and
(iv) the hydrophobicity of the catalyst.

The hydrophobicity of the catalyst, was crucial to prevent the hydrolysis of solketal. Numerous references have reported that mesoporous silica catalysts have the advantage of high stability in the conversion of glycerol to solketal, resulting in processes with a relatively higher percentage of conversion (95%) and selectivity to solketal (98%). A sulfonic acid-functionalized mesoporous polymer (MP-SO_3H) contains a high acidity surface (1.88 mmol/g). The surface acidity of catalytic materials can accelerate the formation products of solketal via ketalization reactions (Fatimah et al., 2019).

3.4 *Ketalization of glycerol over clay minerals*

Studies of different clay-based catalysts with different acid strengths ranging from 0.12 to 5.7 meq/g show that a stronger acidity improved the conversion of glycerol up to ca. 80%. The use of formaldehyde as the major source of solketal production has a lower conversion value (only 83% glycerol conversion), with the K10 montmorillonite used as a catalyst. The reaction between glycerol and acetone is preferred as it produces a more stable intermediate hemicetal compound, with a tertiary carbenium ion. While, in the reaction between glycerol with formaldehyde, the produced hemiacetal formation is not a stable carbenium ion. Thus the conversion value for the glycerol–formaldehyde system is relatively small as compared to the reaction where acetone is used as a co-reactant, as shown in Figure 7 (Fatimah et al., 2019).

The stability of catalysts is one of the main hurdles for the commercialization of glycerol to solketal. Even though the reaction temperature was considered as mild, the stability of most of the solid catalysts decayed in the presence of water as a by-product and other impurities (NaCl, methanol) from the glycerol source. The deactivation rate is even higher when the raw glycerol (contaminated with water) was fed to the reactor. Therefore, the viability of the commercial plant depends on (Fatimah et al., 2019):

(i) the source of feeds,
(ii) availability of glycerol and other feeds, and
(iii) cost of glycerol as the feed.

Three main challenges for production of solketal were identified (Fatimah et al., 2019):

(a) The presence of water and impurities in the feed.
(b) The shift from the batch reactor to the fixed bed reactor.
(c) The presence of equilibrium offers other difficulties as higher acetone demand is expected. However, higher acetone to glycerol will lead to destructive instruments.

Protonic zeolites seem to be the most promising and widely applied acid catalysts in the chemical industry. In addition to presenting the apparent benefits of heterogeneous catalysts like reusability and simple recovery, protonic zeolites exemplify eco-friendly solid acid catalysts unlike conventional liquid acids. These are aluminosilicates characterized by uniform microporous crystalline structure, shape selectivity, hydrophobicity, and high thermal stability. It is ascertained that the stronger acidity of protonic zeolites is consistent with the bridging hydroxyl groups Al(OH)Si located in the cavities of zeolite. H-Mordenite is one such protonic zeolite found to hold excellent structural and textural properties that are not seen in other catalytic materials. Unlike other zeolites, even though mordenite is a microporous material, it possesses a special multiple pore channel system where its elliptical pores are wide enough to allow many reactions to be attained with high selectivity (Priya et al., 2017).

The formation of water during glycerol acetalization reactions is problematic and has to be addressed in order not to trigger the reverse reaction. The hydrophobic property of the zeolite assists in inhibiting the reverse reaction of acetalization by diffusion of water in to the pores of zeolite. The modification of the physiochemical properties of zeolites to further improve their efficiency in reactions can be achieved by incorporating metal atoms into the framework or extra framework of zeolite structure (Priya et al., 2017).

The ketalization reaction proceeds via an acid catalyzed mechanism, hence catalysts with stronger acidity might lead to higher glycerol conversion. The influence of catalyst acidity on the solketal yield is shown in Table 1 below. It is clear that the catalyst acidity is a crucial parameter influencing the catalytic performance (Nanda et al., 2016).

The majority of studies on the synthesis of solketal were carried out in batch reactors with the use of heterogeneous catalysts such as zeolites, amberlyst, montmorillonite, silica-induced heterolpolyacids, and Nafion. However, batch processes have several limitations including the long time their reactions take (often exceeding 2 hours), reducing their efficiency, and the difficulty associated with scaling them up. Using a continuous-flow reactor with heterogenous catalysts is, therefore, more strategic because of the advantages it brings to heat and mass transfer management, and the ease it brings to scaling up from laboratory setups to industrial scale setup. Using a continuous process also leads to more environmental and economical benefits offering a constant quality of the end product (Nanda et al., 2016).

3.5 *Microwave heating*

Most of the liquid phase reactions over heterogeneous catalysts were conducted by direct heating or in oil bath which is an inefficient and slow process in terms of energy transfer. The use of microwave-assisted chemical synthesis has become increasingly popular and has proven to be more efficient than conventional heating methods due to the high-speed synthesis involving rapid internal heating, with a direct transfer of microwave energy from source to the molecules in the reaction. Microwave processing brings about uniform heating that can regulate the temperature and brings about considerable energy savings. The chemical syntheses by microwave irradiation as a heating method has received significant interest over traditional heating methods offering high conversion and selectivity (Priya et al., 2017).

Thermo-chemical conversion methods can be used to convert glycerol into useful secondary products. Common methods include combustion, gasification, and pyrolysis. Combustion is a process which involves the complete oxidation of a product, gasification involves high temperatures but only partial oxidation, while pyrolysis is the thermal degradation process that occurs in the absence of oxygen, typically performed by using conventional furnace heating and microwave heating techniques. The latter method heats more effectively and consumes lower energy than the former. Microwave pyrolysis utilizes electromagnetic waves to transfer heat, and by heating the material internally at the molecular level, heat loss is minimized when compared to conventional heating methods. During the pyrolysis process, primary and secondary pyrolysis can occur. Primary pyrolysis involves the process of creating dehydration, dehydrogenation, decarboxylation or decarbonization reactions. Secondary pyrolysis involves thermal or catalytic cracking where heavy compounds are further broken down into gases. The

Table 1. Influence of catalyst acidity on solketal yield (Nanda et al., 2016)

Active phase	Reaction conditions (°C, A/G, T_r)	Acidity (meq/g)	BET (m^2/g)	Pore size (nm)	Yield (%)	Ref.
H-β Zeolite	40, 6:1, 0.25	5.7	480	2	84	54
Amberlyst-36 wet	40, 6:1, 0.25	5.6	33	24	88	54
Amberlyst 35	40, 6:1, 0.25	5.4	35	16.8	86	54
$ZrSO_4$	40, 6:1, 0.25	–	–	–	77	54
Polymax	40, 6:1, 0.25	–	–	–	60	54
Montmorillonite K10	40, 6:1, 0.25	4.6	264	5.5	68	54
Amberlyst 36	38–40, 1.5:1, 8	5.4	19	20	88	24
Pr-SBA-15	70, 6:1, 0.5	0.94	721	8	79	42
Ar-SBA-15	70, 6:1, 0.5	1.06	712	9	83	42
HAr-SBA-15	70, 6:1, 0.5	1.04	533	8	80	42
Amberlyst 15	70, 6:1, 0.5	4.8	53	30	85	42
$Pr\text{-}SO_3H\text{-}SiO_2$	70, 6:1, 0.5	1.04	301	2–20	77	42
$T\text{-}SiO_2$	70, 6:1, 0.5	0.78	279	2–20	73	42
SAC-13	70, 6:1, 0.5	0.12	>200	>10	74	42

Pr-SBA-15: Propsylsulfonic acid-functionalized mesostructured silica; Ar-SBA-15: Arenesulfonic acid-functionalized mesostructured silica; HAr-SBA-15: Hydrophobised arenesulfonic acid-functionalized mesostructured silica; SAC-13: Nafion silica composite; $T\text{-}SiO_2$: Silica bonded tosic acid; $Pr\text{-}SO_3H\text{-}SiO_2$:Silica bonded propylsulfonic acid.

pyrolysis temperature is one of the main parameters that can affect product yields. A previous study has shown that a higher pyrolysis temperature increases char and gas yields. Pyrolyzed liquid shows a maximum yield at an intermediate temperature but it decreases at higher temperatures due to the thermal cracking of heavy compounds into small-chain products (Leong, Lam, Ani, Ng, & Chong, 2016).

The pyrolysis of crude glycerol using the microwave heating technique to produce pyrolyzed liquid, which can be used as fuels in combustion systems has been investigated. Glycerol could also be converted to glycerol carbonate, a green organic solvent, which has a high boiling point. Therefore, the ketalization of glycerol with acetone to produce an oxygenated compound such as solketal is an interesting question to consider (Sulistyo, Priadana, Fitriandini, Ariyanto, & Azis, 2020).

Temperatures higher than 313 K are needed to obtain sufficient conversion to force water removal and to drive the forward reaction to produce solketal. Acetalization using crude glycerol with an SBA15 catalyst was investigated and found high glycerol conversion (Pandian Manjunathan et al., 2014). To drive the reaction equilibrium, a new method for water removal by refluxing the flask followed by water vaporization under vacuum pressure was proposed. Manjunathan et al. (2014), also produced solketal by reacting glycerol with acetone at ambient temperatures by using a modified beta catalyst. They obtained a glycerol conversion of 87.1% by using a catalyst loading of 7.5%. The pseudo homogeneous kinetics model for the ketalization of glycerol using H-BEA as a catalyst was proposed by Rosa at al. in 2017. At a temperature range of 313 K to 353 K, the activation energies for forward and reverse reactions were 44.77 kJ/mol and 41.40 kJ/mol, respectively (Rossa et al., 2017). The investigations into the kinetics of ketalization of glycerol at 293-323K with Amberlyst 35 as a catalyst. They proposed a Langmuir-Hinshelwood kinetic model and obtained an activation energy of 55.6 kJ/mol (Nanda et al., 2014b). Overall, these studies show that the temperatures between 300-350 K are needed for glycerol conversion. As a result, kinetics studies with solid catalyst should be investigated within that range of temperatures (Sulistyo et al., 2020).

The use of a continuous microwave reactor (CMR) for the synthesis of solketal was reported in which, a solution of acetone, glycerol and pTSA as a homogeneous catalyst was mixed and pumped into the reaction coil (inside the microwave cavity) to react at a desired temperature (process similar to Figure 3). The authors reported a maximum 84% yield of solketal at A/G of 13.5, in the presence of pTSA under the reaction conditions of 132 ° C, 1175 kPa, 1.2 min residence time and of 20 mL/min feeding rate. However, the system was restricted only to homogeneous catalysts. Moreover, this technique would not be appropriate for conducting the reaction at a low temperature or for reactants that are not compatible with microwave energy (Nanda et al., 2016; Nanda et al., 2014b).

3.6 *Kinetics modeling*

Establishing reaction paths for any process is very crucial in the design of a catalyst. In addition, establishing reaction rate equations also helps in designing the reactor. The relative acidity of the catalysts has significant effects on glycerol conversion and product yield. The condensation reaction of glycerol with acetone leads to the formation of both five-membered and six-membered ringed molecules (ketals). However, the six-membered ring ketal is less favorable because one of the methyl groups in the final product

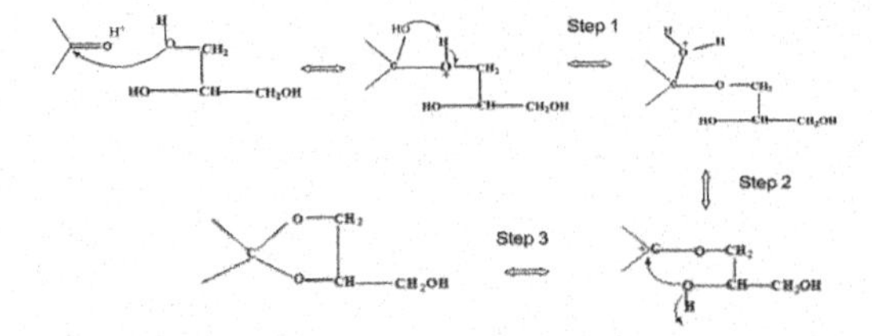

Figure 6. Mechanism used by (Nanda et al., 2014a) for the reaction of acetone and glycerol over acid catalyst (Nanda et al., 2016).

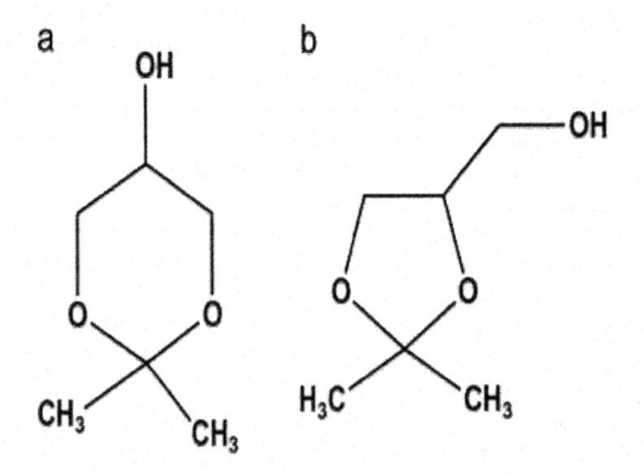

Figure 7. The cyclic acetals from the reaction between glycerol and acetone: (a) 5-hydroxy-2,2-dimethyl-1,3-dioxane (b) solketal i.e 4-hydroxymethyl-2,2-dimethyl-1,3-dioxane (Nanda et al., 2016).

is in axial position of the chair conformation (Figure 8). The resulting product has a ratio of 99:1 for five membered ring (4-hydroxymethyl-2,2-dimethyl-1,3-dioxolane, or solketal) to six-membered ring (5-hydroxy-2, 2-dimethyl-1,3-dioxane). For the ketalization reaction catalyzed by Brønsted acids, the five-membered ring solketal is formed dominantly through a mechanism involving a short-lived carbenium ion as an intermediate (Nanda et al., 2016).

Nanda et al. (2014b) also designed and used the reaction framework in Figure 9 for the ketalization reaction proceeding via acidic catalytic mechanism. The first step involves the surface reaction between the adsorbed acetone and glycerol over the catalyst surface to form the hemi-acetal. The next step is the removal of water leading to the formation of a carbocation on the carbonyl carbon atom, and the last step is the removal of the proton to form solketal (Nanda et al., 2016).

The general reaction rate for the ketalization reaction has been expressed in form of Langmuir–Hinshelwood model with the surface reaction as the rate determining step. The key reaction steps of this model are given as follows (Nanda et al., 2016):

a) The surface reaction between the adsorbed species of glycerol (GF) and acetone (AF) to give adsorbed hemiacetal (HF)

$$GF + AF \leftrightarrow HF + F \tag{1}$$

b) Surface reaction for formation of adsorbed water (WF)

$$HF + F \leftrightarrow IF + WF \tag{2}$$

c) Formation of adsorbed solketal (SF)

$$IF + GF \leftrightarrow SF + F \tag{3}$$

The simplified rate expression for the reaction is given as

$$r = \frac{k[G][A] - [S][W]/K_c[G]}{\{1 + K_w[W]\}^2} \tag{4}$$

where K_W is the equilibrium constant for water adsorption on the catalyst surface.

According to the above kinetic model, three parameters (kinetic constant, k, and water adsorption constant, K_W, and ketalization equilibrium constant, K_c) are to be estimated at each temperature to find the rate of the reaction. The estimated values of these parameters are given in Table 2. Based on the variation of kinetic constant with temperature, the activation energy (E_a) of the reaction has been reported to be 55.673.1 kJ mol^{-1}.

The mechanism of reaction rate for the acetalization reaction has also been alternatively derived through the Eley–Rideal mechanism. The reaction steps of this model can be expressed as follows (Sulistyo et al., 2020);

1. In the first step, the carbonyl group in acetone was activated by acid sites of the catalyst.

$$\mathrm{Ac + s \rightarrow Acs} \tag{5}$$

2. The OH group of glycerol attacks the adsorbed carbonyl to form an intermediate product such as hemiacetal.

$$\mathrm{Acs + G \rightarrow Hs} \tag{6}$$

3. The hemiacetal undergoes cyclization to facilitate formation of adsorbed solketal and water.

$$\mathrm{Hs \rightarrow Ss + W} \tag{7}$$

4. Adsorbed solketal is desorbed from the surface active of the catalyst.

$$\mathrm{Ss \rightarrow S + s} \tag{8}$$

The result of the reaction rate mechanism can be expressed as follows:

$$r_s = \frac{k \cdot \left(C_A \cdot C_G - \frac{C_S \cdot C_W}{K_{eq}}\right)}{1 + K_A \cdot C_A + K_S \cdot C_S} \tag{9}$$

$$K_{eq} = \exp\left(3.6154 \cdot 10^3 \cdot \frac{1}{T(K)} - 11.308\right) \tag{10}$$

$$k = A \cdot \exp\left(\frac{-Ea}{R \cdot T(K)}\right) \tag{11}$$

Equation 9 presents the reaction rate for solketal formation from acetone and glycerol. This equation was

Table 2. Kinetic model parameters at different temperatures (Nanda et al., 2016).

Temperature (K)	K_c	k (L mol^{-1} min^{-1})	K_w
298	2.66	0.11	2.65
303	1.82	0.16	1.50
308	1.51	0.24	1.09
313	1.30	0.33	0.73
323	0.98	0.63	0.34

K_c: Equilibrium constant for the reaction; k: Kinetic constant; K_w: Equilibrium constant for water adsorption on the catalyst surface.

then incorporated with the batch reactor model. There were four parameters: the pre-exponential factor (A), activation energy (Ea), acetone adsorption constant (KA), and solketal desorption constant (KS). In addition, the equilibrium constant is presented as a function of temperature, as shown in Equation 10. Parameters A, Ea, KA, and KS in Equation 10 were estimated by non-linear regression to minimize the SSE of glycerol conversion based on simulation and experimental data. Parameter fitting and simulation were conducted in MATLAB with gradient search methods (lsqnonlin) and ODE solver. The initial concentrations of acetone and glycerol used for the simulation were 11.25 and 2.25 M, respectively (Sulistyo et al., 2020).

It is well-known that the ketalization reaction has a very low equilibrium constant. Therefore, to get high conversions of glycerol it is necessary to shift the equilibrium towards the formation of solketal, which can be achieved by either feeding an excess amount of acetone or by removing the water generated during the reaction continuously. Removing the water by-product is an effective way to break the resulting thermodynamic barriers and several processes have used entrainers, such as petroleum ethers and chloroform, successfully. Benzene is not a preferable entrainer as acetone is removed by distillation before benzene. However, the efficiency of the other mentioned entrainers is not great either as their boiling points are still higher than that of acetone. Acetone co-distillation creates the problem of low efficiency in azeotropic water removal. This phenomenon was evident from its very long reaction time when using petroleum ether as an entrainer. The use of phosphorus pentoxide and sodium sulfate as catalysts as well as desiccants for the removal of water generated from the system has also been reported, but high consumption of the catalysts in this case increased the operation costs (Nanda et al., 2016).

More recently, molecular sieves have been used for this purpose. All these processes are not economical on an industrial scale. These could be addressed more effectively by using excess acetone, which not only acts as a reactant but also as an entrainer for the system. The excess acetone could be distilled off and reused in the same or other processes. In the work by (Nanda et al., 2016) a batch reactor was modified to a membrane batch reactor to remove the water from the reaction system. The authors conducted

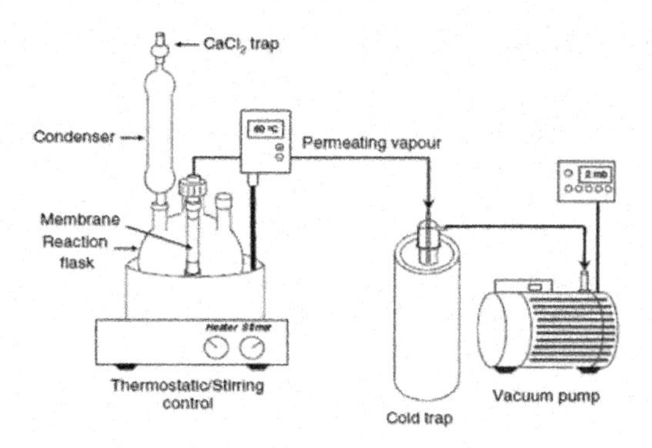

Figure 8. Membrane reactor for synthesis of solketal (Nanda et al., 2016).

the experiment by refluxing a mixture of glycerol, anhydrous acetone and heterogeneous acid catalyst, Montmorillonite K-10 (total weight 1 g) in a three-neck flask (250 mL) equipped with a reflux condenser, a septum cap and a zeolite membrane fixed in the central mouth (Figure 10 below). The membrane allowed permeation of small sized water vapor instead of pervaporation. A maximum solketal yield of 82% was achieved by the authors using a very high A/G (20:1) for 2 h of reaction. As expected, a negligible effect of the catalyst on the solketal yield was observed in this work (Nanda et al., 2016).

Applications of integrated processes, such as reactive distillation and reactive coupling, could be less expensive than conventional reactions for biodiesel production, and this technique could lead to a reduction in the required equipment and energy. Solketal co-production through reactive coupling of triacetin transesterification with condensation of the glycerol by-product with acetone has been demonstrated in a recent study. Although triacetin, a short chain triglyceride, was used, the study suggests that this could be of potential application to the long chain (typically C14-C20) fatty acids found in naturally occurring vegetable oils. It is envisaged that a modification of the biodiesel processing method via in situ reactive coupling of glycerol condensation with acetone during triglyceride transesterification could be applied to produce biodiesel and solketal, reducing the glycerol production in biodiesel plants (Luma Sh. Al-Saadi et al., 2019).

4 CONCLUSIONS

(1) Conversion of glycerol to solketal can proceed either using a homogeneous or heterogeneous catalyst; nevertheless, the use of heterogeneous catalysts is preferred, as there are many shortcomings for using homogeneous catalysts.
(2) Ketalization reactions have a very low equilibrium constant. In order to reach high conversions of glycerol it is necessary to shift the equilibrium towards the formation of solketal, by either feeding excess amount of acetone or by removing

the water generated during the reaction continuously.

(3) All batch processes have common limitation in terms of the difficulty in scaling up for production of solketal on a large scale. Compared with operation in a batch reactor, a continuous-flow process produces similar product yields with relatively shorter reaction times.

(4) The best yields of solketal were achieved by catalysts like Amberlyst-15, Amberlyst-35, Amberlyst-36, Ar-SBA-15, Zeolites, and $SnCl_2$. The preferred reaction conditions are: catalysts with higher acidity, higher A/G, and lower temperature (<70°C). Using Amberlyst-36 wet catalyst, a very high yield of solketal (9472 wt.%) was obtained at 25°C, 500 psi, A/G of 4, and WHSV of 2 h^{-1}.

(5) The ketalization reaction proceeds via acidic catalytic mechanism, hence catalysts with strong acidity might lead to high glycerol conversion.

(6) Heterogeneous catalysts for glycerol ketalization in a continuous-flow reactor can be deactivated, attributed to the loss of active acidic sites during the reaction, not due to fouling. For continuous operation (ranging from a couple of days or months), however, reactor clogging might occur, caused by fine particles of disintegrated catalysts.

(7) Direct use of crude glycerol as feedstock may cause problems such as deactivation of catalyst (by poisoning the active sites by the impurities) or plugging of the reactor (due to deposition of high boiling organic compounds or inorganic salts).

5 RECOMMENDATIONS

The purpose of this review paper is to investigate the production of solketal through microwave heating and the use of catalysis with an aim of finding an optimal production process of solketal from glycerol through the reduced cost of production.

6 LIMITATIONS OF THE STUDY

The retrospective nature of the study limits it to the information it generates from the secondary sources it uses and it cannot verify the results and arguments put forward by their authors using experimental designs because of time constraints.

REFERENCES

Fatimah, I., Sahroni, I., Fadillah, G., Musawwa, M. M., Mahlia, T. M. I., & Muraza, O. (2019). Glycerol to Solketal for Fuel Additive: Recent Progress in Heterogeneous Catalysts. *Energies, 12*. doi:10.3390/en12152872

Ilgen, O., Yerlikaya, S., & Akyurek, F. O. (2017). Synthesis of Solketal from Glycerol and Acetone over Amberlyst-46 to Produce an Oxygenated Fuel Additive. *Periodica Polytechnica Chemical Engineering, 61*(2), 144–148. doi:https://doi.org/10.3311/PPch.8895

Jorge Sepúlveda, Mariana Busto, Carlos Vera, Maraisa Gonçalves, and, W. C., & Mandelli, D. (2015). Synthesis of Oxygenated Fuel Additives from Glycerol *Biofuels – Status and Perspective*: IntechOpen.

Leong, S. K., Lam, S. S., Ani, F. N., Ng, J.-H., & Chong, C. T. (2016). Production of Pyrolyzed Oil from Crude Glycerol Using a Microwave Heating Technique. *International Journal of Technology, 2*, 323–331. doi: http://dx.doi.org/10.14716/ijtech.v7i2.2979

Luma Sh. Al-Saadi, Eze, V. C., & Harvey, A. P. (2019). A reactive coupling process for co-production of solketal and biodiesel. *Green Process Synth, 8*, 516–524. doi:https://doi.org/10.1515/gps-2019-0020

Mota, C. J. A., Silva, C. X. A. d., Nilton Rosenbach, J., Costa, J., & Silva, F. d. (2010). Glycerin Derivatives as Fuel Additives: The Addition of Glycerol/Acetone Ketal (Solketal) in Gasolines. *Energy & Fuels, 24*, 2733–2736. doi:10.1021/ef9015735

Nanda, M. R., Ghaziaskar, H. S., Zhang, Y., Yuan, Z., Qin, W., & Xu, C. C. (2016). Catalytic conversion of glycerol for sustainable production of solketal as a fuel additive: A review. *Renewable and Sustainable Energy Reviews, 56*, 1022–1031. doi:http://dx.doi.org/10.1016/j.rser.2015.12.008

Nanda, M. R., Yuan, Z., Qin, W., Ghaziaskar, H. S., Poirier, M.-A., & Xu, C. C. (2014a). A new continuous-flow process for catalytic conversion of glycerol to oxygenated fuel additive: Catalyst screening. *Applied Energy, 123*, 75–81. doi:http://dx.doi.org/10.1016/j.apenergy.2014.02.055

Nanda, M. R., Yuan, Z., Qin, W., Ghaziaskar, H. S., Poirier, M.-A., & Xu, C. C. (2014b). Thermodynamic and kinetic studies of a catalytic process to convert glycerol into solketal as an oxygenated fuel additive. *Fuel, 117*. doi:http://dx.doi.org/10.1016/j.fuel.2013.09.066

Pandian Manjunathan, Sanjeev P. Maradur, A.B. Halgeri, & Shanbhag, G. V. (2014). Room temperature synthesis of solketal from acetalization of glycerol with acetone: Effect of crystallite size and the role of acidity of beta zeolite. *Journal of Molecular Catalysis A: Chemical, 396*. doi:http://dx.doi.org/10.1016/j.molcata.2014.09.028

Priya, S. S., Selvakannan, P. R., Chary, K. V. R., Kantam, M. L., & Bhargava, S. K. (2017). Solvent-free microwave-assisted synthesis of solketal from glycerol using transition metal ions promoted mordenite solid acid catalysts. *Molecular Catalysis, 434*, 184–193. doi:http://dx.doi.org/10.1016/j.mcat.2017.03.001

Rossa, V., Pessanha, Y. d. S. P., Diaz, G. C., Camara, L. D. T., Pergher, S. B. C., & Aranda, D. A. G. (2017). Reaction Kinetic Study of Solketal Production from Glycerol Ketalization with Acetone. *Industial & Engineering Chemistry Research, 56*(2), 479–488.

Sulistyo, H., Priadana, D. P., Fitriandini, Y. W., Ariyanto, T., & Azis, M. M. (2020). Utilization of Glycerol by Ketalization Reactions with Acetone to Produce Solketal using Indion 225 Na as Catalyst. *International Journal of Technology, 11*(1), 190–199. doi:10.14716/ijtech.v11i1.3093

Talebian-Kiakalaieh, A., Amin, N. A. S., Najaafi, N., & Tarighi, S. (2018). A Review on the Catalytic Acetalization of Bio-renewable Glycerol to Fuel Additives. *Frontiers in Chemistry, 6*. doi:10.3389/fchem.2018.00573

Vinicius Rossa, Gisel Chenard Díaz, Germildo Juvenal Muchave, Gomes Aranda, D. A., & Castellã Pergher, S. B. (2019). Production of Solketal Using Acid Zeolites as Catalysts. In M. Frediani (Ed.), *Glycerine Production and Transformation – An Innovative Platform for Sustainable Biorefinery and Energy*: IntechOpen.

Advances in Phytochemistry, Textile and Renewable Energy Research for Industrial Growth – Nzila et al. (Eds)
© 2022 Copyright the Author(s), ISBN: 978-1-032-11871-0

Solar adsorption cooling with focus on using steatite adsorbent: A review

E.M. Nyang'au & K. Kiriamiti
Department of Mechanical, Production and Energy Engineering, Moi University, Eldoret, Kenya

ABSTRACT: The focus on solar cooling systems is driven by the need for efficient and pollution free climate friendly technologies. Such systems are required to meet cooling requirements in medical, post-harvest preservation of food in remote areas and ice-making. Solar radiation within tropics poses an opportunity for exploitation in solar adsorption systems. Research activities in this sector are geared towards finding technical, environmental and economic solutions to existing systems. The purpose of this paper is to review advancement in solar adsorption cooling systems in view of using novel composite adsorbent consisting of impregnated steatite, paired with methanol as adsorbate. The review presents a brief background on the basic adsorption process, working pairs and challenges of current adsorption systems along with published performance data and improvement strategies. It is noted that solar adsorption systems suffer low conversion efficiency and high capital cost in comparison with conventional vapour compression systems.

Keywords: Solar adsorption, Steatite, Adsorbent, Adsorbate, Refrigeration, Coefficient of Performance

NOMENCLATURE

COP coefficient of performance.
K constant.
n constant.
P adsorption pressure, kPa.
q_{st} isosteric heat of adsorption, kJ/kg.
SCP specific cooling power.
T adsorbent temperature, K.
T_{ads} adsorption temperature, °C.
T_d driving temperature, °C.
T_{evp} evaporation temperature, °C.
T_{sat} saturation temperature corresponding to the refrigerant pressure, K.
X refrigerant concentration, kg refrigerant /kg adsorbent.
x_o refrigerant concentration at saturation conditions, kg refrigerant/kg adsorbent.

1 INTRODUCTION

Vapour compression refrigeration cycles consume large amounts of electrical energy which significantly increases consumption of expensive and polluting fossil derived energy. In addition, operation of vapour compression systems use synthetic refrigerants (Chlorofluorocarbons, CFC'S, Hydro chlorofluorocarbons, HCFC's and Hydro Fluorocarbons, HFC's) which contribute to the Green House Gas (GHG) effect. The energy crisis has triggered attention from researchers and engineers to seek for energy efficient, sustainable and environment friendly solutions to address the increasing demand for energy. Solar and low grade thermal energy present an effective way worth exploring to address both environmental pollution and high energy consumption problems associated with conventional vapour compression cycles. One of the applications of solar and low grade thermal energy is in adsorption refrigeration systems in which adsorption of refrigerant liquid or gas takes place on a solid surface. In this process, the adsorbed particles enter the porous solid adsorbent and sit on its surface by adhesion (Wang et al., 2014). Solar adsorption cooling systems are easy to install and maintain, use clean energy sources and therefore pollution free, do not have moving parts and have long life. They are useful for ice-making, air conditioning, medical and food preservation in off-grid areas. However, the major disadvantage of these systems is low performance compared to other refrigeration and cooling systems.

Application of solar energy in refrigeration is attractive because of the coincidence of peak cooling demand with available solar power. Furthermore, energy demand for air conditioning during summer period in developed countries propagates increased consumption of electricity hence high GHG emissions while within the developing countries, the grid is out of reach in most remote areas thus rendering use of solar for provision of refrigeration services advantageous. Post-harvest losses of agricultural products can be improved if storage is done at low temperatures using solar refrigeration technologies. This would eliminate sharp differences in food supplies between harvest seasons.

Solar refrigeration is dependent upon environmental factors such as cooling water temperature, local weather, solar irradiation, air velocity and temperature.

DOI 10.1201/9781003221968-36

Most solar sorption systems that have been developed are not yet economically justified. However, solar thermal adsorption systems are promising and environmental friendly with low maintenance cost requirements. This work details aspects of adsorption refrigeration systems, criteria for choosing an appropriate working pair, challenges with adsorption systems and published performance data and limitations of these systems.

Commercialization of the technology has been hindered due to comparatively bigger sizes of current adsorption based cooling units, their low specific cooling power and high manufacturing cost. Efforts to enhance their performance have focused attention on improvement of heat and mass transfer and improvements in properties of working pairs among other strategies.

Steatite (also called saponite or soapstone) is a soft, magnesium-rich metamorphic rock composed of talc (Huhta & Kärki, 2018) which can be studied for suitability as a co-adsorbent material. It has been used for carvings, sculptures, countertops, and architectural elements, among other uses. However, the cutting and processing of soapstone produces a large amount of waste powder, which poses a disposal problem. For instance, in the production of dimension stones and other items in Brazil, 60 wt. % of soapstone ends up as waste and is disposed to landfills (Rodrigues & Lima, 2012). In Finland, some 110,000 tons of soapstone was produced annually in 2012 (Pokki, 2014) out of which, the generated waste was 60–75 wt. %.In Kenya, waste that remains after carving is not disposed which leads to environmental degradation and pollution (Onyambu, 2013). It is therefore imperative to consider sustainable ways for utilization of waste associated with this industry.

Consequently, there have been attempts to find uses for steatite waste. Some possible uses have been explored including its application as raw material in ceramics manufacturing (Cota et al., 2018; Panzera et al., 2011). Steatite is also a potential oil spill sorbent due to the hydrophobicity of the talc it contains (Souza et al., 2016). Ultrafine steatite powder is a suitable supplementary cementitious material that can replace up to 25% of Portland cement (Kumar et al., 2017). Another application is its use as a dispersed phase material in composites containing Portland cement binder (Strecker et al., 2010). Steatite-containing mortars have been further supplemented with carbon fibers (Panzera et al., 2011) or thermoset polymers (Cota et al., 2018) and used in the restoration of historical buildings. However, all of the above examples rely on either the use of high temperature (approximately 1,000–1,200°C in ceramics production) or CO_2-intensive Portland cement as a co-binder, which decreases the environmental feasibility of these applications.

Adsorption systems utilize low temperature heat, which provides a more suitable condition to use steatite as a co-adsorbent medium in a low carbon footprint application. This will help develop a method to turn currently underexploited soapstone waste into a useful material.

2 LITERATURE REVIEW

The history of solar adsoption system dates back to 1920s when sulfur dioxide and silica gel were used for the air-conditioning of railway carriages in the USA. But with the development of cheap reliable compressors and the introduction of CFCs, heat based sorption systems took a back seat. After the oil crisis in 1970s and with the restrictions imposed by Montreal (1987) and Kyoto (1997) protocol, research on heat driven sorption cooling systems started again. The development of sorption refrigeration systems powered by solar energy emerged in the late 1970s, following the pioneering work of (Tchernev, 1978), which observed that zeolite, adsorbs large amounts of water vapour when cooled and desorbs the water vapour when heated, thus providing a unique opportunity for its utilization in refrigeration applications.

Fast forward, Wang et al. (2000) and Zhang (2002) proposed solar powered continuous solid adsorption refrigeration and heating hybrid system for water heating and an icemaker. The machine used activated carbon–methanol as the working pair and 2 m^2 of evacuated tube collectors. The daily ice production was about 10 kg when the insolation was about 22 MJ/m^2. Critoph (2000) used a heat pipe to heat and cool the adsorber. He concluded that fluids with different physical and chemical properties should be used for cooling and heating purposes. Different fluids would eliminate the occurrence of possible inward air leaks and the utilization of thick material to enclose the working fluid.

Li et al. (2001) performed experiments with a solar-powered icemaker that had activated carbon–methanol as working pair. This icemaker had a COP ranging from 0.12 to 0.14, and produced between 5 and 6 kg of ice per square meter of collector. They concluded that in order to improve the performance of this system, the heat transfer properties of the adsorber could be enhanced by increasing the number of fins or using consolidated adsorbent.

An adsorption icemaker with activated carbon–methanol pair was tested in Burkina Faso by Buchter (2003). The results of this prototype were comparable to those obtained by (Boubakri, 1992) in Morocco, with a similar system.

In order to solve the corrosion problem of sea water in the steel adsorber, a split heat pipe system was designed and constructed in Shanghai Jiao Tong University (SJTU) (Wang, 2004) using composite adsorbent of $CaCl_2$ and activated carbon to improve adsorption performance through incorporating a mass recovery system. The mass recovery process improved SCP and COP for the system by 15.5% and 24.1%, respectively. The heat transfer performance of this system improved by the introduction of the split heat pipe.

3 THE ADSORPTION PROCESS

Adsorption is a process resulting from an interaction between a solid (adsorbent) and a gas (refrigerant), based on a physical or chemical reaction process. Physical adsorption occurs when the molecules of refrigerant (adsorbate) fix themselves at the surface of an adsorbent by means of Vander Waals forces and electrostatic forces. When heated, this process can be reversed in which adsorbate molecules can be released through a desorption process. In chemical adsorption, ionic or covalent bonds are formed between the adsorbate molecules and the adsorbent. The bonding forces are much greater than that of physical adsorption, releasing more heat. The process cannot be easily reversed and this type of bonding promotes chemical alteration of the adsorbed substance, thus the adsorbate and adsorbent molecules never keep their original state after adsorption .Therefore, most of the adsorption refrigeration systems mainly involve physical adsorption (Alghoul et al., 2007; Anyanwu, 2004).

The quantity of heat released during adsorption depends on the magnitude of electrostatic forces, latent heat and bond energy. A comparison of heat of adsorption between various solid adsorbent pairs is presented in Table 1.

Table 1. Heat of adsorption between adsorbent-adsorbate pairs.

Adsorbent	Adsorbate	Heat of Adsorption
Activated alumina	H_2O	3000
Zeolite	H_20	3300–4200
	NH_3	4000–6000
	CH_3OH	2300–2600
Silica gel	CH_3OH	1000–1500
Silica gel	H_2O	2800
Charcoal	C_2H_4	1000–1200
	NH_3	2000–2700
	H_2O	2300–2600
	CH_3OH	1800–2000
	C_2H_5OH	1200–1400
Calcium chloride	CH_3OH	1800–2000
Metal hydrides	Hydrogen	2300–2600
Complex compounds	Salts and ammonia or water	2000–2700

Source: Anirban & Randip (2010).

All adsorption processes are exothermic in nature. The available energy during adsorption is represented in the form $\Delta G = \Delta H - T\Delta S$ where ΔG is the change in free energy, ΔH is the change in enthalpy, T is adsorption temperature and ΔS is the change in entropy. The three terms which are used in adsorption process are: (i) Integral heat of adsorption which is the total heat released from initial stage to final stage of adsorption loading, (ii) Differential heat of adsorption; the change in integral heat of adsorption with change in loading and (iii) Isosteric heat of adsorption,q_{st} defined using adsorption isosters. Differential and isosteric heat of adsorption are usually identical and can be related as a function of adsorption potential, heat of vaporization and the change in adsorption capacity of an adsorbent with temperature.

When fixed adsorbent beds are employed, which is the common practice, adsorption cycles can be operated without any moving parts. The use of fixed beds results in silence, mechanical simplicity, high reliability and a very long lifetime. On the other hand, it also leads to intermittent cycle operation which decreases the COP of the system. Hence, when constant flow of vapour from the evaporator is required for continuous cooling, two or more adsorbent beds must be operated out of phase, which requires availability of the heat source at all times, which is not the case with solar radiation.

There are 4 processes involved in an adsorption cycle out of which 2 are isosteric processes and remaining 2 are isobaric processes [Fernandes, (2014) and Sneha,(2015)]. The cycle is illustrated on Clapeyron diagram in Figure 1a and 1b.

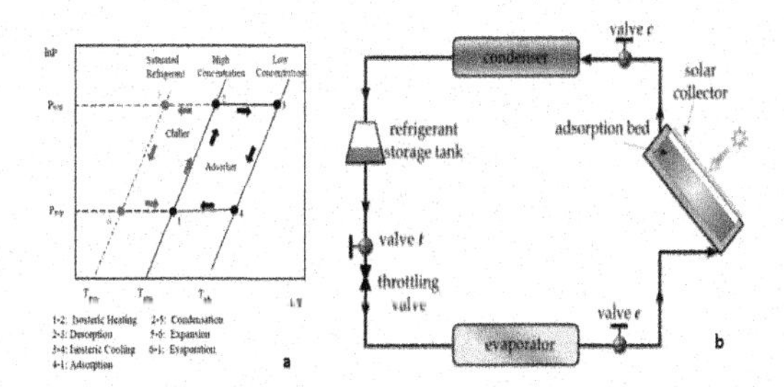

Figure 1. Thermodynamic adsorption cooling cycle-The Clapeyron diagram.

Process I- Isosteric Heating Process 1-2

Heating and pressurization process starts at point 1, when the adsorbent is at adsorption temperature, T_{ads} and low pressure (evaporation pressure, P_{evp}), and adsorbate is at high concentration. The valve which isolates the condenser from the evaporator is closed and, as heat is applied to the adsorbent, both temperature and pressure increase along the isosteric line 1–2, while the mass of adsorbed refrigerant remains constant at maximum value.

Process II- Isobaric Heating Process 2-3

At point 2, the adsorbate attains condenser pressure, P_{con} and desorption process starts. In this process, progressive temperature rise takes place at constant condenser pressure. The refrigerant vapor is released from the adsorbent and then changes into liquid form in the condenser, releasing heat of condensation at condenser temperature, T_{con} and then gets collected in a receiver tank. This stage ends when the adsorbent reaches its maximum regeneration temperature, and the adsorbate concentration decreases to the minimum value.

Process III- Isosteric Cooling Process 3-4

In this process, the adsorbent cools down along the isosteric line 3–4, while the adsorbed refrigerant remains constant at the lowest concentration. During this phase, the valve is opened, allowing for the refrigerant to flow into the evaporator, and the system pressure decreases until it reaches the evaporator pressure, P_{evp}.

Process IV- Isobaric Cooling Process 4-1

In this process, adsorption–cooling phase 4–1 occurs, producing the cooling effect in the evaporator, at evaporation temperature, T_{evp}. At this stage, the vaporized refrigerant in the evaporator flows to the adsorbent bed where it is adsorbed until the maximum concentration is attained, at point 1. During this phase, the adsorbent is cooled down until it reaches the adsorption temperature, by rejecting the sensible heat and the heat of adsorption. At the end of this phase, the valve is closed and the cycle restarts.

3.1 *Selection of working pair*

The working pair influences system performance which is dependent on temperature of the heat source and the desired characteristics of the refrigeration system. Selection of the pair is based on cost, availability, environmental impact, the constituent properties of the working pair and their mutual affinity (which depend on the chemical, physical and thermodynamic properties of the substances).The commonly used selection criteria is the temperature lift capability of the pair at the adsorption-evaporation temperature and generation-condensation temperature.

3.2 *Choice of adsorbent*

The most important features for choosing a suitable adsorbent according to Sumathy et al. (2003) and Odesola & Adebayo (2010) are:

(i) The ability to adsorb a large amount of adsorbate when cooled to ambient temperature,
(ii) Desorption ability of the adsorbate when heated by the available heat source,
(iii) Higher apparent density; high pore volume; high surface area,
(iv) Low specific heat, good thermal conductivity, high porosity
(v) Chemically and physically compatible with the chosen refrigerant and
(vi) Low cost and wide availability.

When selection is done, there must be compromise between the high porosity required for rapid vapor diffusion and the high density suitable for good thermal conductivity (Wang et al., 2012). The most commonly used adsorbents are activated carbon, zeolite and silica-gel. Activated carbon offers a good compromise between high adsorption and desorption capacities. Natural zeolites need to be present in large quantities since only a small amount of adsorbate is desorbed during the temperature increase. However, the adsorption isotherms of zeolites have extremely non-linear pressure dependence, which is relevant for solar refrigeration applications. Contrarily, activated carbon and silica-gel present almost linear pressure-dependent isotherms. Silica-gel satisfies most of the criteria above but it is expensive and may not be readily available. Besides, the deterioration phenomenon of the adsorption capacity and aging of silica-gel is another current issue as found by a number of researchers (Gordeeva et al., 2007; Ruthven, 1984; Sumathy et al., 2003).

3.3 *Choice of adsorbate*

The adsorbate or refrigerant must fulfil the following requirements according to (Choudhury et al., 2013; Odesola & Adebayo, 2010):

(i) High latent heat of vaporization and low specific volume when in liquid state
(ii) High thermal conductivity
(iii) Low viscosity
(iv) Thermally and chemically stable with the adsorbent in the operating temperature range
(v) Non-toxic, non-corrosive and non-flammable
(vi) Low saturation pressures at normal operating temperature and
(vii) Absence of ecological issues.

According to the basic principle and working characteristics of the adsorption refrigeration cycle, there are no working pairs to completely meet the above requirements in practice. However, commonly used working pairs closely meet these requirements (Wang et al., 2010). The governing equation for the adsorption relationship between an adsorption working pair is the Dubinin–Radushkevich (D-R) equation. (Critoph, 1996):

$$X = x_0 e^{\left[X = x_0 e^{\left[-K\left(\frac{T}{T_{sat}}-1\right)^n\right]}\right]} \tag{1}$$

The most commonly used refrigerants are ammonia, methanol and water, which have relatively high latent heat values (1368, 1160 and 2258kJ/kg, respectively) and low specific volumes. Water and methanol operate at sub-atmospheric saturation pressures at the operating temperatures. Ammonia operates at higher pressures so small leakages can be tolerated in some cases. While it is toxic and corrosive, water and methanol are not, but the latter is flammable. Water is the most thermally stable adsorbate closely followed by methanol and ammonia.

4 CHALLENGES

Adsorption cooling systems have low coefficient of performance COP and specific cooling power SCP, huge volume, weight and elevated cost in comparison with compression systems. Heat and mass performance enhancement of the adsorber and modification of system structure improves cooling efficiency. Further drawbacks of adsorption systems noted by Shahab (2018) include intermittent operation, need for special design to maintain high vacuum and large volume and weight relative to traditional systems.

Approaches undertaken to overcome the drawbacks include enhancement of heat and mass transfer of the adsorber, improving adsorption properties of adsorbents, design changes, adoption of different kinds of cycles, improvement of regenerative heat and mass transfer between beds etc. (Wang et al., 2018). Heat transfer problems in adsorption cycles have been intensively investigated by Cacciola et al. (1993). The main factors affecting heat transfer between the thermal fluid and the adsorbent bed include the physical contact at the metal–adsorbent interface and the effective conductivity of the adsorbent bed. The thickness of the adsorbent bed affects temperature distribution between the surface and bottom layers in a flat plate bed arrangement leading to uneven adsorption and desorption. Evacuated tube type of solar collectors can be used to improve heating efficiency in such systems.

The adsorbents in use in most solar adsorption systems have low thermal conductivity and poor porosity characteristics. The effect is a bulky collector component and thus, excessive heating capacity and consequently a low COP. Heat dissipation to the environment limits solar heating efficiency.

To improve performance of adsorbents, utilization of composite adsorbents has been studied by Patel et al. (2016) from which recommendations for adsorbent-adsorbate working pairs with high cyclic adsorption capacity and thermal conductivity, low resistance to adsorbate flow as well as better heat management during the adsorption cycle have been proposed.

5 IMPROVEMENT STRATEGIES

Adsorption systems must have their size and cost reduced to become more commercially attractive. The most promising alternatives to achieve these goals include the improvement of heat management to increase the COP and enhancement of internal and external heat transfer of the adsorber to increase the SCP. The main technologies to enhance the external heat transfer in the adsorber are related to the increase of the heat exchange area, the use of coated adsorbers and utilization of heat pipe technology. To improve the internal heat transfer, the most suitable option is the deployment of consolidated adsorbents.

5.1 *Optimization of adsorption bed design*

Some of the methods used to expedite improved efficiency of the adsorber bed include improving the heat transfer structure of the adsorption bed, enlarged mass transfer channels, thermal insulation and automation. Xu et al. (2014) developed an enhanced heat and mass transfer finned tube casing in which it was found that in comparison with the metal casing of same dimension, finned tube has 51.4 % more heat transfer per unit length.

In his study of heat pipes, Zhang (2002) contends that selection of proper adsorption pairs, heat and mass transfer enhancement in the adsorption bed, invention of advanced refrigeration cycles are always a focus of research interests.

Dusane & Ghuge (2016) conducted studies to analyse performance of a glass tube and reflector system using domestic charcoal-methanol pair and concluded that system performance depended on adsorption pairs and process parameters in agreement with Khattab (2004), whose study on use of domestic charcoal adsorbent achieved higher COP than that of activated charcoal. They recommended further studies for new low cost adsorption material and high performance adsorbent bed.

5.1.1 *Extended surfaces*

Several types of extended surfaces can be considered, such as finned tubes, plate heat exchangers and plate–fin heat exchangers. The drawback of this technology is that it increases the thermal mass of the adsorber; and therefore requires efficient heat management to produce reasonable COPs. Furthermore, this solution should be avoided if the operation pressure is very low and the Knudsen effect can occur (Meunier, 1998).

5.1.2 *Advanced cycles*

Research on advanced cycles involve cycles with heat and mass recovery. A continuous multiple bed adsorption regenerative system with 32 solar modules has been proposed by Critoph (2000) in which the effect of thermal capacity ratio, number of modules, air temperature from solar heater, generator heat transfer coefficient and evaporator air inlet temperature on COP and CSP were reported. Cycles with heat management have achieved energy recovery of about 35% of the total energy transmitted to the adsorber. Besides utilization of heat management cycles, it is possible to employ refrigerant mass recovery between two adsorbent beds to effectively enhance both the cooling power and the COP. Szarzynski (1997) analysed cycles with refrigerant mass recovery and concluded that the SCP could be increased by about 20%. Wang (2001) compared the COP of adsorption systems with and without mass recovery and found that the former could produce a COP from 10% to 100% higher than the latter. The difference between the COPs is higher at lower generation temperatures. Although advanced cycles can increase the performance of the adsorption systems, the complexity of the system also increases. Therefore, among the studied advanced cycles, the mass recovery cycle seems to be one of the most cost effective ways to improve both COP and SCP.

5.2 *Improvement in adsorption properties of working pairs*

Koyama et al. (2014) investigated three pairs of adsorbent/adsorbate according to Malaysia climate conditions. The selected pairs were activated carbon-methanol; activated carbon fibre –ethanol and silica gel-water. Adsorption was highest in activated carbon fibre –ethanol and lowest in silica gel/water. In a simulation experiment, activated carbon fibre –ethanol exhibited high uptake capacity (Loh et.al., 2009). El-sharkawy et al. (2009) observed that Maxsorb III/methanol pair has superiority among other carbonaceous adsorbent/methanol pair for both air conditioning and ice making with adsorption capacity 1.76 times than that of activated charcoal/methanol. Allouhi et al. (2015) had similar observation for activated carbon fibre/methanol pair in which the optimal uptake was 0.3406 kg/kg whereas the maximum specific COP was about 0.384 for silica gel/water.

5.2.1 *Coated adsorbers*

The utilization of coated adsorbers is particularly suited for applications where high COP is not as important as high SCP. This technology improves the wall heat transfer coefficient by effectively decreasing the contact thermal resistance between the heat exchange surface and the adsorbent. Its main disadvantage is a very high ratio between the inert mass and the adsorbent mass, which spoils the COP. In order to overcome this drawback, very effective heat management is required.

5.2.2 *Consolidated and composite adsorbents*

Consolidated adsorbents with high thermal conductivity can be considered as the most promising alternative to enhance the heat transfer within the adsorber. Wang et al. (2004) developed a consolidated compound made from a mixture of $CaCl_2$ and activated carbon. Their experiments showed that utilization of this compound could lead to a cooling density 35% higher than that obtained by use of pure powder of $CaCl_2$. In general, consolidated adsorbents have lower mass transfer properties than granular adsorbents, which could lead to very low adsorption rates especially for refrigerants evaporating under atmospheric pressure such as water or methanol.

Thus, besides experiments to identify the thermal conductivity and the wall heat transfer coefficient of these compounds, experiments to identify their permeability must also be performed when a new consolidated adsorbent is formulated. By controlling the compression pressure and mass ratio between an adsorbent and the inert material, it is possible to control the density of the final compound and its properties of heat and mass transfer.

6 RESULTS AND DISCUSSION

From the foregoing review, Table 2 shows reported levels of COP achieved for different adsorption pairs under a range of driving temperature, T_d and evaporation temperature, T_{evp}.

Performance of some solar adsorption systems from various studies is summarized in Table 3. To choose

Table 2. Comparison between adsorption working pairs.

Ref	Working pair	COP	T_{evp} (°)	T_d (°)	SCP (w/Kg)
Physical adsorbents	Activated Carbon/Ammonia	0.61	−5	100	2000
	Activated Carbon/Methanol	0.78	15	90	16
	Activated Carbon/Ethanol	0.8	3	80	N.A
	Silica gel/Water	0.61	12	82	208
Chemical adsorbents	Metal chloride/Ammonia	0.6	−10	52	NA
	Metalhydride/Hydrogen	0.83	−50	85	300
	Metal Oxides/Water	N.A	100	200	78
Composite adsorbents	Silica gel and chlorides/water	1.65	7	70	N.A
	Silica gel and chlorides/methanol	0.33	−10	47	N.A
	Chlorides and porous media/ammonia	0.35	−15	117.5	493.5
	Zeolite and foam aluminum/water	0.55	10	250	500

Table 3. Typical research on solar adsorption systems.

Ref.	Yr.	Application	Working pair 7	COP	T_{evp} (°C)	Type
Sumathy	2001	Icemaker	Ethanol/Silica gel	0.12	−5	Experiment
Li et al.	2004	Icemaker	Methanol/Activated Carbon	0.2–0.3	0	Experiment
Khattab N.	2004	Icemaker	Methanol/Activated Carbon	0.18	−0.5	Experiment
Wang S.	2005	Icemaker	NH3/Activated Carbon +CaCl2	0.41	–	Experiment
Watheq	2008	Chiller	Methanol/Activated carbon	0.1	10	Experiment
Hassan et al.	2011	Chiller	Methanol/Charcoal	0.21	16.3	Theoretical
Faeza H. et al.	2011	Chiller	NH3/Activated carbon	0.30	7	Experiment
Naef Q. Maged A.	2013	Icemaker	Methanol/Activated carbon	0.24		Theoretical
Berdja et al.	2014	Chiller	Methanol/Olive Waste	0.49	4	Experiment
Fadhil A.	2014	Chiller	Methanol/Activated carbon	0.34	9	Experiment
Khalifa et al.	2015	Hybrid	Water/Silicagel	–	–	Theoretical
Hadj A. et al.	2015	Icemaker	Methanol/Activated carbon	0.21	−1	Theoretical

Source: Author.

the appropriate pair in adsorption cooling application, the running conditions and the aim of the application should be taken into account considering different effects of factors in Table 2. For solar cooling the important parameter governing the choice is the driving temperature as it should be the minimum allowable temperature. The most important running conditions are the driving temperature, evaporation temperature and COP.

The highest COP value reported in the composite adsorbents category is between silica gel with chlorides and water pair attributed to the ionic nature of chlorides which enhance conductivity and the high heat capacity of water. The minimum value for COP occurs in the physical adsorbents category between zeolite and water pair. Metal hydrides and hydrogen show the lowest value for evaporation temperature because hydrogen has a very low normal boiling point (−252.87°C). For the case of driving temperature, silica gel and chlorides with methanol pair has the lowest driving temperature while zeolite and water pair has the highest driving temperature.

Research results on solar adsorption systems from Table 3 indicate attainment of maximum COP value of 0.49 and cooler temperature of 4°C from methanol/olive waste pair compared to COP of 1.65 achieved from composite adsorbents.

7 CONCLUSION AND RECOMMENDATIONS

Although investment costs for adsorption chillers are still high, the environmental benefits are impressive, when compared to conventional compression chillers. The absence of harmful or hazardous products such as CFCs, together with a substantial reduction of CO2 emissions due to very low consumption of electricity, creates an environmentally safe technology. Low temperature waste heat or solar energy can be converted into a chilling capacity as low as 5°C with minor maintenance costs. Nevertheless, some crucial points in the development of sorption systems still exist and those are closely connected to the low specific power of the machine and the investment costs. Recently, more close attention was paid to the development of combined systems of solar cooling and heating in order to make use of all types of energies rationally. All these works will be of great favour to the development of the solar sorption refrigeration system.

More research on adsorbent materials, improved heat and mass transfer, advanced cycles, etc. to make adsorption technology competitive is necessary. This work will help to understand basics and research progress on adsorption systems.

From this research, it is recommended to assess the suitability of steatite as a co-adsorbent material for purposes of improving adsorption capacity of various working pairs to enhance operation with low driving temperature hence boost deployment of solar energy usage.

ACKNOWLEDGMENT

This work was supported by the Africa Centre of Excellence II in Phytochemicals, Textile and Renewable Energy (ACE II – PTRE) at Moi University of the Eastern and Southern Africa Higher Education Centers of Excellence Project through the Association of Energy Professionals in Eastern Africa, AEPEA.

REFERENCES

Abu-Hamdeh N.H., A. K. (2013). Design and performance characteristics of solar adsorption refrigeration system using parabolic trough collector: Experimental and statistical optimization technique. Energy Conversion and Management, Vollume 74, 162–170.

Alghoul MA, Sulaiman MY, Azmi BZ, & Wahab MA. (2007). Advances on multi-purpose solar Adsorption systems for domestic refrigeration and water heating. Applied Thermal Engineering, Volume 27(5–6), 813–822, http://dx.doi.org/10.1016/j.applthermaleng.

Al-Hemiri, A., & Nasiaf, A. (2010). The use of direct solar energy in absorption refrigeration employing Ammonia – water system. Iraqi Journal of Chemical and Petroleum Engineering, Volume 11, No. 4, 13–21.

Allouhia, A., Kousksoub, T., Jamila, A., El Rhafikic T., E., Mourada Y., & Zeraoulib Y. (2015). Optimal working pairs for solar adsorption cooling applications. Energy, Volume 79, 235–247.

Al-Mousawi, F., Al-Dadah, R., & Mahmoud, S. (2016). Low grade heat driven adsorption system for cooling and power generation with small-scale radial inflow turbine. Applied Energy, Volume 183, 1302–1316.

Amar Sadoon A. (2006). Design, built and performance study on adsorption ice maker", M.Sc. thesis. Iraq.: University of Technology.

Ammar, M. H., Benhaoua, B., & Balghouthi, M. (2015). Simulation of tubular adsorber for adsorption refrigeration system powered by solar energy in sub-Sahara region of Algeria. Energy conversion and management, Volume 106, 31–40.

Anyanwu E.E. (2004). Review of solid adsorption solar refrigeration II: an over view of the principles and theory. Energy Conversion and Management, Volume 45(7–8), 1279–1295; http://dx.doi.org/10.1016/j.enconman.2003.08.003.

Anyanwu, E. E. (2003). Design, construction and test run of a solid adsorption solar refrigerator using activated carbon/methanol, as adsorbent/adsorbate pair. Energy Conversion and Management, Volume 44(18), 2879–2892.

Askalany, A., Ismail, I., Salema, M., Ahmed, H., & Morsy, M. (2012). A review on adsorption cooling systems with adsorbent carbon. Renewable and Sustainable Energy Reviews, Volume16, 493–500.

Bajpai V. K. (2012). Design of solar powered vapour absorption system", proceeding of the world congress on engineering, Vol.3. London, U.K. World congress on engineering, Volume 3. London, U.K.

Boubakri A., Arsalane M., Yous B., Ali-Moussa L., Pons M., & F., M. (1992). Experimental study of adsorptive solar powered ice makers in Agadir (Morocco)-1. Performance in actual site. Renewable Energy, Volume 2(1), 7–13.

Boubakri, A., Arsalane, M., Yous, B., Ali-Moussa, L., Pons, M., & Meunier, F. (1992). Experimental study of adsorptive solar powered ice makers in Agadir (Morocco)-2, influences of meteorological parameters. Renewable Energy Volume 2(1), 15–21.

Boubakri, A., Guilleminot, J., & Meunier, F. (2000). Adsorptive solar powered ice maker: experiments and model. Solar Energy, Volume 69, No. 3, 249–263.

Buchter, F., Dind, P., & Pons, M. (2003). An experimental solar powered adsorptive refrigerator tested in Burkina-Faso. International Journal for Refrigeration, 79–86.

Cacciola, G., Guilleminot, J., Chalfen, J., & Choisier, A. (1993). Heat and mass transfer characteristics of composites for adsorption heat pumps. International absorption heat pump conference, American Society of Mechanical Engineers, AES volume 31.

Choudhury B, Saha BB, Chatterjee PK, & Sarkar JP. (2013). An over view of developments In adsorption refrigeration systems towards a sustainable way of cooling. Applied Energy, volume 104, 554–67, http://dx.doi.org/10.1016/j.apenergy.2012.11.042.

Cota T.G., Reis E.L., Lima R.M.F., L., & Cipriano R.A.S. (2018). Incorporation of waste from ferromanganese alloy manufacture and soapstone powder in red ceramic production. Applied Clay Science , Volume 161, 274–281.

Critoph RE. (1999). Rapid cycling solar/biomass powered adsorption refrigeration system. Renewable Energy, Volume 16, 673–678.

Critoph RE. (2000). The use of thermosyphon heat pipes to improve the performance of a carbon–ammonia adsorption refrigerator. Proceedings of IV Minsk international seminar – heat pipes, heat pumps, refrigerators.

Critoph, R. (1992). A forced convection regenerative cycle using the ammonia-carbon pair. In proceedings of Solid Sorption Refrigeration, Paris,Volume 11R, 80–85.

Critoph, R. (1999). Forced convection adsorption cycle with packed bed heat regeneration. International Journal of Refrigeration, Volume 22, 38–46.

Critoph, R. (2000). Multiple bed regenerative adsorption cycle using the monolithic carbon–ammonia pair. Applied Thermal Engineering, Volume 22, 667–677.

Critoph, R. E. (2001). Simulation of a continuous multiple-bed regenerative adsorption cycle. International Journal of Refrigeration, Volume 24 (5), 428–437.

Critoph, R. E., & Z.Tamainot-Telto. (1997). Solar Adsorption Refrigeration. Renewable Energy, volume 12 (4), 409–417.

Critoph, R., Tamainot-Telto, Z., & Davies, G. (2000). A prototype of a fast cycle adsorption refrigerator utilizing a novel carbon – aluminium laminate, Proceedings of the Institution of Mechanical Engineers, Part A. Journal of Power and Energy, Volume 214 (5), 439–448.

De Francisco A., I. R. (2002). Development and testing of a prototype of low power water–ammonia absorption equipment for solar energy applications. Renewable Energy, Volume 25, 537–544.

Demirocak, E. (2008). Thermodynamic and Economic Analysis of a Solar Thermal Powered Adsorption Cooling System. Unpublished M.Sc. Thesis. Turkey: Middle East Technical University.

Dusane N.C., & Ghuge. (2016). A Review on Solar Adsorption Refrigeration System. IOSR Journal of Engineering, Volume 06, Issue 11, 07-13.

El-Sharkawy I.I., Hassan M., Saha B.B., Koyama S., & Nasr M.M., N. (2009). Study on adsorption of methanol onto carbon based adsorbents. International Journal of Refrigeration, Volume 32 (7), 1579–1586.

Faeza M.H., Fawziea M.H., & H.N.K., A. (2011). Experimental study on two beds adsorption chiller with regeneration. Modern Applied Science, Volume 5 (4).

Glaznev, I., Ponomarenko, S., Kirik, Y., & Aristov. (2011). Composites CaCl2/SBA-15 for adsorptive transformation of low temperature heat: Pore size effect. International. Journal for Refrigeration, Volume 34, 1244–1250. doi:10.1016/j.ijrefrig.2011.02.007.

Gordeeva LG, G., Freni A, F., Restuccia G, R., & Aristov YI. (2007). "Influence of characteristics of methanol sorbents "salts in mesoporous silica" on the performance of adsorptive air conditioning cycle. Industrial Engineering and Chemistry, Volume 46(9), 2747–2752, http://dx.doi.org/10.1021/ie060666n.

Guilleminot, J., Choisier, A., Chalfen, J., Nicolas, S., & Reymoney, J. (1993). Heat transfer intensification in fixed bed adsorbers. Heat Recovery Systems, Volume 13(4), 297–300.

Gupta, A., Anand, Y., Anand, S., & Tyagi, S. (2015). Thermodynamic optimization and chemical exergy quantification for absorption-based refrigeration system. Akade'miai Kiado', Budapest, Hungary., DOI 10.1007/s10973-015-4795-6.

H.N. Souza, Reis, E., Lima, R., & Cipriano, R. (2016). Using soapstone waste with diesel oil adsorbed as raw material for red ceramic products. Ceramics International, Volume 42, 16205–16211.

Hai-Ming., L. (2000). An enhanced adsorption cycle operated by periodic reversal forced convection. Applied Thermal Engineering, Volume 20, 595–617.

Hassan Fadiel. (2008). Theoretical and experimental study of a hybrid adsorption refrigeration system", M.Sc. thesis. Iraq: University of Technology.

Hassan, H. M.-A. (2012). Development of a continuously operating solar-driven adsorption cooling system: Thermodynamic analysis and parametric study. Applied Thermal Engineering, doi:10.1016/j.applthermaleng.2012.04.040.

Hassan, H., Mohamad, A., & Bennacer, R. (2011). Simulation of an adsorption solar cooling system. Energy, Volume 36, 530–537.

Hildbrand, C., Dind, P., Pons, M., & Buchter, F. (2004). A new solar powered adsorption refrigerator with high Performance. Solar Energy, Volume 77, 311–318.

Huhta A., & Kärki A. (2018). A proposal for the definition, nomenclature, and classification of soapstones. GFF Volume 140, 38–43.

IEA. (2010). World Energy Outlook Executive Summary. Retrieved from International Energy Agency.

Khalifa, A., Ahmed, Q. M., & Hassan, J. F. (2015). Theoretical study on the effect of operating parameters on the performance of adsorption refrigerator. Journal of kerbala university, Volume 13 (1).

Khattab N.M. (2004). A novel solar-powered adsorption refrigeration module. Applied Thermal Engineering, Volume 24, 2747–2760.

Khattab, N. (2006). Simulation and optimization of a novel solar-powered adsorption refrigeration module. Solar Energy, Volume 80, 823–833.

Koyama K.H, & Bidyut B.S.S. (2014). Study of various adsorbent refrigerant pairs for the application of solar driven adsorption cooling in tropical climates. Applied Thermal Engineering, 266–274.

Kumar P., Sudalaimani K., & Shanmugasundaram M. (2017). An investigation on selfcompacting concrete using ultra-fine natural steatite powder as replacement to cement. Advanced Materials Science and Engineering.

Leite, A., & Daguenet, M. (2000). Performance of a new solid adsorption ice maker with solar energy regeneration. Energy Conversion Management, Volume 41, 1625–1647.

Leite, P. (1998). Thermodynamic analysis and modeling of an adsorption–cycle system for refrigeration from low grade energy sources. Journal of Brazilian Society of Mechanical Science, Volume 20(3), 301–324.

Leite, P., Grilo, R., Andrade, F., Belo, F., & Meunier, F. (2007). Experimental thermodynamic cycles and performance analysis of a solar-powered adsorptive icemaker in hot humid climate. Renewable Energy, Volume 32, 697–712.

Li M. , Wang R.Z., Xu Y.X., Wu J.Y., & Dieng A.O.,. (2002). Experimental study on dynamic performance analysis of a flat-plate solar solid-adsorption refrigeration for ice maker", Renewable Energy 27 (2002) 211–221.

Li. (2003). Experimental study on the adsorption performance of consolidated activated carbon-methanol pair. Unpublished Msc Thesis. Shanghai, China: Shanghai Jiao Tong University.

Li, M., & Wang, R. (2003). Heat and Mass Transfer in a Flat Plate Solar Solid Adsorption Refrigeration Ice Maker. Renewable Energy, Volume 28, 613–622.

Li, M., Huang, H., R.Z.Wang, Wang, L., Cai, W., & Yang, W. (2004). Experimental study on adsorbent of activated carbon with refrigerant of methanol and ethanol for solar icemaker. Renewable Energy, Volume 29, 2235–2244.

Li, M., Sun, C., Wang, R., & Cai, W. (2004). Development of no valve solar ice maker. Applied Thermal Engineering, 865–872.

Li, M., Wang, R., Xu, Y., Wu, J., & Dieng, A. (2001). Experimental study on dynamic performance analysis of a flatplate solar solid-adsorption refrigeration for ice maker. Renew Energy 2001;27:211–21.

Lithoxoos G.P., L. A. (2010). Adsorption of N2, CH4, CO and CO2 gases in single walled carbon nanotubes: A combined experimental and Monte Carlo molecular simulation study. Journal of Supercritical Fluids, Volume 55, 510–523. doi:10.1016/j.supflu.2010.09.017.

Liu, Y., & K C., L. (2006). Numerical study of a novel Cascaded adsorption cycle. International Journal of Refrigeration, Volume 29 (2), 250–259.

Liu, Y., Wang, Z., & Xia, Z. (2005). Experimental study on a continuous adsorption water chiller with novel design. International Journal of Refrigeration, Volume 28 (92), 218–230.

Loh W.S, El-Sharkawy I.I., Ng K.C., & Saha B.B. (2009). Adsorption cooling cycles for alternative adsorbent/adsorbate pairs working at partial vacuum and pressurized conditions. Applied thermal Engineering, Volume 29, 793–798.

Meunier F, K. S. (1996). Comparative thermodynamic study of sorption systems: second law analysis. International Journal for Refrigeration Volume 19(6), 414–421.

Meunier F. (1998). Solid sorption heat powered cycles for cooling and heat pumping applications. Applied Thermal Engineering, Volume 18(9–10), 715–729.

Meunier, F. (1985). Second law analysis of a solid adsorption heat pump operating on reversible cascaded cycles; Application to the Zeolite-water pair. Journal of Heat Recovery System, Volume 5 (2), 133–141.

Miyazaki, T. A. (2010). The performance analysis of novel dual evaporator type three-bed adsorption chiller. International Journal of Heat Recovery Systems, Volume 5 (2), 133–141.

Mohand, B., Brahim, A., Ferhat, Y., Fateh, B., & Maamar, O. (2014). Design and realization of a solar adsorption refrigeration machine powered by solar energy. Energy Procedia, Volume 48, 1226–1235.

Naef A.A.Q., & Maged Al E.S. (2013). Improving ice productivity and performance for an activated carbon/methanol solar adsorption ice maker. Solar Energy, Volume 98, 523–542.

Nidal H., Abu-Hamdeh, A., Khaled A., Alnefaie, A., Khalid H., & Almitani. (2013). Design and performance characteristics of solar adsorption refrigeration system using parabolic trough collector: Experimental and statistical optimization technique. Energy Conversion and Management, Volume 74, 162–170.

Nilesh Pawar, D. P. (2014). Theoretical investigation of refrigeration system for rapid cooling applications. International Journal of Engineering, Business and Enterprise Applications, Volume 7(1), 95–98.

Nunez, T., Mittelbach, W., & Henning, H. (2004). Development of an adsorption chiller and heat pump for domestic heating and air-conditioning applications. Proceedings of the third international conference on heat powered cycles. Cyprus.

Nwamba, J., Meyer, C., & Mbarawa, M. (2008). The Design and Evaluation of a Solarpowered Adsorption Refrigerator for African Conditions. Tshwane University of Technology.

Nwosu M. C., U. S. (2013). Design, construction and characterization of a sliding –plate-evaporator freezer (Spef). Innovative Systems Design and Engineering, Volume 4 (15), ISSN 2222–1727.

Odesola, I., & Adebayo, J. (2010). Solar adsorption technologies for ice-making and recent Developments in solar technologies: a review. International Journal of Advanced Engineering and Technolology, Volume 1(3), 284–303.

Oleg, B. T., & Yury, A. L. (2002). Thermo physical aspects of environmental problems of modern refrigerating engineering. Chemistry and Computational Simulation Butlerov Communications, Volume 3, No.10.

Onyambu M.K. (2013). The development and evolution of soapstone industry in Kisii area of Kenya: 1895-2010 Masters thesis. Eldoret: Moi university.

Panzera T.H., Strecker K., S., Miranda J.D.S., Christoforo A.L., & Borges P.H.R. (2011). Cement – steatite composites reinforced with carbon fibres: an alternative for restoration of Brazilian historical buildings. Materials Restoration, Volume 14, 118–123.

Patel, J. (2016). An Overview of Developments in Adsorption Refrigeration Systems: Principle and Optimization Techniques. International Conference on Advances in Robotic, Mechanical Engineering and Design - ARMED.

Pokki, J. (2014). Summary: geological resources in Finland, production data and annual report 2012, Geological Survey of Finland, Report of Investigation 2014, p. 27. Helsinki: Geological Survey of Finland.

Pongsid Srikhirin, S. A. (2001). A review of absorption refrigeration technologies. Renewable and Sustainable Energy Reviews, Volume 5, 343–372.

Pons M, P. F. (1999). Adsorptive machines with advanced cycles for heat pumping or cooling applications. International Journal for Refrigeration, Volume 22(1), 27–37.

Poyelle, F., Guilleminot, J., & Meunier, F. (1999). Experimental tests and predictive model of an adsorptive air conditioning unit. Industrial Engineering and Chemistry, Volume 38(1), 298–309.

Pridasawas, W., & Lundqvist, P. (2007). A year-round dynamic simulation of a solar driven ejector refrigeration system with iso-butane as a refrigerant. International Journal of Refrigeration, Volume 30(5), 840–850.

Qasem, N., & El-Shaarawi, M. (2013). Improving ice productivity and performance for an activated carbon/methanol solar adsorption ice-maker. Solar Energy, Volume 98, 523–542.

Restuccia, G., Freni, A., Vasta, S., & Aristov, Y. (2004). Selective water sorbent for solid sorption chiller: experimental results and modelling. International Journal of Refrigeration, Volume 27, 284–293.

Rhayt, F., Dezairi, A., Ouaskit, S., Loulijat, H., Zerradi, H., & Mizani, S. (2015). Study of the absorption refrigerating cycle, NH3-H2O coupled with the solar absorption heat transformer, H2O-LiBr by solar data from the City of Oujda (Morocco). Revue des Energies Renouvelables, Volume 18 (1), 39–48.

Rodrigues, M., & Lima, R. (2012). Cleaner production of soapstone in the Ouro Preto region of Brazil: a case study. Journal of Clean Production, Volume 32, 149–156.

Saha B.B., Akisawa A., & Kashiwagi T. (2001). Solar/waste heat driven two-stage adsorption chiller: the prototype. Renewable Energy, Volume 23, 93–101.

Said, W. K. (2008). Solar energy refrigeration by liquid-solid adsorption technique. Nablus – Palestine: An-Najah University.

Shahab, H. (2018). Experimental Study of the Performance of a Continues Solar Adsorption Chiller using Nano-activated Carbon/Methanol as Working Pair". Solar Energy 173(2018), 920-927.

Sierra, Z., Best, R., & Holland, A. (1993). Experiments on an absorption refrigeration system powered by a solar pond. Heat Recovery Systems & Combined Heat and Power, Volume 13(5), 401–408.

Srivastava, N., & Eames, I. (1998). A review of adsobents and adsorbates in solid–vapor adsorption heat pump systems. Applied Thermal Engineering, Volume 18, 704–714.

Strecker K., Panzera T.H., Sabariz A.L.R., & Miranda J.S. (2010). The effect of incorporation of steatite wastes on the mechanical properties of cementitious composites. Materials Structure, Volume 43, 923–932.

Sumathy K, Yeung KH, & Yong L. (2003). Technology development in the solar adsorption refrigeration systems. Progress in Energy Combustion Science, Volume 29(4), 301–327, http://dx.doi.org/10.1016/S03601285(03)000285.

Sumathy K. (2001). An energy efficient solar ice-maker. Department of Mechanical Engineering, University of Hong Kong, Hong Kong.

Sumathy, K., & Yeung, K. (2003). Thermodynamic analysis and optimization of a combined adsorption heating and cooling system. International Journal of Energy Research, Volume 27, 1299–1315.

Sur, A., & Das, R. K. (2010). Review on Solar Adsorption Refrigeration cycle. International Journal of Mechanical Engineering and Technology. Volume 1, 190–226.

Sward, K., Levan, D., & Meunier, F. (2000). Adsorption heat pump modeling: the thermal wave process with local equilibrium. Applied Thermal Engineering, Volume 20 (8), 759–780.

Szarzynski, S., Feng, Y., & Pons, M. (1997). Study of different internal vapour transports for adsorption cycles with heat regeneration. International Journal for Refrigeration, Volume 20(6), 390–401.

Tamainot-Telto Z. & Critoph R. E. (1997). Adsorption refrigerator using monolithic carbon-ammonia pair. International Journal of Refrigeration, Volume 20(2), 146–155.

Tather, M., Tantekin-Ersolmaz, B., & Erdem-Senatalar, A. (1999). A novel approach to enhance heat and mass transfer in adsorption heat pumps using the zeolite–water pair. Micropore Mesopore Materials, Volume 27, 1–10.

Tchernev D.I. (1979). Solar air conditioning and refrigeration systems utilizing zeolites.In: Proceeding of meetings of Commissions E1-E2, Jerusalem. Proceeding of meetings of Commissions, Volume E1-E2, (pp. 209–215). Jerusalem.

Tchernev DI. (1978). Natural zeolites: occurrence properties and use. London: Pergamon Press.

Tierney, M. J. (2002). Feasibility of driven convective thermal wave chiller with low-grade heat. Renewable Energy, Volume 33 (9), 2097–2108.

Tso, C., & Chao, Y. (2012). Activated carbon, silica-gel and calcium chloride composite adsorbents for energy efficient solar adsorption cooling and dehumidification systems. International Journal for Refrigeration, Volume 35, 1626–1638 doi:10.1016/j.ijrefrig.2012.05.007.

Umar, M., & Aliyu, B. (2008). Design and thermodynamic simulation of a solar absorption icemaker. Continental Journal of Engineering Sciences, Volume 3, 42–49.

Uyan, A. S. (2009). Numerical analysis of an advanced three bed mass recovery adsorption refrigeration cycle. Applied Thermal Engineering, Volume 29 (14–15), 2876–2884.

Wang D, Zhang J, Xia Y, Han Y, & Wang S. (2012). Investigation of adsorption performance Deterioration in silica gel–water adsorption refrigeration. Energy Conversion Management, Volume 58, 157–162, http://dx.doi.org/10.1016/j.enconman.2012.01.013.

Wang R.Z., W. J. (2001). Performance researches and improvements on heat regenerative adsorption refrigerator and heat pump". Energy Conversion & Management, Volume 42, 233–249.

Wang R.Z., Wu J.Y., Xu Y.X., & Wang W. (2001). Performance researches and improvements on heat regenerative adsorption refrigerator and heat pump. Energy Conversion & Management, Volume 42, 233–249.

Wang RZ. (2001). Performance improvement of adsorption cooling by heat and mass recovery operation. International Journal for Refrigeration, Volume 24, 601–611.

Wang, K., Wu, J., Xia, Z., Li, L., & Wang, Z. (2008). Design and performance prediction of a novel double heat pipes type adsorption chiller for fishing boats. Renewable Energy, Volume 33, 780–790.

Wang, L. (2018). Experimental Study of an Adsorption Refrigeration Test Unit. Procedia Engineering, Volume 152, 895–903.

Wang, L., Wang, R., Wu, J., & Wang, K. (2004). Compound adsorbent for adsorption ice maker on fishing boats. International Journal of Refrigeration, Volume 27, 401–408.

Wang, R. Z. (2001). Performance improving of adsorption cooling by heat and mass recovery operation. International Journal of Refrigeration, Volume 24 (7), 602–611.

Wang, R., Li, M., Xu, Y., & Wu, J. (2000). An energy efficient hybrid system of solar powered water heater and adsorption ice maker. Solar Energy, Volume 68, 189–195.

Wang., L. (2018). Experimental Study of an Adsorption Refrigeration Test Unit. Energy Procedia, Volume 152, 895–903.

Xia, Z. Z., Chen, C. J., Kiplagat, J., Wang, R. Z., & Hu, J. Q. (2008). Adsorption Equilibrium of water on silica gel. Journal of Chemical & Engineering Data, Volume 35 (10), 2462–2465.

Xu J., Ming L., Jieqing F., Peng Z., Bin L., & W., L. (2014). Structure optimization and performance experiments of a solar-powered finned-tube adsorption refrigeration system. Applied Energy, Volume 113, 1293–1300.

Zhang, J., & Wang, Z. (2002). A new combined adsorption–ejector refrigeration and heating hybrid system powered by solar energy. Applied Thermal Engineering, Volume 22, 1245–1258.

Zhang, X., & Wang, R. (2002). Design and performance simulation of a new solar continuous solid adsorption refrigeration and heating hybrid system. Renewable Energy, Volume 27, 401–415.

Advances in Phytochemistry, Textile and Renewable Energy Research for Industrial Growth – Nzila et al. (Eds)

Electricity consumption and economic growth in Uganda

G.S. Mutumba & J. Otim
Department of Economics and Statistics, Kyambogo University, Kampala, Uganda

S. Watundu
Department of Management Science, Makerere University Business School, Kampala, Uganda

Muyiwa S. Adaramola
Faculty of Environmental Science and Natural Resource Management, Norwegian University of Life Sciences, Aas, Norway

T. Odongo
Department of Economics, Makerere University Business School, Kampala, Uganda

ABSTRACT: This paper examines the causal relationship between electricity consumption and economic growth for Uganda (2008–2018). Electricity consumption is among the key drivers of economic growth. Studies have conflicting results on the direction of causality and methodology. The hypothesis that explains causality follows growth, conservation, feedback, and neutral. The study uses a vector error correction model within a multivariate data framework. The Johansen Cointegration test was carried out to ascertain if there exists a long-run relationship between electricity consumption, real fixed capital formation, labour force, and real GDP. Data from both World Bank Development Indicators and the Electricity Regulatory Authority of Uganda was used. The results indicate bidirectional causality between electricity consumption and economic growth in both the short run and long run. The study recommends Ugandan authorities expand the electricity infrastructure to increase electricity production only and increase electricity consumption with a focus on efficient energy use to support economic growth.

Keywords: Electricity consumption, Vector Error Correction, Cointegration, Economic growth, Uganda.

1 INTRODUCTION

Electricity is an important driver for economic growth and most production and consumption models use electricity as a vital input. (Bhattacharya 2016; Fotourehchi 2017; Paul & Bhattacharya 2004). The direction of causation between electricity consumption and economic growth has very important implications. The relationship between electricity consumption and economic growth, is of keen interest to energy economists as it is not possible to have high rates in one, without the other keeping pace. However, in Uganda the relationship between electricity consumption and economic growth has not been adequately studied (Asafu–Adjaye 2000). In some earlier studies, Mawejje and Mawejje (2016), found a bidirectional causality using Error Correction Mechanism and Granger causality using data for 1971–2011. Similarly, Sekantsi and Okot (2016) confirmed a two-way long-term relationship using autoregressive distribution lag (ARDL) and Granger causality on data for 1981–2013, confirming the existence of long-run relationship between electricity consumption and economic growth and vice versa. In addition, the Granger causality test results confirm the conservation hypothesis in the short run and feedback hypothesis in the long run. The results are contentious and the debate is inconclusive.

The relationship between electricity consumption and economic growth can be categorized into four testable hypotheses: growth, conservation, feedback, and neutrality. The growth hypothesis asserts that electricity consumption affects the growth process both directly through increasing investment and employment, and indirectly as a complement to labour and capital inputs (Sekantsi & Motlokoa 2015). The growth hypothesis is supported if there is unidirectional causality from electricity consumption to economic growth. Under the growth hypothesis, electricity conservation policies that reduce electricity consumption may have an adverse impact on economic growth. Second, the

DOI 10.1201/9781003221968-37

conservation hypothesis postulates that electricity conservation policies designed to reduce electricity consumption and waste may not have an adverse impact on economic growth. The conservation hypothesis is confirmed if there is unidirectional causality from economic growth to electricity consumption. Third, the feedback hypothesis emphasizes the interdependent relationship between electricity consumption and economic growth and their complementarity. The presence of bidirectional causality between electricity consumption and economic growth lends support for the feedback hypothesis. Finally, the neutrality hypothesis considers electricity consumption to be a small component of an economy's overall output and thus may have little or no impact on economic growth. Similar to the conservation hypothesis, electricity conservation policies may not have an adverse impact on economic growth under the neutrality hypothesis. The neutrality hypothesis is supported by the absence of a causal relationship between electricity consumption and economic growth (Apergis & Payne 2010). The hypothesis for this transmission mechanism is given as in Table 1.

H0: Energy consumption has no causal relationship on GDP growth.

H1: Energy consumption has a causal relationship on GDP growth.

Table 1. Hypothesis testing.

1.Energy consumption has no causal relationship on GDP in Uganda	$H_0:\beta=0$
2. Energy consumption has a causal relationship on GDP in Uganda	$H_1:\beta\neq 0$
3. GDP has no causal relationship on energy consumption in Uganda	$H_0:\beta=0$
4. GDP has a causal relationship on energy consumption in Uganda	$H_1:\beta\neq 0$

The main contribution of this study to energy economics literature is to apply a multivariate framework of analysis which is superior to a bivariate framework that is more prone to bias due to an omitted variables problem. This study includes capital and labour as controls to our model (Lutkepohl 1982, 1999). Secondly, the study uses the neoclassical Solow model to construct a theoretical framework based on endogenous growth, rather than basing it on arbitrary model specification. These results are easily interpreted as they are grounded in economic analysis (Tang et al. 2016). Finally on methodology it uses a vector error correction mechanism and Johansen–Juselius cointegration to test the long-term relationship; it gives a robust and more comprehensive analysis. It provides a suitable and valid basis for policy making. The remainder of this paper discusses in section 2 electricity consumption and economic growth in Uganda, the literature in section 3, the methods in section 4, the findings and discussion in section 5; and the conclusions and recommendations in section 6.

2 ELECTRICITY CONSUMPTION AND ECONOMIC GROWTH IN UGANDA

Uganda's economy has been growing at an estimated average of 4.9% of GDP per year. GDP per capita has been growing at 2.2% while energy consumption has been growing at 7%. The study considered data of GDP at constant prices in order to adjust for the effects of inflation.

Uganda's economy has been generally growing over the period 2008 to 2018 (see Figure 1 in the appendices). Likewise, electricity consumption has been rising consistently over the same period. However, using marginal growth rates of GDP and Electricity consumption there is greater volatility in GDP with a gradient of –.0192 while that of electricity consumption is –0.0053, as shown in Figure 2 in the appendices. The question of the strength of causality between the two variables remains paramount and is a subject of investigation by this paper.

3 LITERATURE

The pioneer work of Kraft and Kraft (1978) triggered one of the most interesting debates in contemporary literature. Their argument that efficient electricity consumption has implications for economic growth is a timeless subject of inquiry. Using data of USA's economy from 1947–1974, they found a unidirectional causality from GDP to electricity consumption. Akarca and Long (1980), Yu and Hwang (1984), and Abosedra and Baghestani (1991) used Granger (1969), causality tests or the related test developed by Sims (1972) to test whether electricity use causes economic growth or whether electricity use is determined by the level of output in the context of a bivariate vector autoregression. The results have been inconclusive. Where significant results were obtained, they indicate that causality runs from output to electricity use.

Erol and Yu (1987) found some indications of a causal relationship between electricity and output in a number of industrialised countries with the most significant relationship being for Japanese data of 1950–1982. However, when the sample was restricted to 1950–1973 the relationship was no longer significant. Yu and Choi (1985) also found a causal relationship running from electricity to GDP in the Philippines, and causality from GDP to electricity in the economy of South Korea. Ammah-Tagoe (1990) found causality from GDP to electricity use in Ghana. Stern (1993) tested for Granger causality in a multivariate setting using a vector autoregression (VAR) model of GDP, electricity use, capital, and labour inputs. He also used a quality-adjusted index of electricity input in place of gross electricity use. The multivariate methodology is important because changes in electricity use are frequently countered by the substitution of other factors of production, resulting in an insignificant overall impact on output. Weighting electricity use for changes in the composition of electricity input is important because a large part of the growth effects of electricity are due to substitution of higher quality electricity sources such as electricity for lower quality electricity sources such as coal (;all et al. 1986; Kaufmann 1994; Jorgensen 1984). When both these innovations are employed, electricity is found to Granger-cause GDP. These results are supported by Hamilton (1983) and Burbridge and Harrison (1984), who found that changes in oil prices Granger-cause changes in GNP and unemployment in VAR models whereas oil prices are exogenous to the system.

Reynolds and Kolodzieji (2008) examined the relationship between oil, coal, and natural gas production and GDP in the former Soviet Union. Using Granger-causality tests within a bivariate framework, they found unidirectional causality from oil production to GDP and unidirectional causality from GDP to coal production and natural gas production, respectively. In an examination of the relationship between electricity consumption and economic growth for 11 Commonwealth of Independent States (CIS) countries using a panel error correction model, Apergis and Payne (2009b) revealed unidirectional causality from electricity consumption to economic growth in the short-run and bidirectional causality in the long-run. In a panel study of 15 transition economies (including Belarus, Estonia, Latvia, Lithuania, Moldova, Russia, and the Ukraine), Acaravci and Ozturk (2010) failed to find a cointegrating relationship between electricity consumption per capita and real GDP per capita. In a panel study of the relationship between emissions, electricity consumption, and growth for 11 CIS countries, Smeich and Papeiz (2014) studied electricity consumption and economic growth in the light of meeting the targets of energy policy in the EU using bootstrap panel Granger-causality approach; their results show that the level of compliance with energy policy targets influences linkages between electricity consumption and economic growth. The results indicate causal relations in the group countries with the greatest reduction of greenhouse gases emissions, the highest reduction of energy intensity, and highest share of electricity consumption in total energy consumption. In the remaining groups it showed a neutrality hypothesis.

It is vital that developing nations invest in electricity infrastructure during their economic growth path. Policy makers devote a lot of effort to the efficient use of electricity in the development process. Electricity supply capacity could be expanded to meet consumption needs to be adequate during the time of economic growth to stimulate the process of all-inclusive growth. Paul and Bhattacharya (2004) studied the causal relationship between energy consumption and economic growth in India using sample data of 1950–1996 using standard Granger causality test (1969) for a linear combination of non-stationary variables remaining non-stationary. If a linear combination of non-stationary variables becomes stationary, then the Error Correction Model (ECM) was adopted. Results showed a bi-directional causal relationship between energy consumption and economic growth. The Johansen multivariate approach (1991) at lag length 4 reveals long-run causality exists from energy consumption to economic growth. The results of a standard Granger causality test and the Engel-Granger approach show only unidirectional causality between energy consumption and economic growth. Earlier in an analysis of the relationship between economic growth and energy consumption, Pachuri (1977) and Tyner (1978), using regression approach, found a strong relation between energy consumption and economic growth in India. Cheng (1999) in his quest for causality established a unidirectional causal relation from economic growth to energy consumption. In contrast, Asafu-Adjaye (2000) estimated a unidirectional Granger causality from energy consumption to income. The findings of Asafu-Adjaye (2000) do not support the findings of Cheng (1999). Empirical results identify a bi-directional causality between energy consumption and economic growth. This study, therefore, is situated in a hotly contested and controversial area; the controversy of the debate revolves around different countries, different data sets, different time periods, and different methods, all of which leave no room for a common position of agreement. It is amidst this controversy that we intend to make a contribution.

Omay et al. (2010) investigated the relationship between energy consumption and economic growth using evidence from nonlinear panel cointegration and causality tests and compared both developing and developed countries. It gave a unidirectional causality that growth increases energy consumption in recess and expansion periods, proposing a weak relationship in the short run. Output does not Granger-cause growth in the short run. Odhiambo (2009) examined energy consumption and economic growth in South Africa in a tri-variate causality test. His empirical results showed a bidirectional causality between energy consumption and economic growth, both in the short and long run. Odhiambo (2010) studied energy consumption and economic growth in Democratic Republic of Congo (DRC): using autoregressive distribution lag and Granger causality testing; his findings confirm the existence of a long-run relationship between electricity consumption and economic growth. Odhiambo (2010) investigated energy consumption and economic growth in Tanzania using autoregressive distribution lag and Vector Error Correction Mechanism (VECM); his empirical results showed causality between electricity consumption and economic growth.

Tang et al. (2016) studied energy consumption and economic growth in Vietnam using data from 1971–2011 using a multivariate analysis and neoclassical Solow growth model; his findings were a unidirectional relationship running from energy consumption to GDP while Nguyen and Ngochi in (2019) used data of 1980–2016 using the ARDL and Toda Yamamoto test, and found no relationship. Molem and Ndifor (2016) looked at Cameroon using the generalised method of moments (GMM); they found a unidirectional relationship running from energy consumption to GDP. Bekun and Agbola (2019) used data for Nigeria 1971–2014 using both a dynamic modified and fully modified ordinary least squares (DMOL and FMOL), Maki cointegration, Toda-Yamamota, and Wald tests; they found the relationship running from energy consumption to GDP. While Ogundipe and Oyomide (2013) used data from 1980–2008 and used VECM, Johansen and Juselius cointegration and a Cobb Douglas (CD) production function to find a bidirectional hypothesis. Therefore, a proper and sufficient supply of electricity is of utmost importance for assisting economic growth of a nation (Ogundipe 2016). Bah and Azam (2017) studied energy consumption and economic growth in South Africa using data from 1971–2012, using an ARDL and CD production function and found no causality. Weng and Cheng Lu (2017) studied the economy of Taiwan using data from 1984–2014 and using Granger causality found a bidirectional relationship. Ibrahiem (2018) studied Egypt using VECM and a Johansen test, he found a relationship running from electricity consumption to GDP. Muhammad and Nur-Syazwari (2018) looked at electricity consumption and the economic growth of Malaysia and using ARDL found a relationship running from electricity consumption to GDP. Mukhtarov et al. (2018) studied the electricity consumption and economic growth of Azerbaijan using data of 1992–2015; using VECM and Johansen test they found a unidirectional relationship running from electricity consumption to GDP, while Humbatova (2020), who used data of Azerbaijan of 1995–2017 using ARDL, augmented Dickey fuller (ADF), Phillips Peron (PP), and pairwise Granger causality test found no relationship between the variables. Ozturk et al. (2019) investigated Denmark using data from 1970–2012; they used ARDL and Granger causality test and found no relationship between electricity and GDP. Nepal and Paija (2019) studied Nepal using data of 1974–2014; they used ARDL, Toda Yamamoto, and Granger causality tests and found no relationship. Earlier Bastoola and Sapkoota (2015) had studied the same and found a unidirectional relationship running from electricity to economic growth. Saint Akadiri et al. (2019) studied Turkey using ARDL, Toda Yamamoto and Granger causality test; they used the Environmental Kuznet Curve (EKC) and Long-range Energy Alternative Program (LEAP) and found a bidirectional relationship. Salauddin and Gow (2019) studied Qatar using ARDL and Toda Yamamoto; they used the Environmental Kuznet Curve (EKC) and data of 1980–2016, and they found a bidirectional relationship. Zhang et al. (2020) studied electricity consumption and economic growth in Pakistan using data from 1960–2014; they used a VECM and Average Neural Network (ANN) method and found a unidirectional relationship running from electricity consumption to economic growth, which is in agreement with the earlier work of Agee and Butt (2015). Ridzual et al. (2020) studied the electricity consumption and economic growth of Malaysia using data of 1970–2016; they adopted a multivariate framework using the ARDL and cumulative sum of squares and adopted the Solow growth model. Lin and Zu (2020) studied China and found a relationship running from GDP to electricity, while Junsheg et al. (2018) using Toda Yamamoto and Granger found the relationship running from electricity consumption to GDP.

4 METHODS

4.1 *Data source*

The study used quarterly time series data on aggregate electricity consumption from the Electricity Regulatory Authority (ERA) for series from 2008 to 2018. ERA is an institution with the mandate to regulate the electricity subsector in Uganda. Additional data used was Uganda's real GDP, real gross fixed capital formation, and labour force, using 2010 as the base year from World Bank statistics.

4.2 *The theoretical model—Solow growth model*

The paper adopted a neoclassical growth theory by Solow (1956, 1987)—the Solow growth model—to analyse the relationship between electricity consumption and economic growth. Unlike the Solow-neutral model which augments capital as an input, this study will provide electricity as a separate input into the production model; and investigates its relationship with growth. Output (Y), labour force (L), capital (C), and electricity consumption as a component of renewable energy, i.e., electricity (E), will be measured. Thus the model will be specified as:

$$Y_t = \alpha + \delta_t + \beta_1 E_t + \beta_2 K_t + \beta_3 L_t + u_t \qquad (1)$$

where Y_t is output/real Gross GDP

The parameters α and δ allow for the possibility of specific fixed effects and deterministic trend, respectively, E_t is electricity consumption, K_t is capital formation, L_t is labour force, and u_t is the error term.

The choice of the Solow model production is essentially because of its suitability as a production function in explaining economic growth. The Solow-neutral model augments capital with technological progress and as such this model brings in domestic electricity demand as a variable of study. Economic growth as a matrix of goods and services that an economy can produce is best represented by a production function. The basic inputs are capital, labour, and the composite energy good, electricity. Capital formation and electricity consumption are studied in the same model because capital is treated as a stock and electricity consumption is a flow of resources.

4.3 *The econometric model: vector error correction model*

Economic growth (Y) is modelled as a function of electricity function (E), capital (c), labour (L); which can then be transformed and rewritten by specifying an error-correction representative inclusive vector autoregressive model as follows;

$$(1-\delta)\begin{bmatrix} LogY_t \\ LogE_t \end{bmatrix} = \begin{pmatrix} \alpha_1 \ \gamma 1 \\ \alpha_2 \ \gamma 2 \end{pmatrix}\begin{pmatrix} I \\ X_{t-i} \end{pmatrix} + \sum_{i=1}^{p}$$

$$(1-\delta)\begin{bmatrix} \beta 11 \ \beta 12 \ \beta 13 \\ \beta 21 \ \beta 22 \ \beta 23 \end{bmatrix}\begin{bmatrix} logE \\ logK \\ logL \end{bmatrix} + \begin{pmatrix} V_{1t} \\ V_{2t} \end{pmatrix} \quad (2)$$

Note that all the variables are expressed in logs.

X = the Error-Correction Model (ECM), which is the lagged value of the error term from the following cointegration equation below:

$$Y = \alpha_0 + \gamma E_t + \alpha_1 K_t + \beta L_t + e_t \quad (3)$$

$$(1-\delta)Y_t = \alpha_0 + X_{t-1} + \mathrm{Y}_{t-1} + V_{1t} \quad (4)$$

$$(1-\delta)Y_t = \alpha_0 + X_{t-1} + \sum_{i=1}^{m}(1-\delta)Y_{t-1} + \mathrm{E}_{t-j} + V_{1t} \quad (5)$$

$$(1-\delta)Y_t = \alpha_0 + X_{t-1} + \sum_{i=1}^{m}(1-\delta)Y_{t-1} + \sum_{j=1}^{n}(1-\delta)\mathrm{E}_{t-j} + V_{3t} \quad (6)$$

$$(1-\delta)Y_t = \alpha_0 + X_{t-1} + \sum_{i=1}^{m}(1-\delta)Y_{t-1} + \sum_{j=1}^{n}(1-\delta)E_{t-j} + \sum_{p=1}^{p}(1-\delta)E_{t-p} + V_{3t} \quad (7)$$

While applying Vector error-correction modelling, we follow Miller (1991) by using different variables as the dependent variable and choosing the conditioning (left-hand-side) variable with the highest adjusted R-square. Further, in testing for causality between electricity consumption and economic growth, we used a Granger causality test. We proceeded to estimate this long-run relationship in a vector error correction framework. The normalised cointegrating relationship was between GDP and electricity consumption. These statistics are based on averages of the individual autoregressive coefficients associated with the unit root tests. All tests are distributed asymptotically as standard normal. The results indicate that there is a long-run equilibrium relationship between real GDP and electricity consumption, real gross fixed capital formation, and the labour force. Coefficients for real fixed gross capital, and labour force are positive and statistically significant at the 5% significance level, and given the variables are expressed in natural logarithms, the coefficients can be interpreted as elasticity estimates.

$$\Delta Y_t = \omega_1 + \sum_{k=1}^{q}\theta_{11k}\Delta Y_{t-k} + \sum_{k}^{q}\theta_{12k}\Delta E_{t-k} + \sum_{k}^{q}\theta_{13k}\Delta K_{t-k} + \sum_{k}^{q}\theta_{14k}\Delta L_{t-k} + \lambda_1\varepsilon_{t-1} + u_{1t} \quad (8a)$$

$$\Delta E_t = \omega 2 + \sum_{k=1}^{q}\theta_{21k}\Delta Y_{t-k} + \sum_{k=1}^{q}\theta_{22}\mathrm{k}\Delta E_{t-k} + \sum_{k}^{q}\theta_{23k}\Delta K_{t-k} + \sum_{k}^{q}\theta_{24k}\Delta L_{t-k} + \lambda_2\varepsilon_{t-1} + u_{2t} \quad (8b)$$

$$\Delta K_t = \omega 3 + \sum_{k=1}^{q}\theta_{31k}Y_{t-k} + \sum_{k=1}^{q}\theta_{32}\mathrm{k}\theta_{22k}\Delta E_{t-k} + \sum_{k}^{q}\theta_{33k}\Delta K_{t-k} + \sum_{k}^{q}\theta_{34k}\Delta L_{t-k} + \lambda_2\varepsilon_{t-1} + u_{3t} \quad (8c)$$

$$\Delta L_t = \omega 4 + \sum_{k=1}^{q}\theta_{41k}\Delta Y_{t-k} + \sum_{k=1}^{q}\theta_{42}\mathrm{k}\Delta E_{t-k} + \sum_{k}^{q}\theta_{43k}\Delta K_{t-k} + \sum_{k}^{q}\theta_{44k}\Delta L_{t-k} + \lambda_4\varepsilon_{t-1} + u_{4t} \quad (8d)$$

where Δ is the first-difference operator, q is the lag length set at one based on likelihood ratio tests, and u is the serially uncorrelated error term.

4.4 *Diagnostic tests*

This subsection mainly looks at post-estimation tests particularly the test from stationarity and unit root, serial correlation, functional form, cointegration, heteroscedasticity, and normality as explained.

4.4.1 *Test for stationarity and unit root*

According to Granger's (1969) approach, a variable Y is caused by a variable E if Y can be predicted better from past values of both Y and E than from past values of Y alone. For a simple bivariate model, we tested if E Granger-caused Y by estimating Eq. (9) and then tested the null hypothesis in Eq. (10)

$$Y_t = \mu + \sum_{j-1}^{n}\gamma Y_{t-j} + \sum_{j-1}^{n}\alpha_1\mathrm{E}_{t-j} + \mu_t \quad (9)$$

H_0: $\gamma = 0$ for $j = 1, \ldots,$ n

$$H_1 : \alpha \neq 0 \text{ for at least one j} \quad (10)$$

where μ is a constant and u_t is a white noise process. To test for unit roots in our variables, we used the Augmented Dickey Fuller (ADF) test. Using the results of ADF, the null hypothesis showed that all variables have unit roots. Electricity consumption (E) is non-stationary (therefore the null hypothesis is rejected).

4.4.2 *Test for serial correlation*

According to Samson (2015), autocorrelation is the correlation of a time series with its own past and future values. We used the Breusch-Godfrey LM test (Damodar 2004) for both AR(p) and MA (q) error structures as well as for the presence of lagged regression and explanatory variables. The null hypothesis (Ho) is that there is no serial correlation of any order. If the sample size is large enough, Breusch and Godfrey have shown that:

$$(n-p)R^2 \sim X_p^2. \quad (11)$$

Implying that asymptotically, n-p times the R^2 follows the chi-square distribution with PDF. If in an application, (n-p) R^2 exceeds the critical chi-square value at a chosen level of significance, we reject the null hypothesis. Thus the null hypothesis is rejected if p-value is less than 5%, in our case it is 0.00 so we reject the null hypothesis.

4.4.3 *Determining the appropriate lag length for VECM model*

The need for the lags arises because values in the past affect today's values for a given variable. This is to say the variable in question is persistent. There are various methods to determine how many lags to use. The AIC was used to determine the appropriate lag length given the large sample size of 44 observations. The appropriate lag length is 3.

4.4.4 *Cointegration test*

This test is used to check if there exists a long-run relationship between the study variables. Generally, a set of variables is said to be cointegrated if a linear combination of the individual series, which are I(d), is stationary. Intuitively, if xt $\sim$ I(d) and yt $\sim$ I(d), a regression is run, If the residuals, εt, are I(0), then Et and Y_t are cointegrated. We use Johansen's (1988) approach, which allows us to estimate and test for the presence of multiple cointegration relationships. The choice of lag length is made according to the AIC criterion. In conclusion there is one cointegration rank (long-run relationship). When determining lag structures of the data-generating processes (DGP), we use the Johansen (1988) procedure to test the existence of long-run equilibrium relations using the trace statistic test for cointegration, because our data are based on rather small samples, the estimation procedure that we adopt accounts for the Bartlett correction following Johansen (2000). The Johansen cointegration procedure does not reject the null hypothesis of one cointegrating equation. The Johansen trace and max test statistics suggest the existence of at least 1 cointegrating relationship between GDP and electricity consumption. Hence, we estimated the Vector Error Correction Model (VECM) analysis. Why the VECM? According to Hendry and Mizon (1978), this choice of the econometric model is because of existence of a cointegration problem. Since the error structure in non-stationary in levels, the problem is estimated in a first difference formulation. The Dickey Fuller test is used to determine whether the remainder is stationary Dickey Fuller, and the Augmented Dickey Fuller test is applied on the least squares residual to implement the Engel and Granger procedure.

4.4.5 *Test for functional form*

We may have a model that is correctly specified, in terms of including the appropriate explanatory variables, yet commit functional form misspecification. In this case, the model does not properly account for the form of the relationship between dependent and observed explanatory variables.

4.4.6 *Test for heteroscedasticity*

The error term is found to be homoscedastic using the Breush Pagan test; this shows the stability of the parameters using residual diagnostics to minimize errors (or residuals). The error term is be independently and identically distributed (i.i.d). Using the correlogram, the error term of the estimated model is identified. This procedure of log transformation is important because it stabilises the means, however the means are also found to be nonstationary.

4.4.7 *Test for normality*

The Jacque Bera normality test was used to test for normality, which variable is relevant to be expressed as linear combination among other variables. Using the Maximum Likelihood- Autoregressive Conditional Heteroscedasticity (ML ARCH) the residuals were normally distributed as shown in Figure 3 in the appendices.

5 FINDINGS AND DISCUSSION

5.1 *Findings*

With respect to Eqs. (8a)–(8d), short-run causality is determined by the statistical significance of the partial F-statistics associated with the corresponding right hand side variables. The null hypothesis is of no long-run causality in each equation. Equations. (8a)–(8d) were tested by the statistical significance of the t-statistics for the coefficient on the respective error correction terms represented by λ. In terms of Equation (8b), both economic growth and real gross fixed capital formation each have a positive and statistically significant impact on renewable energy consumption in the short-run, whereas the labour force has a statistically insignificant impact. For Equation (8c), economic growth, renewable energy consumption, and the labour force each have a positive and statistically significant impact on real gross fixed capital formation. Moreover, both renewable energy consumption and the labour force appear to serve as complements to real gross fixed capital formation. With respect to Eq. (8d), at the 95% confidence interval, the results indicate a 1% increase in domestic electricity consumption causes a rise in GDP of 0.032, while a 1% increase in fixed capital formation causes a rise in GDP of 0.757. While a 1% increase in labour causes a drop in GDP of 0.04, both labour and real gross fixed capital formation have a positive and statistically significant impact on the GDP in the short run while electricity consumption is statistically insignificant. The coefficients are exactly identified. Therefore, this makes the VECM an ideal test for this study.

5.2 *GDP and electricity consumption*

5.2.1 *Stationarity*

Using ADF tests, data was found to be non- stationary at level, but all exhibited stationarity at first difference (see Appendices Table 5a and 5b)

5.2.2 *Lag length selection*

Since it has been shown that ADF tests are sensitive to lag lengths (Campbell & Perron 1991), we determine the optimal lag length by using Akaike's information criterion (AIC). The formal testing of the lag structure is based on the maximum likelihood function.

5.2.3 *Normality*

The data was normally distributed (refer to Figure 3 in the appendices).

5.2.4 *Granger causality*

The finding using the Granger causality test showed a bidirectional relationship between log of GDP and log of electricity (as shown in Table 4 in the appendices). This confirmed the feedback hypothesis of the energy–economic growth nexus. However, there was unidirectional relationship running from log of labour force to log of GDP and log of electricity Granger causes log capital formation.

5.2.5 *Cointegration test*

The Johansen trace and max test statistics suggest the existence of at least one cointegrating equation between GDP (LGDP) and electricity consumption (Lelec), So we reject the null hypothesis at the 5% level of significance, thus we go ahead to determine the long run relationship using VECM results that show that a positive and significant relationship exists between electricity consumption and GDP. Specifically, a 1% increase in electricity consumption tends to increase GDP by 0.06% as shown in Table 3 in the appendices.

5.3 *Discussion*

The empirical results from the multivariate framework shows that GDP is explained by 3% electricity consumption, capital 75%, and labour 4%, which is in agreement with Kummel and Lindenberg (2014) and Stern (2004), that electricity consumption has a small cost share yet a very big contribution to the overall GDP, as shown in Table 6. When the model is regressed log electricity consumption alone to log GDP is 37% showing a large contribution of electricity consumption to GDP as shown in Table 7. The results from the error correction models lend support for the feedback hypothesis given that both short-run and long-run bidirectional causality exists between renewable energy consumption and economic growth.

6 CONCLUSIONS AND RECOMMENDATIONS

6.1 *Conclusions*

The paper investigates the causal relationship between electricity consumption and economic growth for Uganda over the period 2008–2018. It is worthwhile to examine electricity consumption as a clean renewable energy. Uganda has performed well in increasing hydroelectric power supply in the last decade with the hope to drive the economy at a much faster rate.

There is a bidirectional causality implying that increasing electricity consumption increases economic growth. And economic growth in turn increases electricity consumption. This explains the critical increase in electricity supply capacity is not followed by increasing electricity consumption in Uganda due to a low economic growth rate.

6.2 *Recommendations*

In this paper, we investigated the causal relationship between electricity demand and economic growth for Uganda and based on the findings make the following recommendations for both energy economists and policy authorities.

Policy makers should encourage a multilateral effort to promote renewable energy and energy efficiency in the region. Regional cooperation on the development of renewable energy markets between public and private sector stakeholders could begin with sharing information across countries with respect to ongoing projects, technologies, as well as the financing and investment strategies. The establishment of partnerships between the public and private sector would also facilitate the technology transfer process of bring renewable energy projects to market.

In addition, policy makers should introduce the appropriate incentive mechanisms for the development and market accessibility of renewable energy. Such incentives could include tax credits and/or subsidies for the production and consumption of renewable energy. The establishment of markets for tradable renewable energy certificates along with the implementation of renewable energy portfolio standards may promote the expansion of the electricity subsector in specific and the renewable sector in general.

Therefore, this paper recommends Ugandan authorities not only to develop energy policies geared towards expanding electricity infrastructure to increase electricity consumption only, but also address efficient energy use to support economic growth.

REFERENCES

Abosedra, S., & Baghestani, H. 1991. New evidence on the causal relationship between United States energy consumption and gross national product. Journal of Energy and Development, 14, 285–292.

Acaravci, A., & Ozturk, I. 2010. *On the Relationship between Energy Consumption, CO2 Emissions and Economic Growth in Europe.* Energy, 35(12), 5412–5420.

Akarca, A.T. & Long, T.V. 1980. *On the relationship between energy and GNP: a re- examination.* The Journal of Energy and Development, pp.326–331.

Ammah-Tagoe, F.A., 1990. *On woodfuel, total energy consumption and GDP in Ghana: a study of trends and causal relations.* Center for Energy and Environmental Studies, Boston University, Boston, MA (mimeo).

Apergis, N. & Payne, J.E. 2009a. *Energy consumption and economic growth in CentralAmerica: evidence from a panel cointegration and error correction model.* Energy Economics 31, 211–216.

Apergis, N. & Payne, J.E. 2009b. *Energy consumption and economic growth: evidence from the commonwealth of independent states.* Energy Economics 31, 641–647.

Apergis, N., & Payne, J.E. 2010. *The Renewable Energy Consumption-Growth Nexus in CentralAmerica,* Working Paper.

Asafu-Adjaye, J. 2000. *The relationship between energy consumption, energy prices and economic growth: time series evidence from Asian developing countries. Energy economics.* 22 (6), 615–625.

Bah, M.M. & Azam, M., 2017. *Investigating the relationship between electricity consumption and economic growth: Evidence from South Africa.* Renewable and Sustainable Energy Reviews, 80, pp.531–537.

Bhattacharya, M. Paramati, S.R, Ozturk, I. & Bhattacharya, S. (2016). The effect of renewable energy consumption on economic growth: Evidence from top 38 countries*Applied Energy* 162 733–741.

Bowden, N. & Payne, J.E. 2010. *Sectoral Analysis of the Causal Relationship between Renewable and Non-Renewable Energy Consumption and Real Output in the U.S.* Energy Sources, Part B: Economics, Planning, and Policy 5 (4), 400– 408.

Burbridge, J. & Harrison, A., 1984. *Testing for the effects of monetary policy shocks: Evidence from the flow of funds.* International Economic Review, 25, pp.459–484.

Campbell, J.Y. & Perron, P., 1991. *Pitfalls and opportunities: what macroeconomists should know about unit roots.* NBER macroeconomics annual, 6, pp.141–201.

Chien, T. & Hu, J.-L. 2007. *Renewable energy and macroeconomic efficiency of OECD and non- OECD economies.* Energy Policy 35, 3606–3615.

Engle, R.F. & Granger, C.W.J. 1987. *Co- integration and error correction: representation, estimation, and testing.* Econometrica 55, 251–276.

Electricity Regulatory Authority annual report (2017, 2019).

Erol, U. & Yu, E.S., 1987. *On the causal relationship between energy and income for industrialized countries.* The Journal of Energy and Development, pp.113–122.

Ewing, B.T. Sari, R., Soytas, U. 2007. *Disaggregate energy consumption and industrial output in the United States.* Energy Policy 35, 1274–1281.

Fotourehchi, Z. (2017), *Renewable energy consumption and economic growth: A case study for developing countries.* International Journal of Energy Economics and Policy, 7(2), 61–64.

Granger, C.W. 1969. *Investigating causal relations by econometric models and cross-spectral methods.* Econometrica 37:424–438.

Hamilton, J.D., 1983. *Oil and the macroeconomy since World War II. Journal of political economy,* 91(2), pp.228–248.

Hendry, D.F. & Mizon, G.E., 1978. *Serial correlation as a convenient simplification, not a nuisance: A comment on a study of the demand for money by the Bank of England.* The Economic Journal, 88(351), pp.549–563.

Holtz-Eakin, D. 1986. *Testing for Individual Effects in Dynamic Models Using Panel Data,* NBER Technical Paper Series, No. 57.

Holtz-Eakin, D. Newey, W. &Rosen, H. 1985. *Implementing Causality Tests with Panel Data with an Example from Local Public Finance,* NBER Technical Working Paper, No. 48.

Humbatova, S. I., Ahmadov, F. S., Seyfullayev, I. Z., & Hajiyev, N. G. O. 2020. *The relationship between electricity consumption and economic growth: evidence from Azerbaijan.* International Journal of Energy Economics and Policy, 10(1), 436.

Ibrahiem, D.M., 2018. *Investigating the causal relationship between electricity consumption and sectoral outputs: evidence from Egypt.* Energy Transitions, 2(1), pp.31–48.

Im, K.S., Pesaran, M.H. & Shin, Y. 2003. *Testing for unit roots in heterogeneous panels.* J. Econometrics 115, 53–74.

Johansen, S. 1988 *Statistical analysis of cointegrating vectors,* Journal of Economic Dynamics and Control, Vol. 12, Nos. 2–3, pp.231–254.

Johansen, S. 1991 *Estimation and hypothesis testing of cointegration vectors in Gaussian vector autoregressive models,* Econometrica, Vol. 59, No. 6, pp.1.551–1.580.

Kaygusuz, K., 2007. *Energy for sustainable development: key issues and challenges.* Energy Sources, Part B: Economics, Planning, and Policy 2, 73–83.

Kaygusuz, K. Yuksek, O. &Sari, A. 2007. *Renewable energy sources in the European Union: markets and capacity.* Energy Sources, Part B: Economics, Planning, and Policy 2, 19–29.

Kraft. J., & Kraft .A, 1978. *On the relationship between energy and GNP' Journal of energy and Development* Vol.3 1978 401–403.

Kummel, R. & Lindenberg,D. 2014. *How energy conversion drives economic growth far from the equilibrium of neoclassical economics.* New J. Physics vol 16:1-31.

Kummel, R. Schmid, J. & Lindenberg, D. 2014. Why production theory and the second law of thermodynamics support high energy taxes. Research gate publications 123-140:

Lütkepohl, H. 1982. *Non-causality due to omitted variables.* Journal of Econometrics, 19(2-3), 367- 378.

Lütkepohl, H. 1999. *Vector auto regression* Unpublished manuscript Institute fur statistic und Okonometrie Humboldt- Universitat Zu Berlin.

Mawejje,J.& Mawejje, N.D. 2016. *Electricity consumption and sectoral output in Uganda: An empirical investigation* J.Econ Structures, 5-21.

Molem, C.S.,& Ndifor, R.T. 2016, *The effect of energy consumption on economic growth in Cameroon.* Asian Economic and Financial Review, 6(9), 510–521.

Nepal, R. & Paija, N., 2019. *A multivariate time series analysis of energy consumption, real output and pollutant emissions in a developing economy: New vidence from Nepal.* Economic Modelling, 77, pp.164-173.

Odhiambo, N.M.2009a. *Energy consumption and economic growth nexus in Tanzania: an ARDL bounds testing approach.* Energy Policy 37:617–622.

Odhiambo, N.M.2009c. *Energy Consumption and economic growth in SouthAfrica: a tri-variate causality test. Energy* Econ 31:635–640.

Odhiambo,N.M.2010. *Energy consumption, prices and economic growth in three SSA countries: a comparative study.* Energy Policy 38:2463–2469.

Okoboi, G., & Mawejje, J. 2016. *The impact of adoption of power factor correction technology on electricity peak demand in Uganda.* J Econ Struct 5(1):1–14.

Omay,T. et al. 2010. *Energy consumption and Economic growth: evidencefrom nonlinear panel co- integration and Causality tests.*

Ozturk, I., 2010. A literature survey on energy-growth nexus. Energy Policy 38, 340–349. Paul, S., & Paul.S & Bhattacharya, R. N. 2004. *Causality between energy consumption and economic growth in India: a note on conflicting results.* Energy economics, 26(6), 977–983.

Pachauri, R.K., 1977. Energy and economic development in India. Praeger, New York.

Payne, J.E., 2009. *On the dynamics of energy consumption and output in the U.S.* Applied Energy 86, 575–577.

Payne, J.E., 2010a. *Survey of the international evidence on the causal relationship between energy consumption and growth.* Journal of Economic Studies 37, 53–95.

Payne, J.E., 2010b. *A survey of the electricity consumption-growth literature.* Applied Energy 87, 723– 731.

Payne, J.E. 2010c. *On Biomass Energy Consumption and Real Output in the U.S.*Energy Sources, Part B:Economics, Planning, and Policy, forthcoming.

Pedroni, P. 1999. *Critical values for cointegration tests in heterogeneous panels with multiple regressors.* Oxford Bulletin of Economics and Statistics 61, 653–670.

Pedroni, P., 2000. *Fully modified OLS for heterogeneous cointegrated panels.* Advanced in Econometrics 15, 93–130.

Pedroni, P. 2004. *Panel cointegration: asymptotic and finite sample properties of pooled time series tests with an application to the ppp hypothesis: new results.* Econometric Theory 20, 597–627.

Pesaran, H.M. Shin, Y. & Smith, R.P. 1999. *Pooled mean group estimation of dynamic heterogeneous panels.* Journal of the American Statistical Association 94, 621–634.

Reynolds, D.B. & Kolodzieji, M., 2008. *Former Soviet Union oil production and GDPdecline: Granger causality and the multi-cycle Hubbert curve.* Energy Economics 30, 271–289.

Ridzuan, A.R., Kamaluddin, M., Ismail, N.A., Razak, M.I.M. & Haron, N.F., 2020. *Macroeconomic indicators for electrical consumption demand model in Malaysia.* International Journal of Energy Economics and Policy, 10(1), p.16.

Sadorsky, P., 2009. *Renewable energy consumption and income in emerging economies.* Energy Policy 37, 4021–4028.

Sari, R.,& Soytas, U. 2004. *Disaggregate energy consumption, employment, and income in Turkey.* Energy Economics 26, 335–344.

Sari, R., Ewing, B.T.,& Soytas, U. 2008. *The relationship between disaggregate energy consumption and industrial production in the United States: an ARDL approach.* Energy Economics 30, 2302–2313.

Śmiech, S. & Papież, M., 2014. *Energy consumption and economic growth in the light of meeting the targets of energy policy in the EU: The bootstrap panel Granger causality approach.* Energy Policy, 71, pp.118–129.

Solow, R.M., 1956. *A contribution to the theory of economic growth.* The quarterly journal of economics, 70(1), pp.65–94.

Solow, J.L., 1987. *The capital-energy complementarity debate revisited.* The American Economic Review, pp.605–614.

Tang, C.F., Tan, B.W. & Ozturk, I., 2016. *Energy consumption and economic growth in Vietnam. Renewable and Sustainable Energy Reviews, 54, pp.1506-1514.*

Tyner, W.A., 1978. *Energy Resources and Economic Development in India.* Boston, Martines Nijhoff Social Sciences Division, Netherlands.

Saint Akadiri, S., Alola, A.A., Akadiri, A.C. & Alola, U.V., 2019. *Renewable energy consumption in EU-28 countries: policy toward pollution mitigation and economic sustainability.* Energy Policy, 132, pp.803–810.

Sekantsi & Motlokoa , M. 2016. *Evidence of the nexus between electricity consumption and economic growth through empirical investigation of Uganda.* Review of Econ & Business studies 8 (1), 149–165.

Sekantsi, L. P., & Okot, N. 2016. *Electricity–economic growth nexus in Uganda.* Energy Sources, Part B: Economics, Planning and Policy 11:1144–1149. doi:10.1080/15567249.2015.1010022

Stern, D.I., 1993. *Energy and economic growth in the USA: a multivariate approach.* Energy economics, 15(2), pp.137–150.

Soytas, U. Sari, R. & Ewing, B.T. 2007. *Energy consumption, income, and carbon emissions in the United States.* Ecological Economics 62, 482–489.

Yu, E.S. & Choi, J.Y., 1985. *The causal relationship between energy and GNP: an international comparison.* The Journal of Energy and Development, pp.249–272.

APPENDICES

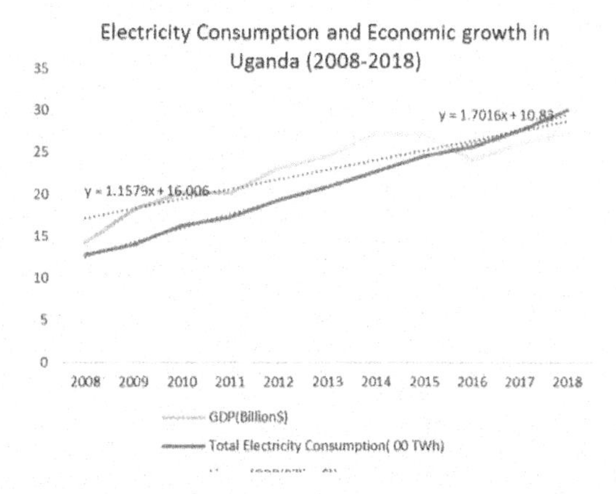

Figure 1. Trend of Uganda's GDP and Electricity consumption (2008–2018). SOURCE: Author's analysis based on data from World Bank and ERA, 2020.

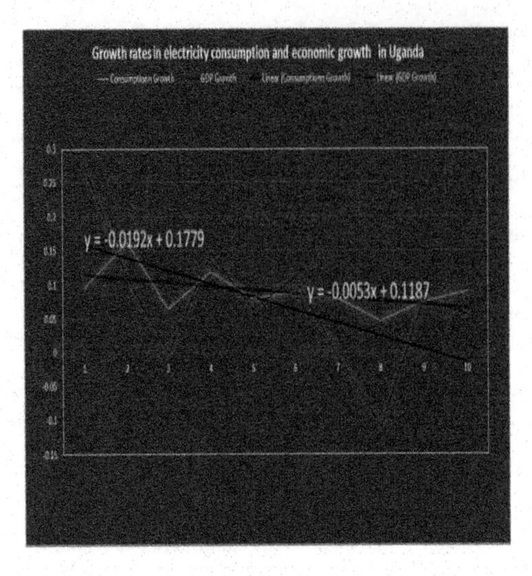

Figure 2. Growth rate of GDP and Electricity Consumption (2008–2018). Source: Author's analysis based on data from World Bank and ERA, 2020.

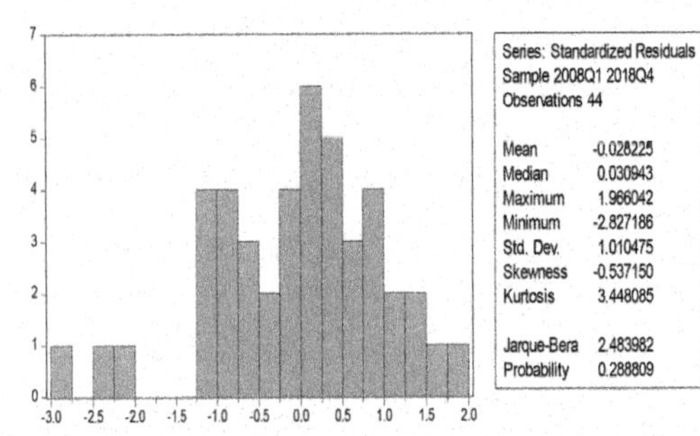

Figure 3. Normality test.

Table 2. Cointegration test.

D log GDP	Variable	Coef	Std.Error	P-value
	Log GDP			
L1.		0.13	0.818	0.25
	LR GDP			
Lag1.LGDP.		−0.165	0.161	−0.13
Lag2 d. LGD		−0.044	0.600	−0.73
	Log elec			
Lag lelec 1.		0.017	0.13	0.894
Lag2 lelec.		0.175	0.06	0.03
	Log cap			
Lag lCap.		0.03	0.19	0.87
Lag12 LCap		−0.19	0.21	0.36
Lag2 lCap		−0.20	0.21	0.35

Table 3. Vector error correction mechanism.

	Coeff	Std. error	P-value
Logelec			
L1.	0.06	0.02	0.01
LR			
Lag Lelec 1.	0.34	0.12	0.06
LagLelec 2.	0.01	0.01	0.34
LagLelec3.	0.01	0.54	
D.log GDP (LY)			
LD.	0.59	0.23	0.01
Lag LY1	0.53	0.19	0.01
Lag LY2	0.11	−1088	
Lag LY3	0.38	0.19	0.04
Lncap			
Lag Lcap	0.15	0.13	0.25

Table 4. Granger causality.

Null Hypothesis:	F-Stat	Prob.
LNELEC does not Granger Cause LNCAP	5.32192	0.0041
LNCAP does not Granger Cause LNELEC	0.22687	0.8770
LNGDP does not Granger Cause LNCAP	1.41890	0.2542
LNCAP does not Granger Cause LNGDP	0.63778	0.5959
LNLAB does not Granger Cause LNCAP	0.82253	0.4906
LNCAP does not Granger Cause LNLAB	0.39181	0.7597
LNGDP does not Granger Cause LNELEC	3.88549	0.0172
LNELEC does not Granger Cause LNGDP	3.43835	0.0275
LNLAB does not Granger Cause LNEEC	0.37452	0.7719
LNELEC does not Granger Cause LNLAB	0.23738	0.8697
LNLAB does not Granger Cause LNGDP	3.85600	0.0178
LNGDP does not Granger Cause LNLAB	0.27577	0.8425

Table 5a. ADF test for stationarity at first difference for log of electricity consumption.

Test critical values	t-statistic	P-value
	−15.24945	0.0000
1% level	−3.610453	
5% level	−2.938987	
10% level	−2.607932	

Table 5b. ADF test for stationarity at first difference for of capital formation.

	t-statistic	P-value
Test critical values	−13.58911	0.0000
1% level	-3.615588	
5% level	-2.941145	
10% level	-2.609066	

Table 6. Regression of a multivariate regression.

Log GDP	coefficient	t-statistic	P-value
Log Elec	0.032	1.26	0.214
Log Cap	0.757	14.22	0.000*
Log Lab	0.04	4.13	0.000*

R-squared = 0.978, Adj.R-squared = 0.976
*shows is statistically significant at 1 %

Table 7. Regression of electricity consumption vs. GDP.

Log GDP	coefficient	t-statistic	P-value
Log Elec	0.37	15.04	0.000*

R-squared = 0.844, Adj. R-squared = 0.840
*shows is statistically significant at 1 %

Advances in Phytochemistry, Textile and Renewable Energy Research for Industrial Growth – Nzila et al. (Eds)

Purification and upgrade of biogas using biomass-derived adsorbents: Review

Elshaday Mulu*
Department of Mechanical, Production and Energy Engineering, School of Engineering, Moi University, Eldoret, Kenya
Faculty of Mechanical and Production Engineering, Arba-minch University, Arba-minch, Ethiopia
African Centre of Excellence in Phytochemicals, Textile and Renewable Energy. Moi University, Eldoret, Kenya (ACE II PTRE)

Milton M. M'Arimi
African Centre of Excellence in Phytochemicals, Textile and Renewable Energy. Moi University, Eldoret, Kenya (ACE II PTRE)
Department of Chemical and Process Engineering, School of Engineering, Moi University, Eldoret, Kenya

Ramkat C. Rose
African Centre of Excellence in Phytochemicals, Textile and Renewable Energy, Moi University, Eldoret, Kenya (ACE II PTRE)
Department of Biological Science, School of Sciences and Aerospace Studies, Moi University, Eldoret, Kenya

ABSTRACT: The contaminants in biogas, which include carbon dioxide and hydrogen sulfide, limit its application as engine fuel because they reduce its energy content and cause corrosion to metals. The aim of this study was to review the applications of biomass materials in the purification and upgrading of biogas. The CO_2 adsorption capacity of activated carbon is dependent on the surface area and pore size of the adsorbent. Biochar has a high adsorption capacity of H_2S, which is dependent on the media alkalinity and the surface chemistry. The capture of CO_2 and H_2S by biomass adsorbent occurs through both physisorption and chemisorption. An increase in adsorption temperature decreases the adsorption capacity of CO_2 but increases the adsorption capacity of H_2S for biomass adsorption. Published data indicate that modification with KOH adsorbents improves CO_2 uptake significantly. Furthermore, impregnation of biomass adsorbents with agents like amine compounds can increase the adsorption capacity of H_2S.

1 INTRODUCTION

Fossil fuel has a high contribution to global warming. Therefore, the use of renewable energy is an attractive alternative source. Biogas is among the renewable energy sources that are currently used as replacements for fossil fuels. In Europe, the biogas potential is estimated at 18 billion M^3 produced per year from animal manure with an average power capacity between 315–515 kWe (Scarlat et al. 2018). Biogas is one of the best alternative renewable energies which can be used for generating electricity, heating, and fueling engines. In addition, it is cost-effective and reduces the environmental effects of waste materials.

Biogas can be produced from various organic compounds by anaerobic digestion. Anaerobic digestion is a process that breaks down organic matter in the absence of oxygen. It consists of four main steps: hydrolysis, acidogenesis, acetogenesis, and methanogens. Hydrolysis is the breakdown of complex organics material (carbohydrate, protein and lipid) into simple monomers (monosaccharide, amino acid, long chain fatty acid) by hydrolytic enzyme. Acidogenesis bacteria convert simple monomers into volatile fatty acids. Acetogenesis bacteria convert the volatile fatty acid into acetic acid, carbon dioxide, and hydrogen. The methanogenesis bacteria convert acetates into methane and carbon dioxide (Murphy J. & Thamsiriroj 2017).

Biogas is composed of methane, carbon dioxide, hydrogen sulfide, nitrogen, siloxanes, oxygen and other minor components. The composition is mainly determined by the substrate and the conditions in the anaerobic digester, such as temperature and media pH. The microbes in the culture may also influence the composition of biogas. The main contaminants in biogas include carbon dioxide and hydrogen sulfide, which limit the application of biogas, especially

*Corresponding author

 DOI 10.1201/9781003221968-38

as engine fuel. The high content of carbon dioxide contaminants reduces the energy density of biogas. Furthermore, the contaminants like hydrogen sulfide affect the internal parts of engine by causing corrosion (Magomnang et al. 2014) In addition, biogas contaminants affect the environment and human health.

The significance of biogas as a bioenergy can be increased by purification and upgrading in a process that entails the removal of contaminants by various methods. Commercial processes for the upgrade of biogas include membrane separation, cryogenic separation, pressure swing adsorption, water scrubbing, physical scrubbing, chemical absorption, hydrate formation, and biological conversion process (Khan et al. 2017; Q. Sun et al. 2015;. Though these methods have high adsorption capacity, they have limitations, which include high-energy requirements and initial investment cost. Biogas can also be purified using natural minerals like clay (Chen et al. 2014), zeolite (Paolini et al. 2015), and coal fly ash (Ferella et al. 2017). They are low cost and have simple operation methods. However, natural minerals adsorbents are not available everywhere. Therefore, biomass adsorbent materials have become an alternative choice for biogas cleaning. Biomass is the most abundant material on the Earth's surface. In addition, it is environmentally friendly and cheap. Furthermore, biomass materials can be converted to adsorbents, such as activated carbon and biochar.

Activated carbon can be derived from any rich carbon organic materials like agricultural waste, plant biomass, industrial waste, and household wastes (Rodriguez-reinoso et al. 2008). It can also be prepared from fossil carbon (Hsu & Teng 2000). However, fossil sources are the main cause of global warming. Therefore, the preparation of activated carbon from plants, agriculture waste, or industrial waste is a more sustainable and preferable method of preparing carbon adsorbents. Biochar is another cheap and widely available material from biomass that can be used for biogas cleaning. It can be produced from carbon-rich materials like agricultural wastes. Biomass adsorbents have been used for various applications, including air pollution control (Nor et al. 2013), battery electrodes and capacitors (Kalyani & Anitha 2013), water and wastewater treatment (Bhatnagar et al. 2015; Wong et al. 2018), remediation of heavy metals like arsenic (Asadullah, Jahan, & Boshir 2014), and adsorbent for carbon dioxide capture (Rashidi et al. 2013).

The process of producing adsorbents from biomass materials may entail physical or chemical treatments. The characteristics of biomass adsorbents, which include the surface area and pore size, are determined by the substrates used and the process conditions (Jung et al. 2019). This paper reviews the documented studies on the removal of biogas contaminations; carbon and hydrogen sulfide using biochar and activated carbon derived from biomass. The summary of characteristics and potential of biomass adsorbents in cleaning of biogas are also given.

2 EFFECT OF BIOGAS CONTAMINATIONS

Carbon dioxide is a greenhouse gas which is emitted in large scale from industries, households, vehicles, and other fuel machines. Furthermore, it is the main contaminant in biogas and has the highest contribution to global warming (Na et al. 2002). The process of global warming starts with the short-wave length solar radiations hitting the Earth's surface. Some radiations are absorbed by the Earth while others are reflected to the atmosphere. Those reflected are in infrared form and are absorbed by greenhouse gases like carbon dioxide and cause an increase in atmospheric temperature, thereby causing global warming. Further, the carbon dioxide content in biogas affects the internal metallic parts when used as fuel for the engine. Basically, the high content of carbon dioxide reduces biogas calorific value, flammability range, and flame velocity (Porpatham et al. 2008).

The hydrogen sulfide contaminants in biogas affect the internal engine parts. A low content of hydrogen sulfide, 50–10,000 ppm is sufficient to cause corrosion and wear on metallic components (Al Mamun et al. 2015; Sevimoğlu & Tansel 2013). Hydrogen sulfide reacts with engine metal components and forms sulfur dioxide and water. Sulfur dioxide and sulfur trioxide react with water to form corrosive sulfurous acid and sulfuric acid, respectively. Therefore, traces of hydrogen sulfide in biogas are undesirable and should be eliminated before usage as engine fuel. In addition, the corrosion caused reduces the effectiveness of the heat exchanger (Razbani et al. 2011). Also, hydrogen sulfide affects human health by causing headaches, dizziness, respiratory illness, and poor memory (Bates et al. 2002). Hence, biogas should be purified before application in engines.

3 BIOMASS ADSORBENTS FOR REMOVAL OF CARBON DIOXIDE FROM BIOGAS

3.1 *Biochar*

Biochar is a cheap abundant material with high carbon content. It is produced from biomass materials and has wide applications in wastewater treatment (Qambrani et al. 2017), soil remediation (Saifullah et al. 2018), and the absorption of carbon dioxide (Madzaki et al. 2016). The main elements in biochar include carbon, hydrogen, oxygen, and ash. It may also contain traces of sulfur and hydrogen. However, the actual composition may vary depending on the substrate type and production process. Biochar can be produced by various processes including pyrolysis, gasification, and hydrothermal carbonization (HTC). In the pyrolysis process, biomass materials are decomposed at temperatures between 350–900°C in the absence of oxygen (Méndez et al. 2014). The main product of the process is syngas (CO, H_2, and hydrocarbon gases), while biochar is a by-product. Biochar production through gasification entails partial oxidation of biomass by thermochemical treatment using agents like air, steam, oxygen, or carbon dioxide. The products of gasification include syngas, bio-oil, and biochar (Wang & Wang 2019). Hydrothermal carbonization is carried out on a mixture of biomass and water. Thermal composition is done at high temperature and pressure resulting in hydrochar formation which has higher carbon content than biochar from processes using dry biomass. Biochar can be modified by various methods to enhance its adsorption ability. Alkaline treatment with chemicals like NaOH, KOH, K_2O, and K_2CO_3 is among the most applied biochar modification processes (Jung et al. 2019). Other modification processes include physical treatment with CO_2 and steam. It can also be impregnated with amino compounds or other surface modifying compounds to enhance the effectiveness as an adsorbent (Yang & Jiang 2014). The adsorption properties of biochar depend on the type of substrate and activation process. Biochar has been used as an adsorbent in gases, soil, and water contaminates (Qian et al. 2015). Recently, it has been applied in the removal of carbon dioxide from gases (Ding & Liu 2020). Its main advantages as a gas cleaning agent are that it is cost-effective and easily available compared to commercial adsorbents used in such applications. Table 1 shows the use of biochar in the capture of carbon dioxide.

Unmodified biochar has low adsorption capacity of carbon dioxide with a range of 0.41–1.7 mmol/g. However, this can be improved by the activation and impregnation with metal compounds. Physical adsorptions depend on the surface area pore volume and the carbon dioxide interaction. The surface area and pore size of biomass adsorbents are determined by the feedstock, activation temperature, and residence time. Table 1 shows that biochar modified with monoethanolamine has high carbon dioxide adsorption capacity. This is possibly because biochar adsorption capacity depends on its surface area and surface chemistry. Monoethanolamine treatment of biochar increases the content of nitrogen which improves the alkalinity of the surface. Therefore, the increase in the basicity of the surface of adsorbent improves the adsorption capacity of carbon dioxide. Generally, chemical activation of biochar improves the adsorption capacity by increasing the surface area and total pore size.

3.2 *Activation of biomass plant*

Activated carbon can be prepared from plant biomass by physical and chemical activation. The quality of adsorbent produced from biomass depends on the types of feedstocks, pyrolysis temperature, and residence time (Dias et al. 2007). They affect the surface area, micropore area, and micropore volume, which are important in enhancing the adsorption capacity of carbon dioxide. In addition, it depends on the chemical agents used. The process of activating carbonaceous material into activated carbon leads to an increase in the surface area and the pore volume. Furthermore, it enhances the degree of surface reactivity. Therefore, the material is able to adsorb various adsorbates such as carbon dioxide. Activated carbon adsorbents have high adsorption capacity, despite their low cost. In addition, their adsorption capacity can be improved with the impregnation of bases. Figure 1 illustrates how activated carbon can be processed from biomass materials and impregnated with bases for enhancement of adsorption capacity.

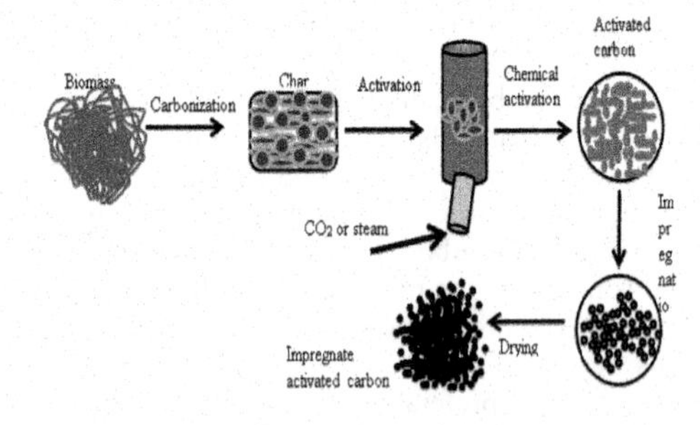

Figure 1. Preparation of activated biomass material by physical and chemical activation of biomass followed by impregnation with bases.

Table 1. The summary of the capture of carbon dioxide using raw and modified biochar adsorbent.

Types of biochar adsorbents	Modification of biochar	Biochar activation temperature	CO_2 uptake mmol/g	Reference
Cotton wood	Mg	600°C	1.4	(Creamer et al. 2016)
Sugarcane bagasse	Unmodified	600°C	1.7	(Creamer et al. 2014)
See-weed porous	KOH	800°C	1.05	(Ding & Liu 2020)
Palm kernel shells	Unmodified	500°C	0.46	(Promraksa & Rakmak 2020)
Rambutan	Mg	–	1.7	(Zubbri et al. 2020)
Sawdust biochar	MEA	850°C	10.7	(Madzaki et al. 2016)
Sewage sludge	Unmodified	500°C	0.41	(Xu et al. 2016)
Wheat straw	Unmodified	500°C	0.78	(Xu et al. 2016)
Pine sawdust	Steam activated	550°C	0.73	(Igalavithana et al. 2020)
Hickory chips	Fe	–	3.64	(Xu et al. 2020)
Walnut shell	Mg	900°C	1.9	(Lahijani et al. 2018)
HF-N-char	Pre-de-ashed	600°C	1.8	(X. Zhang et al. 2015)

Note; MEA: monoethanolamine, HF-N-char: hydrofluoric pre de-ashed rice husk.

3.3 *Physical activation*

Physical activation releases volatile matter from biomass materials. The process depends on the nature of feedstock and activation conditions such as temperature (Thomauske et al. 2007; Williams & Reed 2006). The presence of volatile matter in activated carbon reduces the pore volume and micropores. Therefore, a substrate with high content of volatile matter needs a high activation temperature. The high temperature helps to develop the pore size and creates new pores. Conversely, too high temperature may collapse the pores, which reduces the quality of activated carbon (T. Yang & Lua 2003).

Physical activation is done in two steps: carbonation and activation using steam and carbon dioxide. Carbonation is the first step of physical activation. It entails decomposition of organic matter without oxygen and removes non-carbon components. It converts organic matter to char at temperatures of 600–800°C (Hernandez et al. 2007). Carbonization produces low microporous carbon (Girgis et al. 2002). The second step of the treatment is called activation, in which the char from the carbonation process is activated using steam and carbon dioxide as activating agents. A mixture of char and water is heated to a temperature range of 800–1,000°C (Dobele et al. 2012; Zabaniotou et al. 2008). The steam pyrolysis creates porosity in the range of mesopores. However, the process results in a low surface area (El-hendawy et al. 2001; Girgis et al. 2002). Carbon dioxide activation enhances the microporosity of the product (Yang & Lua 2003). A summary of the use of physically activated biomass materials to remove carbon dioxide is given in Table 2.

The adsorption capacity of carbon dioxide by biomass adsorbents prepared by physical activation depends on the type of feedstock for activation, pyrolysis temperature, and residence time. High pyrolysis temperature decreases the yield of char but produces high-quality adsorbents. The documented data indicate that uptake capacity of carbon dioxide in a range of 0.3-5.6 mmol/g can be achieved using physically activated biomass adsorbents. The quality of the biomass adsorbent is dependent on the substrate. Activated carbon made from coconut shell produced the highest carbon dioxide removal capacity. The adsorbent produced by a pyrolysis step of physical activation has low adsorption capacity. However, after the activation of the adsorbent using either steam or carbon dioxide, the adsorption capacity is improved. Carbon dioxide activation results in adsorbents with higher micropore volume and surface area than steam activation (Pallarés et al. 2018). Therefore, activation using carbon dioxide results in adsorbents with high adsorption capacity.

3.4 *Chemical activation*

Chemical activation is effective in the production of activated carbon. The process uses different chemical agents such as sodium hydroxide, zinc chloride, potassium hydroxide, potassium carbonate, phosphoric acid, and sulfuric acid (Dias et al. 2007). In chemical activation, the prepared biomass material is mixed with chemical agents, which affects its structure. The characteristic of activated carbon depends on type of chemical agent used for treatment and its concentration. After chemical activation, the adsorbent is subjected to heating at high temperature, which creates the porous structure of adsorbent (William & Reed 2006; Zabaniotou et al. 2008).

Activated carbon from biomass can be impregnated with different bases to enhance their capacity to remove carbon dioxide and hydrogen sulfide. The pores of the adsorbent are filled with chemical agents, which increase the adsorption capacity of the adsorbent (Song et al. 2013). However, according to Somy et al. (2009), impregnation of activated carbon with $Fe_2O_3$was not effective. This was probably caused by the blockage of the adsorbent micropores due to large particle size of Fe_2O_3. Therefore, the process decreased the adsorption capacity of carbon dioxide (Somy et al. 2009). Other studies have reported that the carbon dioxide adsorption capacity decreases with an increase in impregnation temperature of solution (Somy et al. 2009). The summary of carbon dioxide removal using chemical activation of biomass is given in in Table 3.

The result from Table 3 shows that an uptake capacity of carbon dioxide range of 0.78–8.0 mmol/g can be achieved using activated biomass. However, NaOH has low adsorption capacity because the particle blocks the surface pores structure. From the Table 3, activation with KOH produces adsorbents with adsorption

Table 2. The removal of carbon dioxide by physically activated of biomass.

Type of adsorbent Biomass	Carbonation temperature	CO_2 uptake mmol/g	Reference
Palm mesocarp fibre	500°C	0.3	(Rashidi et al. 2013)
Coconut shell	900°C	1.8	(Rashidi et al. 2013)
Sugarcane bagasse	600°C	1.7	(Creamer et al. 2014)
Eucalyptus wood	Unmodified	2.9	(Heidari et al. 2014)
Rice husk	600°C	0.46	(Rashidi et al. 2013)
Coconut shells	CO_2 activation	5.6	(Ello et al. 2013)
Coconut shells	CO_2 activation	1.3	(Rashidi et al. 2013)
Whitewood	Steam activation	1.34	(Shahkarami et al. 2015)
Whitewood	CO_2 activation	1.4	(Shahkarami et al. 2015)

Table 3. Removal of carbon dioxide using chemical activation of biomass.

Type of adsorbent	Temperature and pressure	Chemical agent	CO_2 uptake mmol/g	Reference
Rice husk	303K	ZnCl	1.75	(Boonpoke et al. 2011)
Corn cob	–	KOH	1.5	(Song et al. 2013)
Waste celtuce leaves	273K	KOH	6.04	(R. Wang et al. 2012)
Waste celtuce leaves	298K	KOH	4.36	(R. Wang et al. 2012)
Bamboo 3-873	273K	KOH	7.0	(Wei et al. 2012)
Bamboo 3-873	298K	KOH	4.5	(Wei et al. 2012)
Eucalyptus wood based	303K	H_3PO_4	1.1	(Heidari et al. 2014)
Eucalyptus wood based	303K	H_3PO_4	3.22	(Heidari et al. 2014)
GACP48	273K	H_3PO_4	4.16	(Vargas et al. 2012)
GACZn36	273K	$ZnCl_2$	3.3	(Vargas et al. 2012)
GACCa2	273K	$CaCl_2$	3.9	(Vargas et al. 2012)
PC3-780	273K & 0.1 bar	KOH	1.25	(D. Li et al. 2015)
PC3-780	273K	KOH	2.04	(D. Li et al. 2015)
PC3-780	273K & 1 bar	KOH	6.24	(D. Li et al. 2015)
Pomegranate peels	273K	KOH	6.03	(Serafin et al. 2017)
Pomegranate peels	298K	KOH	4.11	(Serafin et al. 2017)
Slash pine	–	KOH	4.93	(Ahmed et al. 2019)
Slash pine	–	$ZnCl_2$	4.32	(Ahmed et al. 2019)
Bamboo 3-873	–	KOH	7.0	(Wei et al. 2012)
Paulownia sawdust	–	KOH	8.0	(Zhu et al. 2014)
Coconut shell	–	NaOH	0.78	(Tan et al. 2014)
Pine nut shell	273K	KOH	7.7	(Deng et al. 2014)
Pine nut shell	298K	KOH	5.0	(Deng et al. 2014)
Lumpy bracket	1bar	KOH	7	(Serafin et al. 2019)
Lumpy bracket	30bar	KOH	14	(Serafin et al. 2019)

Note: GAC- prepared granular in H_3 PO_4 solution 48%w/v, GACZn36- prepared granular in $ZnCl_2$ solution 36% w/v, GACCa2- prepared granular in $CaCl_2$ solution 2%w/v,PC3 – porous carbon prepared from rice husk.

capacity up to 8.0 mmol/g. Therefore, it is the most tried and very effective activation method. According to Vargas et al. (2012), activation with a high concentration of H_3PO_4 produces adsorbents with a high surface area and pore volume. Therefore, its use resulted in high carbon dioxide uptake compared to $ZnCl_2$ and $CaCl_2$ (Vargas et al. 2012). In contrast, the use of low concentration of $CaCl_2$ produces high adsorption capacity of carbon dioxide (Vargas et al. 2012). The process of impregnating adsorbent with chemical agents, such as KOH and K_2CO_3, helps to develop the porosity and surface area which ultimately increases the adsorption capacity of the adsorbent (Tsai et al. 2001).

4 REMOVAL OF HYDROGEN SULFIDE BY BIOMASS DERIVED ADSORBENTS

Hydrogen sulfide can be removed from biogas by the use of adsorbents from biomass material, such as biochar using simple processes. Activated carbon has poor removal of hydrogen sulfide despite its high surface area and porosity. However, the adsorption capacity of hydrogen sulfide by activated biomass can be enhanced through impregnation with ZnO, CuO, Fe_2O_3, and $CaCO_3$. The removal of hydrogen sulfide capacity is enhanced by the presence of moisture content in the gases. In addition, the high pH value of the media results in high hydrogen sulfide adsorption capacity (Xu et al. 2014). The removal of hydrogen sulfide from biogas happens through chemical reactions (chemisorption) with the basic functional groups in the adsorbents or impregnated compounds. The reactions between hydrogen sulfide and ZnO, CuO, Fe_2O_3, and $CaCO_3$ that are commonly impregnated on activated biomass material are given in equations 1, 2, 3, and 4, respectively. The summary of hydrogen sulfide removal using physical and chemical activation of biomass is given in Table 4.

$$C - ZnO + H_2S \longrightarrow C - ZnS + H_2O \tag{1}$$

$$C - CuO + H_2S \longrightarrow C - CuS + H_2O \tag{2}$$

$$C - Fe_2O_3 + 3H_2S \longrightarrow C - Fe_2S_3 + H_2O \tag{3}$$

$$2C - CaCO_3 + 2H_2S \longrightarrow C - Ca(HCO3)_2 + Ca(HS)_2 \tag{4}$$

From Table 4, biochar is very effective as hydrogen sulfide adsorbent and can substitute commercial processes in this regard because it is cheaper. The pyrolysis temperature is important in determining the ability of biochar to adsorb hydrogen sulfide. Shang et al. (2016) studied the effect of pyrolysis temperature on the characteristics of biochar from rice husks,

Table 4. Removal of H_2S using adsorbent from biomass materials.

Type of biomass Adsorbent	Carbonation temperature	Chemical agent	H_2S uptake H_2S mmol/g	Reference
Palm shell	–	H_2SO_4	2.23	(Guo 2007)
Palm shell	–	KOH	2	(Guo 2007)
Palm shell	–	CO_2	1.35	(Guo 2007)
Camphor	400	–	3.21	(Shang et al. 2013)
Bamboo	400	–	9.9	(Shang et al. 2013)
Pig manure	600	–	1.9	(Xu et al. 2014)
Sewage sludge	−600	–	1.4	(Xu et al. 2014)
Coconut shell	–	Fe	3.1	(Locke 2001)
Coconut shell	–	Cu	1.4	(Huang et al. 2006)
Rice hull	400	–	0.6	(Shang et al. 2016)
Rice hull	500	–	11.2	(Shang et al. 2016)
LG700PA	700	–	0.25	(Ortiz et al. 2014)
LG700A	700	–	0.2	(Ortiz et al. 2014)
Potato peel	500	–	1.56	(Y. Sun et al. 2017)
Coffee industry	800	CO_2	8.3	(Nowicki et al. 2014)
Black liquor	900	Steam	0.08	(Ping Zhang et al. 2016)
Wood chips	600	–	8.0	(Kanjanarong et al. 2017)

Note LG: type of sludge.

and observed that high activation temperature promotes the biochar's adsorption capacity of hydrogen sulfide. The nature of the substrate also determines the performance of the adsorbent. Locke (2001) studied adsorbents from sewage sludge and coconut shell and observed that the adsorption capacity of activated carbon from sewage was twice that of activated carbon from coconut shell. Hydrogen sulfide is an acidic gas and therefore and increase in the adsorption capacity can be achieved by increasing the surface pH value. Shang et al. (2013) found a high adsorption capacity of hydrogen sulfide, 11.3mmol/g at pH 9.75 using biochar from rice hulls. Biochar adsorption capacity depends on the temperature and pH.

5 FACTORS AFFECTING ADSORPTION CAPACITY OF BIOMASS MATERIALS

Adsorption of carbon dioxide is a physical process which involves weak bonds between the adsorbates and the adsorbent. Therefore, the process decreases with temperature increase (Rashidi et al. 2013). High temperature increases the interaction due to high kinetic energy of particles at the surface. The increase in temperature therefore causes desorption of carbon dioxide. Serafin et al. (2017) reported that activated carbon from pomegranate peels achieved carbon dioxide uptake of 6.03 mmol/g at 0°C but the same reduced to 4.11 mmol/g at 25°C. Similar observations were reported in an investigation using low temperature by Ello et al. (2013).

Adsorption of carbon dioxide increases with the increase in pressure (Serafin et al. 2019). Li Y. et al. (2014) studied activated carbon derived from starch-based porous carbon and reported carbon dioxide uptake capacity of 21.2 mmol/g at 20 bars. However, the value reduced by 30%, 50%, and 75% at 14, 10, and 5 bar, respectively (Y. Li et al. 2014). The optimum pore volume of micropores for the adsorption of carbon dioxide depends on the carbon dioxide pressure. Presser et al. (2011) observed that the pore sizes for 0.1 and 1 bar were approximately 0.5 nm and 0.8 nm, respectively (Presser et al. 2011).

The gas flow rate affects the adsorption capacity. Yaumi et al. (2018) investigated the effect of flow rate and observed that flow rate increased from 30 to 60 mL/min, while the adsorption capacity decreased from 4.41 to 3.4 mmol/g. For this reason, the contact time between the adsorbent surface area and gases is reduced. In addition, it reduces the external mass transfer (Yaumi et al. 2018). These results are supported by another study where a decrease in total gas feed flow rate increased the adsorption capacity. The total gas flow rate of 90 mL/min produced the adsorption capacity of 27 mg/g which reduced to 25.38 mg/g and 25.73 mg/g at 120 mL/min and 150 mL/min respectively (Tan et al. 2014). The effect of carbon dioxide inlet concentration was studied at 24% and 47%, while the flow rate and breakthrough time were held constant at 50 mL/min and 180 seconds, respectively. It was observed that at high carbon dioxide concentration of 47%, the adsorption capacity was 1.388 mmol/g. In contrast, at 24% concentration of carbon dioxide, the adsorption capacity decreased to 0.713 mmol/g (Munusamy et al. 2012).

6 FUTURE OUTLOOK OF BIOMASS ADSORBENT IN BIOGAS CLEANING

The use of activated biomass materials as adsorption media compares to that of commercial adsorbents and natural mineral adsorbents like zeolite and coal. However, more comparisons should be done based

Table 5. Summary of the prospects of biogas cleaning using adsorbents from biomass.

Activation method	Activating agent	Process conditions	Characteristics of activated carbon produced	CO_2 removal	H_2S removal
Physical activation of biomass	No activation	400-650°C	Low microporous carbon Low surface area Low total volume	Poor	Poor
	CO_2	800-1000°C	Increases the micropore volume High surface area Improved the microporosity	high compared to steam	Poor
	Steam	800-1000°C	Large widening of microporosity Wider pore size distributions Low surface area Low microvolume Develop the mesoporosity	poor	Poor
Chemical activation of biomass	Alkaline	400-800°C	High surface area and pore volume in addition, it creates micropores	Very high	High
Biochar	Unmodified	350-900°C	Low surface area and pore size compare with chemical activation	Poor	Very high
	Modified with amine group		High surface area, creates new pores and improve the width of pores	Very high	High
	Modified with alkaline metal oxide		High surface area and better micropore structure	High	High

on factors such as cost-effectiveness, availability, and environmental sustainability. Hsu and Teng (2000) studied activated carbon from bituminous coal with KOH activating agent. They found that high surface area and porosity are necessary for high capture of carbon dioxide (Hsu & Teng 2000). However, the main disadvantage of coal is that it is not a renewable source and its utilization increases greenhouse emissions in the atmosphere. Furthermore, its emissions can cause health problems. In another study, Alonso- Vicaro et al. (2010) observed that natural clinoptilolite zeolite had 4 mmol/g and 0.04 mmol/g adsorption capacity of carbon dioxide and hydrogen sulfide, respectively (Alonso-Vicario et al. 2010). The impregnated activated carbon (IAC) adsorbents from coal and zeolite 13X have high hydrogen sulfide adsorption capacity of 23 mmol/g and 5 mmol/g, respectively (Sigot et al. 2016). However, considering the cost and availability of natural and activated minerals in the removal of carbon dioxide and hydrogen sulfide, the use of activated biomass adsorbents is preferable. The limitation of zeolite mineral is that it is not abundantly available.

Raw biomass adsorbents have low adsorption capacity of biogas contaminants because of low; surface area, pore size, and mesoporous structure. However, this can be improved by physical or chemical activation. Physical activation improves the creation of micropore, surface area and pore volume. Further enhancement of the surface area, micropore and pore volume can be done using chemical activating agents. The use of biomass adsorbents for removal of carbon dioxide and hydrogen sulfide from biogas is effective and sustainable. Biomass materials, which are the substrates, are cost-effective, and are universally abundant. In addition, the process of adsorbent making and application is an easy operation, with low initial investment and maintenance costs. Furthermore, the production of adsorbents from biomass reduces environmental waste.

The usage of activated biomass materials for purification and upgrade of biogas can help diversify the applications of biogas. It is possible to use upgraded biogas or biomethane to power engines as a replacement of fossil fuels. Therefore, these biomass adsorbents can help in reducing the greenhouse gas emissions significantly as envisaged in Kyoto protocol agreement, which encourages usage of renewable energy instead of fossil fuels. Table 5 indicates the prospects of using biomass adsorbents in the upgrade and purification of biogas. From Table 5, chemical activation of biomass adsorbents with alkaline agents can produce high uptake capacity of biogas contaminants. It is also possible to use unmodified biochar to capture hydrogen sulfide because of its high uptakes.

7 CONCLUSIONS

Biomass derived adsorbents are cost-effective, sustainable, and easily available for purification and upgrading of biogas. Raw biomass materials have low adsorption capacity for carbon dioxide uptake. However, it can be enhanced by physical and chemical activation. The adsorption capacity depends on the surface area and pore volume especially on the micropore. Therefore, chemical activation is more effective than physical activation. Chemical activation increases the surface area, micropore and pore volume. Although in physical activation, carbon dioxide agent is better than steam activation, activated carbon derived from biomass with KOH chemical agent enhances the mesoporosity which increases the carbon dioxide adsorption capacity. The adsorbent was able to achieve uptake of 8 mmol/g of carbon dioxide. The adsorption capacity of carbon dioxide is high at 0°C than at higher temperatures. The low flow rate of the gas through the adsorption bed, such as 30 mL/min, results in greater adsorption capacity compared to higher gas flow rates. The use of high pressure of adsorption can significantly increase the adsorption capacity. Activated biomass adsorbent activated with KOH at 30 bars was able to achieve 14mmol/g carbon dioxide uptake. Raw biochar has high adsorption capacity of hydrogen sulfide but poor uptake of carbon dioxide. The uptake is determined by the substrate and surface modifications. Biochar adsorbents derived from rice hull can adsorb 11 mmol/g of hydrogen sulfide. Modifications with alkaline chemicals and aniline compounds can greatly enhance the uptake of carbon dioxide. Biochar derived from sawdust and activated with MED achieved 10.7 mmol/g of adsorption capacity of carbon dioxide. Hydrogen sulfide adsorption capacity depends on the temperature and pH of the adsorption process. The use of low temperature and pressure processes favors the adsorption capacity of biogas contaminants with biochar. However, there are limited studies on the cost and the methane loss of biomass adsorbents, this should be part of future investigations.

ACKNOWLEDGMENTS

EM is grateful to the World Bank and the Inter-University Council of East Africa (IUCEA) for the scholarship awarded to her through the Africa Center of Excellence II in Phytochemicals, Textiles, and Renewable Energy (ACE II PTRE) at Moi University, Kenya.

REFERENCES

Ahmed, M. B., Johir, M. A. H., Zhou, J. L., Ngo, H. H., Nghiem, L. D., Richardson, C., & Bryant, M. R. (2019). Activated carbon preparation from biomass feedstock: Clean production and carbon dioxide adsorption. J Clean Prod. 225, 405–413.

Al Mamun, M. R., & Torii, S. (2015). Removal of hydrogen sulfide (H_2S) from biogas using zero-valent iron. J. Clean Energy Technol. 3(6), 428–432.

Alonso-Vicario, A., Ochoa-Gómez, J. R., Gil-Río, S., Gómez-Jiménez-Aberasturi, O., Ramírez-López, C. A., Torrecilla-Soria, J., & Domínguez, A. (2010). Purification and upgrading of biogas by pressure swing adsorption on synthetic and natural zeolites. Micropor Mesopor Mater. 134(1–3), 100–107.

Asadullah, M., Jahan, I., & Boshir, M. (2014). Preparation of microporous activated carbon and its modification for arsenic removal from water. J. Indu. Eng. Chem. 20(3), 887–896.

Bates, M. N., Garrett, N., & Shoemack, P. (2002). Investigation of Health Effects of Hydrogen Sulfide from a Geothermal Source. Archiv. Environ. Health. 57(5), 405–411.

Bhatnagar, A., Sillanpää, M., & Witek-krowiak, A. (2015). Agricultural waste peels as versatile biomass for water purification – A review. Chem. Eng. J. 270, 244–271.

Boonpoke, A., Chiarakorn, S., Laosiripojana, N., Towprayoon, S., & Chidthaisong, A. (2011). Synthesis of Activated Carbon and MCM-41 from Bagasse and Rice Husk and their Carbon Dioxide Adsorption Capacity. J Sustain Energy Environ. 2(2), 77–81.

Chen, C., Park, D. W., & Ahn, W. S. (2014). CO_2 capture using zeolite 13X prepared from bentonite. Appl. Surf. Sci. 292, 63–67.

Creamer, A. E., Gao, B., & Wang, S. (2016). Carbon dioxide capture using various metal oxyhydroxide-biochar composites. Chem. Eng. J. 283, 826–832.

Creamer, A. E., Gao, B., & Zhang, M. (2014). Carbon dioxide capture using biochar produced from sugarcane bagasse and hickory wood. Chem. Eng. J. 11964, 105–132.

Deng, S., Wei, H., Chen, T., Wang, B., Huang, J., & Yu, G. (2014). Superior CO_2 adsorption on pine nut shell-derived activated carbons and the effective micropores at different temperatures. Chem. Eng. J. 12097, 115–153.

Dias, J. M., Alvim-ferraz, M. C. M., Almeida, M. F., & Sa, M. (2007). Waste materials for activated carbon preparation and its use in aqueous-phase treatment: A review. J. Environ. Manage. 85, 833–846. 85.

Ding, S., & Liu, Y. (2020). Adsorption of CO_2 from flue gas by novel seaweed based KOH activated porous biochars. Fuel 260 (116382).

Dobele, G., Dizhbite, T., Gil, M. V, Volperts, A., & Centeno, T. A. (2012). Production of nanoporous carbons from wood processing wastes and their use in supercapacitors and CO_2 capture. Biom. Bioenergy. 46, 145–154.

El-Hendawy, A. N. A., Samra, S. E., & Girgis, B. S. (2001). Adsorption characteristics of activated carbons obtained from corncobs. Colloid Sur A: Physicochem. Eng. Aspects. 180(3), 209–221.

Ello, A. S., de Souza, L. K., Trokourey, A., & Jaroniec, M. (2013). Coconut shell-based microporous carbons for CO_2 capture. Micropor Mesopor Mater. 180, 280–283.

Ferella, F., Puca, A., Taglieri, G., Rossi, L., & Gallucci, K. (2017). Separation of carbon dioxide for biogas upgrading to biomethane. J. Clean Prod. (10039).

Sigot, L., Obis, M. F., Benbelkacem, H., Germain, P., & Ducom, G. (2016). Comparing the performance of a 13X zeolite and an impregnated activated carbon for H2S removal from biogas to fuel an SOFC: Influence of water. Int. J. Hydrogen Energy. 41(41), 18533–18541.

Girgis, B. S., Yunis, S. S., & Soliman, A. M. (2002). Characteristics of activated carbon from pea nut hulls in relation to conditions of preparation. Mater. Lett. 57(1), 164–172.

Guo, J., Luo, Y., Lua, A. C., Chi, R. A., Chen, Y. L., Bao, X. T., & Xiang, S. X. (2007). Adsorption of hydrogen sulphide (H_2S) by activated carbons de rived from oil-palm shell. Carbon 45(2), 330–336.

Heidari, A., Younesi, H., Rashidi, A., & Asghar, A. (2014). Evaluation of CO_2 adsorption with eucalyptus wood based activated carbon modified by ammonia solution through heat treatment. Chem. Eng. J. 254, 503–513.

Hernandez, J. R., Capareda, S. C., & Ph, D. (2007). Activated Carbon Production from Pyrolysis and Steam Activation of Cotton Gin Trash. ASAE Annual Meeting (p. 1). Amer. Society. Agri. Biol. Eng. 0300(07).

Hsu, L. Y., & Teng, H. (2000). Influence of different chemical reagents on the preparation of activated carbons from bituminous coal. Fuel Process. Tecnol. 64(1–3), 155–166.

Huang, C., Chen, C., & Chu, S. (2006). Effect of moisture on H_2S adsorption by copper impregnated activated carbon. J. Hazard. Mater. 136, 866–873.

Igalavithana, A. D., Choi, S. W., Shang, J., Hanif, A., Dissanayake, P. D., Tsang, D. C. W., Kwon, J.-H., Lee, K. B., & Ok, Y. S. (2020). Carbon dioxide capture in biochar produced from pine sawdust and paper mill sludge: Effect of porous structure and surface chemistry. Sci. Tot. Environ. 139845.

Jung, S., Park, Y., & Kwon, E. E. (2019). Strategic use of biochar for CO_2 capture and sequestration. J. CO_2 Utilizatio 32, 128–139.

Kalyani, P., & Anitha, A. (2013). Biomass carbon & its prospects in electrochemical energy systems. Int. J. Hydrogen Energy 38(10), 4034–4045.

Kanjanarong, J., Giri, B. S., Jaisi, D. P., Oliveira, F. R., Boonsawang, P., Chaiprapat, S., Singh, R. S., Balakrishna, A., & Khanal, S. K. (2017). Removal of hydrogen sulfide generated during anaerobic treatment of sulfate-laden wastewater using biochar: Evaluation of efficiency and mechanisms. Bioresour. Technol. 234, 115–121.

Khan, I. U., Othman, M. H. D., Hashim, H., Matsuura, T., Ismail, A. F., Rezaei-DashtArzhandi, M., & Azelee, I. W. (2017). Biogas as a renewable energy fuel – A review of biogas upgrading , utilisation and storage. Energy Conver. Manage. 150, 277–294.

Lahijani, P., Mohammadi, M., & Mohamed, A. R. (2018). Metal incorporated biochar as a potential adsorbent for high capacity CO_2 capture at ambient condition. J. CO_2 Utilization 26, 281–293.

Li, D., Ma, T., Zhang, R., Tian, Y., & Qiao, Y. (2015). Preparation of porous carbons with high low-pressure CO_2 uptake by KOH activation of rice husk char. Fuel. 139, 68–70.

Li, Y., Li, D., Rao, Y., Zhao, X., & Wu, M. (2016). Superior CO_2, CH_4, and H_2 uptakes over ultrahigh-surface-area carbon spheres prepared from sustainable biomass-derived char by CO_2 activation. Carbon 105, 454–462.

Locke, D. C. (2001). Sewage Sludge-Derived Materials as Efficient Adsorbents for Removal of Hydrogen Sulfide. Environ. Sci. Technol. 1537–1543.

Madzaki, H., Karimghani, W. A. W. A. B., Nurzalikharebitanim, & Azilbaharialias. (2016). Carbon Dioxide Adsorption on Sawdust Biochar. Procedia Eng. 148, 718–725.

Magomnang, A. S. M., Villanueva, P. E. P., & Ph, D. (2014). Removal of Hydrogen Sulfide from Biogas Using a Fixed Bed of Regenerated Steel Wool. Int. Conf. Agri. Biol. Environ. Sci. 14–17.

Méndez, A., Paz-Ferreiro, J., Araujo, F., & Gascó, G. (2014). Biochar from pyrolysis of deinking paper sludge and its use in the treatment of a nickel polluted soil. J. Anal. Appl. Pyrol. 107, 46–52.

Munusamy, K., Sethia, G., Patil, D. V, Rallapalli, P. B. S., Somani, R. S., & Bajaj, H. C. (2012). Sorption of carbon dioxide , methane , nitrogen and carbon monoxide on MIL-101 (Cr): Volumetric measurements and dynamic adsorption studies. Chem. Eng. J. 196, 359–368.

Murphy, J. D., & Thamsiriroj, T. (2013). Fundamental science and engineering of the anaerobic digestion process for biogas production. In Th biogas handbook (pp. 104-130). Woodhead Pulishing.

Na, B., Lee, H., Koo, K., & Song, H. K. (2002). Effect of Rinse and Recycle Methods on the Pressure Swing Adsorption Process To Recover CO_2 from Power Plant Flue Gas Using Activated Carbon. Ind. Eng. Chem. Res. 5498–5503.

Nor, N. M., Campus, P. P., Lau, L. C., Tunku, U., Rahman, A., Lee, K. T., & Mohamed, A. R. (2013). Synthesis of activated carbon from lignocellulosic biomass and its applications in air pollution control - A review. J. Environ. Chem. Eng. 1(4), 658–666.

Nowicki, P., Skibiszewska, P., & Pietrzak, R. (2014). Hydrogen sulphide removal on carbonaceous adsorbents prepared from coffee industry waste materials. Chem. Eng. J. 248, 208–215.

Ortiz, F. J. G., Aguilera, P. G., & Ollero, P. (2014). Biogas desulfurization by adsorption on thermally treated. Sep. Purif. Technol. 123, 200–213.

Pallarés, J., González-cencerrado, A., & Arauzo, I. (2018). Biomass and Bioenergy Production and characterization of activated carbon from barley straw by physical activation with carbon dioxide and steam. Biomass Bioenergy. 115, 64–73.

Paolini, V., Petracchini, F., Guerriero, E., Bencini, A., & Drigo, S. (2016). Biogas cleaning and upgrading with natural zeolites from tuffs. Environ. Technol. 37(11), 1418–1427.

Ping Zhang, J., Sun, Y., Woo, M. W., Zhang, L., & Xu, K. Z. (2016). Preparation of steam activated carbon from black liquor by flue gas precipitation and its performance in hydrogen sulfide removal: Experimental and simulation works. J. Taiwan Inst. Chem. Eng. 59, 395–404.

Porpatham, E., Ramesh, A., & Nagalingam, B. (2008). Investigation on the effect of concentration of methane in biogas when used as a fuel for a spark ignition engine. Fuel 87, 1651–1659.

Presser, V., McDonough, J., Yeon, S. H., & Gogotsi, Y. (2011). Effect of pore size on carbon dioxide sorption by carbide derived carbon. Energy Environ. Sci. 4(8), 3059–3066.

Promraksa, A., & Rakmak, N. (2020). Biochar production from palm oil mill residues and application of the biochar to adsorb carbon dioxide. Heliyon 6(5).

Qambrani, N. A., Rahman, M. M., Won, S., Shim, S., & Ra, C. (2017). Biochar properties and eco-friendly applications for climate change mitigation, waste management, and wastewater treatment: A review. Renew. Sustain Energy Rev. 79, 255–273.

Qian, K., Kumar, A., Zhang, H., Bellmer, D., & Huhnke, R. (2015). Recent advances in utilization of biochar. Renew. Sustain. Energy Rev. 42, 1055–1064.

Rashidi, N. A., Yusup, S., & Hameed, B. H. (2013). Kinetic studies on carbon dioxide capture using lignocellulosic based activated carbon. Energy 61, 440–446.

Razbani, O., Mirzamohammad, N., & Assadi, M. (2011). Literature review and road map for using biogas in internal combustion engines. Third Int. Conf. Appl. Energy. 10.

Rodriguez-reinoso, F., Prauchner, M. J., & Rodríguez-reinoso, F. (2008). Preparation of Granular Activated Carbons for Adsorption of Natural Gas Microporous and Mesoporous Materials Chemical versus physical activation of coconut shell: A comparative study. Micropor Mesopor Material. 152, 163–171.

Saifullah, Dahlawi, S., Naeem, A., Rengel, Z., & Naidu, R. (2018). Biochar application for the remediation of salt-affected soils: Challenges and opportunities. Sci. Tot. Environ. 625, 320–335.

Scarlat, N., Dallemand, J., & Fahl, F. (2018). Biogas: Developments and perspectives in Europe. Renew. Energy. 129, 457–472.

Schröder, E., Thomauske, K., Weber, C., Hornung, A., & Tumiatti, V. (2007). Experiments on the generation of activated carbon from biomass. J. Anal. Appl. Pyrol. 79(1-2), 106-111.
Serafin, J., Baca, M., Biegun, M., Mijowska, E., Kaleńczuk, R. J., Sreńscek-Nazzal, J., & Michalkiewicz, B. (2019). Direct conversion of biomass to nanoporous activated biocarbons for high CO_2 adsorption and supercapacitor applications. Appl. Surf. Sci. 497, 143722.
Serafin, J., Narkiewicz, U., Morawski, A. W., Wróbel, R. J., & Michalkiewicz, B. (2017). Highly microporous activated carbons from biomass for CO_2 capture and effective micropores at different conditions. J. CO_2 Utilization 187, 73–79.
Sevimoğlu, O., & Tansel, B. (2013). Effect of persistent trace compounds in landfill gas on engine performance during energy recovery: a case study. Waste Manage. 33(1), 74-80.
Shahkarami, S., Azargohar, R., Dalai, A. K., & Soltan, J. (2015). ScienceDirect Breakthrough CO_2 adsorption in bio-based activated carbons. J. Environ. Sci. 34, 68–76.
Shang, G., Liu, L., Chen, P., Shen, G., & Li, Q. (2016). Kinetics and the mass transfer mechanism of hydrogen sulfide removal by biochar derived from rice hull. J. Air Waste Manage. Assoc. 66(5), 439–445.
Shang, G., Shen, G., Liu, L., Chen, Q., & Xu, Z. (2013). Bioresource Technology Kinetics and mechanisms of hydrogen sulfide adsorption by biochars. Bioresour. Technol. 133, 495–499.
Somy, A., Mehrnia, M. R., Amrei, H. D., Ghanizadeh, A., & Safari, M. (2009). Adsorption of carbon dioxide using impregnated activated carbon promoted by Zinc. Int. J. Greenhouse Gas Control 3, 249–254.
Song, M., Jin, B., Xiao, R., Yang, L., Wu, Y., Zhong, Z., & Huang, Y. (2013). The comparison of two activation techniques to prepare activated carbon from corn cob. Biomass Bioenergy. 48, 250–256.
Sun, Q., Li, H., Yan, J., Liu, L., Yu, Z., & Yu, X. (2015). Selection of appropriate biogas upgrading technology-a review of biogas cleaning , upgrading and utilisation. Renew. Sustain. Energy Rev. 51, 521–532.
Sun, Y., Yang, G., Zhang, L., & Sun, Z. (2017). Preparation of high performance H_2S removal biochar by direct fluidized bed carbonization using potato peel waste. Proc. Safety Environ Protection 107, 281–288.
Tsai, W. T., Chang, C. Y., Wang, S. Y., & Chang, C. F. (2001). Cleaner production of carbon adsorbents by utilizing agricultural waste corn cob. Resour. Conserv. Recyc. 32, 43–53.
Vargas, D. P., Giraldo, L., & Moreno-piraján, J. C. (2012). CO_2 adsorption on granular and monolith carbonaceous materials. J. Anal. Appl. Pyrol. 96, 146–152.
Wang, J., & Wang, S. (2019). Preparation, modification and environmental application of biochar: A review. J. Clean Prod. 227, 1002–1022.
Wang, R., Wang, P., Yan, X., Lang, J., Peng, C., & Xue, Q. (2012). Promising Porous Carbon Derived from Celtuce Leaves with Outstanding Supercapacitance and CO_2 Capture Performance. ACS appl. Material Interf. 11 (2012): 5800-5806.
Wei, H., Deng, S., Hu, B., Chen, Z., Wang, B., & Huang, J. (2012). Granular Bamboo-Derived Activated Carbon for High CO_2 Adsorption: The Dominant Role of Narrow Micropores. Chem. Sus. Chem. 5(12), 2354-2360.
Williams, P. T., & Reed, A. R. (2006). Development of activated carbon pore structure via physical and chemical activation of biomass fiber waste. Biomass Bioenergy 30(2), 144–152.
Wong, S., Ngadi, N., Inuwa, I. M., & Hassan, O. (2018). Recent advances in applications of activated carbon from biowaste for wastewater treatment: A short review. J Clean Prod. 175, 361–375.
Xu, X., Cao, X., Zhao, L., & Sun, T. (2014). Comparison of sewage sludge- and pig manure-derived biochars for hydrogen sulfide removal. Chemosphere 111, 296–303.
Xu, X., Kan, Y., Zhao, L., & Cao, X. (2016). Chemical transformation of CO_2 during its capture by waste biomass derived biochars. Environ. Poll. 213, 533–540.
Xu, X., Xu, Z., Gao, B., Zhao, L., Zheng, Y., Huang, J., Tsang, D. C. W., Ok, Y. S., & Cao, X. (2020). New insights into CO_2 sorption on biochar/Fe oxyhydroxide composites: Kinetics, mechanisms, and in situ characterization. Chem. Eng. J. 384, 123289.
Yang, G. X., & Jiang, H. (2014). Amino modification of biochar for enhanced adsorption of copper ions from synthetic wastewater. Water Res. 48(1), 396–405.
Yang, T., & Lua, A. C. (2003). Characteristics of activated carbons prepared from pistachio-nut shells by physical activation. J. Colloid Interface Sci. 267, 408–417.
Yaumi, A. L., Bakar, M. Z. A., & Hameed, B. H. (2018). Melamine-nitrogenated mesoporous activated carbon derived from rice husk for carbon dioxide adsorption in fixed-bed. Energy J. 155, 46–55.
Zabaniotou, A., Stavropoulos, G., & Skoulou, V. (2008). Activated carbon from olive kernels in a two-stage process: Industrial improvement. Bioresour. Technol. 99, 320–326.
Zhang, X., Zhang, S., Yang, H., Shao, J., Chen, Y., Feng, Y., Wang, X., & Chen, H. (2015). Effects of hydrofluoric acid pre-deashing of rice husk on physicochemical properties and CO_2 adsorption performance of nitrogen-enriched biochar. Energy J. 91, 903–910.
Zhu, X., Wang, P., Peng, C., Yang, J., & Yan, X. (2014). Activated carbon produced from paulownia sawdust for high-performance CO_2 sorbents. Chin. Chem. Lett. 25, 929–932.
Zubbri, N. A., Mohamed, A. R., Kamiuchi, N., & Mohammadi, M. (2020). Enhancement of CO_2 adsorption on biochar sorbent modified by metal incorporation. Environ. Sci. Poll. Res. 27(11), 11809–11829.

Advances in Phytochemistry, Textile and Renewable Energy Research for Industrial Growth – Nzila et al. (Eds)

Evaluation of sugarcane vinasse and maize stalks for anaerobic digestion

Mohamed Kibet Kiplagat, Cleophas Achisa Mecha & Zachary Otara Siagi
Department of Mechanical & Production Engineering, Moi University, Eldoret, Kenya

ABSTRACT: Sugar cane vinasse and maize stalks waste disposal poses serious challenges to the environment. However, anaerobic digestion is an attractive treatment method for these wastes. The current study therefore established the suitability of theses wastes for anaerobic digestion. Specifically, the study aimed at characterizing the substrates and anaerobic digestion of the substrates. The raw materials were collected from Muhoroni Sugar Company and Uasin-Gishu county maize farms, respectively. The pH, moisture content, chemical oxygen demand (COD), total solids (TS), total suspended solids (TSS), total dissolved solids (TDS), total organic carbon (TOC), nitrogen content, and CN ratio were determined based on standard methods. Experiment setups were run depending on batch experiment. The study established that the pH, moisture content, COD,TS, TSS, TDS, TOC, and nitrogen content for vinasse were 4.34, 93.91%, 71.28g/L, 7.05%, 6.04%, 1.01%, 2.23 g/L, and 0.27g/L, respectively. For the maize stalks, the pH, moisture content, TS, TSS, TDS, TOC, and nitrogen content were 7.52, 9.52%, 91.50%, 90.12%, 7.38%, 49.51g/kg, and 1.28g/kg, respectively. The cumulative biogas yield for sugarcane vinasse and maize stalks were 329mL and 385 mL, respectively. In conclusion, this study found that the characteristics of vinasse and maize stalks are suitable for biogas production.

1 INTRODUCTION

High oil prices, increasing population, industrialization, and the decline in energy security have all led to an increase in global interest in biofuels. However, studies on biomass availability have concluded that by 2050, the possible contribution of biomass to global energy supply could vary from 100 EJ/year to 400 EJ/year, which represents 21%–85% of the world's current total energy consumption, estimated at 470 EJ.Although biofuels are only a fraction of total biomass, biofuels still have the potential to play a significant role in meeting future global energy demand, if developed through the appropriate channels (Bailis et al. 2015).

Furthermore, the disposal of some biomass waste has proved to be a problem in developing countries. The disposal of vinasse and maize stalks in Kenya poses serious challenges to the environment. Iqbal (2016) stated that vinasse has a dark colour, the concentration of total solid and COD value are very high, and the pH condition of vinasse is 3.25-4.97, while the total solid (TS) value in vinasse is 63,000–79,000 mg/L. If vinasse is discharged directly into the rivers without treatment, water biota will be killed. Dissolved oxygen in the rivers is used by oxidation bacteria to degrade COD and BOD, hence the availability of dissolved oxygen is depleted, so the water biota cannot breath and finally die (Summardiono et al. 2013). The same applies for maize stalks. Here in Uasin-Gishu county maize stalks are abundant since they are left in the farms and even sometimes burned. These disposal methods of maize stalks cause harm to the environment.

Attempts have been made to establish ways to dispose of vinasse with minimum negative effect on the environment. Iqbal (2016) in his review further supported and concluded that it is more effective to degrade organic materials through anaerobic digestion than aerobic treatment. He further stated that anaerobic digestion is a viable option for sugarcane vinasse processing and enables energy recovery as biogas production.Studies have established that maize stalks could be used as a raw material for anaerobic fermentation (Adebayo et al. 2014; Bruni 2010; Zhong et al. 2011). A pretreatment process is required to decompose cellulose to reduce the volume of material and increase production of biogas (Antognoni et al. 2013).

Based on these facts and problems associated with disposal of these wastes, this study evaluated the potential of these wastes (vinasse and maize stalks) as substrates for co-digestion. Co-digestion is the simultaneous digestion of a homogenous mixture of two or more substrates. The most common situation is when a major amount of a main basic substrate (such as manure or sewage sludge) is mixed and digested together with minor amounts of a single, or a variety of additional substrates.

*Corresponding author

DOI 10.1201/9781003221968-39

2 LITREATURE REVIEW

2.1 *Biogas*

High oil prices, increasing population, industrialization, and the decline in energy security have all led to an increase in global interest in biofuels. However, despite this growth, the global market for biofuels is still in its infancy. The future global potential for biofuel production is also difficult to estimate, due to a number of factors including the limits of natural resources and the need for food security above biofuel use. However, studies on biomass availability have concluded that by 2050, the possible contribution of biomass to global energy supply could vary from 100 EJ/year to 400 EJ/year, which represents 21%–85% of the world's current total energy consumption, estimated at 470 EJ.Although biofuels are only a fraction of total biomass, biofuels still have the potential to play a significant role in meeting future global energy demand, if developed through appropriate channels (Bailis et al 2015).

Many developing countries are faced with the dilemma of finding alternative energy sources with reduced environmental impact. Wind energy, hydro, solar, and biofuels energy sources are potential solutions. According to Earley et al. (2015), global ethanol and biodiesel production increased from about 4.8 billion gallons in 2000 to about 21 billion gallons in 2008. The goal of replacing fossil fuels with biofuels has resulted in a high production of ethanol (Gil 2011).

Ethanol is a prominent biofuel because of its advantages; however, there is a challenge in disposing of vinasse, which is the effluent from the distillation columns of ethanol industries (Wilkie et al. 2000). It is said that for each litre of ethanol produced, between 0.8 and 3.0 litres of vinasse are obtained (Asocaña 2011). Vinasse has a dark colour and a low pH (Iqbal S. 2016). The concentration of total solid and COD value are very high. The pH condition of vinasse is 3.25–4.97, while the total solid (TS) value in vinasse is 63,000–79,000 mg/L (Iqbal S. 2016). Budiyonoet al.(2014) reported that vinasse contains COD content of299,250 mg/L. Wilkie et al.(2000) further stated that vinasse is an organic liquid residue comprising about 93% water, 5% organic matter mainly unfermented sugars and other carbohydrates, and about 2%inorganic dissolved solids. Vinasse contains many kinds of organic compounds, such as acetic acids, lactic acids, glycerol, phenols, polyphenols, and melanoidins (Budiyono et al. 2014).

Vinasse is wastewater that is a by-product of distillation from the production of ethanol by fermentation. Besides containing high COD, vinasse has a strongly acidic character (pH 3.67–4.98) (Lutoslawski et al. 2011; Siles et al. 2011). Vinasse characteristics pose serious challenges to the environment in terms of disposal. The strongly acidic pH of vinasse causes the remobilization of heavy metals in soil (Kafle et al. 2012). The dark colour in vinasse is not good for the environment, being unsightly. Besides that, it also can hamper penetration of sunlight into the rivers, so water plants cannot photosynthesise (Iqbal S. 2016). Soluble salts in vinasse can cause soil salinity and sodicity. It can cause poor soil structure and reduce fertility. Vlyssideset al. (1997) stated that high concentration of P and N nutrients cause eutrophication in water bodies. The temperature of fresh vinasse from a distillation unit is 65–105°C. If vinasse is disposed of in water bodies, without cooling, the temperature of water bodies can increase. It can also disturb fish activity (Siles et al. 2011).

Attempts have been made to establish ways to dispose of vinasse with a minimum negative effect on the environment. In Brazil, vinasse is applied directly to the soil because of its organic matter and nutrient content; potassium, nitrogen and phosphorus make it a good organic fertilizer for sugarcane farms (Ferraz et al. 2015).From an economic perspective, the soil application of vinasse represents the simplest and cheapest solution. However, continuous application of vinasse to the soil results in soil and groundwater contamination, leaching and salinization, and seed germination inhibition (Ferraz et al. 2015).Dávila et al. (2009) evaluated the electro-flotation/oxidation process for vinasse, obtaining reductions in COD of 58%. Similarly, Yavuz (2007) achieved a 90% reduction in total organic carbon in vinasse through electro-coagulation with the use of a supporting electrolyte and the gradual addition of hydrogen peroxide. Goncalves (2006) performed research for the treatment of the vinasse by utilizing coagulation and flocculation. The study evaluated several variables influencing COD removal which was used to develop a model. The model demonstrated that the COD removal varied as a function of the pH and mixing. The best results were achieved when calcium oxide and ferrous sulfate were used, with pH values of 12.4 the removal efficiencies were 52% and 44%, respectively. The study established that the resulting sludge could be used as a fertilizer because it was rich in nutrient content. The major challenge in utilizing sludge as organic fertilizer from vinasse is the high pH of 12.4 which causes soil pollution. Tang et al (2007) and Íñiguez-Covarrubias and Peraza-Luna (2007) found that biological treatment such as active sludge is expensive and it produces poison. According to Satyawali et al. (2007), anaerobic treatment is the most attractive primary treatment of vinasse due to the BOD and COD removal being over 80%, and the energy recovery in the form of biogas. Ribas (2006) stated that the anaerobic reactors are a promising alternative because they accomplish a high rate of organic load removal and produce biogas. Iqbal (2016) in his review concluded that it is more effective to degrade organic materials through anaerobic digestion than aerobic treatment. However, the value of COD removal is not maximal. That is caused bythe presence of phenolic compounds in vinasse. He further stated that anaerobic digestion is a viable option for sugarcane vinasse processing and enables energy recovery as biogas production. To further support and minimize these challenges, the current study will digest vinasse.Wilkie et al (2000) also stated that anaerobic treatment can reduce organic matter to biogas that can be used for heating for an evaporation and distillation unit or can be saved for aerobic–anaerobic treatment.

Studies have established that maize stalks could be used as raw material for anaerobic fermentation (Adebayo et al. 2014; Bruni 2010; Zhong et al. 2011). A pretreatment process is required to decompose cellulose to reduce the volume of material and increase the production of biogas (Antognoni et al. 2013). In Uasin-Gishu county maize stalks are abundant since they are left in the farms and even sometimes burned. These practices cause harm to the environment. When the maize stalks are left or burnt they cause air pollution. The current study will also digest maize stalks.

3 MATERIALS AND METHODS

The test substrates consisted of maize stalks and sugarcane vinasse. Maize stalks was collected from the farms in Uasin-Gishu County (Indany Kipsum farm) during harvesting season and dried. The maize stalks samples were milled and dried at ambient conditions to average equilibrium moisture content of 10% (±1.5).

The vinasse was obtained from Muhoroni Sugar Company, ethanol plant, collected in 20 litre plastic containers and transported to the laboratory.

Table 1. Materials, reagents, and apparatus.

Raw materials	Reagents	Equipment/apparatus	
			Model
Maize	Sodium	Weighing balance –	Ohaus-
stalks	hydroxide	Digital	Scout Pro
Sugarcane	Calcium	Milling machine	Alfa
Vinasse	hydroxide	Digital pH meter	Machines
	Hydrochloric	Thermo balance	HANNA
	acid	Beaker	(HI9812)
	Distilled water	Clamps	HANNA
	$(NH_4)_2SO_4$	Measuring	
	Copper sulfate	cylinders	
	Potassium	Thermometer	
	dichromate	Conical Flask	
	solution	Gas deliveryr tubes	
	($K_2Cr_2O_7$)	Water bath	
	Silver	Kjeldahlr apparatus	Beckman
	sulfate	& condenser	DU640
	(Ag_2SO_4)	UV-Vis	AGRIPRO
	$HgSO_4$	Spectrometer	605
	Iron sulfate	Moisture analyser	HANNA
	heptahydrate	Photometer	(HI83300)
	Potassium	COD reactor	HANNA
	sulfate		(HI839800)

3.1 *Determination of pH of the substrates*

The pH of vinasse was measured using a Hanna pH-meter directly. The maize stalks were sun-dried for 15 days, and then milled and sieved to obtain different substrate particle sizes: 0.5 mm, 1 mm, and 2 mm (Elely et al. 2018). 5g of crushed (1mm) maize stalks was placed in a glass test tube, 150 Ml of distilled water was added and stirred, after 3 hours the pH readings were taken. The samples were made in three replicates.

3.2 *Determination moisture content*

The moisture content of the substrates was measured using moisture analysers. The moisture content for vinasse was determine using the MAX50 moisture analyser while for maize stalks the AGRIPRO 6095 moisture meter was used. The procedure was done in three replicates

3.3 *Determination of nitrogen*

Nitrogen content for the substrates was determined based on the persulfate standard method—4,500-N_{org} D (LAMNDA 900). The method determines total nitrogen content by oxidation of all nitrogenous compounds to nitrate. Alkaline oxidation at 100–110°C converts organic and inorganic nitrogen to nitrates. The total nitrogen is determined by analysing the nitrate in the digestate. A standard curve was constructed by plotting the standard sample absorbance due to NO_3against strength/concentration. The table below shows the results for absorbance against strength for the standard samples

3.4 *Determination of carbon content*

The carbon content for the substrates was determined from Kenya Agricultural & Livestock's Research Organization (KALRO), Soil laboratory, Nairobi centre.

3.5 *Determination of C/N ratio*

The C/N ratio was determined by dividing the total organic carbon content by the total nitrogen content, according to Xiaojiaoet al. (2014). The carbon content and the nitrogen content of the respective substrates (VN and MS) were determined based on the standard methods highlighted in 3.3 and 3.4 above. The equation used to determine the ratio was as follows;

$$C/N = \frac{W1C1}{W1 \times N1} \tag{1}$$

$$C/N = \frac{(W1 \times C) + (W2 \times C2)}{(W1 \times N1) + (W2 \times N2)} \tag{2}$$

Where W1, W2, and W3 were the TS weight in a single substrate in the mixture; C1, C2, and C3 were the organic carbon content (g kg^{-1}VS) in each substrate; and N1, N2, and N3 were the nitrogen content (g kg^{-1}VS) in each substrate.

3.6 *Determination of COD*

COD was determined based on closed reflux, Colorimetric standard method-5220D. The standard and substrate samples were treated in standard potassium dichromate solution (K_2Cr_2O) and sulfuric acid reagent. The reduction of dichromate absorbance was measured using a UVs spectrometer. The standard samples were prepared using potassium hydrogen phthalate with COD equivalence.

3.7 *Determination of total solids (TS), total suspended solids (TSS), and total dissolved solids (TDS)*

TS, TSS, and TDS for the substrates was determined based on standard methods (2,540B, 2,540C, and 2,540D) for the examination of water and wastewater by Eaton et al. (1995). Vinasse was well mixed and evaporated in a weighed dish and dried to constant weight in an oven at 103 – 105°C, while maize stalks were dried directly. The decrease in weight over that of the empty dish represents the solids. The relationship between TS, TSS, and TDS is given by the following expression;

$$TS = TSS + TDS \tag{3}$$

3.8 *Biogas production*

3.8.1 *Experiment setup*

Biogas production from sugar cane vinasse (SV) and maize stalks (MS) was analysed in a batch anaerobic test at 37.5°C according to Linke and Schelle (2000). A constant mesophilic temperature of 37.5°C was maintained through a water bath. Fresh cow-dung from Moi university farm was used as inocula in this study. 20mL of inocula was mixed with 40mL of vinasse and 40mL of pretreated maize stalks in separate digestion vessels (250mL conical flasks) and labelled runs 1 and 2, respectively. In each run, three conical flasks we rearranged in a way that the first flask contained substrate; the middle contained water and the last was for collecting water that was expelled out of the second container (Budiyono et al 2010a, 2010b). The cocks of all digesters were sealed tightly using clear silicon glue in order to control the entry of air and loss of biogas as indicated in Figure 1. Shaking of digesters was done manually on a daily basis to ensure contact between the substrate molecules and microbial cells.

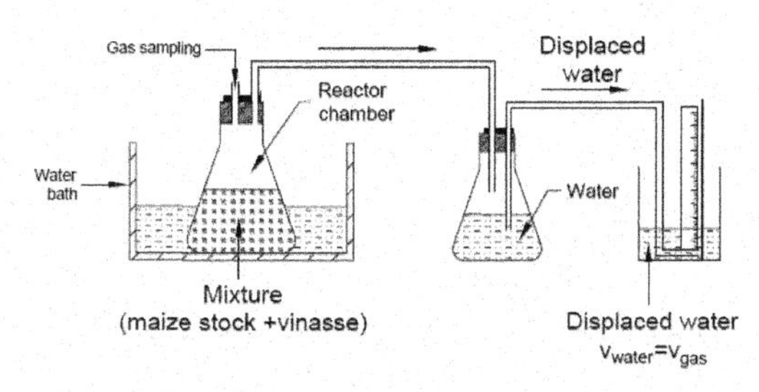

Figure 1. Experiment setup.

4 RESULTS AND DISCUSSION

4.1 *Substrates*

The characteristics of the substrates were analysed before digestion. The pH, moisture %, COD g/L, total solids %, total suspended solids %, total dissolved solids %, carbon content (TOC), nitrogen content, and C/N ration were determined based on standard procedures outlined in the study. The results were recorded in the table below.

Table 2. Composition of various components of maize stalks and vinasse.

Parameters	Maize stalks	Vinasse
pH	7.52	4.34
Moisture %	9.52	93.91
COD g/L	**	71.28
Total solids (%)	91.50	7.05
Total suspended solids %	90.12	6.04
Total dissolved solids %	7.38	1.01
Carbon content (TOC)	49.51g/kg	2.23g/l
Nitrogen content	1.28g/kg	0.27g/l
C/N ratio	38.68	8.25

Key: Number of repetition (n) =3, except for pH and TOC which has n = 1. **Not applicable.

The characteristics of the substrates are suitable for biogas production as supported by the result posted by Kirchgebner(1997) and Mahnert et al. (2002).

4.2 *Biogas production*

The parameter measured was biogas production daily. The tested samples of sugarcane vinasse and maize stalks showed a normal curve of accumulated biogas production. The curves had a steep increase followed by a decrease of biogas production. The maximum biogas yields were obtained in the first 2weeks of digestion experiment (Figures 2 and 3). The two lines represent the two substrates.

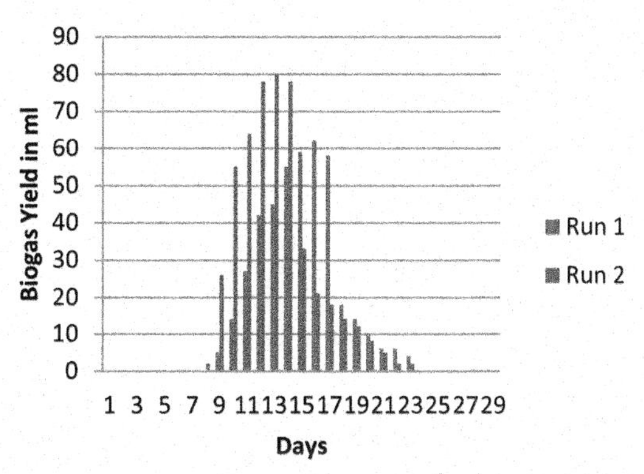

Figure 2. Daily biogas yield (*Run 1, 100% sugarcane vinasse and temperature of* 37.5°C; *Run 2, 100% maize stalks and temperature of* 37.5°C).

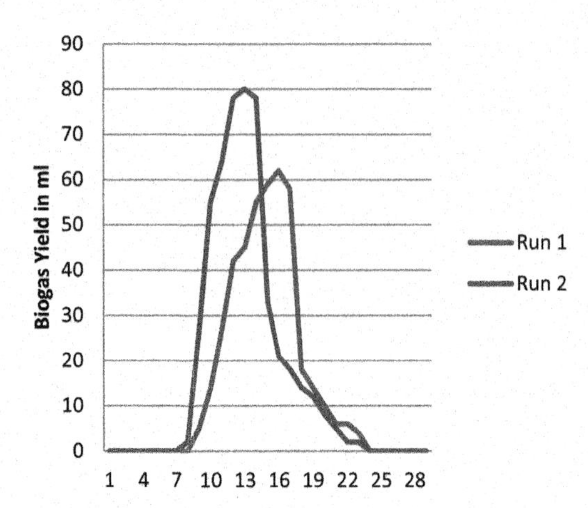

Figure 3. Daily biogas yield (*Run 1, 100% sugarcane vinasse and temperature of 37.5° C; Run 2, 100% maize stalks and temperature of* 37.5°C).

A maximum production of biogas was produced by 100% maize stalks (Run 2) followed by 100% sugarcane vinasse (Run 1). The cumulative biogas yields for sugarcane vinasse and maize stalks were 8.225mL/g and 38.5mL/g respectively. Some researchers have also worked on various residues for the production of biogas. Anunputtikul and Rodtong (2007) reported a biogas yield of $0.36m^3/kg$ from 1.00% (w/v) TS when a single stage digester of 5L is used. Similar observations have been recorded by Somayaji and Khanna (1994). Reports have also showed that retention time for biogas production depends on the type of substrate (Ezekoye et al. 2006) Similarly, in this study biogas production from Run 1 (100% sugarcane vinasse) and Run 2 (100% maize stalks)was different, as shown in Figure 2. This could also depend on the amount and growth phase of the added cow-dung (20 mL inocula) that might create a prolonged lag phase of the methanogenic bacteria. According, to Wilkie (2008) the quality and quantity of inocula are critical to the performance, time required, and stability of bio-methanogenesis for the commencement of the anaerobic digester.

5 CONCLUSION AND RECOMMENDATION

The biogas potential of sugarcane vinasse and maize stalks was investigated by batch experiment under mesophilic conditions (37.5°C). It can be concluded that the characteristics of the test substrates (sugarcane vinasse and maize stalks) are suitable for biogas production. These characteristics made favourable conditions for the multiplication of bacteria. The results for anaerobic digestion strongly support the potential of these wastes to produce biogas.

The study recommends further research to be done to determine the optimum conditions for maximum yield (pH, temperature, quality and quantity of inocula used), co-digestion of these wastes, and to analyse the methane content of the biogas.

ACKNOWLEDGMENTS

The authors are grateful to the ACE II centre and Moi University for financial support through the award of research scholarship to carry out this work at the Engineering School-Moi University, Kenya.

REFERENCES

Adebayo A. O., Jekayinfa S. O., & Linke B., 2014. Anaerobic co-digestion of cattle slurry with maize stalks at mesophilic temperature. American Journal of Engineering Research.

Al Seadi, T.; Rutz, D.; Prassl, H.; Köttner, M.;& Finsterwalder, T. (2008) *Biogas Handbook;* University of Southern Denmark: Funen, Denmark.

Anthony M., 2006. Effect of particle size on biogas yield from sisal fibre waste. Journal of Renewable Energy 31: 23 85–2392.

Antognoni S., (2013). Potential effects of mechanical pretreatments on methane yield from solid waste anaerobically digested. Science and Education Publishing.

Anunputtikul W.,& Rodtong S. (2004). Laboratory scale experiment for biogas production from cassava tubers. The joint international conference on sustainable energy and environmental (SEE), Hua Hin, Thailand.

Bailis R., Drigo R., Ghilardi A., & Masera O. 2015. The carbon footprint of traditional woodfuels. *Nat. Clim. Change*

Bruna S. M, Marcelo Z, & Antonio B, 2015. Anaerobic digestion of vinasse from sugarcane ethanol production in Brazil: Challenges and perspectives, Elsevier Journal.

Bruni E., 2010. Improved anaerobic digestion of energy crops and agricultural residues, Ph.D. thesis, Technical University of Denmark.

Budiyono, Syaichurrozi, I. & Sumardiono, S. 2013. Biogas production from bioethanol waste: the effect of pH and urea addition to biogas production rate. Waste Technology.

Budiyono, Widiasa I. N., Johari S.,& Sunarso. 2010b. The Kinetic of Biogas Production rate from Cattle Manure in Batch Mode. International Journal of Chemical and Biological Engineering.

Budiyono, Widiasa I. N., Johari S., & Sunarso.2010a. Increasing Biogas Production Rate from Cattle Manure Using Rumen Fluid as Inoculums. International Journal of Basic & Applied Sciences.

Earley, J.H., Bourne, R.A., Watson, M.J., & Poliakoff M. (2015) Continuous catalytic upgrading of ethanol to n-butanol and >C_4 products over Cu/CeO_2 catalyst in supercritical CO_2.

Eaton, D.W., Clesceri, L.S., Greenberg, A.E. & Franson, M.A.H. (1995) Standard methods for examination of water and wastewater, American public Health Association, Washington DC.

Ezekoye VA, Okeke CE (2006) Design, construction and performance evaluation of plastic bio-digester and the storage of biogas. The Pacific J Sci Technol.

Fachagentur, N. R. (2010). *Guide to Biogas. From Production to Use;* Gülzow, Brazil, p. 24.

Girmaye, K., & Ebsa, K. 2019. Optimization of Biogas Production from Avocado Fruit Peel Wastes Co-digestion with Animal Manure Collected from Juice Vending House in Gimbi Town, Ethiopia. Fermentation Technology.

Kayhanian, M.; & Rich, D. (1995).Pilot-scale high solids thermophilic anaerobic digestion of municipal solid waste with an emphasis on nutrient requirements. *Biomass Bioenergy.*

Lauterböck, B.; Nikolausz, M.; Lv, Z.; Baumgartner, M.; Liebhard, G.; & Fuchs, W.(2014) Improvement of anaerobic digestion performance by continuous nitrogen removal with a membrane contactor treating a substrate rich in ammonia and sulfide. *Bioresour. Technol.*

Lv, Z.; Hu, M.; Harms, H.; Richnow, H.H.; Liebetrau, J.; & Nikolausz, M. (2014) Stable isotope composition of biogas allows early warning of complete process failure as a result of ammonia inhibition in anaerobic digesters. *Bioresour. Technol.*

Mshandete A. L., Bjornsson A. K., & Kivaisi M. S. T., 2006. Effect of particle size on biogas yield from sisal fibre waste. Journal of Renew Energy.

Siles, J.A., I. García-García, A. Martín, & M.A. Martín. 2011. Integrated ozonation and biomethanization treatments of vinasse derived from ethanol manufacturing. Journal of Hazardous Materials.

Somayaji, D., & Khanna, S. (1994) Biomethanation of rice and wheat straws. World J microbial Biotechnol.

Tang Y-Q, Fujimura Y, Shigematsu T, Morimura S, & Kida K. 2007. Anaerobic treatment performance and microbial population of thermophilic up flow anaerobic filter reactor treating awamori distillery wastewater.

Wang X, Lu X, Li F, & Yang G. 2014 Effects of Temperature and Carbon-Nitrogen (C/N) ratio on the performance of Anaerobic Co-Digestion of dairy manure, Chicken manure and rice straws.

Wilkie AC (2008) Biomethane from biomass, Biowaste and Biofuel, Bioenergy. Washington DC.

Wilkie AC, Riedesel KJ, & Owens JM. 2000. Stillage characterization and anaerobic treatment of ethanol stillage from conventional and cellulosic feedstocks. Biomass Bioenergy.

Yavuz, Y., 2007. EC and EF processes for the treatment of alcohol distillery wastewater. Sep. Purif. Technol. 53, 135–140.

Zhong, W., Zhang, Z., Luo, Y., Sun, S., Qiao, W., Xiao, & M., 2011. Effect of biological pretreatments in enhancing corn straw biogas production. Bioresour. Technol.

Advances in Phytochemistry, Textile and Renewable Energy Research for Industrial Growth – Nzila et al. (Eds)

Redox potential advances of quinone derivatives for energy storage applications

C.K. Kosgei & H. Kirimi

ABSTRACT: Natural quinone's electron transfer role is an important aspect in a number of areas like biochemistry, medicine, and electrochemical redox reactions for energy storage applications. This electroactive nature of quinones has placed them as of interest for energy storage and energy harvesting applications. Recent rechargeable energy storage systems which have been advanced are redox flow batteries (RFB), pseudocapacitors, and Li-ion batteries made up of reversible quinone redox couples. Quinone and its derivatives are preferred as redox active compounds used to fabricate rechargeable batteries due to their relative high energy density, fast charging rate, solubility in electrolytes, abundance, and cyclic stability. This review paper summarizes quinone's molecular structure, its electrochemical behavior, quinone redox predictions, and the strides made in predicting its redox potential computationally. These recent advances in the functionalization of quinone hybrid materials based on their redox properties applications can provide solutions to the engineering of bio-inspired energy storage systems such as rechargeable batteries.

Keywords: Quinone derivatives, rechargeable batteries, quinone, electron-withdrawing functional groups, electron-donating functional groups

1 INTRODUCTION

There is pressure to focus on renewable energy sources like wind and photovoltaics. The mismatch between this intermittent supply and demand has increased the need for rechargeable batteries. Storage of this energy with a growing need for the development of renewable energy sources is inevitable. A major challenge for exploratory research is to design electrical energy storage (EES) batteries for both mobile and stationary applications. Lithiumbased rechargeable battery anode materials has resulted in an advanced battery technology with high theoretical capacity. However, these materials are expensive, limited, and are a threat to the environment.

Here we discuss a category of energy storage electrode materials that exhibits favorable electrochemical and chemical properties of a class of molecules called quinones (Hukinston et al., 2014). These molecules are organic carbonyl compounds whose role is to transport electrons in biological processes like respiration and photosynthesis. In respiration they transfer electrons from enzyme complex I to enzyme complex II, while in photosynthesis quinones' role is to transfer electrons from photosystem II to photo-system I. Due to these energy conversion involvements and their stable redox chemistry quinones have been valued in electrical energy storage buildup. Therefore, it makes it necessary to investigate further the factors regulating the reaction pathways and the potentials of different species of quinone–hydroquinone biological systems for energy storage applications. Electron spin resonance (Dan & Neta, 1975) (Meisei & Czapski, 1975; Meisei & Fessenden, 1976), Pulse radiolysis techniques and electrochemical methods like cyclic voltammetry, polarography, square wave voltammetry, DFT, and so forth, have been employed in the investigation of the redox behavior of varied quinone systems.[1]

[1] (Rao, Lown, & Plambeck, 1978) (S.I & M.I, 1985) (Molinier-Jumel, Malfoy, Reynaud, & Aubel-Sadron, 1978) (Baldwin, Packett, & Woodcock, 1981) (Chaney Jr & Baldwin, 1982) (Kano, Konse, Nishimura, & Kubota, 1984) (Guin, Das, & Mandal, 2008) (Mairanovskii, 1968) (Tachi, 1954) (Hahn & Lee, 2004) (El-Hady, Abde-Hamid, Seliem, & E-Maali, 2004) (Zhang, Wu, & Hu, 2002) (Hu & Li, 1999) (Kertesz, Chambers, & Mullenix, 1999) (Yi & Gratzl, 1998) (Gill & Stonehill, 1952) (Furman & Stone, 1948) (Guin, Das, & Mandal, 2010) (Anson & Epstein, 1968) (O'Kelly & Forester, 1998) (Berg, 1961) (Shiu & Song, 1996) (Kahlert, 2008) (Shi & Shiu, 2004) (Bechtold, Gutmann, Burtscher, & Bobleter, 1997) (Soriaga & Hubbard, 1982) (He, Crooks, & Faulkner, 1990) (Turner & Elying, 1965) (Koshy, Sawayambunathan, & Periasamy, 1980) (Soriaga, White, & Hubbard, Orientation of aromatic compounds adsorbed on platinum electrodes. The effect of temperature, 1983) (Soriaga, Stickney, & Hubrard, Electrochemical oxidation of aromatic compounds adsorbed on platinum electrodes. The influence of molecular orientation, 1983) (Hubbard, Stickney, & Soriaga, 1984) (Soriaga, Binamira-Soriaga, Benziger,

DOI 10.1201/9781003221968-40

This review paper will cover areas of research on the redox behavior of quinones from a simulation point of view. The aim of this paper is to provide an indepth understanding of specific aspects of the reduction of quinones on different electrodes using different electrolytes and solvents by using different computational approaches. We will make an introduction on the chemistry of quinones, then the core background for arriving at redox potentials, followed by the description of the thermodynamic cycle in varied redox potential prediction approaches. This will focus on how these computational research results are applied to redox potentials predictions of quinones derivatives for energy storage applications.

2 THEORY OF QUINONES

Quinone comprises organic carbonyl redox-active compounds. In nature they play an important role in a number of electrochemical reactions for energy storage and transduction, which includes photosynthesis and respiration processes. For example, in the chloroplasts of green plants the electron transfer between secondary and primary quinones triggers the conversion of solar energy to chemical energy. Also respiration involves the redox-active quinones where the transfer of protons and electrons results in the production of energetic molecules like ATP. This process has been utilized to develop energy storage batteries like redox flow batteries.

An intense research attempt is ongoing to replace inorganic battery couples like V, Cr, or Fe, which are expensive and toxic (Ding, Zhang, Li, Liu, & Xing, 2013; Dunn, Kamath, & Tarascon, 2011), with quinone-based electrode materials, e.g., anthraquinone (AQ), and benzoquinone (BQ) (Huskinsom et al., 2014; Lin et al., 2015; ;erry & Weber, 2016; Yang et al., 2014). Furthermore, a number of quinones are being applied to develop new lithium battery types through simulation and computational studies (Ding & Yu, 2016; Kim, Liu, Lee, & Jang, 2016; Ma, Zhao, Wang, & Pan, 2016; Nokami et al., 2012; Park et al., 2015). Recent research studies has proved quinone's potential to be well-utilized in energy applications and its hybridization with a number of inorganic and organic materials (Hu et al., 2016; Oyaizu et al., 2012; Schmidt, Hager, & Schubert, 2016; Wang et al., 2016; ;Yang, Wang, & Guo, 2016). Among the methods discovered to hybridize quinone with these nanomaterials is the use of polydopamine (PDA), which is quinone-rich and mimics mussel adhesive proteins containing 3,4-dihydroxylphenylalanine (DOPA). This is achieved through self-polymerization of dopamine molecules in a slightly alkaline environment (Lee et al., 2007). The redox reaction property of quinone ligands in PDA has been known to be maintained after polymerization (Yang et al., 2015; Yu et al., 2010), thus recently, materials coated with PDA have been applied in various energy systems, such as fuel cells, artificial photosynthetic systems and Li-ion batteries (Liu, Ai, & Lu, 2014).

2.1 *Structural pattern of quinones*

Quinones represent a class of compounds known to be extensively distributed in nature. In excess of 1200 quinones have been documented (Dey P & Harborne, 1989). The three basic quinone structural patterns are benzoquinone (BQs), napthoquinone (NQs), and anthraquinones (AQs), and they have classes of 1-, 2-, and 3-ring quinone isomers respectively. They are enumerated as per the position of ketone groups on pure quinones. These pure basic quinones have 2, 6, and 9 different classes for BQs, NQs, and AQs, respectively (Figure 1; shown in the left white column).

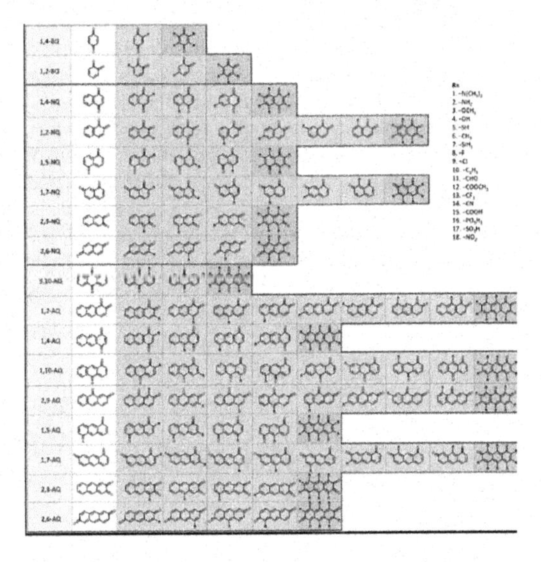

Figure 1. A schematic representation of the molecular screening library. The parent BQ, NQ, and AQ isomers are shown on left (white). These quinone isomers are functionalized with 18 different R-groups singly (gray) and fully (green) to generate a total of 1710 quinone molecules (Er et al. 2015).

In an attempt to design high quinone-based electrode materials and lower the cost of electrical energy, a virtual library of molecules has been proposed computationally by substituting the main structures of quinones and branding quinone cores with interesting chemical substituents. Among these R-groups substituents, as listed in Figure 1 are: $-NH_2$,$-N(CH_3)_2$, $-OCH_3$,$-CH_3$, -OH, -SH , $-SiH_3$, -CL, -F, $-C_2H_3$, -CHO, $-CF_3$, $-COOCH_3$-CN, $-PO_3H_2$, $-NO_2$, and -SO_3H (Figure 1) (Er et al., 2015). The position of these substituents has been found to significantly alter the electrochemistry of quinones. (;Ajloo, Yoonesi, & Soleymanpour, 2010; Guin, Das, & Mandal, 2011; Manisankar & Valarselvan, 2012; Song & Buettner, 2010). This electrochemical effect (redox potential E_0) of the substituted R-groups was computationally

& Pang, 1985) (Han, Ha, Kim, & Kim, 1998) (Han, Joo, Ha, & Kim, 2000) (Bard & Faulkner, 2000)

investigated in aqueous media and this was done only on the full and single substitutions as the solvation-free energy (DG_{0solv}) and E_0 of the intermediate substitutions are assumed to be in-between the two ends.

2.2 *Redox properties*

The biological nature of quinones can be summarized by their two properties. First is its ability to undergo nucleophilic attack leading to either increased toxicity or detoxification. Secondly is that quinone and its derivatives undergo a reversible oxido-reduction process. From Laviron's theory, (Laviron, 1984) this can sequentially be coupled with electrons and protons through a number of oxidation and reduction reactions. Studies on the electrochemistry of quinones in aprotic, buffed, and unbuffered aqueous solutions have previously been carried out to shed light on their redox behavior (Guin, Das, & Mandal, 2011; Hawley, Piekarski, & Adams, 1967; Rich, 1982; Sasaki et al., 1990). Quinones undergo one-step two-proton–electron reduction in aqueous buffer at neutral and acidic pH displaying a reversible reduction CV wave. This corresponds to the oxidation and reduction of quinone/hydroquinone couple. The redox potential has perfect linear pH dependence with –59 pH^{-1} as its slope. (Laitinen & Harris, 1975). The two protons and two electrons involved in the redox reaction have three redox states and three protonations which results in the generation of nine species (Figure 2)

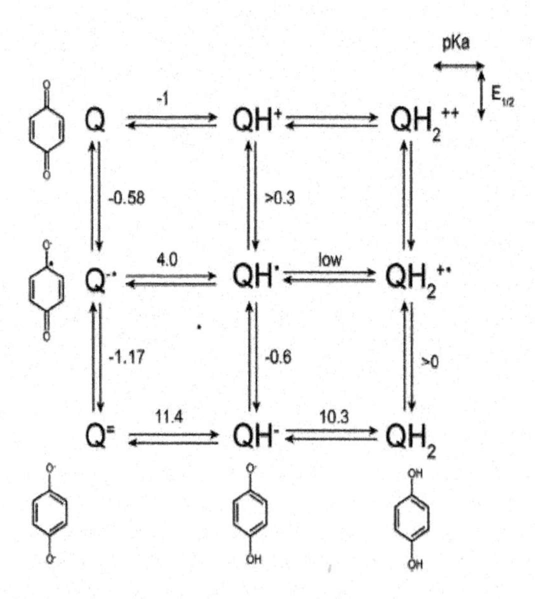

Figure 2. 1,4-benzoquinone thermodynamic cycles. The systematic transfer of protons electrons and makes quinone suitable as a redox mediator. (Monks et al., 1992) Copyright 1992, Elsevier Inc.

(Monks et al., 1992; Sasaki et al., 1990. The electron and proton sequence of transfer at pH 7 is: quinone (Q), quinone anion radical ($Q^{\cdot -}$), semiquinone radical (HQ), semiquinone anion (QH^-), and hydroquinone (HQ_2), (Kim, Chung, & Bull, 2014; Laviron, 1984; Zhang & Burgess, 2011). In the unbuffered media, reduction is a single-step two-proton two-electron process if the concentration of protons is similar to that of quinone but intermediate species are generated in the case of lower proton concentration. In nonaqueous media quinone's reduction is a two-step successive one-electron process, where step one is completely reversible and the second step is a quasi-reversible process at customized scan rates. These two reduction channels are majorly influenced by intra-molecular hydrogen bonding, nature of supporting electrolyte, protonation and deprotonation equilibrium, ion pair formation, addition of basic and acidic additives, nature of the solvents, polarity of solvents, and other factors.

Research has shown that quinone together with its derivatives are promising redox active radicals for aqueous redox flow batteries due to their high energy density as a result of their two electron redox reaction property in contrast to conventional inorganic redox couples (e.g., chromium, iron and vanadium) (Yang et al., 2014). By introducing electronegative substituents like nitro, carbonyl groups, and halogens, quinone is made to be a stronger oxidant, while electron-donating groups like the amine, methoxy group, and hydroxyl make it a weaker oxidant (Chambers, 2010.). Due to quinones' electrochemical and structural properties, quinones functionalized with electronegative or electron donating groups can be attracted to the positive (or negative) side of the redox flow battery. Er (2015) and co-workers screened redox potential of 18 varied substituents of quinones, a crucial factor to achieve high-cell voltage in a redox flow battery (Er, 2015). From density functional theory (DFT) results, the electron donating groups, e.g., amine and hydroxyl, showed a decrease in electron affinity, hence low redox potentials. Comparatively, quinone electron withdrawing substituents, e.g., nitro, potassium, and sulfonate showed high redox potentials. Furthermore, another interesting factor in determining the redox potential is the position of the substituent groups on the quinone ring. Flores and co-workers found that a big change in quinone redox potential is seen when two functionals are placed on the opposite rings, and the smallest is seen on placing them on the same ring, e.g., thiophenoquinone-based derivatives (Pineda, Martin-Noble, & Phillips, 2015). In their computational analysis they suggested common patterns with high or low redox potentials as per the site and type of functional groups located in the quinone ring.

3 THEORY OF REDOX PREDICTIONS

A one-electron redox process of any redox reaction can simply be defined in the form of a half-cell reaction as follows:

$$Oxd^m + e - Red^{m-1} \tag{1}$$

where Red is a reduced species and Oxd is an oxidized species. During a redox reaction, the reduced species Red is formed when an electron is gained by Oxd species. This overall reaction is called a redox half-cell reaction. Such two half-cell reactions combine to get a complete redox reaction.

In this section we will describe how reference electrodes are treated computationally for non-aqueous and aqueous solutions. A brief explanation of how the calculation of redox potentials relates to the thermodynamic cycle will then be given. It will also be prudent to illustrate the existing methods used to predict the Gibbs free energy of a redox reaction and then highlight the values of absolute potentials of common reference electrodes in non-aqueous and aqueous solutions.

3.1 *Reference electrodes on aqueous solutions*

Reduction potential of a setup of chemical species measures their tendency or ease to undergo oxidation or reduction. Standard reduction potential which is basically referenced relative to the standard hydrogen electrode (SHE) is a potential measured under standard conditions (1 atm, 298 K, and 1 M). When used its value is arbitrarily given as 0.00 volts. In practice there exists other reference electrodes that are easier to use in aqueous solution than the SHE such as the Ag^{+}/AgCl electrode and saturated calomel electrode (SCE). Procedural experimentalists normally uses any practical and suitable reference electrode to examine the redox potentials of particular complexes, thereafter this redox potential would then be converted with respect to the SHE, SCE or Ag^{+}/AgCl. However during conversion, the liquid junction is normally problematic and therefore caution should be taken when converting the determined experimental redox values with respect to the two chosen reference electrode versus another. An example is when converting redox potential obtained against SCE to SHE, a value of 0.24 V have to be subtracted from the values obtained.

Typically, in an experiment the redox potential values obtained are referenced with respect to the reference electrode values in literature. Although experimentally the SHE value being considered as zero, its absolute reference values have been reported to range from 4.24 to 4.73 eV (Fawcett & Acta, 2008; Hansen & Kolb, 1979; Kelly, Cramer, & Truhlar, 2007; Reiss, 1985; Trasatti, 1986). SHE absolute value has been debated for a number of years and in 1986 IUPAC recommended a SHE absolute value of 4.44 eV (Trasatti, 1986), a value recently confirmed experimentally (Fawcett & Acta, 2008).

3.2 *Reference electrodes on non-aqueous solutions*

Other reactants may require the use of non-aqueous solutions if they are insoluble or unstable in water. In such a case ferrocene redox couple has been considered suitable for non-aqueous solutions. The redox system of ferrocene/ferrocenium ion (Fc/Fc^{+}) was tested in 22 non-aqueous solvents where it showed solvent independent redox results. Hence IUPAC has recommended it as a reference electrode for electrochemical experiments involving non-aqueous solutions (Gritzner & Kuta, 1982), and since then this electrode has been widely accepted and used as a suitable reference electrode for non-aqueous solutions. The Fc/Fc+ reference electrode half-cell reaction system is as below:

$$Fc + +e- > Fc$$

3.3 *Quinone computational studies for battery materials*

Intensive research on quinones from various disciplines is approached through DFT modeling, electrochemical tests, and organic synthesis, as depicted in Figure 3. Although quinones and its derivatives have been identified as promising materials (organic electrodes), only a few computational research studies to date have examined their redox properties. Recently Er et al (2015) used a high-throughput computational approach and identified redox potentials of about 300 quinone derivatives with varied functionalities and backbone lengths (Er, 2015). They predicted a reduction potential of above 0.7V, and refined a list of feasible quinones possible to be synthesized (Table 1).

Table 1. List of molecules of benzoquinone computationally predicted showing interesting redox properties vs. SHE, based on HT screening.

	R-group substituted		ΔG^0 solv solv (kJ/mol)	$E0$ (V vs. SHE)
1,4-BQ	NH_2	Full	–70.68	0.03
1,4-BQ	OH	Full	–52.90	0.17
1,2-BQ	C l	R6	–24.76	0.90
1,2-BQ	$COOCH_3$	R5	–29.30	0.92
1,2-BQ	CF_3	R5	–21.62	0.92
1,2-BQ	PO_3H_2	R5	–83.76	0.92
1,2-BQ	CF_3	R6	–26.25	0.93
1,4-BQ	$COOCH_3$	R2	–28.89	0.93
1,2-BQ	COOH	R5	–44.99	0.93
1,2-BQ	SO_3H	R5	–55.33	0.94
1,2-BQ	CHO	R5	–32.21	0.95
1,2-BQ	CN	R5	–25.49	0.95
1,2-BQ	SO_3H	R6	–45.23	0.95
1,2-BQ	CN	R6	–19.04	0.97
1,4-BQ	NO_2	R2	–24.91	0.98
1,2-BQ	SO_3H	Full	–132.64	0.98
1,4-BQ	COOH	Full	–105.46	0.99
1,2-BQ	NO_2	R5	–26.41	1.00
1,4-BQ	CN	Full	–32.01	1.02
1,4-BQ	PO_3H_2	Full	–142.99	1.02
1,2-BQ	COOH	R6	–45.00	1.03
1,4-BQ	$COOCH_3$	Full	–45.18	1.04
1,4-BQ	CN	Full	-32.01	1.02

1,4-BQ	PO_3H_2	Full	–142.99	1.02
1,2-BQ	COOH	R6	-45.00	1.03
1,4-BQ	$COOCH_3$	Full	–45.18	1.04
1,2-BQ	CF_3	Full	–8.19	1.11
1,4-BQ	NO_2	Full	–14.08	1.11
1,2-BQ	CHO	R6	–33.91	1.11
1,2-BQ	COOH	Full	–107.66	1.12
1,2-BQ	$COOCH_3$	R6	–34.49	1.13
1,2-BQ	CN	Full	-39.96	1.14
1,2-BQ	NO_2	R6	–33.43	1.17
1,2-BQ	CHO	Full	–47.99	1.19
1,2-BQ	NO_2	Full	–16.68	1.19
1,2-BQ	$COOCH_3$	Full	–46.14	1.21
1,2-BQ	PO_3H_2	Full	–168.34	1.23
1,4-BQ	SO_3H	Full	–96.55	1.32
1,4-BQ	CHO	Full	–46.79	1.34

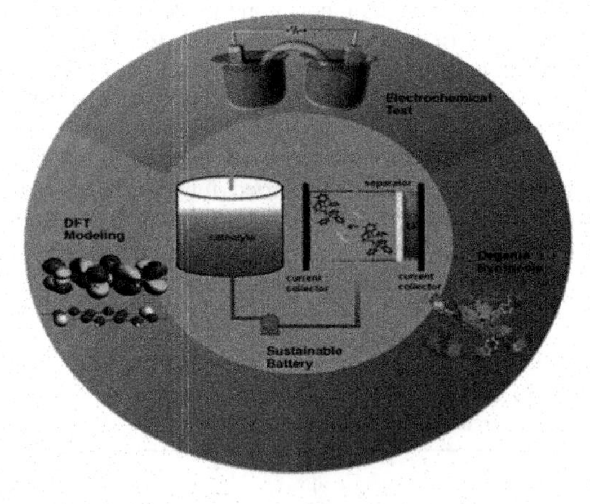

Figure 3. Quinone chart on an integrated approach with organic synthesis, theoretical calculation and electrochemical analysis. (Ding & Li, 2016).

Furthermore, they also concluded that functionalization near the ketone affects the reduction potential, and away from the ketone enhances the solubility, which is key in improving new quinone electrolytes. Assary and co-workers also computationally investigated the first and the second anthraquinone's (AQ) redox reactions of its derivatives. (Bachman & Curtis, 2014). Their computations suggested that lithium ions would increase athraquinone's redox potential by ~0.4, caused by lithium ions pairing on forming a complex of Lewis base–Lewis acid. Their suggestion was that to come up with new active redox species, AQ is substituted with electron donating groups in order to improve its reduction window having sufficient oxidative stability. They also suggested that when oxy-methyl dioxolane is incorporated as a substituent in the AQ framework it can improve its solubility and raise its interaction ability with non-aqueous solvent.

In their study Lee and the group carried out DFT redox potential investigation on the oxygen functional group in the hydrothermally reduced graphene oxides. (Liu et al., 2015). Quinone derivatives can exothermically react with Li atoms and concomitantly form a chemical bond of Li-oxygen. Therefore, it is expected that quinone electronic properties would be altered by the attached Li atoms. In their other DFT calculations Jang and co-workers (Kim, Liu, & Lee, 2016) found that on their interaction of quinone derivatives and Li atoms, Li atoms initially bind in the test molecules with the carbonyl groups. They further observed that quinone derivatives' redox properties can be tuned as needed by systematically introducing electron-with drawing functional groups to modify their chemical structure.

Further, DFT investigations on Li and quinone interaction gave an insight on the changes introduced during discharging process on their redox properties. As Li's atom number is increased there is a decrease in the redox potential. Though Jang's group further suggested that to improve their charge capacity as well as their redox potential quinone derivatives can be functionalized with carboxylic acids. They also established that on discharging quinone derivatives, its cathodic activity strongly relies on the number of carbonyl groups still available for further Li binding as well as the solvation effect.

4 APPLICATION

4.1 *Quinone derivatives for energy storage*

Rechargeable batteries as well as supercapacitors are two promising energy storage devices due to their high power density, high energy density and a reasonable life cycle. (Shi et al., 2018). the reversible and faradic reaction in Lithium ion batteries (LIB) endows LIB with high energy density of about 150–250 Wh/kg, high output voltage (3.5V), relatively good cycle stability, and high energy density. What inspires research on quinones is its electron transfer properties in biological systems such as photosynthesis and respiration. Due to this electroactive nature of findings have led top interest in quinones for energy storage and energy harvesting applications (Figure 4). Furthermore, quinone derivatives have been identified as promising candidates both for positive and negative electrode components sides for rechargeable batteries. In this sub-section we give an overview of recent research milestones of functionalized quinone materials in rechargeable batteries.

Recent rechargeable energy storage systems such as redox flow batteries (Chen, Eisenach, & Aziz, 2016; Chen et al., 2016; Huskinsom et al., 2014; Huskinson et al., 2013; Saraf Nawar and Aziz, 2013) and Li-ion batteries (Park et al., 2015; Song, Zhan, & Zhou, 2009; Song et al., 2014; Yokoji, Matsubara, & Satoh, 2014; Zhu, Guo, Shi, Tao, & Chen, 2014) are made up of reversible quinone redox couples. They are particularly attractive because of their relative high redox potential versus Li-intercalated or Li anode ranging from 1.8 to 3.1 V (vs. Li/L^{+}). (Aupler, Wild, & Schubert,

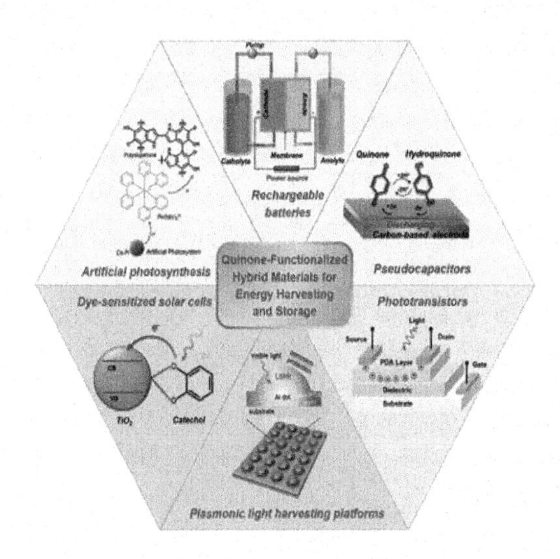

Figure 4. An illustration of applications of quinone functionalized materials for energy storage and harvesting. (Kim, Lee, & Park, 2014).

2015). Quinone's lower molar mass and its ability to undergo one-electron redox activity in organic electrolytes results in a high theoretical charge capacity of up to 500 mAh/g (Aupler, Wild, & Schubert, 2015), and displaying energy densities which can be compared to those inorganic cathode materials, such as $LiFePO_4$. (Choi, Harada, Oyaizu, & Nishide, 2011; Chen, Armand et al., 2008; Lee et al., 2013; Wang, Wang, Zhang, Zhu, Tao, & Chen, 2013;. Of more interest, is that quinone redox properties can be tuned to highly performing Li-ion batteries by introducing the electron-withdrawing groups (Lee et al., 2013). For instance, introduction of carboxylic groups introduces electrophilic quinones whereby the redox potential and the bound Li atoms increased in some quinones, as confirmed in anthraquinone-2,6-dicaboxylic acid and anthraquinone-2-carboxilic acid, resulting in high charge capacities.

4.2 *Electrolyte factor*

From literature it is clear that quinone redox electrochemistry is significantly affected by the nature of the supporting electrolyte. For example, quinone redox process with a tetrabutylammonium hexafluorophosphate ($TBAPF_6$) in acetonitrile (MeCN) is a two-step reversible one-electron redox reaction that peaks at –1.7V and –0.9V (vs. Fc^+/Fc) (Gamboa-Valero et al., 2016; Sereda et al., 2006). To the contrary, a one-step reversible two electron process is observed at –0.2 V (vs. Fc^+/Fc^0) with the cycling ion being H^+ and at –0.7V (vs. Fc^+/Fc^0) with the cycling ion being Li^+. (Emanuelsson et al., 2017; Emanuelsson et al., 2016). Table 2 gives a summary of a series of quinone derivatives characterized in different electrolytes. It is clear that the redox potential ($E^{0'}$) in an acetonitrile (MeCN) solvent follows the trend $H^+>Li^+>TBA^+$. The $E^{0'}$ of quinone positively shifted with the ability of the electron withdrawing subtituents (Wang et al., 2020). The potential shift of $E^{0'}$ surprisingly was lost in aqueous solution to a large extent. Wang et al. (2020) attributed this to being from electron donation due to the effect of substitution being counteracted by water solvent molecules.

The need for an electrolyte with both low temperature operation and high ionic conductivity is not achievable by the known organicbased electrolytes (Schmitz et al., 2014; Xu, 2004). Nitriles with acetonitrile (AN) being the most studied, are a class of solvents which offer good ionic conductivity of a given electrolyte component (Isken et al., 2011). AN is known to have decent dielectric permittivity and low viscosity (Yamada et al., 2013 & Han et al., 2013), with the best AN-based ionic conductivity of 30mS/cm and above (Wakihara, 2001; Xu K., 2014). However, the limitation of AN is its inability to form a proper solid electrolyte interface on both graphite electrodes and lithium metal (Peled, 1979; Peled, Golodnitsky et al., 1998; Winter, 2009; Zhang et al., 2001). This is ineffective solid electrolyte interphase electrolyte decomposition and exfoliation of graphite (Lee et al., 2013; Wagner et al., 2005; Winter et al., 1998).

Quinone dissolution in organic electrolytes happens due to the low molecular weight and hydrophobicity and this mostly limits the life cycle of Li-ion batteries, resulting in capacity loss during repeated steps of charging and discharging (Song & Zhou, 2013; Shimizu et al., 2014; Yao et al., 2010). In their work Shimizu et al. (2014) observed that dissolution diminishes in organic electrolytes on introducing two lithiooxycarbonyl groups into quinones, e.g., pyrene-4,5,9,10-tetraone, 9,10-anthraquinone, and 9,10-phenanthrenequinone. This happens without the influence of their redox potentials. Their suggestion was that when the above substitution is made the Li atom of the lithiooxycarbonyl group caused the decrease in solubility through the formation of a network by Li coordination, making lithiooxycarbonyl-functionalized quinones behave like polymers. Yao et al. (2010) demonstrated that the functionalization of 1,4-benzoquinone with methoxy groups can lower its solubility because this observed dissolution would be suppressed by the intermolecular forces, e.g., $\pi - \pi$ interaction and hydrogen bonding between the molecules. Other recent research studies suggest ways to overcome dissolution by the attachment of the molecules of quinone to carbon-based conducting nanomaterials, e.g., carbon nanotubes and grapheme through covalent anchoring (Genorio et al., 2010; Pirnat et al., 2012), $\pi - \pi$ interaction (Lee et al., 2015), the use of solid electrolytes (Zhu et al., 2014), or quinone-based polymers (Zhao et al.,2013).

As compared to Li-ion batteries which operate in solid or organic electrolytes, redox flow batteries operate (store electrical energy) through reversible electrochemical reaction between the cell electrodes and redox active aqueous electrolytes. Due to the possibility to scale up the energy storage density of redox

Table 2. Series of quinone derivertives characterized in different electrolytes.

Quinone molecule component	Name	Structure	Redox potential	Specific Capacity	Electrolyte	Reference	Method
1,4-BQ	*para*-quinone	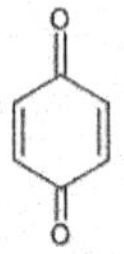	2.8 V vs. Li/Li+	496	Tetra-n-butylammonium perchlorate/ acetonitrile	Yokoji 2016	Experimental
1,4-BQ	*para*-quinone	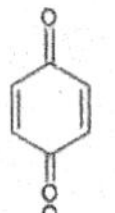	3.1 V vs. Li/Li+)			Kim *et., al*	DFT
1,4-BQ	*para*-quinone	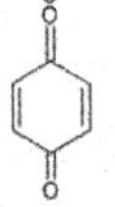	2.55V		Electrolyte solvent (1,2-dimethoxymethane (DME)	Miao et al	DFT
1,4-BQ	*para*-quinone	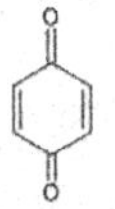	0.19 V vs. Fc^+/Fc^0)		Acetonitrile in the presence of pyridine/pyridinium acid base	(Emanuelsson *et. al.*, 2016)	Exp
1,2-BQ	Ortho-quinone	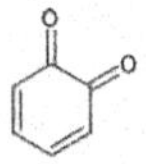	3.08V		Electrolyte solvent (1,2-dimethoxymethane (DME)	Miao et al	DFT
Electron withdrawing groups (increases potential)							
fluoro (–F)	2-flouro-1,4-benzoquinone	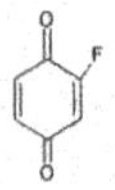	0.86 V vs. $Fc+/Fc^0$)		In TBAPF6/MeCN	Wang et al 2020	EXP
"	"	"	0 .07 V vs. $Fc+/Fc^0$		In LiClO4/MeCN	"	"
"	"	"	–0.33 V vs. $Fc+/Fc^0$		In Pyrid TMF/ Pyridine/MeCN	"	"
			0.27 V vs. $Fc+/Fc^0$		In H2SO4/H2O	"	"
fluoro (–2F)	5,6-flouro-1,4-benzoquinone	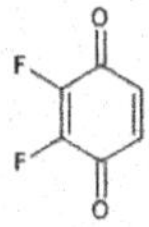	0.88 V vs. $Fc+/Fc^0$)		In TBAPF6/MeCN	"	"
"	"	"	0 .08 V vs. $Fc+/Fc^0$)		In LiClO4/MeCN	"	"
"	"	"	–0.26 V vs. $Fc+/Fc^0$)		In Pyrid TMF/Pyridine/MeCN	"	"
"	"	"	0.27 V vs. $Fc+/Fc^0$)		In H2SO4/H2O	"	"
chloro (–Cl)	2,3-dichloro-5,6-dicyanobenzoquinone (DDQ)	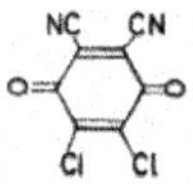	1.19		a mixture of ethylene carbonate (EC, 30 vol%) and diethyl carbonate (DEC, 70 vol%) containing 1.0 M LiPF6.	Yokoji 2014	"

(continued)

Table 2. (Continued).

Quinone molecule component	Name	Structure	Redox potential	Specific Capacity	Electrolyte	Reference	Method
trifluoromethyl (–CF3)	2,5-bis(trifluoromethyl)-1,4-benzoquinone (CF3-BQ)	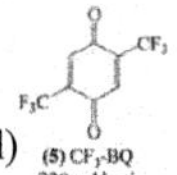 (5) CF_3-BQ 220 mAh g^{-1}	3.0 V versus Li/Li+.	220 mAh/g	a mixture of ethylene carbonate (EC, 30 vol%) and diethyl carbonate (DEC, 70 vol%) containing 1.0 M LiPF6.	Yokoji 2014	Exp
"	"	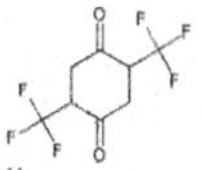	0.94 V vs. Fc+/Fc0		In TBAPF6/MeCN	Wang et al 2020	"
"	"	"	0 .24 V vs. Fc^+/Fc^0		In LiClO4/MeCN	"	"
"	"	"	- 0.20 V vs. Fc^+/Fc^0		In Pyrid TMF/ Pyridine/MeCN	"	"
'	"	"	0.28 V vs. Fc^+/Fc^0		InH2SO4/H2O	"	"
per- fluorobutyl (–C(CF3)3)	2,5-bis (perfluorob -1,4-benzoquinone (Rf4-BQ	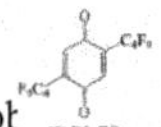 (6) Rf_4-BQ 99 mAh g^{-1}	"	99	"	Yokoji 2014	"
perfluorohexyl (–C6F13)	2,5-bis (perfluorohexyl)- 3,6-dichloro- 1,4-benzoquinone (Rf6-Cl-BQ)	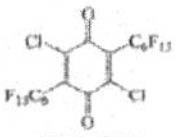 (7) Rf_6-Cl-BQ 66 mAh g^{-1}	3.1	66	"	"	Exp
Electron donating groups (decreases potential)							
methyl (Me, –CH3)	2,5-dimethyl-1,4-benzoquinone (CH3-BQ)	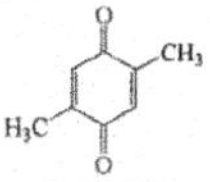 (3) CH_3-BQ 394 mAh g^{-1}	2.7 V vs. Li/Li+	394 $mAhg^{-1}$	mixture of ethylene carbonate (EC, 30 vol%) and diethyl carbonate (DEC, 70 vol%) containing 1.0 M LiPF6.	Yokoji 2014 (pg 2)	Exp
methoxy (OMe, –OCH3)	2,5-dimethoxy-1,4-benzoquinone (DMBQ)	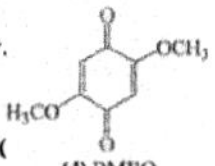 (4) DMBQ 319 mAh g^{-1}	2.6 V vs. Li/Li+	312 mAh/g	acetonitrile/tetra-n-butylammonium perchlorate system (0.1 mol L-1)	Yao 2010	Exp

flow batteries, they are suited for largescale applications. This is achieved by adjusting the size of its external tank containing the electrolytes (Huang & Wang, 2015; Perry & Weber, 2016; Soloveichik, 2015). The solubility and redox potential of reversible redox active components governs the energy density of the redox flow batteries. This necessitates a proper design and selection of the redox couple.

The redox couples of quinone/hydroquinone have been identified as candidates of interest for metal-free redox flow rechargeable batteries due to their good solubility (greater than 1 M in a pH 0 electrolyte for 9,10-anthraquinone-2,7-disulphonic acid; AQDS) in aqueous solvent, earth abundance (Huskinsom et al., 2014; Saraf Nawar & Aziz, 2013), and fast kinetic rate constants of charge transfer (e.g., for AQDS it is 7.2×10^{-3} cm s^{-1}). Solubility and redox potentials of quinones can further be tailored on introducing functional groups such as hydrophilic (Chen, Eisenach, & Aziz, 2016; Chen et al., 2016; Er, 2015; Huskinson et al., 2014; Huskinson et al., 2013; Saraf Nawar & Aziz, 2013; Yang et al., 2014), electron-withdrawing, and electron-donating groups. Generally solubility in water improves with an increase in the number of functional groups substituted, moreso on attaching the hydrophilic groups away from the ketone units of the quinones. Stability of quinones is essential for the long life cycle of redox flow batteries. Er et al. found that quinone oxidation results in a decomposition reaction, e.g., carbon backbone oxidation and polymerization, which is essential to lasting device performance. They suggested a way to improve the stability by introducing substituents, such as sulfonyl, hydroxyl, carboxyl, and amino substituents to quinone's C-H groups adjacent to C=O groups, which are prone to oxidation. A summary of the quinone's dependency on the substituents redox potential's is given in Table 3.

5 CONCLUSION

Quinone being the main redox mediator in cellular respiration and natural photosynthesis for delivering protons and electrons with low recombination has inspired the advancement of manmade energy devices. Quinone's unique redox properties have been utilized to fabricate energy storage materials. Its fast redox and reversible reactions can support extraction/ion insertion that is good for rechargeable energy storage systems like Li-ion batteries, pseudocapacitors and redox flow batteries. Quinone and its derivatives are preferred as redox active compounds used to fabricate rechargeable batteries due to their relative high energy density, fast charging rate, solubility in electrolytes, abundance, and cyclic stability.

Progress has been made to utilize quinone and its derivatives in fabricating energy material. However a number of issues have to be addressed in order to fabricate more efficient energy storage systems. Quinone redox couple's longterm stability is a crucial issue to consider for it to be used as a component in energy storage devices like Li-ion batteries. Often, the aromatic nature of quinonebased molecules causes problems by dissolving into electrolytes, decreasing the cyclic stability and the rate of performance of energy storage devices. With regard to this a number of strategies are now underway to enhance longterm stability which includes the anchoring/grafting of quinone through polymerization (Zhou et al., 2015), covalent-bonded hybridization of quinone to electrode materials with conductive property (Genorio et al., 2010) and the use of solid electrolytes (Zhu et al., 2014) Further studies also are required on the relationship between quinone's electrochemical behavior and its molecular structure, geared towards optimizing the solubility of quinone molecules (Er, 2015; Hong et al., 2014; Morita

Table 3. List of quinone functionalized materials for energy storage applications.

Quinone Moiety	Properties	Functions	Applications	Quinone-based hybrid materials	Ref
1,4-Benzoquinone	Redox	Redox couple for electrochemical charge storage	Redox flow battery	1,2-Benzoquinone-3,5disulfonic acid catholyte	(Yang et al., 2014)
1,4-Naphthoquinone	Redox	Redox pair for electrochemical charge storage	Redox flow battery	9,10-athraquinone-2,7-disulhonic acid anolyte 2,6-Dihy droxyanthraquinone anolyte 9,10-Athraqu inone-2-sulfonic acid anolyte	(Yang et al., 2014)

et al., 2011). Such efforts hopefully will allow the engineering of practical and sustainable energy storage batteries with better and more feasible cycle lives.

ACKNOWLEDGEMENTS

I thank Professor Henry Kiriamiti at Moi University for the fruitful input. I can't forget my colleagues Mulei, Alice, Ken, and Erick for the comrade motivation. This review study was supported by The Center of Excellence in Phytochemicals, Textile and Renewable Energy ACE II PTRE).

REFERENCES

Ajloo, D., Yoonesi, B., & Soleymanpour, A. (2010). Int. J. Electrochem. Sci. (5), 459–477.

Anson, F. C., & Epstein, B. (1968). A chronocoulometric study of the adsorption of anthraquinone monosulfonate on mercury. *Journal of The Electrochemical Society, 115*(11), 1155–1158.

Aupler, B. H., Wild, A., & Schubert, U. S. (2015). Adv. Energy Mater., (5), 1402034.

Bachman, J. E., & Curtis, L. A. (2014). Investigation of the Redox Chemistry of Anthraquinone Derivatives Using Density Functional Theory. *J.Phy.Chem*(118), 2374–2482.

Baldwin, R. P., Packett, D., & Woodcock, T. M. (1981). "Electrochemical behaviour of adriamycin at carbon paste electrodes. *Analytical Chemistry, 53*(3), 540–542.

Bard, A. J., & Faulkner, L. R. (2000). *Electrochemical Methods* (2nd ed.). John Wiley & Sons.

Bechtold, T., Gutmann, R., Burtscher, E., & Bobleter, O. (1997). Cyclic voltammetric study of anthraquinone-2-sulfonate in the presence of Acid Yellow 9. *Electrochimica Acta, 42*(23-24), 3483–3487.

Berg, H. (1961). "Irreversible" anthrachinone,". *Die Natur wissenschaften, 48*(23), 714.

Chambers, J. (2010.). *Quinonoid Compounds,* .Hoboken, NJ, John Wiley& Sons, Ltd,.

Chaney Jr, E. N., & Baldwin, R. P. (1982). "Electrochemical determination of adriamycin in urine by preconcentration at carbon electrodes. *Analytical Chemistry, 54*(14), 2556–2560.

Chen, H., Armand, M., Demailly, G., Dolhem, F., Poizot, P., & Tarascon, J. -M. (2008). ChemSusChem, . (1), 348–355.

Chen, Q., Eisenach, L., & Aziz, M. J. (2016,). J. Electrochem. Soc. (163), A5057–A5063.

Chen, Q., Gerhardt, M. R., Hartle, L., & Aziz, M. J. (2016). J. Electrochem. Soc.,. (163), A5010–A5013.

Choi, W., Harada, D., Oyaizu, K., & Nishide, H. (2011). J. Am. Chem. Soc. (133,), 19839–19843.

D, M., & Czapski. (1975). One-electron transfer equilibria and redox potentials of radicals studied by pulse radiolysis. *Journal of physical chemistry, 79*(15), 1503–2509.

Dan, M., & Neta, P. (1975). One-electron redox potentials of nitro compounds and radiosensitizers. Correlation with spin densities of their radical anions. *Journal of the American Chemical Society, 97*(18), 5198–5203.

Determination of the orientation of aromatic molecules adsorbed on platinum electrodes: the influence of iodide, a surface-active anion. (1982). *Journal of the American Chemical Society, 105*(10), 2742–2747.

Dey P, M., & Harborne, J. B. (1989). *Methods in plant biochemistry. In: Harborne JB (ed) Plant phenolics.* London: Academic Press.

Ding, C., Zhang, H., Li, X., Liu, T., & Xing, F. (2013). J. Phys. Chem. (4), 1281–1294.

Ding, Y., & Li, Y. Y. (2016). Exploring Bio-inspired Quinone-Based Organic Redox Flow Batteries: A Combined Experimental and Computational Study. (1), 790–801.

Ding, Y., & Yu, G. (2016). Angew. Chem. *Int. Ed* (128), 4850–4854.

Dunn, B., Kamath, H., & Tarascon, -M. J. (2011). *Science* (334), 928–935.

El-Hady, D. A., Abde-Hamid, M. I., Seliem, M. M., & E-Maali, N. A. (2004). "Voltammetric studies of Cu-adriblastina complex and its effect on ssDNA-adriblastina interaction at in situ mercury film electrode,". *Archives of Pharmacal Research*, 1161–1167.

Emanuelsson, R., Huang, H., Gogoll, A., Strømme, M., & Sjödin, M. (2016). Enthalpic versus Entropic Contribution to the Quinone Formal Potential in a Polypyrrole-Based Conducting Redox Polymer. *J. Phys. Chem. C., 120*, 21178–21183.

Emanuelsson, R., Karlsson, C., Huang, H., Kosgei, C., Strømme, M., & Sjödin, M. (2016). Quinone Based Conducting Redox Polymers for Electrical Energy Storage. *Russian Journal of Electrochemistry, 53*(1), 8–15.

Emanuelsson, R., Sterby, M., Strømme, M., & Sjödin, M. (2017). An All-Organic Proton Battery. *J. Am. Chem. Soc, 139*, 4828–4834.

Er, S. S.-G. (2015). Computational design of molecules for an all-quinone redox flow battery. *Chemical Science* (6), 885–893.

Er, s., Changwon, S., Michae, l. P., Marshak, & Al'an, A. (2015). Computational design of molecules for an all-quinone redox flow battery. *Chem. Sci* (6), 885.

Fawcett, W. R., & Acta, S. E. (2008). The ionic work function and its role in estimating absolute electrode potentials. In *Langmuir* (pp. 9865–9875).

Furman, N. H., & Stone, K. (1948). A polarographic study of certain anthraquinones. *Journal of the American Chemical Society, 70*(9), 3055–3061.

Gamboa-Valero, N., Astudillo, P., Gonzalez-Fuentes, M., Leyva, M., Rosales-Hoz, M. d., & Gonzalez, F. (2016). Hydrogen Bonding Complexes in the Quinone-Hydroquinone System and the Transition to a Reversible Two-Electron Transfer Mechanism. *Electrochimica Acta., 188*, 602–610.

Genorio, B., Pirnat, K., Cerc-Korosec, R., Dominko, R., & Gaberscek, M. (2010). Angew.Chem., Int. Ed., (49), 7222–7224.

Gill, R., & Stonehill, H. I. (1952). A polarographic investigation of the tautomerism of 2-hydroxy- and 2: 6-dihydroxyanthraquinol. *Journal of the Chemical Society (Resumed)*, 1845-1857.

Gritzner, G., & Kuta, J. (1982). Recommendations on reporting electrode potenitals in nonaqueous solvents.. *Pure Appl. Chem.* (54), 1527–1532.

Guin, P. S., Das, S., & Mandal, P. (2011). Int. J. Electrochem., 816202.

Guin, P. S., Das, S., & Mandal, P. C. (2008)., "Electrochemicalreduction of sodium 1,4-dihydroxy-9,10-anthraquinone-2-sulphonate in aqueous and aqueous dimethyl formamide mixed solvent: a cyclic voltammetric study,". *International Journal of Electrochemical Science*, 1016–1028.

Guin, P. S., Das, S., & Mandal, P. C. (2010). Sodium 1, 4-dihydroxy- 9, 10-anthraquinone- 2-sulphonate interacts with calf thymus DNA in a way that mimics anthracycline antibiotics: an electrochemical and spectroscopic study. *Journal of Physical Organic Chemistry, 23*(6), 447–482.

Hahn, Y., & Lee, H. Y. (2004). "Electrochemical behavior and square wave voltammetric determination of doxorubicin hydrochloride,". *Archives of Pharmacal Research, 27*(1), 31–34.
Han, S. -D., Borodin, O., Allen, J. L., Seo, D. M., Mc Owen, D. W., Yun, S.-H. et al. (2013). J. Am. Chem. Soc,. *11*(160), A2100–A2110.
Han, S. W., Ha, T. H., Kim, C. H., & Kim, K. (1998). Self-assembly of anthraquinone-2-carboxylic acid on silver: Fourier transform infrared spectroscopy, ellipsometry, quartz. *Langmuir, 14*(21), 6113–6120.
Han, S. W., Joo, S. W., Ha, T. H., & Kim, Y. K. (2000). Adsorption characteristics of anthraquinone-2-carboxylic acid on gold. *Journal of Physical Chemistry, 104*(50), 11987–11995.
Hansen, W. N., & Kolb, D. M. (1979). The work function of emersed electrodes. *J. Electroanal. Chem.*(100), 493–500.
Hawley, M. D., Piekarski, S., & Adams, R. N. (1967). J. Am. Chem. Soc.,. (89), 447–450.
He, P., Crooks, R. M., & Faulkner, L. R. (1990). Adsorption and electrode reactions of disulfonated anthraquinones at mercury electrodes. *Journal of Physical Chemistry, 94*(3), 1135–1141.
Hong, J., Lee, M., Lee, B., Seo, D. -H., Park, C. B., & Kang, K. (2014). *Commun.*(5), 1–6.
Hu, J., & Li, Q. (1999). Voltammetric behavior of adriamycin and its determination at Ni ion-implanted electrode. *Analytical Sciences, 15*(12), 1215–1218.
Hu, P., Wang, H., Yang, Y., Yang, J., J, L., & Guo, L. (2016). Adv. Mater. (28), 3486–3492.
Huang, Q., & Wang, Q. (2015). ChemPlusChem,. (80), 312–322.
Hubbard, A. T., Stickney, J. L., & Soriaga, M. P. (1984). Electrochemical processes at well-defined surfaces. *Journal of Electroanalytical Chemistry, 168*(1–2), 43–66.
Hukinston, B., Michael, P. M., Changwon, S., Leyman, E., Michael, R. G., Cooper, J. G. et al. (2014). A metal-free organic-inorganic aqueous flow battery. *Nature, 505*(7482), 195–198.
Huskinsom, B., Marshak, M. P., Suh, C., Er, S., Gerhardt, M. R., Galvin, C. J. et al. (2014). *Nature* (505), 195–194.
Huskinson, B., Marshak, M. P., Gerhardt, M. R., & Aziz, M. J. (2014). *ECS Trans* (61), 27–30.
Huskinson, B., Nawar, S., Gerhardt, M. R., & Aziz, M. J. (2013). ECS Trans. (53), 101–105.
Isken, P., Dippel, C., Schmitz, R., Schmitz, R. W., Kunze, M., & Passerini, S. et al. (2011). Electrochim. Acta,. 22(56), 7530–7535.
Kahlert, H. (2008). Functionalized carbon electrodes for pH determination. *Journal of Solid State Electrochemistry, 12*(10), 1255–1266.
Kano, K., Konse, T., Nishimura, N., & Kubota, T. (1984). "Electrochemical properties of adriamycin adsorbed on a mercury electrode surface,". *Bulletin of the Chemical Society of Japan, 57*(9), 2383–2390.
Kelly, C. P., Cramer, C. J., & Truhlar, D. G. (2007). Single-ion solvation free energies and the normal hydrogen electrode potential in methanol, acetonitrile and dimethyl sulfoxide. *Journal of physical Chemistry* (111), 408–422.
Kertesz, V., Chambers, J. Q., & Mullenix, A. N. (1999). Chronoamperometry of surface-confined redox couples. Application to daunomycin adsorbed on hanging mercury drop electrodes. *Electrochimica Acta, 45*(7), 1095–1104.
Kim, J. H., Lee, M., & Park, C. B. (2014). Angew. Chem., Int. Ed .(53), 6364–6368.
Kim, K. C., Liu, T., & Lee, S. W. (2016). J. Am. Chem. Soc. (138), 2374–2382.
Kim, K. C., Liu, T., Lee, S. W., & Jang, S. ,. (2016). J.Am. Chem. Soc. (138), 2374–2382.
Kim, R. S., Chung, T. D., & Bull, K. (2014). orean Chem. Soc., (35), 3143–3155.
Koshy, V. J., Sawayambunathan, V., & Periasamy, N. (1980). A reversible redox couple in quinone-hydroquinone system in nonaqueous medium. *Journal of the Electrochemical Society, 127*(12), 2761–2763.
Laitinen, H. A., & Harris, W. E. (1975). *Chemical Analysis: an Advanced Text and Reference.* Tokyo: McGraw-Hill.
Laviron, E. (1984). Interfacial Electrochem.,. *J. Electroanal. Chem.* (164), 213–227.
Lee, H., Dellatore, S. M., Miller, W. M., & Messersmith, P. B. (2007). Science. (318), 426-430.
Lee, M., Hong, J., Kim, J., Lim, H. -D., Cho, B., Kang, K. et al. (2015). Coord. Chem. Rev., . 102–108.
Lee, m., Hong, J., Seo, D. -H., Nam, D. -H., Nam, K. T., Kang, K. et al. (2013). Angew. Chem., Int.Ed. (52), 8322–8328.
Lee, M., Hong, J., Seo, D. -H., Nam, D. H., Nam, K. T., Kang, K. et al. (2013,). Angew. Chem., Int.Ed.,. (52,), 8322–8328.
Liang, Y., Zhang, P., & Chen, J. (2013). Chem. Sci., . (4), 1330–1337.
Lin, K., Chen, Q., Gerhardt, M. R., Tong, L., Kim, S. B., Eisenach, L. et al. (2015). *Science* (349), 1529–1532.
Liu, T., Kim, K. C., Kavian, R., Jang, S. S., & Lee, S. W. (2015). Chem. Mater. (27), 3291.
LIU, Y., Ai, K., & Lu, L. (2014). Chem. Rev. (114), 5057–5115.
Ma, T., Zhao, Q., Wang, J., & Pan, Z. (2016). J.Chem. *Int Ed*, 6428–6432.
Mairanovskii, S. G. (1968). *Catalytic and kinetic Waves in Polarography.* New York, NY, USA: Plenum Press.
Manisankar, P., & Valarselvan, S. (2012). Ionics. (18), 679–686.
Molinier-Jumel, C., Malfoy, B., Reynaud, J. A., & Aubel-Sadron, G. (1978). "Electrochemical study of DNA-Athracyclines interaction", *Biochemical and Biophysical Research Communications, 84*(2), 441–449.
Monks, T. J., Hanzlik, R. P., Cohen, G. M., Ross, D., & Graham, D. G. (1992). Toxicol. Appl. Pharmacol. (112), 2–16.
Morita, Y., Nishida, S., Murata, T., Moriguchi, M., Ueda, A., Satoh, M. et al. (2011). *Nat. Mater* (10), 947–951.
Nokami, T., Matsuo, Y., Hojo, N., Tsukagoshi, T., Yoshizawa, H., Shimizi, A. et al. (2012). J. am. Chem. *Soc*, 19694-19700.
O'Kelly, J. P., & Forester, R. J. (1998). Potential dependent adsorption of anthraquinone-2,7-disulfonate on mercury. *Analyst, 123*, 1987–1993.
Oyaizu, K., Niibori, Y., Takahashi, A., & Nishide, H. (2012). J. Inorg. Organomet. Polym. (23), 243–250.
Park, M., Shin, D. -S., Ryu, J., Choi, M., Park, N., Hong S, Y. et al. (2015). Adv. Mater. (27), 51415146.
Peled, E. (1979). J. Am. Chem. Soc. *12*(126), 2047–2051.
Peled, E., Golodnitsky, D., Menachem, C., & Bar-Tow, D. (1998). J. Am. Chem. Soc. *10*(145), 3482–3486.
Perry, M. L., & Weber, Z. (2016). J.Electrochem. Soc. (163), A5064–A5067.
Pineda, S. D., Martin-Noble, R. L., & Phillips, J. S. (2015). Bio-inspired electroactive organic molecules for aqueous redox flow batteries, I. Theophenoquinones. *J. Phys.Chem.C*(119), 21800–21809.
Pirnat, K., Dominko, R., Cerc-Korosec, R., Mali, G., Genorio, B., & Gaberscek, M. (2012). J. Power Sources. (199), 308–314.

Rao, G. M., Lown, J. W., & Plambeck, J. A. (1978). "Electrochemical studies of antitumor antibiotics, III, Daunorubicin and adriamycin. *Journal of Electrochemical Society, 125*(4), 534–539.

Reiss, H. (1985). The absolute potential of the standard hydrogen electrode: A new estimate. *J. Phys.Chem*(89), 4207–4213.

Rich, P. R. (1982). Farad. Discuss. (74), 349–364.

S.I, B., & M.I, R. (1985). A cyclic voltametric study of the aqueous electrochemistry of some quinones. *Electrochimica Acta, 30*(1), 3-12.

Saraf Nawar & Aziz, M. (2013). Mater. Res. Soc. Symp. Proc. (1491,), 1–6.

Saraf Nawar, B. H., & Aziz, M. (2013). *Mater. Res. Soc. Symp. Proc.*, (1491), 1–6.

Sasaki, k., Kashimura, T., Ohura, M., Ohsaki, Y., & Ohta, N. (1990). J. Electrochem. Soc. (137), 2437–2443.

Sereda, G., Van Heukelom, J., Koppang, M., Ramreddy, S., & Collins, N. (2006). Effect of Transannular Interaction on the Redox-Potentials in a Series of Bicyclic Quinones. *J. Org. Chem.*, 2.

Schmidt, D., Hager, M. D., & Schubert. (2016). adv Energy. Mater. (6), 1500369.

Schmitz, R. W., Murmann, P., Schmitz, R., Müller, R., Krämer, L., Kasnatscheew, J. et al. (2014). Prog. Solid State Chem. *4*(42), 65-84.

Shi, K., & Shiu, -K. K. (2004). Adsorption of some quinone derivatives at electrochemically activated glassy carbon electrodes. *Journal of Electroanalytical Chemistry, 574*(1), 63-70.

Shimizu, A., Kuramoto, H., sujii, Y. T., Nokami, T., Inatomi, Y., Hojo, N. et al. (2014). J.PowerSources. (260), 211–217.

Shiu, -K. K., & Song, F. D.-P. (1996). Potentiometric pH sensor with anthraquinone sulfonate adsorbed on glassy carbon electrodes. *Electroanalysis, 8*(12), 1160–1164.

Soloveichik, G. L. (2015). Chem. Rev., . (115), 11533–11558.

Song, Y., & Buettner, G. R. (2010). Free Radical. *Biol. Med* (49), 919–962.

Song, Z., & Zhou, H. (2013). Energy Environ. Sci. (2280–2301.), 2280–2301.

Song, Z., Qian, Y., Liu, X., Zhang, T., Zhu, Y., Yu, H. O. et al. (2014). Energy Environ. Sci. (7), 4077–4086.

Song, Z., Zhan, H., & Zhou, Y. (2009). Chem.Commun. 448–450.

Soriaga, M. P., & Hubbard, A. T. (1982). Determination of the orientation of adsorbed molecules at solid-liquid interfaces by thin-layer electrochemistry: aromatic compounds at platinum electrodes. *Journal of the American Chemical Society, 104*(10), 2735–2742.

Soriaga, M. P., Binamira-Soriaga, H. A., Benziger, J. B., & Pang, k. W. (1985). Surface coordination chemistry of platinum studied by thin-layer electrodes. Adsorption, orientation, and mode of binding of aromatic and quinonoid compounds. *Inorganic Chemistry, 24*(1), 65–73.

Soriaga, M. P., Stickney, J. L., & Hubrard, A. T. (1983). Electrochemical oxidation of aromatic compounds adsorbed on platinum electrodes. The influence of molecular orientation. *Journal of Electroanalytical Chemistry, 144*(1–2), 207–215.

Soriaga, M. P., White, J. H., & Hubbard, A. T. (1983). Orientation of aromatic compounds adsorbed on platinum electrodes. The effect of temperature. *Journal of Physical Chemistry, 87*(16), 3048–3054.

Tachi, I. (1954). *Polarography.* Tokyo, Japan: Iswanami.

Trasatti, S. (1986). The absolute electrode potential: An explanatory note. *Pure Appl. Chem.*(58), 955–966.

Turner, W. E., & Elying, P. T. (1965). Electrochemical behavior of the quinone-hydroquinone system in pyridine. *Journal of The Electrochemical Society, 112*(12), 1215–1217.

Wagner, M. R., Albering, J. H., Möller, K. C., Besenhard, J. O., & Winter, M. (2005). Electrochem. Commun.,. *9*(7), 947–952.

Wakihara, M. (2001). Mater. Sci. Eng. R,. *4*(33), 109–134.

Wang, H., Emanuelsson, R., Banerjee, A., Ahuja, R., Strømme, M., & Sjödin, M. (2020). Effect of Cycling Ion and Solvent on the Redox Chemistry of Substituted Quinones and Solvent-Induced Breakdown of the Correlation Between Redox Potential and Electron Withdrawing Power of Substituents. *J. Phys. Chem*, 2–18.

Wang, H., Li, F., Zhu, B., Guo, L., Yang, Y., Hao, R. et al. (2016). Adv. Funct. Mate. (26), 3472–3479.

Wang, S., Wang, L., Zhang, K., Zhu, Z., Tao, Z., & Chen, J. (2013). Nano Lett., . (13), 4404–4409.

Winter, M. (2009). Z. Phys. Chem. *10-11*(223), 1395–1406.

Winter, M., Besenhard, J. O., Spahr, M. E., & Novak, P. (1998). Adv. Mater. *10*(10), 725–763.

Xu, K. (2004,). Chem. Rev. *10*(104), 4303–4417.

Xu, K. (2014). Chem. Rev. *23*(114), 11503–11618.

Yamada, Y., Furukawa, K., Sodeyama, K., Kikuchi, K., Yaegashi, M., Tateyama, Y. et al. (2013). J. Am. Chem. Soc,. *13*(136), 5039–5046.

Yang, B., Hoober-Burkhardt, L., Wang, F., Surya Prakash, G. K., & Narayanan, S. R. (2014). J. Electrochem. Soc. (161), A1371–A1380.

Yang, J., Niu, L., Zhang, Z., Zhao, J., & Chou, L. (2015). Anal. Lett. (48), 2031–2039.

Yang, Y., Wang, H., R, H., & Guo, L. (2016). Small.

Yao, M., Senoh, H., Yamazaki, S. -i., Siroma, Z., Sakai, T., & Yasuda, K. (2010). J. Power Sources. (195), 8336–8340.

Yi, C., & Gratzl. (1998). Continuous in situ electrochemical monitoring of doxorubicin efflux from sensitive and drug resistant cancer cells. *Biophysical Journal, 75*(5), 2255–2261.

Yokoji, T., Matsubara, H., & Satoh, M. (2014). J. Mater. Chem . (2), 19347–19354.

Yu, B., Liu, J., Liu, S., & Zhou, F. (2010). Chem. Commun. (46), 5900–5902.

Zhang, S. S., Ding, M. S., Xu, K., Allen, J., & Jow, T. R. (2001). Electrochem. Solid-State Lett. *12*(4), A206–A208.

Zhang, S., Wu, K., & Hu, S. (2002) , "Carbon paste electrode based on surface activation for trace adriamycin determination by a preconcentration and voltammetric method,". *Analytical Sciences, 18*(10), 1089–1092.

Zhang, W., & Burgess, I. J. (2011). Phys.Chem.Chem. Phys., (13), 2151–2159.

Zhao, L., Wang, W., Wang, A., Yuan, K., Chen, S., & Yang, Y. (2013). J. Power Sources. (233), 23–27.

Zhou, Y., Wang, B., Liu, C., Han, N., Xu, X., Zhao, F. et al. (2015). *Nano Energy* (15), 654–661.

Zhu, Z. H., Guo, D., Shi, J., Tao, Z., & Chen, J. (2014). J. Am. Chem. (136), 16461–16464.

Zhu, Z., Hong, M., Guo, D., Shi, J., Tao, Z., & Chen, J. (2014). J. Am. Colloid Interface Sci., U lle Chem. Soc. (136), 16461–16464.

Advances in Phytochemistry, Textile and Renewable Energy Research for Industrial Growth – Nzila et al. (Eds)
© 2022 Copyright the Editor(s), ISBN: 978-1-032-11871-0
Open Access: www.taylorfrancis.com, CC BY-NC-ND 4.0 license

Author index